Lecture Notes in Computer Science 8591

Commenced Publication in 1973
Founding and Former Series Editors:
Gerhard Goos, Juris Hartmanis, and Jan van Leeuwen

T0213758

Zhipeng Cai   Alex Zelikovsky
Anu Bourgeois (Eds.)

Computing
and Combinatorics

20th International Conference, COCOON 2014
Atlanta, GA, USA, August 4-6, 2014
Proceedings

 Springer

Volume Editors

Zhipeng Cai
Alex Zelikovsky
Anu Bourgeois

Georgia State University, Department of Computer Science
34 Peachtree Street, Atlanta, GA 30303, USA
E-mail: zcai@gsu.edu; {alexz, anu}@cs.gsu.edu

ISSN 0302-9743   e-ISSN 1611-3349
ISBN 978-3-319-08782-5   e-ISBN 978-3-319-08783-2
DOI 10.1007/978-3-319-08783-2
Springer Cham Heidelberg New York Dordrecht London

Library of Congress Control Number: 2014942318

LNCS Sublibrary: SL 1 – Theoretical Computer Science and General Issues

Typesetting: Camera-ready by author, data conversion by Scientific Publishing Services, Chennai, India

Printed on acid-free paper

Springer is part of Springer Science+Business Media (www.springer.com)

Preface

The 20th International Computing and Combinatorics Conference (COCOON 2014) was held during August 4–6, 2014 in Atlanta, Georgia, USA. COCOON 2014 provided a forum for researchers working in the area of theoretical computer science and combinatorics.

The technical program of the conference includes 51 contributed papers selected by the Program Committee from 110 full submissions received in response to the call for papers. All the papers were peer reviewed by at least three Program Committee members or external reviewers. The papers cover various topics, including algorithms and data structures, algorithmic game theory, approximation algorithms and online algorithms, automata, languages, logic, and computability, complexity theory, computational learning theory, cryptography, reliability and security, database theory, computational biology and bioinformatics, computational algebra, geometry, number theory, graph drawing and information visualization, graph theory, communication networks, optimization, and parallel and distributed computing. Some of the papers will be selected for publication in special issues of Algorithmica, Theoretical Computer Science (TCS), and Journal of Combinatorial Optimization (JOCO). It is expected that the journal version papers will appear in a more complete form.

The proceeding also includes 8 papers selected from a workshop on computational social networks (CSoNet 2014) co-located with COCOON 2014, held on August 6th, 2014. An independent Program Committee was chaired by Dr. Yingshu Li and Dr. Yu Wang. We appreciate the work by the CSoNet Program Committee that helped with enriching the conference topics.

We would like to thank the Program Committee members and external reviewers for volunteering their time to review conference papers. We would like to extend special thanks to the publication, publicity, and local organization chairs for their hard work in making COCOON 2014 a successful event. Last but not least, we would like to thank all the authors for presenting their works at the conference.

August 2014

Zhipeng Cai
Alex Zelikovsky
Anu Bourgeois

Conference Organization

Program Chairs

Zhipeng Cai   Georgia State University, USA
Alex Zelikovsky   Georgia State University, USA

Publication Chairs

Chunyu Ai   University of South Carolina Upstate, USA
Meng Han   Georgia State University, USA

Publicity Chairs

Yingshu Li   Georgia State University, USA
Siyao Cheng   Harbin Institute of Technology, China
Mingyuan Yan   Georgia State University, USA

Local Organization Chair

Anu Bourgeois   Georgia State University, USA

Program Committee

Yossi Azar   Tel-Aviv University, Israel
Sang Won Bae   Kyonggi University, South Korea
Yixin Cao   Hungarian Academy of Sciences, Hungary
Zhixiang Chen   University of Texas Pan American, USA
Xi Chen   Columbia University, USA
Siyao Cheng   Harbin Institute of Technology, China
Yongxi Cheng   Xi'an Jiaotong University, China
Janos Csirik   University of Szeged, Hungary
Bhaskar Dasgupta   University of Illinois at Chicago, USA
Ling Ding   University of Washington Tacoma, USA
Dingzhu Du   The University of Texas at Dallas, USA
Zachary Friggstad   University of Alberta, Canada
Xiaofeng Gao   Shanghai Jiao Tong University, China
Juraj Hromkovic   ETH Zurich, Switzerland
Tsan-Sheng Hsu   Academia Sinica, Taiwan

Xiao-Dong Hu	Chinese Academy of Sciences, China
Klaus Jansen	University of Kiel, Germany
Valentine Kabanets	Simon Fraser University, Canada
Iyad Kanj	DePaul University, USA
Ming-Yang Kao	Northwestern University, USA
Donghyun Kim	North Carolina Central University, USA
Piotr Krysta	The University of Liverpool, UK
Minming Li	City University of Hong Kong, China
Jian Li	Tsinghua University, China
Julian Mestre	University of Sydney, Australia
Benjamin Moseley	Toyota Technological Institute, Japan
Mitsunori Ogihara	University of Miami, USA
Desh Ranjan	Old Dominion University, USA
He Sun	Max Planck Institute for Informatics, Germany
Yilin Shen	Samsung Research America, USA
Marc Uetz	University of Twente, The Netherlands
Yitong Yin	Nanjing University, China
Guochuan Zhang	Zhejiang University, China

CSoNet Program Chairs

Yingshu Li	Georgia State University, USA
Yu Wang	University of North Carolina at Charlotte, USA

CSoNet Program Committee

Mihaela Cardei	Florida Atlantic University, USA
Ionut Cardei	Florida Atlantic University, USA
Thang N. Dinh	Virginia Commonwealth University, USA
Xiaolin Fang	Harbin Institute of Technology, USA
Xiaohua Jia	City University of Hong Kong, USA
Timothy Killingback	University of Massachusetts Boston, USA
Donghyun Kim	North Carolina Central University, USA
Fan Li	Beijing Institute of Technology, China
Zaixin Lu	University of Southern Texas, USA
Panos Pardalos	University of Florida, USA
Meirui Ren	Heilongjiang University, China
Shaojie Tang	Temple University, USA
Selcuk Uluagac	Georgia Institute of Technology, USA
Chaokun Wang	Tsinghua University, China
Jie Wang	University of Massachusetts Lowell, USA
Xiaoming Wang	Shaanxi Normal University, China
Lichen Zhang	Shaanxi Normal University, China
Hong Zhao	Lanzhou University, China

Additional Reviewers

Adamaszek, Anna
Ahn, Hee-Kap
Allender, Eric
An, Hyung-Chan
Antoniadis, Antonios
Bansal, Nikhil
Barequet, Gill
Barhum, Kfir
Bodlaender, Hans L.
Boeckenhauer, Hans-Joachim
Bogdanov, Andrej
Bonamy, Marthe
Cada, Roman
Calamoneri, Tiziana
Cao, Bo
Chan, Siu Man
Chang, Hsien-Chih
Chen, Lin
Chen, Zongchen
Cornelissen, Kamiel
De Haan, Ronald
de Jong, Jasper
Epstein, Leah
Fernau, Henning
Flatland, Robin
Fotakis, Dimitris
Gaspers, Serge
Gu, Zhangxu
Guo, Jiong
Gurski, Frank
Haitner, Iftach
Hoefer, Martin
Hoffmann, Michael
Jeffery, Stacey
Jeż, Łukasz
Jin, Kai
Jin, Yifei
Jukna, Stasys
Kliemann, Lasse
Komm, Dennis
Korman, Matias
Kortsarz, Guy
Kothari, Robin

Kraft, Stefan
Krivelevich, Michael
Krug, Sacha
Li, Jun
Li, Liang
Li, Xianyue
Liu, Wenqing
Liu, Zhen
Lu, Hsueh-I
Manthey, Bodo
Mastrolilli, Monaldo
Megow, Nicole
Meulemans, Wouter
Nakano, Shin-Ichi
Okamoto, Yoshio
Perkovic, Ljubomir
Pferschy, Ulrich
Qiu, Xian
Rahman, Md. Saidur
Rahn, Mona
Rao B.V., Raghavendra
Reidl, Felix
Rutter, Ignaz
Saldamli, Gokay
Sauerwald, Thomas
Schmid, Markus L.
Shaw, Peter
Shin, Chan-Su
Sivan, Balu
Skopalik, Alexander
Smula, Jasmin
Son, Junggab
Son, Wanbin
Steinová, Monika
Sun, Xiaoming
Tackmann, Björn
Tanigawa, Shin-Ichi
van Leeuwen, Erik Jan
Vardi, Adi
Wang, Haitao
Wang, Wei
Wu, Chenggang
Wu, Weiwei

Xia, Ge

Yakaryilmaz, Abuzer

Ye, Deshi

Zanetti, Luca

Zhang, Chihao

Zhang, Qiang

Zhang, Qin

Zhang, Zhao

Zhou, Yuan

Zhu, Xudong

Table of Contents

Sampling and Randomized Methods

Logic, Algebra and Automata

Database and Data Structures

Parameterized Complexity and Algorithms(I)

Computational Complexity

Computational Biology and Computational Geometry

Parameterized Complexity and Algorithms(II)

Approximation Algorithm(I)

Approximation Algorithm(II)

Graph Theory and Algorithms(I)

Approximation Algorithm(III)

Graph Theory and Algorithms(II)

Game Theory and Cryptography

Scheduling Algorithms and Circuit Complexity

CSoNet I

CSoNet II

Building above Read-once Polynomials: Identity Testing and Hardness of Representation

Meena Mahajan[1], B.V. Raghavendra Rao[2], and Karteek Sreenivasaiah[1]

[1] The Institute of Mathematical Sciences, Chennai, India
{meena,karteek}@imsc.res.in
[2] Indian Institute of Technology Madras, Chennai, India
bvrr@cse.iitm.ac.in

Abstract. Polynomial Identity Testing (PIT) algorithms have focussed on polynomials computed either by small alternation-depth arithmetic circuits, or by read-restricted formulas. Read-once polynomials (ROPs) are computed by read-once formulas (ROFs) and are the simplest of read-restricted polynomials. Building structures above these, we show:

1. A deterministic polynomial-time non-black-box PIT algorithm for $\sum^{(2)} \cdot \prod \cdot \mathsf{ROF}$.
2. Weak hardness of representation theorems for sums of powers of constant-free ROPs and for 0-justified alternation-depth-3 ROPs.

1 Introduction

The Polynomial Identity Testing (PIT) problem is the most fundamental computational question that can be asked about polynomials: is the polynomial given by some implicit representation identically zero? The implicit representations of the polynomials can be arithmetic circuits, branching programs etc., or the polynomial could be presented as a black-box, where the black-box takes a query in the form of an assignment to the variables and outputs the evaluation of the polynomial on the assignment. PIT has a randomized polynomial time algorithm on almost all input representations, independently discovered by Schwartz and Zippel [Sch80, Zip79]. However, obtaining deterministic polynomial time algorithms for PIT remained open since then. In 2004, Impagliazzo and Kabanets [KI04] showed that a deterministic polynomial time algorithm for PIT implies lower bounds (either $\mathsf{NEXP} \not\subset \mathsf{P/poly}$ or permanent does not have polynomial size arithmetic circuits), thus making it one of the central problems in algebraic complexity. Following [KI04], intense efforts over the last decade have been directed towards de-randomizing PIT (see for instance [SY10, Sax14]). The attempts fall into two categories: considering special cases ([Sax14]), and optimizing the random bits used in the Schwartz-Zippel test [BHS08, BE11].

The recent progress on PIT mainly focusses on special cases where the polynomials are computed by restricted forms of arithmetic circuits. They can be seen as following one of the two main lines of restrictions: 1. Shallow circuits based on alternation depth of circuits computing the polynomial. 2. Restriction

Z. Cai et al. (Eds.): COCOON 2014, LNCS 8591, pp. 1–12, 2014.

on the number of times a variable is read by formulas (circuits with fanout 1) computing the polynomial.

The study of PIT on shallow circuits began with depth two circuits, where deterministic polynomial time algorithms are known even when the polynomial is given as a black-box [BOT88, KS01]. Further, there were several interesting approaches that lead to deterministic PIT algorithms on depth three circuits with bounded top fan-in [DS07, KS07]. However, progressing from bounded fan-in depth three circuits seemed to be a big challenge. In 2008, Agrawal and Vinay [AV08] explained this difficulty, showing that deterministic polynomial time algorithms for PIT on depth four circuits implies sub-exponential time deterministic algorithms for general circuits. There have been several interesting approaches towards obtaining black-box algorithms for PIT on restricted classes of depth three and four circuits, see [Sax14, SY10] for further details. Recently, Kamath, Kayal and Saptharishi [GKKS13] showed that, over infinite fields, deterministic polynomial time algorithms for PIT on depth three circuits would also imply lower bounds for the permanent.

A formula computing a polynomial that depends on all of its variables must read each variable at least once (count each leaf labeled x as reading the variable x). The simplest such formulas read each variable exactly once; these are Read-Once Formulas ROFs, and the polynomials computed by such formulas are known as read-once polynomials (ROP). In the case of an ROP f presented by a read-once formula computing it, a simple reachability algorithm on formulas can be applied to test if $f \equiv 0$. Shpilka and Volkovich [SV08] gave a deterministic polynomial time algorithm for PIT on ROPs given as a black-box. Generalizing this to formulas that read a variable more than once, they obtained a deterministic polynomial time algorithm for polynomials presented as a sum of $O(1)$ ROFs. Anderson et. al [AvMV11] showed that if a read-k formula, with $k \in O(1)$, is additionally restricted to compute multilinear polynomials at every gate, then PIT on such formulas can be done in deterministic polynomial time. The result by [AvMV11] subsumes the result in [SV08] since a k-sum of read-once formulas is read-k and computes multilinear polynomials at every gate. However, both [SV08] and [AvMV11] crucially exploit the multilinearity property of the polynomials computed under the respective models. In [MRS14], the authors explored eliminating the multilinear-at-each-gate restriction, and gave a non-blackbox deterministic polynomial time algorithm for read-3 formulas. However for the case of Read-k formulas for $k \geq 4$, even the non-blackbox version of the problem is open. Note that multilinearity checking itself is equivalent to PIT on general circuits [FMM12].

Our Results: In this paper, we explore further structural properties of ROPs and polynomials that can be expressed as polynomial functions of a small number of ROPs. Our structural observations lead to efficient algorithms on special classes of bounded-read formulas.

We attempt to extend the class considered in [SV08] (namely, formulas of the form $\sum_i f_i$ where each f_i is an ROF) to the class of polynomials of the form $\sum_{i=1}^{k} f_i g_i$ where the f_is and g_is are presented as ROFs and k is some constant. These are read-$2k$ polynomials, not necessarily multilinear. Over the ring of integers and the field of rationals, we can give an efficient deterministic non-blackbox PIT algorithm for the case $k = 2$; the polynomial is $f_1 f_2 + g_1 g_2$ where f_1, f_2, g_1, g_2 are all read-once polynomials presented by ROFs. This class can also be seen as a special case of read-4 polynomials. Our algorithm exploits the structural decomposition properties of ROPs and combines this with an algorithm that extracts greatest common divisors of the coefficients in an ROP. The algorithm easily generalises to polynomials of the form $f_1 f_2 f_3 \cdots f_m + g_1 g_2 \cdots g_s$ where f_is and g_is are presented as ROFs, but m, s can be unbounded; that is, the class $\sum^{(2)} \cdot \prod \cdot \text{ROF}$. Note that this class of polynomials includes non-multilinear polynomials and also polynomials with no bound on the number of times variables are read. Thus it is incomparable with the classes considered in [SV08], [AvMV11] and [MRS14]. This result is presented in Section 3, Theorem 1.

(At a recent Dasgtuhl seminar 14121, Amir Shpilka pointed out to the first author that this method can be adapted to work over any field. That is, over any field, identity testing for polynomials of the form $\sum^{(2)} \cdot \prod \cdot \text{ROF}$ can be done deterministically and efficiently. Details will appear in the full version.)

Central to the PIT algorithm in [SV08] is a "hardness of representation" lemma showing that the polynomial $\mathcal{M}_n = x_1 x_2 \cdots x_n$, consisting of just a single monomial, cannot be represented as a sum of less than $n/3$ ROPs of a particular form (weakly 0-justified). More recently, a similar hardness of representation result appeared in [Kay12]: if $\mathcal{M}_n$ is represented as a sum of powers of low-degree (at most d) polynomials, then the number of summands is $\exp(\Omega(n/d))$. As is implicit in [Kay12], such a hardness of representation statement can be used to give a PIT algorithm. We analyze this connection explicitly, and show that the results in [Kay12] lead to a deterministic sub-exponential time algorithm for black-box PIT for sums of powers of polynomials with appropriate size and degree (Section 4, Theorem 2).

A minor drawback of both these statements is that they consider a model that cannot even individually compute all monomials. One would expect any reasonable model of representing polynomials to be able to compute $\mathcal{M}_n$. In Section 5, we consider the restriction of read-once formulas to *constant-free* formulas that are only allowed leaf labels ax, where x is a variable and a is a field element. This model can compute any single monomial. We show (Theorem 3) that the elementary symmetric polynomial $\text{Sym}_{n,d}$ of degree d cannot be written as a sum of powers of such formulas unless the number of summands is $\Omega(\log(n/d))$. This appears weak compared to the $n/3$ bound from [SV08], but this is to be expected since unlike in [SV08] where the ROPs could only be added, we allow sums of powers. We also consider $0 - \text{justified}$ read-once formulas with alternation depth (between $+$ and $\times$) 3, and obtain a similar hardness-of-representation result for the polynomial $\mathcal{M}_n$ against sums of powers of polynomials computed by such formulas, showing that $n^{\frac{1}{2}-\epsilon}$ summands are needed (Theorem 4). Again, this appears weak compared to the $\exp(\Omega(n/d))$ bound from [Kay12], but unlike in [Kay12] where the degree of the inner functions is a parameter, our inner ROPs could have arbitrarily high degree.

2   Preliminaries

An arithmetic formula on n variables $X = \{x_1, \ldots, x_n\}$ is a rooted binary tree with leaves labeled from $\mathbb{F} \cup X$ and internal nodes labeled by $\circ \in \{+, \times\}$. Each node computes a polynomial in the obvious way, and the formula computes the polynomial computed at the root gate. An arithmetic formula is said to be read-once (ROF) if each $x \in X$ appears at most once at a leaf. Polynomials computed by ROFs are called read-once polynomials ROPs.

It is more convenient for us to allow leaf labels $ax + b$ for some $x \in X$ and some $a, b \in \mathbb{F}$. This does not change the class of polynomials computed, even when restricted to ROFs. Henceforth we assume that ROFs are of this form.

The alternation depth of the formula is the maximum number of maximal blocks of $+$ and $\times$ gates on any root-to-leaf path in the formula.

We say that an ROF is constant-free (denoted CF-ROF) if the labels at the leaves are of the form ax for $x \in X$ and $a \in \mathbb{F} \setminus \{0\}$. We call polynomials computed by such formulas constant-free ROPs, denoted CF-ROP.

For a polynomial $f \in \mathbb{F}[x_1, x_2, \cdots, x_n]$, a set $S \subseteq [n]$ and an assignment a, let $f_{S \to a_S}$ denote the polynomial on variables $\{x_i : i \notin S\}$ obtained from f by setting $x_j = a_j$ for $j \in S$. Using notation from [SV08], for a polynomial f, $\mathsf{var}(f)$ denotes the set of variables that f depends on non-trivially. We say that f is 0-justified if for all $S \subseteq \mathsf{var}(f)$, $\mathsf{var}(f|_{S \to 0}) = \mathsf{var}(f) \setminus S$. Equivalently, f is 0-justified if and only if $\forall x \in \mathsf{var}(f)$, the monomial x has a non-zero coefficient.

3   Identity Testing for $\sum^{(2)} \cdot \prod \cdot$ROPs over $\mathbb{Z}$ or $\mathbb{Q}$

In this section we show that PIT can be solved efficiently for formulas of the form $f_1 f_2 \cdots f_m + g_1 g_2 \cdots g_s$, where each f_i, g_j is an ROF over the field of rationals.

Theorem 1. *Given Read-Once Formulas computing each of the polynomials $f_1, f_2, \cdots, f_r, g_1, g_2, \ldots, g_s \in \mathbb{Q}[x_1, \ldots, x_n]$, checking if $f_1 \cdot f_2 \cdots f_r \equiv g_1 \cdot g_2 \cdots g_s$ can be done in deterministic polynomial time.*

A crucial ingredient in our proof is the following structural characterization from [RS11, RS13] and its constructive version; this is a direct consequence of the characterisation of ROPs given in [SV08].

Lemma 1 ([RS13], follows from [SV08]). *Let f be an ROP. Then exactly one of the following holds:*

1. *$k \geq 1$, there exist ROPs $f_1, \ldots, f_k$, with $\mathsf{var}(f_i) \cap \mathsf{var}(f_j) = \emptyset$ for all distinct $i, j \in [k]$, such that $f = a + f_1 + \cdots + f_k$, for some $a \in \mathbb{F}$, and each f_i is either uni-variate or decomposes into variable-disjoint factors.*
2. *$k \geq 2$, there exist ROPs $f_1, \ldots, f_k$, with $\mathsf{var}(f_i) \cap \mathsf{var}(f_j) = \emptyset$ for all distinct $i, j \in [k]$, such that $f = a \times f_1 \times f_2 \times \cdots \times f_k$ for some $a \in \mathbb{F} \setminus \{0\}$, and none of the f_is can be factorised into variable-disjoint factors.*

Furthermore, ROFs computing such f_is can be constructed from an ROF computing f in polynomial time.

Given an ROF over $\mathbb{Q}$, we can clear all denominators to get an ROF over $\mathbb{Z}$, without changing the status of the $? \equiv 0$? question. So we now assume that all the numbers a, b appearing in the ROF (recall, leaf labels are of the form $ax+b$) are integers. For a polynomial $p(X)$, let $\mathsf{content}(p(X))$ denote the greatest common divisor (gcd) of the non-zero coefficients of p. The next crucial ingredient in our proof is that for an ROF f, we can efficiently compute its content.

Lemma 2. *There is a polynomial-time algorithm that, given an ROF f in $\mathbb{Z}[X]$, computes $\mathsf{content}(f)$ and constructs an ROF f' in $\mathbb{Q}[X]$ such that $f = \mathsf{content}(f) \cdot f'$.*

Proof. It suffices to show how to compute $\mathsf{content}(f)$; then the ROF f' is just $\frac{1}{\mathsf{content}(f)} \times f$. We proceed bottom-up, or alternatively, we prove this by induction on the structure of f.

For a polynomial $p \in \mathbb{Z}[X]$, let $\hat{p} = p - p(0)$, where $p(0) = p(0, \ldots, 0)$, and let $\hat{p}'$ be the polynomial such that $\hat{p} = \mathsf{content}(\hat{p})\hat{p}'$.

If f is a single leaf node, then computing $\mathsf{content}(f)$ and $\mathsf{content}(\hat{f})$ is trivial. Otherwise, say $f = g \circ h$. Since f is an ROF, $\mathsf{var}(g) \cap \mathsf{var}(h) = \emptyset$.

Case $f = g + h$: Then $\hat{f} = \hat{g} + \hat{h}$, and $f(0) = g(0) + h(0)$. So

$$\mathsf{content}(f) := \gcd(\mathsf{content}(\hat{g}), \mathsf{content}(\hat{h}), g(0) + h(0)),$$
$$\mathsf{content}(\hat{f}) := \gcd(\mathsf{content}(\hat{g}), \mathsf{content}(\hat{h})).$$

Case $f = g \times h$: Then $\hat{f} = \hat{g}\hat{h} + h(0)\hat{g} + g(0)\hat{h}$, and $f(0) = g(0)h(0)$. We can show that

Claim. For any two variable-disjoint polynomials $p, q \in \mathbb{Z}[X]$, $\mathsf{content}(pq) = \mathsf{content}(p)\mathsf{content}(q)$.

Proof. Let $p = \mathsf{content}(p)(a_1 M_1 + a_2 M_2 + \cdots + a_k M_k)$ and $q = \mathsf{content}(q)(b_1 N_1 + b_2 N_2 + \cdots + b_\ell N_\ell)$, where M_i, N_j are monomials. By definition of content, $\gcd(\ldots, a_i, \ldots) = \gcd(\ldots, b_j, \ldots) = 1$. Since p and q are variable-disjoint, every monomial of the form $\mathsf{content}(p)\mathsf{content}(q)(a_i b_j M_i N_j)$ appears in the polynomial $p \times q$, and there are no other monomials. Hence $\mathsf{content}(p)\mathsf{content}(q) | \mathsf{content}(p \times q)$. For the converse, we need to show that $\gcd(S) = 1$, where $S = \{a_i b_j \mid i \in [k], j \in [\ell]\}$. Suppose not. Let c be the largest prime that divides all numbers in S. Then, $\forall i \in [k]$,

$$c | a_i b_1 \text{ and } c | a_i b_2 \text{ and } \ldots \text{ and } c | a_i b_k.$$
$$\text{Hence } c | a_i \text{ or } (c | b_1, c | b_2, \cdots, c | b_\ell).$$
$$\text{Hence } c | a_i \text{ or } c = 1, \text{ since } \gcd(b_1, \ldots, b_\ell) = 1.$$

Thus we conclude that c divides $\gcd(a_1, \ldots, a_k) = 1$, a contradiction. $\square$

Using this claim, we see that

$$\mathsf{content}(f) := \mathsf{content}(g) \times \mathsf{content}(h),$$
$$\mathsf{content}(\hat{f}) := \gcd(\mathsf{content}(\hat{g})\mathsf{content}(\hat{h}), h(0)\mathsf{content}(\hat{g}), g(0)\mathsf{content}(\hat{h})).$$

$\square$

Now we have all the ingredients for proving Theorem 1.

Proof (of Theorem 1). Let $f = f_1 \cdot f_2 \cdots f_r$ and $g = g_1 \cdot g_2 \cdots g_s$ As discussed above, without loss of generality, each f_i, g_i is in $\mathbb{Z}[X]$. Using Lemma 1 and 2, we can compute the irreducible variable-disjoint factors of each f_i and each g_i, and also pull out the content for each factor. That is, we express each f_i as $\alpha_i f_{i,1} \cdots f_{i,k_i}$, and each g_i as $\beta_i g_{i,1} \cdots g_{i,\ell_i}$ where the $f_{i,j}$s, $g_{i,j}$s are irreducible and have content 1. We obtain ROFs in $\mathbb{Q}[X]$ for each of the $f_{i,j}$s and $g_{i,j}$s. Note that if $\sum_i k_i \neq \sum_j \ell_j$, then there cannot be a component-wise matching between the factors of f and g, and hence we conclude $f \not\equiv g$. Otherwise, $\sum_i k_i = \sum_j \ell_j$. We now form multisets of the factors of f and of g, and we knock off equivalent factors one by one. (See Algorithm 1.) Detecting equivalent factors (the condition

---

Algorithm 1. Test if $\prod_{i=1}^{r} \alpha_i \prod_{j=1}^{k_i} f_{i,j} \equiv \prod_{i=1}^{s} \beta_i \prod_{j=1}^{\ell_i} g_{i,j}$

1: $S \leftarrow \{f_{1,1}, \cdots, f_{1,k_1}, f_{2,1}, \cdots, f_{2,k_2}, \ldots, f_{r,1}, \cdots, f_{r,k_r}\}$
2: $T \leftarrow \{g_{1,1}, \cdots, g_{1,\ell_1}, g_{2,1}, \cdots, g_{2,\ell_2}, \ldots, g_{s,1}, \cdots, g_{s,\ell_s}\}$
3: (Both S and T are multisets; repeated factors are retained with multiplicity.)
4: **for** $p \in S$ **do**
5:     **for** $q \in T$ **do**
6:         **if** $p \equiv q$ **then**
7:             **if** S and T have unequal number of copies of p and q **then**
8:                 Return No
9:             **else**
10:                 $S \leftarrow S \setminus \{p\}$. (Remove all copies).
11:                 $T \leftarrow T \setminus \{q\}$. (Remove all copies).
12:             **end if**
13:         **end if**
14:     **end for**
15: **end for**
16: **if** $(\alpha_1 \alpha_2 \cdots \alpha_r = \beta_1 \beta_2 \cdots \beta_s) \wedge (S = T = \emptyset)$ **then**
17:     Return Yes
18: **else**
19:     Return No
20: **end if**

---

in Step 6) requires an identity test $p \equiv q$?, or $p - q \equiv 0$?, for ROFs in $\mathbb{Q}[X]$. Since we have explicit ROFs computing p and q, this can be done using [SV08]. □

4   PIT for Sums of Powers of Low Degree Polynomials

In this section, we give a blackbox identity testing algorithm for multilinear sums of powers of low-degree polynomials.

We say that a polynomial f has a sum-powers representation of degree d and size s if there are polynomials f_i each of degree at most d, and a set of positive integers e_i, such that $f = f_1^{e_1} + \ldots + f_s^{e_s}$. In [Kay12], it is shown that computing the full multilinear monomial $\mathcal{M}_n = x_1 x_2 \cdots x_n$ using sums of powers of low-degree polynomials requires exponentially many summands:

Proposition 1. *[Kay12] There is a constant c such that for the polynomial $x_1 x_2 \cdots x_n$, any sum-powers representation of degree d requires size $s \geq 2^{\frac{cn}{d}}$.*

Shpilka and Volkovich [SV08] proved that sum of less than $n/3$ 0-justified ROPs cannot equal $\mathcal{M}_n$, and used it to obtain a black-box PIT algorithm for bounded sums of ROPs. Using these ideas along with Proposition 1, we note that such a hardness of representation for sums of powers of low-degree polynomials, where the final sum is multilinear, gives sub-exponential time algorithms for black-box PIT for this class.

Let $R = \{0, 1\} \subseteq \mathbb{F}$ be a finite set that contains 0. For any $k > 0$, define

$$W_k^n(R) \triangleq \{\boldsymbol{a} \in R^n \mid \boldsymbol{a} \text{ has at most } k \text{ non-zero coordinates}\}.$$

In Theorem 7.4 of [SV10], it is shown that for a certain kind of formula F (k-sum of degree-d 0-justified preprocessed ROP), and for any $R \subseteq \mathbb{F}$ containing 0 and of size at least $d + 1$, $F \equiv 0$ if and only if $F|_{W_{3k}^n(R)} \equiv 0$. The proof uses the Combinatorial Nullstellensatz [Alo99], see also Lemma 2.13 in [SV10]. We re-state it here for convenience:

Proposition 2 (Combinatorial Nullstellensatz, [Alo99]). *Let $P \in \mathbb{F}[x_1, \ldots, x_n]$ be a polynomial where for every $i \in [n]$, the degree of x_i is bounded by t. Let $R \subseteq \mathbb{F}$ have size at least $t + 1$, and $S = R^n$. Then $P \equiv 0 \Leftrightarrow P|_S \equiv 0$.*

Along similar lines, using Propositions 1,2, we show that

Lemma 3. *Let $C(n, s, d)$ be the class of all n-variate multilinear polynomials that have a sum-powers representation of degree d and size s. Let c be the constant from Proposition 1. For $f \in C(n, s, d)$, $R = \{0, 1\}$, and $k = (d \log s)/c$, $f|_{W_k^n(R)} \equiv 0 \iff f \equiv 0$.*

Proof. The $\Leftarrow$ direction in the claim is trivial. To prove the $\Rightarrow$ direction, we proceed by induction on n.

Base Case: $n \leq k$. Then $W_k^n(R) = R^n$. Using Proposition 2 (since f is multilinear, R is large enough), we conclude that $f \equiv 0$.

Induction Step: $n > k$. Suppose $f \not\equiv 0$. Consider any $i \in [n]$, and let $f' = f|_{x_i = 0}$. Then $f' \in C(n - 1, s, d)$. Since $f|_{W_k^n(R)} \equiv 0$, we have $f'|_{W_k^{n-1}(R)} \equiv 0$. So by the induction hypothesis, $f' \equiv 0$. Hence $x_i | f$. Since this holds for every $i \in [n]$, the monomial $x_1 \cdots x_n$ must divide f. Since f is multilinear, it must be that $f = x_1 \cdots x_n$. But $n > k = (d \log s)/c$, so $s < 2^{cn/d}$. This contradicts Proposition 1. Hence we conclude $f \equiv 0$. $\qquad\square$

This gives the required black-box PIT algorithm, since for our choice of k in the above lemma, $|W_k^n(\{0, 1\})| \in n^{O(k)} \in 2^{O(d \log s \log n)}$. Thus

Theorem 2. *Let $C(n, s, d)$ be the class of all n-variate multilinear polynomials that have a sum-powers representation of degree d and size s. There is a deterministic black-box PIT algorithm for $C(n, s, d)$ running in time $2^{O(d \log n \log s)}$.*

Remark 1. Though f is multilinear in Lemma 3 (and hence Theorem 2), the polynomials f_i in the sum-powers representation of f need not be multilinear.

5   Hardness of Representation for Sum of Powers of CF-ROPs

The hardness of representation result from [Kay12], stated in Proposition 1, and its precursor from [SV08],[SV10], are both for $\mathcal{M}_n$, the former using low-degree polynomials and the latter using a kind of ROPs called 0-justified ROPs. Note that ROPs, even when 0-justified, can have high degree, so these results are incomparable. Here we extend such a hardness result in two ways.

Our first hardness result is for elementary symmetric polynomials $\mathsf{Sym}_{n,d}$, not just for $d = n$. It works against another subclass of ROPs, CF-ROF; as is the case in [SV08, SV10], this class too can have high-degree polynomials. Recall that this class consists of polynomials computed by read-once formulas that have $+$ and $\times$ gates, and labels ax at leaves ($a \neq 0$). Hence for any f in this class, $f(0) = 0$. We show that powers of such polynomials cannot add up to elementary symmetric polynomials of arbitrary degree $d \leq n$ unless there are many such summands. First, we establish a useful property of this class.

Lemma 4. *For every* CF-ROP $f \in \mathbb{F}[x_1, \ldots, x_n]$, *there is a set* $S \subseteq [n]$ *with* $|S| \leq |\mathsf{var}(f)|/2$ *such that* $\deg(f|_{S \to 0}) \leq 1$.

Proof. Consider a CF-ROF F computing f. If F has a single node, then f is already linear, so $S = \emptyset$. Otherwise, $F = G_1 \circ G_2$, where G_1, G_2 are variable-disjoint CF-ROFs computing CF-ROPs g_1, g_2, respectively.
Case 1: $\circ = \times$. Without loss of generality, assume $|\mathsf{var}(g_1)| \leq |\mathsf{var}(f)|/2$. For $S = \{i : x_i \in \mathsf{var}(g_1)\}$, $g_1|_{S \to 0} \equiv f|_{S \to 0} \equiv 0$.
Case 2: $\circ = +$. Inductively, we can find sets S_i of at most half the variables of each g_i, such that $g_i|_{S_i \to 0}$ has degree at most 1. Define $S = S_1 \cup S_2$. Since G_1, G_2 are variable-disjoint, $|S| \leq |\mathsf{var}(f)|/2$, and $f|_{S \to 0}$ has degree at most 1. □

We use this to get our hardness-of-representation result for CF-ROPs, irrespective of degree.

Theorem 3. *Fix any* $d \in [n]$. *Suppose there are* CF-ROPs $f_1, \ldots, f_k$, *and positive integers* $e_1, \ldots, e_k$ *such that*

$$\sum_{i=1}^{k} f_i^{e_i} = \mathsf{Sym}_{n,d}.$$

Then $k \geq \min\{\log \frac{n}{d}, 2^{\Omega(d)}\}$.

Proof. Let $f = \mathsf{Sym}_{n,d}$.

We repeatedly apply Lemma 4 to restrictions of the f_i's obtain a formula of degree at most 1. Let $S_0 = T_0 = \emptyset$, and let S_{i+1} be the set obtained by applying the Lemma to $f_{i+1}|_{T_i \to 0}$, where each $T_i = S_1 \cup \ldots \cup S_i$. Define $S = T_k$. Since at least half the variables survive at each stage, we see that $r \triangleq |\mathsf{var}(f|_{S \to 0})| \geq |\mathsf{var}(f)|/2^k = n/2^k$.

  – If $r \geq d$, then $f|_{S \to 0} = \mathsf{Sym}_{r,d} \not\equiv 0$. Add any $r - d$ surviving variables to the set S to obtain the expression $\mathsf{Sym}_{d,d} = f|_{S \to 0} = \sum_{i=1}^{k} (f_i|_{S \to 0})^{e_i}$ where each f_i is either linear or identically 0. Let k' be the number of non-zero polynomials $f_i|_{S \to 0}$. By Proposition 1, $k' \in 2^{\Omega(d)}$, and $k \geq k'$.
  – If $r < d$, then $n/2^k \leq r < d$. So $k > \log(\frac{n}{d})$.

Thus if $k \leq \log \frac{n}{d}$, then $k \in 2^{\Omega(d)}$. $\square$

What this tells us is that there is a threshold $r \sim \log \log n$ such that any sum-powers representation of $\mathsf{Sym}_{n,d}$ using CF-ROPs needs size $2^{\Omega(d)}$ for $d \leq r$, and size $\geq \log \frac{n}{d}$ for $d \geq r$.

Our second hardness result is for $\mathcal{M}_n$, but works against a different class of ROFs. These ROFs may not be constant-free, but they have bounded alternation-depth, and are also 0-justified. Again, first we establish a useful property of the class.

Lemma 5. *Let $f \in \mathbb{F}[x_1, \ldots, x_n]$ be computed by an ROF with alternation depth 3. For any degree bound $1 \leq d \leq n$, there is an $S \subseteq [n]$ of size at most $|\mathsf{var}(f)|/d$, and an assignment of values A_S to the variables x_i for $i \in S$, such that $\deg(f|_{S \to A}) \leq d$. Moreover, if f is 0-justified, then we can find an A_S with all non-zero values.*

Proof. Let f be computed by the ROF F with alternation depth 3, where no gate computes the 0 polynomial.

If the top gate in F is a $+$, then $F = \sum_{i=1}^{r} f_i$, where each summand f_i is of the form $\prod_{j=1}^{t_i} \ell_{i,j}$ and the factors $\ell_{i,j}$'s are linear forms on disjoint variable sets. We find a partial assignment that kills all summand of degree more than d. For each such summand f_i, identify the factor with fewest variables, and assign values to the variables in it to make it 0. We assign values to at most $|\mathsf{var}(f_i)|/d$ variables, so overall no more than $|\mathsf{var}(f)|/d$ variables are set.

Further, if f is 0-justified and read-once, then each f_i is also a 0-justified ROF. Hence no factor of f_i vanishes at 0; each factor $\ell_{i,j}$ is of the form $\sum_{k=1}^{p} a_{i,j,k} x_{i,j,k} - c_{i,j}$ where $c_{i,j} \neq 0$. We can kill such a factor with an assignment avoiding 0s (eg set $x_{i,j,k} = c_{i,j}/pa_{i,j,k}$.)

If the top gate in F is a $\times$, then $F = \prod_{i=1}^{r} F_i$, where the F_i have alternation depth 2 and are on disjoint variables. If f has degree more than d, it suffices to kill any one factor F_i to make the polynomial 0. Choosing the factor with fewest variables, and proceeding as above, we set no more than $|\mathsf{var}(f)|/d$ variables. Again, since F is an ROF, if F is 0-justified, then so are the F_i. So A_S can be chosen avoiding 0s. $\square$

Using this, we get a hardness of representation result for 0-justified alternation-depth 3 ROPs.

Theorem 4. *Let $\epsilon \in (0, \frac{1}{2})$. If there are 0-justified, alternation-depth-3 ROPs $f_1, \ldots, f_s$, and non-negative integers $e_1, \ldots, e_s$ such that*

$$\sum_{i=1}^{s} f_i^{e_i} = x_1 \cdots x_n$$

then $s \geq n^{\frac{1}{2} - \epsilon}$.

Proof. Let d be a parameter to be chosen later. We identify a subset of variables S and an assignment A avoiding zeroes to variables of S, such that under this partial assignment, all the f_i's are reduced to degree at most d. We show that for any $d \in [n]$, this is possible with $|S| = t \le \frac{s^2 n}{d}$. This gives a sum-powers representation of degree d and size s for $\prod_{x_i \notin S} x_i = M_{n-t}$. Invoking Kayal's result from Proposition 1, we see that $s \ge 2^{c(n-t)/d}$, and hence $\log s + \frac{cns^2}{d^2} \ge \frac{cn}{d}$. Choosing $d = 4n^{1-2\epsilon}$, we conclude that $s \ge n^{\frac{1}{2}-\epsilon}$.

The construction of S proceeds in stages. At the kth stage, polynomials $f_1, \ldots, f_{i-1}$ have already been reduced to low-degree polynomials, and we consider f_i. We want to use Lemma 5 at each stage. This requires that each polynomial f_i, **after all the substitutions from the previous stages**, is still a 0-justified ROF with alternation-depth 3. The alternation-depth-3 ROF is obvious; it is only maintaining 0-justified that is a bit tricky. We describe the construction for stage 1; the other stages are similar.

Applying Lemma 5 to f_1 with d as the parameter, we obtain a set R_1 of variables with $|R_1| \le n/d$ and an assignment A_{R_1} avoiding 0, such that $\deg(f_1|_{R_1 \to A_{R_1}}) \le d$. It may be the case that for some $i > 1$, the polynomial $f_i|_{R_1 \to A_{R_1}}$ is no longer 0-justified. We fix this by augmenting R_1 as follows.

Assume first that the ROFs for all the f_i's have top-gate $+$; we will discuss top-gate $\times$ later. So, as discussed in the proof of Lemma 5, each f_i has the form $\sum \prod \ell_{j,k}$ where each $\ell_{j,k}$ is a linear form. If $f_i|_{R_1 \to A_{R_1}}$ is not 0-justified, then some of the linear forms in it are homogeneous linear (no constant term). We identify such linear forms in each f_i, $i \ge 2$. Call this set L_1. That is,

$$L_1 = \left\{ \ell \mid \begin{array}{l} \ell \text{ is a linear form at level-2 of some } f_i; \\ \ell|_{R_1 \to A_{R_1}} \text{ is homogeneous linear but not} \\ \text{identically 0.} \end{array} \right\}$$

Since each f_i is a ROF, it contributes at most $|R_1|$ linear forms to L_1. Hence $|L_1| \le (s-1)|R_1|$. Now pick a minimal set T_1 of variables from $X \setminus R_1$ that intersects each of the linear forms in L_1. By minimality, $|T| \le |L_1| \le (s-1)|R_1|$. We want to assign non-zero values A_{T_1} to variables in T_1 in such a way that for all $i \ge 2$, the $f_i|_{R_1 \to A_{R_1}; T_1 \to A_{T_1}}$ are 0-justified. We must ensure that the linear forms in L_1 become homogeneous (or vanish altogether), and we must also ensure that previously non-homogeneous forms do not become homogeneous. To achieve this, consider

$$L_2 = \left\{ \ell \mid \begin{array}{l} \ell \text{ is a linear form at level-2 of some } f_i; \\ \ell|_{R_1 \to A_{R_1}} \ne 0; \ell|_{R_1 \to A_{R_1}} \text{ contains a variable from } T_1. \end{array} \right\}$$

Clearly, $L_1 \subseteq L_2$. It suffices to find an assignment A_{T_1} to variables in T_1, avoiding zeroes, such that for each $\ell \in L_2$, either $\ell|_{R_1 \to A_{R_1}; T_1 \to A_{T_1}} \equiv 0$ or $\ell|_{R_1 \to A_{R_1}; T_1 \to A_{T_1}}(0) \ne 0$. For sufficiently large fields, such an assignment can always be found.

If some of the f_i's have top-gate $\times$, we need only a minor modification. We use this fact:

Observation 1. *If $F = \prod F_r$ is a read-once formula, then F is* 0-justified *if and only if for each r, F_r is* 0-justified *and satisfies $F_r(0) \neq 0$.*

Treat each factor of the polynomials with top-gate $\times$ exactly as we dealt with the other polynomials. Add their level-2 linear factors to L_1. Note that each such f_i can have many factors, but since it is read-once, any one variable can occur in at most one of these factors. So f_i still contributes no more than R_1 linear forms to L_1. Also modify the definition of L_2 to include also all linear forms at level 3 of such f_i's, containing a variable of T_1. Finally, look for an assignment also satisfying the additional condition that the factors do not vanish at 0. Again, over sufficiently large fields, it is possible to find such an assignment.

Now we set $S_1 = R_1 \cup T_1$, and $A_1 = A_{R_1} \cup A_{T_1}$. We have ensured the following:

1. $\deg(f_1|_{S_1 \to A_1}) \leq d$; and
2. for $i \geq 2$, $f_i|_{S_1 \to A_1}$ is 0-justified.

Furthermore, $|S_1| = |R_1| + |T_1| \leq |R_1|(1 + (s - 1)) \leq sn/d$.

Other stages are identical, working on the polynomials restricted by the already-chosen assignments. Finally, $S = S_1 \cup \ldots \cup S_s$, and so $|S| \leq s^2 n/d$, as required.   □

6   Further Questions

- Can the results of [SV08] be extended to the case $\sum_{i=1}^{k} f_i^{r_i}$, where f_i's are ROFs?
- Can a hardness of representation for $\mathsf{Sym}_{n,d}$ be transformed into a polynomial identity test for a related model?
- Can the bound given by Theorem 3 be improved? We conjecture:

  Conjecture 1. There is a constant $\epsilon > 0$ such that if there are CF-ROPs $f_1, \ldots, f_k$, and integers $e_1, \ldots e_k \geq 0$ satisfying

  $$\sum_{i=1}^{k} f_i^{e_i} = \mathsf{Sym}_{n,n/2},$$

  then $k = \Omega(n^\epsilon)$.

- Do the results of [AvMV11] extend to read-k-multilinear branching programs?

References

[Alo99]   Alon, N.: Combinatorial nullstellensatz. Combinatorics, Problem and Computing 8 (1999)

[AV08]   Agrawal, M., Vinay, V.: Arithmetic circuits: A chasm at depth four. In: FOCS, pp. 67–75 (2008)

[AvMV11] Anderson, M., van Melkebeek, D., Volkovich, I.: Derandomizing polynomial identity testing for multilinear constant-read formulae. In: CCC, pp. 273–282 (2011)

[BE11] Bläser, M., Engels, C.: Randomness efficient testing of sparse black box identities of unbounded degree over the reals. In: STACS, pp. 555–566 (2011)

[BHS08] Bläser, M., Hardt, M., Steurer, D.: Asymptotically optimal hitting sets against polynomials. In: Aceto, L., Damgård, I., Goldberg, L.A., Halldórsson, M.M., Ingólfsdóttir, A., Walukiewicz, I. (eds.) ICALP 2008, Part I. LNCS, vol. 5125, pp. 345–356. Springer, Heidelberg (2008)

[BOT88] Ben-Or, M., Tiwari, P.: A deterministic algorithm for sparse multivariate polynominal interpolation (extended abstract). In: STOC, pp. 301–309 (1988)

[DS07] Dvir, Z., Shpilka, A.: Locally decodable codes with two queries and polynomial identity testing for depth 3 circuits. SIAM J. Comput. 36(5), 1404–1434 (2007)

[FMM12] Fournier, H., Malod, G., Mengel, S.: Monomials in arithmetic circuits: Complete problems in the counting hierarchy. In: STACS, pp. 362–373 (2012)

[GKKS13] Gupta, A., Kamath, P., Kayal, N., Saptharishi, R.: Arithmetic circuits: A chasm at depth three. In: FOCS, pp. 578–587 (2013)

[Kay12] Kayal, N.: An exponential lower bound for the sum of powers of bounded degree polynomials. ECCC 19(TR12-081), 81 (2012)

[KI04] Kabanets, V., Impagliazzo, R.: Derandomizing polynomial identity tests means proving circuit lower bounds. Computational Complexity 13(1-2), 1–46 (2004)

[KS01] Klivans, A., Spielman, D.A.: Randomness efficient identity testing of multivariate polynomials. In: STOC, pp. 216–223 (2001)

[KS07] Kayal, N., Saxena, N.: Polynomial identity testing for depth 3 circuits. Computational Complexity 16(2), 115–138 (2007)

[MRS14] Mahajan, M., Raghavendra Rao, B.V., Sreenivasaiah, K.: Monomials, multilinearity and identity testing in simple read-restricted circuits. Theoretical Computer Science 524, 90–102 (2014), preliminary version in MFCS 2012

[RS11] Raghavendra Rao, B.V., Sarma, J.M.N.: Isomorphism testing of read-once functions and polynomials. In: FSTTCS, pp. 115–126 (2011)

[RS13] Raghavendra Rao, B.V., Sarma, J.M.N.: Isomorphism testing of read-once functions and polynomials (2013) (Submitted Manuscript)

[Sax14] Saxena, N.: Progress on polynomial identity testing - ii. CoRR, abs/1401.0976 (2014)

[Sch80] Schwartz, J.T.: Fast probabilistic algorithms for verification of polynomial identities. J. ACM 27(4), 701–717 (1980)

[SV08] Shpilka, A., Volkovich, I.: Read-once polynomial identity testing. STOC, pp. 507–516 (2010), See also ECCC TR-2010-011

[SV10] Shpilka, A., Volkovich, I.: Read-once polynomial identity testing. In: ECCC, p. 011 (2010), Preliminary version in STOC 2010

[SY10] Shpilka, A., Yehudayoff, A.: Arithmetic circuits: A survey of recent results and open questions. Found. Trends Theor. Comput. Sci. 5(3), 207–388 (2010)

[Zip79] Zippel, R.: Probabilistic algorithms for sparse polynomials. In: Ng, K.W. (ed.) EUROSAM 1979 and ISSAC 1979. LNCS, vol. 72, pp. 216–226. Springer, Heidelberg (1979)

Sampling from Dense Streams without Penalty:
Improved Bounds for Frequency Moments and Heavy Hitters

Vladimir Braverman* and Gregory Vorsanger

Department of Computer Science
Johns Hopkins University
Baltimore, Maryland, USA
vova@cs.jhu.edu, gregvorsanger@jhu.edu

Abstract. We investigate the ability to sample relatively small amounts of data from a stream and approximately calculate statistics on the original stream. McGregor et al. [29] provide worst case theoretical bounds that show space costs for sampling that are inversely correlated with the sampling rate. Indeed, while the lower bound of McGregor et al. cannot be improved in the general case, we show it is possible to improve the space bound for stream D of domain n, when the average positive frequency $\mu = F_1/F_0$ is sufficiently large. We consider the following range of parameters: $\mu \geq \log(n)$ and sample rate $p \geq C_k \mu^{-1} \log(n)$, where C_k is a constant. On these streams we improve the bound from $\tilde{O}(\frac{1}{p}n^{1-2/k})$ to $\tilde{O}(n^{1-2/k})$ thus giving polynomial improvement in space for sufficiently large μ and p^{-1}.

Keywords: Streaming Algorithms, Sampling, Frequency Moments, Heavy Hitters.

1 Introduction

An exciting topic of current algorithms research is evaluating the ability to sample relatively small amounts of data from a stream and to be able to approximately calculate statistics on the stream as a whole. In a recent paper [29], McGregor, Pavan, Tithrapura, and Woodruff provided worst case theoretical bounds that show space costs for sampling that are inversely correlated with the sampling rate. [1] This implies it is not possible to sample effectively on the stream without a cost tradeoff. However, experimental work has shown that sampling can be performed on the stream without sacrificing additional space for accuracy [30]. Let us define the following terms:

Definition 1. *Let m,n be positive integers. A **stream** $D = D(n,m)$ is a sequence of size m of integers $a_1, a_2, \ldots, a_m$, where $a_i \in \{1, \ldots, n\}$. A **frequency vector** is a vector of dimensionality n with non-negative entries $f_i, i \in [n]$ defined as:*

---

* This work was supported in part by DARPA grant N660001-1-2-4014. Its contents are solely the responsibility of the authors and do not represent the official view of DARPA or the Department of Defense.

[1] We have recently learned from an anonymous reviewer that these lower bounds may not hold. We stress that our techniques are independent of this result, and thus they hold regardless of the correctness of this work.

Z. Cai et al. (Eds.): COCOON 2014, LNCS 8591, pp. 13–24, 2014.
© Springer International Publishing Switzerland 2014

$$f_i = |\{j : 1 \leq j \leq m, a_j = i\}|$$

Definition 2. *A k-th frequency moment of a stream D is defined by $F_k(D) = \sum_{i \in [n]} f_i^k$. F_0 is the number of distinct elements in the stream and $F_\infty = max_{i \in [n]} f_i$.*

Definition 3. *A dense stream is any stream D s.t $F_1(D)/F_0(D) \geq \log(n)$*

While the lower bound of McGregor et al. cannot be improved in the general case, we show it is possible to improve the space bound for a stream D of domain n and length m, when the the average positive frequency $\mu = F_1/F_0$ is sufficiently large. Specifically, we consider the following range of parameters: $\mu \geq \log(n)$ and $p \geq C_k \mu^{-1} \log(n)$, where C_k is a constant (defined in (6)).

As our main technical claim, we show in Theorem 1 that the frequency moment on the sampled stream, D_p, is a $1 + \epsilon$ approximation for the frequency moment on the entire stream with high probability. As a result, we show the problem of computing F_k on D is reducible to the problem of computing F_k on D_p and the reduction preserves the space bounds up to a constant factor. In particular, the space bounds are independent of the sample rate, p. We stress that for our range of parameters the problem of approximating F_k is as hard as the problem of approximating F_k on the set of all streams. In this case, the lower bound from [15] still applies. However, the lower bound from [29] does not apply, as this bound is proven for streams with average frequency bounded by a constant. On these streams we improve the bound[2] from $\tilde{O}(\frac{1}{p} n^{1-2/k})$ to $\tilde{O}(n^{1-2/k})$ thus giving polynomial improvement in space for sufficiently large μ and p^{-1}. Additionally, we provide proof that the same result is applicable for finding heavy elements (heavy hitters) in the stream. Specifically, we show that heavy elements in the original stream are heavy elements in the sampled stream. Thus, techniques to analyze heavy elements are also unaffected by the sampling rate. We also describe several practical applications where streams have high average frequency.

1.1 Related Work

For many applications it is practical to consider sampling data instead of attempting to process the entire data set. This is especially true as data sets grow larger and larger. The concept of accurately calculating statistics using small portions of a stream is not new, and sampling algorithms in the streaming setting have been studied for a long time. Sampling algorithms in the streaming setting have been studied for a long time [2], [3], [12], [13,14], [16], [32].

Calculating frequency moments is one of the central problems for streaming algorithms, see, e.g., [1], [4], [5], [7,8], [10,11], [15], [17,18,19,20,21], [24,25], [27], and the references therein.

Further, computing frequency moments and other functions using sampling has been an intriguing question for a long time [9]. For example, Bar-Yossef showed[3] [3] that

---

[2] The $\tilde{O}$ notation suppresses factors polynomial to $\frac{1}{\epsilon}$ and factors logarithmic in m and n.

[3] While Bar-Yossef showed his results in a slightly different model the lower bounds are applicable for the sampled streams as well. See also Theorem 3.1 from [29].

the complexity of sampling from streams differs from the complexity of sketching by a polynomial factor in the worst case. Specifically computing F_2 is possible using $\tilde{O}(1)$ bits, but $\tilde{\Omega}(n^{0.5})$ samples are still needed.

In 2007 Bhattacharyya, Madeira, Muthukrishnan, and Ye [6] considered skipping certain portions of the stream and only examining every Nth item deterministically.

Following this, in [28] Problem 13, Matias asked about the effects of subsampling on the streaming data. His question addresses the issue of very fast streams, ones that cannot be analyzed effectively even if each element can be processed in O(1) time. In addition to asking questions regarding [6], he also asked about how subsampling effects the accuracy of standard calculations, such as frequency moments.

Recent work provided by McGregor, Pavan, Tirthapura, and Woodruff [29] considered sampling streams and addressed several fundamental problems, including frequency moments, heavy hitters, entropy, and distinct elements. In particular, they provide a matching (up to a polylogarithmic factor) upper and lower bound for the problem of frequency moments for $k > 2$ with lower bound of $\tilde{\Omega}(\frac{1}{p}n^{1-2/k})$. However, if we can observe the entire stream, then we can apply the well known upper bound of $\tilde{O}(n^{1-2/k})$ from Indyk and Woodruff [22]. Thus, the bound of [29] shows that it is not possible to obtain approximations without increasing the space required by a factor of p^{-1}, in the worst case.

However, in 2009, Rusu and Dobra [30] experimentally showed that when 10% of the original stream is sampled then the second frequency moment is still preserved. This provides intuition that there may be certain inputs that allow for an improvement over the bound from [29].

1.2 Relation to Existing Work on Lower Bounds

We now explain why the lower bound of $\tilde{\Omega}(\frac{1}{p}n^{1-2/k})$ does not apply to our analysis.

The lower bound in question only applies when $n = \Theta(m)$. Consider streams such that $F_0 = \Omega(n)$ and for all i either $f_i = 0$ or $f_i > n$. Clearly in this case $n = o(m)$ and thus the lower bound of [29] do not necessarily apply. Indeed, if we sample with probability $p = n^{-0.5}$ then, with high probability, all sampled frequencies will be in the range $[(1 - \epsilon)n^{-0.5}f_i, (1 + \epsilon)n^{-0.5}f_i]$ for constant ϵ and sufficiently large n. Thus, it is not hard to show that F_k on the entire stream can be approximated by computing the frequency moment on the sampled stream, $\tilde{F}_k$. In this paper we investigate the range of parameters for which sampled streams possess these properties.

Consider Theorem 4.33 from [3]. Let us consider the case when $k = 2$. To prove the lower bound of Bar-Yossef considers the following example. Either (1) the stream represents a frequency vector with all frequencies bounded by 1 or (2) the stream represents a frequency vector with all frequencies bounded by 1 and one frequency is $O(n^{1/2})$. Observe that in both cases the average non-zero frequency $\mu = O(1)$. Since we require $\mu = \Omega(C_k \log(n))$ the lower bound from [3] is not applicable directly to our range of parameters. Is it possible to increase μ by repeating the same element many times. However, the bound from [3] is for algorithms that are based solely on sampled data. In our model, we first sample and then we can apply an arbitrary algorithm, including the sketching algorithm for F_2 from [1]. In this case the lower bound on the number of samples from [3] becomes the lower bound on the length of the sampled stream D_p which is $\tilde{F}_1$.

In the same way consider Section 3.3 from [29]. The authors explicitly state that their bound is for the case when $m = \Theta(n)$. (It is important to note that the implicit (and standard) assumption in [29] is that $F_0 = \Theta(n)$. Otherwise, better bounds are possible, even on the original stream. E.g., if $F_0 = O(1)$ then we can compute any F_k precisely). In the proof of Theorem 3.3 in [29] the construction requires each element to be included at most once in the stream (except for one special element).

We give polynomial improvements over previous methods for the case when the non-zero average frequency is polynomial. Consider the stream where the average non-zero frequency is n^3. For sampling rate $p = \frac{1}{n^2}$ the bound [29] is $\tilde{\Omega}(min(n, \frac{1}{p}n^{1-2/k})) = \tilde{\Omega}(n)$. Our improvement for such streams can be as large as $\tilde{\Omega}(n^{2/k})$. Consider streams with the average frequency n^{ζ} where $0 < \zeta < 1$. If the sampling rate is $p = C_k \log(n)\frac{1}{n^{\zeta}}$ then our improvement is of order $\tilde{\Omega}(\frac{1}{p})$.

1.3 Results

We show in this paper that the space requirement bound in [29] can be improved on a sufficiently long stream, given input with specific characteristics such that the stream is a dense stream. Specifically, we improve these results for stream D of domain n, when the average frequency of all elements in D is greater than $C_k \log(n)$.

To the best of our knowledge, this is the first theoretical bound that shows strict improvement for sampling (no time/space trade-off) and thus gives justification for practical observations such as [30]. Note that our results do not contradict the lower bounds of [29]. In [29], the lower bound is given for the case when $F_1 = \Theta(F_0)$; this is not the case for the streams we analyze, and thus does not effect correctness of the upper bounds in this paper.

All of our results are applicable for the following range of parameters: $\mu \geq \log(n)$ and $p \geq C_k \mu^{-1} \log(n)$, where C_k is a constant defined in (6). Our contributions are:

- As our main technical claim, we show in Theorem 1 that the frequency moment on the sampled stream is a $1 + \epsilon$ approximation for the frequency moment on the entire stream with high probability.
- As a result, we show the problem of computing F_k on D is reducible to the problem of computing F_k on D_p and the reduction preserves the space bounds up to a constant factor. In particular, the space bounds are independent of the sample rate, p.
- We provide the bound of $\tilde{O}(n^{1-2/k})$ for $k > 2$. On our range of parameters we improve the bounds of [29] by a factor of $1/p$. In fact, our recent result [7] implies a bound of $O(n^{1-2/k})$ bits.
- We provide the bound of $\tilde{O}(1)$ for $1 \leq k < 2$ for F_k approximation. To the best of our knowledge this is the first theoretical bound for this range of k on sampled streams.
- We provide proof that our result is also applicable for finding heavy elements (heavy hitters) in a stream. See Section 4. To the best of our knowledge this is the first theoretical bound for heavy hitters in sampled streams.

– We give a concenatration bound on the sum of k-th powers of binomial random variables using inequalities for Sterling numbers of the 2nd kind, Bell numbers, and the Hölder Inequality.

It is important to note that the space lower bound $\tilde{\Omega}(n^{1-2/k})$ holds for streams with arbitrary large μ. To see this, consider a stream D with the average non-zero frequency smaller than some parameter t. Replace stream D with stream D', where every element of D is repeated exactly t times. In this case the average non-zero frequency in D' is increased exactly by factor of t. Since the μ is always at least one, we conclude that $\mu(D') \geq t$. It is not hard to see that the lower bound from [15] will be applicable for such D'. Thus, our restrictions do not make the problem of approximating F_k easier.

1.4   Intuition

Given a stream D, of length m with domain n, we assume that $m = \theta(n)$. However, as datasets get large, it is often the case that the expected frequency of a given element increases significantly. If this is the case then we can sample the stream without losing much precision (at least for the F_k approximation). As a result, we can improve the space bounds for frequency moments on sampled streams.

Our main claim is that $\tilde{F}_k$ is approximately $p^{-k}F_k$ if the *expected* frequency μ is sufficiently large and $p \geq \mu^{-1}\log(n)$. Specifically, we prove that the value of the frequency moment will be preserved (up to a multiplicative error) with high probability. It is easy to see[4] that the sampled frequency $\tilde{f}_i$ is a random variable with binomial distribution. Thus, the frequency moment on the sampled stream is $\tilde{F}_k = \sum_{i=1}^{n} \tilde{f}_i^k$ where $\tilde{f}_i \sim B(f_i, p)$.[5] Note that $\tilde{f}_i$s are independent but not identically distributed since the numbers of trials are different.[6] To obtain our result, we use the relation between the the moments of $\tilde{f}_i$, the Stirling numbers of the second kind and the Bell numbers.

Intuitively, when sampling datasets with large average frequency, we can divide all elements into one of three categories: A_1, the category of all elements with frequency greater than the sampling rate multiplied by an $O(\log(n))$ factor, A_2, elements with frequency greater than the sampling rate but less than A_1, and A_3, elements with frequency smaller than the sampling rate. With this, we can prove that the group of elements in A_1 dominates the frequency moment of a dense stream. In this paper, we prove that the contribution of the sampled frequencies from the second two groups is negligible, with high probability. This allows us to accurately estimate the frequency moment of the sampled stream using only elements in A_1. We also prove that the frequency of each element f_i in A_1 is preserved within $1 \pm \Theta(\epsilon)$, while sampling with rate $p \geq C_k\mu^{-1}\log(n)$, for sufficiently large constant C_k. Thus, $\tilde{F}_k$ is a $(1 \pm \epsilon)$-approximation of F_k, and we can accurately perform our computations on D_p instead of D.

---

[4] Similar observation has been made in [29].

[5] We denote $B(0, p)$ as the degenerate distribution concentrated at 0.

[6] A slightly different case is well studied, when $Y = \sum_{i=1}^{n} Y_i^k$ where $Y_i \sim B(n, p_i)$, i.e., the number of trials is the same, but success probabilities are different. See e.g., [23] for more details.

2   Definitions and Facts

The average positive frequency is defined as

$$\mu = \mu(D) = F_1/F_0. \tag{1}$$

Note that $\mu \geq 1$. Let us prove the following simple fact.

Fact 1. $\mu^k F_0 \leq F_k$

Proof. By Hölder inequality $F_1 \leq F_0^{1-1/k} F_k^{1/k}$. Thus, $\mu^k F_0 = (F_1/F_0)^k F_0 \leq F_k$.

Definition 4. *Given data stream* $D = \{a_1, a_2, \ldots, a_m\}$ *and a fixed real* $p \in (0, 1)$, *let* D_p *be a random sub-stream of* D *obtained as follows. Let* $Z_1, \ldots, Z_m$ *be independent random variables such that* $Z_i = a_i$ *with probability* p *and* $Z_i = -1$ *with probability* $(1-p)$. *Denote* D' *to be the sequence* $Z_1, \ldots, Z_m$. *Next let* D_p *be the subsequence of* D' *obtained by deleting all* -1s. *Define*[7]

$$\tilde{f}_i = frequency\ of\ i\ in\ D_p. \tag{2}$$

$$\tilde{F}_k = \sum_{i=1}^{n} \tilde{f}_i^k. \tag{3}$$

$B(N, p)$ is the binomial distribution with N trials and success probability p, where N is a positive integer and $p \in [0, 1]$. For completeness, define $B(0, p)$ to be the degenerate distribution concentrated at 0.

3   Frequency Moments on Sampled Streams

Define:

$$\alpha_k = 64(k/\epsilon)^2, \tag{4}$$

$$\beta_k = (k+1)B_k, \tag{5}$$

where B_k is the k-th Bell number (see [26] for the definition).

$$C_k = \epsilon^{-1/k}(10\beta_k)^{1/k}\alpha_k. \tag{6}$$

Consider stream D such that:

$$\mu \geq C_k \log(n). \tag{7}$$

Let p be such that:

$$1 \geq p \geq \mu^{-1}C_k \log(n). \tag{8}$$

Let $k > 1$ and ϵ be arbitrary constants. We now divide elements by frequency. Define:

$$S_1 = \{i : f_i \geq \alpha_k p^{-1} \log(n)\}, \tag{9}$$

---

[7] Note that we make "two passes" on D to define D_p but our algorithms will only need one pass on D_p.

$$S_2 = \{i : p^{-1} \le f_i < \alpha_k p^{-1} \log(n)\}, \tag{10}$$

$$S_3 = \{i : f_i < p^{-1}\}, \tag{11}$$

Denote random variables $X_j, j \in \{1, 2, 3\}$:

$$X_j = p^{-k} \sum_{i \in S_j} \tilde{f}_i^k. \tag{12}$$

Denote numbers $A_j, j \in \{1, 2, 3\}$:

$$A_j = \sum_{i \in S_j} f_i^k. \tag{13}$$

For completeness define $A_j = X_j = 0$ if $S_j = \emptyset$ for $j = 1, 2, 3$. It follows that $p^{-k} \tilde{F}_k = X_1 + X_2 + X_3$ and $F_k = A_1 + A_2 + A_3$. We will show that, with high probability: A_1 is very close to X_1, $A_3 + A_2$ is negligible in terms of F_k, and $X_2 + X_3$ is bounded by $c(A_3 + A_2)$ for some constant c. As a result, we will prove that $X_1 \approx p^{-k} \tilde{F}_k$ is a good approximation of F_k. Define

$$\gamma = \epsilon/2k \tag{14}$$

Fact 2. *For any i the following is true. If*

$$|p^{-1} \tilde{f}_i - f_i| \le \gamma f_i \tag{15}$$

then

$$|p^{-k} \tilde{f}_i^k - f_i^k| \le \epsilon f_i^k. \tag{16}$$

Lemma 1. $A_2 + A_3 \le 0.1 \beta_k^{-1} \epsilon F_k < \epsilon F_k$.

Proof. Recall that $i \in (S_2 \cup S_3)$ implies $f_i < \alpha_k p^{-1} \log(n)$. Thus,

$$A_2 + A_3 = \sum_{i \in S_2 \cup S_3} f_i^k \le (\alpha_k p^{-1} \log(n))^k F_0. \tag{17}$$

Recall that $p \ge C_k \mu^{-1} \log(n)$. Thus,

$$A_2 + A_3 \le F_0 (\alpha_k C_k^{-1} \mu)^k. \tag{18}$$

Equation (6) yields

$$A_2 + A_3 \le F_k (\alpha_k C_k^{-1})^k. \tag{19}$$

The first inequality of the lemma follows from the definition (6) of C_k. The second inequality follows since $\beta_k > 1$.

Lemma 2. *Let $X \sim B(N, p)$. There exists a constant β_k that depends only on k and such that if $Np \ge 1$ then*

$$E(X^k) \le \beta_k (Np)^k, \tag{20}$$

and if $Np < 1$ then

$$E(X^k) \le \beta_k. \tag{21}$$

Due to a lack of space, we omit the proof. The proof is included in the full version of the paper.

Lemma 3. $P(X_2 \geq \epsilon F_k) \leq 0.1$

Proof. To bound X_2 we observe that $\tilde{f}_i \sim B(f_i, p)$. Also $i \in S_2$ implies that $1/p \leq f_i$. Thus, we can apply Lemma 2. In particular, the case (20) gives:

$$E(\tilde{f}_i^k) \leq \beta_k (f_i p)^k, \tag{22}$$

which in turn gives

$$E(X_2) = \frac{1}{p^k} \sum_{i \in S_2} E(\tilde{f}_i^k) \leq \beta_k \sum_{i \in S_2} f_i^k = \beta_k A_2. \tag{23}$$

Combining (23) with Lemma (1) we obtain $E(X_2) \leq 0.1\epsilon F_k$. Note that X_2 is non-negative. Thus, the lemma follows from Markov inequality.

Lemma 4. $P(X_3 \geq \epsilon F_k) \leq 0.1$

Proof. To bound X_3 we observe that $i \in S_3$ implies $1/p > f_i$. Thus we can apply Lemma 2. In particular (21) gives us:

$$E(X_3) = \frac{1}{p^k} \sum_{i \in S_3} E(\tilde{f}_i^k) \leq \frac{1}{p^k} \beta_k F_0. \tag{24}$$

Recall that $p \geq C_k \mu^{-1} \log(n)$ (see (8)). Thus, Fact 1 gives us:

$$E(X_3) \leq \frac{1}{p^k} \beta_k F_0 \leq \frac{C_k^{-k} \beta_k \mu^k F_0}{\log^k(n)} \leq 0.1\epsilon F_k. \tag{25}$$

The lemma follows.

Lemma 5. *If $i \in \{2, 3\}$ and $|X_i - A_i| > \epsilon F_k$ then $X_i > \epsilon F_k$.*

Proof. If $|X_i - A_i| > \epsilon F_k$ then either

$$X_i > A_i + \epsilon F_k \tag{26}$$

or

$$X_i < A_i - \epsilon F_k. \tag{27}$$

Note that $0 \leq A_i < \epsilon F_k$ (by the definition and Lemma 1) and $X_i \geq 0$ (by the definition). Thus (27) is not possible and (26) implies $X_i > \epsilon F_k$.

Lemma 6. $P(|X_1 - A_1| > \epsilon F_k) \leq 0.1$.

Proof. Note that if $S_1 = \emptyset$ then $X_1 = A_1 = 0$ and thus the lemma is correct. Otherwise, let $i \in S_1$ be fixed. First, we will show that

$$P(|\tilde{f}_i - pf_i| > \gamma pf_i) \leq \frac{1}{10n}. \tag{28}$$

Indeed, $\tilde{f}_i = \sum_{j=1}^{f_i} Y_{i,j}$ where $Y_{i,j}$ are i.i.d. indicators with mean p. Thus $E(\tilde{f}_i) = pf_i$, and by Chernoff bound (see e.g., [31], B.2) we have:

$$P(|\tilde{f}_i - pf_i|) > \gamma pf_i) \leq 2e^{(-\gamma^2 pf_i)/4}. \tag{29}$$

Direct computations and the definitions (4) and (14) imply that $\gamma^2 \alpha_k = 16$. Since $i \in S_1$, it follows that $f_i \geq p^{-1}\alpha_k \log(n)$. Thus, $\gamma^2 pf_i \geq \gamma^2 \alpha_k \log(n) = 16 \log(n)$. Substituting this bound into (29) we obtain (for sufficiently large n):

$$P(|\tilde{f}_i - pf_i|) > \gamma pf_i) \leq 2e^{-4\log(n)} \leq \frac{1}{10n},$$

and thus (28) holds. Further, Fact 2 and (28) imply

$$P(|p^{-k}\tilde{f}_i^k - f_i^k| > \epsilon f_i^k) \leq \frac{1}{10n}. \tag{30}$$

If we apply (30) to every $i \in S_1$ and use the union bound and the fact that $|S_1| \leq n$ then the lemma follows immediately. Indeed,

$$P(|X_1 - A_1| > \epsilon F_k) \leq P(|X_1 - A_1| > \epsilon A_1) = \tag{31}$$

$$P(|\sum_{i \in S_1} p^{-k}\tilde{f}_i^k - \sum_{i \in S_1} f_i^k| > \epsilon(\sum_{i \in S_1} f_i^k)) \leq$$

$$P(\cup_{i \in S_1}(|p^{-k}\tilde{f}_i^k - f_i^k| > \epsilon f_i^k)) \leq \sum_{i \in S_1} P(|p^{-k}\tilde{f}_i^k - f_i^k| > \epsilon f_i^k) \leq 0.1.$$

Theorem 1. *Let D be a stream such that $\mu = \mu(D) \geq C_k \log(n)$ and let p be a number such that $1 \geq p \geq \mu^{-1}C_k \log(n)$. Let D_p be the sampled stream (see Definition 4). Let $k > 1$ and ϵ be constants. Then the following bound holds for sufficiently large n.*

$$P(|\tilde{F}_k - F_k| > 3\epsilon F_k) \leq 0.3.$$

Proof. Indeed,
$$P(|\tilde{F}_k - F_k| > 3\epsilon F_k) \leq \tag{32}$$

$$P(|X_1 - A_1| > \epsilon F_k) + P(|X_2 - A_2| > \epsilon F_k) + P(|X_3 - A_3| > \epsilon F_k).$$

Applying Lemma 5 we obtain:

$$P(|\tilde{F}_k - F_k| > 3\epsilon F_k) \leq \tag{33}$$

$$P(|X_1 - A_1| > \epsilon F_k) + P(X_2 > \epsilon F_k) + P(X_3 > \epsilon F_k).$$

The theorem follows from the union bound and Lemmas 6, 4, 3.

Theorem 2. *Let D be a stream such that $\mu = \mu(D) \geq C_k \log(n)$ and let p be a number such that $1 \geq p \geq \mu^{-1} C_k \log(n)$. Let D_p be the sampled stream. Let $k > 1$ and ϵ be constants. Then it is possible to output the $(1 \pm \epsilon)$-approximation of F_k by making a single pass over D_p and computing $\tilde{F}_k$. Thus, the problem of computing F_k on D is reducible to the problem of computing F_k on D_p and the reduction preserves the space bounds. In particular, the space bounds are independent of p. Current best bounds for F_k include:*

1. *$\tilde{O}(n^{1-2/k})$ memory bits for $k > 2$.*
2. *$\tilde{O}(1)$ memory bits for $1 \leq k < 2$.*

4   Finding Heavy Elements

Definition 5. *Let D be a stream and ρ be a parameter. The index $i \in [n]$ is a ρ-heavy element if $f_i^k \geq \rho F_k$.*

In this section, we show that a heavy element in the original stream remains a heavy element in the sampled stream, and therefore we can apply existing techniques for heavy hitters. The frequency of the found heavy element is $(1 \pm \epsilon) p f_i$, with high probability, by Chernoff bound.

Theorem 3. *Let D be a stream and i be a heavy element w.r.t. F_k on D. Let $k \geq 1$ and let $p \geq \mu^{-1} = F_0/F_1$. Then there exists a constant c_k such that with a constant probability, i is a c_k-heavy element w.r.t. F_k on D_p.*

Proof. By Chernoff bound, the frequency of i in D_p is at least $(1-\epsilon) p f_i$ with high probability. By Fact 3, the k-th frequency moment of D_p is bounded by $\alpha_k \mu^{-k} \sum_{i=1}^{n} v_i^k$. Thus, i is a heavy element.

Fact 3. *Let $V \in (Z^+)^n$ be a vector with strictly positive integer entries v_i. Let $\mu = \frac{1}{n} \sum_{i=1}^{n} v_i$. Note that $\mu \geq 1$. Let $X_i \sim B(v_i, \mu^{-1})$ and $X = \sum_{i=1}^{n} X_i^k$. Then there exists a constant α_k that depends only on k such that $P(X > \alpha_k \mu^{-k} \sum_{i=1}^{n} v_i^k) < 0.1$.*

Proof. By Lemma 2

$$E(X_i^k) \leq \beta_k((\mu^{-1} v_i)^k + 1)$$

Thus,

$$E(X) < \beta_k(\mu^{-k} \sum_{i=1}^{n} v_i^k) + \beta_k n.$$

Also, by the Hölder inequality

$$\frac{\sum_{i=1}^{n} v_i}{n^{1-1/k}} \leq (\sum_{i=1}^{n} v_i^k)^{1/k}$$

Thus,

$$n^{1/k} = \mu^{-1} \frac{\sum_{i=1}^{n} v_i}{n^{1-1/k}} \leq \mu^{-1} (\sum_{i=1}^{n} v_i^k)^{1/k}$$

Finally, $n < (\mu^{-k} \sum_{i=1}^{n} v_i^k)$. We conclude the proof by putting $\alpha_k = 200 \beta_k$ and applying Markov's inequality.

5   Discussion, Open Questions, and Appendix

Due to space constraints, we have omitted the discussion, open questions, and appendix. These are available in the full version of our paper on arXiv.

References

1. Alon, N., Matias, Y., Szegedy, M.: The space complexity of approximating the frequency moments. J. Comput. Syst. Sci. 58(1), 137–147 (1999)
2. Babcock, B., Datar, M., Motwani, R.: Sampling from a moving window over streaming data. In: SODA, pp. 633–634 (2002)
3. Bar-Yossef, Z.: The complexity of massive data set computations. PhD thesis, Berkeley, CA, USA, AAI3183783 (2002)
4. Bar-Yossef, Z., Jayram, T.S., Kumar, R., Sivakumar, D.: An information statistics approach to data stream and communication complexity. J. Comput. Syst. Sci. 68(4), 702–732 (2004)
5. Bar-Yossef, Z., Jayram, T.S., Kumar, R., Sivakumar, D., Trevisan, L.: Counting distinct elements in a data stream. In: Rolim, J.D.P., Vadhan, S.P. (eds.) RANDOM 2002. LNCS, vol. 2483, pp. 1–10. Springer, Heidelberg (2002)
6. Bhattacharyya, S., Madeira, A., Muthukrishnan, S., Ye, T.: How to scalably and accurately skip past streams. In: Proceedings of the 2007 IEEE 23rd International Conference on Data Engineering Workshop, ICDEW 2007, pp. 654–663. IEEE Computer Society, Washington, DC (2007)
7. Braverman, V., Katzman, J., Seidell, C., Vorsanger, G.: Approximating large frequency moments with $o(n^{1-2/k})$ bits. CoRR, abs/1401.1763 (2014)
8. Braverman, V., Ostrovsky, R.: Smooth histograms for sliding windows. In: Proceedings of the 48th Annual IEEE Symposium on Foundations of Computer Science, FOCS 2007, pp. 283–293. IEEE Computer Society, Washington, DC (2007)
9. Braverman, V., Ostrovsky, R.: Zero-one frequency laws. In: Proceedings of the 42nd ACM Symposium on Theory of Computing, STOC 2010, pp. 281–290. ACM, New York (2010)
10. Braverman, V., Ostrovsky, R.: Approximating large frequency moments with pick-and-drop sampling. In: Raghavendra, P., Raskhodnikova, S., Jansen, K., Rolim, J.D.P. (eds.) APPROX/RANDOM 2013. LNCS, vol. 8096, pp. 42–57. Springer, Heidelberg (2013)
11. Braverman, V., Ostrovsky, R.: Generalizing the layering method of Indyk and Woodruff: Recursive sketches for frequency-based vectors on streams. In: Raghavendra, P., Raskhodnikova, S., Jansen, K., Rolim, J.D.P. (eds.) APPROX/RANDOM 2013. LNCS, vol. 8096, pp. 58–70. Springer, Heidelberg (2013)
12. Braverman, V., Ostrovsky, R., Vilenchik, D.: How hard is counting triangles in the streaming model? In: Fomin, F.V., Freivalds, R., Kwiatkowska, M., Peleg, D. (eds.) ICALP 2013, Part I. LNCS, vol. 7965, pp. 244–254. Springer, Heidelberg (2013)
13. Braverman, V., Ostrovsky, R., Vorsanger, G.: Weighted sampling without replacement from data streams (2013) (submitted)
14. Braverman, V., Ostrovsky, R., Zaniolo, C.: Optimal sampling from sliding windows. In: PODS, pp. 147–156 (2009)
15. Chakrabarti, A., Khot, S., Sun, X.: Near-optimal lower bounds on the multi-party communication complexity of set disjointness. In: IEEE Conference on Computational Complexity, pp. 107–117 (2003)
16. Chaudhuri, S., Motwani, R., Narasayya, V.: On random sampling over joins. In: Proceedings of the 1999 ACM SIGMOD International Conference on Management of Data, SIGMOD 1999, pp. 263–274. ACM, New York (1999)

17. Coppersmith, D., Kumar, R.: An improved data stream algorithm for frequency moments. In: SODA, pp. 151–156 (2004)
18. Cormode, G., Datar, M., Indyk, P., Muthukrishnan, S.: Comparing data streams using hamming norms (how to zero in). IEEE Trans. on Knowl. and Data Eng. 15(3), 529–540 (2003)
19. Feigenbaum, J., Kannan, S., Strauss, M., Viswanathan, M.: An approximate l1-difference algorithm for massive data streams. In: FOCS 1999: Proceedings of the 40th Annual Symposium on Foundations of Computer Science, FOCS 1999, p. 501. IEEE Computer Society, Washington, DC (1999)
20. Ganguly, S.: Estimating frequency moments of data streams using random linear combinations. In: Jansen, K., Khanna, S., Rolim, J.D.P., Ron, D. (eds.) APPROX and RANDOM 2004. LNCS, vol. 3122, pp. 369–380. Springer, Heidelberg (2004)
21. Ganguly, S., Cormode, G.: On estimating frequency moments of data streams. In: Charikar, M., Jansen, K., Reingold, O., Rolim, J.D.P. (eds.) APPROX and RANDOM. LNCS, vol. 4627, pp. 479–493. Springer, Heidelberg (2007)
22. Indyk, P., Woodruff, D.: Optimal approximations of the frequency moments of data streams. In: STOC 2005: Proceedings of the Thirty-Seventh Annual ACM Symposium on Theory of Computing, pp. 202–208. ACM, New York (2005)
23. Johnson, N.L., Kemp, A.W., Kotz, S.: Univariate discrete distributions. Wiley-Interscience (2005)
24. Kane, D.M., Nelson, J., Woodruff, D.P.: On the exact space complexity of sketching and streaming small norms. In: Proceedings of the 21st Annual ACM-SIAM Symposium on Discrete Algorithms, SODA 2010 (2010)
25. Kane, D.M., Nelson, J., Woodruff, D.P.: An optimal algorithm for the distinct elements problem. In: PODS 2010: Proceedings of the Twenty-ninth ACM SIGMOD-SIGACT-SIGART Symposium on Principles of Database Systems of Data, pp. 41–52. ACM, New York (2010)
26. Knuth, D.E.: The art of computer programming, fundamental algorithms, 3rd edn., vol. 1. Addison Wesley Longman Publishing Co., Inc., Redwood City (1997)
27. Li, P.: Compressed counting. In: SODA 2009: Proceedings of the Nineteenth Annual ACM -SIAM Symposium on Discrete Algorithms, pp. 412–421. Society for Industrial and Applied Mathematics, Philadelphia (2009)
28. McGregor, A.: Open problems in data streams and related topics. In: IITK Workshop on Algorithms for Data Streams (2006), http://www.cse.iitk.ac.in/users/sganguly/data-stream-probs.pdf (2007)
29. McGregor, A., Pavan, A., Tirthapura, S., Woodruff, D.: Space-efficient estimation of statistics over sub-sampled streams. In: Proceedings of the 31st Symposium on Principles of Database Systems, PODS 2012, pp. 273–282. ACM, New York (2012)
30. Rusu, F., Dobra, A.: Sketching sampled data streams. In: Proceedings of the 2009 IEEE International Conference on Data Engineering, ICDE 2009, pp. 381–392. IEEE Computer Society, Washington, DC (2009)
31. Vazirani, V.V.: Approximation algorithms. Springer-Verlag New York, Inc., New York (2001)
32. Vitter, J.S.: ACM Transactions on Mathematical Software, 11(1), 37–57

L_∞-Discrepancy Analysis
of Polynomial-Time Deterministic Samplers
Emulating Rapidly Mixing Chains*

Takeharu Shiraga, Yukiko Yamauchi, Shuji Kijima, and Masafumi Yamashita

Graduate School of Information Science and Electrical Engineering,
Kyushu University, Japan
{takeharu.shiraga,yamauchi,kijima,mak}@inf.kyushu-u.ac.jp

Abstract. *Markov chain Monte Carlo* (*MCMC*) is a standard technique
to sample from a target distribution by simulating Markov chains. In an
analogous fashion to MCMC, this paper proposes a *deterministic sampling* algorithm based on *deterministic random walk*, such as the rotor-
router model (a.k.a. Propp machine). For the algorithm, we give an upper
bound of the point-wise distance (i.e., infinity norm) between the "distributions" of a deterministic random walk and its corresponding Markov
chain in terms of the *mixing time* of the Markov chain. As a result, for
uniform sampling of #P-complete problems, such as 0-1 knapsack solutions, linear extensions, matchings, etc., for which rapidly mixing chains
are known, our deterministic algorithm provides samples with a "distribution" with a point-wise distance at most ε from the target distribution,
in time polynomial in the input size and ε^{-1}.

Keywords: rotor-router model, #P-complete, Markov chain Monte
Carlo, mixing time.

1 Introduction

Motivated by a new general scheme for a derandomization of randomized algorithms, this paper proposes a *deterministic sampling* algorithm, that is a deterministic algorithm to provide samples. Our approach is an analogy of *Markov
chain Monte Carlo* (*MCMC*), and uses the idea of *deterministic random walk*.

Background: Sampling and approximate counting. Counting is a fundamental
topic in Combinatorics, and it is highly related to sampling, a fundamental topic
in Probability Theory. #P, a computational class of the counting version of NP,
is an important class of polynomial-time complexity theory. Several counting
problems are known to be #P-complete.

A number of randomized approximate counting (cf. FPRAS[1]) based on MCMC
sampling are devised for #P complete problems, such as knapsack solutions [17],

---

* This manuscript is an extended abstract version of [19].
[1] An algorithm (for counting, for simplicity) is called *fully polynomial-time approximation scheme* (FPRAS) if the output approximate value $Z \in \mathbb{R}$ for the exact number
$A \in \mathbb{Z}$ satisfies that $\Pr(|Z - A|/A \le \varepsilon) \ge 1 - \delta$ for any $\varepsilon \in (0, 1)$ and $\delta \in (0, 1)$, and
the algorithm terminates in time polynomial in the input size, ε^{-1} and $\log(\delta^{-1})$.

Z. Cai et al. (Eds.): COCOON 2014, LNCS 8591, pp. 25–36, 2014.
© Springer International Publishing Switzerland 2014

linear extensions [14,3], matchings [12,13], etc. (see Section 5). The idea of MCMC is simple; design an *ergodic* Markov chain with a desired limit distribution, and sample from the limit distribution simulating the chain (see Section 2). It is easy to design Markov chain with a desired limit distribution based on *reversible* chains, and a major issue is *mixing time* of chains; "How long shall we simulate the chain to obtain samples from an approximately limit distribution?" Several techniques are developed for estimating mixing times concerning total variation distance or relative point-wise distance (see e.g., [20,16]).

Recently, *deterministic* approximation for #P-hard problems is a major challenge. For instance, deterministic approximation algorithm based on the dynamic programing was proposed for counting knapsack solutions [9].

Deterministic random walk. The *rotor-router model*, also known as the *Propp machine*, is a deterministic process analogous to random walk on a graph [6,15]. Instead of distributing tokens to randomly chosen neighbors, the rotor-router model deterministically serves the neighbors in a fixed order by associating to each vertex a "rotor-router" pointing to one of its neighbors. Doerr et al. [5,7] first called the rotor-router model *deterministic random walk*, meaning a "derandomized, hence *deterministic*, version of a *random walk*."

The first remarkable result on the rotor-router model is due to Cooper and Spencer [6]. They are concerned with the model of multiple tokens (*multiple-walk*) on $\mathbb{Z}^n$, and investigated the discrepancy on a single vertex: they gave a bound that $|\chi_v^{(t)} - \mu_v^{(t)}| \leq c_n$ where $\chi_v^{(t)}$ (resp. $\mu_v^{(t)}$) denotes the number (resp. expected number) of tokens on vertex $v \in \mathbb{Z}^n$ in a rotor-router model (resp. corresponding random walk) at time t on the condition that $\mu_v^{(0)} = \chi_v^{(0)}$ for any v, and c_n is a constant depending only on n but independent of the total number of tokens in the system. Cooper et al. [5] showed $c_1 \simeq 2.29$, and Doerr and Friedrich [7] showed that c_2 is 7.29 or 7.83 depending on the routing rules. On the other hand, Cooper et al. [4] gave an example of $|\chi_v^{(t)} - \mu_v^{(t)}| = \Omega(\sqrt{kt})$ on infinite k-regular trees, the example implies that the discrepancy can get infinitely large as increasing the total number of tokens.

Motivated by a derandomization of Markov chains, Kijima et al. [15] are concerned with the multiple-walks on general finite *multidigraphs* $(V, \mathcal{A})$, and gave a bound $|\chi_v^{(t)} - \mu_v^{(t)}| = O(|V||\mathcal{A}|)$ in case that corresponding Markov chain is ergodic, reversible and lazy. They also gave some examples of $|\chi_v^{(t)} - \mu_v^{(t)}| = \Omega(|\mathcal{A}|)$. In the context of load balancing, Rabani et al. [18] are concerned with a deterministic algorithm similar to the rotor-router model corresponding to Markov chains with *symmetric* transition matrices, and gave a bound $O(\Delta \log(|V|)/(1 - \lambda_*))$ where Δ denotes the maximum degree of the transition diagram and λ_* denotes the second largest eigenvalue of the transition matrix.

For some specific finite graphs, namely hypercubes and tori, some bounds in terms of logarithm of the size of transition diagram are known [15,8,1]. For instance, Akbari and Berenbrink [1] gave a bound $O(n^{1.5})$ for n-dimensional hypercube. Those analyses highly depend on the structures of the specific graphs, and it is difficult to extend the technique to other combinatorial graphs.

Kijima et al. [15] gave rise to a question if there is a deterministic random walk for #P-complete problems, such as 0-1 knapsack solutions, bipartite matchings, etc., such that $|\chi_v^{(t)} - \mu_v^{(t)}|$ is bounded by a polynomial in the input size.

There are a number of results related to deterministic random walk. Here, we briefly refer some of them. Holroyd and Propp [10] analyzed "hitting time" of the rotor-router machine with a *single token* (*single-walk*) on finite simple graphs, and gave a bound $|\nu_v^{(t)} - t\pi_v| = O(|V||\mathcal{A}|)$ where $\nu_v^{(t)}$ denotes the frequency of visits of the token at vertex v in t steps, and π denotes the stationary distribution of the corresponding random walk. Holroyd and Propp [10] also proposed a generalized model called *stack walk*, which is the first model of deterministic random walk for irrational transition probabilities, as far as we know. While Holroyd and Propp [10] showed the existence of routers which approximates irrational transition probabilities well, Angel et al. [2] gave a simple routing algorithm, which serves tokens in a greedy manner based on the "shortest remaining time" strategy.

Our Results. This paper proposes a deterministic algorithm for sampling from a finite set $V = \{1, \ldots, N\}$. Our algorithm is based on a version of *deterministic random walk* which emulates a Markov chain with a transition matrix P. In the algorithm, a configuration of M tokens over V is deterministically updated; let $\chi^{(t)} = (\chi_1^{(t)}, \ldots, \chi_N^{(t)}) \in \mathbb{Z}_{\geq 0}^N$ denote the configuration at time $t = 0, 1, 2, \ldots$, i.e., $\chi_v^{(t)}$ denotes the number of tokens on $v \in V$, and hence $\sum_{v \in V} \chi_v^{(t)} = M$. For comparison, let $\mu^{(0)} = \chi^{(0)}$, and let $\mu^{(t)} = \mu^{(0)} P^t$, then $\mu^{(t)} \in \mathbb{R}_{\geq 0}^N$ denotes the expected configuration of M tokens independently according to P for t steps. A main contribution of the paper is to show that $|\chi_v^{(t)} - \mu_v^{(t)}| \leq 3(\pi_{\max}/\pi_{\min})t^*\Delta$ holds for any $v \in V$ at any time t in case that P is *ergodic* and *reversible*, where $\pi_{\max}$ and $\pi_{\min}$ are maximum/minimum values of π respectively, t^* is the *mixing rate* of the corresponding Markov chain, and Δ is the maximum degree of the transition diagram.

This result suggests polynomial-time deterministic algorithms (with polynomial space) for uniform sampling for #P-complete problems, such as knapsack solutions, linear extensions, matchings, etc., for which rapidly mixing chains exist. Thus, our result affirmatively answers the question by Kijima et al. [15]. Setting the number of tokens $M \geq 3\varepsilon^{-1}t^*\Delta$ for an arbitrary ε ($0 < \varepsilon < 1$), our algorithm provides M samples with a "distribution" $\widetilde{\chi}^{(t)} := \chi^{(t)}/M$, of which the *point-wise distance* $\|\widetilde{\chi}^{(t)} - \pi\|_\infty$ is at most ε from the uniform distribution π over the target set. For instance, our algorithm runs in $O^*(n^{11.1}\varepsilon^{-1})$ time for n-dimensional 0-1 knapsack solutions, in $O^*(n^8\varepsilon^{-1})$ time for linear extensions of n elements poset, in $O^*(m^4n^4\varepsilon^{-1})$ time for all matchings in a graph with n vertices and m edges, where O^* notation ignores $\mathrm{poly}(\log(\varepsilon^{-1}), \log m, \log n)$ factors. Note that those orders of magnitude are not optimized, for simplicity of the main arguments. Unfortunately, these running times are the best possible in terms of ε^{-1} for any deterministic sampler, because of the integrality gap concerning the number of tokens. See also the full-paper version [19] for a relationship to the previous deterministic random walks.

Organization. This paper is organized as follows. In Section 2, we briefly reviews MCMC, as a preliminary of our algorithm and analysis. In Section 3, we describe out algorithm, and explain a summary of our main result. In Section 4, we prove the main theorem. In Section 5, we show examples of polynomial-time uniform samplers, namely for knapsack solutions, linear extensions, and matchings.

2   Preliminaries: Markov Chain Monte Carlo

As a preliminary step of our deterministic sampling, this section briefly reviews the Markov chain Monte Carlo (MCMC). See e.g., [20,16] for detail of MCMC.

Let $V \stackrel{\text{def.}}{=} \{1, \ldots, N\}$ be a finite set, and suppose that we wish to sample from V with a probability proportional to a given positive vector $f = (f_1, \ldots, f_N) \in \mathbb{R}_{\geq 0}^N$; for example, we are concerned with *uniform* sampling of 0-1 knapsack solutions in Section 5.1, where V denotes the set of 0-1 knapsack solutions and $f_v = 1$ for each $v \in V$. The idea of a Markov chain Monte Carlo (MCMC) is to sample from a limit distribution of a Markov chain which is equal to the target distribution $f/\|f\|_1$ where $\|f\|_1 = \sum_{v \in V} f_v$ is the normalizing constant.

Let $P \in \mathbb{R}_{\geq 0}^{N \times N}$ be a transition matrix of a Markov chain with the state space V, where $P_{u,v}$ denotes the transition probability from u to v ($u, v \in V$). A transition matrix P is *irreducible* if $P_{u,v}^t > 0$ for any u and v in V, and is *aperiodic* if $\text{GCD}\{t \in \mathbb{Z}_{>0} \mid P_{x,x}^t > 0\} = 1$ holds for any $x \in V$, where $P_{u,v}^t$ denotes the (u,v) entry of P^t, the t-th power of P. An irreducible and aperiodic transition matrix is called *ergodic*. It is well-known for a ergodic P, there is a unique *stationary distribution* $\pi \in \mathbb{R}_{\geq 0}^N$, i.e., $\pi P = \pi$, and the limit distribution is π, i.e., $\xi P^\infty = \pi$ for any probability distribution $\xi \in \mathbb{R}_{\geq 0}^N$ on V.

An ergodic Markov chain defined by a transition matrix $P \in \mathbb{R}_{\geq 0}^{N \times N}$ is *reversible* if the *detailed balance equation*

$$f_u P_{u,v} = f_v P_{v,u} \tag{1}$$

holds for any $u, v \in V$. When P satisfies the detailed balance equation, it is not difficult to see that $fP = f$ holds, meaning that $f/\|f\|_1$ is the limit distribution (see e.g., [16]). Let ξ and ζ be a distribution on V, then the *total variation distance* $\mathcal{D}_{\text{tv}}$ between ξ and ζ is defined by

$$\mathcal{D}_{\text{tv}}(\xi, \zeta) \stackrel{\text{def.}}{=} \max_{A \subseteq V} \sum_{v \in A} (\xi_v - \zeta_v) = \frac{1}{2} \|\xi - \zeta\|_1. \tag{2}$$

Note that $\mathcal{D}_{\text{tv}}(\xi, \zeta) \leq 1$, since $\|\xi\|_1$ and $\|\zeta\|_1$ are equal to one, respectively. The *mixing time* of a Markov chain is defined by

$$\tau(\varepsilon) \stackrel{\text{def.}}{=} \max_{v \in V} \min \left\{ t \in \mathbb{Z}_{\geq 0} \mid \mathcal{D}_{\text{tv}}(P_{v,\cdot}^t, \pi) \leq \varepsilon \right\} \tag{3}$$

for any $\varepsilon > 0$, where $P_{v,\cdot}^t$ denotes the v-th row vector of P^t; i.e., $P_{v,\cdot}^t$ denotes the distribution of a Markov chain at time t stating from the initial state $v \in V$.

In other words, the distribution $P_{v,\cdot}^t$ of the Markov chain after $\tau(\varepsilon)$ transition satisfies $\mathcal{D}_{\text{tv}}(P_{v,\cdot}^t, \pi) \leq \varepsilon$, meaning that we obtain an approximate sample from the target distribution.

For convenience, let $h(t) \stackrel{\text{def.}}{=} \max_{w \in V} \mathcal{D}_{\text{tv}}(P_{w,\cdot}^t, \pi)$ for $t \geq 0$, then it is well-known that h satisfies a kind of *submultiplicativity*. We will use the following proposition in the analysis of our algorithm in Section 4.2. See [16] or [19] for the proof.

Proposition 1. *For any integers ℓ $(\ell \geq 1)$ and k $(0 \leq k < \tau(\gamma))$,*

$$h(\ell \cdot \tau(\gamma) + k) \leq \frac{1}{2}(2\gamma)^\ell$$

holds for any γ $(0 < \gamma < 1/2)$.   ∎

By the submultiplicativity, $t^* \stackrel{\text{def.}}{=} \tau(1/4)$, called *mixing rate*, is often used as a characterization of P.

3 Deterministic Sampling Algorithm

Now, we explain our algorithm in Section 3.1, and exhibit a summary of our main theorem in Section 3.2. Our algorithm is based on the idea of *deterministic random walks*, such as the rotor-router model (see e.g., [6,15]) or the stack walk (greedy-routing) [10,2,21], but a major difference is that our algorithm is *oblivious*; while the rotor-router model and greedy-routing model memorizes the configurations of tokens and routers, our algorithm memorizes the configuration of tokens only. It makes the description of the algorithm simple, compared with other deterministic random walks.

See Section 5 for the detailed description of deterministic sampling algorithms for particular applications, such as 0-1 knapsack solutions (Section 5.1), linear extensions (Section 5.2), and matchings (Section 5.3), where we also discusses the computational complexities of our algorithm for the applications. See also the full paper version [19] for a relationship to other deterministic random walk, including the rotor-router model [6,15], greedy-routing [10,2,21], etc.

3.1 Algorithm

Let $P \in \mathbb{R}_{\geq 0}^{N \times N}$ be a transition matrix of an *ergodic* Markov chain with the state space V. Let $\mu^{(0)} = (\mu_1^{(0)}, \ldots, \mu_N^{(0)}) \in \mathbb{Z}_{\geq 0}^N$ denote an initial configuration of M tokens over V, and let $\mu^{(t)} \in \mathbb{R}_{\geq 0}^N$ denote the *expected* configuration of tokens independently according to P at time $t \in \mathbb{Z}_{\geq 0}$, i.e., $\|\mu^{(t)}\|_1 = M$ and $\mu^{(t)} = \mu^{(0)} P^t$. Let $\widetilde{\mu}^{(t)} = \mu^{(t)}/M$, for simplicity, then clearly $\widetilde{\mu}^{(\infty)} = \pi$ holds, since P is ergodic (see Section 2).

The idea of our algorithm is to simulate $\mu^{(t)}$ in a deterministic way. Let $\mathcal{G} = (V, \mathcal{E})$ be the transition digram of P, meaning that $\mathcal{E} = \{(u, v) \in V^2 \mid P_{u,v} > 0\}$.

Note that $\mathcal{E}$ may contain self-loop edges such as (v,v). Let $\mathcal{N}^+(v)$ and $\mathcal{N}^-(v)$ respectively denote the out-neighborhood and in-neighborhood of $v \in V$, i.e., $\mathcal{N}^+(v) = \{u \in V \mid P_{v,u} > 0\}$ and $\mathcal{N}^-(v) = \{u \in V \mid P_{u,v} > 0\}$. Note that v is a member of both $\mathcal{N}^+(v)$ and $\mathcal{N}^-(v)$ if $(v,v) \in \mathcal{E}$. For convenience, let $\delta^+(v) = |\mathcal{N}^+(v)|$ and $\delta^-(v) = |\mathcal{N}^-(v)|$. In case that P is reversible, $\mathcal{N}^+(v) = \mathcal{N}^-(v)$ holds, and let $\mathcal{N}(v)$ denote them and let $\delta(v) = |\mathcal{N}(v)|$, for simplicity.

Let $\chi^{(0)} = \mu^{(0)}$, and let $\chi^{(t)} \in \mathbb{Z}_{\geq 0}^N$ denote the configuration of tokens at time $t \in \mathbb{Z}_{\geq 0}$ in our algorithm. A configuration $\chi^{(t)}$ is updated, imitating $P_{v,u}$, as follows. Without loss of generality, we may assume that an arbitrary ordering $u_1, \ldots, u_{\delta^+(v)}$ is defined on $\mathcal{N}^+(v)$ for each $v \in V$. Then, we define the number of tokens $Z_{v,u}^{(t)}$ sent from v to u during the time interval from t to $t+1$ by

$$Z_{v,u_i}^{(t)} = \begin{cases} \left\lfloor \chi_v^{(t)} P_{v,u_i} \right\rfloor + 1 & (i \leq i^*) \\ \left\lfloor \chi_v^{(t)} P_{v,u_i} \right\rfloor & \text{(otherwise)} \end{cases} \tag{4}$$

where

$$i^* = \chi_v^{(t)} - \sum_{i=1}^{\delta^+(v)} \left\lfloor \chi_v^{(t)} P_{v,u_i} \right\rfloor$$

denoting the number of "surplus" tokens. Then, $\chi^{(t+1)}$ is defined by

$$\chi_u^{(t+1)} \stackrel{\text{def.}}{=} \sum_{v \in V} Z_{v,u}^{(t)} \tag{5}$$

for each $u \in V$.

Remark that $\mu_u^{(t+1)} = \sum_{v \in V} \mu_v^{(t)} P_{v,u}$ holds for each $u \in V$ and $t \geq 0$, in the multiple random walk, meaning that if $\chi^{(t)}$ approximates $\mu^{(t)}$ well, then we can expect that $Z_{v,u}^{(t)}$ approximates the "*expected* flow of tokens" $\mu_v^{(t)} P_{v,u}$ and hence that $\chi^{(t+1)}$ approximates $\mu^{(t+1)}$ well. In fact, it is not difficult to see the following observation, which we will use in the analysis in Section 4.2.

Lemma 2. *For the above algorithm,*

$$\left| Z_{v,u}^{(t)} - \chi_v^{(t)} P_{v,u} \right| \leq 1$$

holds for any $u, v, \in V$ *and* $t \geq 0$.

3.2   Main Results

By the definition of the mixing time, $\mathcal{D}_{\text{tv}}(\widetilde{\mu}^{(\tau(\varepsilon))}, \pi) \leq \varepsilon$ holds where $\tau(\varepsilon)$ denotes the mixing time of P, meaning that $\widetilde{\mu}$ approximates the target distribution π well. Thus, we hope our deterministic sampler that the "distribution" $\widetilde{\chi}^{(T)} \stackrel{\text{def.}}{=} \chi^{(T)}/M$ approximates the target distribution π well. We define a *point-wise distance* $\mathcal{D}_{\text{pw}}(\xi, \zeta)$ between $\xi \in \mathbb{R}_{\geq 0}^N$ and $\zeta \in \mathbb{R}_{\geq 0}^N$ satisfying $\|\xi\|_1 = \|\zeta\|_1 = 1$ by

$$\mathcal{D}_{\text{pw}}(\xi, \zeta) \stackrel{\text{def.}}{=} \max_{v \in V} |\xi_v - \zeta_v| = \|\xi - \zeta\|_\infty. \tag{6}$$

Theorem 3. *Let $P \in \mathbb{R}_{\geq 0}^{N \times N}$ be a reversible transition matrix with a stationary distribution π, then*

$$\mathcal{D}_{\mathrm{pw}}\left(\widetilde{\chi}^{(T)}, \widetilde{\mu}^{(T)}\right) \leq \frac{\pi_{\max}}{\pi_{\min}} \cdot \frac{3t^* \Delta}{M}$$

holds for any $T \geq 0$, where $\pi_{\max} = \max\{\pi_v \mid v \in V\}$ and $\pi_{\min} = \min\{\pi_v \mid v \in V\}$.

In a special case that the stationary distribution is uniform, we obtain the following.

Corollary 4. *Let $P \in \mathbb{R}_{\geq 0}^{N \times N}$ be an ergodic and reversible transition matrix with a uniform stationary stationary distribution π. Set $M \geq 6\varepsilon^{-1}t^* \Delta$, then the "distribution" $\widetilde{\chi}^{(T)}$ of the deterministic sampler after $T \geq \tau(\varepsilon/2)$ steps satisfies that $\mathcal{D}_{\mathrm{pw}}\left(\widetilde{\chi}^{(T)}, \pi\right) \leq \varepsilon$.*

4 Analysis of the Point-Wise Distance

This section proves Theorem 3. Some of basic techniques in our proof are based on or similar to previous works [6,15,18].

4.1 Framework

To begin with, we establish the following key lemma. See [19] for the proof.

Lemma 5. *Let $P \in \mathbb{R}_{\geq 0}^{N \times N}$ be a transition matrix of an ergodic Markov chain with a state space V, and let π be the stationary distribution of P. Then, the configurations $\chi^{(T)}$ and $\mu^{(T)}$ of tokens in the algorithm and in corresponding random walk satisfy*

$$\chi_w^{(T)} - \mu_w^{(T)} = \sum_{t=0}^{T-1} \sum_{u \in V} \sum_{v \in \mathcal{N}^-(u)} \left(Z_{v,u}^{(t)} - \chi_v^{(t)} P_{v,u}\right)\left(P_{u,w}^{T-t-1} - \pi_w\right)$$

for any $w \in V$ and for any $T \geq 0$. ∎

4.2 Analysis for Reversible Chains

Now, we are concerned with *reversible* Markov chains, and show the following theorem.

Theorem 6. *Let $P \in \mathbb{R}_{\geq 0}^{N \times N}$ be a transition matrix of a reversible and ergodic Markov chain with a state space V, and let π be the stationary distribution of P. Then, the configurations $\chi^{(T)}$ and $\mu^{(T)}$ of tokens in the algorithm and in its corresponding random walk satisfy*

$$\left|\chi_w^{(T)} - \mu_w^{(T)}\right| \leq \frac{2(1-\gamma)}{1-2\gamma}\tau(\gamma)\frac{\pi_w}{\pi_{\min}}\Delta \tag{7}$$

for any $w \in V$, $T \geq 0$ and γ $(0 < \gamma < 1/2)$.

Remark that our main Theorem 3 is immediate from Theorem 6 by setting $\gamma = 1/4$ and dividing (7) by the total number of tokens M.

Proof. By Lemma 5 and Observation 2, we obtain that

$$\left| \chi_w^{(T)} - \mu_w^{(T)} \right| \leq \sum_{t=0}^{T-1} \sum_{u \in V} \sum_{v \in \mathcal{N}(u)} \left| Z_{v,u}^{(t)} - \chi_v^{(t)} P_{v,u} \right| \left| P_{u,w}^{T-t-1} - \pi_w \right|$$

$$\leq \sum_{t=0}^{T-1} \sum_{u \in V} \sum_{v \in \mathcal{N}(u)} \left| P_{u,w}^{T-t-1} - \pi_w \right| = \sum_{t=0}^{T-1} \sum_{u \in V} \delta(u) \left| P_{u,w}^t - \pi_w \right| \qquad (8)$$

holds. Since P is reversible, $P_{u,w}^t = \frac{\pi_w}{\pi_u} P_{w,u}^t$ holds for any w and u in V. Thus

$$(8) = \sum_{t=0}^{T-1} \sum_{u \in V} \delta(u) \left| \frac{\pi_w}{\pi_u} \left(P_{w,u}^t - \pi_u \right) \right|$$

$$\leq \Delta \frac{\pi_w}{\pi_{\min}} \sum_{t=0}^{T-1} \sum_{u \in V} \left| P_{w,u}^t - \pi_u \right| = 2\Delta \frac{\pi_w}{\pi_{\min}} \sum_{t=0}^{T-1} \mathcal{D}_{\mathrm{tv}} \left(P_{w,\cdot}^t, \pi \right) \qquad (9)$$

where the last equality follows the fact that $\sum_{u \in V} \left| P_{w,u}^t - \pi_u \right| = 2\mathcal{D}_{\mathrm{tv}} \left(P_{w,\cdot}^t, \pi \right)$, by the definition of the total variation distance (2). By Proposition 1, we obtain the following.

Lemma 7. *For any* $v \in V$ *and for any* $T > 0$,

$$\sum_{t=0}^{T-1} \mathcal{D}_{\mathrm{tv}} \left(P_{v,\cdot}^t, \pi \right) \leq \frac{1-\gamma}{1-2\gamma} \tau(\gamma)$$

holds for any γ $(0 < \gamma < 1/2)$.

Proof. Let $h(t) = \mathcal{D}_{\mathrm{tv}} \left(P_{w,\cdot}^t, \pi \right)$, for convenience. Then, $h(t)$ is at most 1 for any $t \geq 0$, by the definition of the total variation distance (2). By Proposition 1,

$$\sum_{t=0}^{T-1} \mathcal{D}_{\mathrm{tv}} \left(P_{w,\cdot}^t, \pi \right) = \sum_{t=0}^{T-1} h(t) \leq \sum_{t=0}^{\infty} h(t) = \sum_{\ell=0}^{\infty} \sum_{k=0}^{\tau(\gamma)-1} h(\ell \cdot \tau(\gamma) + k)$$

$$= \sum_{k=0}^{\tau(\gamma)-1} h(k) + \sum_{\ell=1}^{\infty} \sum_{k=0}^{\tau(\gamma)-1} h(\ell \cdot \tau(\gamma) + k) \leq \sum_{k=0}^{\tau(\gamma)-1} 1 + \sum_{\ell=1}^{\infty} \sum_{k=0}^{\tau(\gamma)-1} \frac{1}{2} (2\gamma)^\ell$$

$$= \tau(\gamma) + \sum_{\ell=1}^{\infty} \tau(\gamma) \frac{1}{2} (2\gamma)^\ell = \tau(\gamma) + \frac{\gamma}{1-2\gamma} \tau(\gamma) = \frac{1-\gamma}{1-2\gamma} \tau(\gamma)$$

holds, and we obtain the claim. □

Now we obtain Theorem 6 from (9) and Lemma 7. □

5   Applications to Rapidly Mixing Chains

In this section, we show some examples of polynomial-time deterministic samplers for uniform sampling of combinatorial objects, whose counting is known to be #P-complete.

5.1   0-1 Knapsack Solutions

Given $a \in \mathbb{R}_{\geq 0}^n$ and $b \in \mathbb{R}_{>0}$, the set of the 0-1 knapsack solutions is defined by $\Omega_{\mathrm{Kna}} = \{x \in \{0,1\}^n \mid \sum_{i=1}^n a_i x_i \leq b\}$. We define a transition matrix $P_{\mathrm{Kna}} \in \mathbb{R}^{|\Omega_{\mathrm{Kna}}| \times |\Omega_{\mathrm{Kna}}|}$ by

$$P_{\mathrm{Kna}}(x, y) = \begin{cases} 1/2n & (\text{if } y \in \mathcal{N}_{\mathrm{Kna}}(x)) \\ 1 - |\mathcal{N}_{\mathrm{Kna}}(x)|/2n & (\text{if } y = x) \\ 0 & (\text{otherwise}) \end{cases}$$

for $x, y \in \Omega_{\mathrm{Kna}}$, where $\mathcal{N}_{\mathrm{Kna}}(x) = \{y \in \Omega_{\mathrm{Kna}} \mid \|x - y\|_1 = 1\}$. Note that the stationary distribution of P_{Kna} is uniform on Ω_{Kna} since P_{Kna} is symmetric. The following theorem is due to Morris and Sinclair [17].

Theorem 8. *[17] The mixing time $\tau(\gamma)$ of P_{Kna} is $O(n^{\frac{9}{2}+\alpha} \log \gamma^{-1})$ for any $\alpha > 0$ and for any $\gamma > 0$.*

For the Markov chain defined by P_{Kna}, our deterministic sampler is described as follows. Note that the following implementation does not optimize the time and space complexity, for simplicity of the arguments.

Algorithm 1
Step 0. Set $W^0[i] := \mathbf{0}$ for each $i = 1, \ldots, M$.
  /* $W^t[i]$ stores a solution in Ω_{Kna}, where token i is. */
Step 1. For $(t = 0 \text{ to } T - 1)\{$
  (a). **Set** list $S_x^{(t)} := \{i \in \{1, \ldots, M\} \mid W^t[i] = x\}$ for each $x \in \Omega_{\mathrm{Kna}}$ as long as $S_x^{(t)} \neq \emptyset$.
  (b). **Serve** tokens in $S_x^{(t)}$ to neighboring vertices according to (4) for each $x \in \Omega_{\mathrm{Kna}}$ satisfying that $S_x^{(t)} \neq \emptyset$, and **set** $W^{t+1}[i]$ be the solution in Ω_{Kna} at which token i arrived.
  $\}$
Step 2. Output $W^T[i]$ for each $i = 1, \ldots, M$.

Theorem 9. *For an arbitrary ε $(0 < \varepsilon < 1)$, set $M := c_1 n^{\frac{11}{2}+\alpha} \varepsilon^{-1}$ and $T := c_2 n^{\frac{9}{2}+\alpha} \log \varepsilon^{-1}$ with appropriate constants c_1, c_2 and α, then Algorithm 1 outputs M samples over Ω_{Kna} satisfying that*

$$\mathcal{D}_{\mathrm{pw}}\left(\widetilde{\chi}^{(T)}, \pi\right) \leq \varepsilon \tag{10}$$

where π is the uniform distribution over Ω_{Kna}. The running time of Algorithm 1 is

$$O(TM \log(M) \, n \, \mathrm{poly}(\log a, \log b)) = O^*(n^{11+2\alpha} \varepsilon^{-1})$$

where O^* ignores poly log term.

Proof. We check the complexity of Algorithm 1 for each Step. Step 0 sets all M tokens on $\mathbf{0} \in \Omega_{\mathrm{Kna}}$, which takes $O(Mn)$ time. Step 1(a) constructs the configuration $\chi^{(t)}$ of M tokens over Ω_{Kna}. Note that the number of lists is at most M, since Step 1(a) constructs a list for $v \in \Omega_{\mathrm{Kna}}$ only when at least one token exists on v. Step 1(a) takes $O(M \log(M) n)$ time, by heapifying $W^t[i]$ $(i = 1, \ldots, M)$ with the lexicographic order on Ω_{Kna}. Step 1(b) updates a configuration according to our deterministic sampling algorithm described in Section 3.1. It takes $O(n \operatorname{poly}(\log a, \log b))$ time to find all feasible solutions neighboring to $\mathbf{x}$. Once the algorithm finds all feasible solutions neighboring to $\mathbf{x}$, then it is easy to let every token of $\chi_{\mathbf{x}}^{(t)}$ go to the neighboring vertex according to (4), in $O(n \chi_{\mathbf{x}}^{(t)})$ time, like the rotor-router. Since we repeat Step 1 T times, then we obtain the time complexity $O(TM \log(M) n \operatorname{poly}(\log a, \log b))$.

Now, (10) is clear from Corollary 4, since Algorithm 1 is an implementation of the deterministic sampler described in Section 3.1.   □

5.2   Linear Extensions of a Poset

Let $S = \{1, 2, \ldots, n\}$, and $Q = (S, \preceq)$ be a partial order. A linear extension of Q is a total order $X = (S, \sqsubseteq)$ which respects Q, i.e., for all $i, j \in S$, $i \preceq j$ implies $i \sqsubseteq j$. Let Ω_{Lin} denote the set of all linear extensions of Q. We define a relationship $X \sim_p X'$ $(p \in \{1, \ldots, n\})$ for a pair of linear extensions X and X' $\in \Omega_{\mathrm{Lin}}$ satisfying that $x_p = x'_{p+1}$, $x_{p+1} = x'_p$, and $x_i = x'_i$ for all $i \neq p, p+1$, i.e.,

$$X = (x_1, x_2, \ldots, x_{p-1}, x_p, x_{p+1}, x_{p+2}, \ldots, x_n)$$
$$X' = (x_1, x_2, \ldots, x_{p-1}, x_{p+1}, x_p, x_{p+2}, \ldots, x_n)$$

holds. Then, we define a transition matrix $P_{\mathrm{Lin}} \in \mathbb{R}^{|\Omega_{\mathrm{Lin}}| \times |\Omega_{\mathrm{Lin}}|}$ by

$$P_{\mathrm{Lin}}(X, X') = \begin{cases} F(p)/2 & \text{(if } X' \sim_p X) \\ 1 - \sum_{I \in \mathcal{N}_{\mathrm{Lin}}(X)} P_{\mathrm{Lin}}(X, I) & \text{(if } X' = X) \\ 0 & \text{(otherwise)} \end{cases}$$

for $X, X' \in \Omega_{\mathrm{Lin}}$, where $\mathcal{N}_{\mathrm{Lin}}(X) = \{Y \in \Omega_{\mathrm{Lin}} \mid X \sim_p Y (p \in \{1, \ldots, n-1\})\}$ and $F(p) = \frac{p(n-p)}{\frac{1}{6}(n^3-n)}$. Note that P_{Lin} is ergodic and reversible, and its stationary distribution is uniform on Ω_{Lin} [3]. The following theorem is due to Bubley and Dyer [3].

Theorem 10. *[3] The mixing time $\tau(\gamma)$ of P_{Lin} is $O\left(n^3 \log n \gamma^{-1}\right)$ for any $\gamma > 0$.*

Thus, we obtain a deterministic algorithm running in $O^*(n^8 \varepsilon^{-1})$ time, in a similar way as 0-1 knapsack in Section 5.1. See [19] for detail.

5.3   Matchings in a Graph

Counting all matchings in a graph, related to the *Hosoya index* [11], is known to be #P-complete [22]. Jerrum and Sinclair [12] gave a rapidly mixing chain.

This Section is concerned with sampling of all matchings in a graph. Remark that counting all perfect matchings in a bipartite graph, related to the *permanent*, is also well-known #P-complete problem, and Jerrum, Sinclair, and Vigoda [13] gave a celebrated FPRAS based on an MCMC method using annealing. To apply our algorithm to sampling perfect matchings, we need some assumptions on the input graph (see e.g., [20,12,13]).

Let $H = (U, F)$ be an undirected graph. A matching in H is a subset $\mathcal{M} \subseteq F$ such that no edges in $\mathcal{M}$ share an endpoint. Let $N_C(\mathcal{M}) = \{e = \{u, v\} \mid e \notin \mathcal{M}, \text{both } u \text{ and } v \text{ are matched in } \mathcal{M}\}$ and let $\mathcal{N}_{\mathrm{Mat}}(\mathcal{M}) = \{e \mid e \notin N_C(\mathcal{M})\}$. Then, for $e = \{u, v\} \in \mathcal{N}_{\mathrm{Mat}}(\mathcal{M})$, we define $\mathcal{M}(e)$ by

$$\mathcal{M}(e) = \begin{cases} \mathcal{M} - e & \text{(if } e \in \mathcal{M}) \\ \mathcal{M} + e & \text{(if } u \text{ and } v \text{ are unmatched in } \mathcal{M}) \\ \mathcal{M} + e - e' & \text{(if exactly one of } u \text{ and } v \text{ is matched in } M, \text{ and} \\ & \quad e' \text{ is the matching edge).} \end{cases}$$

Let Ω_{Mat} denote the set of all possible matchings of H. The we define the transition matrix $P_{\mathrm{Mat}} \in \mathbb{R}^{|\Omega_{\mathrm{Mat}}| \times |\Omega_{\mathrm{Mat}}|}$ by

$$P_{\mathrm{Mat}}(\mathcal{M}, \mathcal{M}') = \begin{cases} 1/2m & \text{(if } \mathcal{M}' = \mathcal{M}(e)) \\ 1 - |\mathcal{N}_{\mathrm{Mat}}(\mathcal{M})|/2m & \text{(if } \mathcal{M}' = \mathcal{M}) \\ 0 & \text{(otherwise)} \end{cases}$$

for any $\mathcal{M}, \mathcal{M}' \in \Omega_{\mathrm{Mat}}$, where $m = |F|$. Note that P_{Mat} is ergodic and reversible, and its stationary distribution is uniform on Ω_{Mat} [12]. The following theorem is due to Jerrum and Sinclair [12].

Theorem 11. *[12] The mixing time $\tau(\gamma)$ of P_{Mat} is $\mathrm{O}\left(mn^2 \log n\gamma^{-1}\right)$ for any $\gamma > 0$.*

Thus, we obtain a deterministic algorithm running in $\mathrm{O}^*(m^4 n^4 \varepsilon^{-1})$ time, in a similar way as 0-1 knapsack in Section 5.1. See [19] for detail.

6   Concluding Remarks

This paper proposed an algorithm for deterministic sampling, and gave an upper bound of the point-wise distance $\mathcal{D}_{\mathrm{pw}}(\widetilde{\chi}^{(t)}, \widetilde{\mu}^{(t)})$. Using the algorithm, we obtain polynomial-time deterministic algorithms for uniform sampling of #P-complete problems, such as knapsack solutions, linear extensions and matchings. A bound of the point-wise distance independent of $\pi_{\max}/\pi_{\min}$ is a future work. Development of deterministic approximation algorithms based on a deterministic sampler for #P-hard problems is a challenge.

Acknowledgment. This work was supported by JSPS KAKENHI Grant Number 25700002, 24650008, and MEXT Grant-in-Aid for Scientific Research on Innovative Areas "Molecular Robotics" (No. 25104519).

References

1. Akbari, H., Berenbrink, P.: Parallel rotor walks on finite graphs and applications in discrete load balancing. In: Proc. SPAA 2013, pp. 186–195 (2013)
2. Angel, O., Holroyd, A.E., Martin, J., Propp, J.: Discrete low discrepancy sequences, arXiv:0910.1077
3. Bubley, R., Dyer, M.: Faster random generation of linear extensions. Discrete Mathematics 201, 81–88 (1999)
4. Cooper, J., Doerr, B., Friedrich, T., Spencer, J.: Deterministic random walks on regular trees. Random Structures & Algorithms 37, 353–366 (2010)
5. Cooper, J., Doerr, B., Spencer, J., Tardos, G.: Deterministic random walks on the integers. European Journal of Combinatorics 28, 2072–2090 (2007)
6. Cooper, J., Spencer, J.: Simulating a random walk with constant error. Combinatorics, Probability and Computing 15, 815–822 (2006)
7. Doerr, B., Friedrich, T.: Deterministic random walks on the two-dimensional grid. Combinatorics, Probability and Computing 18, 123–144 (2009)
8. Friedrich, T., Gairing, M., Sauerwald, T.: Quasirandom load balancing. SIAM Journal on Computing 41, 747–771 (2012)
9. Gopalan, P., Klivans, A., Meka, R., Stefankovic, D., Vempala, S., Vigoda, E.: An FPTAS for #knapsack and related counting problems. In: Proc. FOCS 2011, pp. 817–826 (2011)
10. Holroyd, A.E., Propp, J.: Rotor walks and Markov chains. In: Lladser, M., Maier, R.S., Mishna, M., Rechnitzer, A. (eds.) Algorithmic Probability and Combinatorics, pp. 105–126. The American Mathematical Society (2010)
11. Hosoya, H.: Topological index. A newly proposed quantity characterizing the topological nature of structural isomers of saturated hydrocarbons. Bulletin of the Chemical Society of Japan 44, 2332–2339 (1971)
12. Jerrum, M., Sinclair, A.: Approximation algorithms for NP-hard problems. In: Hochbaum, D.S. (ed.) The Markov Chain Monte Carlo Method: An Approach to Approximate Counting and Integration. PWS Publishing (1996)
13. Jerrum, M., Sinclair, A., Vigoda, E.: A polynomial-time approximation algorithm for the permanent of a matrix with nonnegative entries. Journal of the ACM 51, 671–697 (2004)
14. Karzanov, A., Khachiyan, L.: On the conductance of order Markov chains. Order 8, 7–15 (1991)
15. Kijima, S., Koga, K., Makino, K.: Deterministic random walks on finite graphs. In: Proc. ANALCO 2012, pp. 16–25 (2012)
16. Levine, D.A., Peres, Y., Wilmer, E.L.: Markov Chain and Mixing Times. American Mathematical Society (2008)
17. Morris, B., Sinclair, A.: Random walks on truncated cubes and sampling 0-1 knapsack solutions. SIAM Journal on Computing 34, 195–226 (2004)
18. Rabani, Y., Sinclair, A., Wanka, R.: Local divergence of Markov chains and analysis of iterative load balancing schemes. In: Proc. FOCS 1998, pp. 694–705 (1998)
19. Shiraga, T., Yamauchi, Y., Kijima, S., Yamashita, M.: Deterministic random walks for rapidly mixing chains, arXiv:1311.3749
20. Sinclair, A.: Algorithms for Random Generation & Counting, A Markov chain approach. Birkhäuser (1993)
21. Tijdeman, R.: The chairman assignment problem. Discrete Math. 32, 323–330 (1980)
22. Valiant, L.G.: The complexity of enumeration and reliability problems. SIAM Journal on Computing 8, 410–421 (1979)

Sampling Query Feedback Restricted Repairs of Functional Dependency Violations: Complexity and Algorithm*

Dongjing Miao, Xianmin Liu, and Jianzhong Li**

Harbin Institute of Technology, Harbin, Heilongjiang, 150001, China
{miaodongjing,xianmliu,lijzh}@hit.edu.cn

Abstract. An inconsistent database is a database instance violating integrity constraints. A repair of an inconsistent database is a maximal consistent subset. Sampling from the repair space is an alternative approach meeting the needs of many applications. In this paper, we introduce a new class of repair, query feedback restricted repair, based on the feedback on user's witness query. We first map out a complete picture of both data and combined complexities of repair existence problems under different cases to identify the intractable cases. Especially, we show that if the query is a projection or a union query, then the decision problem is NP-*complete*; Even worse, if the query is a conjunctive query, the decision problem becomes Σ_2^P-*complete*. At last, we provide a random repair sampling algorithm when the witness query is a selection-join query, and it is still polynomial even under the combined complexity.

Keywords: repair sampling, database, complexity.

1 Introduction

In many novel database applications, violations of integrity constrains cannot be avoided. For example, two consistent data sources will contribute conflicting information in data integration [1]. Integrity constrains (such as FDs [2]/CFDs [3]) can be used to identify conflicts in the database. A resolution of a conflict is the deletion of one of the tuples resulting in the conflict. Generally, there are many nondeterministic choices to resolve the conflicts when repairing the database, because integrity constrains can not be able to further determine which tuple should be deleted. Due to the exponential space of possible subset-repairs, we may not be able to, or may not want to, generate all repairs. Therefore, repair sampling is proposed as an alternative approach which aims to provide repairs sampled from the original data to user in some strategies in order to help

* This work was supported in part by the National Grand Fundamental Research 973 Program of China under grant 2012CB316200, the Major Program of National Natural Science Foundation of China under grant 61190115, the Key Program of the National Natural Science Foundation of China under grant 61033015, 60933001.
** Corresponding author.

Z. Cai et al. (Eds.): COCOON 2014, LNCS 8591, pp. 37–48, 2014.

user find a more reasonable repair. This approach will meet the needs of some applications such as interactive data cleaning, data integration and uncertain query answering, see [4].

Consider an example about the cargo information of a retail store. Suppose a schema R(Item: string, Type: string, Brand: string, Origin: string, Price: numerical), an database instance I_0 over schema R is shown as follow,

$\{t_1$(Tea, Green, China, 100), t_2(Tea, Red, India, 110),

t_3(Tea, Green, China, 120), t_4(Tea, Red, India, 130)$\}$

Additionally, an FD φ is defined over R as follow, φ: *Item, Type, Origin* $\rightarrow$ *Price*. The semantic of φ is that if two tuples have the same value on the attributes "Item", "Type", "Origin", the values of attribute "Price" must be the same. According to such FD, tuple "t_1" is inconsistent with "t_2", and tuple "t_3" is inconsistent with "t_4". Usually, inconsistencies imply that the data has errors. However, φ cannot further guide how to repair such two inconsistencies in order to recover the correct information. A subset-repair is the maximal consistent subset of the inconsistent database, thus there are exponential possible repairs with respect to the number of inconsistencies such as the repair set of I_0 includes four possible subset-repairs as follow,

I'_1: $\{$(Tea, Green, China, 100), (Tea, Red, India, 110)$\}$

I'_2: $\{$(Tea, Green, China, 100), (Tea, Red, India, 130)$\}$

I'_3: $\{$(Tea, Green, China, 120), (Tea, Red, India, 110)$\}$

I'_4: $\{$(Tea, Green, China, 120), (Tea, Red, India, 130)$\}$

Sampling algorithm is to generate a sample of possible repairs of the input database under some repair semantic, moreover, it will return empty when no repairs are found.

In real life, there are many users having useful knowledge which can be used to guide how to resolve the conflicts, for example, they are able to provide a preference on how conflicts should be resolved which has been investigated in [1]. In this paper, we consider another way of exploiting users' knowledge, in which user will give a feedback on a witness query result to guide how to resolve the conflicts. Comparing with preference on conflicts in [1], query feedback has a stronger expression ability, and this implies that user knowledge can be exploited as more as possible. Often, users have the knowledge about the result of his witness query, but they may not be permitted to modify the database directly due to some reasons, since they may not have the complete knowledge about the whole database. We restrict that the feedback can be only specified on the witness query result, rather than the modifications directly on the original database. Obviously, the guide from user's feedback on the witness query will narrow down the set of repairs to a set of repairs consistent with the users' knowledge at least.

Continue the example above, user A wants to know "the information of all the tea produced in China that this retail store sailed". Motivated by this, user A submits a witness database query $\mathbf{Q}$: $\sigma_{Item=Tea, Origin=China}$ (I_0).[1] And an result of $\mathbf{Q}$ is returned as $\{$(Tea, Green, China, 100), (Tea, Green, China, 120)$\}$. In fact, user A has the useful knowledge about the cargo information of this

---

[1] Selection symbol σ. See [2]

retail store, and make sure that *"The prices of green tea this store sailed are no more than 100 !"* Consequently, for **Q**, user A will provide a feedback that Δ =(Tea, Green, China, 120) should be eliminated. Then, these feedbacks are the belief that we are allowed to eliminate tuple t_2 and t_5 from I_0. Because user query feedback can be used to eliminate incorrect repairs, the sampling space of possible repairs gets even better and smaller as follows, and we call the two repairs *query feedback restricted repair*,

I_1'': {(Tea, Green, China, 100), (Tea, Red, India, 110)},

I_2'': {(Tea, Green, China, 100), (Tea, Red, India, 130)}.

In this paper, we want to make a theoretical and algorithmic study on sampling query feedback restricted repair. We first investigate the repair existence problem. This is motivated that the repair existence problem is the fundamental problem of repair sampling. We set the integrity constrain as the simple case and set the witness query as a single relation algebraic query (not multiple queries), then give the complexity analysis of the decision problems under different query classes. After identifying all the intractable cases, we will provide polynomial sampling algorithm for the tractable case.

Contributions. We summarize our contributions as follows. First, we formally introduce query feedback restricted repair. Second, we give the thorough complexity analysis of its existence decision problem, *qfr*-RE. On data complexity aspect, *qfr*-RE is at least NP-*hard* if the witness query includes projection or union; On combined complexity aspect, *qfr*-RE is also intractable if projection or union included, and it turns to Σ_2^P-*complete* if the witness query is a SPJ (selection-projection-join) query (i.e., conjunctive query) or a SPJU (selection-projection-join-union) query. In brief, we map out a complete picture of the data and combined complexities of the three problems. Finally, we provide a random repair sampling algorithm when the witness query is a selection-join query and user has the complete knowledge about the witness query, the algorithm is still polynomial even under the combined complexity.

2   Related Works

Optimal data repairing and consistent query answer are the most popular approaches to deal with violations of FDs and other integrity constraints. The former aim to find a repair with a minimum modifications on the given database, including minimally differs from the original one (e.g., [5], [6], [7], [8]), minimize the description length (e.g., [9]) and so on. The limitation of them is that there may be many different optimal repairs. The latter aims to find answer of a query that are true in every possible repair. It usually employ techniques of condensed representation of possible repairs(e.g., [10], [11]) or query rewriting (e.g., [12], [13]) to obtain consistent answer. Unfortunately, there are lots of classes of queries have to be answered approximately. Sampling repairs is an alternative approach proposed to overcome several drawbacks of optimal repairing and consistent query answering. It is to generate a sample of possible repairs of the input database under some repair semantic, moreover, it will return empty query results when no consistent answers are found, such as [4] proposed three classes of repairs and the corresponding sampling algorithm.

Different from the existing work, this paper consider using user query feedback to guide how to resolve conflicts so that a more reasonable sampling space could be obtained. Comparing with the preferred repair in [1], query feedback restricted repair defined in this paper has a stronger expressive ability. Moreover, we focus on sampling repairs, not the consistent query answering. We study the complexity of repair existence problem not the repair checking problem that whether a given instance is a repair of the input instance, see [14].

Another related problems is view update problem that given a view and an update against a view, the problem is to translate the update into a corresponding update against the base data, see [2]. There are several complexity bounds are known on relational view updates, [15], [16], [17], and [18] give out the tractability and intractability results of finding a minimal view complement for relational views. [19] gave out the complexity of view update analysis under key preserving condition which can not be extended to ours. There are lots of works on the algorithms for translating view update to base table update, such as [20], [21], [22] and so on. Especially, [23] and [18] both considered the presence of certain functional dependencies and provided algorithms for translating restricted view updates to base table updates without side effects or with minimum side effects. Their goal was to define correctness properties of these translations and to characterize precisely the conditions for the existence of translations possessing these properties. Different from our work, the database they considered is a consistent one not a inconsistent database so that the presence of FD and other integrity constrains simplifies view update problem which it is in contrary to this paper.

Dependency propagation is another related problem, it is to determine that given a view defined on data sources and a set of dependencies on the sources, whether another dependency is guaranteed to hold on the view, e.g., [24], [25] which are the first to investigate dependency propagation. [26] extended [24], [25] by providing complexity bounds for FD propagation in the general setting, and for CFD propagation. However, this is a problem different from ours.

3   Notations and Definitions

A *schema* is a finite sequence $\mathbf{R} = \langle R_1, \ldots, R_m \rangle$ of distinct relation symbols, where each R_i has an arity $r_i > 0$ and includes several attributes, denoted by $R_i = (A_1, \ldots, A_{r_i})$. Each attribute A_j has a corresponding set $dom(A_j)$ which is the domain of values appearing in A_j. An database instance I (over $\mathbf{R}$) is a sequence $\langle R_1^I, \ldots, R_m^I \rangle$, such that each R_i^I is a finite set of tuples $\{t_1, \ldots, t_N\}$, each tuple t_k belongs to the set $dom(A_1) \times \cdots \times dom(A_{r_i})$. We use $I[R_i]$ to represent the relation R_i in database I. If I and J are two instances over $\mathbf{R} = \langle R_1, \ldots, R_m \rangle$, then J is a sub-instance of I denoted $J \subseteq I$ if $R_i^J \subseteq R_i^I$, for all $i = 1, \ldots, m$.

An FD (Functional Dependency[2]) φ over a relation R can be represented by $\varphi : (X \to A)$, where both X and A are a set of attributes from R. Such dependency means the values of any two tuples' attributes A should be same if they have same value in attributes X. Given a database instance I and an FD φ, if there is no tuple pair violate the FD rule, we denote that $I \models \varphi$. Usually, we use Σ to denote the set of FDs. Given a database I and an FD set Σ, for every FD $\varphi \in \Sigma$, if $I \models \varphi$, we call I consistent, denoted as $I \models \Sigma$. Otherwise, we call it *inconsistent*.

In this paper, we suppose that the user has complete knowledge about the answer of his witness query, he is able to guarantee what should be preserved in the result and what should not be. Therefore, we define the query feedback restricted repair as follow.

Definition 1 (Query feedback restricted repair). *Given a database I, FD set Σ, a query Q, its result $Q(I)$ and a subset of result need to delete specified by users Δ (i.e., $\Delta \subseteq Q(I)$), for any instance I_r, it is called a subset-repair of I such that (1) $I_r \models \Sigma$, and (2) $Q(I_r) = Q(I) \backslash \Delta$, (3) I_r is a maximal sub-instance of I.*

Before we give the sampling algorithm, we first study the complexity of the basic decision problem of repair existence, called **qfr-RE** . Here, the feedback Δ specified by user is a subset of the query result $Q(I)$ where the witness query Q is written by operations in relational algebra including **S** (selection), **P** (projection), **J** (join), **U** (union), **RA** (Relation algebra). Recall the example in section 1, both I_1'' and I_2'' are repairs restricted by the feedback $\Delta = \{(Tea, Green, China, 120)\}$.

4   Intractable Cases

In this section, we list the intractable cases on two aspects including both data and combined complexity. We remark that *data complexity* is the complexity expressed in terms of the size of the database only, while *combined complexity* is the complexity expressed in terms of both the size of the database and the query expression [27].

4.1   Data Complexity Aspect

Theorem 1. *qfr-RE is NP-hard for P query.*

Proof Sketch: We construct a PTIME reduction from 3SAT to this problem. Given a boolean variable set $X = \{x_1, \ldots, x_n\}$, the input of 3SAT problem is a formula $\phi = C_1 \wedge \ldots \wedge C_m$ where $C_i = \{l_1, l_2, l_3\}$ and l_j is either x_k or $\overline{x}_k$ for $k \in 1, \ldots, n$, reduction can be described as follows. (1)*Base instance.* Let I contains only one relation R including three attributes (L, X, C). For each clause $C_i \in \phi$ and each literal $l_j \in C_i$ ($j \in \{1, 2, 3\}$), a tuple t_{ij} is built and inserted into R as follows. If l_j is x_k, let $t_{ij} = (+, X_k, c_i)$. If l_j is $\overline{x}_k$, let $t_{ij} = (-, X_k, c_i)$. (2)*FD set.* Let Σ be $\{X \rightarrow L\}$. (3) *Witness query.* Let Q be $\pi_C(R)$. (4)*Query result.* Let $Q(I) = \{(c_1), \ldots, (c_m)\}$. (5)*Feedback.* Let Δ be $\emptyset$. One can verify the ϕ is satisfied if and only if there is a valid repair of I.

Theorem 2. *qfr-RE is NP-hard for U query.*

Proof Sketch: We construct a PTIME reduction from *Monotone* 3SAT problem. Similar with 3SAT problem, an instance of *Monotone* 3SAT problem is a formula $\phi = C_1 \wedge \cdots \wedge C_m$, where each clause C_i includes only *positive* or *negative* literals. The reduction can be made as follows. (1) *Base relations*. First, suppose there are n variables $x_1, \ldots, x_n$, then let I contains n relations $R_1, \ldots, R_n$. Each relation R_i has attribute set $\{X, C\}$. Second, for each C_j, if $x_i \in C_j$, add $(+, C_j)$ to R_i, if $\overline{x}_i \in C_j$, add $(-, C_j)$ to R_i; (2) *FD set*. For each R_i, add rule $\emptyset \rightarrow X$ to the rule set Σ; (3) *Witness query*. Let Q be $R_1 \cup \cdots \cup R_n$; (4) *Query result*. Initially, $Q(I)$ includes m tuples where tuple t_i is $(+, C_j)$ if C_j contains positive; otherwise, it is $(-, C_j)$; (5) *Feedback*. Let Δ be $\emptyset$. One can verify that ϕ is satisfied if and only if there is a valid repair of I.

Theorem 3. *qfr-RE is NP for RA query.*

4.2   Combined Complexity Aspect

To analyze the SPJ query, we will use the term *fact*. Given a database instance I and a SPJ query Q in a form of $\pi_{\mathbf{A}} (\sigma_{con}(R_1 \times \cdots \times R_q))$. A *fact* μ of I is a tuple sequence $(t_1, t_2, \ldots, t_q) \in R_1 \times \cdots \times R_q$, where $t_i \in R_i^I$ for each $1 \leq i \leq q$. If $(t_1, t_2, \ldots, t_q)$ satisfies the selection condition *con*, then we denote it as $Q(\mu) \in Q(I)$.

Theorem 4. *qfr-RE is Σ_2^P-complete for SPJ query and SPJU query.*

Proof. We will prove the following two statements to show the correctness of the theorem. (i) We prove the upper bound of *qfr*-RE for SPJU query is Σ_2^P by giving a Σ_2^P algorithm as follows. First, guess a sub-instance I_r of I satisfying Σ. Then, determining whether $Q(I_r) \cap \Delta = \emptyset$ and $Q(I) \backslash \Delta \subseteq Q(I_r)$. The former question is in coNP, because any SPJU query has a form of $q_1 \cup \cdots \cup q_z$ where each q_i is a SPJ query, so that its complement can be solved by determining whether for there is a fact μ of I_r such that $Q(\mu) \in Q(I_r)$ and $Q(\mu) \in \Delta$. The latter question is also in coNP, because its complement can be solved by determining whether there is a fact μ of I such that $Q(\mu) \in Q(I) \backslash \Delta$ but μ is not a fact of I_r (then it must be $Q(\mu) \notin Q(I_r)$, because $Q(\mu) \in Q(I)$).

(ii) We prove the lower bound of *qfr*-RE for SPJ query is Σ_2^P-*hard* by a reduction from QSAT$_2$ problem. An instance of QSAT$_2$ problem includes two variable sets $X_1 = \{x_1, ..., x_{n'}\}$ and $X_2 = \{x_{n'+1}, ..., x_{n'+n''}\}$, and a 3-DNF boolean expression ϕ with m clauses $\{C_1, \ldots, C_m\}$, the task is to determine whether there is an assignment τ for X_1 such that ϕ is satisfied by all assignments for X_2. Let $n = n' + n''$, that is $|X_1| + |X_2| = n$, we show the reduction as follows. (An example of the reduction for a QSAT$_2$ instance $\phi = \exists x_1 x_2 \forall x_3 x_4 (x_1 \wedge x_2 \wedge x_3) \vee (x_1 \wedge \overline{x}_2 \wedge x_3) \vee (\overline{x}_1 \wedge \overline{x}_3 \wedge x_4)$ is shown in Fig. 1.)
Base instance I. We build I including $n + m + 3$ relations $\{S_i, i = 1, \ldots, n\} \cup \{R_k, k = 1, \ldots, m\} \cup \{G_p, p = 1, 2, 3\}$, where S_i simulates x_i, R_k simulates clause C_k and G_1, G_2, G_3 are three auxiliary relations. Concretely, (1) For each variable

x_i $(1 \leq i \leq n)$, construct relation $S_i = \{A_1, A_2\}$ and add three tuples $(X, 1)$ and $(X, 0)$ and (Y, B) to S_i. (2) For each clause C_i $(1 \leq i < m)$, build a quintuple relation R_i $(A_1, A_2, A_3, A_4, A_5)$. We add 8 tuples into $I[R_i]$. In the first 7 tuples, values of A_1, A_2, A_3 refer to the 7 false value assignments of the 3 variables, values of A_4 are always '$-$', and the values of A_5 are the ids of these 7 tuples. The last two tuples are auxiliary tuples $(Z, Z, Z, -, 8)$ and $(B, B, B, B, 9)$. (3) G_1 includes two tuples $(\underbrace{X \cdots X}_{n'})$, $(\underbrace{Y \cdots Y}_{n'})$; G_2 includes three tuples $(\underbrace{0 \cdots 0}_{n''})$, $(\underbrace{1 \cdots 1}_{n''})$, $(\underbrace{B \cdots B}_{n''})$; G_3 includes eight tuples $(\underbrace{1 \cdots 1}_{m})$, $(\underbrace{2 \cdots 2}_{m})$, ..., $(\underbrace{9 \cdots 9}_{m})$.

FD set Σ. For each relation S_i, $i \in [1, n']$, add FD: $S_i.A_1 \to S_i.A_2$ into Σ.

Witness query Q. Construct the query Q as follows. We denote $R_1 \times \cdots \times R_m$, $S_1 \times \cdots \times S_{n'}$ and $S_{n'+1} \times \cdots \times S_n$ as $\mathbf{R}$, $\mathbf{S}^1$, $\mathbf{S}^2$. For each clause $C_k \in \phi$, without loss of generality, it is assumed that $C_k = x_{k1} \wedge \overline{x}_{k2} \wedge x_{k3}$, let the condition con_k be $(S_{k1}.A_2 = R_k.A_1) \wedge (S_{k2}.A_2 = R_k.A_2) \wedge (S_{k3}.A_2 = R_k.A_3)$. Let the condition con_{A4} be $R_1.A_4 = R_2.A_4 = \cdots = R_m.A_4$. Then, let the witness query $Q = Q_0 \times Q_1 \times Q_2 \times Q_3$, where

$$Q_0 = \pi_{R_1.A_4,...,R_m.A_4}(\sigma_{con_1 \wedge \cdots \wedge con_m \wedge con_{A4}} \mathbf{S}^1 \times \mathbf{S}^2 \times \mathbf{R}),$$
$$Q_1 = \pi_{G_1.A_1,...,G_1.A_{n'}}(\sigma_{S_1.A_1=G_1.A_1 \wedge \cdots \wedge S_{n'}.A_1=G_1.A_{n'}}(\mathbf{S}^1 \times G_1)),$$
$$Q_2 = \pi_{G_2.A_1,...,G_2.A_{n''}}(\sigma_{S_{n'+1}.A_2=G_2.A_1 \wedge \cdots \wedge S_n.A_2=G_2.A_{n''}}(\mathbf{S}^2 \times G_2)),$$
$$Q_3 = \pi_{G_3.A_1,...,G_3.A_m}(\sigma_{R_1.A_5=G_3.A_1 \wedge \cdots \wedge R_m.A_5=G_3.A_m}(\mathbf{R} \times G_3)).$$

Query result $Q(I)$. Initially, let $Q(I) = \{t, t'\} \times G_1 \times G_2 \times G_3$, where $t = (\underbrace{-, \ldots, -}_{m})$, $t' = (\underbrace{B, \ldots, B}_{m})$.

Feedback Δ. Let $\Delta = \{t\} \times G_1 \times G_2 \times G_3$.

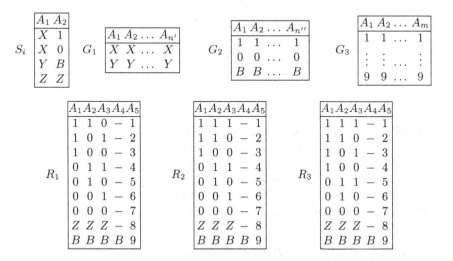

Fig. 1. Example for the reduction of Theorem 4

Some key properties are introduced before the correctness proof.

P1. A solution I_r, $Q(I_r) = Q(I) \backslash \Delta$ if and only if, $Q_0(I_r) = \{t'\}$, $Q_1(I_r) = G_1$, $Q_2(I_r) = G_2$ and $Q_3(I_r) = G_3$.

P2. $Q_1(I_r) = G_1$ where $I_r \models \Sigma$ if and only if, either $S_i = \{(X,0), (Y,B)\}$ or $S_i = \{(X,1), (Y,B)\}$ holds, for each S_i of I_r where $1 \leq i \leq n'$. It simulates that each variable in X_1 has one and only one assignments.

P3. $Q_2(I_r) = G_2$ if and only if each S_i of I_r is the same as it is in I where $n' + 1 \leq i \leq n$. It simulates that each variable in X_2 can be assigned arbitrarily.

P4. $Q_3(I_r) = G_3$ if and only if each R_i of I_r is the same as it is in I where $1 \leq i \leq m$.

Next, we show the correctness of the reduction by following two statements.

$\Rightarrow$ If the answer of QSAT$_2$ instance ϕ is 'yes', there must be a sub-instance I_r obtained from I by deleting some tuples such that $I_r \models \Sigma$ and $Q(I_r) = Q(I) \backslash \Delta$. Suppose ϕ is satisfiable for all assignments of X_2 under the assignment $\tau(X_1)$. Given τ, we will construct a repair I_r satisfying the conditions of qfr-RE. First, delete (Z,Z) from each S_i ($1 \leq i \leq n$); Second, for each relation S_i satisfying $1 \leq i \leq n'$ of I_r, let the corresponding variable of S_i is $x_i \in X_1$. If $\tau(x_i) = 1$, the tuple $(X,0)$ will be deleted, otherwise, the tuple $(X,1)$ will be deleted. Obviously, $I_r \models \Sigma$ and $Q_i(I_r) = G_i$ ($1 \leq i \leq 3$), since the RHS of **P2**, **P3**, **P4** are satisfied. One can verify that $Q_0(I_r) = t'$ since that ϕ is a tautology under $\tau(X_1)$.

$\Leftarrow$ One the other hand, if there is a sub-instance $I_r \models \Sigma$ and $Q(I_r) = Q(I) \backslash \Delta$, then we can construct an assignment $\tau(X_1)$ such that ϕ is *true* under any assignment of X_2. In I_r, each R_k related with clause C_k is the same as it is in I where $1 \leq k \leq m$, each S_i related with variable in X_2 is the same as it is in I where $n' + 1 \leq i \leq n$, and each S_i related with variable in X_1 excludes either $\{(X,0)\}$ or $\{(X,1)\}$, because of S_i should satisfy $S_i.A_1 \rightarrow S_i.A_2$, where $1 \leq i \leq n'$. Note that, in I_r, we do not care about that whether (Z,Z) is preserved in each S_i. For each variable $x_i \in X_1$, let S_i be the corresponding relation in I_r, then the assignment τ can be built as follows,

$$\tau(x_i) = \begin{cases} 1, & \text{if } (X,1) \text{ is in } S_i; \qquad\qquad (1) \\ 0, & \text{otherwise.} \qquad\qquad\qquad\qquad (2) \end{cases}$$

To show that τ is a valid assignment for ϕ, consider any tuple $\tilde{t} \in \mathbf{S}^1 \times \mathbf{S}^2 \times \mathbf{R}^1$ in I_r such that $\tilde{t}[S_i.A_2] = \tau(x)$ ($1 \leq i \leq n'$), $\tilde{t}[S_i.A_2] \neq B$ ($n' + 1 \leq i \leq n$). It is not hard to verify that such $\tilde{t}$ always violates at least one condition defined in Q_0 due to **P1** where $Q_0(I_r) = \{t'\}$. That is, some condition con_k such that t does not satisfy con_k. Because $\tilde{t}$ satisfies the conditions $t[R_k.A_4] = '-'$, then we have that $\tilde{t}$ refer to the assignments on $\{A_1, A_2, A_3\}$ which does not appear in R_k, it means that clause C_k is satisfied under such assignment. Because for each assignment on X_2, there is at least one clause is true, we have that ϕ is tautology under assignment τ.

Theorem 5. *qfr-RE is PSPACE-complete for RA query.*

Proof. qfr-RE is in PSPACE obviously, since the Relation algebraic *Query Evaluation* problem for *RA* is in PSPACE-*complete* [2]. We next prove that qfr-RE is PSPACE-*hard* by reduction from query evaluation problem. Given an instance of *Query Evaluation* problem $\langle I, q, t \rangle$ where q is a relational algebraic query, it is to decide if tuple $t \in q(I)$. Without loss of generality, we assume that I consists of n relations, $R_1, \ldots, R_n$, each relation R_i contains c_i columns, and t is a d-dimension tuple. Then, by means of the technique similar with the previous proof, an instance of *qfr*-RE can be built as follows.

Base instance. Let I' contain relations $R'_1, \ldots, R'_n$ and two auxiliary relations R'_x, R'_y. Each R'_i $(1 \leq i \leq n)$ is obtained by adding an addition column c_{i+1} to R_i and filling the additional column using integer numbers in $[\Sigma_{1 \leq j \leq i-1}|R_j| + 1, \Sigma_{1 \leq j \leq i}|R_j|]$ to identify each tuple uniquely in I'. For convenience, let $N = |R_1| + \cdots + |R_n|$. Then let $R'_x = \{t\}$ and $R'_y = \{(t', -), (t'', +)\}$ where t' and t'' are two different d-dimension tuples as long as they are different from t.

Witness query Q. We denote $q(\pi_I(I'))$ as A_q, construct the query Q as follows,

$$Q_a(I') = \pi_{A_{c_1+1}}(R'_1) \bigcup \cdots \bigcup \pi_{A_{c_n+1}}(R'_n),$$
$$Q(I') = [Q_a(I') \times \{(-), (+)\} - \pi_{Q_a}(Q_a(I') \times ((R'_x - A_q) \bigcup R'_y))] \times R'_x.$$

Here, the operator π_{Q_a} extracts all attributes in the scheme of Q_a, and the operator π_I extracts all attributes in I.

Query result $Q(I)$. Initially, $Q(I) = \{(1), \ldots, (N)\} \times \{(-), (+)\} \times \{t\}$.

Feedback Δ. Let $\Delta = \{(1, -), \ldots, (N, -)\} \times \{t\}$.

Obviously, the reduction described above can be finished in polynomial time. The correctness of the reduction can be obtained by observing that $Q(I_r) = Q(I) \backslash \Delta$ if and only if (1) no tuple in I' disappears in I_r since $Q(I) \backslash \Delta$ includes every unique identification number, and (2) $t \in q(I)$.

5   Repair Sampling Algorithm for FD and SJ Query

In the previous section, we conclude that there is no polynomial sampling algorithm if the witness query include Projection or Union operation. In this section, we propose a polynomial algorithm 1 to sample the subset-repair restricted by SJ query. We first give a overview of the sampling algorithm and its intuitive idea, then we show its polynomial implementation.

Given any *SJ* query Q, an equivalent standard form $\sigma_C(R_1 \times \cdots \times R_m)$ can be built in PTIME. Here, each R_i $(1 \leq i \leq m)$ is a relation included in I, and two relations in $\{R_1, \ldots, R_m\}$ may be same relation in the database. Next, we provide a polynomial repair sampling algorithm only considering the self-join-free case, i.e. each relation is allowed to appear at most once in the query Q. Then, we extend such algorithm with general setting.

Algorithm works as follows. Build an anchor I_r which is a sub-instance of the instance I. Specifically, for each relation $R_i \in I$, $I_r[R_i]$ is set to be $\pi_{R_i}(Q(I) \backslash \Delta)$, if R_i appears in the query Q; Let $I_r[R_i]$ be empty, otherwise. Once I_r is obtained, we first test that whether $I_r \models \Sigma$. If not, return ; else, we should test if $Q(I_r) = Q(I) \backslash \Delta$. If $Q(I_r) \neq Q(I) \backslash \Delta$, return $\emptyset$, which means that there is no repair satisfying the user feedback restriction. Pick a tuple $a \in I \backslash I_r$ randomly, let $a \in R_i$, if there is a result $t \in \Delta$ such that $t[R_i] = a$ and each $t[R_j] \in \Delta$ $(i \neq j)$, then discard a. Otherwise if $I_r \cup \{a\} \models \Sigma$, add a into I_r. Loop this step until there is no new tuple can be added into I_r. At last, output I_r as a sample repair. The pseudo-code is given as below.

Algorithm 1. Sampling Algorithm

---

1. **for all** i such that $1 \leq i \leq m$ **do**
2.   **if** R_i appears in the query Q **then**
3. $I_r[R_i] \leftarrow \pi_{R_i}(Q(I)\backslash\Delta)$
4.   **else**
5. $I_r[R_i] \leftarrow \emptyset$
6. **if** $I_r \nvDash \Sigma$ **or** $Q(I_r) \neq Q(I)\backslash\Delta$ **then**
7.   **return** $\emptyset$
8. **while** there is no new tuple can be added into I_r **do**
9.   Pick a tuple $a \in I\backslash I_r$ randomly where $a \in R_i$
10.   **if** $\exists t$ of Δ, $t[R_i] = a$ **and** $\forall j \neq i, t[R_j] \in \Delta$ **then**
11.   discard a
12.   **else if** $I_r \cup \{a\} \models \Sigma$ **then**
13.   add a into I_r
14. **return** I_r

The intuitive idea behind such algorithm as follows. All valid repair for I must be a superset of the anchor I_r, since that in order to guarantee that $Q(I)\backslash\Delta$ is part of $Q(I_r)$, the tuples in I_r must be preserved when deleting tuples. An important observation about FD is that if database instance I does not satisfy some FD rule and $I \subseteq J$, we have J will not satisfy the rule also. Therefore, I_r is the only candidate solution needed to be considered for the given qfr-RE problem, since there will be no valid repair if I_r is invalid.

Now, we give a polynomial implementation of **line 6** in algorithm 1. Obviously, whether I_r satisfies the Σ can be determined polynomially trivially. It is a little complicated to determine whether $Q(I_r) = Q(I)\backslash\Delta$, since a trivial solution will take exponential time cost. A polynomial implementation for deciding whether or not $Q(I_r) = Q(I)\backslash\Delta$ can be designed as follows.

First, in I_r, compute the transitive closure of attributes in the database as follows. Build a node for each attribute and constant appearing in the condition, then, build an edge between attribute A and constant c if $A = c$ is in the condition, and an edge between two attributes A and A' if $A = A'$ is in the condition. For each connected component in the graph, build a group, if there is a group containing two different constants, then return *false*, since such query condition is unsatisfiable. Otherwise, there must be two kinds of groups, one is that containing one constant, the other one is that containing all variables. We call each group with some constant c is '*constant group*' and use $[c]$ to denote it, and call each group without constant is '*variable group*'. Then, all attributes in $\bigcup\{R_i\}$ can be divided into three parts, first part is the *constant* group, the second part is the *variable* group, and the last part is others.

Then, partition tuples in each relation R_i into several equivalent classes according to the following steps. (i) For attributes in '*constant group*', obviously, only the corresponding *constant* value can appear in the table. (ii) For the attributes in the '*variable group*', partition the tuples according to different value combinations. Therefore, all tuples in each R_i are divided into several equivalent classes $\{q_{i1}, \ldots, q_{iy}\}$, each class represents a unique value combination for the 'variable' attributes. We use $|q_{ij}|$ to represent the size of the class.

For each variable group g, let $g \cap R_i$ be the 'variable' attributes in R_i. Let $R_i[g \cap R_i]$ be the corresponding value set. Compute the intersections of all such value set, $V_g = \bigcap \{R_i[g \cap R_i]\}$. Delete those tuples of R_i whose values on attributes $g \cap R_i$ are not in V_g.

Compute join result $R_{\bowtie}$ of all R_i relations as follows. For each two relations R_i and R_j, compute the join and filter the conditions C. Notice that there are at most $m - 1$ joins needed to be computed. Each join operation involves two relations R_i and a temporal result. The size of each temporal result will not exceed $|Q(I) \backslash \Delta|$. It should be noticed that the join operations are executed on the equivalent classes and the 'free' attributes are not included in the result.

If $R_{\bowtie}$ contains a tuple not included by $Q(I) \backslash \Delta$, returns *false*. Otherwise, for each tuple t in $R_{\bowtie}$, compute size $size(t)$ as follows. For each relation R_i, let $[t[R_i]]$ be the class in R_i containing the same values as t, $size(t)$ is $\times_{R_i} |[t[R_i]]|$.

Compare $\Sigma_{t \in R_{\bowtie}} size(t)$ with $|Q(I) \backslash \Delta|$, if they are not equal, return *false*, else return *true*.

References

1. Staworko, S., Chomicki, J., Marcinkowski, J.: Prioritized repairing and consistent query answering in relational databases. Annals of Mathematics and Artificial Intelligence 64(2-3), 209–246 (2012)
2. Abiteboul, S., Hull, R., Vianu, V.: Foundations of Databases. Addison-Wesley (1995)
3. Bohannon, P., Fan, W., Geerts, F., Jia, X., Kementsietsidis, A.: Conditional functional dependencies for data cleaning. In: IEEE 23rd International Conference on Data Engineering, ICDE 2007, pp. 746–755. IEEE (2007)
4. Beskales, G., Ilyas, I.F., Golab, L., Galiullin, A.: Sampling from repairs of conditional functional dependency violations. The VLDB Journal 23(1), 103–128 (2014)
5. Bohannon, P., Fan, W., Flaster, M., Rastogi, R.: A cost-based model and effective heuristic for repairing constraints by value modification. In: Proceedings of the 2005 ACM SIGMOD International Conference on Management of Data, SIGMOD 2005, pp. 143–154. ACM, New York (2005)
6. Cong, G., Fan, W., Geerts, F., Jia, X., Ma, S.: Improving data quality: Consistency and accuracy. In: Proceedings of the 33rd International Conference on Very Large Data Bases, VLDB 2007, pp. 315–326. VLDB Endowment (2007)
7. Lopatenko, A., Bravo, L.: Efficient approximation algorithms for repairing inconsistent databases. In: IEEE 23rd International Conference on Data Engineering, ICDE 2007, pp. 216–225. IEEE (2007)
8. Chomicki, J., Marcinkowski, J.: Minimal-change integrity maintenance using tuple deletions. Inf. Comput. 197(1/2), 90–121 (2005)
9. Chiang, F., Miller, R.J.: A unified model for data and constraint repair. In: IEEE 27th International Conference on Data Engineering, ICDE 2011, pp. 446–457. IEEE (2011)
10. Lopatenko, A., Bertossi, L.: Complexity of consistent query answering in databases under cardinality-based and incremental repair semantics. In: Schwentick, T., Suciu, D. (eds.) ICDT 2007. LNCS, vol. 4353, pp. 179–193. Springer, Heidelberg (2007)

11. Wijsen, J.: Condensed representation of database repairs for consistent query answering. In: Calvanese, D., Lenzerini, M., Motwani, R. (eds.) ICDT 2003. LNCS, vol. 2572, pp. 378–393. Springer, Heidelberg (2003)

12. Arenas, M., Bertossi, L., Chomicki, J.: Consistent query answers in inconsistent databases. In: Proceedings of the Eighteenth ACM SIGMOD-SIGACT-SIGART Symposium on Principles of Database Systems, PODS 1999, pp. 68–79. ACM, New York (1999)

13. Fuxman, A., Miller, R.J.: First-order query rewriting for inconsistent databases. J. Comput. Syst. Sci. 73(4), 610–635 (2007)

14. Afrati, F.N., Kolaitis, P.G.: Repair checking in inconsistent databases: Algorithms and complexity. In: Proceedings of the 12th International Conference on Database Theory, ICDT 2009, pp. 31–41. ACM, New York (2009)

15. Cosmadakis, S.S., Papadimitriou, C.H.: Updates of relational views. J. ACM 31(4), 742–760 (1984)

16. Lechtenbörger, J., Vossen, G.: On the computation of relational view complements. ACM Trans. Database Syst. 28(2), 175–208 (2003)

17. Bancilhon, F., Spyratos, N.: Update semantics of relational views. ACM Trans. Database Syst. 6(4), 557–575 (1981)

18. Kimelfeld, B., Vondrák, J., Woodruff, D.P.: Multi-tuple deletion propagation: Approximations and complexity. Proc. VLDB Endow., pp. 1558–1569 (2013)

19. Cong, G., Fan, W., Geerts, F., Li, J., Luo, J.: On the complexity of view update analysis and its application to annotation propagation. IEEE Trans. on Knowl. and Data Eng. 24(3), 506–519 (2012)

20. Keller, A.M.: Algorithms for translating view updates to database updates for views involving selections, projections, and joins. In: Proceedings of the Fourth ACM SIGACT-SIGMOD Symposium on Principles of Database Systems, PODS 1985, pp. 154–163. ACM, New York (1985)

21. Bohannon, A., Pierce, B.C., Vaughan, J.A.: Relational lenses: A language for updatable views. In: Proceedings of the Twenty-fifth ACM SIGMOD-SIGACT-SIGART Symposium on Principles of Database Systems, PODS 2006, pp. 338–347. ACM, New York (2006)

22. Cui, Y., Widom, J.: Run-time translation of view tuple deletions using data lineage. Technique report (2001)

23. Dayal, U., Bernstein, P.A.: On the correct translation of update operations on relational views. ACM Trans. Database Syst. 7(3), 381–416 (1982)

24. Klug, A.C.: Calculating constraints on relational expression. ACM Trans. Database Syst. 5(3), 260–290 (1980)

25. Klug, A.C., Price, R.: Determining view dependencies using tableaux. ACM Trans. Database Syst. 7(3), 361–380 (1982)

26. Fan, W., Ma, S., Hu, Y., Liu, J., Wu, Y.: Propagating functional dependencies with conditions. Proc. VLDB Endow. 1(1), 391–407 (2008)

27. Vardi, M.Y.: The complexity of relational query languages (extended abstract). In: Proceedings of the Fourteenth Annual ACM Symposium on Theory of Computing, STOC 1982, pp. 137–146. ACM, New York (1982)

Online Prediction Problems with Variation

Chia-Jung Lee, Shi-Chun Tsai, and Ming-Chuan Yang

Department of Computer Science,
National Chiao-Tung University, Hsinchu, Taiwan
leecj@nctu.edu.tw, sctsai@cs.nctu.edu.tw, mingchuan.cs96g@g2.nctu.edu.tw

Abstract. We study the prediction with expert advice problem, where in each round, the player selects one of N actions and incurs the corresponding loss according to an N-dimensional linear loss vector, and aim to minimize the regret. In this paper, we consider a new measure of the loss functions, which we call L_∞-*variation*. Consider the loss functions with small L_∞-variation, if the player is allowed to have some information related to the variation in each round, we can obtain an online bandit algorithm for the problem without using the self-concordance methodology, which conditionally answers an open problem in [8]. Another related problem is the combinatorial prediction game, in which the set of actions is a subset of $\{0,1\}^d$, and the loss function is in $[-1,1]^d$. We provide an online algorithm in the semi-bandit setting when the loss functions have small L_∞-variation.

Keywords: prediction with expert advice problem, combinational prediction game, semi-bandit setting, bandit setting, variation.

1 Introduction

We study the prediction with expert advice problem, in which the player has to make repeated decisions for T rounds in the following way. Suppose that there are N actions. In each round t, the adversary chooses a loss vector $f_t = (f_{t,1}, \cdots, f_{t,N}) \in [-1,1]^N$, and simultaneously, the player chooses an action I_t to play. After the choice, the player suffers the corresponding loss f_{t,I_t} and obtains some feedback. In the full information setting, the player obtains the entire loss function f_t, while in the bandit setting, the player only knows the corresponding loss f_{t,I_t}. The goal of the player is to minimize the *expected regret*:

$$\mathbb{E}\left[\sum_{t=1}^{T} f_{t,I_t}\right] - \min_{i \in \{1,\cdots,N\}} \sum_{t=1}^{T} f_{t,i},$$

which is the difference between the expected total loss of the player and the total loss of the best fixed action.

In the full information setting, one can achieve a regret of $O\left(\sqrt{T \log N}\right)$ using the multiplicative update algorithm [9,6]. Note that they considered arbitrary sequence of loss functions, that is, they only considered the worst case scenario.

Z. Cai et al. (Eds.): COCOON 2014, LNCS 8591, pp. 49–60, 2014.
© Springer International Publishing Switzerland 2014

When the loss functions have some restriction, a smaller regret can be achieved. Hazan and Kale [7] considered the measure of the loss functions, called *variation* $\text{VAR}_T^{\max}$, which is defined to be

$$\text{VAR}_T^{\max} = \max_{t \leq T} \{\text{VAR}_t(\ell_t)\},$$

where ℓ_t is the best action till the t-th round, and $\text{VAR}_t(i) = \sum_{\tau=1}^{t}(f_{\tau,i} - u_{t,i})^2$ with $u_t = \frac{1}{t}\sum_{\tau=1}^{t}f_\tau = (u_{t,1}, \ldots, u_{t,N})$, and they provided an online algorithm with a regret bound of $O\left(\sqrt{\text{VAR}_T^{\max} \log N}\right)$. Note that since $\text{VAR}_T^{\max} \leq O(T)$, this result recovers that in [9,6]. Chiang et al. considered another measure of the loss functions, called L_∞-*deviation*, which is defined as

$$D_\infty = \sum_{t=1}^{T} \|f_t - f_{t-1}\|_\infty^2,$$

where f_0 is the all-0 function, and showed that a regret bound of $O\left(\sqrt{D_\infty \log N}\right)$ can be achieved [4]. Note that since $D_\infty \leq O(T)$, this result also recovers that in [9,6], but it is incomparable to that of [7].

In the bandit setting, one can obtain a regret bound of $O(\sqrt{NT})$ [2,3]. When the loss functions have the quadratic variation $Q = \sum_{t=1}^{T}\|f_t - \mu\|_2^2$, where $\mu = \frac{1}{T}\sum_{t=1}^{T}f_t$ is the mean of the loss functions, Hazan and Kale [8] achieved a regret of $\tilde{O}\left(N^2\sqrt{Q}\right)$[1]. However, this algorithm used some methodology based on self-concordant barrier functions, which were first used in online learning by Abernethy et al. [1]. Besides, no bound is known for the loss functions with a small L_∞-deviation.

Another related problem is the combinatorial prediction game. Suppose that the set of actions is $\mathcal{A} = \{A_1, \cdots, A_N\}$, where $A_i \in \{0,1\}^d$. In each round t, the adversary secretely chooses a loss function $f_t \in [-1,1]^d$. Simultaneously, the player chooses an action A_{I_t} and suffers a loss of $\langle f_t, A_{I_t} \rangle$, where $\langle x, y \rangle$ is the inner product of $x, y \in \mathbb{R}^d$. In the combinatorial prediction game, there are three kinds of feedback. (1) In the full information setting, the player knows the entire loss function f_t. (2) In the bandit setting, the player only obtains the corresponding loss value $\langle f_t, A_{I_t} \rangle$. (3) In the semi-bandit setting, the player can know the loss values $f_{t,j}$ for any $j \in [d]$ satisfying $A_{I_t,j} = 1$. The target of the player is also to minimize the *expected regret*:

$$\mathbb{E}\left[\sum_{t=1}^{T}\langle f_t, A_{It}\rangle\right] - \min_{i \leq N}\sum_{t=1}^{T}\langle f_t, A_i\rangle.$$

In the full information setting and the semi-bandit setting, Audibert et al. obtained a regret bound of $O\left(d\sqrt{T}\right)$ [3]. In the bandit setting, a regret of $O\left(d^{5/2}\sqrt{T}\right)$ can be achieved [5].

---

[1] We use the notation $\tilde{O}(\cdot)$ to hide the dependence on $\text{poly}(\log T)$ factor.

In this paper, we consider a new measure of the loss functions, which we call L_∞-*variation*, defined to be

$$V_\infty = \sum_{t=1}^{T} \|f_t - \mu\|_\infty^2 .$$

It is easy to see that the L_∞-variation V_∞ is bounded by $O(T)$, and the L_∞-deviation D_∞ in [4] is bounded by $O(V_\infty)^2$, but the L_∞-variation is incomparable to the variation $\text{VAR}_T^{\max}$ in [7]. However, our definition of L_∞-variation is simple and more intuitive. Besides, in each round, after the player makes his choice, he is allowed to receive some information related to $\|f_t - \mu\|_\infty$. Note that he may not know the true value of $\|f_t - \mu\|_\infty$, and he is not able to know which choice has the biggest difference from the mean. Consider the loss functions with a small L_∞-variation, we can obtain an online bandit algorithm for prediction with expert advice problem without using the self-concordance methodology, which conditionally answers an open problem in [8]. Nevertheless, there is an additional factor $\sqrt{\log T}$ in our regret bound in the full information setting. More precisely, when the loss functions have L_∞-variation V_∞, we can prove a regret of $\tilde{O}\left(\sqrt{V_\infty \log N}\right)$ in the full information setting, and a regret of $\tilde{O}\left(\sqrt{N V_\infty \log N}\right)$ in the bandit setting.

We also consider the combinatorial prediction game in which the loss functions have a small L_∞-variation. Note that many situations in daily life can be modeled as a combinatorial prediction game, for example, the commuting problem. Each morning, one has to choose one of N routes to work where each route may contain several roads. The environment will decide the commuting time for each road. Observe that for each road, the commuting time between different days may be very similar, which implies that in this problem, the L_∞-variation of the loss functions may be small. Since the combinatorial prediction game is a special case of the online linear optimization problem[3], when the loss functions are with the quadratic variation $Q = \sum_{t=1}^{T} \|f_t - \mu\|_2^2$, one can achieve a regret of $O\left(\sqrt{Q}\right)$ in the full information setting [7], and a regret of $\tilde{O}\left(d^{3/2}\sqrt{Q}\right)$ in the bandit setting [8]. For loss functions with L_2-deviation $D_2 = \sum_{t=1}^{T} \|f_t - f_{t-1}\|_2^2$, Chiang et al. showed a regret bound of $O\left(\sqrt{D_2}\right)$ in the full information setting [4]. However, no result exists for the loss functions with a small variation in the semi-bandit setting. None is known either for loss functions with a small deviation in the semi-bandit setting and the bandit setting. Our final contribution is to provide an online algorithm and obtain a regret bound of $\tilde{O}\left(d\sqrt{V_\infty \log N}\right)$ in the semi-bandit setting, when the loss functions have the L_∞-variation V_∞.

---

[2] By the triangle inequality and the fact that for any two real numbers a, b, $(a+b)^2 \le 2a^2 + 2b^2$, we have $D_\infty = \sum_t \|f_t - f_{t-1}\|_\infty^2 \le \sum_t \left(\|f_t - \mu\|_\infty + \|\mu - f_{t-1}\|_\infty\right)^2 \le 4V_\infty$.

[3] In the online linear optimization problem, the set of actions can be arbitrary subset of $\mathbb{R}^d$, while in the combinatorial prediction game, the set of actions is a subset of $\{0, 1\}^d$.

2   Preliminaries

Let $\mathbb{N}$ be the set of positive integers and $\mathbb{R}$ the set of real numbers. For $n \in \mathbb{N}$, let $\mathbb{R}^n$ be the set of n-dimensional vectors over $\mathbb{R}$ and $[n]$ be $\{1, 2, \cdots, n\}$. We denote the inner product of $x, y \in \mathbb{R}^n$ by $\langle x, y \rangle$ and the infinity norm of x by $\|x\|_\infty$. Let $\mathbf{RE}(x\|y) = \sum_{i=1}^N x_i \ln(x_i/y_i)$ be the relative entropy of x with respect to y for non-negative x, $y \in \mathbb{R}^N$. For some event A, $\mathbb{1}[A]$ is the indicator variable of A. Let $\mathbb{E}_t$ denote the expectation conditioned on the randomness up to round $t - 1$. Let $\{e_1, \cdots, e_N\}$ be the set of standard basis of $\mathbb{R}^N$. We denote $\mathbf{1}$ to be the all-1 vector.

2.1   Problems

In this work, we address two categories of prediction problems; one is a prediction with expert advice and another is a combinatorial prediction game. Generally, in a prediction game a player makes repeated decisions and then suffers the corresponding loss, and the goal is to minimize the player's total loss of T rounds with respect to an optimal fixed decision. We formally define the problems as follows:

Prediction with Expert Advice: In round $t = 1, 2, \cdots, T$

- The player chooses $I_t \in [N]$ and reveals his estimator $\tilde{u}_{t-1}$ of μ
- Simultaneously, the adversary chooses a loss function
  $f_t = (f_{t,1}, \cdots, f_{t,N}) \in [-1, 1]^N$
- The player incurs the loss f_{t,I_t}, and observes $\|f_t - \tilde{u}_{t-1}\|_\infty$ [4] and either
  - the whole loss function f_t [**the full information version**], or
  - the loss value f_{t,I_t} [**the bandit version**]

Target: minimize the expected regret: $\mathbb{E}\left[\sum_{t=1}^T f_{t,I_t}\right] - \min_{i \in [N]} \mathbb{E}\left[\sum_{t=1}^T f_{t,i}\right]$

Combinatorial Prediction Game: Let $\mathcal{A} = \{A_1, \cdots, A_N\} \subset \{0, 1\}^d$ be a set of actions.
In round $t = 1, 2, \cdots, T$

- The player chooses $I_t \in [N]$ and reveals his estimator $\tilde{u}_{t-1}$ of μ
- Simultaneously, the adversary chooses a loss function
  $f_t = (f_{t,1}, \cdots, f_{t,d}) \in [-1, 1]^d$
- The player incurs the loss $\langle f_t, A_{I_t} \rangle$, and observes $\|f_t - \tilde{u}_{t-1}\|_\infty$ and either
  - the whole loss function f_t [**the full information version**],
  - the coordinates $f_{t,j} \mathbb{1}[A_{I_t,j} = 1]$ [**the semi-bandit version**], or
  - the loss $\langle f_t, A_{I_t} \rangle$ [**the bandit version**]

Target: minimize the expected regret: $\mathbb{E}\left[\sum_{t=1}^T \langle f_t, A_{I_t} \rangle\right] - \min_{i \in [N]} \mathbb{E}\left[\sum_{t=1}^T \langle f_t, A_i \rangle\right]$

---

[4] In fact, the player only needs the information $\|f_t - \tilde{u}_{t-1}\|_\infty + \varepsilon$ for some $0 \leq \varepsilon \leq O(1/\sqrt{T})$.

2.2   Tools

We need the following simple fact, whose proof is omitted.

Proposition 1. *Let $f_1, \cdots, f_n$ be a sequence of bounded d-dimensional vectors over $\mathbb{R}$. Then we have $\|f_1 + \cdots + f_n\|_\infty^2 \leq n \cdot \left(\|f_1\|_\infty^2 + \cdots + \|f_n\|_\infty^2 \right)$*

Reservoir Sampling is a procedure that can obtain an unbiased estimator of the mean of a stream of data [10]. Consider a situation that the data in the stream can only be seen in one pass and can not be visited again. The problem is how to sample this stream such that in any time the empirical mean is good enough? The idea of RESERVOIR SAMPLING is to maintain a randomly chosen subset S of size k, without replacement, from the stream of real numbers $l_1, l_2, \cdots$, and the average of the sample is used as an estimator of the current mean $u_t = \frac{1}{t} \sum_{\tau=1}^{t} l_\tau$. Algorithm 1 [8] is a formal statements of this sampling method.

Algorithm 1. Reservoir Sampling

---

1. **Initialize** S by including the first k elements $l_1, l_2, \cdots$ in the stream.
2. **for** $t = k+1, k+2, \cdots$ **do**
3.   include l_t in S with probability k/t;
4.   if decide to include l_t, then a random element of S is replaced by l_t.
5. **end for**

2.3   Meta Algorithm

All of our algorithms in this paper is based on the META algorithm, as shown in Algorithm 2. The parameter m in the META algorithm is the dimension of the loss functions, and for different types of problems, we will choose different v_i's.

The META algorithm is inspired by the full information algorithm modified from the multiplicative update algorithm in [4] for the loss functions with a small deviation $D_\infty = \sum_{t \in [T]} \|f_t - f_{t-1}\|_\infty^2$. In round t, the multiplicative update algorithm suggests the player should choose the action according to a distribution p_t. It can be shown that if one can select the action according to p_{t+1} in round t, then the regret will be small. However, to compute p_{t+1}, we need the loss function f_t, which is not available before the round t. Nevertheless, since the loss functions have a small deviation, f_{t-1} may be close to f_t. Hence, in round t, it may be a good idea to choose the action according to the distribution computed by f_{t-1} instead of f_t, and this indeed obtains a small regret in terms of deviation. Here, we consider the measure L_∞-variation $V_\infty = \sum_{t \in [T]} \|f_t - \mu\|_\infty^2$, where $\mu = \frac{1}{T} \sum_{t \in [T]} f_t$. Similarly, when the loss functions have a small variation, μ may be a good approximation of f_t. However, we do not know μ in round t. In the full information setting, it is natural to use $u_{t-1} = \frac{1}{t-1} \sum_{\tau=1}^{t-1} f_\tau$ as an approximation of μ in round t. Therefore, we choose the action according to the distribution $\hat{p}_t$ computed by u_{t-1} instead of f_t. While in the semi-bandit setting and the bandit setting, we cannot know u_{t-1}, so we borrow the idea of [8] to compute an estimator $\tilde{u}_{t-1}$ of u_{t-1} by reservoir sampling, and choose the action according to the distribution $\hat{p}_t$ computed by $\tilde{u}_{t-1}$ instead of f_t. In the META algorithm, the **SAMPLE STEP** is used to maintain a good estimator $\tilde{u}_t$ of u_t, and the **UPDATE STEP** is used to update the distributions p_t and $\hat{p}_t$. Note that in the full information setting, the META algorithm always executes the **UPDATE STEP**.

Algorithm 2. META algorithm

---

1. Initially, let $p_1 = \hat{p}_1$ be the uniform distribution over $[N]$, $S = (0)_{i,j} \in \mathbb{R}^{m \times k}$, and $S' = (0)_{i,j} \in \mathbb{R}^{m \times Nk}$
2. Let $\pi : [Nk] \to [Nk]$ be a random permutation
3. **for** $t = 1$ to T **do**
4.   Toss the biased coin whose outcome is $r_t \in \{0, 1\}$ with $r_t = 0$ for the full-information setting; otherwise, $\Pr[r_t = 1] = \min\left\{\frac{Nk}{t}, 1\right\}$
5.   **if** $r_t = 1$ **then**
6.   // **SAMPLE STEP** to maintain S
7.   **if** $t \leq Nk$ **then**
8.   Choose $I_t = (\pi(t) \mod N) + 1$, and reveal $\tilde{u}_{t-1}$
9.   For each $j \in [m]$, if we observe $f_{t,j}$, then put it into an empty bucket in the jth row of S'
10.   If $t = Nk$, for each $j \in [m]$, randomly select k elements without replacement from the nonempty buckets in the jth row of S', and put them into the jth row of S
11.   **else** $\{// \text{ i.e. } t > Nk\}$
12.   Choose $I_t \in [N]$ uniformly at random, and reveal $\tilde{u}_{t-1}$
13.   Update S by an additional rule // defined in later sections
14.   **end if**
15.   Estimate $\tilde{u}_{t,i} = \frac{1}{k} \sum_{j=1}^{k} S_{i,j}$
16.   Let $p_{t+1} = p_t$, and $\hat{p}_{t+1} = \hat{p}_t$
17.   **else** $\{// \text{ i.e. } r_t = 0\}$
18.   // **UPDATE STEP**
19.   Choose I_t according to the distribution $\hat{p}_t$, and reveal $\tilde{u}_{t-1}$
20.   Compute the estimated loss $\tilde{f}_t = \tilde{g}_t + \tilde{u}_{t-1}$ // defined in later sections
21.   Compute $\tilde{u}_t$ // defined in later sections
22.   Update $\forall i \in [N]$,
23. $p_{t+1,i} = \frac{p_{t,i} \exp\left(-\eta\langle \tilde{f}_t, v_i \rangle\right)}{Z_{t+1}}$, where $Z_{t+1} = \sum_{j=1}^{N} p_{t,j} e^{\left(-\eta\langle \tilde{f}_t, v_j \rangle\right)}$
24. $\hat{p}_{t+1,i} = \frac{p_{t+1,i} \exp\left(-\eta\langle \tilde{u}_t, v_i \rangle\right)}{\hat{Z}_{t+1}}$, where $\hat{Z}_{t+1} = \sum_{j=1}^{N} p_{t+1,j} e^{\left(-\eta\langle \tilde{u}_t, v_j \rangle\right)}$
25.   **end if**
26. **end for**

Next, we bound the regret of META algorithm. Let p^* be the distribution over $[N]$ such that $p_{i^*}^* = 1$ for the best fixed action $i^* \in [N]$ and $p_j^* = 0 \; \forall j \in [N] \setminus \{i^*\}$.

Moreover, let $\tilde{F}_{t,i} = \left\langle \tilde{f}_t, v_i \right\rangle$, $\tilde{U}_{t,i} = \langle \tilde{u}_t, v_i \rangle$, and $\tilde{G}_{t,i} = \tilde{F}_{t,i} - \tilde{U}_{t-1,i} = \langle \tilde{g}_t, v_i \rangle$. We will need the following lemma, whose proof is omitted due to the page limit.

Lemma 1. *Let T_u include the rounds that run the* **UPDATE STEP**. *Then, for any $\eta > 0$, if $\tilde{G}_{t,i} \geq 0$ for all $t \in T_u$, and $i \in [N]$, we have* $\mathbb{E}\left[\sum_{t \in T_u} \tilde{F}_{t,I_t}\right] -$ $\mathbb{E}\left[\sum_{t \in T_u} \tilde{F}_{t,i^*}\right] \leq 2\eta \mathbb{E}\left[\sum_{t \in T_u} \left\|\tilde{G}_t\right\|_\infty \mathbb{E}_{i \sim \hat{p}_t}\left[\tilde{G}_{t,i}\right]\right] + \frac{\log N}{\eta}$.

3 Prediction with Expert Advice

In this section, we consider the prediction with expert advice problem when the loss functions have small L_∞-variation, defined to be $V_\infty = \sum_{t=1}^T \|f_t - \mu\|_\infty^2$, where $\mu = \frac{1}{T} \sum_{t=1}^T f_t$ is the mean of f_t's.

3.1 Full Information Setting

In this subsection, we consider the full information setting. As mentioned, the **UPDATE STEP** of META algorithm can deal with the full information game. Therefore, we do not need to choose the parameter k. We instantiate the META algorithm with $v_i = \mathbf{e}_i$ for each $i \in [N]$, and:

- (step 20) Compute $\tilde{f}_t = \tilde{g}_t + \tilde{u}_{t-1}$, with $\tilde{u}_{t-1} = u_{t-1}$, and $\tilde{g}_t = f_t - u_{t-1} + \|f_t - \tilde{u}_{t-1}\|_\infty \cdot \mathbf{1}$.
- (step 21) Compute $u_t = \frac{1}{t} \sum_{\tau=1}^t f_\tau$.

Then, the updates in step 23 and 24 of META algorithm are:

- $p_{t+1,i} = p_{t,i} \cdot \exp\left(-\eta \tilde{f}_{t,i}\right) / Z_{t+1}$, where $Z_{t+1} = \sum_{j \in [N]} p_{t,j} \cdot \exp\left(-\eta \tilde{f}_{t,j}\right)$;
- $\hat{p}_{t+1,i} = p_{t+1,i} \cdot \exp\left(-\eta u_{t,i}\right) / \hat{Z}_{t+1}$, where $\hat{Z}_{t+1} = \sum_{j \in [N]} p_{t+1,j} \cdot \exp\left(-\eta u_{t,j}\right)$.

The following theorem shows the regret bound of our algorithm.

Theorem 1. *When the L_∞-variation of the loss functions is V_∞, the regret of our algorithm is at most $O\left(\sqrt{V_\infty \log T \log N}\right)$.*

Proof. By the choices of $v_i = \mathbf{e}_i$ for each $i \in [N]$, and the fact that $\tilde{f}_t = f_t + \|f_t - \tilde{u}_{t-1}\|_\infty \cdot \mathbf{1}$, the expected regret of our algorithm is at most

$$2\eta \mathbb{E}\left[\sum_{t \in [T]} \|\tilde{g}_t\|_\infty \mathbb{E}_{i \sim \hat{p}_t}\left[\tilde{g}_{t,i}\right]\right] + \frac{\log N}{\eta} \leq 2\eta \mathbb{E}\left[\sum_{t \in [T]} \|\tilde{g}_t\|_\infty^2\right] + \frac{\log N}{\eta},$$

by Lemma 1. It remains to bound the term $\|\tilde{g}_t\|_\infty^2$, which is at most $4\|f_t - u_{t-1}\|_\infty^2$ by Proposition 1. Let $f'_t = f_t - \mu$, which implies $u'_t = \frac{1}{t} \sum_{\tau=1}^t f'_\tau = u_t - \mu$ and $\|f_t - u_{t-1}\|_\infty^2 = \|f'_t - u'_{t-1}\|_\infty^2$. By Proposition 1 and the definition of V_∞, we have $\sum_t \|f'_t - u'_{t-1}\|_\infty^2 \leq O(1) + \sum_{t=2}^T \|f'_t - u'_{t-1}\|_\infty^2$ is at most

$$O(1) + \sum_{t=2}^{T} \left(2 \|f'_t\|_\infty^2 + 2 \|u'_{t-1}\|_\infty^2\right) \leq O(V_\infty) + 2 \sum_{t=2}^{T} \|u'_{t-1}\|_\infty^2 \,.$$

Note that by Proposition 1, $\sum_{t=2}^{T} \|u'_{t-1}\|_\infty^2$ is

$$\sum_{t=2}^{T} \left\| \frac{1}{t-1} \sum_{\tau=1}^{t-1} f'_\tau \right\|_\infty^2 \leq \sum_{t=2}^{T} \frac{1}{t-1} \sum_{\tau=1}^{t-1} \|f'_\tau\|_\infty^2 = \sum_{t=1}^{T-1} \|f'_t\|_\infty^2 \cdot \left(\sum_{\tau=t}^{T-1} \frac{1}{\tau}\right),$$

which implies $\sum_{t=2}^{T} \|u'_{t-1}\|_\infty^2 \leq \sum_{t=1}^{T} \|f'_t\|_\infty^2 \cdot \log T = V_\infty \cdot \log T$.

Combining these bounds together, the expected regret of our algorithm is at most

$$O\left(\eta V_\infty \cdot \log T\right) + \frac{\log N}{\eta} \leq O\left(\sqrt{V_\infty \log T \log N}\right),$$

by setting $\eta = \sqrt{\frac{\log N}{V_\infty \log T}}$.

3.2   Bandit Setting

In this subsection, we consider the bandit setting. Recall that in the bandit setting, we can only obtain the corresponding loss value and $\|f_t - \tilde{u}_{t-1}\|_\infty$. To obtain a small regret in terms of L_∞-variation, we instantiate the META algorithm with parameters $m = N$, $k = \log T$, $v_i = \mathbf{e}_i$ for each $i \in [N]$, and:

- (step 13) Choose j uniformly from $[k]$, and update $S_{I_t,j} = f_{t,I_t}$.
- (step 20) Compute $\tilde{f}_t = \tilde{g}_t + \tilde{u}_{t-1}$, where
  $\tilde{g}_t = \frac{1}{\hat{p}_{t,I_t}} \left(f_{t,I_t} - \tilde{u}_{t-1,I_t} + \|f_t - \tilde{u}_{t-1}\|_\infty\right) \mathbf{e}_{I_t}$.
- (step 21) Compute $\tilde{u}_t = \tilde{u}_{t-1}$.

Then, the updates in step 23 and 24 of META algorithm are:

- $p_{t+1,i} = p_{t,i} \cdot \exp\left(-\eta \tilde{f}_{t,i}\right) / Z_{t+1}$, where $Z_{t+1} = \sum_{j \in [N]} p_{t,j} \cdot \exp\left(-\eta \tilde{f}_{t,j}\right)$;
- $\hat{p}_{t+1,i} = p_{t+1,i} \cdot \exp\left(-\eta \tilde{u}_{t,i}\right) / \hat{Z}_{t+1}$, where $\hat{Z}_{t+1} = \sum_{j \in [N]} p_{t+1,j} \cdot \exp\left(-\eta \tilde{u}_{t,j}\right)$.

It is easy to verify that for each **UPDATE STEP** t,

$$\mathbb{E}_t\left[\tilde{f}_t\right] = \mathbb{E}_t\left[\tilde{g}_t\right] + \tilde{u}_{t-1}$$

$$= \sum_{I_t \in [N]} \hat{p}_{t,I_t} \frac{1}{\hat{p}_{t,I_t}} \left(f_{t,I_t} - \tilde{u}_{t-1,I_t} + \|f_t - \tilde{u}_{t-1}\|_\infty\right) \mathbf{e}_{I_t} + \tilde{u}_{t-1}$$

$$= f_t + \|f_t - \tilde{u}_{t-1}\|_\infty \cdot \mathbf{1},$$

where the first equality is due to the fact that $\tilde{u}_{t-1}$ is fixed when conditioned on the randomness up to round $t - 1$. Moreover, for each $i \in [N]$, $\tilde{g}_{t,i} \geq 0$.

In the bandit setting, we need to estimate the function u_t. The following lemma shows that $\tilde{u}_t$ is an unbiased estimator of u_t, for any $t \geq Nk$. We omit the proof due to the page limit.

Lemma 2. For each $t \geq Nk$, $\mathbb{E}\left[\tilde{u}_t\right] = u_t$, and $\mathbb{E}\left[\left\|u_t - \tilde{u}_t\right\|_\infty^2\right] \leq \frac{V_\infty}{kt}$.

The main result of this subsection is the following.

Theorem 2. Let $k = \log T$. When the L_∞-variation of the loss functions is V_∞, the regret of our algorithm is at most $O\left(N \log^2 T + \sqrt{N V_\infty \log N \log T}\right)$.

Proof. Let T_s include the rounds that run the **SAMPLE STEP**, that is, $T_s = \{t \in [T] : r_t = 1\}$, and T_u include the rounds that run the **UPDATE STEP**, that is, $T_u = [T] \setminus T_s$. Note that $\mathbb{E}\left[T_s\right] \leq O(N \log^2 T)$. Since for any distribution q over $[N]$, $\langle f_t, q - p^* \rangle \leq 2$, the expected regret of our algorithm is at most

$$O(N \log^2 T) + \mathbb{E}\left[\sum_{t \in T_u} \langle f_t, \hat{p}_t - p^* \rangle\right]$$

$$= O(N \log^2 T) + \mathbb{E}\left[\sum_{t \in T_u} \left\langle \tilde{f}_t, \hat{p}_t - p^* \right\rangle\right],$$

$$\leq O(N \log^2 T) + \mathbb{E}\left[2\eta \sum_{t \in T_u} \left(\|\tilde{g}_t\|_\infty \cdot \mathbb{E}_{i \sim \hat{p}_t}\left[\tilde{g}_{t,i}\right]\right)\right] + \frac{\log N}{\eta} \quad (1)$$

where the equality is due to the fact that $\mathbb{E}_t\left[\tilde{f}_t\right] = f_t + \|f_t - \tilde{u}_{t-1}\|_\infty \cdot \mathbf{1}$, and the last inequality follows from Lemma 1.

It remains to bound the second term of (1). First, note that by the definition of $\tilde{g}_t$, we have

$$\mathbb{E}_{i \sim \hat{p}_t}\left[\tilde{g}_{t,i}\right] = f_{t,I_t} - \tilde{u}_{t-1,I_t} + \|f_t - \tilde{u}_{t-1}\|_\infty \leq 2\|f_t - \tilde{u}_{t-1}\|_\infty. \quad (2)$$

On the other hand, since for each $t \in T_u$, when conditioned on the randomness up to round $t - 1$, $\tilde{u}_{t-1}$ is fixed, we obtain

$$\mathbb{E}_t\left[\|\tilde{g}_t\|_\infty\right] = \mathbb{E}_t\left[\left\|\frac{1}{\hat{p}_{t,I_t}}\left(f_{t,I_t} - \tilde{u}_{t-1,I_t} + \|f_t - \tilde{u}_{t-1}\|_\infty\right)\mathbf{e}_{I_t}\right\|_\infty\right]$$

$$= \sum_{I_t \in [N]} \hat{p}_{t,I_t} \left\|\frac{1}{\hat{p}_{t,I_t}}\left(f_{t,I_t} - \tilde{u}_{t-1,I_t} + \|f_t - \tilde{u}_{t-1}\|_\infty\right)\mathbf{e}_{I_t}\right\|_\infty$$

$$\leq 2N\|f_t - \tilde{u}_{t-1}\|_\infty \quad (3)$$

Therefore, using (2), (3), and Proposition 1, the second term of (1) is at most

$$8\eta N \cdot \mathbb{E}\left[\sum_{t \in T_u} \|f_t - \tilde{u}_{t-1}\|_\infty^2\right]$$

$$\leq 24\eta N \cdot \left(\sum_{t \in T_u} \|f_t - \mu\|_\infty^2 + \sum_{t \in T_u} \|\mu - u_{t-1}\|_\infty^2 + \mathbb{E}\left[\sum_{t \in T_u} \|u_{t-1} - \tilde{u}_{t-1}\|_\infty^2\right]\right)$$

where the first term in the parenthesis is bounded by V_∞, and the second term is at most $V_\infty \log T$ as in the proof of Theorem 1. Moreover, the last term is at most $\sum_t V_\infty/(kt) \le V_\infty \log T/k$ by Lemma 2. Therefore, we can bound

$$\mathbb{E}\left[2\eta \sum_{t \in T_u} \left(\|\tilde{g}_t\|_\infty \cdot \mathbb{E}_{i \sim \hat{p}_t}\left[\tilde{g}_{t,i}\right]\right)\right] \le O\left(\eta N V_\infty \log T\right). \tag{4}$$

Finally, by plugging (4) into (1) and setting $\eta = \sqrt{\frac{\log N}{NV_\infty \log T}}$, the expected regret of our algorithm is at most

$$O(N \log^2 T) + O\left(\eta N V_\infty \log T\right) + \frac{\log N}{\eta} \le O\left(N \log^2 T + \sqrt{N V_\infty \log N \log T}\right).$$

4   Combinatorial Prediction Game

In this section, we consider the combinatorial prediction game in which the set of actions is $\mathcal{A} = \{A_1, \cdots, A_N\} \subset \{0,1\}^d$, and the loss function $f_t \in [-1,1]^d$. Moreover, we consider the semi-bandit setting, that is, in round t, we can receive the values $f_{t,j}$ for any $j \in [d]$ satisfying $A_{I_t,j} = 1$ and $\|f_t - \tilde{u}_{t-1}\|_\infty$. To obtain a small regret in terms of L_∞-variation, we instantiate the META algorithm with parameters $m = d$, $k = \log T$, $v_i = A_i$, and:

- (step 13) For each $j \in [d]$ satisfying $A_{I_t,j} = 1$, update, with probability $1/n_j$, where $n_j = \sum_{i \in [N]} A_{i,j}$, $S_{j,a} = f_{t,j}$ for a random index $a \in [k]$.
- (step 20) Compute $\tilde{f}_t = \tilde{g}_t + \tilde{u}_{t-1}$, where
  $\tilde{g}_{t,j} = \frac{1}{\sum_{i;A_{i,j}=1}\hat{p}_{t,i}}\left(f_{t,j} - \tilde{u}_{t-1,j} + \|f_t - \tilde{u}_{t-1}\|_\infty\right)A_{I_t,j}$.
- (step 21) Compute $\tilde{u}_t = \tilde{u}_{t-1}$.

Then, with the choice of $v_i = A_i$, the updates in step 23 and 24 are

- $p_{t+1,i} = \frac{p_{t,i} \cdot \exp\left(-\eta\langle \tilde{f}_t, A_i\rangle\right)}{Z_{t+1}}$, where $Z_{t+1} = \sum_{j \in [N]} p_{t,j} \cdot \exp\left(-\eta\left\langle \tilde{f}_t, A_j\right\rangle\right)$;
- $\hat{p}_{t+1,i} = \frac{p_{t+1,i} \cdot \exp(-\eta\langle \tilde{u}_t, A_i\rangle)}{\hat{Z}_{t+1}}$, where $\hat{Z}_{t+1} = \sum_{j \in [N]} p_{t+1,j} \cdot \exp\left(-\eta\langle \tilde{u}_t, A_j\rangle\right)$.

Note that for each **UPDATE STEP** t, and for each $j \in [d]$, $\tilde{g}_{t,j} \ge 0$, and

$$\mathbb{E}_t\left[\tilde{g}_{t,j}\right] = \sum_{i;A_{i,j}=1} \hat{p}_{t,i}\frac{1}{\sum_{i;A_{i,j}=1}\hat{p}_{t,i}}\left(f_{t,j} - \tilde{u}_{t-1,j} + \|f_t - \tilde{u}_{t-1}\|_\infty\right)$$
$$= f_{t,j} - \tilde{u}_{t-1,j} + \|f_t - \tilde{u}_{t-1}\|_\infty,$$

which implies that $\mathbb{E}_t\left[\tilde{f}_t\right] = f_t + \|f_t - \tilde{u}_{t-1}\|_\infty \cdot \mathbf{1}$.

Moreover, $\tilde{u}_t$ is an unbiased estimator of u_t, whose proof is omitted.

Lemma 3. *For each $t \ge Nk$, $\mathbb{E}\left[\tilde{u}_t\right] = u_t$.*

The regret bound of our algorithm is guaranteed by the following.

Theorem 3. *When the L_∞-variation of the loss functions is V_∞, the regret of our algorithm is at most $O\left(N \log^2 T + d\sqrt{V_\infty \log N \log T}\right)$.*

Proof. As in the proof of Theorem 2, let T_s include the rounds that run the **SAMPLE STEP**, and T_u include the rounds that run the **UPDATE STEP**. Note that $\mathbb{E}[T_s] = O(N \log^2 T)$.

By the choices of $v_i = A_i$ for each $i \in [N]$ and Lemma 1, the expected regret of algorithm $\mathbb{E}\left[\sum_{t \in [T]} \left\langle \tilde{F}_t, \hat{p}_t - p^* \right\rangle\right]$ is at most

$$O(N \log^2 T) + \mathbb{E}\left[\sum_{t \in T_u} 2\eta \left\|\tilde{G}_t\right\|_\infty \mathbb{E}_{i \sim \hat{p}_t}\left[\tilde{G}_{t,i}\right]\right] + \frac{\log N}{\eta} \tag{5}$$

Next, we bound the second term of (5). For convenience, let $q_{t,j} = \sum_{i; A_{i,j}=1} \hat{p}_{t,i} = \mathbb{E}_{i \sim \hat{p}_t}[A_{i,j}]$. Then, since $A_{i,j} \in \{0, 1\}$, the term $\mathbb{E}_{i \sim \hat{p}_t}\left[\tilde{G}_{t,i}\right]$ is

$$\mathbb{E}_{i \sim \hat{p}_t}\left[\sum_{j \in [d]} \tilde{g}_{t,j} A_{i,j}\right] \le \sum_{j \in [d]} \frac{1}{q_{t,j}} (f_{t,j} - \tilde{u}_{t-1,j} + \|f_t - \tilde{u}_{t-1}\|_\infty) \mathbb{E}_{i \sim \hat{p}_t}[A_{i,j}]$$

$$\le 2d \|f_t - \tilde{u}_{t-1}\|_\infty . \tag{6}$$

On the other hand, observe that for each $t \in T_u$,

$$\mathbb{E}_t\left[\left\|\tilde{G}_t\right\|_\infty\right] \le \mathbb{E}_t\left[\sum_{j \in [d]} \frac{2}{q_{t,j}} \|f_t - \tilde{u}_{t-1}\|_\infty A_{I_t,j}\right]$$

$$\le \sum_{j \in [d]} \frac{2}{q_{t,j}} \|f_t - \tilde{u}_{t-1}\|_\infty \mathbb{E}_{I_t \sim \hat{p}_t}[A_{I_t,j}]$$

$$= 2d \|f_t - \tilde{u}_{t-1}\|_\infty , \tag{7}$$

where the first inequality is due to the fact that for each $i \in [N]$, $\tilde{G}_{t,i} = \sum_{j \in [d]} \tilde{g}_{t,j} A_{i,j} \le \sum_{j \in [d]} \frac{2}{q_{t,j}} \|f_t - \tilde{u}_{t-1}\|_\infty A_{I_t,j}$.

Therefore, the second term in (5) is at most $8\eta d^2 \mathbb{E}\left[\sum_{t \in T_u} \|f_t - \tilde{u}_{t-1}\|_\infty^2\right]$ by (6) and (7). By Lemma 3, $\tilde{u}_t$ is an unbiased estimator of u_t, and hence, we can follow the same argument in the proof of Theorem 2 to bound

$$8\eta d^2 \mathbb{E}\left[\sum_{t \in T_u} \|f_t - \tilde{u}_{t-1}\|_\infty^2\right] \le O\left(\eta d^2 V_\infty \log T\right). \tag{8}$$

Finally, by plugging (8) into (5), the regret bound is at most

$$O(N \log^2 T) + O\left(\eta d^2 V_\infty \log T\right) + \frac{\log N}{\eta}$$

$$\le O\left(N \log^2 T + d\sqrt{V_\infty \log T \log N}\right)$$

when $\eta = \sqrt{\log N / d^2 V_\infty \log T}$.

5   Conclusion and Open Problems

By introducing a new measure of the loss functions and with the help of some additional information related to the variation, we obtain a new bound for the prediction with expert advice problem on the bandit setting without using the self-concordance methodology and thus conditionally answer an open problem raised by Hazan and Kale [8]. We also prove a new regret bound for the combinatorial prediction game under the semi-bandit setting.

For future work, we provide the following open problems. (1) Obtain the regret bounds without the additional information. (2) Remove the terms poly($\log T$) in the regret bounds of this paper. (3) Obtain the regret bounds in terms of the deviation for the prediction problems in the bandit setting.

References

1. Abernethy, J., Hazan, E., Rakhlin, A.: Competing in the dark: An efficient algorithm for bandit linear optimization. In: COLT, pp. 263–274 (2008)
2. Audibert, J.-Y., Bubeck, S.: Regret Bounds and Minimax Policies under Partial Monitoring. Journal of Machine Learning Research 11, 2635–2686 (2010)
3. Audibert, J.-Y., Bubeck, S., Lugosi, G.: Minimax Policies for Combinatorial Prediction Games. In: COLT, pp. 107–132 (2011)
4. Chiang, C.-K., Yang, T., Lee, C.-J., Mahdavi, M., Lu, C.-J., Jin, R., Zhu, S.: Online optimization with gradual variations. In: COLT, pp. 6.1–6.20 (2012)
5. Dani, V., Hayes, T., Kakade, S.M.: The Price of Bandit Information for Online Optimization. In: NIPS, pp. 345–352 (2008)
6. Freund, Y., Schapire, R.E.: A Decision-Theoretic Generalization of On-Line Learning and an Application to Boosting. J. Comput. Syst. Sci. 55(1), 119–139 (1997)
7. Hazan, E., Kale, S.: Extracting certainty from uncertainty: Regret bounded by variation in costs. Machine Learning 80(2-3), 165–188 (2010)
8. Hazan, E., Kale, S.: Better Algorithms for Benign Bandits. Journal of Machine Learning Research 12, 1287–1311 (2011)
9. Littlestone, N., Warmuth, M.K.: The Weighted Majority Algorithm. Inf. Comput. 108(2), 212–261 (1994)
10. Vitter, J.S.: Random sampling with a reservoir. ACM Trans. Math. Softw. 11(1), 37–57 (1985)

Nondeterministic Automatic Complexity
of Almost Square-Free
and Strongly Cube-Free Words

Kayleigh K. Hyde and Bjørn Kjos-Hanssen

[1] University of Hawai'i at Mānoa, Honolulu, HI 96822, USA
bjoernkh@hawaii.edu
http://math.hawaii.edu/wordpress/bjoern/
[2] University of Hawai'i at Mānoa, Honolulu, HI 96822, USA
kkhyde@hawaii.edu
http://math.hawaii.edu/wordpress/graduate-alumni/kkhyde/

Abstract. Shallit and Wang studied deterministic automatic complexity of words. They showed that the automatic Hausdorff dimension $I(\mathbf{t})$ of the infinite Thue word satisfies $1/3 \le I(\mathbf{t}) \le 2/3$. We improve that result by showing that $I(\mathbf{t}) \ge 1/2$. For nondeterministic automatic complexity we show $I(\mathbf{t}) = 1/2$. We prove that such complexity A_N of a word x of length n satisfies $A_N(x) \le b(n) := \lfloor n/2 \rfloor + 1$. This enables us to define the complexity deficiency $D(x) = b(n) - A_N(x)$. If x is square-free then $D(x) = 0$. If x almost square-free in the sense of Fraenkel and Simpson, or if x is a strongly cube-free binary word such as the infinite Thue word, then $D(x) \le 1$. On the other hand, there is no constant upper bound on D for strongly cube-free words in a ternary alphabet, nor for cube-free words in a binary alphabet.

The decision problem whether $D(x) \ge d$ for given x, d belongs to $\mathrm{NP} \cap \mathrm{E}$.

1 Introduction

The Kolmogorov complexity of a finite word w is roughly speaking the length of the shortest description w^* of w in a fixed formal language. The description w^* can be thought of as an optimally compressed version of w. Motivated by the non-computability of Kolmogorov complexity, Shallit and Wang studied a deterministic finite automaton analogue.

Definition 1 (Shallit and Wang [3]). *The* automatic complexity *of a finite binary string* $x = x_1 \ldots x_n$ *is the least number* $A_D(x)$ *of states of a deterministic finite automaton* M *such that* x *is the only string of length* n *in the language accepted by* M.

This complexity notion has two minor deficiencies:

1. Most of the relevant automata end up having a "dead state" whose sole purpose is to absorb any irrelevant or unacceptable transitions.

Z. Cai et al. (Eds.): COCOON 2014, LNCS 8591, pp. 61–70, 2014.

2. The complexity of a string can be changed by reversing it. For instance,

$$A_D(011100) = 4 < 5 = A_D(001110).$$

If we replace deterministic finite automata by nondeterministic ones, these deficiencies disappear. The NFA complexity turns out to have other pleasant properties, such as a sharp computable upper bound.

Technical Ideas and Results. In this paper we develop some of the properties of NFA complexity. As a corollary we get a strengthening of a result of Shallit and Wang on the complexity of the infinite Thue word **t**. Moreover, viewed through an NFA lens we can, in a sense, characterize exactly the complexity of **t**. A main technical idea is to extend Shallit and Wang's Theorem 9 which said that not only do squares, cubes and higher powers of a word have low complexity, but a word completely free of such powers must conversely have high complexity. The way we strengthen their results is by considering a variation on square-freeness and cube-freeness, *strong cube-freeness*. This notion also goes by the names of *irreducibility* and *overlap-freeness* in the combinatorial literature. We also take up an idea from Shallit and Wang's Theorem 8 and use it to show that the natural decision problem associated with NFA complexity is in $E = DTIME(2^{O(n)})$. This result is a theoretical complement to the practical fact that the NFA complexity can be computed reasonably fast; to see it in action, for strings of length up to 23 one can view automaton witnesses and check complexity using the following URL format

http://math.hawaii.edu/wordpress/bjoern/complexity-of-110101101/

and check one's comprehension by playing a Complexity Guessing Game at

http://math.hawaii.edu/wordpress/bjoern/software/web/
complexity-guessing-game/

Let us now define our central notion and get started on developing its properties.

Definition 2. *The nondeterministic automatic complexity $A_N(w)$ of a word w is the minimum number of states of an NFA M, having no ϵ-transitions, accepting w such that there is only one accepting path in M of length $|w|$.*

The minimum complexity $A_N(w) = 1$ is only achieved by words of the form a^n where a is a single letter.

Theorem 3 (Hyde [2]). *The nondeterministic automatic complexity $A_N(x)$ of a string x of length n satisfies*

$$A_N(x) \leq b(n) := \lfloor n/2 \rfloor + 1.$$

Proof (Proof sketch.). If x has odd length, it suffices to carefully consider the automaton in Figure 1. If x has even length, a slightly modified automaton can be used.

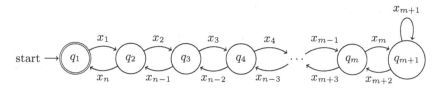

Fig. 1. A nondeterministic finite automaton that only accepts one string $x = x_1x_2x_3x_4 \ldots x_n$ of length $n = 2m + 1$

Definition 4. *The* complexity deficiency *of a word x of length n is*

$$D_n(x) = D(x) = b(n) - A_N(x).$$

Table 1. Probability of strings of having positive complexity deficiency D_n, truncated to 3 decimal digits

Length n	$\mathbb{P}(D_n > 0)$
0	0.000
2	0.500
4	0.500
6	0.531
8	0.617
10	0.664
12	0.600
14	0.687
16	0.657
18	0.658
20	0.641
22	0.633
24	0.593

Length n	$\mathbb{P}(D_n > 0)$
1	0.000
3	0.250
5	0.250
7	0.234
9	0.207
11	0.317
13	0.295
15	0.297
17	0.342
19	0.330
21	0.303
23	0.322
25	0.283

(a) Even lengths.   (b) Odd lengths.

The notion of deficiency is motivated by the experimental observation that about half of all strings have deficiency 0; see Table 1.

2   Time Complexity

Definition 5. *Let* DEFICIENCY *be the following decision problem.*
    Given a binary word w and an integer $d \geq 0$, is $D(w) > d$?

2.1   NP

Theorem 6 is not surprising; we do not know whether DEFICIENCY is NP-complete.

Theorem 6. DEFICIENCY *is in NP.*

Proof. Shallit and Wang's Theorem 2 showed that one can efficiently determine whether a given DFA uniquely accepts w among string of length $|w|$. Hyde [2], Theorem 2.2, extended that result to NFAs, from which the result easily follows.

2.2   E

Definition 7. *Suppose M is an NFA with q states that uniquely accepts a word x of length n. Throughout this paper we may assume that M contains no edges except those traversed on input x. Consider the almost unlabeled transition diagram of M, which is a directed graph whose vertices are the states of M and whose edges correspond to transitions. Each edge is labeled with a 0 except for an edge entering the initial state as described below.*

We define the accepting path P *for x to be the sequence of $n+1$ edges traversed in this graph, where we include as first element an edge labeled with the empty string ε that enters the initial state q_0 of M.*

We define the abbreviated accepting path P' *to be the sequence of edges obtained from P by considering each edge in order and deleting it if it has previously been traversed.*

Lemma 8. *Let v be a vertex visited by an abbreviated accepting path $P' = (e_0, \ldots, e_t)$. Then v is of one of the following five types.*

1. *In-degree 1 (edge e_i), out-degree 1 (edge e_{i+1}).*
2. *In-degree 2 (edges e_i and e_j with $j > i$), out-degree 1 (e_{i+1}).*
3. *In-degree 1 (edge e_i), out-degree 2 (edges e_{i+1} and e_j, $j > i+1$).*
4. *In-degree 2 (edges e_i and e_j with $j > i$), out-degree 2 (e_{i+1} and e_{j+1}).*
5. *In-degree 1 (edge e_t), out-degree 0.*[1]

Proof. The out-degree and in-degree of each vertex encountered along P' are both ≤ 2, since failure of this would imply non-uniqueness of accepting path. Since all the edges of M are included in P, the list includes all the possible in-degree, out-degree combinations. We can define i by the rule that e_i is the first edge in P' entering v. Again, since all the edges of M are included in P, e_{i+1} must be one of the edges contributing to the out-degree of v, if any, and e_j must also be as specified in the types.

Lemma 8 implies that Definition 9 makes sense.

Definition 9. *For $0 \leq i \leq t+1$ and $0 \leq n \leq t+1$ we let $E(i, n)$ be a string representing the edges $(e_i, \ldots, e_n)$. The meaning of the symbols is as follows: 0 represents an edge. A left bracket [represents a vertex that is the target of a backedge. A right bracket] represents a backedge. The symbol + represents a vertex of out-degree 2. When $i > n$, we set $E(i, n) = \varepsilon$. Next, assuming we have defined $E(j, m)$ for all m and all $j > i$, we can define $E(i, n)$ by considering the type of the vertex reached by the edge e_i. Let $a_i \in \{0, \varepsilon\}$ be the label of e_i.*

---

[1] This type was omitted by Shallit and Wang.

1. $E(i, n) := a_i E(i + 1, n)$.
2. $E(i, n) := a_i [E(i + 1, j - 1)] E(j + 1, n)$.
3. $E(i, n) := a_i + E(i + 1, n)$.
4. $E(i, n) := a_i [+E(i + 1, j - 1)] E(j + 1, n)$.
5. $E(i, n) := a_i E(i + 1, n)$.

Lemma 10. *The abbreviated accepting path P' can be reconstructed from $E(0, t)$.*

Lemma 11

$$|E(a, b)| \leq 2(b - a + 1).$$

Theorem 12. DEFICIENCY *is in E.*

Proof. Let w be a word of a length n, and let $d \geq 0$. To determine whether $D(w) > d$, we must determine whether there exists an NFA M with at most $\lfloor \frac{n}{2} \rfloor - d$ states which accepts w, and accepts no other word of length n. Since there are *prima facie* more than single-exponentially many automata to consider, we consider instead codes $E(0, t)$ as in Definition 9. By Lemma 10 we can recover the abbreviated accepting path P' and hence M from such a code. The number of edges t is bounded by the string length n, so by Lemma 11

$$|E(0, t)| \leq 2(t + 1) \leq 2(n + 1);$$

since there are four symbols this gives

$$4^{2(n+1)} = O(16^n)$$

many codes to consider. Finally, to check whether a given M accepts uniquely takes only polynomially many steps, as in Theorem 6.

Remark 13. *The bound 16^n counts many automata that are not uniquely accepting; the actual number may be closer to 3^n based on computational evidence.*

3 Powers and Complexity

In this section we shall exhibit infinite words all of whose prefixes have complexity deficiency bounded by 1. We say that such a word has a hereditary deficiency bound of 1.

3.1 Square-Free Words

Lemma 14. *Let a and b be strings in an arbitrary alphabet with $ab = ba$. Then there is a string c and integers k and ℓ such that $a = c^k$ and $b = c^\ell$.*

We will use the following simple strengthening from DFAs to NFAs of a fact used in Shallit and Wang's Theorem 9 [3].

Theorem 15. *If an NFA M uniquely accepts w of length n, and visits a state p as many as $k + 1$ times, where $k \geq 2$, during its computation on input w, then w contains a kth power.*

Proof. Let $w = w_0 w_1 \cdots w_k w_{k+1}$ where w_i is the portion of w read between visits number i and $i + 1$ to the state p. Since one bit must be read in one unit of automaton time, $|w_i| \geq 1$ for each $1 \leq i \leq k$ (w_0 and/or w_{k+1} may be empty since the initial and/or final state of M may be p). For any permutation π on $1, \ldots, k$, M accepts $w_0 w_{\pi(1)} \cdots w_{\pi(k)} w_{k+1}$. Let $1 \leq j \leq k$ be such that w_j has minimal length and let $\hat{w}_j = w_1 \cdots w_{j-1} w_{j+1} \cdots w_k$. Then M also accepts

$$w_0 w_j \hat{w}_j w_{k+1} \quad \text{and} \quad w_0 \hat{w}_j w_j w_{k+1}.$$

By uniqueness,

$$w_0 w_j \hat{w}_j w_{k+1} = w = w_0 \hat{w}_j w_j w_{k+1}$$

and so

$$w_j \hat{w}_j = \hat{w}_j w_j$$

By Lemma 14, w_j and $\hat{w}_j$ are both powers of a string c. Since $|\hat{w}_j| \geq (k-1)|w_j|$, $w_j \hat{w}_j$ is at least a kth power of c, so w contains a kth power of c. $\quad\blacksquare$

We next strengthen a particular case of Shallit and Wang's Theorem 9 to NFAs.

Theorem 16. *A square-free word has deficiency 0.*

Corollary 17. *There exists an infinite word of hereditary deficiency 0.*

Proof. There is an infinite square-free word over the alphabet $\{0, 1, 2\}$ as shown by Thue [5][6]. The result follows from Theorem 16. $\quad\blacksquare$

3.2   Cube-Free Words

Definition 18. *For a word u, let $first(u)$ and $last(u)$ denote the first and last letters of u, respectively. A weak cube is a word of the form $uu\,first(u)$ (or equivalently, $last(u)\,uu$). A word w is strongly cube-free if it does not contain any weak cubes.*

Theorem 19 (Shelton and Soni [4]). *The set of all numbers that occur as lengths of squares within strongly cube-free binary words is equal to*

$$\{2^a : a \geq 1\} \cup \{3 \cdot 2^a : a \geq 1\}.$$

Lemma 20. *If a cube www contains another cube xxx then either $|x| = |w|$, or $xx\,first(x)$ is contained in the first two consecutive occurrences of w, or $last(x)\,xx$ is contained in the last two occurrences of w.*

Theorem 21. *The deficiency of cube-free binary words is unbounded.*

Proof. Given k, we shall find a cube-free word x with $D(x) \geq k$. Pick a number n such that $2^n \geq 2k + 1$. By Theorem 19, there is a strongly cube-free binary word that contains a square of length 2^{n+1}; equivalently, there is a strongly cube-free square of length 2^{n+1}. Thus, we may choose w of length $\ell = 2^n$ such that ww

is strongly cube-free. Let $x = ww\hat{w}$ where $\hat{w}$ is the proper prefix of w of length $|w| - 1$. By Lemma 20, x is cube-free. The complexity of x is at most $|w|$ as we can just make one loop of length w, with code (Theorem 12)

$$[w_1 \ldots w_{\ell-1}]_{w_\ell}.$$

And so

$$D(x) \geq \lfloor |x|/2 + 1 \rfloor - |w| \geq |x|/2 - |w| = \frac{3|w| - 1}{2} - |w|$$

$$= |w|/2 - 1/2 \geq k.$$

3.3 Strongly Cube-Free Words

Theorem 22 (Thue [5][6]). *The infinite Thue word*

$$\mathbf{t} = t_0 t_1 \ldots = 0110\,1001\,1001\,0110 \ldots$$

given by

$$b = \sum b_i 2^i, \quad b_i \in \{0,1\} \quad \Longrightarrow \quad t_b = \sum b_i \quad \bmod 2,$$

is strongly cube-free.

Lemma 23. *For each $k \geq 1$ there is a sequence $x_{1,k}, \ldots, x_{k,k}$ of positive integers such that*

$$\sum_{i=1}^{k} a_i x_{i,k} = 2 \sum_{i=1}^{k} x_{i,k} \quad \Longrightarrow \quad a_1 = \cdots = a_k = 2$$

Let t_j denote bit j of the infinite Thue word. Then we can ensure that

1. $x_{i,k} + 1 < x_{i+1,k}$ *and*
2. $t_{x_{i,k}} \neq t_{x_{i+1,k}}$ *for each $1 \leq i < k$.*

Theorem 24. *For an alphabet of size three, the complexity deficiency of strongly cube-free words is unbounded.*

Proof. Let $d \geq 1$. We will show that there is a word w of deficiency $D(w) \geq d$. Let $k = 2d - 1$. For each $1 \leq i \leq k$ let $x_i = x_{k+1-i,k}$ where the $x_{j,k}$ are as in Lemma 23. Note that since $x_{i,k} + 1 < x_{i+1,k}$, we have $x_i > x_{i+1} + 1$. Let

$$w = \left(2 \prod_{i=1}^{x_1-1} t_i \right)^2 t_{x_1} \left(2 \prod_{i=1}^{x_2-1} t_i \right)^2 t_{x_2} \left(2 \prod_{i=1}^{x_3-1} t_i \right)^2 \cdots t_{x_{k-1}} \left(2 \prod_{i=1}^{x_k-1} t_i \right)^2$$

$$= \lambda_1 t_{x_1} \lambda_2 \cdots t_{x_{k-1}} \lambda_k$$

where $\lambda_i = (2\tau_i)^2$, $\tau_i = \prod_{j=1}^{x_i-1} t_j$, and where t_i is the ith bit of the infinite Thue word on $\{0,1\}$, which is strongly cube-free (Theorem 22). Let M be the NFA with code (Theorem 12)

$$[+0^{x_1-1}]0[+0^{x_2-1}]0 \cdots 0 * [+0^{x_k-1}]$$

(where $*$ indicates the accept state). Let $X = \sum_{i=1}^{k} x_i$. Then M has $k - 1 + X$ many edges but only $q = X$ many states; and w has length

$$n = k - 1 + 2X = 2(d - 1) + 2X$$

giving $n/2 + 1 = d + X$.

Suppose v is a word accepted by M. Then M on input v goes through each loop of length x_i some number of times $a_i \geq 0$, where

$$k - 1 + \sum_{i=1}^{k} a_i x_i = |v|.$$

If additionally $|v| = |w|$, then by Lemma 23 we have $a_1 = a_2 = \cdots = a_k$, and hence $v = w$. Thus

$$D(w) \geq \lfloor n/2 + 1 \rfloor - q = d + X - X = d.$$

We omit the proof that w is strongly cube-free in this version of the paper.

Definition 2 yields the following lemma.

Lemma 25. *Let* $(q_0, q_1, \ldots)$ *be the sequence of states visited by an NFA* M *given an input word* w. *For any* t, t_1, t_2, *and* r_i, s_i *with*

$$(p_1, r_1, \ldots, r_{t-2}, p_2) = (q_{t_1}, \ldots, q_{t_1+t})$$

and

$$(p_1, s_1, \ldots, s_{t-2}, p_2) = (q_{t_2}, \ldots, q_{t_2+t}),$$

we have $r_i = s_i$ *for each* i.

Note that in Lemma 25, it may very well be that $t_1 \neq t_2$.

Theorem 26. *Strongly cube-free binary words have deficiency bound 1.*

Proof. Suppose w is a word satisfying $D(w) \geq 2$ and consider the sequence of states visited in a witnessing computation. As in the proof of Theorem 32, either there is a state that is visited four times, and hence there is a cube in w, or there are three *state cubes* (states that are visited three times each), and hence there are three squares in w. By Theorem 19, a strongly cube-free binary word can only contain squares of length 2^a, $3 \cdot 2^a$, and hence can only contain powers u^i where $|u|$ is of the form 2^a, $3 \cdot 2^a$, and $i \leq 2$.

In particular, the length of one of the squares in the three state cubes must divide the length of another. So if these two state cubes are disjoint then the shorter one repeated can replace one occurrence of the longer one, contradicting Lemma 25.

So suppose we have two state cubes, at states p_1 and p_2, that overlap. At p_1 then we read consecutive words ab that are powers $a = u^i$, $b = u^j$ of a word u, and since there are no cubes in w it must be that $i = j = 1$ and so actually

$a = b$. And at p_2 we have words c, d that are powers of a word v and again the exponents are 1 and $c = d$.

The overlap means that in one of the two excursions of the same length starting and ending at p_1, we visit p_2. By uniqueness of the accepting path we then visit p_2 in both of these excursions. If we suppose the state cubes are chosen to be of minimal length then we only visit p_2 once in each excursion. If we write $a = rs$ where r is the word read when going from p_1 to p_2, and s is the word going from p_2 to p_1, then $c = sr$ and w contains $rsrsr$. In particular, w contains a weak cube.

Definition 27. *For an infinite word* **u** *define the* deterministic automatic Hausdorff dimension *of* **u** *by*

$$I(\mathbf{u}) = \liminf_{u \ prefix \ of \ \mathbf{u}} A_D(u)/|u|.$$

and the deterministic automatic packing dimension *of* **u** *by*[2]

$$S(\mathbf{u}) = \limsup_{u \ prefix \ of \ \mathbf{u}} A_D(u)/|u|.$$

For nondeterministic complexity, in light of Theorem 3 it is natural to make the following definition.

Definition 28. *Define the* nondeterministic automatic Hausdorff dimension *of* **u** *by*

$$I_N(\mathbf{u}) = \liminf_{u \ prefix \ of \ \mathbf{u}} \frac{A_N(u)}{|u|/2}$$

and define S_N analogously.

Theorem 29 (Shallit and Wang's Theorem 18). $\frac{1}{3} \le I(\mathbf{t}) \le \frac{2}{3}$.

Here we strengthen Theorem 29.

Theorem 30. $I(\mathbf{t}) \ge \frac{1}{2}$. *Moreover* $I_N(\mathbf{t}) = S_N(\mathbf{t}) = 1$.

Proof. This follows from the observation that the proof of Theorem 26 applies equally for deterministic complexity.

3.4   Almost Square-Free Words

Definition 31 (Fraenkel and Simpson [1]). *A word all of whose contained squares belong to $\{00, 11, 0101\}$ is called* almost square-free.

Theorem 32. *A word that is almost square-free has a deficiency bound of 1.*

---

[2]   There is some connection with Hausdorff dimension and packing dimension. For instance, if the effective Hausdorff dimension of an infinite word **x** is positive then so is its automatic Hausdorff dimension, by a Kolmogorov complexity calculation in Shallit and Wang's Theorem 9.

Corollary 33. *There is an infinite binary word having hereditary deficiency bound of 1.*

Proof. We have two distinct proofs. On the one hand, Fraenkel and Simpson [1] show there is an infinite almost square-free binary word, and the result follows from Theorem 32. On the other hand, the infinite Thue word is strongly cube-free (Theorem 22) and the result follows from Theorem 26.

Conjecture 34. *There is an infinite binary word having hereditary deficiency 0.*

We have some numerical evidence for Conjecture 34, for instance there are 108 strings of length 18 with this property.

References

1. Fraenkel, A.S., Simpson, R.J.: How many squares must a binary sequence contain? Electron. J. Combin. 2: Research Paper 2, approx., 9 p. (electronic) (1995)
2. Hyde, K.: Nondeterministic finite state complexity. Master's thesis, University of Hawaii at Manoa, U.S.A. (2013)
3. Shallit, J., Wang, M.-W.: Automatic complexity of strings. J. Autom. Lang. Comb. 6(4), 537–554 (2001), 2nd Workshop on Descriptional Complexity of Automata, Grammars and Related Structures (London, ON, 2000)
4. Shelton, R.O., Soni, R.P.: Chains and fixing blocks in irreducible binary sequences. Discrete Math. 54(1), 93–99 (1985)
5. Thue, A.: Über unendliche zeichenreihen. Norske Vid. Skrifter I Mat.-Nat. Kl., Christiania 7, 1–22 (1906)
6. Thue, A.: Über die gegenseitige lage gleicher teile gewisser zeichenreihen. Norske Vid. Skrifter I Mat.-Nat. Kl., Christiania 1, 1–67 (1912)

An Axiomatization for Cylinder Computation Model*

Nan Zhang, Zhenhua Duan**, and Cong Tian

Institute of Computing Theory and Technology, and ISN Laboratory
Xidian University, Xi'an 710071, China
zhhduan@mail.xidian.edu.cn

Abstract. To model and verify multi-core parallel programs, the paper proposes an axiom system for Propositional Projection Temporal Logic with Cylinder Computation Model (CCM-PPTL). To do so, the syntax and semantics of CCM-PPTL are presented. Further, based on the logical laws of PPTL, some algebraic laws of sequence expressions and logical laws regarding CCM operators are proved. Moreover, the axiom system of CCM-PPTL is established by extending that of PPTL with some axioms and inference rules of CCM operators. In addition, the soundness and completeness of the system are proved.

Keywords: Axiom System, Multi-core, Parallel, Formal Method.

1 Introduction

With the rapid development of integrated circuits technology and the demand for higher performance, on-chip multi-core processors (CMP) have been brought into being. The reality of multi-core processor has made parallel programs pervasive. Creating a correct parallel program is not a straightforward process even for a considerable small system, because programmers are forced to consider that the program will always yield to a correct result no matter what order the instructions are executed in. To improve the reliability of parallel programs, formal verification is an important viable approach. Modeling multi-core parallel programs is a crucial step for formal verification of correctness and reliability of many core parallel programs.

Model checking and theorem proving are two key verification methods. With model checking, the system is often modeled as a finite transition system or automaton M, and the property is specified using a temporal logic formula P. Then a model checking procedure is employed to check whether or not $M \models P$. If so, the property is verified otherwise a counterexample can be found. The advantage of model checking is that the verification can be done automatically. However, model checking suffers from the state explosion problem [10]. Further, most of web applications are data-intensive which are not suitable to be verified by means of model checking since the treatment of the data usually leads to a huge, even infinite state space. Some successful model checking tools are SPIN [9], SMV [10] and so on. By contrast, theorem proving can

---

* The research is supported by the National Program on Key Basic Research Project of China (973 Program) Grant No.2010CB328102, National Natural Science Foundation of China under Grant No. 61133001, 61202038, 61272117, 61272118, 61322202 and 91218301.
** Corresponding author.

handle many complex structures abstractly without state space explosion but requires more human intervention and are often time-consuming. With theorem proving, both the system behavior and the desired property are specified as formulas, say S and P, in some appropriate logic. To demonstrate that the system satisfies the property amounts to proving that $\vdash S \to P$ is a theorem within the proof system of the logic. Some famous theorem prover are PVS [11], ACL2 [2], Coq [1], Isabella [12], HOL [8] and so on. Verification of multi-core parallel programs raises a great challenge for theorem proving since it requires that the logic for modeling multi-core systems and specifying the expected properties has a powerful expressiveness. However, the widely used Propositional Linear Temporal Logic (PLTL) and Computational Tree Logic (CTL) are not powerful enough. In fact, they are not full regular. Further, Quantified Linear time Temporal Logic (QLTL) [13], Extended Temporal Logic (ETL) [16] and Linear mu-Calculus [15] have a more powerful expressiveness of full regular language. However, these logics are not practical since they are not intuitive or too complicated. Propositional Projection Temporal Logic (PPTL) [3] allows us to specify ω full regular properties [14]. Further, a decision procedure [4,7] and a complete proof system for PPTL [6] have been established. A model checker based on SPIN [5] and a theorem prover based on PVS have also been developed. Cylinder Computation Model (CCM) [17] is a concurrent semantic model which is defined based on PPTL and has been implemented in the interpreter of MSVL, which is an executable subset of Projection Temporal Logic. CCM can be employed to model multi-core parallel programs since the sequence expressions in it have the nature of regular expressions. With CCM, the autonomy and parallelism of the processes occupying different cores on one chip can be described neatly and concisely. In [6], we have proposed an axiom system for PPTL, and proved its soundness and completeness. To specify and verify multi-core parallel programs in a uniform framework, this paper proposes an axiom system for CCM-PPTL which extends that of PPTL by including transformation rules for sequence expressions and axioms and inference rules on CCM operators. Furthermore, the soundness and completeness of the extended axiom system are also proved.

The paper is organized as follows. In the next section, the underlying logic PPTL and the semantic model CCM are reviewed, including their syntax, semantics and PPTL axiom system. Based on PPTL, CCM-PPTL is proposed in Section 3, including its syntax and semantics. In Section 3, we further give an axiom system for CCM-PPTL and prove the soundness and completeness of the system. Finally, conclusions are drawn in Section 4.

2   Preliminaries

2.1   Propositional Projection Temporal Logic

Our underlying logic is Propositional Projection Temporal Logic. The formula P of PPTL is given by the following grammar.

$$P ::= p \mid \bigcirc P \mid \neg P \mid P_1 \vee P_2 \mid (P_1, \ldots, P_m)\ prj\ P$$
$$\mid (P_1, \ldots, (P_i, \ldots, P_l)^{\oplus}, \ldots, P_m)\ prj\ P$$

where $p \in Prop$, $P_i(1 \le i \le m)$ and P are well-formed PPTL formulas, and $\bigcirc$, prj and $prj\oplus$ (projection-plus) are primitive temporal operators. A formula is a state formula if it contains no temporal operators, otherwise it is a temporal formula.

We define a state s over $Prop$ to be a mapping from $Prop$ to B. $s[p]$ denotes the valuation of p at state s. An interval σ is a non-empty finite or infinite sequence of states. The length, $|\sigma|$, of σ is ω if σ is infinite, and the number of states minus 1 if σ is finite. We consider the set N_0 of non-negative integers and ω, $N_\omega = N_0 \cup \{\omega\}$ and extend the comparison operators, $=$, $<$, $\le$, to N_ω by considering $\omega = \omega$, and for all $i \in N_0$, $i < \omega$. Furthermore, we define $\preceq$ as $\le -\{(\omega, \omega)\}$. σ is often denoted by $\langle s_0, \ldots, s_{|\sigma|} \rangle$, where $s_{|\sigma|}$ is undefined if σ is infinite. With such a notation, $\sigma_{(i...j)}$ $(0 \le i \preceq j \le |\sigma|)$ denotes the sub-interval $\langle s_i, \ldots, s_j \rangle$ and σ^i $(0 \le i \le |\sigma|)$ denotes the prefix interval $\langle s_0, \ldots, s_i \rangle$. The concatenation of a finite σ with another interval (or empty string) σ' is denoted by $\sigma \cdot \sigma'$ (not sharing any states). Let $\sigma = \langle s_0, s_1, \ldots, s_{|\sigma|} \rangle$ be an interval and $r_1, \ldots, r_h$ be integers $(h \ge 1)$ such that $0 \le r_1 \le r_2 \le \ldots \le r_h \preceq |\sigma|$. The projection of σ onto $r_1, \ldots, r_h$ is the interval (called projected interval) $\sigma \downarrow (r_1, \ldots, r_h) = \langle s_{t_1}, s_{t_2}, \ldots, s_{t_l} \rangle$ where $t_1, \ldots, t_l$ are obtained from $r_1, \ldots, r_h$ by deleting all duplicates. That is, $t_1, \ldots, t_l$ is the longest strictly increasing subsequence of $r_1, \ldots, r_h$. An interpretation is a triple $\mathcal{I} = (\sigma, k, j)$, where σ is an interval, k an integer, and j an integer or ω such that $0 \le k \preceq j \le |\sigma|$. We use the notation $(\sigma, k, j) \models P$ to indicate that some formula P is interpreted and satisfied over the subinterval $\langle s_k, \ldots, s_j \rangle$ of σ with the current state being s_k. The satisfaction relation ($\models$) is inductively defined as follows.

$\mathcal{I} \models p$ iff $s_k[p] = true$, for any atomic proposition p.

$\mathcal{I} \models \neg P$ iff $\mathcal{I} \not\models P$.

$\mathcal{I} \models \bigcirc P$ iff $k < j$ and $(\sigma, k+1, j) \models P$.

$\mathcal{I} \models P \vee Q$ iff $\mathcal{I} \models P$ or $\mathcal{I} \models Q$.

$\mathcal{I} \models (P_1, \ldots, P_m)$ prj Q iff there exist integers $k = r_0 \le \cdots \le r_{m-1} \preceq r_m \le j$; for all $1 \le l \le m$, $(\sigma, r_{l-1}, r_l) \models P_l$; $(\sigma', 0, |\sigma'|) \models Q$ for one of the following σ':
- $r_m < j$ and $\sigma' = \sigma \downarrow (r_0, \ldots, r_m) \cdot \sigma_{(r_m+1..j)}$, or
- $r_m = j$ and $\sigma' = \sigma \downarrow (r_0, \ldots, r_h)$ for some $0 \le h \le m$.

$\mathcal{I} \models (P_1, \ldots, (P_u, \ldots, P_l)^{\oplus}, \ldots, P_m)$ prj Q iff one of following cases holds:
- $1 \le u \le l \le m$ and there exists an integer $n \ge 1$ and $\mathcal{I} \models (P_1, \ldots, (P_u, \ldots, P_l)^{(n)}, \ldots, P_m)$ prj Q, or
- $1 \le u \le l = m$, $j = \omega$ and there exist infinitely many integers $k = r_0 \le r_1 \le \cdots \le r_n \preceq \omega$ such that for all $1 \le x \le u-1$, $(\sigma, r_{x-1}, r_x) \models P_x$, and $\lim_{n \to \infty} r_n = \omega$
and $(\sigma, r_{u+t(l-u+1)+n-1}, r_{u+t(l-u+1)+n}) \models P_{u+n}$, for all $t \ge 0$ and $0 \le n \le l-u$, and $\sigma \downarrow (r_0, r_1, \ldots, r_h, \omega) \models Q$ for some $h \in N_\omega$.

The axiom system for CCM-PPTL presented later is based on that of PPTL, which has been proved to be sound and complete. For more detail, please refer to [6].

2.2 Cylinder Computation Model

In this section, Cylinder Computation Model (CCM) is reviewed [17], including its syntax and semantics which are based on sequence expressions. Then the logical laws on CCM are drawn. First, sequence expressions are defined as follows.

$$l ::= \emptyset \mid \epsilon \mid n \mid l_1, l_2 \mid l_1 \otimes l_2 \mid l^*$$

From the syntax of the sequence expression given above, we see that it is an analogue of regular expressions where $\emptyset$ denotes empty set, ϵ empty sequence expression and n any non-negative integer. The concatenation $(\cdot)$, sum $(\otimes)$ or Kleene closure $(*)$ of any two sequence expressions is also a sequence expression. The semantics is also defined by a satisfaction relation, $\Vdash$, by means of interpretation $\mathcal{I} = (\sigma, k, j)$.

1. $\mathcal{I} \nVdash \emptyset$ for all $\mathcal{I}$.
2. $\mathcal{I} \Vdash \epsilon$ iff $j = k$.
3. $\mathcal{I} \Vdash n$ iff $j - k = n$.
4. $\mathcal{I} \Vdash l_1 \otimes l_2$ iff $\mathcal{I} \Vdash l_1$ or $\mathcal{I} \Vdash l_2$.
5. $\mathcal{I} \Vdash l_1, l_2$ iff there exists r, $k \leq r \preceq j$, such that $\mathcal{I}_1 = (\sigma, k, r) \Vdash l_1$ and $\mathcal{I}_2 = (\sigma, r, j) \Vdash l_2$.
6. $\mathcal{I} \Vdash l^*$ iff $j = k$ or there exist finitely many integers $k = r_0 \leq r_1 \leq \ldots \preceq r_n = j$ such that for all h, $1 \leq h \leq n$, $(\sigma, r_{h-1}, r_h) \Vdash l$.

In the semantics of sequence expressions, $\emptyset$ cannot be satisfied by any interpretation. The empty sequence expression ϵ is equivalent to the sequence expression 0. In order to avoid an excessive number of parentheses, the precedence rules are given from high to low: (1) $*$ (iteration); (2) $\cdot$ (concatenation); (3) $\otimes$ (selection). Some algebraic laws of sequence expressions are summarized in [17]. Then, the syntax of CCM is defined as follows.

$$CCM ::= P \text{ ov } (l) \mid CCM_1 \parallel CCM_2$$

where P is a PPTL formula, l a sequence expression and the parallel ($\parallel$) composition of any two CCM formulas is also a CCM formula. CCM operators " ov " and " $\parallel$ " are temporal. So all the CCM formulas are temporal formulas. With CCM formulas, the interpretation of P is controlled by the sequence expression l. The beginning and ending points generated by the interpretation of l make up of the coarse-grained interval of P. Therefore, to give the semantics of CCM formulas, it is necessary to define the set of ending point lists denoted by $S_l^{\mathcal{I}}$. First, some notations are defined. For any two strings $X \overset{\text{def}}{=} (x_1, \ldots, x_m)$ and $Y \overset{\text{def}}{=} (y_1, \ldots, y_n)$, the concatenation of X and Y is defined as: $X, Y \overset{\text{def}}{=} (x_1, \ldots, x_m), (y_1, \ldots, y_n) \overset{\text{def}}{=} (x_1, \ldots, x_m, y_1, \ldots, y_n)$. For any two sets S_1 and S_2, of strings, the concatenation of S_1 and S_2 is defined as: $S_1, S_2 \overset{\text{def}}{=} \{X, Y \mid X \in S_1 \text{ and } Y \in S_2\}$.

Definition 1 (set of ending point lists $S_l^{\mathcal{I}}$) Let $\mathcal{I}$ be an arbitrary interpretation (σ, i, k, j). $S_l^{\mathcal{I}}$ is inductively defined as follows:

1. $S_{\emptyset}^{\mathcal{I}} = \emptyset$.
2. If $\mathcal{I} \Vdash \epsilon$, then $S_{\epsilon}^{\mathcal{I}} = \{(k, j)\}$.
3. If $\mathcal{I} \Vdash n$, then $S_n^{\mathcal{I}} = \{(k, j)\}$.
4. If $\mathcal{I} \Vdash l_1, l_2$, then $S_{l_1, l_2}^{\mathcal{I}} = \left\{ \tau \left| \begin{array}{l} \text{there exists } r, k \leq r \preceq j, \text{ such that } \mathcal{I}_1 = \\ (\sigma, k, r) \Vdash l_1 \text{ and } \mathcal{I}_2 = (\sigma, r, j) \Vdash l_2 \text{ and} \\ \tau \in S_{l_1}^{\mathcal{I}_1}, S_{l_2}^{\mathcal{I}_2} \end{array} \right. \right\}$
5. If $\mathcal{I} \Vdash l_1 \otimes l_2$, then $S_{l_1 \otimes l_2}^{\mathcal{I}} = S_{l_1}^{\mathcal{I}} \bigcup S_{l_2}^{\mathcal{I}}$.

6. If $\mathcal{I} \Vdash l^*$, then $S_{l^*}^{\mathcal{I}} = S_\epsilon^{\mathcal{I}} \cup \left\{ \tau \left| \begin{array}{l} \text{there exist finitely many integers } k = r_0 \le r_1 \ldots \preceq \\ r_n = j \text{ such that for all } 1 \le h \le n, \mathcal{I}_h = \\ (\sigma, r_{h-1}, r_h) \Vdash l \text{ and } \tau \in S_l^{\mathcal{I}_1}, S_l^{\mathcal{I}_2}, \ldots, S_l^{\mathcal{I}_n} \end{array} \right. \right\}$

Then, the semantics of Cylinder Computation Model is defined by a satisfaction relation $\models$ by means of the interpretation $\mathcal{I} = (\sigma, k, j)$.

$\mathcal{I} \models P$ ov (l) iff one of the following cases holds:

(a) $\mathcal{I} \Vdash l$ and there exists $(r_0, r_1, \ldots, r_n) \in S_l^{\mathcal{I}}$, $n \in N_0$ such that P is satisfied by $\sigma \downarrow (r_0, r_1, \ldots, r_h)$ for some $0 \le h \le n$;

(b) there exists $r, k \le r \preceq j$ such that $\mathcal{I}_1 = (\sigma, k, r) \Vdash l$ and there exists $(r_0, r_1, \ldots, r_n) \in S_l^{\mathcal{I}_1}$, $n \in N_0$ and P is satisfied by $\sigma \downarrow (r_0, r_1, \ldots, r_n) \cdot \sigma_{(r_n+1..j)}$.

$\mathcal{I} \models CCM_1 \parallel CCM_2$ iff one of the following cases holds:

(a) $\mathcal{I} \models CCM_1$ and there exists $r, k \le r \preceq j, (\sigma, k, r) \models CCM_2$

(b) $\mathcal{I} \models CCM_2$ and there exists $r, k \le r \preceq j, (\sigma, k, r) \models CCM_1$

In fact, an element in $S_l^{\mathcal{I}}$ is a sequence of non-negative integers, and a sequence expression can be satisfied by an interpretation in more than one way. Each element in $S_l^{\mathcal{I}}$ records one particular way in which $\mathcal{I}$ satisfies l, containing all the beginning and ending points. The definition of $S_l^{\mathcal{I}}$ is necessary since the PPTL formula P in CCM is interpreted over a coarse-grained interval composed of the points from one of the elements in $S_l^{\mathcal{I}}$.

3   CCM-PPTL and Axiom System

To model and verify multi-core parallel programs, CCM is included in PPTL. Then an axiom system is proposed in this section. The syntax of CCM-PPTL is inductively defined as follows:

$$\beta ::= p \mid \neg \beta \mid \bigcirc \beta \mid \beta_1 \vee \beta_2 \mid (\beta_1, \ldots, \beta_m) \, prj \, \beta$$
$$\mid (\beta_1, \ldots, (\beta_u, \ldots, \beta_l)^\oplus, \ldots, \beta_m) \, prj \, \beta \mid CCM$$

where p is an arbitrary atomic proposition; β, β_i are arbitrary CCM-PPTL formulas; CCM an arbitrary CCM formula defined in section 2. The semantics of CCM-PPTL is also defined as a satisfaction relation $\models$ by means of the interpretation $\mathcal{I} = (\sigma, k, j)$.

$\mathcal{I} \models p$ iff $s_k[p] = true$ for any atomic proposition p.

$\mathcal{I} \models \neg \beta$ iff $\mathcal{I} \not\models \beta$.

$\mathcal{I} \models \bigcirc \beta$ iff $(\sigma, k+1, j) \models \beta$.

$\mathcal{I} \models \beta_1 \vee \beta_2$ iff $\mathcal{I} \models \beta_1$ or $\mathcal{I} \models \beta_2$.

$\mathcal{I} \models (\beta_1, \ldots, \beta_m) \, prj \, \beta$ iff

there exist $k = r_0 \le r_1 \le \ldots \le r_{m-1} \preceq r_m \le j$ such that for all $1 \le l \le m$, $(\sigma, r_{l-1}, r_l) \models \beta_l$, and one of the following cases holds:

(a) $r_m < j$ and β is satisfied by $\sigma \downarrow (r_0, \ldots, r_m) \cdot \sigma_{(r_m+1..j)}$;

(b) $r_m = j$ and β is satisfied by $\sigma \downarrow (r_0, \ldots, r_h)$ for some $0 \le h \le m$.

$\mathcal{I} \models (\beta_1, \ldots, (\beta_u, \ldots, \beta_l)^\oplus, \ldots, \beta_m) \, prj \, \beta$ iff one of the following cases holds:

(a) $\mathcal{I} \models (\beta_1, \ldots, (\beta_u, \ldots, \beta_l)^{(n)}, \ldots, \beta_m) \, prj \, \beta$ for some $n \ge 1$ and $n \in N_0$;

(b) $l = m$ and $j = \omega$ and there exist infinitely many integers $k = r_0 \le r_1 \le \cdots$ and $\lim\limits_{x \to \infty} r_x = \omega$, such that

- $(\sigma, r_{x-1}, r_x) \models \beta_x$ for $1 \le x \le u - 1$, and
- $(\sigma, r_{u+t(l-u+1)+n-1}, r_{u+t(l-u+1)+n}) \models \beta_{u+n}$, for $t \ge 0$ and $0 \le n \le l - u$, and
- β is satisfied by $\sigma \downarrow (r_0, r_1, \ldots, r_h)$ for some $h \in N_\omega$.

$\mathcal{I} \models CCM$ see the semantics of CCM in Section 2.

It should be noted that the formula P appearing in P ov (l) is a PPTL formula and doesn't contain the ov operator. That is why the syntax and semantics of PPTL and CCM-PPTL have to be separated to define. CCM is of a typical form of P_1 ov $(l_1) \parallel \cdots \parallel P_m$ ov (l_m) where each P_i is a PPTL formula and each l_i is a sequence expression. With this parallelism, a main time interval is the sequence of fine-grained unit subintervals with length one while several coarse-grained projected intervals over which processes are interpreted are in parallel with the main time interval. This computation model can be viewed as m processes that share one processor and each occupies an execution core cooperating to complete their tasks in a parallel way. Each process progresses in its own speed and communicates with each other at some global states which indicates the coordination among these processes. Sequence expression l_i is used to control and determine the execution points (states) of P_i. $\parallel$ is the main operator in CCM. Thus P_1 ov $(l_1) \parallel \cdots \parallel P_m$ ov (l_m) is endowned with the semantics of many-core parallel computing. For example, the interval satisfying CCM formula P_1 ov $(2, 3, 3, 4) \parallel P_2$ ov $(3, 5, 3, 6) \parallel P_3$ ov $(2, 1, 2, 3, 3, 1, 5)$ is given in Fig.1.

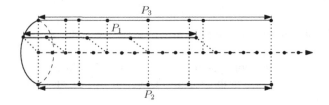

Fig. 1. P_1 ov $(2, 3, 3, 4) \parallel P_2$ ov $(3, 5, 3, 6) \parallel P_3$ ov $(2, 1, 2, 3, 3, 1, 5)$

All the logical laws in PPTL also hold in CCM-PPTL. In addition, we also prove some logical laws on CCM, for more details, refer to [3]. Some of these laws are chosen to be axioms later and are used to transform any CCM formula into its normal form. Now we introduce a normal form for CCM-PPTL formulas upon which the completeness proof of the axiom system is based.

Definition 2 (normal form of CCM-PPTL). A CCM-PPTL formula β is in normal form if it conforms to the following syntax: $\alpha_e \wedge \varepsilon \vee \bigvee_{i=1}^{r}(\alpha_i \wedge \bigcirc \beta_i)$, where $r \ge 1$, α_e and the α_i's are state formulas, whereas the β_i's are general CCM-PPTL formulas. Moreover, β is in complete normal form if $\bigvee_{i=1}^{r} \alpha_i \equiv true$ and $\bigvee_{i \ne j}(\alpha_i \wedge \alpha_j) \equiv false$.

Now, an axiom system for CCM-PPTL is formalized based on that of PPTL. Axioms and inference rules on CCM operators are included. Since the deduction of CCM formulas depends on the nature of sequence expressions, some essential transformation rules on sequence expressions need to be included in the proof system.

S1	$\epsilon \simeq \epsilon^* \simeq 0 \simeq 0^*$	S7	$l^*, l^* \simeq l^*$
S2	$0, l \simeq l, 0 \simeq l$	S8	$(l^*)^* \simeq l^*$
S3	$l_1, (l_2, l_3) \simeq (l_1, l_2), l_3 \simeq l_1, l_2, l_3$	S9	$l_1, (l_2, l_1)^* \simeq (l_1, l_2)^*, l_1$
S4	$l_1, (l_2 \otimes l_3), l_4 \simeq (l_1, l_2, l_4) \otimes (l_1, l_3, l_4)$	S10	$(0 \otimes l)^* \simeq l^*$
S5	$l^* \simeq \epsilon \otimes (l, l^*) \simeq (\epsilon \otimes l)^*$	S11	$l_1 \simeq l_2 \Longrightarrow l \simeq l[l_2/l_1]$
S6	$l, l^* \simeq l^*, l$	S12	$l \simeq (l_1, l) \otimes l_2 \Longrightarrow l \simeq l_1^*, l_2$

Then, the axioms on CCM formulas are given as follows based on the transformation rules of sequence expressions.

A1	$P \text{ ov } (l_1, \emptyset, l_2) \cong false$
A2	$P \text{ ov } (0) \cong P$
A3	$\varepsilon \text{ ov } (m, l) \cong \bigcirc(\varepsilon \text{ ov } (m - 1, l)) \ (m > 0)$
A4	$\bigcirc P \text{ ov } (m, l) \cong \bigcirc^m \varepsilon; (P \text{ ov } (l)) \ (m > 0)$
A5	$(w \wedge P) \text{ ov } (l) \cong w \wedge (P \text{ ov } (l))$
A6	$P \text{ ov } (l_1 \otimes l_2) \cong (P \text{ ov } (l_1)) \vee (P \text{ ov } (l_2))$
A7	$(P_1 \vee P_2) \text{ ov } (l) \supset (P_1 \text{ ov } (l)) \vee (P_2 \text{ ov } (l))$
A8	$CCM_1 \parallel CCM_2 \cong (CCM_1; true) \wedge CCM_2 \vee (CCM_2; true) \wedge CCM_1$

The inference rules are presented in the following:

I1	$P \supset P' \Longrightarrow P \text{ ov } (l) \supset P' \text{ ov } (l)$
I2	$l_1 \simeq l_2 \Longrightarrow P \text{ ov } (l_1) \cong P \text{ ov } (l_2)$

Some explanations are needed. A1 means that any CCM formula P ov (l) with an unsatisfiable sequence expression l is also unsatisfiable. A2 means that CCM formula P ov (0) is deduced to the PPTL formula P. A3 means that if a sequence expression begins with a positive integer m and the PPTL formula is ε, we can extract one *next* operator directly with m decreasing by one. A4 means that if a sequence expression begins with a positive integer m and the major operator of the PPTL formula is the *next* operator ($\bigcirc$), we can extract m *next* operators directly with deleting m from the sequence expression and the next operator from $\bigcirc P$. A5 means that if the PPTL formula contains a conjunction being a state formula w, then w can be extracted from the PPTL formula and treated as a conjunction of the whole formula. A6 indicates the distributivity of the sum operator $\otimes$ over the ov operator. A7 describes the distributivity of disjunction over the ov operator. A8 presents the semantics of the parallel operator, that is, CCM_1 and CCM_2 are interpreted in parallel and may specify their own lengths. I1 means that the implication between any two PPTL formulas is preserved by the ov operator. I2 means that if l_1 is deduced to be the equivalent of l_2, the two CCM formulas with l_1 and l_2 being sequence expressions respectively are also deduced to be equivalent. Then the soundness and completeness of the axiom system of CCM-PPTL are demonstrated.

Theorem 1 (Soundness). *For any CCM-PPTL formula β, if $\vdash \beta$, then $\models \beta$.*

Proof. We need to prove that each axiom in the proof system of CCM-PPTL is valid in the model theory of CCM-PPTL and each inference rule preserves the validity of premises. Since the proof system of PPTL is sound, we only need to consider the axioms and inference rules on CCM operators. We can prove that each transformation rule of sequence expressions is an algebraic law, each axiom on CCM is also a logical law. Two inference rules I1 and I2 are also easy to understand, which formalize the idea of substitution. All the above ensures the soundness of the axiom system of CCM-PPTL. Since the proof is not difficult, we omit it here.

To prove the completeness of the axiom system given in Theorem 3, ten lemmas are proved in advance. In general, the set of CCM-PPTL formulas are partitioned into terminable and non-terminable formulas. We will prove that any terminable formula is satisfiable (Lemma 7), and that for any non-terminable formula β, if $\not\vdash \beta \rightarrow false$, then β is satisfiable (Lemma 10). Lemma 10 is based on a fact that any CCM-PPTL formula can be deduced into a normal form which is proved in Theorem 2. The proof of Theorem 2 depends on Lemma 1-6. Because of space limitations, most of the details of proofs is omitted here.

Lemma 1. *For any CCM-PPTL formula β, if $\beta \cong \beta'$ where β' is in normal form, there exists a CCM-PPTL formula β_c in complete normal form satisfying $\beta \cong \beta_c$.*

Lemma 2. *Let $\alpha_1, \ldots, \alpha_n$ be state formulas, and β_i a general CCM-PPTL formula. If $\bigvee_{i=1}^n \alpha_i \cong true$ and $\bigvee_{i \neq j} \alpha_i \wedge \alpha_j \cong false$, then $\neg(\bigvee_{i=1}^n \alpha_i \wedge \beta_i) \cong \bigvee_{i=1}^n (\alpha_i \wedge \neg\beta_i)$.*

Lemma 1 indicates that any normal form can be deduced into a complete normal form. In the deduction of $\neg\beta$ into a normal form as we will see later on in Theorem 2, β is deduced into its normal form first, then further deduced into its complete normal form by means of Lemma 1. Finally, we deduce $\neg\beta$ into its normal form using Lemma 2 based on β's complete normal form.

Lemma 3. *If $\beta_i \cong \beta_i'$ $(0 \leq i \leq m)$, where β_i''s are CCM-PPTL formulas in normal form, then there exists a formula β in normal form such that $(\beta_1, \ldots, \beta_m)$ prj $\beta_0 \cong \beta$.*

Lemma 4. *If $\beta_i \cong \beta_i'$ $(0 \leq i \leq m)$, where β_i''s are CCM-PPTL formulas in normal form, then (1) there exists a formula β in normal form such that*
$$((\beta_1, \ldots, \beta_l)^\oplus, \ldots, \beta_m) \text{ prj } \beta_0 \cong \beta \ (1 \leq l \leq m);$$
(2) there exists a formula β in normal form such that
$$(\beta_1, \ldots, (\beta_i, \ldots, \beta_l)^\oplus, \ldots, \beta_m) \text{ prj } \beta_0 \cong \beta \ (1 < i \leq l \leq m).$$

Lemma 3 and 4 show that the projection and projection plus operators can be deduced into normal forms. They are integral parts of the proof of Theorem 2. To deduce CCM formulas into their normal forms, first we need to deduce the sequence expression into one of the three following forms using transformation rules, which are formalized in the following lemma.

Lemma 5. *For any sequence expression l, l can be deduced using the transformation rules into one of the following forms:*
(Form 1) $\emptyset$
(Form 2) $0 \otimes \bigotimes_{i=1}^m (n_i, l_i)$ where $m \geq 0$ and $n_i \in N$ for all $1 \leq i \leq m$.
(Form 3) $\bigotimes_{i=1}^m (n_i, l_i)$ where $m \geq 1$ and $n_i \in N$ for all $1 \leq i \leq m$.

Proof. The proof proceeds by induction on the structure of sequence expressions.

Base:

(1) l is $\emptyset$, then it is already in Form 1.

(2) l is ϵ, then $l \simeq 0$ according to the algebraic laws of sequence expressions, which is in Form 2 under the condition $m = 0$.

(3) l is n, if n is zero, it is the same as case (2); if n is a positive integer, according to the algebraic laws we have $n \simeq (n, 0)$, which is in Form 3 under the condition $m = 1$, $n_1 = n$ and $l_1 = 0$.

Induction:

(4) l is (l_1, l_2), with the hypothesis that both of l_1 and l_2 can be transformed into one of the three forms, then there are 3×3 possible combinations.

If l_1 or l_2 is transformed into $\emptyset$, which covers 5 possible combinations, l can be equivalently transformed into $\emptyset$ which is in Form 1.

If both of the transformations of l_1 and l_2 are in Form 2, l is transformed into Form 2.

$$
\begin{aligned}
l &\simeq (l_1, l_2) & \\
&\simeq (0 \otimes \bigotimes_{i=1}^{m}(n_i, l_i), 0 \otimes \bigotimes_{j=1}^{n}(n_j', l_j')) & \text{Hypothesis} \\
&\simeq (0, 0 \otimes \bigotimes_{j=1}^{n}(n_j', l_j')) \otimes (\bigotimes_{i=1}^{m}(n_i, l_i), 0 \otimes \bigotimes_{j=1}^{n}(n_j', l_j')) & \text{S4} \\
&\simeq (0, 0) \otimes (0, \bigotimes_{j=1}^{n}(n_j', l_j')) \otimes (\bigotimes_{i=1}^{m}(n_i, l_i), 0) \otimes (\bigotimes_{i=1}^{m}(n_i, l_i), \bigotimes_{j=1}^{n}(n_j', l_j')) & \text{S4} \\
&\simeq 0 \otimes \bigotimes_{j=1}^{n}(n_j', l_j') \otimes \bigotimes_{i=1}^{m}(n_i, l_i) \otimes \bigotimes_{i=1}^{m} \bigotimes_{j=1}^{n}(n_i, l_i, n_j', l_j') & \text{S2, S4}
\end{aligned}
$$

If l_1 is in Form 2 and l_2 Form 3, then l is in Form 3.

$$
\begin{aligned}
l &\simeq (l_1, l_2) \simeq (0 \otimes \bigotimes_{i=1}^{m}(n_i, l_i), \bigotimes_{j=1}^{n}(n_j', l_j')) & \text{Hypothesis} \\
&\simeq (0, \bigotimes_{j=1}^{n}(n_j', l_j')) \otimes (\bigotimes_{i=1}^{m}(n_i, l_i), \bigotimes_{j=1}^{n}(n_j', l_j')) & \text{S4} \\
&\simeq \bigotimes_{j=1}^{n}(n_j', l_j') \otimes \bigotimes_{i=1}^{m} \bigotimes_{j=1}^{n}(n_i, l_i, n_j', l_j') & \text{S2, S4}
\end{aligned}
$$

If l_1 is in Form 3 and l_2 Form 2, then l will be transformed into Form 3, and if both l_1 and l_2 are in Form 3, then l will be in Form 3. The proofs of these two cases are similar as the proof given above, so they are omitted here.

(5) l is $l_1 \otimes l_2$, with the hypothesis that l_1 and l_2 are transformed into l_1' and l_2', then $l_1' \otimes l_2'$ is already in one of the three forms.

(6) l is $(l')^*$, using the transformation rule S5, we have $l \simeq \epsilon \otimes (l', (l')^*)$; then using S1, we have $l \simeq 0 \otimes (l', (l')^*)$. Suppose that l' has been transformed into l''. If $l'' \simeq \emptyset$, then $l = 0$ which is in Form 2. If $l'' \simeq 0 \otimes \bigotimes_{i=1}^{m}(n_i, l_i)$, then l is in Form 2.

$$
\begin{aligned}
l &\simeq (l')^* & \\
&\simeq (0 \otimes \bigotimes_{i=1}^{m}(n_i, l_i))^* & \text{Hypothesis} \\
&\simeq (\bigotimes_{i=1}^{m}(n_i, l_i))^* & \text{S10} \\
&\simeq 0 \otimes (\bigotimes_{i=1}^{m}(n_i, l_i), (\bigotimes_{i=1}^{m}(n_i, l_i))^*) & \text{S5} \\
&\simeq 0 \otimes \bigotimes_{i=1}^{m}(n_i, l_i, (\bigotimes_{j=1}^{m}(n_j, l_j))^*) & \text{S4}
\end{aligned}
$$

If l'' is in Form 3, that is, $l'' \simeq \bigotimes_{i=1}^{m}(n_i, l_i)$, then l is in Form 2.

$$
\begin{aligned}
l &\simeq (l')^* & \\
&\simeq 0 \otimes (l', (l')^*) & \text{S5, S1} \\
&\simeq 0 \otimes (\bigotimes_{i=1}^{m}(n_i, l_i), (l')^*) & \text{Hypothesis} \\
&\simeq 0 \otimes \bigotimes_{i=1}^{m}(n_i, l_i, (l')^*) & \text{S4}
\end{aligned}
$$

Lemma 6. *For any CCM formula β, there exists a formula β' in normal form such that $\beta \cong \beta'$.*

Proof. The proof proceeds by induction on the syntax of CCM.

Base: For P ov (l), by Lemma 5, if $l \simeq \emptyset$, then P ov $(l) \cong false$ (A1).

If $l \simeq 0 \otimes \bigotimes_{i=1}^{m}(n_i, l_i)$, by the theorem that any PPTL formula P can be deduced into its normal form, we have

$$
\begin{aligned}
&P \text{ ov } (l) \\
&\cong P \text{ ov } (0 \otimes \bigotimes_{i=1}^{m}(n_i, l_i)) &&\text{I2} \\
&\cong P \text{ ov } (0) \vee \bigvee_{i=1}^{m} P \text{ ov } (n_i, l_i) &&\text{A6} \\
&\cong P \vee \bigvee_{i=1}^{m} P \text{ ov } (n_i, l_i) &&\text{A2} \\
&\cong P_e \wedge \varepsilon \vee \bigvee_{j=1}^{n}(P_{cj} \wedge \bigcirc P'_{cj}) \vee \bigvee_{i=1}^{m}(P_e \wedge \varepsilon \vee \bigvee_{j=1}^{n}(P_{cj} \wedge \bigcirc P'_{cj})) \text{ ov } (n_i, l_i) &&\text{Hypothesis} \\
&\cong P_e \wedge \varepsilon \vee \bigvee_{j=1}^{n}(P_{cj} \wedge \bigcirc P'_{cj}) \vee \bigvee_{i=1}^{m} \bigvee_{j=1}^{n}(P_{cj} \wedge \bigcirc P'_{cj}) \text{ ov } (n_i, l_i) &&\text{A7} \\
&\cong P_e \wedge \varepsilon \vee \bigvee_{j=1}^{n}(P_{cj} \wedge \bigcirc P'_{cj}) \\
&\quad \vee \bigvee_{i=1}^{m} P_e \wedge (\varepsilon \text{ ov } (n_i, l_i)) \vee \bigvee_{i=1}^{m} \bigvee_{j=1}^{n} P_{cj} \wedge (\bigcirc P'_{cj} \text{ ov } (n_i, l_i)) &&\text{A5} \\
&\cong P_e \wedge \varepsilon \vee \bigvee_{j=1}^{n}(P_{cj} \wedge \bigcirc P'_{cj}) \\
&\quad \vee \bigvee_{i=1}^{m} P_e \wedge \bigcirc(\varepsilon \text{ ov } (n_i - 1, l_i)) \vee \bigvee_{i=1}^{m} \bigvee_{j=1}^{n} P_{cj} \wedge \bigcirc^{n_i}(P'_{cj} \text{ ov } (l_i)) &&\text{A3, A4}
\end{aligned}
$$

If $l \simeq \bigotimes_{i=1}^{m}(n_i, l_i)$, P ov (l) also can be deduced into its normal form in a similar way. Induction: For $CCM_1 \parallel CCM_2$, suppose that CCM_1 and CCM_2 have been deduced into their normals, then

$$
\begin{aligned}
&CCM_1 \parallel CCM_2 \\
&\cong (CCM_1; true) \wedge CCM_2 \vee (CCM_2; true) \wedge CCM_1 &&\text{A8} \\
&\cong ((\alpha_{1e} \wedge \varepsilon \vee \bigvee_{i=1}^{m} \alpha_{1i} \wedge \bigcirc \beta_{1i}); true) \wedge (\alpha_{2e} \wedge \varepsilon \vee \bigvee_{j=1}^{n} \alpha_{2j} \wedge \bigcirc \beta_{2j}) \\
&\quad \vee ((\alpha_{2e} \wedge \varepsilon \vee \bigvee_{j=1}^{n} \alpha_{2j} \wedge \bigcirc \beta_{2j}); true) \wedge (\alpha_{1e} \wedge \varepsilon \vee \bigvee_{i=1}^{m} \alpha_{1i} \wedge \bigcirc \beta_{1i}) &&\text{Hypothesis} \\
&\cong ((\alpha_{1e} \wedge \varepsilon; true) \vee \bigvee_{i=1}^{m}(\alpha_{1i} \wedge \bigcirc \beta_{1i}; true)) \wedge (\alpha_{2e} \wedge \varepsilon \vee \bigvee_{j=1}^{n} \alpha_{2j} \wedge \bigcirc \beta_{2j}) \\
&\quad \vee ((\alpha_{2e} \wedge \varepsilon; true) \vee \bigvee_{j=1}^{n}(\alpha_{2j} \wedge \bigcirc \beta_{2j}; true)) \wedge (\alpha_{1e} \wedge \varepsilon \vee \bigvee_{i=1}^{m} \alpha_{1i} \wedge \bigcirc \beta_{1i}) &&\text{PDF, PEB} \\
&\cong (\alpha_{1e} \wedge \varepsilon; true) \wedge (\alpha_{2e} \wedge \varepsilon) \\
&\quad \vee \bigvee_{j=1}^{n}(\alpha_{1e} \wedge \varepsilon; true) \wedge (\alpha_{2j} \wedge \bigcirc \beta_{2j}) \vee \bigvee_{i=1}^{m} \bigvee_{j=1}^{n}(\alpha_{1i} \wedge \bigcirc \beta_{1i}; true) \wedge (\alpha_{2j} \wedge \bigcirc \beta_{2j}) \\
&\quad \vee \bigvee_{i=1}^{m}(\alpha_{2e} \wedge \varepsilon; true) \wedge (\alpha_{1i} \wedge \bigcirc \beta_{1i}) \vee \bigvee_{j=1}^{n} \bigvee_{i=1}^{m}(\alpha_{2j} \wedge \bigcirc \beta_{2j}; true) \wedge (\alpha_{1i} \wedge \bigcirc \beta_{1i}) &&\text{TAU} \\
&\cong \alpha_{1e} \wedge \alpha_{2e} \wedge \varepsilon \\
&\quad \vee \bigvee_{j=1}^{n} \alpha_{1e} \wedge \alpha_{2j} \wedge \bigcirc \beta_{2j} \vee \bigvee_{i=1}^{m} \alpha_{2e} \wedge \alpha_{1i} \wedge \bigcirc \beta_{1i} \\
&\quad \vee \bigvee_{i=1}^{m} \bigvee_{j=1}^{n} \alpha_{1i} \wedge \alpha_{2j} \wedge \bigcirc((\beta_{1i}; true) \wedge \beta_{2j} \vee (\beta_{2j}; true) \wedge \beta_{1i}) &&\text{PSM, PEB}
\end{aligned}
$$

Lemma 6 tells us that any CCM formula can be deduced into a normal form after its sequence expression having been deduced into one of three forms, which is also an integral part of the proof of Theorem 2.

Theorem 2. *For any CCM-PPTL formula β, there exists a formula β' in normal form such that $\beta \cong \beta'$.*

Proof. The proof proceeds by induction on the syntax of CCM-PPTL.

Base:

(1) For any atomic proposition p, $p \cong p \wedge \varepsilon \vee p \wedge \bigcirc true$.

(2) For $\bigcirc \beta$, $\bigcirc \beta \cong true \wedge \bigcirc \beta$.

Induction:

(3) For $\neg \beta$, suppose that β can be deduced into its normal form, from Lemma 1, β also can be deduced into its complete normal form $\alpha_e \wedge \varepsilon \vee \bigvee_{i=1}^{r}(\alpha_i \wedge \bigcirc \beta_i)$ where $\bigvee_{i=1}^{r} \alpha_i \cong true$ and $\bigvee_{i \neq j} \alpha_i \wedge \alpha_j \cong false$. Then we have

$$\neg\beta \cong \neg(\alpha_e \wedge \varepsilon \vee \bigvee_{i=1}^{r}(\alpha_i \wedge \bigcirc\beta_i)) \cong \neg\alpha_e \wedge \varepsilon \vee \bigvee_{i=1}^{r}(\alpha_i \wedge \bigcirc\neg\beta_i)$$

(4) For $\beta_1 \vee \beta_2$, suppose β_1 and β_2 have been transformed into their normal form, then

$$\beta_1 \vee \beta_2 \cong \alpha_{1e} \wedge \varepsilon \vee \bigvee_{i=1}^{m}\alpha_{1i} \wedge \bigcirc\beta_{1i} \vee \alpha_{2e} \wedge \varepsilon \vee \bigvee_{j=1}^{n}\alpha_{2j} \wedge \bigcirc\beta_{2j}$$
$$\cong (\alpha_{1e} \vee \alpha_{2e}) \wedge \varepsilon \vee \bigvee_{i=1}^{m}\alpha_{1i} \wedge \bigcirc\beta_{1i} \vee \bigvee_{j=1}^{n}\alpha_{2j} \wedge \bigcirc\beta_{2j}$$

(5) For $(\beta_1, \ldots, \beta_m)$ prj β_0, refer to the proof of Lemma 3.
(6) For $(\beta_1, \ldots, (\beta_i, \ldots, \beta_l)^{\oplus}, \ldots, \beta_m)$ prj β_0, refer to the proof of Lemma 4.
(7) For CCM, refer to the proof of Lemma 6.

Definition 3 (terminable formula and non-terminable formula) For any CCM-PPTL formula β, if $\beta \wedge \Diamond\varepsilon \not\equiv false$, then β is a terminable formula. Otherwise, it is a non-terminable formula.

We can easily prove that any terminable CCM-PPTL formula β, β is satisfiable. Since β is terminable, by Definition 3, $\beta \wedge \Diamond\varepsilon \not\equiv false$, which means that there exists a model σ satisfies $\beta \wedge \Diamond\varepsilon$. Then σ is also a model of β and so β is satisfiable. Then we derive the following conclusion.

Lemma 7. *For any terminable CCM-PPTL formula β, if $\not\vdash \beta \rightarrow false$, then β is satisfiable.*

We can prove by contradiction that for any CCM-PPTL formulas β and β', if $\beta \equiv \beta'$ where β is non-terminable and β' in normal form, then β' is of the form $\bigvee_{i=1}^{n} \alpha_i \wedge \bigcirc\beta_i$ with each α_i being a state formula and each β_i being non-terminable. From hypothesis, we can derive a contradiction to the premise that β is non-terminable. Then we have the following conclusion.

Lemma 8. *For any CCM-PPTL formula β, if $\beta \equiv \beta'$ where β' is in normal form of $\alpha_e \wedge \varepsilon \vee \bigvee_{i=1}^{n} \alpha_i \wedge \bigcirc\beta_i$, then*
(1) If $\alpha_e \not\equiv false$, then β is terminable.
(2) If there exists some β_i being terminable, then β is terminable.

With Lemma 8 which is a conclusion on model theory of CCM-PPTL, we can derive a similar conclusion on the axiom system by using contradiction proof.

Lemma 9. *For any CCM-PPTL formula β and β', if $\beta \cong \beta'$ where β is non-terminable and β' in normal form, then β' must be of the form $\bigvee_{i=1}^{n} \beta_i \wedge \bigcirc\beta_i$ with each β_i being non-terminable.*

Lemma 9 means that if a non-terminable formula β has been deduced into its normal form β' using the axiom system, then we can infer that there is no terminal product $\alpha_e \wedge \varepsilon$ in β', and each future product in normal form be also non-terminable. Otherwise, there will be a contradiction to the premise that β is non-terminable.

Lemma 10. *For any non-terminable CCM-PPTL formula β, if $\not\vdash \beta \rightarrow false$, then β is satisfiable.*

The proof of Lemma 10 is with intricacy. It involves constructing an interval for β and then prove the interval is indeed a model of β. Two famous theorems on fix-point are used in the proof, one is Taski's fix-point theorem and the other is Scott's fix-point induction. The proof is omitted here. From Lemma 7 and Lemma 10, we can derive that any CCM-PPTL formula β, no matter whether it is terminable or non-terminable, if $\nvdash \beta \rightarrow false$, then β is satisfiable. Then we have the following corollary.

Corollary 1. *For any CCM-PPTL formula β, if β is unsatisfiable, then $\vdash \beta \rightarrow false$.*

Theorem 3 (Completeness). *For any CCM-PPTL formula β, if $\models \beta$, then $\vdash \beta$.*

Proof. From the premise of β is valid, we can derive the duality that $\neg\beta$ is unsatisfiable. By Corollary 1, we have $\neg\beta \rightarrow false$ is a theorem in the proof system of CCM-PPTL, which means that β is a theorem in the proof system.

4   Conclusion

We introduce a Cylinder Computation Model into Propositional Projection Temporal Logic and propose an axiom system for CCM-PPTL, which can be employed to model and verify many-core computation systems. In the future, we need to do some further case studies for more complex many-core computation. Further, to provide a highly automatical verification approach, the existing tool for PPTL theorem proving will be extended to support CCM operators. Moreover, we will explore the verification methodology which combines model checking and theorem proving.

References

1. Bertot, Y., Castéran, P.: Interactive Theorem Proving and Program Development, Heidelberg (2004)
2. Brock, B., Kaufmann, M., Moore, J.: ACL2 theorems about commercial micro-processors. In: Srivas, M., Camilleri, A. (eds.) FMCAD 1996. LNCS, vol. 1166, pp. 275–293. Springer, Heidelberg (1996)
3. Duan, Z.: Temporal Logic and Temporal Logic Programming. Science Press, Beijing (2006)
4. Duan, Z., Tian, C., Zhang, L.: A decision procedure for propositional projection temporal logic with infinite models. Acta Informatica 45, 43–78 (2008)
5. Duan, Z., Tian, C.: A unified model checking approach with projection temporal logic. In: Liu, S., Araki, K., Maibaum, T. (eds.) ICFEM 2008. LNCS, vol. 5256, pp. 167–186. Springer, Heidelberg (2008)
6. Duan, Z., Zhang, N., Koutny, M.: A complete proof system for propositional projection temporal logic. Theoretical Computer Science 497, 84–107 (2013)
7. Duan, Z., Tian, C.: A practical decision procedure for propositional projection temporal logic with infinite models. Theoretical Computer Science (2014) doi:10.1016/j.tcs.2014.02.011
8. Gordon, M., Melham, T.: Introduction to HOL: A Theorem Proving Environment for Higher Order Logic. Cambridge University Press (1993)
9. Holzmann, G.: The model checker SPIN. IEEE Trans. Softw. Eng. 23(5), 279–295 (1997)
10. McMillan, K.: Symbolic Model Checking: An Approach to the State Explosion Problem, Dordrecht (1993)
11. Owre, S., Rushby, J., Shankar, N.: PVS: A prototype verification system. In: Kapur, D. (ed.) CADE 1992. LNCS (LNAI), vol. 607, pp. 748–752. Springer, Heidelberg (1992)
12. Paulson, L.C.: Isabelle. LNCS, vol. 828. Springer, Heidelberg (1994)

13. Sistla, A.: Theoretical issues in the design and verification of distributed systems, Ph.D. Thesis. Harvard University (1983)
14. Tian, C., Duan, Z.: Expressiveness of propositional projection temporal logic with star. Theoretical Computer Science 412(18), 1729–1744 (2011)
15. Vardi, M.: A temporal fixpoint calculus. In: POPL 1988, pp. 250–259 (1988)
16. Wolper, P.: Temporal logic can be more expressive. Information and Control 56, 72–99 (1983)
17. Zhang, N., Duan, Z., Tian, C.: A cylinder computation model for many-core parallel computing. Theoretical Computer Science 497, 68–83 (2013)

Normal Form Expressions of Propositional Projection Temporal Logic[*]

Zhenhua Duan, Cong Tian[**], and Nan Zhang

Institute of Computing Theory and Technology, and ISN Laboratory
Xidian University, Xi'an 710071, China
ctian@mail.xidian.edu.cn

Abstract. This paper presents normal form expressions of Propositional Projection Temporal Logic (PPTL). For doing so, a PPTL formula is represented as the disjunction of formulas in form of $e_\varepsilon^k = \bigwedge_{0 \le i \le k \in N_0} \bigcirc^i S_i \wedge \bigcirc^k \varepsilon$ or $e_\omega^{(k,l)} = \bigwedge_{0 \le i \le k \in N_0} \bigcirc^i S_i \wedge \bigwedge_{k \le j \in N_\omega} \bigcirc^j (\bigcirc S_{k+1} \wedge \bigcirc^2 S_{k+2} \wedge \cdots \wedge \bigcirc^l S_{k+l}), 1 \le l \in N_0$. Here e_ε^k denotes a finite model with length being k while $e_\omega^{(k,l)}$ indicates an infinite model. We show that any PPTL formula can be expressed as a normal form expression. As a consequence, satisfiability of PPTL formulas can easily be achieved.

Keywords: Propositional projection temporal logic, normal form expression, normal form, specification, satisfiability.

1 Introduction

Temporal logics are popular formalisations that can express properties about the temporal order of events. They are widely used in model checking for specifying desired properties of a system to be verified. The family of temporal logics has grown over the years, containing linear [9] and branching time logics [4,2], and, more recently, game, alternating time, and coordination logics [1,10]. While linear time temporal logics are concerned with properties of paths, branching time logics describe properties that depend on the branching of computational tree structures.

Interval-based temporal logics such as Interval Temporal Logic (ITL) [11] and Projection Temporal Logic (PTL) [5,6,7] which extends ITL with infinite models and a new projection construct, $(P_1, \ldots, P_m)$ prj Q, are a more recent branch of temporal logics with their own niche of interesting applications. Propositional PTL (PPTL) is a propositional subset of PTL with a usual next construct and the projection construct that is able to express chop construct, often denoted by the symbol '; ', by $P ; Q \stackrel{\text{def}}{=} (P, Q)$ prj ε. Compared with classic temporal logics, interval-based temporal logics greatly simplify the formulation of certain correctness properties [12], which underlines the usefulness of these logics for specification and formal reasoning about concurrent systems. Interval-based temporal logics lend themselves particularly well to reasoning about

---

[*] The research is supported by the National Program on Key Basic Research Project of China (973 Program) Grant No.2010CB328102, National Natural Science Foundation of China under Grant No. 61133001, 61202038, 61272117, 61272118, 61322202 and 91218301.

[**] Corresponding author.

Z. Cai et al. (Eds.): COCOON 2014, LNCS 8591, pp. 84–93, 2014.
© Springer International Publishing Switzerland 2014

properties with a 'scope'; such properties are quite common in most programming languages. Further, with *chop* operators, sequential behaviours can be described elegantly and succinctly; and full regular expressiveness is achieved by projection construct.

In this paper, we present normal form expression that represents a PPTL formula as the disjunction of formulas in form of

$$e_\varepsilon^k = \bigwedge_{0 \le i \le k \in N_0} \bigcirc^i S_i \wedge \bigcirc^k \varepsilon$$

or

$$e_\omega^{(k,l)} = \bigwedge_{0 \le i \le k \in N_0} \bigcirc^i S_i \wedge \bigwedge_{k \le j \in N_\omega} \bigcirc^j (\bigcirc S_{k+1} \wedge \bigcirc^2 S_{k+2} \wedge \cdots \wedge \bigcirc^l S_{k+l})$$

$1 \le l \in N_0$, that implicitly depicts a finite or an infinite model of the corresponding PPTL formula, respectively. We prove that any PPTL formula can be expressed in a normal form expression. As a consequence, satisfiability of PPTL formulas can easily be achieved.

The rest of the paper is organized as follows. The following section presents syntax and semantics of PPTL. Normal form expressions are defined in Section 3. We then show that any PPTL formula can be represented as a normal form expression in Section 4. As a consequence, a decision procedure for checking the satisfiability of PPTL formulas based on normal form expressions is presented in Section 5. Finally, conclusions are drawn in Section 6.

2   Propositional Projection Temporal Logic

Propositional Projection Temporal Logic (PPTL) [5,13] is an extension of Propositional ITL (PITL) [14] with infinite models and a new projection construct [6,15].

Let *Prop* be a countable set of atomic propositions and $B = \{true, false\}$ the boolean domain. We use small letters, possibly with subscripts, like p,q,r to denote atomic propositions, and capital letters, possibly with subscripts, for instance P, Q, R to indicate general PPTL formulas. Formulas of PPTL are defined by the following grammar:

$$P ::= p \mid \neg P \mid P_1 \vee P_2 \mid \bigcirc P \mid (P_1, \ldots, P_m) \text{ prj } P$$

where $p \in Prop$, $\bigcirc$ (next), and prj (projection) are temporal operations.

We define a *state* s over *Prop* to be a mapping from *Prop* to B, $s : Prop \rightarrow B$. We write $s[p]$ to denote the valuation of p at state s. An *interval* $\sigma = \langle s_0, s_1, \ldots \rangle$ is a non-empty sequence of states, which can be finite or infinite. The length of σ, $|\sigma|$, is the number of states in σ minus one if σ is finite; otherwise it is ω. Let N_0 denote the set of non-negative integers. To have a uniform notation for both finite and infinite intervals, we will use *extended integers* as indices, that is $N_\omega = N_0 \cup \{\omega\}$, and extend the comparison operators, $=, <, \le$, to N_ω by considering $\omega = \omega$ and for all $i \in N_0, i < \omega$. Moreover, we write $\le$ as $\le -\{(\omega, \omega)\}$.

To formalize the semantics of the projection construct, we need an auxiliary operator $\downarrow$. Let $\sigma = \langle s_0, s_1, \ldots \rangle$ be an interval and $r_1, \ldots, r_h$ be integers ($h \ge 1$) such that

$0 \leq r_1 \leq \ldots \leq r_h \leq |\sigma|$. The projection of σ onto $r_1, \ldots, r_h$ is the *projected interval*, $\sigma \downarrow (r_1, \ldots, r_h) \stackrel{def}{=} \langle s_{t_1}, s_{t_2}, \ldots, s_{t_l} \rangle$, where $t_1, \ldots, t_l$ are attained from $r_1, \ldots, r_h$ by deleting all duplicates. In other words, $t_1, \ldots, t_l$ is the longest strictly increasing subsequence of $r_1, \ldots, r_h$. For instance, $\langle s_0, s_1, s_2, s_3 \rangle \downarrow (0, 2, 2, 3) = \langle s_0, s_2, s_3 \rangle$. The concatenation($\cdot$) of an interval $\sigma = \langle s_0, s_1, \ldots, s_{|\sigma|} \rangle$ with another interval $\sigma' = \langle s'_0, s'_1, \ldots, s'_{|\sigma|} \rangle$ is represented by $\sigma \cdot \sigma' = \langle s_0, s_1, \ldots, s_{|\sigma|}, s'_0, s'_1, \ldots, s'_{|\sigma|} \rangle$ (not sharing any states).

An *interpretation* is a tuple $\mathcal{I} = (\sigma, k, j)$, where $\sigma = \langle s_0, s_1, \ldots \rangle$ is an interval, k is a non-negative integer, and j is an integer or ω, such that $0 \leq k \leq j \leq |\sigma|$. We write (σ, k, j) to mean that a formula is interpreted over a subinterval $\sigma_{k, \ldots, j}$ with the current state being s_k. We utilize I^k_{prop} to stand for the state interpretation at state s_k. The satisfaction relation $\models$ for formulas is given as follows:

$$\mathcal{I} = (\sigma, k, j) \models p \qquad \text{iff } s_k[p] = I^k_{prop}[p] = true$$
$$\mathcal{I} = (\sigma, k, j) \models \neg P \qquad \text{iff } \mathcal{I} \not\models P$$
$$\mathcal{I} = (\sigma, k, j) \models P_1 \wedge P_2 \qquad \text{iff } \mathcal{I} \models P_1 \text{ and } \mathcal{I} \models P_2$$
$$\mathcal{I} = (\sigma, k, j) \models \bigcirc P \qquad \text{iff } k < j \text{ and } (\sigma, k+1, j) \models P$$
$$\mathcal{I} = (\sigma, k, j) \models (P_1, \ldots, P_m) \text{ prj } P \qquad \text{iff there exist integers } r_0, \ldots, r_m, \text{ and } k = r_0 \leq \ldots \leq$$
$$r_{m-1} \leq r_m \leq j \text{ such that } (\sigma, r_{l-1}, r_l) \models P_l \text{ for all}$$
$$1 \leq l \leq m \text{ and } (\sigma', 0, |\sigma'|) \models P \text{ for } \sigma' \text{ given by :}$$
$$(1) \ r_m < j \text{ and } \sigma' = \sigma \downarrow (r_0, \ldots, r_m) \cdot \sigma_{(r_m+1, \ldots, j)}$$
$$(2) \ r_m = j \text{ and } \sigma' = \sigma \downarrow (r_0, \ldots, r_h) \text{ for some } 0 \leq h \leq m$$

For convenience, some derived formulas from elementary PPTL formulas are presented below. The abbreviations true, false, $\vee$, $\rightarrow$ and $\leftrightarrow$ are defined as usual.

ε	$\stackrel{def}{=} \neg \bigcirc true$		more	$\stackrel{def}{=} \neg \varepsilon$
$\Diamond P$	$\stackrel{def}{=} (true, P) \text{ prj } \varepsilon$		$\Box P$	$\stackrel{def}{=} \neg \Diamond \neg P$
$fin(P)$	$\stackrel{def}{=} \Box(\varepsilon \rightarrow P)$		$halt(P)$	$\stackrel{def}{=} \Box(\varepsilon \leftrightarrow P)$
$keep(P)$	$\stackrel{def}{=} \Box(\neg\varepsilon \rightarrow P)$		$rem(P)$	$\stackrel{def}{=} \Box(more \rightarrow \bigcirc P)$
$P ; Q$	$\stackrel{def}{=} (P, Q) \text{ prj } \varepsilon$		$P ;_w Q$	$\stackrel{def}{=} (P ; Q) \vee (P \wedge \Box more)$
fin	$\stackrel{def}{=} \Diamond \varepsilon$		inf	$\stackrel{def}{=} \Box more$
$len(0)$	$\stackrel{def}{=} \varepsilon$		$len(n)$	$\stackrel{def}{=} \bigcirc len(n-1), n \geq 1$
$\odot P$	$\stackrel{def}{=} \varepsilon \vee \bigcirc P$		$skip$	$\stackrel{def}{=} len(1)$

A PPTL formula containing no temporal operators is called a state formula.

Further, we have the following useful logic laws, whose proofs can be found in [13]:

(L1) $\Diamond P$	$\equiv P \vee \bigcirc \Diamond P$	(L2) $\Box P$	$\equiv P \wedge \varepsilon \vee P \wedge \bigcirc \Box P$
(L3) $\neg \bigcirc P$	$\equiv \bigodot \neg P$	(L4) $Q \,; (P_1 \vee P_2)$	$\equiv (Q \,; P_1) \vee (Q \,; P_2)$
(L5) $\Box \neg Q$	$\equiv \neg \Diamond Q$	(L6) keep(P)	$\equiv \varepsilon \vee P \wedge \bigcirc$keep$(P)$
(L7) $\Box P \vee \Box Q$	$\supset \Box (P \vee Q)$	(L8) halt(P)	$\equiv P \wedge \varepsilon \vee \neg P \wedge \bigcirc$halt$(P)$
(L9) $\Box (P \wedge Q)$	$\equiv \Box P \wedge \Box Q$	(L10) fin(P)	$\equiv P \wedge \varepsilon \vee \bigcirc$fin$(P)$
(L11) true	$\equiv \Diamond \varepsilon \vee \Box$more	(L12) $P_1 \,; (P_2 \,; P_3)$	$\equiv (P_1 \,; P_2) \,; P_3$
(L13) $\bigcirc (P \vee Q) \equiv \bigcirc P \vee \bigcirc Q$		(L14) $\Box (P \wedge$ more$)$	$\equiv P \wedge \bigcirc \Box (P \wedge$ more$)$
(L15) $\bigcirc (P \wedge Q) \equiv \bigcirc P \wedge \bigcirc Q$		(L16) $\Box (P \rightarrow Q)$	$\supset (\Box P \rightarrow \Box Q)$
(L17) true	$\equiv \varepsilon \vee \bigcirc$true	(L18) more $\wedge \bigcirc \neg P \equiv$ more $\wedge \neg \bigcirc P$

3 Normal Form Expressions

Now we define normal form expressions that implicitly express models of temporal logic formulas.

Definition 1 (Normal Form Expressions). Let

$$E_\varepsilon ::= \{e_\varepsilon^k \mid e_\varepsilon^k = \bigwedge_{0 \leq i \leq k \in N_0} \bigcirc^i S_i \wedge \bigcirc^k \varepsilon\}$$

$$E_\omega ::= \{e_\omega^{(k,l)} \mid e_\omega^{(k,l)} = \bigwedge_{0 \leq i \leq k \in N_0} \bigcirc^i S_i \wedge \bigwedge_{k \leq j \in N_\omega} \bigcirc^j (\bigcirc S_{k+1} \wedge \bigcirc^2 S_{k+2} \wedge \cdots \wedge \bigcirc^l S_{k+l}), 1 \leq l \in N_0\}$$

Here, each S (possibly with subscripts) is a state formula. The set of normal form expressions are defined by:

$$E ::= \{e \mid e = \bigvee_{1 \leq m \in N_0} e^m, e^m \in E_\varepsilon \cup E_\omega\}$$

Every $e \in E$ is a normal form expression. □

Intuitively, in a normal form expression, each

$$e_\varepsilon^k = \bigwedge_{0 \leq i \leq k \in N_0} \bigcirc^i S_i \wedge \bigcirc^k \varepsilon$$

in E_ε denotes a finite interval with length being k, where for each $0 \leq i \leq k$, state formula S_i holds at state i as illustrated in Fig. 1 (1). Whereas each

$$e_\omega^{(k,l)} = \bigwedge_{0 \leq i \leq k \in N_0} \bigcirc^i S_i \wedge \bigwedge_{k \leq j \in N_\omega} \bigcirc^j (\bigcirc S_{k+1} \wedge \bigcirc^2 S_{k+2} \wedge \cdots \wedge \bigcirc^l S_{k+l})$$

in E_ω depicts an infinite model with a loop suffix where S_i holds at state i in case $0 \leq i \leq k$, and $S_{k+j}, 1 \leq j \leq l$, holds at state $k + m \times j$ for all $m \geq 1$ as shown in Fig. 1 (2). Further, let E be the set of all normal form expressions, *true* (or T) and *false* (or F) can be expressed by *true* $\overset{\text{def}}{=} E$ and *false* $\overset{\text{def}}{=} \emptyset$, respectively.

The merits of normal form expressions are twofold: (1) compared with temporal logic formulas, they are much more intuitive in acquiring the underlying meaning of the formula; (2) in contrast to graphical models of temporal logic formulas, they are more compact and convenient in logic operations. In the following, we show two simple examples of normal form expressions.

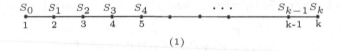

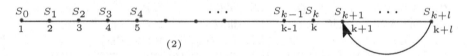

Fig. 1. Intervals expressed by e_ε^k and $e_\omega^{(k,l)}$

Example 1. Examples of normal form expressions:

(1) Normal form expression of a proposition p:

$$p \equiv \bigvee_{0 \leq i \in N_0} p \wedge \bigcirc^i \varepsilon \vee p \wedge \bigcirc^\omega true$$

It hints that two kinds of models will satisfy p. The first kind of models contains the finite ones with an arbitrary length such that p holds at the first state as shown in Fig. 2 (1), while the second kind includes only one infinite model where p holds at the first states as illustrated in Fig. 2 (2), here T denotes true.

$$(1) \quad \underset{0 \quad\quad 1 \quad\quad 2 \quad\quad\quad\quad k\text{-}1 \quad\quad k}{\overset{p \quad\quad T \quad\quad T \quad \cdots \quad T \quad\quad T}{\bullet\!\!-\!\!\bullet\!\!-\!\!\bullet\!\!-\!\!\!-\!\!\bullet\!\!-\!\!\bullet}}$$

$$(2) \quad \overset{p \quad\quad T}{\bullet\!\!-\!\!\bullet}\!\!\bigcirc$$
$$\quad\quad\quad\ 0 \quad\ 1$$

Fig. 2. Models of proposition p

(2) Normal form expression of formula $\Diamond p$:

$$\Diamond p \equiv \bigwedge_{0 \leq i < k \in N_0, 0 \leq j} \bigcirc^i true \wedge \bigcirc^k p \wedge \bigcirc^{k+j} \varepsilon \vee \bigwedge_{0 \leq i < k \in N_0} \bigcirc^i true \wedge \bigcirc^k p \wedge \bigcirc^\omega true$$

It indicates that the models that satisfy $\Diamond p$ are finite or infinite ones where p holds at some state throughout the intervals as shown in Fig. 3 (1) and (2), respectively.

4   Normal Form Expressions of PPTL

In this section, we show that any PPTL formula can be equivalently transformed to a normal form expression. We first show some results useful in the transformation.

Lemma 1 shows that the negation of a normal form expression will still be a normal form expression.

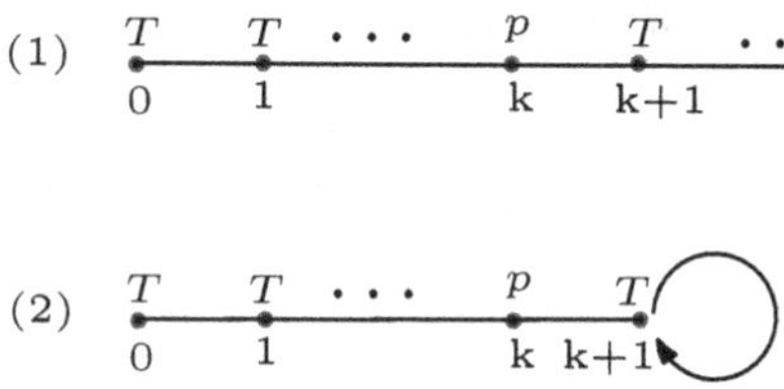

Fig. 3. Models of $\Diamond p$

Lemma 1. *For any* $e \in E_\varepsilon \cup E_\omega$, $\neg e$ *can be transformed to normal form expression.*

Proof: In case $e = \bigwedge_{0 \le i \le k \in N_0} \bigcirc^i S_i \wedge \bigcirc^k \varepsilon$. We have:

$$\neg e \equiv \neg(\bigwedge_{0 \le i \le k \in N_0} \bigcirc^i S_i \wedge \bigcirc^k \varepsilon)$$

$$\equiv \bigvee_{0 \le i \le k \in N_0} \neg \bigcirc^i S_i \vee \neg \bigcirc^k \varepsilon$$

$$\equiv \bigvee_{0 \le i \le k \in N_0} \bigodot^i \neg S_i \vee \bigodot^k \neg \varepsilon$$

$$\equiv \bigvee_{0 \le i \le k \in N_0} \bigodot^i \neg S_i \vee \bigodot^k more$$

$$\equiv \bigvee_{0 \le i \le k \in N_0} \bigcirc^i \neg S_i \vee \bigcirc^k more \vee \bigvee_{0 \le i \le k-1 \in N_0} \bigcirc^i \varepsilon$$

$$\equiv \bigvee_{0 \le i \le k \in N_0} \bigcirc^i(\bigvee_{0 \le j \in N_0} \neg S_i \wedge \bigcirc^j \varepsilon \vee \neg S_i \wedge \bigcirc^\omega true) \vee \bigcirc^k more \vee \bigvee_{0 \le i \le k-1 \in N_0} \bigcirc^i \varepsilon$$

$$\equiv \bigvee_{0 \le i \le k \in N_0} \bigvee_{0 \le j \in N_0} \bigcirc^i \neg S_i \wedge \bigcirc^{i+j} \varepsilon \vee \bigcirc^i \neg S_i \wedge \bigcirc^\omega true \vee \bigcirc^k more \vee \bigvee_{0 \le i \le k-1 \in N_0} \bigcirc^i \varepsilon$$

So in this case $\neg e$ has been represented as a normal form expression. Further, if $e = \bigwedge_{0 \le i \le k \in N_0} \bigcirc^i S_i \wedge \bigwedge_{k \le j \in N_\omega} \bigcirc^j(\bigcirc S_{k+1} \wedge \bigcirc^2 S_{k+2} \wedge \cdots \wedge \bigcirc^l S_{k+l})$, $1 \le l \in N_0$. We have:

$$\neg e \equiv \neg(\bigwedge_{0 \le i \le k \in N_0} \bigcirc^i S_i \wedge \bigwedge_{k \le j \in N_\omega} \bigcirc^j(\bigcirc S_{k+1} \wedge \bigcirc^2 S_{k+2} \wedge \cdots \wedge \bigcirc^l S_{k+l}))$$

$$\equiv \bigvee_{0 \le i \le k \in N_0} \bigcirc^i \neg S_i \vee \bigvee_{k \le j \in N_\omega} \bigcirc^j(\bigcirc \neg S_{k+1} \vee \bigcirc^2 \neg S_{k+2} \vee \cdots \vee \bigcirc^l \neg S_{k+l}) \vee \bigvee_{0 \le i \in N_\omega} \bigcirc^i \varepsilon$$

$$\equiv \bigvee_{0 \le i \in N_\omega} \bigcirc^i \neg S_i \vee \bigvee_{0 \le i \in N_\omega} \bigcirc^i \varepsilon$$

$$\equiv \bigvee_{0 \le i \in N_\omega} \bigcirc^i(\bigvee_{0 \le j \in N_0} \neg S_i \wedge \bigcirc^j \varepsilon \vee \neg S_i \wedge \bigcirc^\omega true) \vee \bigvee_{0 \le i \in N_\omega} \bigcirc^i \varepsilon$$

$$\equiv \bigvee_{0 \le i \le j \in N_\omega} \bigcirc^i \neg S_i \wedge \bigcirc^{i+j} \varepsilon \vee \bigvee_{0 \le i \in N_\omega} \bigcirc^i \neg S_i \wedge \bigcirc^{i+\omega} true \vee \bigvee_{0 \le i \in N_\omega} \bigcirc^i \varepsilon$$

$$\equiv \bigvee_{0 \le i \le j \in N_0} \bigcirc^i \neg S_i \wedge \bigcirc^{i+j} \varepsilon \vee \bigvee_{0 \le i \in N_0} \bigcirc^i \neg S_i \wedge \bigcirc^\omega true \vee \bigvee_{0 \le i \in N_0} \bigcirc^i \varepsilon$$

Hence the lemma holds. □

Lemma 2 indicates that the conjunction of normal form expressions is still a normal form expression.

Lemma 2. *Let e_1 and e_2 be normal form expressions. $e_1 \wedge e_2$ can be expressed by a normal form expression.*

Proof: Suppose $e_1 \equiv \bigvee_{0 \leq k_1 \in N_0} e_\varepsilon^{k_1} \vee \bigvee_{0 \leq k_1' \in N_0} e_\omega^{(k_1', l_1)}$, $1 \leq l_1 \in N_0$, and $e_2 \equiv \bigvee_{0 \leq k_2 \in N_0} e_\varepsilon^{k_2} \vee \bigvee_{0 \leq k_2' \in N_0} e_\omega^{(k_2', l_2)}$, $1 \leq l_2 \in N_0$. We have,

$$
e_1 \wedge e_2 \equiv \left(\bigvee_{0 \leq k_1 \in N_0} e_\varepsilon^{k_1} \vee \bigvee_{0 \leq k_1' \in N_0} e_\omega^{(k_1', l_1)} \right) \wedge \left(\bigvee_{0 \leq k_2 \in N_0} e_\varepsilon^{k_2} \vee \bigvee_{0 \leq k_2' \in N_0} e_\omega^{(k_2', l_2)} \right)
$$

$$
\equiv \bigvee_{0 \leq k_1 \in N_0} \bigvee_{0 \leq k_2 \in N_0} e_\varepsilon^{k_1} \wedge e_\varepsilon^{k_2} \vee e_\omega^{(k_1', l_1)} \wedge e_\omega^{(k_2', l_2)}
$$

It is ready that $e_1 \wedge e_2$ can be expressed by a normal form expression. $\square$

Lemma 3 presents how normal form expression of a chop construct is obtained.

Lemma 3. *$e_\varepsilon^{k_1} \, ; e_\varepsilon^{k_2}$ and $e_\varepsilon^{k_1} \, ; e_\omega^{k,l}$ can be expressed by normal form expressions.*

Proof: Suppose,

$$
e_\varepsilon^{k_1} \equiv \bigwedge_{0 \leq i \leq k_1 \in N_0} \bigcirc^i S_i \wedge \bigcirc^{k_1} \varepsilon
$$

$$
e_\varepsilon^{k_2} \equiv \bigwedge_{0 \leq i \leq k_2 \in N_0} \bigcirc^i S_i \wedge \bigcirc^{k_2} \varepsilon
$$

$$
e_\omega^{(k,l)} \equiv \bigwedge_{0 \leq i \leq k \in N_0} \bigcirc^i S_i \wedge \bigwedge_{k \leq j \in N_\omega} \bigcirc^j (\bigcirc S_{k+1} \wedge \bigcirc^2 S_{k+2} \wedge \cdots \wedge \bigcirc^l S_{k+l})
$$

We have,

$$
e_\varepsilon^{k_1} \, ; e_\varepsilon^{k_2} \equiv \left(\bigwedge_{0 \leq i \leq k_1 \in N_0} \bigcirc^i S_i \wedge \bigcirc^{k_1} \varepsilon \right) ; \left(\bigwedge_{0 \leq i \leq k_2 \in N_0} \bigcirc^i S_i \wedge \bigcirc^{k_2} \varepsilon \right)
$$

$$
\equiv \bigwedge_{0 \leq i \leq k_1 + k_2 \in N_0} \bigcirc^i S_i \wedge \bigcirc^{k_1 + k_2} \varepsilon
$$

$$
e_\varepsilon^{k_1} \, ; e_\omega^{k,l} \equiv \left(\bigwedge_{0 \leq i \leq k_1 \in N_0} \bigcirc^i S_i \wedge \bigcirc^{k_1} \varepsilon \right) ; \left(\bigwedge_{0 \leq i \leq k \in N_0} \bigcirc^i S_i \wedge \bigwedge_{k \leq j \in N_\omega} \bigcirc^j (\bigcirc S_{k+1} \wedge \bigcirc^2 S_{k+2} \wedge \cdots \wedge \bigcirc^l S_{k+l}) \right)
$$

$$
\equiv \bigwedge_{0 \leq i \leq k_1 + k \in N_0} \bigcirc^i S_i \wedge \bigwedge_{k_1 + k \leq j \in N_\omega} \bigcirc^j (\bigcirc S_{k+1} \wedge \bigcirc^2 S_{k+2} \wedge \cdots \wedge \bigcirc^l S_{k+l})
$$

Thus, the lemma holds. $\square$

Lemma 4 shows that projection construct can be expressed by a normal form expression.

Lemma 4. *If $P_0, \cdots, P_m$, and Q can be expressed by normal form expressions, so does $(P_0, \cdots, P_m) \, prj \, Q$.*

Proof: Without loss of generality, suppose

$$
E_Q \equiv e_\varepsilon^k \vee e_\omega^{(k', l)}
$$

$$
E_{P_0} \equiv e_\varepsilon^{k_0} \vee e_\omega^{(k_0', l_0)}
$$

$$
\cdots
$$

$$
E_{P_m} \equiv e_\varepsilon^{k_m} \vee e_\omega^{(k_m', l_m)}
$$

Here for convenience, we represent E_Q as $\bigwedge\limits_{0\leq i\in N_\omega} \bigcirc^i S_i$. It has:

$$(P_0,\cdots,P_m)\,prj\,Q \equiv (E_{P_0},\cdots,E_{P_m})\,prj\,E_Q$$

By the semantics of projection construction, $P_1,\cdots$, and P_{m-1} are confined to finite models. In case $i \leq m$, we have:

$$(E_{P_0},\cdots,E_{P_m})\,prj\,E_Q \equiv e_\varepsilon^{k_0} \wedge \mathsf{fin}(S_0)\,;\,\cdots\,;\,e_\varepsilon^{k_i} \wedge \mathsf{fin}(S_i)\,;\,e_\varepsilon^{k_{i+1}}\cdots\,;\,e_\varepsilon^{k_{m-1}}\,;\,(e_\varepsilon^{k_m} \vee e_\omega^{(k_m',l_m)})$$

In case $i > m$, we have:

$$(E_{P_0},\cdots,E_{P_m})\,prj\,E_Q \equiv e_\varepsilon^{k_0} \wedge \mathsf{fin}(S_0)\,;\,\cdots\,;\,e_\varepsilon^{k_m} \wedge \mathsf{fin}(S_m)\,;\,\bigwedge\limits_{m<i} \bigcirc^i S_i$$

Note that each $e_\varepsilon^{k_i} \wedge \mathsf{fin}(S_i)$ means that S_i is conjuncted with the state formula holding at the last state of the interval specified by $e_\varepsilon^{k_i}$. Thus, by Lemma 3, the lemma is already proved.   □

Now the main theorem is presented.

Theorem 5. *Any PPTL formula can be equivalently expressed by a normal form expression.*

Proof: The proof proceeds by induction on the structure of PPTL formulas. As the base case, we have shown that a proposition p can be expressed as a normal form expression. Suppose PPTL formulas P (or P with subscripts) and Q have been expressed as normal form expressions E_P and E_Q, respectively.

1. Formula $\neg P$ can be expressed as normal form expression by:

$$\begin{aligned}\neg P &\equiv \neg E_p \\ &\equiv \neg(\bigvee\limits_{1\leq i\in N_0} e^i),\, e^i \in E_\varepsilon \cup E_\omega \\ &\equiv \bigwedge\limits_{1\leq i\in N_0} \neg e^i\end{aligned}$$

   By Lemma 1 and 2, $\neg P \equiv \bigwedge\limits_{1\leq i\in N_0} \neg e^i$ can be further expressed in a normal form expression.

2. Formula $P_1 \vee P_2$ can be expressed as a normal form expression by:

$$P_1 \vee P_2 \equiv E_{P_1} \vee E_{P_2}$$

3. Formula $\bigcirc P$ can be expressed as a normal form expression by:

$$\bigcirc P \equiv \bigcirc E_P$$

   $\bigcirc E_P$ is already in normal form expression.

4. Formula $P\,;\,Q$ can be expressed as a normal form expression by:

$$P_1 \, ; P_2 \equiv E_{P_1} \, ; E_{P_2}$$

$$\equiv (\bigvee_{0 \le k_1 \in N_0} e_\varepsilon^{k_1} \vee \bigvee_{0 \le k_1' \in N_0} e_\omega^{(k_1', l_1)}) \, ; (\bigvee_{0 \le k_2 \in N_0} e_\varepsilon^{k_2} \vee \bigvee_{0 \le k_2' \in N_0} e_\omega^{(k_2', l_2)})$$

$$\equiv (\bigvee_{0 \le k_1 \in N_0} e_\varepsilon^{k_1}) \, ; (\bigvee_{0 \le k_2 \in N_0} e_\varepsilon^{k_2} \vee \bigvee_{0 \le k_2' \in N_0} e_\omega^{(k_2', l_2)})$$

$$\equiv \bigvee_{0 \le k_1 \in N_0} \bigvee_{0 \le k_2 \in N_0} (e_\varepsilon^{k_1} \, ; e_\varepsilon^{k_2} \vee e_\varepsilon^{k_1} \, ; e_\omega^{(k_2', l_2)})$$

By Lemma 3, $P_1 \, ; P_2 \equiv \bigvee_{0 \le k_1 \in N_0} \bigvee_{0 \le k_2 \in N_0} (e_\varepsilon^{k_1} \, ; e_\varepsilon^{k_2} \vee e_\varepsilon^{k_1} \, ; e_\omega^{(k_2', l_2)})$ can be further expressed in a normal form expression.

5. By Lemma 4, formula $(P_1, \cdots, P_m) \, prj \, Q$ can be expressed in a normal form expression.

Accordingly, any PPTL formula can be equivalently expressed by a normal form expression.   □

The above proofs also provide an approach for transforming a PPTL formula to a normal form expression.

5   Decision Procedure of PPTL

Based on normal form expressions, how to check the satisfiability of PPTL formulas becomes simple. Give a PPTL formula P, we first transform P to its normal form expression:

$$P \equiv \bigvee_{0 \le k \in N_0} e_\varepsilon^k \vee \bigvee_{0 \le k' \in N_0} e_\omega^{(k', l)}$$

where

$$e_\varepsilon^k = \bigwedge_{0 \le i \le k \in N_0} \bigcirc^i S_i \wedge \bigcirc^k \varepsilon$$

$$e_\omega^{(k', l)} = \bigwedge_{0 \le i \le k' \in N_0} \bigcirc^i S_i \wedge \bigwedge_{k' \le j \in N_\omega} \bigcirc^j (\bigcirc S_{k'+1} \wedge \bigcirc^2 S_{k'+2} \wedge \cdots \wedge \bigcirc^l S_{k'+l})$$

$1 \le l \in N_0$. Then for all k and k', if there exists an S (or with subscript) such that S is unsatisfiable, P is unsatisfiable; otherwise, P is satisfiable. As a matter of fact, each S (or with subscript) is a state formula without any temporal operators, i.e. a typical propositional logic formula, a SAT solver can be employed to check its satisfiability automatically.

Example 2. Satisfiability of PPTL formula $(p \wedge \square \bigcirc p) \, ; q$.

We first present $p \wedge \square \bigcirc p$ and q in normal form expressions:

$$p \wedge \square \bigcirc p \equiv p \wedge \bigwedge_{0 \le j \in N_\omega} \bigcirc^j (\bigcirc p)$$

$$q \equiv \bigvee_{0 \le i \in N_0} q \wedge \bigcirc^i \varepsilon \vee q \wedge \bigcirc^\omega true$$

There are no $e \in E_\varepsilon$ occuring in the normal form expression of $p \wedge \Box \bigcirc p$. Thus,

$$p \wedge \Box \bigcirc p ; q \equiv p \wedge \bigwedge_{0 \leq j \in N_\omega} \bigcirc^j(\bigcirc p) ; (\bigvee_{0 \leq i \in N_0} q \wedge \bigcirc^i \varepsilon \vee q \wedge \bigcirc^\omega true)$$
$$\equiv false$$

So, PPTL formula $(p \wedge \Box \bigcirc p) ; q$ is unsatisfiable.

6   Conclusion

In this paper, we present normal form expressions and show that any PPTL formula can be represented as a normal form expression. When presented as a normal form expression, the underlying models of a PPTL formula is easy to be acquired that leads to a simple decision procedure for checking the satisfiability of PPTL formulas.

References

1. Alur, R., Henzinger, T.A., Kupferman, O.: Alternating-Time Temporal Logic. Journal of the ACM 49(5), 672–713 (2002)
2. Ben-Ari, M., Manna, Z., Pnueli, A.: The temporal logic of branching time. Acta Informatica 20, 207–226 (1983)
3. Chandra, A., Halpern, J., Meyer, A., Parikh, R.: Equations between regular terms and an application to process logic. In: Proceedings of the Thirteenth Annual ACM Symposium on Theory of Computing (STOC 1981), pp. 384–390 (1981)
4. Clarke, E.M., Emerson, E.A.: Design and synthesis of synchronization skeletons using branching time temporal logic. In: Kozen, D. (ed.) Logic of Programs 1981. LNCS, vol. 131, pp. 52–71. Springer, Heidelberg (1982)
5. Duan, Z.: An Extended Interval Temporal Logic and A Framing Technique for Temporal Logic Programming. PhD thesis. University of Newcastle Upon Tyne (May 1996)
6. Duan, Z., Koutny, M., Holt, C.: Projection in Temporal Logic Programming. In: Pfenning, F. (ed.) LPAR 1994. LNCS, vol. 822, pp. 333–344. Springer, Heidelberg (1994)
7. Duan, Z., Tian, C., Zhang, L.: A Decision Procedure for Propositional Projection Temporal Logic with Infinite Models. Acta Informatica 45(1), 43–78 (2008)
8. Clark, M., Gremberg, O., Peled, A.: Model Checking. The MIT Press (2000)
9. Pnueli, A.: The temporal logic of programs. In: Proceedings of the 18th IEEE Symposium on Foundations of Computer Science (FOCS 1977), pp. 46–57 (1977)
10. Finkbeiner, B., Schewe, S.: Coordination Logic. In: Dawar, A., Veith, H. (eds.) CSL 2010. LNCS, vol. 6247, pp. 305–319. Springer, Heidelberg (2010)
11. Halpern, J., Manna, Z., Moszkowski, B.: A hardware semantics based on temporal intervals. In: Díaz, J. (ed.) ICALP 1983. LNCS, vol. 154, pp. 278–291. Springer, Heidelberg (1983)
12. Emerson, E.A.: Temporal and Modal Logic, Computer Science Department. University of Texas at Austin, USA (1995)
13. Duan, Z.: Temporal Logic and Temporal Logic Programming. Science Press, Beijing (2006)
14. Moszkowski, B.: Executing temporal logic programs. Cambridge University Press (1986)
15. Duan, Z., Tian, C.: A practical decision procedure for propositional projection temporal logic with infinite models. Theoretical Computer Science (2014), doi:10.1016/j.tcs.2014.02.011

On the Smoothed Heights of Trie and Patricia Index Trees

Weitian Tong, Randy Goebel, and Guohui Lin*

Department of Computing Science, University of Alberta
Edmonton, Alberta T6G 2E8, Canada
{weitian,rgoebel,guohui}@ualberta.ca

Abstract. Two of the most popular data structures for storing strings are the Trie and the Patricia index trees. Let H_n denote the height of the Trie (the Patricia, respectively) on a set of n strings. It is well known that under the uniform distribution model on the strings, for Trie $H_n/\log n \to 2$ and for Patricia $H_n/\log n \to 1$, when n approaches infinity. Nevertheless, in the worst case, the height of the Trie on n strings is unbounded, and the height of the Patricia on n strings is in $\Theta(n)$. To better understand the practical performance of both the Trie and Patricia index trees, we investigate these two classical data structures in a smoothed analysis model. Given a set $\mathcal{S} = \{s_1, s_2, \ldots, s_n\}$ of n binary strings, we perturb the set by adding an *i.i.d* Bernoulli random noise to each bit of every string. We show that the resulting smoothed heights of Trie and Patricia trees are both $\Theta(\log n)$.

1   Introduction

A *Trie*, also known as a *digital tree*, is an ordered tree data structure for storing strings over an alphabet Σ. It was initially developed and analyzed by Fredkin [6] in 1960, and is one of the first collected in "The art of computer programming" by Knuth [7] in 1973. Such a data structure is used for storing a dynamic set to be exploited as an associative array, where keys are strings. There has been much recent exploitation of such index trees for processing genomic data.

In the simplest form, let the alphabet be $\Sigma = \{0, 1\}$ and consider a set $\mathcal{S} = \{s_1, s_2, \ldots, s_n\}$ of n binary strings over Σ, where each s_i is a countable string of 0's and 1's. The Trie for storing these n binary strings is an ordered binary tree $T_{\mathcal{S}}$: first, each s_i defines a path (infinite if its length $|s_i|$ is infinite) in the tree, starting from the root, such that a 0 forces a move to the left and a 1 indicates a move to the right; if one node is the highest in the tree that is passed through by only one string $s_i \in \mathcal{S}$, then the path defined by s_i is truncated at this node, which becomes a leaf in the tree and is associated (i.e., labelled) with s_i. The *height* of the Trie $T_{\mathcal{S}}$ built over $\mathcal{S}$ is defined as the number of edges on the longest root-to-leaf path. Fig. 1 shows the Trie constructed for a set of six strings. The strings can be long or even infinite, but only the first 5 bits are shown, which are those used in the example construction.

---

* Correspondence author.

Z. Cai et al. (Eds.): COCOON 2014, LNCS 8591, pp. 94–103, 2014.

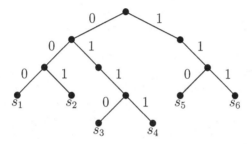

Fig. 1. The Trie constructed for $\{s_1 = 00001\ldots, s_2 = 00111\ldots, s_3 = 01100\ldots, s_4 = 01111\ldots, s_5 = 11010\ldots, s_6 = 11111\ldots\}$

Let H_n denote the height of the Trie on a set of n binary strings. It is not hard to see that in the worst case H_n is unbounded, because any two of the strings can have an arbitrary long common prefix. In the uniform distribution model, bits of s_i are *independent and identically distributed* (*i.i.d.*) Bernoulli random variables each of which takes 1 with probability $p = 0.5$. The asymptotic behavior of Trie height H_n under the uniform distribution model had been well studied in the 1980s [13,8,5,4,3,11,12,15,16], and it is known that *asymptotically almost surely* (*a.a.s.*)

$$H_n/\log_2 n \to 2, \quad \text{when } n \to \infty.$$

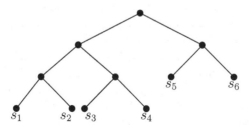

Fig. 2. The Patricia constructed for $\{s_1 = 00001\ldots, s_2 = 00111\ldots, s_3 = 01100\ldots, s_4 = 01111\ldots, s_5 = 11010\ldots, s_6 = 11111\ldots\}$

A Patricia index tree is a space-optimized variant of the Trie data structure, in which every node with only one child is merged with its child. Such a data structure was firstly discovered by Morrison [9] in 1968, and then well analyzed in "The art of computer programming" by Knuth [7] in 1973. Fig. 2 shows the Patricia tree constructed for the same set of six strings used in Fig. 1. Again let H_n denote the height of the Patricia tree on a set of n binary strings. In the worst case, $H_n \in \Theta(n)$. Under the same uniform distribution model assumed for an average case analysis on Trie height, Pittel showed that a.a.s. the height of Patricia is only 50% of the height of Trie [11], that is,

$$H_n / \log_2 n \to 1, \text{ when } n \to \infty.$$

The average case analysis is intended to provide insight on the practical performance as a string indexing structure. Recently, Nilsson and Tikkanen experimentally investigated the height of Patricia trees and other search structures [10]. In particular, they showed that the heights of the Patricia trees on sets of $50,000$ random uniformly distributed strings are 15.9 on average and 20 at most. For real datasets consisting of $19,461$ strings from geometric data on drill holes, $16,542$ ASCII character strings from a book, and $38,367$ strings from Internet routing tables, the heights of the Patricia trees are on average 20.8, 20.2, 18.6, respectively, and at most 30, 41, 24, respectively.

Theoretically speaking, these experimental results suggest that worst-case instances are perhaps only isolated peaks in the instance space. This hypothesis is partially supported by the average case analysis on the heights of Trie and Patricia structures, under the uniform distribution model, that suggests the heights are a.a.s. logarithmic. Nevertheless, these average case analysis results on the specific random instances generated under the uniform distribution model could be inconclusive, because the specific random instances have very special properties inherited from the model, and thus would distinguish themselves from real-world instances. To overcome the fact that real-world instances are not captured by a single probabilistic distribution, Spielman and Teng introduced the idea of *smoothed analysis* [14], which can be considered as a hybrid of the worst-case and the average-case analyses, and inherits the advantages of both. In brief, an given string instance is perturbed by adding a slight random noise to generate an instance neighborhood and the average performance on this neighbor is evaluated; the smoothed performance is then taken as the worst among all these local average performances. One can image that when the magnitude of random noise approaches 0, the smoothed analysis becomes the worst case analysis; when the magnitude of random noise approaches infinity, the smoothed analysis becomes the average case analysis under the probabilistic distribution assumed on the random noise. In practice, such a magnitude is set to be small; then a good smoothed analysis result under certain reasonable probabilistic distribution assumed on the random noise generally implies a good practical performance in real world applications. One key reason underlying this hypothesis is that real world instances are often subject to a slight amount of noise, especially when they are obtained from measurements of real world phenomena. The classic example is the Simplex method for solving linear programming. The Simplex method is one kind of practical algorithm for solving linear programming, all of which have worst case exponential running time. Spielman and Teng showed that Simplex algorithms have polynomial smoothed running time [14], which explained their practical performance.

In this paper, we conduct the smoothed analysis on the height of Trie and Patricia structures, to reveal certain essential properties of these two data structures. In the next section, we first introduce the string perturbation model, and we show an a.a.s. upper bound $O(\log n)$ and an a.a.s. lower bound $\Omega(\log n)$ on the Trie height H_n. The consequence is that the smoothed height of the Trie on n strings is in $\Theta(\log n)$. In Section 3, we achieve similar results for the smoothed height of the Patricia tree on n strings.

2   The Smoothed Height of Trie

We consider an arbitrary set $\mathcal{S} = \{s_1, s_2, \ldots, s_n\}$ of n strings over alphabet $\{0,1\}$, where each string may be infinitely long. Let $s_i(\ell)$ denote the ℓ-th bit in string s_i, for $i = 1, 2, \ldots, n$ and $\ell = 1, 2, 3, \ldots$. Every string s_i is perturbed by adding a noise string ν_i, giving rise to the perturbed string $\tilde{s}_i = s_i + \nu_i$, where $\tilde{s}_i(\ell) = s_i(\ell)$ if and only if $\nu_i(\ell) = 0$. The noise string ν_i is independently generated by a memoryless source, which assigns 1 to every bit of string ν_i independently and with a small probability $\epsilon \in [0, 0.5]$. More formally, $Pr\{\nu_i(\ell) = 1\} = \epsilon$ for each $\ell = 1, 2, 3, \ldots$. Essentially the perturbation flips each bit of every string independently and with a probability ϵ. Let $\tilde{\mathcal{S}} = \{\tilde{s}_1, \tilde{s}_2, \ldots, \tilde{s}_n\}$ denote the set of perturbed strings.

Let p_{ij}^ℓ be the probability of the event $\{\tilde{s}_i(\ell) = \tilde{s}_j(\ell)\}$. We have

$$p_{ij}^\ell = \begin{cases} 2\epsilon(1-\epsilon) \overset{\triangle}{=} p, & \text{if } s_i(\ell) \neq s_j(\ell), \\ \epsilon^2 + (1-\epsilon)^2 = 1 - p \overset{\triangle}{=} q, & \text{if } s_i(\ell) = s_j(\ell). \end{cases} \tag{1}$$

We can clearly note that $q \geq p$, since $\epsilon \leq 0.5$. Let C_{ij} denote the length of the longest common prefix between $\tilde{s}_i$ and $\tilde{s}_j$. Since $C_{ij} = k$ if and only if $\tilde{s}_i(\ell) = \tilde{s}_j(\ell)$ for $\ell = 1, 2, \ldots, k$ but not for $\ell = k+1$, the probability of $\{C_{ij} = k\}$ for any $k \geq 0$ is

$$Pr\{C_{ij} = k\} = \left(\prod_{\ell=1}^{k} p_{ij}^\ell \right) (1 - p_{ij}^{k+1}).$$

From the fact that $\{C_{ij} = k\}$ and $\{C_{ij} = m\}$ are disjoint events when $k \neq m$, we have for any $k \geq 1$

$$Pr\{C_{ij} < k\} = \sum_{m=0}^{k-1} \left(\prod_{\ell=1}^{m} p_{ij}^\ell - \prod_{\ell=1}^{m+1} p_{ij}^\ell \right) = 1 - \prod_{\ell=1}^{k} p_{ij}^\ell.$$

Consequently, the probability that the longest common prefix between $\tilde{s}_i$ and $\tilde{s}_j$ is at least k long is

$$Pr\{C_{ij} \geq k\} = 1 - Pr\{C_{ij} < k\} = \prod_{\ell=1}^{k} p_{ij}^\ell. \tag{2}$$

2.1   An a.a.s. Upper Bound

We use a slight abuse of notation H_n to also denote the height of the Trie constructed for $\tilde{\mathcal{S}}$. We can express H_n in terms of C_{ij} as

$$H_n = \max_{1 \leq i < j \leq n} C_{ij} + 1.$$

By Boole inequality [2], we have

$$Pr\{H_n > k\} = Pr\{\max_{1 \leq i < j \leq n} C_{ij} \geq k\} \leq \binom{n}{2} \prod_{\ell=1}^{k} p_{ij}^{\ell} \leq \binom{n}{2} q^k,$$

where the last equality holds when all the n strings $\{s_1, s_2, \ldots, s_n\}$ have the same prefix of length k. By setting $k = 2(1 + \delta) \log_{1/q} n$ for a constant $\delta > 0$, we have

$$Pr\{H_n > k\} \leq \binom{n}{2} q^{2(1+\delta) \log_{1/q} n} \leq n^{-2\delta} \to 0,$$

as $n \to \infty$. Therefore, $H_n \leq 2 \log_{1/q} n$ with high probability, when n approaches infinity.

2.2   An a.a.s. Lower Bound

To estimate a lower bound, we will use the following Chunge-Erdös formulation of the second moment method on a set of events:

Lemma 1. (Chunge-Erdös) [1] *For any set of events $E_1, E_2, \ldots, E_n$,*

$$Pr\{\cup_{i=1}^{n} E_i\} \geq \frac{\left(\sum_{i=1}^{n} Pr\{E_i\}\right)^2}{\sum_{i=1}^{n} Pr\{E_i\} + \sum_{i \neq j} Pr\{E_i \cap E_j\}}.$$

Let A_{ij} denote the event $\{C_{ij} \geq k\}$, for every pair $\{i, j\}$ such that $1 \leq i < j \leq n$; also define the following two sums:

$$S_1 \triangleq \sum_{1 \leq i < j \leq n} Pr\{A_{ij}\}, \text{ and } S_2 \triangleq \sum_{\{i,j\} \neq \{s,t\}} Pr\{A_{ij} \cap A_{st}\}.$$

Then by Chunge-Erdös formulation (Lemma 1), we have

$$Pr\{H_n > k\} = Pr\{\cup_{1 \leq i < j \leq n} A_{ij}\} \geq \frac{S_1^2}{S_1 + S_2}. \tag{3}$$

Let's first estimate S_1. From Eq. (2), one clearly sees that

$$S_1 = \sum_{1 \leq i < j \leq n} Pr\{A_{ij}\} = \sum_{1 \leq i < j \leq n} \prod_{\ell=1}^{k} p_{ij}^{\ell}. \tag{4}$$

Recall the definition of p_{ij}^{ℓ} and its value in Eq. (1). The following Lemma 2 is then straight-forward:

Lemma 2. *For any $\ell \geq 1$ and any three perturbed strings $\tilde{s}_i, \tilde{s}_j, \tilde{s}_t$, if $p_{ij}^{\ell} = p_{it}^{\ell}$, then $p_{jt}^{\ell} = q$.*

Lemma 3. *For any three perturbed strings* $\tilde{s}_i, \tilde{s}_j, \tilde{s}_t,$

$$S_0 \overset{\triangle}{=} \prod_{\ell=1}^{k} p_{ij}^{\ell} + \prod_{\ell=1}^{k} p_{it}^{\ell} + \prod_{\ell=1}^{k} p_{jt}^{\ell} \geq 3p^{\frac{2}{3}k}q^{\frac{1}{3}k}.$$

Proof. For the string pair (s_i, s_j), let Z_{ij} denote the number of $(0,1)$-pairs and $(1,0)$-pairs in $\{(s_i(\ell), s_j(\ell)), 1 \leq \ell \leq k\}$, that is, the number of bits where s_i and s_j have different values among the first k bits. Clearly from Eq. (1),

$$\prod_{\ell=1}^{k} p_{ij}^{\ell} = p^{Z_{ij}} q^{k-Z_{ij}}.$$

For the string triple (s_i, s_j, s_t), let x_{ij} denote the number of $(0,0,1)$-triples and $(1,1,0)$-triples in $\{(s_i(\ell), s_j(\ell), s_t(\ell)), 1 \leq \ell \leq k\}$; likewise, x_{it} and x_{jt} are similarly defined. Also let y denote the number of $(0,0,0)$-triples and $(1,1,1)$-triples in $\{(s_i(\ell), s_j(\ell), s_t(\ell)), 1 \leq \ell \leq k\}$. The following relationships are direct consequences of the definitions:

$$\begin{aligned}
Z_{ij} &= x_{it} + x_{jt}, \\
Z_{it} &= x_{ij} + x_{jt}, \\
Z_{jt} &= x_{ij} + x_{it}, \\
k &= x_{ij} + x_{it} + x_{jt} + y.
\end{aligned}$$

It follows that

$$\begin{aligned}
S_0 \overset{\triangle}{=} &\prod_{\ell=1}^{k} p_{ij}^{\ell} + \prod_{\ell=1}^{k} p_{it}^{\ell} + \prod_{\ell=1}^{k} p_{jt}^{\ell} \\
= &\, p^{x_{it}+x_{jt}} q^{x_{ij}+y} + p^{x_{ij}+x_{jt}} q^{x_{it}+y} + p^{x_{ij}+x_{it}} q^{x_{jt}+y} \\
= &\, p^k \left[\left(\frac{q}{p}\right)^{x_{ij}+y} + \left(\frac{q}{p}\right)^{x_{it}+y} + \left(\frac{q}{p}\right)^{x_{jt}+y} \right].
\end{aligned}$$

One can check that, since $q \geq p$, the quantity in the last line reaches the minimum when $x_{ij} = x_{it} = x_{jt} = k/3$ and $y = 0$. That is,

$$S_0 \overset{\triangle}{=} \prod_{\ell=1}^{k} p_{ij}^{\ell} + \prod_{\ell=1}^{k} p_{it}^{\ell} + \prod_{\ell=1}^{k} p_{jt}^{\ell} \geq 3p^{\frac{2}{3}k}q^{\frac{1}{3}k}.$$

This proves the lemma. $\qquad\qquad\qquad\qquad\qquad\qquad\qquad\qquad\qquad\qquad\square$

Note that each string pair (s_i, s_j) is involved in exactly $n-2$ string triples (s_i, s_j, s_t), for $t \neq i, j$. By Lemma 3, Eq. (4) becomes

$$S_1 = \sum_{1 \leq i < j \leq n} \prod_{\ell=1}^{k} p_{ij}^{\ell} \geq \frac{1}{n-2} \binom{n}{3} 3p^{\frac{2}{3}k} q^{\frac{1}{3}k} = \binom{n}{2} p^{\frac{2}{3}k} q^{\frac{1}{3}k}. \qquad (5)$$

We next estimate S_2, which is a bit harder because two events A_{ij} and A_{st} may not be independent. We split S_2 into two parts: $S_2 = S_2' + S_2''$, where

$$S_2' \triangleq \sum_{\{i,j\} \cap \{s,t\} = \emptyset} Pr\{A_{ij} \cap A_{st}\}, \text{ and}$$

$$S_2'' \triangleq \sum_{\{i,j\} \cap \{s,t\} \neq \emptyset} Pr\{A_{ij} \cap A_{st}\}.$$

Since two events C_{ij} and C_{st} are independent when $\{i,j\} \cap \{s,t\} = \emptyset$, we can estimate S_2' as follows:

$$S_2' = \sum_{\{i,j\} \cap \{s,t\} = \emptyset} \left(Pr\{A_{ij}\} Pr\{A_{st}\} \right) \leq \left(\sum_{\{i,j\}} Pr\{A_{ij}\} \right)^2 = S_1^2.$$

Event $\{A_{ij} \cap A_{it}\}$ is equivalent to the event in which the first k bits of all three perturbed strings $\tilde{s}_i, \tilde{s}_j$, and $\tilde{s}_t$ are identical. Using $\epsilon \leq 0.5$, we have

$$Pr\{A_{ij} \cap A_{it}\} = Pr\{\tilde{s}_i(\ell) = \tilde{s}_j(\ell) = \tilde{s}_t(\ell), 1 \leq \ell \leq k\} \leq \left(\epsilon^3 + (1 - \epsilon)^3 \right)^k.$$

It follows that

$$S_2'' = \sum_{\{i,j\} \cap \{s,t\} \neq \emptyset} Pr\{A_{ij} \cap A_{st}\} \leq 3 \binom{n}{3} \left(\epsilon^3 + (1 - \epsilon)^3 \right)^k \leq 3 \binom{n}{3},$$

where the factor 3 arises because a string triple $\{\tilde{s}_i, \tilde{s}_j, \tilde{s}_t\}$ gives rise to three events $\{A_{ij} \cap A_{it}\}$, $\{A_{ij} \cap A_{jt}\}$, and $\{A_{it} \cap A_{jt}\}$.

Putting S_2' and S_2'' together, we can upper bound S_2 by

$$S_2 = S_2' + S_2'' \leq S_1^2 + 3 \binom{n}{3}. \tag{6}$$

Using the estimates of S_1 and S_2 in Eqs. (5) and (6) respectively, Eq. (3) becomes

$$Pr\{H_n > k\} \geq \frac{S_1^2}{S_1 + S_2}$$

$$= \frac{1}{1/S_1 + (S_2' + S_2'')/S_1^2}$$

$$\geq \frac{1}{1/S_1 + 1 + S_2''/S_1^2}$$

$$\geq \frac{1}{1 + \dfrac{1}{\binom{n}{2} p^{\frac{2}{3}k} q^{\frac{1}{3}k}} + \dfrac{3\binom{n}{3}}{\left(\binom{n}{2} p^{\frac{2}{3}k} q^{\frac{1}{3}k} \right)^2}}$$

$$\geq \frac{1}{1 + 4n^{-2} p^{-\frac{2}{3}k} q^{-\frac{1}{3}k} + 2n^{-1} p^{-\frac{4}{3}k} q^{-\frac{2}{3}k}}$$

$$\geq \frac{1}{1 + 4n^{-2}n^{2(1-\delta)} + 2n^{-1}n^{1-\delta}} \tag{7}$$

$$= \frac{1}{1 + 4n^{-2\delta} + 2n^{-\delta}}$$

$$\geq 1 - O(n^{-\delta}) \to 1,$$

where the inequality Eq. (7) is achieved by setting

$$k = 2(1-\delta)\log_{p^{-2/3}q^{-1/3}} n, \text{ that is, } p^{-\frac{2}{3}k}q^{-\frac{1}{3}k} = n^{2(1-\delta)},$$

for a constant $\delta > 0$. Therefore, H_n is larger than $2\log_{p^{-2/3}q^{-1/3}} n$ with a high probability when n approaches infinity.

Theorem 1. *The smoothed height of the Trie on n strings is in $\Theta(\log n)$, where the bit perturbation model is i.i.d. Bernoulli distribution.*

3   The Smoothed Height of Patricia

Here we briefly do the smoothed analysis on the height of the Patricia tree on a set of n binary strings. We adopt the same *i.i.d.* Bernoulli bit perturbation model as in the last section. Again, we present an a.a.s. upper bound and an a.a.s. lower bound for the smoothed height.

3.1   An a.a.s. Upper Bound

Following Pittel [11], on the set of n perturbed strings $\tilde{S} = \{\tilde{s}_1, \tilde{s}_2, \ldots, \tilde{s}_n\}$, we claim that for any fixed integers $k \geq 0$ and $b \geq 2$, the event $\{H_n \geq k + b - 1\}$ implies the event that there exist b strings $\tilde{s}_{i_1}, \tilde{s}_{i_2}, \ldots, \tilde{s}_{i_b}$ such that their common prefix is of length at least k (denoted as $C_{i_1 i_2 \ldots i_b} \geq k$). The correctness of the above claim follows from because, in Patricia trees, there are no degree-2 nodes (except for the root), and thus a path of length $k + b - 1$ hints at least b leaves in the subtree rooted at the node at distance k from the Patricia root.

Similar to the definition of p_{ij}^{ℓ} in Eq. (1), $p_{i_1 i_2 \ldots i_b}^{\ell}$ denotes the probability of the event $\{\tilde{s}_{i_1}^{\ell} = \tilde{s}_{i_2}^{\ell} = \ldots = \tilde{s}_{i_b}^{\ell}\}$, for any $b \geq 2$, which is calculated as follows:

$$p_{i_1 i_2 \ldots i_b}^{\ell} = (1-\epsilon)^{k_0}\epsilon^{k_1} + (1-\epsilon)^{k_1}\epsilon^{k_0},$$

where k_0 and k_1 are the number of 0's and 1's among the b bit values $\tilde{s}_{i_1}(\ell), \tilde{s}_{i_2}(\ell), \ldots, \tilde{s}_{i_b}(\ell)$, respectively. By a similar argument as presented for $Pr\{A_{ij}\}$ in Section 2, we have

$$Pr\{C_{i_1 i_2 \ldots i_b} \geq k\} = \prod_{\ell=1}^{k} p_{i_1 i_2 \ldots i_b}^{\ell}.$$

For a fixed $b \geq 2$, let $q_b = \epsilon^b + (1-\epsilon)^b$ and $k = k_b = b(1 + \delta/2)\log_{1/q_b} n$. We have

$$k = b(1 + \delta/2)\log_{1/q_b} n$$

$$= (1 + \delta/2)\frac{\ln n}{\ln q_b^{-1/b}}$$

$$= (1 + \delta/2)\frac{\ln n}{\ln\left(\epsilon^b + (1-\epsilon)^b\right)^{-1/b}}$$

$$\leq (1 + \delta/2)\frac{\ln n}{\ln\left(\epsilon^2 + (1-\epsilon)^2\right)^{-1/2}} \tag{8}$$

$$= 2(1 + \delta/2)\log_{1/q} n,$$

where the inequality in Eq. (8) holds for any $b \geq 2$. Setting $b = \delta \log_{1/q} n$, it follows that

$$Pr\{H_n \geq 2(1+\delta)\log_{1/q} n\} \leq Pr\{H_n \geq k + b - 1\}$$

$$\leq Pr\{\max_{i_1, i_2, \ldots, i_b} C_{i_1 i_2 \ldots i_b} \geq k\}$$

$$\leq n^b \prod_{\ell=1}^{k} p_{i_1 i_2 \ldots i_b}^{\ell}$$

$$\leq n^b q_b^k$$

$$\in O(n^{-b\delta}) \to 0,$$

when $n \to \infty$.

In summary, for any $\delta > 0$, we have

$$Pr\{H_n \geq 2(1+\delta)\log_{1/q} n\} \in O(n^{-b\delta}) \to 0,$$

when n approaches infinity, and thus a.a.s. $H_n \leq 2(1+\delta)\log_{1/q} n$.

3.2   An a.a.s. Lower Bound

Let D_i be the depth of node labelled $\tilde{s}_i$ in the Patricia tree.

Clearly, $H_n = \max_{i=1}^{n} D_i$ and the $\tilde{s}_{i*}$ reaching the maximum depth must be a leaf node. It follows that if $H_n < k$, then at least one of the 2^k possible length-k strings does not appear as a prefix of any perturbed strings $\tilde{s}_1, \tilde{s}_2, \ldots, \tilde{s}_n$.

Let $\mathbb{L}n = \log_{1/\epsilon} n$ and $k = \mathbb{L}\frac{n}{\mathbb{L}\ln n}$. We have

$$Pr\{H_n < k\} \leq 2^k Pr\{\text{no } \tilde{s}_i \text{ starts with } k \text{ 0's}\}$$

$$\leq 2^k(1 - \epsilon^k)^n$$

$$\leq 2^k e^{-\epsilon^k n}$$

$$= \exp\{k \ln 2 - \epsilon^k n\}$$

$$= \exp\{\ln 2 \cdot \mathbb{L}\frac{n}{\mathbb{L}\ln n} - \mathbb{L}\ln n\} \to 0,$$

when n approaches infinity, and thus a.a.s. $H_n \geq \mathbb{L}\frac{n}{\mathbb{L}\ln n}$.

In summary, we have the following theorem.

Theorem 2. *The smoothed height of the Patricia on n strings is in $\Theta(\log n)$, where the bit perturbation model is i.i.d. Bernoulli distribution.*

4   Conclusion

Under the *i.i.d.* Bernoulli bit perturbation model, we have shown that the smoothed heights of both Trie and Patricia index trees on n strings are in the order of $\log n$. These theoretical results explain the typical probabilistic behavior of these two important data structures on real-world applications.

Acknowledgement. This research was supported in part by NSERC, AITF and iCORE.

References

1. Chung, K.L., Erdös, P.: On the application of the Borel-Cantelli lemma. Transactions of the American Mathematical Society 72, 179–186 (1952)
2. Comtet, L.: Advanced Combinatorics: The Art of Finite and Infinite Expansions. Springer (1974)
3. Devroye, L.: A probabilistic analysis of the height of tries and of the complexity of triesort. Acta Informatica 21, 229–237 (1984)
4. Flajolet, P.: On the performance evaluation of extendible hashing and trie search. Acta Informatica 20, 345–369 (1983)
5. Flajolet, P., Steyaert, J.M.: A branching process arising in dynamic hashing, trie searching and polynomial factorization. In: Nielsen, M., Schmidt, E.M. (eds.) ICALP 1982. LNCS, vol. 140, pp. 239–251. Springer, Heidelberg (1982)
6. Fredkin, E.: Trie memory. Communications of the ACM 3, 490–499 (1960)
7. Knuth, D.E.: The Art of Computer Programming. Sorting and Searching, vol. III. Addison-Wesley (1973)
8. Mendelson, H.: Analysis of extendible hashing. IEEE Transactions on Software Engineering 8, 611–619 (1982)
9. Morrison, D.R.: Patricia — practical algorithm to retrieve information coded in alphanumeric. Journal of the ACM 15, 514–534 (1968)
10. Nilsson, S., Tikkanen, M.: An experimental study of compression methods for dynamic tries. Algorithmica 33, 19–33 (2002)
11. Pittel, B.: Asymptotical growth of a class of random trees. Annals of Probability 13, 414–427 (1985)
12. Pittel, B.: Path in a random digital tree: limiting distributions. Advances in Applied Probability 18, 139–155 (1986)
13. Régnier, M.: On the average height of trees in digital searching and dynamic hashing. Information Processing Letters 13, 64–66 (1981)
14. Spielman, D.A., Teng, S.-H.: Smoothed analysis of algorithms: Why the simplex algorithm usually takes polynomial time. Journal of the ACM 51, 385–463 (2004)
15. Szpankowski, W.: Some results on V-ary asymmetric tries. Journal of Algorithms 9, 224–244 (1988)
16. Szpankowski, W.: Digital data structures and order statistics. In: Dehne, F., Santoro, N., Sack, J.-R. (eds.) WADS 1989. LNCS, vol. 382, pp. 206–217. Springer, Heidelberg (1989)

One-Dimensional k-Center on Uncertain Data*

Haitao Wang and Jingru Zhang

Department of Computer Science
Utah State University, Logan, UT 84322, USA
haitao.wang@usu.edu, jingruzhang@aggiemail.usu.edu

Abstract. Problems on uncertain data have attracted significant attention due to the imprecise nature of many measurement data. In this paper, we consider the k-center problem on one-dimensional uncertain data. The input is a set $\mathcal{P}$ of (weighted) uncertain points on a real line, and each uncertain point is specified by its probability density function (pdf) which is a piecewise-uniform function (i.e., a histogram). The goal is to find a set of Q of k points on the line to minimize the maximum expected distance from the uncertain points of $\mathcal{P}$ to their expected closest points in Q. We present efficient algorithms for this uncertain k-center problem and their running times almost match those for the "deterministic" k-center problem. The techniques proposed in the paper may also be useful for solving other related problems on uncertain data.

1 Introduction

A large amount of work has been done on *deterministic* data, e.g., points with exact positions. Recently, due to the observation that many real-world measurements are inherently accompanied with uncertainty, problems on uncertain data have attracted dramatically increasing amount of attention. Two models are commonly used for data uncertainty: the *existential* model (or tuple model) [23,24,37] and the *locational* model (or attribute model) [1,2,16,34]. In the existential model, each uncertain point has a specific location but its existence is uncertain, following a given probability density function. In the locational model, each uncertain point always exists but its location is uncertain and follows a probability density function. In this paper, we consider the k-center problem on one-dimensional uncertain data under the locational model.

1.1 Problem Definitions and Our Results

Let $\mathcal{P} = \{P_1, P_2, \ldots, P_n\}$ be a set of n uncertain points on the x-axis, where each uncertain point P_i is specified by its probability density function (pdf) $f_i \colon \mathbb{R} \to \mathbb{R}^+ \cup \{0\}$, which is a piecewise-uniform function (i.e., a histogram), consisting of at most $m + 1$ pieces (e.g., see Fig. 1). More specifically, for each uncertain point P_i, there are m x-coordinates $x_{i1} < x_{i2} < \ldots < x_{im}$ and $m - 1$

---

* This research was supported in part by NSF under Grant CCF-1317143.

Z. Cai et al. (Eds.): COCOON 2014, LNCS 8591, pp. 104–115, 2014.

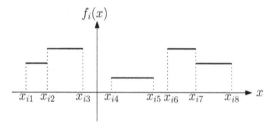

Fig. 1. Illustrating the pdf f_i of an uncertain point P_i with $m = 8$

nonnegative values $y_{i1}, y_{i2}, \ldots, y_{i,m-1}$ such that $f_i(x) = y_{ij}$ ($y_{ij} = 0$ is possible) for $x_{ij} \leq x < x_{i,j+1}$ with $0 \leq j \leq m$, and we set $x_{i0} = -\infty$, $x_{i,m+1} = +\infty$, and $y_{i0} = y_{im} = 0$. In addition, the uncertain points of $\mathcal{P}$ are independent.

Note that in some applications each uncertain point has a *discrete* pdf, that is, it could appear at one of a few locations, each with a probability. This discrete case can also be represented by the above histogram model using infinitesimal pieces around these locations, and thus the histogram model also incorporate the discrete case. In other words, the discrete case is a special case of our model.

Let L denote the x-axis. For any certain point $p \in L$, we let x_p denote its x-coordinate. The *expected distance* between p and any uncertain point P_i is defined as

$$\mathsf{Ed}(p, P_i) = \int_{-\infty}^{+\infty} f_i(x)|x - x_p|dx.$$

Let Q be a set of (certain) points on L, called *facilities*. For any uncertain point P_i, we use $\mathsf{Ed}(Q, P_i)$ to denote the smallest expected distance from P_i to all points of Q, i.e., $\mathsf{Ed}(Q, P_i) = \min_{q \in Q} \mathsf{Ed}(q, P_i)$. The facility q with $\mathsf{Ed}(q, P_i) = \mathsf{Ed}(Q, P_i)$ is called the *expected closest* facility of P_i in Q, and we also say P_i is "served" by the facility q or P_i is "assigned" to q. The k-*center* problem is to find a set Q of k points on L to minimize the maximum expected distance from the uncertain points of $\mathcal{P}$ to their expected closest facilities in Q, i.e., the value $\max_{P_i \in \mathcal{P}} \mathsf{Ed}(Q, P_i)$.

In a *realization*, each uncertain point will appear at a deterministic location abiding by its pdf. We should point out that our problem definitions imply that we always assign each uncertain point P_i to its expected closest facility and we never change the assignment in any realization even through the actual location of P_i in a realization may be closer to a different facility.

For differentiation, we refer to the traditional k-center problem where each point is given in an exact position as the *deterministic version*.

In this paper, we present an algorithm for the uncertain k-center problem and the running time is $O(mn \log mn + n \log k \log n \log mn)$. Further, for the *discrete* case where the pdf of each uncertain point of $\mathcal{P}$ is discrete, i.e., each uncertain point P_i has m possible locations, each with a probability, we have a more efficient algorithm with running time $O(mn \log mn + n \log k \log n)$. Since mn is the input size, as will be seen soon, our results almost match those for the corresponding deterministic k-center problems.

Note that our algorithms can also solve the *weighted* case where each uncertain point P_i has a nonnegative weight w_i and we consider the *weighted expected distance*, i.e., $w_i \cdot \text{Ed}(q, P_i)$, from P_i to each facility q in Q. To solve the weighted problems (for both the general and the discrete cases), we only need to change each value y_{ij} to $w_i \cdot y_{ij}$ for $1 \leq j \leq m - 1$, and then simply apply our algorithms for the corresponding unweighted problems. The time complexities do not change asymptotically.

1.2 Related Work

The deterministic k-center (and k-median) problems are classical problems that have been extensively studied. It is well-known that the k-center problem is NP-hard even in the plane [28] and approximation algorithms have been proposed (e.g., see [3,5,8,22]). Efficient algorithm were also given for some special cases, e.g., the smallest enclosing circle [27], k-center on trees [10,20,29]. Refer to [19] for other variations of facility location problems. The deterministic k-center in one-dimensional space is solvable in $O(n \log n)$ time [14,15,20,30].

The k-center problems on uncertain data in high-dimensional space have been proposed. For example, approximation algorithms were given in [18] for different problem models, e.g., the assigned model that is similar to our problem model and the unassigned model which was relatively easy because it can be reduced to the corresponding deterministic problem, as shown in [18]. Other problems on clustering uncertain data were also studied and heuristic algorithms were proposed [4,11,31]. Other facility location problems on uncertain data under various models, e.g., the minmax regret [6,7,36,38], have also been studied (see [33] for a survey).

To the best of our knowledge, the uncertain k-center problem proposed in this paper has not been particularly studied before.

1.3 Our Approach

For the deterministic one-dimensional k-center problem, there is an observation that there exists an optimal facility set Q such that the input points served by each facility are consecutive if we order them from left to right on L; this observation is crucial for designing the algorithms [14,15,17,30]. In our uncertain problem, however, since the input points of $\mathcal{P}$ are uncertain, it is not clear how to "sort" them; consequently, the algorithmic techniques used before for solving the deterministic problems are not applicable here.

As explained above, one main difficulty for solving the uncertain k-center problem is that we do not have an "order" for the uncertain points of $\mathcal{P}$ to help us design algorithms. Instead, we use the following approach. We first solve the *decision* problem which is to determine whether the minimized value $\max_{P_i \in \mathcal{P}} \text{Ed}(Q, P_i)$ in the optimal solution is less than or equal to a given value ϵ, and if yes, ϵ is called a *feasible value*. We solve the decision problem with the following result: with $O(mn)$ time preprocessing, for any given ϵ, we can determine whether ϵ is a feasible value in $O(\log m + n \log k)$ time.

By using the above result for the decision problem, we solve the k-center problem by using parametric search [17,26]; however, there are some issues that do not allow us to use the parametric search in [17,26] directly and we have to make certain modifications. A useful observation discovered in the paper is that the expected distance $\mathsf{Ed}(p, P_i)$ is a unimodal function (i.e., first monotonically decreasing and then increasing) as p moves from left to right on L.

For the discrete case, we reduce the problem to finding a particular vertex in a line arrangement. By using the arrangement searching technique in [13], we can solve the discrete case in a faster way, in $O(mn \log mn + n \log k \log n)$ time.

The rest of the paper is organized as follows. Section 2 introduces some observations. In Section 3, we present our results for the decision algorithms. In Section 4, we solve the k-center problem, which is referred to as the *optimization* problem. Section 5 presents our algorithm for the discrete case.

2 Observations

Consider any uncertain point P_i of $\mathcal{P}$. For any point p, its expected distance to P_i is $\mathsf{Ed}(p, P_i) = \int_{-\infty}^{+\infty} f_i(x)|x - x_p| dx$. With a little abuse of notation, we also use $\mathsf{Ed}(x_p, P_i)$ to denote $\mathsf{Ed}(p, P_i)$, but we normally consider $\mathsf{Ed}(x_p, P_i)$ as a function of x_p for $x_p \in \mathbb{R} = (-\infty, +\infty)$ as p moves on L.

A function $g : \mathbb{R} \to \mathbb{R}$ is a *unimodal function* if there exists a value x' such that $g(x)$ is monotonically decreasing on $x \in (-\infty, x']$ and monotonically increasing on $x \in [x', +\infty)$, i.e., for any $x_1 < x_2$, $g(x_1) \geq g(x_2)$ holds if $x_2 \leq x'$ and $g(x_1) \leq g(x_2)$ holds if $x' \leq x_1$.

We assume the m coordinates $x_{i1}, \ldots, x_{im}$ of P_i are given sorted. We have the following lemma, which is crucial to our algorithm. The proof is omitted.

Lemma 1. *The function* $\mathsf{Ed}(x_p, P_i)$ *for* $x_p \in \mathbb{R}$ *is a unimodal function and can be explicitly computed in* $O(m)$. *More specifically,* $\mathsf{Ed}(x_p, P_i)$ *is a parabola (of constant complexity) on the interval* $[x_k, x_{k+1})$ *for each* $0 \leq k \leq m$.

By using Lemma 1, we can obtain the following corollary.

Corollary 1. *For each uncertain point* P_i, *with* $O(m)$ *time preprocessing, we can compute the value* $\mathsf{Ed}(x_p, P_i)$ *in* $O(\log m)$ *time for any query point* p *on* L.

Consider any uncertain point $P_i \in \mathcal{P}$. According to Lemma 1, there is a point $p \in L$ that minimizes the value $\mathsf{Ed}(p, P_i)$ and let p_i denote such a point; note that such a point may not be unique, in which case we let p_i denote any such point. We refer to p_i as the *centroid* of P_i. By Lemma 1, we can compute the centroids for all uncertain points of $\mathcal{P}$ in $O(nm)$ time, by explicitly computing the functions $\mathsf{Ed}(x_p, P_i)$ for all uncertain points P_i of $\mathcal{P}$.

3 The Decision k-Center Problem

In order to solve our k-center problem, we first solve the decision version of the problem in this section.

Recall that our goal for the k-center problem is to find a set Q of k points such that $\max_{P_i \in \mathcal{P}} \mathsf{Ed}(Q, P_i)$ is minimized, where $\mathsf{Ed}(Q, P_i) = \min_{q \in Q} \mathsf{Ed}(q, P_i)$. Below, for any set Q of points on L, let $\psi(Q) = \max_{P_i \in \mathcal{P}} \mathsf{Ed}(Q, P_i)$. Denote by ϵ^* the value $\psi(Q)$ in an optimal solution for the k-center problem.

Given any real value ϵ, the *decision k-center* problem is to determine whether there exist a set Q of k points on L such that $\psi(Q) \leq \epsilon$ (i.e., determine whether $\epsilon^* \leq \epsilon$), and if yes, then we say the decision problem is *feasible* and ϵ is a *feasible value*. To distinguish from the decision problem, we refer to our original k-center problem the *optimization problem*. Clearly, ϵ^* is the smallest feasible value.

Consider any value ϵ and any uncertain point $P_i \in \mathcal{P}$. Let Q be any set of k points on L. If $\mathsf{Ed}(Q, P_i) \leq \epsilon$, then there is at least one point q in Q with $\mathsf{Ed}(q, P_i) \leq \epsilon$. Let $\alpha(P_i, \epsilon)$ be the set of points p of L such that $\mathsf{Ed}(x_p, P_i) \leq \epsilon$. A line segment on L is also called an *interval* of L. By using Lemma 1, we obtain the following result, whose proof is omitted.

Lemma 2. *For any uncertain point P_i and any value ϵ, $\alpha(P_i, \epsilon)$ is an interval of L ($\alpha(P_i, \epsilon) = \emptyset$ is possible); with $O(m)$ time preprocessing, we can compute $\alpha(P_i, \epsilon)$ in $O(\log m)$ time for any given ϵ.*

We say that a point on L *covers* an interval of L if the interval contains the point. Let $\alpha(\mathcal{P}, \epsilon)$ is the set of all intervals $\alpha(P_i, \epsilon)$ for $i = 1 \ldots n$. We have the following observation.

Observation 1 *The value ϵ is a feasible value if and only if there exist a set Q of k points on L such that each interval of $\alpha(\mathcal{P}, \epsilon)$ is covered by at least one point in Q.*

Hence, to determine whether ϵ is feasible, it is sufficient to solve the following *interval covering problem*: determine whether there exist a set Q of k points on L such that each interval of $\alpha(\mathcal{P}, \epsilon)$ is covered by at least one point in Q.

To solve the interval covering problem, we can compute the minimum number k^* of points that can cover all intervals of $\alpha(\mathcal{P}, \epsilon)$, and the problem can be solved in $O(n)$ time by a simple greedy algorithm after the endpoints of all intervals of $\alpha(\mathcal{P}, \epsilon)$ are sorted [21]. Specifically, we scan the sorted endpoints of intervals of $\alpha(\mathcal{P}, \epsilon)$ from left to right until we first encounter a right endpoint of an interval. We add this right endpoint into Q and removes all intervals that contain this point. This process is repeated until no intervals remain. However, due to the sorting procedure, the total time for computing k^* is $O(n \log n)$.

Snoeyink [32] gave an $O(n \log k^*)$ time algorithm for computing k^* without sorting. If $k^* \leq k$, then we have $n \log k^* = O(n \log k)$, which means that we can solve the interval covering problem in $O(n \log k)$ time. However, if $k^* > k$, since it is possible that $n \log k = o(n \log k^*)$ (e.g., $k = O(1)$ and $k^* = \Theta(n)$), we cannot bound the time by $O(n \log k)$. To ensure that the interval covering algorithm can still be solved in $O(n \log k)$ time even if $k^* > k$, we modify Snoeyink's algorithm [32] in the following way.

Observe that to solve the interval covering algorithm, it is sufficient to know whether $k^* \geq k$ holds. Snoeyink's algorithm finds a set Q of points one by one in $O(n \log k^*)$ time, with $k^* = |Q|$. Since the points of Q are computed one by one, when we run the algorithm, we simply stop the algorithm when there are $k + 1$ points in the current Q. In this way, the recursion tree of the algorithm has $k + 1$ leaves (instead of k^* leaves) and thus the running time is $O(n \log k)$ according to Lemma 1 in [32].

As a summary, given any ϵ, we solve the decision k-center as follows. First, we compute all intervals $\alpha(\mathcal{P}, \epsilon)$, in $O(n \log m)$ time by Lemma 2. Then, by modifying the algorithm in [32] as discussed above, we can solve the interval covering problem in $O(n \log k)$ time. The decision problem is thus solved. The total time is $O(n \log m + n \log k)$, after the $O(mn)$ time preprocessing in Lemma 2 for all uncertain points. By using fractional cascading [12], we further reduce the running time in Lemma 3, whose proof is omitted.

Lemma 3. *With $O(mn)$ time preprocessing, we can determine whether ϵ is a feasible value in $O(\log m + n \log k)$ time for any given ϵ.*

4   The Optimization Problem

In this section, we present our algorithm for the original k-center problem, which we refer to as the *optimization* problem, and our goal is to find the smallest feasible value ϵ^* and the corresponding optimal facility set Q. Based on some observations and our decision algorithm in Lemma 3, we finally compute ϵ^* by modifying the parametric search technique [17,26].

For any $\epsilon > 0$, for each $1 \le i \le n$, let $\alpha(P_i, \epsilon) = [l_i(\epsilon), r_i(\epsilon)]$, i.e., $l_i(\epsilon)$ is the x-coordinate of the left endpoint of $\alpha(P_i, \epsilon)$ and $r_i(\epsilon)$ is the x-coordinate of the right endpoint of $\alpha(P_i, \epsilon)$; below we will consider $l_i(\epsilon)$ and $r_i(\epsilon)$ as functions of ϵ. With a little abuse of notation, we also use $l_i(\epsilon)$ and $r_i(\epsilon)$ to denote the left and right endpoints of $\alpha(P_i, \epsilon)$, respectively. Define $E(\epsilon)$ to be the set of the endpoints of all intervals in $\alpha(\mathcal{P}, \epsilon)$. Notice that if we know the sorted order of the endpoints of $E(\epsilon^*)$ at the value ϵ^*, we can easily find an optimal facility set Q, e.g., by using the greedy algorithm mentioned before. Although we do not know ϵ^*, but we can still sort the values in $E(\epsilon^*)$ by making use of our decision algorithm to resolve comparisons, which is the key idea of parametric search [17,26]. However, our problem does not allow us to apply the parametric search approaches in [17,26] directly, because in our problem we cannot resolve each "comparison" by a single call on the decision algorithm (since a comparison may have multiple "roots"). The details are given below.

Suppose in our sorting algorithm we want to resolve a comparison between two values in $E(\epsilon^*)$. Depending on whether the two values are left endpoints or right endpoints, there are two cases.

1. If a value is a left endpoint, say $l_i(\epsilon^*)$, and the other value is a right endpoint, say $r_j(\epsilon^*)$, then the comparison between them is called a *type-1* comparison. We resolve this type of comparison in the following way.
   Recall that p_k is the centroid for each uncertain point $P_k \in \mathcal{P}$. We denote the function $\mathsf{Ed}(x_p, P_k)$ on $x_p \in (-\infty, x_{p_k}]$ by $\mathsf{Ed}_L(x_p, P_k)$ and denote $\mathsf{Ed}(x_p, P_k)$

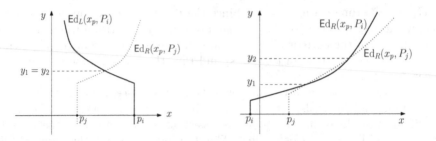

Fig. 2. Illustrating the intersection of $\mathsf{Ed}_L(x_p, P_i)$ and $\mathsf{Ed}_R(x_p, P_j)$, where the intersection is a single point and thus $y_1 = y_2$. $l_i(\epsilon^*) \geq r_j(\epsilon^*)$ if and only if $\epsilon^* \leq y_1$.

Fig. 3. Illustrating two intersections of $\mathsf{Ed}_R(x_p, P_i)$ and $\mathsf{Ed}_R(x_p, P_j)$. For example, if $\epsilon^* \in [y_1, y_2]$, then $r_i(\epsilon^*) \geq r_j(\epsilon^*)$.

on $x \in [x_{p_k}, +\infty)$ by $\mathsf{Ed}_R(x_p, P_k)$. By Lemma 1, $\mathsf{Ed}_L(x_p, P_k)$ is monotonically decreasing and $\mathsf{Ed}_R(x_p, P_k)$ is monotonically increasing. Further, $l_k(\epsilon) \leq x_{p_k} \leq r_k(\epsilon)$ holds. To simplify the discussion below, for each $P_k \in \mathcal{P}$, we add a vertical half-line on the function $\mathsf{Ed}_L(x_p, P_k)$ from the point p_k downwards to $-\infty$ and we also add the same half-line to $\mathsf{Ed}_R(x_p, P_k)$. Note that each new $\mathsf{Ed}_L(x_p, P_k)$ is still monotonically decreasing and each new $\mathsf{Ed}_R(x_p, P_k)$ is still monotonically increasing.

To resolve the comparison between $l_i(\epsilon^*)$ and $r_j(\epsilon^*)$, our goal is to determine whether $l_i(\epsilon^*) \leq r_j(\epsilon^*)$ or $l_i(\epsilon^*) \geq r_j(\epsilon^*)$ holds. To this end, we first determine whether $\mathsf{Ed}_L(x_p, P_i)$ intersects $\mathsf{Ed}_R(x_p, P_j)$.

If $x_{p_i} < x_{p_j}$, then since $\mathsf{Ed}_L(x_p, P_i)$ is to the left of p_i and $\mathsf{Ed}_R(x_p, P_j)$ is to the right of p_j, the two functions do not intersect and $l_i(\epsilon^*) \leq r_j(\epsilon^*)$ always holds.

Otherwise, since $\mathsf{Ed}_L(x_p, P_i)$ is monotonically decreasing and $\mathsf{Ed}_R(x_p, P_j)$ is monotonically increasing, $\mathsf{Ed}_L(x_p, P_i)$ must intersect $\mathsf{Ed}_R(x_p, P_j)$ and the intersection is a line segment (may be degenerated into a single point) that spans an interval $[y_1, y_2]$ on y-coordinates (e.g., see Fig. 2). Observe that $l_i(\epsilon^*) < r_j(\epsilon^*)$ if $\epsilon^* > y_2$, $l_i(\epsilon^*) = r_j(\epsilon^*)$ if $\epsilon^* \in [y_1, y_2]$, and $l_i(\epsilon^*) > r_j(\epsilon^*)$ if $\epsilon^* < y_1$.

Hence, to resolve the comparison between $l_i(\epsilon^*)$ and $r_j(\epsilon^*)$, it sufficient to resolve the comparisons among ϵ^*, y_1, and y_2, which can be done by calling the decision algorithm to determine whether y_1 and y_2 are feasible values. Specifically, if $\epsilon = y_2$ is not feasible, then $\epsilon^* > y_2$ and we obtain $l_i(\epsilon^*) < r_j(\epsilon^*)$. If $\epsilon = y_2$ is feasible, then $\epsilon^* \leq y_2$. We further check whether $\epsilon = y_1$ is feasible. If y_1 is not feasible, then we have $\epsilon^* \in (y_1, y_2]$ and thus obtain $l_i(\epsilon^*) = r_j(\epsilon^*)$; otherwise, we have $\epsilon^* \leq y_1$ and obtain $l_i(\epsilon^*) \geq r_j(\epsilon^*)$. In summary, we can resolve the comparison between $l_i(\epsilon^*)$ and $r_j(\epsilon^*)$ by first finding the intersection of $\mathsf{Ed}_L(x_p, P_i)$ and $\mathsf{Ed}_R(x_p, P_j)$ and subsequently at most two calls on the decision algorithm. The intersection of $\mathsf{Ed}_L(x_p, P_i)$ and $\mathsf{Ed}_R(x_p, P_j)$ can be found by binary search in $O(\log m)$ time and the two calls on the decision algorithm takes $O(\log m + n \log k)$ time.

Hence, we can resolve each type-1 comparison in $O(\log m + n \log k)$ time.

2. If the two values involved in the comparison are both left endpoints or both right endpoints, then we call it a *type-2* comparison. It becomes more complex to resolve this type of comparison. Assume both values are two right endpoints, say $r_i(\epsilon^*)$ and $r_j(\epsilon^*)$, and the case where both values are two left endpoints can be handled similarly. In the sequel, we resolve the comparison in the following way.

   As in the type-1 case, we first compute the intersections between the two functions $\mathsf{Ed}_R(x_p, P_i)$ and $\mathsf{Ed}_R(s, P_j)$. Although both functions are monotonically increasing, there may be $\Theta(m)$ intersections as their complexities are $\Theta(m)$ in the worst case (e.g., see Fig. 3). All intersections can be computed in $O(m)$ time. If there is no intersection, then $r_i(\epsilon^*) \leq r_j(\epsilon^*)$ if and only if $x_{p_i} \leq x_{p_j}$, where p_i and p_j are the centroids.

   Otherwise, let $y_1, y_2, \ldots, y_h$ be the y-coordinates of all intersections, sorted in ascending order, with $h = O(m)$. We can compute this sorted list in $O(m)$ time as we compute the intersections. Using our decision algorithm, we can determine an interval $(y_k, y_{k+1}]$ that contains ϵ^*, by binary search with $O(\log m)$ calls on the decision algorithm. After finding the interval $(y_k, y_{k+1}]$, we can easily determine whether $r_i(\epsilon^*) \leq r_j(\epsilon^*)$ or $r_i(\epsilon^*) \geq r_j(\epsilon^*)$ in the similar way as in the type-1 case (e.g., see Fig. 3).

   Hence, we can resolve each type-2 comparison in $O(m + n \log k \log m)$ time.

The above shows that we can resolve each comparison in $O(m + n \log k \log m)$ time, which is dominated by the type-2 comparisons.

Now we apply the parametric search scheme to our problem by resolving comparisons in the above ways. We first consider Megiddo's approach [26]. We can use n processors to do the soring in $O(\log n)$ parallel steps. For each parallel step, we need to resolve n "independent" comparisons. Our problem is different from other problems in the sense that each type-2 comparison can have $O(m)$ "roots" (i.e., the y-coordinates of the intersections). Nevertheless, we can still be able to resolve all these comparisons in a simultaneous way, as follows.

First, for each comparison, we compute the coordinates of the $O(m)$ intersections as discussed above. The intersections of all n comparisons can be computed in $O(mn)$ time. Then, we have $O(mn)$ roots. Suppose $y_1, y_2, \ldots, y_h$ are the list of all $O(mn)$ roots sorted in ascending order, with $h = O(mn)$. Note that we only use this sorted list to explain our approach and our algorithm do not compute this sorted list. By using our decision algorithm, we determine the interval $(y_k, y_{k+1}]$ that contains ϵ, which can be done in $O(mn)$ time plus $O(\log mn)$ calls on the decision algorithm by using the linear time selection algorithm and binary search (without computing the above sorted list). Further, all n comparisons are resolved on the interval $(y_k, y_{k+1}]$. Therefore, we can resolve all these n independent comparisons in $O(mn + n \log k \log mn)$ time. Since there are $O(\log n)$ parallel steps, we can resolve all comparisons and compute the order for $E(\epsilon^*)$ in $O(mn \log n + n \log k \log n \log mn)$ time.

Once the order for $E(\epsilon^*)$ is determined, we can easily compute ϵ^* and obtain an optimal facility set Q by using the greedy algorithm discussed in Section 3.

In fact, we can immediately determine ϵ^* after the above parametric search finishes. Specifically, after the parametric search finishes, the algorithm also gives us an interval $(y_k, y_{k+1}]$ that contains ϵ^*. We claim that $\epsilon^* = y_{k+1}$. Indeed, an observation is that ϵ^* is always equal to the y-coordinate of an intersection of two functions $\mathsf{Ed}(x_p, P_i)$ and $\mathsf{Ed}(x_p, P_j)$ since otherwise we would always make ϵ^* smaller without changing the order of $E(\epsilon^*)$. On the other hand, the parametric search essentially finds y_{k+1} as the smallest y-coordinate of such function intersections that is feasible. Therefore, $\epsilon^* = y_{k+1}$.

In summary, we solve the k-center problem in $O(mn \log n + n \log k \log n \log mn)$ time. Note that this result is based on the assumption that x_{ij} for $j = 1, \ldots, m$ are given sorted for each uncertain point $P_i \in \mathcal{P}$. If they are not given sorted, then we need an extra step to sort them first, which takes $O(mn \log m)$ time in total. Therefore, we have the following lemma.

Theorem 1. *The optimization version of the k-center problem can be solved in* $O(mn \log mn + n \log k \log n \log mn)$ *time.*

One may wonder that Cole's parametric search [17] can be used to further reduce the time complexity by a logarithmic factor, i.e., reduce the time to $O(mn \log mn + n \log k \log mn)$. However this is not the case because resolving each type-2 comparison needs to consider $O(m)$ roots. Specifically, in Cole's parametric search, calling the decision algorithm on the weighted median root of all roots in each comparison level can resolve a weighted-half comparisons in the level. However, in our problem, to resolve the each type-2 comparison, calling the decision algorithm once is not enough. Therefore, Cole's approach is not applicable to our problem.

Since even Megiddo's parametric search may not be quite practical, Van Oostrum and Veltkamp [35] showed that one can replace the parallel sorting scheme in Megiddo's parametric search by the randomized quicksort to obtain a practical solution with the same expected running time. By using the randomized quicksort, our algorithm can solve the k-center problem in expected $O(mn \log nm + n \log k \log n \log mn)$ time and the algorithm is practical.

5   The Discrete k-Center Problem

In this section, we present an algorithm for the discrete version of the k-center problem, and due to some special properties of the discrete case, the algorithm is faster than the one in Theorem 1 for the general case.

In the discrete k-center problem, each uncertain point P_i has m possible locations, denoted by $p_{i1}, p_{i2}, \ldots, p_{im}$, each having a probability. Since this is a special case of the general k-center problem, the previous results on the general k-center problem (e.g., Lemma 3 and Theorem 1) are still applicable.

By Lemma 1, the function $\mathsf{Ed}(x_p, P_i)$ for $x_p \in \mathbb{R}$ is still a unimodal function, but in the discrete version, $\mathsf{Ed}(x_p, P_i)$ is a piecewise linear function. After the locations $p_{i1}, p_{i2}, \ldots, p_{im}$ are sorted in $O(m \log m)$ time, the function $\mathsf{Ed}(x_p, P_i)$

can be computed in additional $O(m)$ time by Lemma 1. In the following, we assume all functions $\text{Ed}(x_p, P_i)$ for $i = 1, 2, \ldots, n$ have been computed.

We define the decision problem in the same way as before. Our goal is to find the smallest feasible value ϵ^*. As we discussed in the general k-center problem, ϵ^* is the y-coordinate of the intersection of two functions $\text{Ed}(x_p, P_i)$ and $\text{Ed}(x_p, P_j)$ for some i and j. Let $\mathcal{I}$ be the set of intersections of all functions $\text{Ed}(x_p, P_i)$ for $i = 1, 2, \ldots, n$, and for simplicity of discussion, we assume each such intersection is a single point (the general case can be solved by the same techniques with more tedious discussion). Then, ϵ^* is the smallest feasible value among the y-coordinates of all points of $\mathcal{I}$. The algorithm of Theorem 1 uses parametric search to find ϵ^*. In the discrete version, due to the property that each $\text{Ed}(x_p, P_i)$ is a piecewise linear function, we compute ϵ^* by using a technique for searching line arrangement [13], as follows.

We first define an arrangement $\mathcal{A}$. For each $1 \leq i \leq n$, since $\text{Ed}(x_p, P_i)$ is a piecewise linear function, it consists of $O(m)$ line segments and two half-lines, and we let A_i denote the set of lines containing all line segments and half-lines of $\text{Ed}(x_p, P_i)$. Hence, $|A_i| = O(m)$ for each $1 \leq i \leq n$. We explicitly compute each A_i in $O(m)$ time. Let $\mathcal{A}$ be the arrangement of the lines in $\bigcup_{i=1}^{n} A_i$. Note that our algorithm does not compute $\mathcal{A}$ explicitly. With a little abuse of notation, we also use $\mathcal{A}$ to denote the set of all *vertices* of $\mathcal{A}$ (i.e., all line intersections). Clearly, $\mathcal{I} \subseteq \mathcal{A}$. Hence, ϵ^* is also the smallest feasible value among the y-coordinates of the vertices of $\mathcal{A}$, and in other words, ϵ^* is the y-coordinate of the lowest vertex v^* of $\mathcal{A}$ whose y-coordinate is a feasible value for the decision problem. To search the particular vertex v^*, we use the decision algorithm in Lemma 3 and the following arrangement searching technique given in [13].

Suppose there is a function $g : \mathbb{R} \to \{0, 1\}$, such that the description of g is unknown but it is known that g is monotonically increasing. Further, given any value y, we have a "black-box" that can evaluate $g(y)$ (i.e., determine whether $g(y)$ is 1 or 0) in $O(G)$ time, which we call the g-oracle. Let B be a set of n lines in the plane and let $\mathcal{B}$ denote their arrangement. Note that $\mathcal{B}$ is not computed explicitly. For any vertex v of $\mathcal{B}$, let y_v be the y-coordinate of v. The *arrangement searching* is to find the lowest vertex vertex v of $\mathcal{B}$ such that $g(y_v) = 1$. An $O((n + G) \log n)$ time algorithm is given in [13] to solve the arrangement searching problem by modifying the slope selection algorithm [9,25], without using parametric search.

In our problem, we are searching the vertex v^* in the arrangement $\mathcal{A}$. We can define such a function g as follows. For any value y, $g(y) = 1$ if and only if y is a feasible value. Clearly, g is monotonically increasing since for any feasible value y, any value larger than y is also feasible. Hence, v^* is the lowest point in $\mathcal{A}$ with $g(y_{v^*}) = 1$. We use our decision algorithm in Lemma 3 as the g-oracle with $G = O(\log m + n \log k)$. By the result in [13], after the $O(mn)$ lines of $\bigcup_{i=1}^{n} A_i$ are computed, we can compute v^* in $O((mn + \log m + n \log k) \log mn)$ time. It can be verified that $(mn + \log m + n \log k) \log mn = O(mn \log mn + n \log k \log n)$. Consequently, we can obtain ϵ^*. An optimal solution set Q can be found by using the decision algorithm on ϵ^* in additional $O(\log m + n \log k \log n)$ time.

Theorem 2. *The optimization version of the discrete k-center problem can be solved in $O(mn \log mn + n \log k \log n)$ time.*

References

1. Agarwal, P.K., Cheng, S.-W., Tao, Y., Yi, K.: Indexing uncertain data. In: Proc. of the 28th Symposium on Principles of Database Systems (PODS), pp. 137–146 (2009)
2. Agarwal, P.K., Efrat, A., Sankararaman, S., Zhang, W.: Nearest-neighbor searching under uncertainty. In: Proc. of the 31st Symposium on Principles of Database Systems (PODS), pp. 225–236 (2012)
3. Agarwal, P.K., Sharir, M.: Efficient algorithms for geometric optimization. ACM Computing Surveys 30(4), 412–458 (1998)
4. Aggarwal, C., Yu, P.S.: A framework for clustering uncertain data streams. In: Proc. of the 24th International Conference on Data Engineering (ICDE), pp. 150–159 (2008)
5. Arya, V., Garg, N., Khandekar, R., Meyerson, A., Munagala, K., Pandit, V.: Local search heuristics for k-median and facility location problems. SIAM Journal on Computing 33, 544–562 (2004)
6. Averbakh, I., Bereg, S.: Facility location problems with uncertainty on the plane. Discrete Optimization 2, 3–34 (2005)
7. Averbakh, I., Berman, O.: Minimax regret p-center location on a network with demand uncertainty. Location Science 5, 247–254 (1997)
8. Badoiu, M., Har-Peled, S., Indyk, P.: Approximate clustering via core-sets. In: Proc. of the 34th Annual Symposium on Theory of Computing (STOC), pp. 250–257 (2002)
9. Brönnimann, H., Chazelle, B.: Optimal slope selection via cuttings. Computational Geometry: Theory and Applications 10(1), 23–29 (1998)
10. Chandrasekaran, R., Tamir, A.: Polynomially bounded algorithms for locating p-centers on a tree. Mathematical Programming 22(1), 304–315 (1982)
11. Chau, M., Cheng, R., Kao, B., Ng, J.: Uncertain data mining: An example in clustering location data. In: Ng, W.-K., Kitsuregawa, M., Li, J., Chang, K. (eds.) PAKDD 2006. LNCS (LNAI), vol. 3918, pp. 199–204. Springer, Heidelberg (2006)
12. Chazelle, B., Guibas, L.: Fractional cascading: I. A data structuring technique. Algorithmica 1(1), 133–162 (1986)
13. Chen, D.Z., Wang, H.: A note on searching line arrangements and applications. Information Processing Letters 113, 518–521 (2013)
14. Chen, D.Z., Li, J., Wang, H.: Efficient algorithms for one-dimensional k-center problems. arXiv:1301.7512 (2013)
15. Chen, D.Z., Wang, H.: Efficient algorithms for the weighted k-center problem on a real line. In: Asano, T., Nakano, S., Okamoto, Y., Watanabe, O. (eds.) ISAAC 2011. LNCS, vol. 7074, pp. 584–593. Springer, Heidelberg (2011)
16. Cheng, R., Xia, Y., Prabhakar, S., Shah, R., Vitter, J.S.: Efficient indexing methods for probabilistic threshold queries over uncertain data. In: Proc. of the 30th International Conference on Very Large Data Bases (VLDB), pp. 876–887 (2004)
17. Cole, R.: Slowing down sorting networks to obtain faster sorting algorithms. Journal of the ACM 34(1), 200–208 (1987)
18. Cormode, G., McGregor, A.: Approximation algorithms for clustering uncertain data. In: Proc. of the 27th Symposium on Principles of Database Systems (PODS), pp. 191–200 (2008)

19. Drezner, Z., Hamacher, H.W.: Facility Location: Applications and Theory. Springer, New York (2004)
20. Frederickson, G.N.: Parametric search and locating supply centers in trees. In: Dehne, F., Sack, J.-R., Santoro, N. (eds.) WADS 1991. LNCS, vol. 519, pp. 299–319. Springer, Heidelberg (1991)
21. Gupta, U.I., Lee, D.T., Leung, J.Y.-T.: Efficient algorithms for interval graphs and circular-arc graphs. Networks 12, 459–467 (1982)
22. Hochbaum, D., Shmoys, D.: A best possible heuristic for the k-center problem. Mathematics of Operations Research 10, 180–184 (1985)
23. Kamousi, P., Chan, T.M., Suri, S.: Closest pair and the post office problem for stochastic points. In: Dehne, F., Iacono, J., Sack, J.-R. (eds.) WADS 2011. LNCS, vol. 6844, pp. 548–559. Springer, Heidelberg (2011)
24. Kamousi, P., Chan, T.M., Suri, S.: Stochastic minimum spanning trees in Euclidean spaces. In: Proc. of the 27th Annual Symposium on Computational Geometry (SoCG), pp. 65–74 (2011)
25. Katz, M., Sharir, M.: Optimal slope selection via expanders. Information Processing Letters 47(3), 115–122 (1993)
26. Megiddo, N.: Applying parallel computation algorithms in the design of serial algorithms. Journal of the ACM 30(4), 852–865 (1983)
27. Megiddo, N.: Linear-time algorithms for linear programming in R^3 and related problems. SIAM Journal on Computing 12(4), 759–776 (1983)
28. Megiddo, N., Supowit, K.J.: On the complexity of some common geometric location problems. SIAM Journal on Comuting 13, 182–196 (1984)
29. Megiddo, N., Tamir, A.: New results on the complexity of p-centre problems. SIAM J. on Computing 12(4), 751–758 (1983)
30. Megiddo, N., Tamir, A., Zemel, E., Chandrasekaran, R.: An $O(n \log^2 n)$ algorithm for the k-th longest path in a tree with applications to location problems. SIAM J. on Computing 10, 328–337 (1981)
31. Ngai, W.K., Kao, B., Chui, C.K., Cheng, R., Chau, M., Yip, K.Y.: Efficient clustering of uncertain data. In: Proc. of the 6th International Conference on Data Mining (ICDM), pp. 436–445 (2006)
32. Snoeyink, J.: Maximum independent set for intervals by divide and conquer with pruning. Networks 49, 158–159 (2007)
33. Snyder, L.V.: Facility location under uncertainty: A review. IIE Transactions 38, 537–554 (2006)
34. Tao, Y., Xiao, X., Cheng, R.: Range search on multidimensional uncertain data. ACM Transactions on Database Systems 32 (2007)
35. van Oostrum, R., Veltkamp, R.C.: Parametric search made practical. Computational Geometry: Theory and Applications 28, 75–88 (2004)
36. Wang, H.: Minmax regret 1-facility location on uncertain path networks. In: Cai, L., Cheng, S.-W., Lam, T.-W. (eds.) ISAAC 2013. LNCS, vol. 8283, pp. 733–743. Springer, Heidelberg (2013)
37. Yiu, M.L., Mamoulis, N., Dai, X., Tao, Y., Vaitis, M.: Efficient evaluation of probabilistic advanced spatial queries on existentially uncertain data. IEEE Transactions on Knowledge and Data Engineering 21, 108–122 (2009)
38. Yu, H.-I., Lin, T.-C., Wang, B.-F.: Improved algorithms for the minmax-regret 1-center and 1-median problems. ACM Transactions on Algorithms 4(3), Article No. 36 (2008)

The Range 1 Query (R1Q) Problem*

Michael A. Bender[1,2], Rezaul A. Chowdhury[1], Pramod Ganapathi[1],
Samuel McCauley[1], and Yuan Tang[3]

[1] Department of Computer Science, Stony Brook University, Stony Brook, NY, USA
{bender,rezaul,pganapathi,smccauley}cs.stonybrook.edu
[2] Tokutek, Inc., USA
[3] Software School, Fudan University, Shanghai, China
yuantang@csail.mit.edu

Abstract. We define the *range 1 query* (R1Q) problem as follows. Given
a d-dimensional ($d \geq 1$) input bit matrix A, preprocess A so that for any
given region $\mathcal{R}$ of A, one can efficiently answer queries asking if $\mathcal{R}$ contains a 1 or not. We consider both orthogonal and non-orthogonal shapes
for $\mathcal{R}$ including rectangles, axis-parallel right-triangles, certain types of
polygons, and spheres. We provide space-efficient deterministic and randomized algorithms with constant query times (in constant dimensions)
for solving the problem in the word RAM model. The space usage in
bits is sublinear, linear, or near linear in the size of A, depending on the
algorithm.

Keywords: R1Q, range query, range emptiness, randomized, rectangular, orthogonal, non-orthogonal, triangular, polygonal, circular, spherical.

1 Introduction

Range searching is one of the fundamental problems in computational geometry
[1,19]. It arises in application areas including geographical information systems,
computer graphics, computer aided design, spatial databases, and time series
databases. Range searching encompasses different types of problems, such as
range counting, range reporting, emptiness queries, and optimization queries.

The *range 1 query* (R1Q) problem is defined as follows. Given a d-dimensional
($d \geq 1$) input bit matrix A (consisting of 0's and 1's), preprocess A so that one
can efficiently answer queries asking if any given range $\mathcal{R}$ of A is empty (does
not contain a 1) or not, denoted by $R1Q_A(\mathcal{R})$ or simply $R1Q(\mathcal{R})$. In 2-D, the
range $\mathcal{R}$ can be a rectangle, a right triangle, a polygon or a circle.

In this paper, we investigate solutions in the word RAM model sharing the
following characteristics. First of all, we want queries to run in constant time,
even for $d \geq 2$ dimensions. Second, we are interested in solutions that have space

* Rezaul Chowdhury & Pramod Ganapathi are supported in part by NSF grant CCF-
1162196. Michael A. Bender & Samuel McCauley are supported in part by NSF
grants IIS-1247726, CCF-1217708, CCF-1114809, and CCF-0937822.

Z. Cai et al. (Eds.): COCOON 2014, LNCS 8591, pp. 116–128, 2014.

linear or sublinear in the number of bits in the input grid. Note that while our sublinear bounds are parameterized by the number of 1s in the grid, this is still larger than the information-theoretic lower bounds. For our motivating applications, information-theoretically optimal space is less important than constant query times. Third, we are interested in grid inputs [15,16], viewing the problem in terms of pixels/voxels rather than a set of spatial points. This grid perspective enables constant-time operations such as table lookup and hashing. Finally, we are interested in both orthogonal and nonorthogonal queries, and we require solutions that are concise enough to be implementable.

Previous Results. The R1Q problem can be solved using data structures such as balanced binary search trees, kd-trees, quad trees, range trees, partition trees, and cutting trees (see [9]), which take the positions of the 1-bits as input. It can also be solved using a data structure of Overmars [16], which uses priority search trees, y-fast tries, and q-fast tries and takes the entire grid as input. However, in d-D ($d \geq 2$), in the worst case these data structures have a query time at least polylogarithmic and occupy a near-linear number of bits.

The R1Q problem can also be solved via range partial sum [7, 21] and the range minimum query (RMQ) [2–6, 10–12, 17, 18, 22] problems. Though several efficient algorithms have been developed to solve the problem in 1-D and 2-D, their generalizations to 3-D and higher dimensions occupying a linear number of bits are not known yet. Also, there is little work on space-efficient constant-time RMQ solutions for non-orthogonal ranges.

The R1Q problem can also be solved using rank queries [13, 14]. Again, its generalization to 2-D and higher dimensions has not yet been studied.

Motivation. We encountered the R1Q and R0Q (whether a range contains a 0) problems while trying to optimize stencil computations in the *Pochoir* stencil compiler [20], where we had to answer octagonal R1Q and octagonal R0Q on a static 2-D property grid. Stencil computations have applications in physics, computational biology, computational finance, mechanical engineering, adaptive statistical design, weather forecasting, clinical medicine, image processing, quantum dynamics, oceanic circulation modeling, electromagnetics, multigrid solvers, and many other areas (see the references in [20]).

In Fig. 1, we provide a simplified exposition of the problem encountered in Pochoir. There are two grids of the same size: a static property grid and a dynamic value grid. Each property grid cell is set to 1 if it satisfies property $\mathcal{P}$ and 0 otherwise. When Pochoir needs to update a range $\mathcal{R}$ in the value grid (see Alg. 1), its runtime system checks whether all or none of the points in $\mathcal{R}$ satisfy $\mathcal{P}$ in the property grid, and based on the query result it uses an appropriate precompiled optimized version of the original code (see Algs. 3, 4) to update the range in the value grid. To check if all points in $\mathcal{R}$ satisfy $\mathcal{P}$, Pochoir uses R0Q($\mathcal{R}$), and to check if no points in $\mathcal{R}$ satisfy $\mathcal{P}$, it uses R1Q($\mathcal{R}$).

Pochoir needs time-, space-, and cache-efficient data structures to answer R1Q. It can also tolerate some false-positive errors. The solutions should achieve con-

Algorithm 1. : UPDATERANGE($\mathcal{R}$)

1. **if** !R0Q($\mathcal{R}$) **then**
2.   {all points in $\mathcal{R}$ satisfy $\mathcal{P}$.}
3. $funcptr \leftarrow$ PUPDATEPOINT
4. **else if** !R1Q($\mathcal{R}$) **then**
5.   {no points in $\mathcal{R}$ satisfy $\mathcal{P}$.}
6. $funcptr \leftarrow$ NUPDATEPOINT
7. **else**
8.   {not all points in $\mathcal{R}$ satisfy $\mathcal{P}$.}
9. $funcptr \leftarrow$ UPDATEPOINT

10. **for** each grid point p in $\mathcal{R}$ **do**
11. $funcptr(p)$

Algorithm 2. : UPDATEPOINT(p)

1. {update p only if it satisfies $\mathcal{P}$.}
2. **if** $p.property = 1$ **then**
3. $p.value \leftarrow$ new value
4. **do** some stuff

Algorithm 3. : PUPDATEPOINT(p)

1. {p satisfies $\mathcal{P}$. update p.}
2. $p.value \leftarrow$ new value
3. **do** some stuff

Algorithm 4. : NUPDATEPOINT(p)

1. {p doesn't satisfy $\mathcal{P}$. don't update p.}
2. **do** some stuff

Fig. 1. Examples of the procedures in Pochoir that make use of R1Q and R0Q

stant query time and work in all dimensions. Although it is worth trading off space to achieve constant query times, space is still a scarce resource.

Our Contributions. We solve the R1Q problem for orthogonal and non-orthogonal ranges. Our major contributions as shown in Table 1 are as follows:

1. [Orthogonal Deterministic.] We present a deterministic data structure to answer R1Q for orthogonal ranges in all dimensions and for any data distribution. It occupies linear space in bits and answers queries in constant time for any constant dimension.
2. [Orthogonal Randomized.] We present randomized data structures to answer R1Q for orthogonal ranges. The structures occupy sublinear space in bits and provide a tradeoff between query time and error probability.
3. [Non-Orthogonal Deterministic.] We present deterministic data structures to answer R1Q for non-orthogonal shapes such as axis-parallel right-triangles (for 2-D) and spheres (for all dimensions). The structures occupy near-linear space in bits and answer queries in constant time.

We use techniques such as *power hyperrectangles*, *power right-triangles*, *sketches*, *sampling*, *the four Russians trick*, and *compression* in our data structures. A careful combination of these techniques allows us to solve a large class of R1Q problems. Techniques such as power hyperrectangles, table lookup, and the four Russians trick are already common in RMQ-style operations, while sketches, power right-triangles, and compression are not.

Organization of the Paper. Section 2 presents deterministic and randomized algorithms to answer orthogonal R1Qs on a grid in constant time for constant dimensions. Section 3 presents deterministic algorithms to answer non-orthogonal R1Qs on a grid, for axis-parallel right triangles, some polygons, and spheres.

Table 1. R1Q algorithms in this paper. Here, N = total #bits, N_1 = #nonzero bits, and N_0 = #zero bits in the input bit matrix A, and d = #dimensions. If $|A|$ appears in the space complexity, it means that A must be retained, otherwise it can be discarded.

Shape	Space (in bits)	Time	Comments
Orthogonal (Deterministic)			
d-D	$\mathcal{O}\left((d+1)!\left(\frac{2}{\ln 2}\right)^d N\right) + |A|$	$\mathcal{O}\left(4^d d\right)$	for d dimensions
Orthogonal (Randomized)			
1-D (Sketch)	$\mathcal{O}\left(\sqrt{NN_1}\log N \log \frac{1}{\delta}\right)$	$\mathcal{O}\left(\ln \frac{1}{\delta}\right)$	$\delta \in (0, \frac{1}{4})$; correct for range size $\geq \sqrt{N/N_1}$, otherwise correct with prob $\geq 1 - 4\delta$; extendible to $\geq$ 2-D
1-D (Sketch)	$\mathcal{O}\left(\begin{array}{c} N_1 \log^3 N \log_{1+\gamma} \frac{1}{\delta} \\ +N^{\frac{1}{c}}\log\log N \end{array}\right)$	$\mathcal{O}\left(\log \frac{c}{\delta}\right)$	$\gamma, \delta \in (0, \frac{1}{4})$, integer $c > 1$; with prob $\geq 1 - 4\delta$ at most 4γ fraction of all query results will be wrong; extendible to $\geq$ 2-D
1-D (Sampling)	$\mathcal{O}\left(s\right) + |A|$	$\mathcal{O}\left(\frac{1}{\varepsilon}\ln\frac{1}{\delta}\right)$	$\varepsilon, \delta \in (0, 1)$, $s = \Omega(\log N)$; always correct for range size $\geq (N \log N)/s$, otherwise correct with prob $\geq 1 - \delta$ when $\geq \varepsilon$ fraction of all range entries are 1; extendible to $\geq$ 2-D
Non-Orthogonal (Deterministic)			
Right Triangles	$\mathcal{O}\left(N \log N + N_0 \log^3 N\right)$	$\mathcal{O}(1)$	not extendible to $\geq$ 3-D
2-D Spheres	$\mathcal{O}\left(N\sqrt{\log N}\right)$	$\mathcal{O}(1)$	extendible to $\geq$ 3-D

2 Orthogonal Range 1 Queries (R1Q)

In this section, we present deterministic and randomized algorithms for answering orthogonal R1Qs in constant time and up to linear space.

The algorithms in this paper rely on finding the most significant bit (MSB) of positive integers in constant time and sublinear space as follows:

Theorem 1. *Given integers $N \in [1, 2^w)$ and $r \in [1, w]$ in the word-RAM model with w-bit words, one can construct a table occupying $\mathcal{O}\left(N^{1/r} \log\log N\right)$ bits of space to answer MSB queries for integers in $[1, N]$ in $\mathcal{O}\left(1 + \log r\right)$ time.*

2.1 Preliminaries: Deterministic 1-D Algorithm

The input is a bit vector $A[0 \ldots N - 1]$, where $N \in [1, 2^w)$ and w is the word size. The query $\text{R1Q}_A(i, j)$, where $i \leq j$, asks if there exists a 1 in the subarray $A[i \ldots j]$. For simplicity, assume N is an even power of 2.

Preprocessing. Array A has $M = \frac{N}{w}$ words. For each $p \in [0, \log M]$, we construct arrays: L_p and R_p, of size $\frac{M}{2^p}$ each. Let $W(i)$ denote the ith ($i \in [0, M - 1]$) word in A. Then, L_p is defined as follows: $L_0[i]$ is 0, if $W(i)$ has a 1, 1 otherwise.

$$L_{p(\geq 1)}[i] = \begin{cases} L_{p-1}[2i] & \text{if } L_{p-1}[2i] < 2^{p-1}. \\ 2^{p-1} + L_{p-1}[2i+1] & \text{otherwise.} \end{cases}$$

The R_p array can be computed similarly. The array element $L_p[i]$ (and $R_p[i]$) stores the distance of the leftmost (respectively, rightmost) word that contains a 1 in the ith block of 2^p contiguous words of A, measured from the start (and end) of the block. The value $L_p[i] = 2^p$ ($R_p[i] = 2^p$) means that the ith block of 2^p contiguous words of A does not contain a 1.

Query Execution. To answer $\text{R1Q}_A(i,j)$, we consider two cases: (1) *Intra-word queries:* If (i,j) lies inside one word, we answer R1Q using bit shifts. (2) *Inter-word queries:* If (i,j) spans multiple words, then the query gets split into three subqueries: (a) R1Q from i to the end of its word, (b) R1Q of the words between i's and j's word (both exclusive), and (c) R1Q from the start of j's word to j.

The answer to an inter-word query is 1 if and only if the R1Q for at least one of the three subqueries is 1. The first and third subqueries are intra-word queries and can be answered using bit shifts. Let the words containing indices i and j be I and J, respectively. Then, the second subquery, denoted by $\text{R1Q}_{L_0}(I+1, J-1)$, is answered as follows. Using the MSB of $J - I - 1$, we find the largest integer p such that $2^p \leq J - I - 1$. The query $\text{R1Q}_{L_0}(I+1, J-1)$ is then decomposed into the following two overlapping queries of size 2^p each: $\text{R1Q}_{L_0}(I+1, I+2^p)$ and $\text{R1Q}_{L_0}(J - 2^p, J-1)$. If either of those two ranges contains a 1 then the answer to the original query will be 1, and 0 otherwise. We show below how to answer $\text{R1Q}_{L_0}(I+1, I+2^p)$. Query $\text{R1Q}_{L_0}(J-2^p, J-1)$ is answered similarly.

Split L_0 into blocks of size 2^p. Then, the range $\text{R1Q}_{L_0}(I+1, I+2^p)$ can be covered by one or two consecutive blocks. Let $I+1$ be in the kth block. If the range lies in one block, we find whether a 1 exists in that block by checking whether $L_p[k] < 2^p$ is true. If the range is split across two consecutive blocks, we find whether a 1 exists in at least one of the two blocks by checking whether at least one of $R_p[k] \leq (k+1)2^p - I$ or $L_p[k+1] \leq I + 2^p - (k+1)2^p$ is true.

2.2   Deterministic d-D Algorithm

For d-D ($d \geq 2$) R1Q, the input is a bit matrix A of size $N = n^d$. Here we give an algorithm for a 2-D matrix of size $N = n \times n$, but the algorithm extends to higher dimensions. For simplicity, we assume n is a power of 2. The query $\text{R1Q}([i_1, j_1][i_2, j_2])$ asks if there exists a 1 in the submatrix $A[i_1 \ldots j_1][i_2 \ldots j_2]$.

Preprocessing. For each $p, q \in [0, \log n]$, we partition A into $\frac{n}{2^p} \times \frac{n}{2^q}$ blocks, each of size $2^p \times 2^q$ called a (p,q)-block. For each (p,q) pair, we construct four tables of size $\frac{N}{2^{p+q}} \times \min(2^p, 2^q)$ each:
(i) $TL_{p,q}$: if $p \leq q$, $TL_{p,q}[i,j][k]$ indicates that any rectangle of height $k \in [0, 2^p)$ starting from the top-left corner of the current block must have width at least $TL_{p,q}[i,j][k]$ in order to include at least one 1-bit.
(ii) BL, TR, BR: similar to TL but starts from the bottom-left, top-right and bottom-right corners, respectively.
In all cases, a stored value of $\max(2^p, 2^q)$ indicates that the block has no 1.

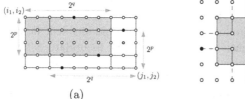
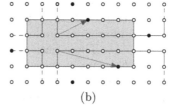

(a)   (b)

Fig. 2. Rectangles: (a) Query rectangle split into four possibly overlapping power rectangles. (b) Power rectangle divided into four regions by four split rectangles.

Query Execution. Given a query $[i_1, j_1][i_2, j_2]$, we find the largest integers p and q such that $2^p \leq j_1 - i_1 + 1$ and $2^q \leq j_2 - i_2 + 1$. The original query range can then be decomposed into four overlapping (p, q)-blocks, which we call *power rectangles*, each with a corner at one of the four corners of the original rectangle, as in Fig. 2(a). If any of these four rectangles contains a 1, the answer to the original query will be 1, and 0 otherwise. We show below how to answer an R1Q for a power rectangle.

We consider the partition of A into preprocessed (p, q)-blocks. It is easy to see that each of the four power rectangles of size $2^p \times 2^q$ will intersect at most four preprocessed (p, q)-blocks. We call each rectangle contained in both the power rectangle and a (p, q)-block a *split rectangle* (see Fig. 2(b)). The R1Q for a split rectangle can be answered using a table lookup, checking if the table values of the appropriate (p, q)-blocks are inside the power rectangle boundary, as shown in Fig. 2(b). The proof of the following theorem will be given in the full paper.

Theorem 2. *Given a d-D input grid of size $N = n^d$, each orthogonal R1Q on the grid can be answered deterministically in $\mathcal{O}\left(4^d d\right)$ time after preprocessing the grid in $\Theta(N)$ time using $\mathcal{O}\left((d+1)! \left(2/\ln 2\right)^d N\right)$ bits of space. In 1-D, the space can be reduced to $\mathcal{O}(N/\log N)$ bits.*

2.3   Randomized Algorithms

In this section, we present randomized algorithms that build on the deterministic algorithms given in Sections 2.1 and 2.2. We describe the algorithms for one dimension only. Extensions to higher dimensions are straightforward.

Sketch Based Algorithms. Our algorithms provide probabilistic guarantees based on the Count-Min (CM) sketch data structure proposed in [8]. Let N_1 be the number of 1-bits in the input bit array $A[0 \ldots N-1]$ for any data distribution. Then, the (preprocessing) time and space complexities depend on N_1 while the query time remains constant.

A CM sketch with parameters $\varepsilon \in (0, 1]$ and $\delta \in (0, 1)$ can store a summary of any given vector $\boldsymbol{a} = \langle a_0, a_1, \ldots, a_{n-1} \rangle$ with $a_i \geq 0$ in only $\lceil \frac{e}{\varepsilon} \rceil \lceil \ln \frac{1}{\delta} \rceil \log \|\boldsymbol{a}\|_1$ bits of space, where $\|\boldsymbol{a}\|_1$ (or $\|\boldsymbol{a}\|$) $= \sum_{i=0}^{n-1} a_i$, and can provide an estimate $\hat{a}_i$

of any a_i with the following guarantees: $a_i \leq \hat{a}_i$, and with probability at least $1 - \delta$, $\hat{a}_i \leq a_i + \varepsilon ||a||_1$. It uses $t = \lceil \ln \frac{1}{\delta} \rceil$ hash functions $h_1 \ldots h_t : \{0 \ldots n - 1\} \to \{1 \ldots b\}$ chosen uniformly at random from a pairwise-independent family, where bucket size $b = \lceil \frac{e}{\varepsilon} \rceil$. These hash functions are used to update a 2-D matrix $c[1 : t][1 : b]$ of bt counters initialized to 0. For each $i \in [0, n - 1]$ and each $j \in [1, t]$ one then updates $c[j][h_j(i)]$ to $c[j][h_j(i)] + a_i$. After the updates, an estimate $\hat{a}_i$ for any given query point a_i is obtained as $\min_{1 \leq j \leq t} c[j][h_j(i)]$.

Preprocessing. In the deterministic algorithms we first compressed the input array by converting each word into a single bit, and then constructed L_0 and R_0 arrays from the compressed array. In the current algorithm we build the L_0 and R_0 arrays directly from the uncompressed input. For $p \in \left[0, \frac{1}{2} \log (N/N_1)\right]$, the L_p and R_p arrays are stored as CM sketches while $p \in \left[\frac{1}{2} \log (N/N_1) + 1, \log N\right]$ the arrays are stored directly as in the deterministic case. Each $L_p[i]$ is added as $(L_p[i] + 1) \mod (2^p + 1)$ to the CM sketch (similarly for $R_p[i]$). Thus a nonzero entry (of value at most 2^p) is added to the CM sketch provided the corresponding block contains a 1, otherwise nothing is added. As a result for any given L_p summation of all entries added to the CM sketch is at most $N_1 \times 2^p$, and we set $\varepsilon = \frac{1}{2 \times N_1 \times 2^p}$ for that sketch.

Query Execution. Given a query $R1Q_A(i, j)$, we use the MSB of $j - i + 1$ to find the largest value of p with $2^p \leq j - i + 1$, and then follow the approach for answering case (b) of inter-word queries described in Section 2.1. If $2^p > \sqrt{N/N_1}$, we use L_p and R_p arrays to answer the query correctly, otherwise we use the L_p and R_p values obtained from the corresponding CM sketches.

Error Bound. If the query range is larger than $\sqrt{N/N_1}$, the answer is always correct. For smaller queries we use CM sketches. Recall that for $p \in \left[0, \frac{1}{2} \ln(N/N_1)\right]$, we store each L_p (and R_p) as a CM sketch with parameter $\varepsilon = \frac{1}{2 \times N_1 \times 2^p}$. Hence, the estimated value $\hat{L}_p[i]$ of an entry $L_p[i]$ returned by the CM sketch is between $L_p[i]$ and $L_p[i] + \varepsilon ||L_p|| \leq L_p[i] + 0.5$ with probability at least $1 - \delta$. In other words, with probability at least $1 - \delta$, the CM sketch returns the correct value. In order to answer an R1Q we need to access at most four CM sketches. Hence, with probability at least $(1 - \delta)^4 \geq 1 - 4\delta$, the query will return the correct answer.

Theorem 3. *Given a 1-D bit array of length N containing N_1 nonzero entries, and a parameter $\delta \in (0, \frac{1}{4})$, one can construct a data structure occupying $\mathcal{O}\left(\sqrt{NN_1} \log N \log \left(\frac{1}{\delta}\right)\right)$ bits (and discard the input array) to answer each R1Q correctly in $\mathcal{O}\left(\ln \frac{1}{\delta}\right)$ worst-case time with probability at least $1 - 4\delta$. For query ranges larger than $\sqrt{N/N_1}$ the query result is always correct.*

By tweaking the algorithm described above slightly, we can reduce the space complexity even further at the cost of providing a weaker correctness guarantee. We assume that we are given an additional parameter $\gamma \in (0, \frac{1}{4})$. The required modification is described below.

For each $p \in [0, \log N]$, we store the L_p and R_p arrays as CM sketches. However, instead of adding a value v directly to a CM sketch, we now add a $(1 + \gamma)$ approximation of v. More precisely, we add $\lceil \log_{1+\gamma} (1 + v) \rceil$ instead of v. Hence, for a given L_p, the summation of all entries added to its CM sketch is at most $N_1 \lceil \log_{1+\gamma} (1 + 2^p) \rceil$, and so we set the parameter ε to $1/ \left(2N_1 \lceil \log_{1+\gamma} (1 + 2^p) \rceil \right)$ for that sketch. The total space used by all CM sketches can be shown to be $\mathcal{O} \left(N_1 \log^3 N \log_{1+\gamma} (1/\delta) \right)$. We store a lookup table of size $\mathcal{O} \left(\log^2 N \right)$ for conversions from $\lceil \log_{1+\gamma} (1 + v) \rceil$ to v, and an MSB table of size $\mathcal{O} \left(N^{1/c} \log \log N \right)$ for some given integer constant $c > 1$.

We first show that for any given $p \in [0, \log N]$ at most 2γ fraction of the queries of size 2^p can return incorrect answers. Consider any two consecutive blocks of size 2^p, say, blocks $i \in [0, \frac{N}{2^p} - 1)$ and $i + 1$. Exactly 2^p different queries of size 2^p will cross the boundary between these two blocks. The answer to each of these queries will depend on the estimates of $R_p[i]$ and $L_p[i+1]$ obtained from the CM sketches. Under our construction the estimates are $\hat{R}_p[i] \leq (1+\gamma)R_p[i] \leq R_p[i] + \gamma \cdot 2^p$ and $\hat{L}_p[i+1] \leq (1+\gamma)L_p[i+1] \leq L_p[i+1] + \gamma \cdot 2^p$. Hence, at most $\gamma \cdot 2^p$ of those 2^p queries will produce incorrect results due to the error in estimating $R_p[i]$, and at most $\gamma \cdot 2^p$ more because of the error in estimating $L_p[i+1]$. Thus with probability at least $(1 - \delta)^2$, at most 2γ fraction of those 2^p queries will return wrong results. Recall from Section 2.1 that we answer given queries by decomposing the query range into two overlapping query ranges. Hence, with probability at least $(1-\delta)^4 \geq 1 - 4\delta$, at most $2\gamma + 2\gamma = 4\gamma$ fraction of all queries can produce wrong answers.

Theorem 4. *Given a 1-D bit array of length N containing N_1 nonzero entries, and two parameters $\gamma \in \left(0, \frac{1}{4}\right)$ and $\delta \in \left(0, \frac{1}{4}\right)$, and an integer constant $c > 1$, one can construct a data structure occupying $\mathcal{O} \left(N_1 \log^3 N \log_{1+\gamma} \left(\frac{1}{\delta}\right) + N^{1/c} \log \log N \right)$ bits (and discard the input array) to answer each R1Q in $\mathcal{O} \left(\log \frac{c}{\delta} \right)$ worst-case time such that with probability at least $1 - 4\delta$ at most 4γ fraction of all query results will be wrong.*

Sampling Based Algorithm. Suppose we are allowed to use only $\mathcal{O}(s)$ bits of space (in addition to the input array A), and $s = \Omega (\log_2 N)$. We are also given two constants $\varepsilon \in (0, 1)$ and $\delta \in (0, 1)$. We build L_p and R_p arrays for each $p \in \left[\log \frac{N}{s} + \log \log N, \log N \right]$, and an MSB lookup table to support constant time MSB queries for integers in $[1, s/ \log N]$. Consider the query $\text{R1Q}_A(i, j)$. If $j - i + 1 \leq w$, we answer the query correctly in constant time by reading at most 2 words from A and using bit shifts. If $j - i + 1 \geq 2^{\log \frac{N}{s} + \log \log N} = \frac{N \log N}{s}$, we use the L_p and R_p arrays to correctly answer the query in constant time. If $w < j - i + 1 < \frac{N \log N}{s}$, we sample $\lceil \frac{1}{\varepsilon} \ln \left(\frac{1}{\delta}\right) \rceil$ entries uniformly at random from $A[i \ldots j]$, and return their bitwise OR. It is easy to show that the L_p and R_p tables use $\mathcal{O}(s)$ bits in total, and the MSB table uses $o(s)$ bits of space. The query time is clearly $\mathcal{O} \left(\frac{1}{\varepsilon} \ln \left(\frac{1}{\delta}\right) \right)$.

Error Bound. If at least an ε fraction of the entries in $A[i \ldots j]$ are nonzero then the probability that a sample of size $\lceil \frac{1}{\varepsilon} \ln \left(\frac{1}{\delta} \right) \rceil$ chosen uniformly at random from the range will pick at least one nonzero entry is $\geq 1 - (1 - \varepsilon)^{\frac{1}{\varepsilon} \ln \left(\frac{1}{\delta} \right)} \approx 1 - \delta$.

Theorem 5. *Given a 1-D bit array of length N, a space bound $s = \Omega \left(\log N \right)$, and two parameters $\varepsilon \in (0, 1)$ and $\delta \in (0, 1)$, one can construct a data structure occupying only $\mathcal{O} \left(s \right)$ bits of space (in addition to the input array) that in $\mathcal{O} \left(\frac{1}{\varepsilon} \ln \left(\frac{1}{\delta} \right) \right)$ time can answer each $R1Q_A(i, j)$ correctly with probability at least $1 - \delta$ provided at least an ε fraction of the entries in $A[i \ldots j]$ are nonzero. If $j - i + 1 \leq w$ or $j - i + 1 \geq \frac{N \log N}{s}$, the query result is always correct.*

3   Non-Orthogonal Range 1 Queries (R1Q)

In this section, we show how to answer R1Q for non-orthogonal ranges, such as axis-parallel right triangles, spheres and certain type of polygons, given an input matrix of size $N = n \times n$.

3.1   Right Triangular R1Q

A right triangular query $R1Q(ABC)$ asks if there exists a 1 in an axis-parallel right triangle ABC defined by three grid points A, B, and C. In the rest of the paper, right triangles will mean axis-aligned right triangles.

Preprocessing. For every grid point (x, y) containing a 0, for each $p \in \left[0, \frac{\log N}{2} \right]$, we store the coordinates of 8 other grid points for 8 different orientations. For example, consider Fig. 3(a) in which each black grid point corresponds to a 1, and each white point corresponds to a 0. For the point $P = (x, y)$ in the figure, we show the eight black points (i.e., L_C, L_{CC}, R_C, R_{CC}, U_C, U_{CC}, D_C and D_{CC}) we store for a given p. For example, L_C is a black point that lies to the left of P within a horizontal distance of 2^p from it (in terms of the number of grid points including P) such that PL_C makes the smallest angle θ_{LC} in the clockwise direction with the horizontal line passing through P. The significance of L_C is that no right triangle with a horizontal base of length 2^p that has one endpoint at (x, y), another endpoint to the left of (x, y), and whose hypotenuse makes a smaller nonnegative angle than θ_{LC} in the clockwise direction with the horizontal line can contain a 1. Similarly, other points are identified.

Query Execution. We show how to answer a right triangular R1Q in $\Theta \left(1 \right)$ time. Say, we want to answer $R1Q(ABC)$ (see Fig. 3(b)). Let 2^p be the largest power of 2 not larger than $|AB|$, and 2^q be the largest power of 2 not larger than $|CB|$. Find grid points D and E on AB and CB, respectively, such that $|AD| = 2^p$ and $|CE| = 2^q$. Suppose the horizontal line passing through D intersects BC at G, and the vertical line passing through E intersects BC at H. Observe that G and H are not necessarily grid points. We assume w.l.o.g. that none of the vertices A, B and C contains a 1 (as otherwise we can answer the query trivially in

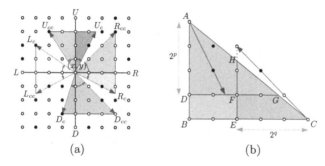

Fig. 3. Right Triangular R1Q. (*a*) Preprocessing. (*b*) Query Execution. Each black grid point contains a 1 while each white grid point contains a 0.

constant time). Observe that we can answer R1Q(ABC) if we can answer R1Q for triangles ADG and CEH, and the rectangle $BDFE$. R1Q for the rectangle can be answered using our deterministic algorithm described in Section 2.2. R1Q for a right triangle of a particular orientation with height or base length equal to a power of two can be answered in constant time. This is done by checking whether the point stored (from preprocessing) with the appropriate endpoint of the hypotenuse for that specific orientation is inside the triangle or not.

Theorem 6. *Given a 2-D bit matrix of size $N = \sqrt{N} \times \sqrt{N}$ containing N_0 zero bits, one can construct a data structure occupying $\mathcal{O}\left(N \log N + N_0 \log^2 N\right)$ bits in $\mathcal{O}\left(N^{1.5}\right)$ time (and discard the input matrix) to answer each axis-aligned right triangular R1Q with the three vertices on the grid points in $\mathcal{O}\left(1\right)$ time.*

3.2 Polygonal R1Q

Consider a simple polygon with its vertices on grid points satisfying the following.

Property 1. For every two adjacent vertices (a, b) and (c, d), one of the two right triangles with the third vertex being either (a, d) or (c, b) is completely inside the polygon.

It can be shown that such a polygon can be decomposed into a set of possibly overlapping right triangles and rectangles with only grid points as vertices that completely covers the polygon (see Fig. 4(*a*)). Examples of polygons that do not satisfy the constraint are given in Fig. 4(*b, c*), but we can still answer R1Q for the polygon in (*c*). A simple polygon with k vertices satisfying propery 1 can be decomposed into $\mathcal{O}\left(k\right)$ right triangles and rectangles and hence can be answered in $\mathcal{O}\left(k\right)$ time.

3.3 Spherical R1Q

The spherical R1Q problem is defined as follows. Given a d-dimensional $(d \geq 2)$ input bit matrix A, preprocess A such that given any grid point p in A and a

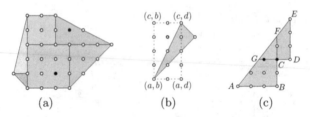

Fig. 4. Black grid points contain 1's and white grid points contain 0's. Polygon in (a) satisfies Prop. 1. Polygons in (b) and (c) do not satisfy Prop. 1. Still, R1Q can be answered for (c).

radius $r \in \mathbb{R}^+$, find efficiently if there exists a 1 in the d-sphere centered at p of radius r. Here, we present the algorithm for 2-D. The approach can be extended to higher dimensions.

A nearest 1-bit of a grid point p is called a *nearest neighbor* (NN) of p. We pre-process A by computing and compresssing the NNs of all grid points in A (only one NN per grid point). We then answer a spherical R1Q by checking whether a NN of the given center point is inside the circle of given radius.

Preprocessing. We store the locations of the NNs of the grid points of A in a temporary NN matrix that occupies $\mathcal{O}(N \log N)$ bits, but can be compressed to occupy $\mathcal{O}(N\sqrt{\log N})$ bits as follows.

We divide the grid into $\sqrt{(\log N)/6} \times \sqrt{(\log N)/6}$ blocks. We store the NN position for all points on the boundary of each block. The interior points will be replaced with arrows $(\rightarrow, \leftarrow, \uparrow, \downarrow)$ and bullets $(\bullet)$ as follows. If a grid point p contains a 1 then p is replaced with a $\bullet$. An arrow at a grid point gives the direction of its NN. If we follow the arrows from any interior point, we end up in either a boundary point or an interior point containing a 1. For any given block the matrix created as above will be called a *symbol matrix* representing the block.

Two blocks are of the same *type* if they have the same symbol matrix. Each symbol can be represented using only three bits. Since each block has $(\log N)/6$ symbols, there are $2^{\frac{3 \log N}{6}} = \sqrt{N}$ possible block types. For each block type we create a *position matrix* that stores, for each grid point within a block, the pointer to its NN if the NN is an interior point, or a pointer to a boundary point if the NN is an exterior or boundary point. The boundary point will have stored its own NN position in the input array.

We can now discard the original input matrix, and replace it with the following compressed representation. For each block in the input matrix we store its block type (i.e., a pointer to the corresponding block type) followed by the NN positions of its boundary points. For each block type we retain its position matrix.

Query Execution. We can answer a spherical R1Q by checking whether the NN position of the center point is inside the query sphere. The approach of finding the NN position is as follows. We find the block to which the given point belongs and follow the pointer to its block type. We check the position stored at the given point in the position matrix. If it points to an internal point, then that point is the correct NN. If it points to a boundary point, we again follow the pointer stored at the boundary point to get the correct NN.

Theorem 7. *Given a 2-D bit array of size N, one can construct a data structure occupying $\mathcal{O}\left(N\sqrt{\log N}\right)$ bits (and discard the input array) to answer each spherical R1Q in $\mathcal{O}\left(1\right)$ time.*

Acknowledgments. We like to thank Michael Biro, Dhruv Matani, Joseph S. B. Mitchell, and anonymous referees for insightful comments and suggestions.

References

1. Agarwal, P.K., Erickson, J.: Geometric range searching and its relatives. Contemporary Mathematics 223, 1–56 (1999)
2. Amir, A., Fischer, J., Lewenstein, M.: Two-dimensional range minimum queries. In: Ma, B., Zhang, K. (eds.) CPM 2007. LNCS, vol. 4580, pp. 286–294. Springer, Heidelberg (2007)
3. Bender, M.A., Farach-Colton, M.: The lca problem revisited. In: Gonnet, G.H., Viola, A. (eds.) LATIN 2000. LNCS, vol. 1776, pp. 88–94. Springer, Heidelberg (2000)
4. Bender, M.A., Farach-Colton, M., Pemmasani, G., Skiena, S., Sumazin, P.: Lowest common ancestors in trees and directed acyclic graphs. Journal of Algorithms 57(2), 75–94 (2005)
5. Berkman, O., Vishkin, U.: Recursive star-tree parallel data structure. SIAM Journal on Computing 22(2), 221–242 (1993)
6. Brodal, G.S., Davoodi, P., Rao, S.S.: On space efficient two dimensional range minimum data structures. Algorithmica 63(4), 815–830 (2012)
7. Chazelle, B., Rosenberg, B.: Computing partial sums in multidimensional arrays. In: SoCG, pp. 131–139. ACM (1989)
8. Cormode, G., Muthukrishnan, S.: An improved data stream summary: The count-min sketch and its applications. Journal of Algorithms 55(1), 58–75 (2005)
9. De Berg, M., Cheong, O., van Kreveld, M., Overmars, M.: Computational geometry. Springer (2008)
10. Fischer, J.: Optimal succinctness for range minimum queries. In: López-Ortiz, A. (ed.) LATIN 2010. LNCS, vol. 6034, pp. 158–169. Springer, Heidelberg (2010)
11. Fischer, J., Heun, V.: A new succinct representation of rmq-information and improvements in the enhanced suffix array. In: Chen, B., Paterson, M., Zhang, G. (eds.) ESCAPE 2007. LNCS, vol. 4614, pp. 459–470. Springer, Heidelberg (2007)
12. Fischer, J., Heun, V., Stühler, H.: Practical entropy-bounded schemes for o(1)-range minimum queries. In: Data Compression Conference, pp. 272–281. IEEE (2008)
13. Golynski, A.: Optimal lower bounds for rank and select indexes. TCS 387(3), 348–359 (2007)
14. González, R., Grabowski, S., Mäkinen, V., Navarro, G.: Practical implementation of rank and select queries. In: Poster Proc. WEA, pp. 27–38 (2005)
15. Navarro, G., Nekrich, Y., Russo, L.: Space-efficient data-analysis queries on grids. TCS (2012)

16. Overmars, M.H.: Efficient data structures for range searching on a grid. Journal of Algorithms 9(2), 254–275 (1988)
17. Sadakane, K.: Compressed suffix trees with full functionality. Theory of Computing Systems 41(4), 589–607 (2007)
18. Sadakane, K.: Succinct data structures for flexible text retrieval systems. JDA 5(1), 12–22 (2007)
19. Sharir, M., Shaul, H.: Semialgebraic range reporting and emptiness searching with applications. SIAM Journal on Computing 40(4), 1045–1074 (2011)
20. Tang, Y., Chowdhury, R., Kuszmaul, B.C., Luk, C.K., Leiserson, C.E.: The Pochoir stencil compiler. In: SPAA, pp. 117–128. ACM (2011)
21. Yao, A.C.: Space-time tradeoff for answering range queries. In: STOC, pp. 128–136. ACM (1982)
22. Yuan, H., Atallah, M.J.: Data structures for range minimum queries in multidimensional arrays. In: SODA, pp. 150–160 (2010)

Competitive Cost-Savings
in Data Stream Management Systems

Christine Chung[1], Shenoda Guirguis[2], and Anastasia Kurdia[3]

[1] Department of Computer Science,
Connecticut College, New London, CT, USA
cchung@conncoll.edu
[2] Oracle Labs, Redwood City, CA, USA
shenoda.guirguis@oracle.com
[3] Department of Computer Science,
Bucknell University, Lewisburg, PA, USA
ak034@bucknell.edu

Abstract. In Continuous Data Analytics and in monitoring applications, hundreds of similar Aggregate Continuous Queries (ACQs) are registered at the Data Stream Management System (DSMS) to continuously monitor the infinite input stream of data tuples. Optimizing the processing of these ACQs is crucial in order for the DSMS to operate at the adequate required scalability. One optimization technique is to share the results of partial aggregation operations between different ACQs on the same data stream. However, finding the query execution plan that attains maximum reduction in total plan cost is computationally expensive. *Weave Share*, a multiple ACQs optimizer that computes query plans in a greedy fashion, was recently shown in experiments to achieve more than an order of magnitude improvement over the best existing alternatives. Maximizing the benefit of sharing, i.e., maximizing the cost-savings achieved by sharing partial aggregation results, is the goal of Weave Share. In this paper we prove that Weave Share approximates the optimal cost-savings to within a factor of 4 for a practical variant of the problem. To the best of our knowledge, this is the first theoretical guarantee provided for this problem. We also provide exact solutions for two natural special cases.

1   Introduction

In Continuous Data Analytics, such as pay-per-click applications, and in monitoring applications, such as network, financial, health and military monitoring, hundreds of similar Aggregate Continuous Queries (ACQs) are typically registered to continuously monitor unbounded input streams of data updates [6, 12]. For example, a stock market monitoring application allows each of its numerous users to register several monitoring queries. Traders interested in a certain stock might register ACQs to monitor the `average` or `maximum` trade volume in a certain period of time, e.g., the last 1, 8, or 24 hours. Meanwhile, decision makers might register monitoring queries for analysis purposes with coarse time granularity over the same data stream, e.g., the `average` trade volume in last week or month. Given the high data arrival rates, optimizing the processing of ACQs is crucial for scalability of the system. Data Stream Management

Z. Cai et al. (Eds.): COCOON 2014, LNCS 8591, pp. 129–140, 2014.

Systems (DSMSs) were developed to be at the heart of every monitoring application, (e.g., [1–3, 8, 13, 14]). DSMSs must efficiently handle the unbounded streams with large volumes of data and large numbers of continuous queries. Thus, devising ways to optimize the processing of multiple continuous queries is imperative for DSMSs to exhibit the scalability required. The commonality of many of the ACQs is what makes optimization possible.

An ACQ is typically defined over a certain window of the input data stream, to bound its computations. (For example, an ACQ that monitors the average trade volume of a stock index could report every hour the average trade volume in the past 24 hours). Partial aggregation has been proposed to optimize the processing of an ACQ [4, 10, 11] by minimizing the repeated processing of overlapping windows. Partial aggregation has also been utilized to share the processing of multiple similar ACQs with different windows [6, 7, 9, 12]. Recently, the concept of *Weaveability* of two sets of ACQs was introduced as an indicator of the potential benefits of sharing their processing [6]. By exploiting weavability, the algorithm *Weave Share* optimizes the shared processing of ACQs. *Weave Share* considers all factors that affect the cost of the shared query plan. It selectively groups ACQs into multiple "execution trees" to minimize the total plan cost. It was shown experimentally that *Weave Share* generates up to 40 times better quality plans compared to the best alternative sharing scheme [6].

Contributions. In this paper we provide formal guarantees on the performance of the Weave Share algorithm. The total cost of a query plan can be represented as the cost of the no-share query plan, in which all partial aggregations are independent, minus the cost-savings achieved by sharing some partial aggregation operations. We provide a lower bound for the cost-savings achieved by Weave Share. Specifically, we show that for a widely applicable variant of the problem, in the worst case, Weave Share is guaranteed to achieve a cost-savings of at least $\frac{1}{4}$ of the maximum possible cost-savings. In contrast with total cost of the query plan, cost-savings is actually a more incisive measure that removes the distraction of the minimum "base-cost" that exists for any given instance, even under the most optimal sharing arrangement. We also remark on two practical special cases of the problem.

2   Background and Definitions

We set the stage by providing the necessary background on ACQ semantics, the paired window technique for partial aggregation [9], and the procedure for "composing" multiple ACQs together [9] so that the results of their partial aggregation can be "shared." We then give a formal definition of the optimization problem at hand and an overview of an efficient practical algorithm for the problem.

2.1   Partial Aggregation for One ACQ

Each ACQ (or *query*) comprises an aggregation operator (sum, max, count, etc.) along with two parameters: the *range* r, the length (in time) of the window of data being aggregated, and the *slide* s, which indicates how frequently the results should be

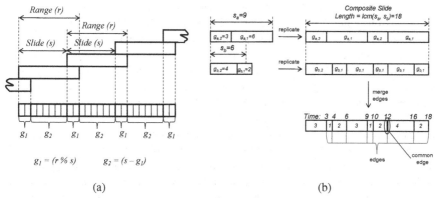

(a)   (b)

Fig. 1. (a) The paired window technique. (b) Creation of a composite slide for 2 ACQs, a and b, with respective ranges of 12 and 10 seconds.

reported. For example, an ACQ may request that the maximum price of a particular stock over the last hour ($r = 60$ min) be reported every 10 minutes ($s = 10$ min). If $r > s$, as is the case in this example, we have a *sliding* or *overlapping* window, where a single data tuple belongs to more than one window. If a new stock price value is generated every minute, then this tuple participates in six aggregation operations.

Rather than aggregate the entire window of tuples from scratch each time, *partial aggregation* [9, 10] first computes sub-aggregations of successive pieces of the window, then applies a *final aggregation* function over these sub-aggregates. For example an aggregate count would be computed by first using a count on each part of the window, then using a sum over the partial counts. This technique can be used over all the distributive functions widely used in database systems. It reduces query processing cost by preventing tuples from having to be aggregated repeatedly. Instead, each input tuple is processed once by the *sub-aggregation operator*, the result of the sub-aggregation gets buffered, and the final aggregate is assembled from those partial aggregates.

To capitalize on the idea of partial aggregation, Krishnamurthy et al. [9] proposed the *paired window* technique, whereby each *slide* is partitioned into at most two slices or *fragments* g_1 and g_2. See Figure 1a for an illustration. As pictured, $g_1 = r \mod s$, and $g_2 = s - g_1$. Thus, since r/s is the number of slides per window, computing each final aggregation uses at most $\lceil 2r/s \rceil$ operations. This paired window approach allows for effective sharing of the partial aggregation results for different queries on the same data, the details of which we describe in the following section.

2.2 Merging Multiple ACQs

To process multiple ACQs with different range and slide parameters, there are two basic strategies [9]: *unshared partial aggregation* (also referred to as *no-sharing*), or *shared partial aggregation*. When unshared partial aggregation is used, each query is simply processed separately using the paired window technique described above. This requires storing multiple copies of the input tuples, as each query is answered using its own individual sub-aggregation results.

For the shared partial aggregation strategy for $k > 1$ ACQs, we need to compute the fragments on which partial aggregation is applied in a different manner, so that the sub-aggregation results can be reused in computing a different final aggregation for each query. These fragments are computed as follows. For k ACQs with slides $s_1, s_2, \ldots, s_k$, we create a *composite slide* of length s' equal to the least common multiple of $(s_1, s_2, \ldots, s_k)$. Copy and repeat each slide $i = 1 \ldots k$ along with its corresponding paired fragments s'/s_i times, to fit the full length of the composite slide s' (see Figure 1b). The end of each fragment is referred to as an *edge*, and it serves as a demarcation of the boundary between two fragments. (We use the term *edge* in this paper to ensure consistency with previous work on query planning).

The fragments (and edges) of the final composite slide are determined by the distinct edges that remain after all k slides have been overlayed. After these new fragments are computed, partial aggregation can be applied on each fragment, and the results can be shared between different ACQs. While shared partial aggregation certainly reduces processing costs at the sub-aggregation level, it increases costs at final aggregation level. Depending on the queries, the overall total cost may be higher than in case of no sharing at all (Example 2 in [6]).

2.3   The Objective Function

An input instance for our optimization problem is comprised of a set of n ACQs. We will say that two or more ACQs are *shared* if their partial aggregation was shared via the shared partial aggregation strategy described above. We define an *execution tree* to be a subset of the n ACQs in which all the ACQs in the subset are shared. A *query plan* Q is then a grouping of the n ACQs into m execution trees, $t_1, t_2, \ldots, t_m$.

For each tree t_i, let E_{t_i} (or, more simply, E_i) be the number of fragments generated per second (also referred to as the *edge rate*). Let Ω_i denote the total number of final-aggregation operations performed on each fragment, which is termed the *tree overlap factor*. If tree i consists of k shared queries, we can compute

$$\Omega_i = \sum_{j=1}^{k} \frac{r_j}{s_j}.$$

The *cost* of processing a single tree t_i, in terms of the total number of aggregate operations per second, is thus

$$C(t_i) = \lambda + E_i \Omega_i \tag{1}$$

where λ is the number of tuples arriving per second (tuple input rate) that represents the cost at sub-aggregation level. The *total cost* of a query plan Q with m trees is simply the sum of the costs of the individual trees:

$$C(Q) = m\lambda + \sum_{i=1}^{m} E_i \Omega_i \tag{2}$$

$C(Q)$ represents here the total number of aggregations per second for all ACQs. The formal problem statement is then as follows.

Given a set of n ACQs, find a query plan (a partitioning of the ACQs) that minimizes total cost.

To recap, the cost of one execution tree comprises two parts: the cost of the sub-aggregations at the intermediate aggregation step (first term in (2)) *plus* the cost of final aggregations at the final aggregation step (second term in (2)). When queries in two trees are merged into one execution tree, the number of necessary intermediate (sub) aggregations decreases, because partial aggregates, which previously needed to be computed for both trees independently, now can be computed just once and reused in answering both queries. However, the number of final aggregation operations increases: first, because the number of edges per second (edge rate) in the resulting execution tree is at least the maximum edge rate of individual trees, and second, because now the final aggregations need to be performed for each edge, for each query. The goal of sharing query processing is to maximize savings during the sub-aggregation step while minimizing the costs at the final-aggregation step.

2.4 Weave Share

Weave Share (WS) is a recently proposed greedy heuristic algorithm for computing query plans [6]. Guirguis et al. first formalized the notion of *weavability* of multiple queries as the ratio of the number of edges common to multiple ACQs to the total number of edges in the composite slide. When several queries share partial aggregation operations, the more common edges between ACQs that exist in their composite slide, the more weavable they are. Naturally, to maximize the benefit of sharing ACQs, either the shared ACQs should exhibit a high degree of weavability or the total overlap factor (number of total final aggregation operations performed on shared fragments) should be low, or both. Weave Share considers both of these factors in optimizing the processing of ACQs. Each iteration of WS involves one *merge* step, where the queries of two separate execution trees are combined into one tree, so that the two previously separate groups of shared queries are now all shared together in one execution tree. There may be up to $n - 1$ iterations, and the total time required by the WS algorithm is $O(n^2)$, where n is the number of ACQs.

Weave Share Algorithm. The input to the algorithm is the original set of n ACQs, and the output is a query plan, or a partition of the ACQs into $m \leq n$ disjoint groups (execution trees).

1. Create n trees, one ACQ per tree.
2. Consider all possible pairs of trees. For each pair of trees, compute the reduction in cost that would be achieved if queries belonging to both trees were merged.
3. Find the maximum reduction in cost over all possible pairs of trees. Ties may be broken arbitrarily.
4. If this value is positive (i.e. it is indeed a cost-reduction), merge these trees and repeat from step 2. Otherwise (the value is not positive), terminate the algorithm.

Experimental Performance of Weave Share. Experimental results by Guirguis et al. [6] demonstrate that the query plans produced by Weave Share outperform query plans generated by other common algorithms, such as *No Share* (in which partial aggregation results are not shared), *Random* (in which random trees are iteratively merged until there is no longer an improvement), *Local Search* (that explores the solution subspace by starting with a random partition of ACQs into trees and iteratively moving single ACQs between trees), *Shared* (in which all queries constitute a single execution tree) and *Insert-then-Weave* (in which each individual query is inserted one-by-one into the tree that it weaves best with). For different input parameters (λ and n) the Weave Share plan had a cost as much as 40 times better than other plans. For small problem instances, exhaustive search was used to find the optimal query plan; Weave Share was able to find these plans in all but one instance.

The evaluation of Weave Share's performance was conducted on a synthetic data stream, to allow control over input parameters and to cover the most likely real scenarios. Although Weave Share is shown to outperform other strategies on a synthetic data stream, there is no guarantee on the quality of the solution produced by the algorithm. Extensive comparison of Weave Share with an optimal solution produced by exhaustive search is not feasible for any practical number of ACQs. In short, Weave Share performs better than the alternative heuristic algorithms but not much is understood about how many more aggregations per second the query plan produced by Weave Share requires compared to the number of aggregations per second in an optimal query plan.

3   A Cost-Savings Approximation

In this section, we give a guarantee on the amount of cost-savings achieved by Weave Share. The outcome of the Weave Share algorithm features a decrease in the total cost of the query plan compared to the *no-share* query plan in which each query is executed by itself. It is this improvement, or *savings*, achieved by the Weave Share algorithm that we seek to bound. We find this measure of maximizing cost-savings appealing, as it focuses on the achievements of the algorithm compared with that of the optimal solution. In contrast, under the umbrella of the objective of minimizing total cost, we would include costs that are inherent and unavoidable, to both WS and OPT.

We first introduce some notation. Recall that a query plan is a partitioning of the queries into execution trees. We refer to an execution tree that consists of more than one query as a *multi-tree*. We will indicate an execution tree by listing its ACQs in square brackets, for example: $[q_1, q_2, \ldots, q_k]$ refers to the multi-tree composed of queries $q_1, \ldots, q_k$ merged together, and $[q_i]$ refers to an execution tree with a single stand-alone query q_i. We can indicate a query plan using a set of such lists. For example, $\{[q_1], [q_2, q_3], [q_4]\}$ is a query plan with three execution trees, one with only query q_1, another with two merged queries q_2 and q_3, and another with only query q_4.

Let Q denote the query plan produced by the Weave Share algorithm, and m denote the number of execution trees in Q. Let Q^* denote an optimal query plan (one that minimizes total cost over all query plans), and m^* denote the number of its execution trees. Let N denote the no-share query plan, which has n execution trees, one for each stand-alone ACQ. Let μ be the number of merges made by WS to reach Q, and μ^* be the number of merges required to reach Q^* from N. Note that

$$n = m + \mu = m^* + \mu^*, \tag{3}$$

since each separate tree represents a potential merge that did not take place.

For any query plan X we denote the difference between the cost of the no-share plan and the cost of X by $R(X) = C(N) - C(X)$. Note that if $R(X)$ is positive, it represents a reduction, or savings, in cost compared to the no-share plan.

We are now ready to state the central theorem of this work.

Theorem 1. *The amount of savings achieved by the Weave Share query plan is at least a quarter of the savings achieved by an optimal query plan, i.e.,*

$$R(Q) \geq R(Q^*)/4,$$

under the following three conditions:

- *the range and slide of each query coincide (i.e., $r_i = s_i$ for each query q_i)*
- *the tuple input rate λ exceeds twice the edge rate of any tree comprised of exactly two ACQs (i.e., $\lambda \geq 2E_{[q_i,q_j]}$ for any $i, j \in \{1, \ldots, n\}, i \neq j$)*
- *the number of merges made by WS in Q is at least half the number of merges made in Q^* (i.e., $\mu \geq \mu^*/2$)*

We note that the special case of the problem enforced by the above three conditions is quite natural and applicable to practical settings. It is common in real-world applications for query windows to "tumble" [9], with disjoint windows that cover the entire input. The assumption on the size of λ is very modest in practice. After applying equation (3), the third condition is equivalent to $m \leq n/2 + m^*/2$, which has held in all previous WS experiments [5, 6]. In fact, a stronger condition $m \leq n/2$ easily held in all experiments involving medium or high tuple input rate λ. For example, for the input rates equal to or exceeding 50 tuples per second, the number of resulting multi-trees produced by WS was at most $0.06n$. In the special case when input rate was set to be ten thousand tuples per second, m did not even exceed $0.01n$ [5]. Problem instances that involve such medium or high tuple input rates appear in the majority of data stream applications and in nearly all monitoring applications (network and stock price monitoring, pay-per-click applications).

To prove this theorem, we first establish two useful lemmas, both of which provide interesting insight into the nature of sharing and merging.

3.1 Savings Dilution

Our first lemma demonstrates that under some practical assumptions, there is some degree of "dilution" in cost-savings when merging a query into a multi-tree compared with merging two stand-alone queries. Specifically, we consider the restricted version of the problem where for each query its range and slide coincide ($r = s$). Moreover, assume that the tuple input rate λ exceeds twice the edge rate of any tree comprised of exactly two individual queries, i.e. $\lambda \geq 2E_{[a,b]}$ for any two queries a and b. Under these circumstances, we show that the savings achieved by merging two individual queries together is at least half the savings achieved by merging an individual query into a multi-tree.

Let $C([a], [b])$ denote the total cost of processing the two queries a and b separately, with no sharing. Let $C([a, b])$ denote the total cost when processing the two queries a and b as one, merged, execution tree (sharing their partial aggregates). Let $r([a], [b]) = C([a], [b]) - C([a, b])$, the cost-savings, or reduction in cost, from merging the two trees $[a]$ and $[b]$. More generally, let $r(t_1, t_2)$ denote the cost-savings from merging the two trees t_1 and t_2, either or both of which may be a multi-tree.

Lemma 1. *Consider an execution tree of k queries $[q_1, q_2, \ldots, q_k]$. Assume $r_i = s_i$ for all queries $i = 1 \ldots k$. Further assume $\lambda \geq 2E_{[q_i, q_j]}$ for any $i, j \in \{1, \ldots, k\}, i \neq j$. Let S be any subset of $\{1, \ldots, k\}$. For any $a, b, c \in S$,*

$$2r([q_a], [q_b]) \geq r([q_{S-c}], [q_c]),$$

where q_{S-c} denotes the multi-tree comprised of all queries in S other than c.

Proof. To improve readability, we will write E_a for $E_{[q_a]}$ and E_{ab} for $E_{[q_a, q_b]}$. We will similarly simplify the Ω terms. Since $r_i = s_i$ for all i, we have $\Omega_a = \Omega_b = \Omega_c = 1$. We begin by noting that the savings achieved by merging q_a and q_b is $r([q_a], [q_b]) = \lambda + E_a + E_b - 2E_{ab}$. The savings achieved by merging q_c and q_{S-c} is $r([q_{S-c}], [q_c]) = \lambda + (|S| - 1)E_{S-c} + E_c - |S| \cdot E_S$. Using our assumption about λ, we have $\lambda + E_S \geq E_c + 2E_{ab}$. Hence, since $E_a + E_b \geq E_{ab}$, we can say $\lambda + 2(E_a + E_b) - 4E_{ab} \geq E_c - E_S$. From this we can conclude, using $E_S \geq E_{S-c}$, that $2(\lambda + E_a + E_b - 2E_{ab}) \geq \lambda + (|S| - 1)E_{S-c} + E_c - |S|E_S$, or equivalently, that $2r([q_a], [q_b]) \geq r([q_{S-c}], [q_c])$. ∎

3.2   The OPT-Sequence

Our second lemma will involve a careful WS-based specification of Q^*, our optimal query plan, which will ultimately allow us to map the merges made by WS to increments of savings in cost achieved by Q^*. We begin with a helpful observation.

Observation 2. *Consider an execution tree consisting of several ACQs. The total cost of this tree, and thus the savings achieved by merging the constituent queries together into the tree, does not depend on the order in which queries are merged into the tree.*

Construction of OPT-lists. Based on the above observation, any valid query plan, including Q^*, can be represented as a specific sequence of steps whereby each step constitutes merging an individual query either with another individual query or into a tree of queries. In other words, each tree of Q^* can be represented as an ordered list of constituent queries, and the order of list elements (from right to left) defines a specific order of merges for that tree. For instance, for four queries a, b, c, and d, we can use the ordered list $[q_a, q_b, q_c, q_d]$ to denote the execution tree that is comprised of all four queries, as well as to indicate the ordered steps of first merging q_c with q_d, then merging q_b with the tree $[q_c, q_d]$, then merging q_a with $[q_b, q_c, q_d]$.

Below, we describe a procedure for constructing an ordered query list for each tree of Q^*. We will refer to each ordered list of queries as an OPT-*list*, and the execution tree to which it refers as an OPT-*tree*. We will refer to them collectively as the set of OPT-*lists*, or just OPT for short. This representation of Q^* will allow us to compare the savings earned by the Weave Share solution Q to that of Q^*. Recall that m (respectively, m^*) is the number of trees in the final Weave Share plan Q (respectively, Q^*).

---

1. initialize all m^* OPT-lists (one list for each tree of Q^*) to empty
   {now consider the steps of Weave Share algorithm on the set of queries, one by one}
2. **for** each merge i of the Weave Share algorithm, $i = 1 \ldots \mu$ **do**
3.   **if** merge i is between two individual queries **then**
4.   let the two queries be called x and y
5.   add query x to the end of the OPT-list it belongs to
6.   add query y to the end of the OPT-list it belongs to
7.   **else if** merge i is between a multi-tree and a single query **then**
8.   add that query to the end of the OPT-list it belongs to
9.   **else** {merge i is between 2 multi-trees}
10.   the queries involved have already been added in a previous iteration
11.   **end if**
12. **end for**

Note that by following this procedure, the OPT-lists may get fully populated before Weave Share finishes (if the final merges of WS are between existing multi-trees). And conversely, Weave Share may finish before the OPT-lists get fully populated, if WS leaves many queries as stand-alone queries. In the latter case, the incomplete OPT-lists may be populated in an arbitrary order.

Construction of OPT-sequence. The sought-after final sequence of merges leading to Q^* can be attained by considering the OPT-lists in arbitrary order and walking through each tree-list from right to left, merging one query at a time into its corresponding final OPT-tree. Recall that μ^* denotes the number of merges in OPT. In executing this OPT sequence of merges, we are effectively starting at the query plan N where there is no sharing—all queries are in their own individual execution trees—and proceeding step by step to the optimal query plan Q^*. After each merge, we are at an intermediary query plan, where some queries that OPT will eventually merge are still individual queries, and some of the final OPT-trees are not complete. We denote this sequence of intermediate query plans $(N = Q_0, Q_1, \ldots, Q_{\mu^*} = Q^*)$, numbered in the order specified by the procedure above, and we refer to it as the OPT-*sequence*. We will abuse this term and also use it to refer to the sequence of merges made by OPT in proceeding from $N = Q_0$ to Q^*.

Consider the change in cost between each adjacent pair of query plans in the OPT-sequence. Let us define for $j = 1 \ldots \mu^*$, each change in cost $r_j^* = C(Q_{j-1}) - C(Q_j)$. We now sort the r_j^*'s in non-increasing order and renumber them so that

$$r_1^* \geq r_2^* \geq r_3^* \geq \ldots \geq r_{\mu^*}^*. \tag{4}$$

A given WS merge (either between two individual queries, a query and a tree, or two trees) is said to *map to* a query in an OPT-list if, upon executing the merge, that query is added to an OPT-list in the above procedure. The following lemma is implicit in the construction procedure of the OPT-lists. For completeness we give a proof.

Lemma 2. *Each merge of Weave Share maps to at most two queries in the OPT-lists defined above. And each query in an OPT-list is only mapped to once.*

Proof. If at merge i, Weave Share merges two existing multi-trees, then merge i of Weave Share adds no new queries to the OPT-lists. If merge i involves one stand-alone query (in the case of merging a stand-alone query into a multi-tree), then it adds one query to an OPT-list. If merge i involves two stand-alone queries then it adds two queries to one or two OPT-lists. Hence each Weave Share merge maps to at most 2 of OPT's merges. Each query is only added once because the procedure specifies that queries are added to an OPT-list only the first time they are "touched" by Weave Share: when they are still stand-alone queries.

Loosely speaking, the construction procedure of the OPT-sequence above accounts for each r_j^* using a WS merge. It effectively ensures that each WS merge precludes at most two of OPT's merges in the OPT-sequence defined above.

3.3   Proof of Main Theorem

We are now ready to prove Theorem 1. The idea of the proof is to break both OPT and Weave Share down into a sequence of merge-steps and then compare the savings achieved by Weave Share at each step to some corresponding savings achieved by OPT. This allows us to compare the total savings achieved by Weave Share with the total savings achieved by OPT.

Proof. First we handle a formality of bookkeeping. Recall that by assumption we have:

$$\mu \geq (1/2)\mu^* \tag{5}$$

Further recall that each merge of WS may map to 2 queries in OPT, which means WS may be making more merges after all of the queries in OPT have been mapped. Hence we simply define $r_j^* = 0$ for all $j = \mu^* + 1 \ldots 2\mu$.

We now proceed with our main argument. We denote the reductions in cost of each WS merge to be $r_1, \ldots, r_\mu$, indexed by the order of merges executed by the Weave Share algorithm. That is, r_i is the savings earned from the ith merge of the Weave Share algorithm. The bulk of the proof will be dedicated to showing that for any $i = 1 \ldots \mu$, we have $2r_i \geq r_{2i-1}^*$.

Savings at the first iteration of Weave Share. We begin with $i = 1$, when Weave Share merges two individual queries. The savings achieved at this step is at least half of the maximum individual savings achieved by OPT, i.e. $2r_1 \geq r_1^*$. Indeed, if r_1^* is achieved by merging an individual query into a multi-tree, the inequality follows from Lemma 1 in Section 3.1. If r_1^* is achieved by merging two stand-alone queries, then from the greedy nature of Weave Share it follows that $r_1 \geq r_1^*$. Note that by definition of the OPT-sequence, r_j^* was not achieved by merging two multi-trees, for any $j = 1 \ldots \mu^*$. Hence, we obtain $2r_1 \geq r_1^*$. This also implies $2r_1 \geq r_2^*$, due to the re-numbering in (4).

Savings at i^{th} iteration of Weave Share We now generalize the argument for savings r_i that is obtained at the ith merge of Weave Share. We will demonstrate that $2r_i \geq r^*_{2i-1}$, and hence that also $2r_i \geq r^*_{2i}$ (since $r^*_{2i-1} \geq r^*_{2i}$). Note that if $2i - 1 > \mu^*$ then the claim is trivially true, since then $r_{2i-1} = 0$. Hence we restrict our attention to the case that $2i - 1 \leq \mu^*$. Recall that each reduction in cost r^*_j is derived from merging a single query into an existing OPT-tree. We will refer to this query as x_j, and the last query that was added to the same OPT-list before x_j will be referred to as y_j. I.e., x_j and y_j are adjacent in some OPT-list. We will call x_j and y_j *the queries of r^*_j*.

Since Weave Share is greedy, r_i is at least the savings that we can get from any merge of two stand-alone queries that are still available just before the ith merge of Weave Share. So let us consider the quality of the merges that are still available to WS at this juncture. Currently, at merge i of Weave Share, either:

1. x_{2i-1} and y_{2i-1} are still stand-alone queries, available to be merged by Weave Share, or
2. Weave Share has already used either x_{2i-1} or y_{2i-1} in one of its earlier $i-1$ merges. In this case another pair of queries (x_1 and y_1, x_2 and y_2,..., or x_{2i-2} and y_{2i-2}) are still available to be merged. We know this because by Lemma 2, a total of at most $2(i - 1)$ OPT queries have been mapped by the first $i - 1$ WS merges. By the pigeon hole principle, we must still have at least one pair of queries (x_1 and y_1, x_2 and y_2, ..., or x_{j-1} and y_{j-1}) that are unmapped, i.e., stand-alone queries thus far untouched by WS.

Let us use x and y to denote the pair of queries that are still available to be merged of these possibilities. By Lemma 1, we know that merging x and y achieves at least as much reduction in cost as half of the reduction achieved when OPT merged x into the OPT-tree that x and y are a part of. Therefore, we know that

$$r(x,y) \geq \frac{1}{2}\min(r^*_1,\ldots,r^*_{2i-1}) \geq \frac{1}{2}(r^*_{2i-1}).$$

We also know by definition of WS that $r_i \geq r(x,y)$. Hence, for $i = 1 \ldots \mu$ we have:

$$2r_i \geq r^*_{2i-1}, \text{ and} \tag{6}$$
$$2r_i \geq r^*_{2i}$$

We sum up the inequalities in (6) to get $2 \cdot 2(r_1 + r_2 + \ldots + r_\mu) \geq (r^*_1 + r^*_2 + r^*_3 + \ldots + r^*_{2\mu})$. And then, using inequality (5), we have $(r^*_1 + r^*_2 + r^*_3 + \ldots + r^*_{2\mu}) \geq (r^*_1 + r^*_2 + r^*_3 + \ldots + r^*_{\mu^*})$. Hence we obtain $4R(Q) \geq R(Q^*)$.

4   Summary and Open Problems

In this paper we have studied an important optimization problem for efficient sharing of ACQs for DSMSs. We analyzed a previously proposed greedy algorithm, Weave Share, which performed extremely well in an experimental study. We show that under some practical assumptions this algorithm guarantees a 4-approximation to the optimal cost-savings. Our analysis technique allowed us to elucidate some properties of the effect of sharing partial aggregates on total cost, and also required an exploration into the structural properties of an optimal solution. We remark that a restricted version of the problem, in which each tree of the resulting query plan is allowed to have at most two queries in it, can be solved exactly and in polynomial time by reducing it to maximum weighted matching in a graph, in which the vertices represent the ACQs and the edges represent the reduction in cost achieved by sharing the two queries. Another interesting variation that occurs when the ACQs are ordered (say, by expiration time) and the trees in the query plan are formed by contiguous queries in order, can be solved exactly and in polynomial time via a dynamic programming solution.

Several open questions remain. Determining whether there is a tighter analysis of WS, or an algorithm with a better approximation, are probably two the most immediate. Removing the assumptions required for proving our approximation may also be possible. An analysis of a follow-up Tri-Weave [7] algorithm would also be interesting.

References

1. Abadi, D.J., Ahmad, Y., Balazinska, M., Çetintemel, U., Cherniack, M., Hwang, J.-H., Lindner, W., Maskey, A., Rasin, A., Ryvkina, E., Tatbul, N., Xing, Y., Zdonik, S.B.: The design of the Borealis stream processing engine. In: CIDR (2005)
2. Abadi, D.J., Carney, D., Çetintemel, U., Cherniack, M., Convey, C., Lee, S., Stonebraker, M., Tatbul, N., Zdonik, S.: Aurora: A new model and architecture for data stream management. VLDB Journal (2003)
3. Arasu, A., Babcock, B., Babu, S., Datar, M., Ito, K., Nishizawa, I., Rosenstein, J., Widom, J.: STREAM: The Stanford stream data manager. In: SIGMOD (2003)
4. Ghanem, T.M., Hammad, M.A., Mokbel, M.F., Aref, W.G., Elmagarmid, A.K.: Incremental evaluation of sliding-window queries over data streams. IEEE TKDE (2007)
5. Guirguis, S.: Scalable Processing of Multiple Aggregate Continuous Queries. PhD thesis. University of Pittsburgh (2011)
6. Guirguis, S., Sharaf, M.A., Chrysanthis, P.K., Labrinidis, A.: Optimized processing of multiple aggregate continuous queries. In: CIKM (2011)
7. Guirguis, S., Sharaf, M.A., Chrysanthis, P.K., Labrinidis, A.: Three-level processing of multiple aggregate continuous queries. In: ICDE, pp. 929–940 (2012)
8. Hammad, M.A., Mokbel, M.F., Ali, M.H., Aref, W.G., Catlin, A.C., Elmagarmid, A.K., Eltabakh, M.Y., Elfeky, M.G., Ghanem, T.M., Gwadera, R., Ilyas, I.F., Marzouk, M.S., Xiong, X.: Nile: A query processing engine for data streams. In: ICDE (2004)
9. Krishnamurthy, S., Wu, C., Franklin, M.: On-the-fly sharing for streamed aggregation. In: SIGMOD (2006)
10. Li, J., Maier, D., Tufte, K., Papadimos, V., Tucker, P.A.: No pane, no gain: Efficient evaluation of sliding-window aggregates over data streams. SIGMOD Record (2005)
11. Li, J., Maier, D., Tufte, K., Papadimos, V., Tucker, P.A.: Semantics and evaluation techniques for window aggregates in data streams. In: SIGMOD (2005)
12. Naidu, K.V.M., Rastogi, R., Satkin, S., Srinivasan, A.: Memory-constrained aggregate computation over data streams. In: ICDE (2011)
13. Streambase (2006), http://www.streambase.com
14. System S (2008), http://domino.research.ibm.com

On the Kernelization Complexity
of String Problems

Manu Basavaraju, Fahad Panolan, Ashutosh Rai, M.S. Ramanujan,
and Saket Saurabh

The Institute of Mathematical Sciences, Chennai, India
{manub,fahad,ashutosh,msramanujan,saket}@imsc.res.in

Abstract. In CLOSEST STRING problem we are given an alphabet Σ,
a set of strings $S = \{s_1, s_2, \ldots, s_k\}$ over Σ such that $|s_i| = n$ and an
integer d. The objective is to check whether there exists a string s over
Σ such that $d_H(s, s_i) \leq d$, $i \in \{1, \ldots, k\}$, where $d_H(x, y)$ denotes the
number of places strings x and y differ at. CLOSEST STRING is a prototype string problem. This problem together with several of its variants
such as DISTINGUISHING STRING SELECTION and CLOSEST SUBSTRING
have been extensively studied from parameterized complexity perspective. These problems have been studied with respect to parameters that
are combinations of k, d, $|\Sigma|$ and n. However, surprisingly the kernel-
ization question for these problems (for the versions when they admit
fixed parameter tractable algorithms) is not studied at all. In this paper
we fill this gap in the literature and do a comprehensive study of these
problems from kernelization complexity perspective. We almost settle all
the problems by either obtaining a polynomial kernel or showing that
the problem does not admit a polynomial kernel assuming a complexity
theoretic assumption.

Keywords: Closest String, Kernelization, Cross-Composition.

1 Introduction

String matching problems generally involve finding a string which is at a small
(or large) distance from given set(s) of strings. The distance here is measured
in terms of Hamming distance. These problems arise in a variety of fields including computational biology [6, 19] and linguistics. These are also interesting
combinatorial and algorithmic problems. String problems have been an active
area of research in algorithmics and as always most of the interesting problems
in the area are NP-complete [12, 14, 15]. Thus, these problems have been studied extensively from algorithmic approaches that cope with NP-hardness, like
approximation [1, 14–17] and parameterized complexity [5, 9, 10, 13, 18]. In this
paper we study a variety of string problems from kernelization perspective – a
subarea in parmeterized complexity – which is the first such approach towards
these problems to the best of our knowledge.

In the parameterized complexity setting, an instance comes with an integer
parameter k – formally, a parameterized problem Q is a subset of $\Sigma^* \times \mathbb{N}$ for some

Z. Cai et al. (Eds.): COCOON 2014, LNCS 8591, pp. 141–153, 2014.
© Springer International Publishing Switzerland 2014

finite alphabet Σ. We say that the problem is *fixed parameter tractable* (*FPT*) if there exists an algorithm solving any instance (x, k) in time $f(k)(|x|^{O(1)})$ for some computable function f. It is known that a decidable problem is FPT if and only if it is kernelizable: a kernelization algorithm for a problem Q takes an instance (x, k) and in time polynomial in $|x| + k$ produces an equivalent instance (x', k') (i.e., $(x, k) \in Q$ iff $(x', k') \in Q$) such that $|x'| + k' \leq g(k)$ for some computable function g. The function g is the *size of the kernel*, and if it is polynomial, we say that Q admits a polynomial kernel. The study of kernelization is a major research frontier of parameterized complexity and many important recent advances in the area are on kernelization. The recent development of a framework for ruling out polynomial kernels under certain complexity-theoretic assumptions [2, 7, 11] has added a new dimension to the field and strengthened its connections to classical complexity.

In what follows, we formally define the problems we study in this paper, give a short overview of previous work on each of them and describe our results. For an ease of presentation we also give our results in Table 1.

Closest String and Distinguishing String Selection. The first problem we look at is the following CLOSEST STRING problem.

CLOSEST STRING

Input: An alphabet Σ, a set of strings $S = \{s_1, s_2, \ldots, s_k\}$ over Σ such that $|s_i| = n$, $i \in [k] = \{1, \ldots, k\}$ and an integer d.

Question: Does there exist a string $s, |s| = n$ such that $\forall i \in [k], d_H(s, s_i) \leq d$?

The problem is NP-complete because of results of Frances and Litman [12]. Another related problem, which is a generalization of CLOSEST STRING is DISTINGUISHING STRING SELECTION (DSS).

DISTINGUISHING STRING SELECTION (DSS)

Input: An alphabet Σ, two sets of strings $S_1 = \{s_1, s_2, \ldots, s_{k_1}\}, S_2 = \{s'_1, s'_2, \ldots, s'_{k_2}\}$ over Σ such that $|s_i| = |s'_j| = n \; \forall i \in [k_1]$ and $\forall j \in [k_2]$ and integers d_1 and d_2.

Question: Does there exist a string $s, |s| = n$ such that $d_H(s, s_i) \leq d_1 \; \forall i \in [k_1]$ and $d_H(s, s'_j) \geq n - d_2 \; \forall j \in [k_2]$?

Clearly, DSS is also NP-complete via a trivial reduction from CLOSEST STRING. Gramm et al. [13] studied both of these problems and proved that the problems are FPT when parameterized by the distance d (in case of DSS, we take $d = d_1 + d_2$). They also show that CLOSEST STRING is FPT when parameterized by the number of strings k.

We show that both these problems are not likely to have polynomial kernels when parameterized by the distance d. We arrive at the results by showing a polynomial parameter transformation from CNF-SAT parameterized by the number of variables. Our reduction also shows as a by-product that both the problems do not have polynomial kernels when parameterized by both the distance d and the length of the strings n. We also obtain a simple kernel for DSS when parameterized by d_1, d_2 and k_1, k_2 together.

Table 1. A catalogue of string problems and their polynomial kernelization status. For DSS, we take $k = k_1 + k_2$ and $d = d_1 + d_2$. All the no polynomial kernel results and the $\mathcal{O}(k^2 d \log k)$ kernel for DSS are new.

			Parameters			
	d	k	(d,k)	(d,n)	l	(k,l,d)
CLOSEST STRING	None	open	$\mathcal{O}(k^2 d \log k)$ [13]	None	-	-
DSS	None	FPT not known	$\mathcal{O}(k^2 d \log k)$	None	-	-
CLOSEST SUBSTRING	W[1]-hard	W[1]-hard	W[1]-hard	-	None	None

Closest Substring. Another problem, which we study is a generalization of CLOSEST STRING, namely CLOSEST SUBSTRING.

CLOSEST SUBSTRING

Input: An alphabet Σ, a set of strings $S = \{s_1, s_2, \ldots, s_k\}$ (where s_i's can have different lengths) over Σ, and integers l and d.

Question: Does there exist strings $s, s_1', s_2', \ldots, s_k', |s| = |s_i'| = l \ \forall i \in [k]$ such that $d_H(s, s_i') \le d$ and s_i' is a substring of $s_i \ \forall i \in [k]$?

Clearly, when $|s_i| = n = l \ \forall i \in [k]$, the above problem is equivalent to CLOSEST STRING. Fellows et al. [10] proved that this problem is strictly harder than CLOSEST STRING in the sense that even for constant alphabet size, the problem is W[1]-hard when parameterized by the number of strings k. Marx [18] showed that the problem is W[1]-hard parameterized by the distance d, even when the alphabet is of a constant size. He also proved that the problem is FPT when parameterized by the target string length l and hence when parameterized by a combination of all three parameters, d, k and l. We prove that CLOSEST SUBSTRING does not have a polynomial kernel when parameterized by k, l and d together unless coNP $\subseteq$ NP/poly. We arrive at the result by giving a cross-composition from a restricted version of the same problem [3].

2   Preliminaries

In this section we define some concepts and mention some theorems which will be used to derive results in the later sections.

Strings. For a string s, we use $s[i]$ to denote the symbol occurring at the ith position for $1 \le i \le |s|$, where $|s|$ denotes the size of the string s. A *substring* of a string s is a continuous sequence of symbols from a string, while a *subsequence* of a string is a sequence of symbols which appear in the same order (not necessarily consecutively) in the original string. We denote the Hamming distance between two equal length strings s_i and s_j by $d_H(s_i, s_j)$. Formally, $d_H(s_i, s_j) = |\{p \in [|s_i|] \mid s_i[p] \ne s_j[p]\}|$. We use $[k]$ for denoting the set $\{1, 2, 3, \ldots, k\}$.

Kernelization Hardness Framework. The two main ways of showing kernelization lower bounds are to either give some variant of composition or to transfer a lower bound from another problem to the target problem by an appropriate reduction. Next we provide the essential definitions for both strategies.

Definition 1 (polynomial equivalence relation [3]). *An equivalence relation $\mathcal{R}$ on Σ^* is called a* polynomial equivalence relation *if the following holds:*
(i) Equivalence of any $x, y \in \Sigma^*$ can be checked in time polynomial in $|x| + |y|$.
(ii) Any finite set $S \subseteq \Sigma^*$ has at most $(\max_{x \in S} |x|)^{\mathcal{O}(1)}$ $\mathcal{R}$-equivalence classes.

The idea behind polynomial equivalence relations is that it suffices to give cross-compositions that work for any single equivalence class.

Definition 2 (cross-composition [3]). *Let $L \subseteq \Sigma^*$ and let $Q \subseteq \Sigma^* \times \mathbb{N}$ be a parameterized problem. We say that L* cross-composes *into Q if there is a polynomial equivalence relation $\mathcal{R}$ and an algorithm which, given t strings $x_1, x_2, \ldots, x_t$ belonging to the same equivalence class of $\mathcal{R}$, computes an instance $(x^*, k^*) \in \Sigma^* \times \mathbb{N}$ in time polynomial in $\sum_{i=1}^{t} |x_i|$ such that: (i) $(x^*, k^*) \in Q \Leftrightarrow x_i \in L$ for some $1 \leq i \leq t$ and (ii) k^* is bounded by a polynomial in $(\max_{1 \leq i \leq t} |x_i| + \log t)$.*

This definition of cross composition can be generalized to accommodate multiple parameters. In that case, we require each of them to be bounded by a polynomial in $(\max_{1 \leq i \leq t} |x_i| + \log t)$. The following theorem relates cross composition to kernelization lower bounds and holds true for multiple parameterizations.

Theorem 1 ([3]). *If an NP-hard problem $L \subseteq \Sigma^*$ has a cross-composition into a parameterized problem Q and Q has a polynomial kernel then coNP $\subseteq$ NP/poly.*

The second lower bound strategy, instead of a direct proof via compositions, is to provide a polynomial parameter transformation from a hard problem.

Definition 3 ([4]). *Let $P, Q \subseteq \Sigma^* \times \mathbb{N}$ be parameterized problems. We say that a polynomially computable function $f \colon \Sigma^* \times \mathbb{N} \to \Sigma^* \times \mathbb{N}$ is a* polynomial parameter transformation *(PPT) from P to Q if for all $(x, k) \in \Sigma^* \times \mathbb{N}$ the following holds: $(x, k) \in P$ if and only if $(x', k') = f(x, k) \in Q$ and $k' \leq k^{\mathcal{O}(1)}$.*

Again, to have a PPT reduction from a problem P having multiple parameters to another problem Q, we require each of the parameters of Q to be polynomial in parameters of P. The following theorem holds true for this case also.

Theorem 2 ([4, 8]). *Let P and Q be parameterized problems and $\tilde{P}$ and $\tilde{Q}$ be the unparameterized versions of P and Q respectively. Suppose that $\tilde{P}$ is NP-hard and $\tilde{Q}$ is in NP. Assume that there is a polynomial parameter transformation from P to Q. Then if Q admits a polynomial kernel, so does P. (Hence if P admits no polynomial kernel under some assumption then neither does Q.)*

For all the parameterized problems considered, we precede the problem name with respective parameters. For an example, (d, n)CLOSEST STRING is CLOSEST STRING parameterized by the distance d and the length of the strings n.

3   Kernel Lower Bounds for CLOSEST STRING

We prove that (d, n)CLOSEST STRING does not admit a polynomial kernel unless coNP⊆NP/poly through a PPT reduction from CNF-SAT parameterized by the number of variables, following an approach which is an extension of the proof on NP-completeness of CLOSEST STRING by Frances and Litman [12]. Given a CNF-SAT formula $F = C_1 \wedge C_2 \wedge \ldots \wedge C_m$ with variables $x_1, x_2 \ldots, x_n$, we obtain an instance of (d, n)CLOSEST STRING as follows.

Step 1. We first transform the formula F to F' with $2n$ variables such that each of the clauses in F' have length either n or 1. We do so by adding n new variables, namely $y_1, y_2, \ldots, y_n$. We first add n clauses to the formula which are all single literals and contain negation of the new variables. Also, for any clause having less than n (say k) variables in F, we add $(n - k)$ new literals, namely $y_1, y_2, \ldots, y_{n-k}$ to it. Formally,

$$F' = C'_1 \wedge C'_2 \wedge \ldots \wedge C'_m \wedge \neg y_1 \wedge \neg y_2 \wedge \ldots \wedge \neg y_n$$

$$C'_i = C_i \vee y_1 \vee y_2 \vee \ldots \vee y_{n-k} \text{ where } |C_i| = k$$

It is easy to see that F is satisfiable if and only if F' is satisfiable. We name each singleton clause $\neg y_i$ as C'_{m+i} for all $i \leq n$ for the sake of clarity.

Step 2. In this step, we obtain an instance of (d, n)CLOSEST STRING from F'. The instance has $m + 13n - 8$ strings each of length $6n - 4$. The set S is a disjoint union of sets S_1 and S_2. In S_1, there is a string corresponding to each of the clauses in F' accounting for a total of $(m + n)$ strings. S_2 will consists of $12n - 8$ strings, four for each $i \in [3n - 2]$. For a variable x_j (or y_j) and a clause C'_i, we define a two bit string $X_{i,j}$ (or $Y_{i,j}$) as follows.

$$X_{i,j}(\text{or } Y_{i,j}) = \begin{cases} 11, & \text{if } x_j(\text{or } y_j) \text{ occurs positively in } C'_i \\ 00, & \text{if } x_j(\text{or } y_j) \text{ occurs negatively in } C'_i \\ 10, & \text{if } x_j(\text{or } y_j) \text{ does not occur in } C'_i \end{cases}$$

1. For each of the C'_i's, $1 \leq i \leq m$, we define a string s_i and add it to S_1, where $s_i = X_{i,1}X_{i,2}\ldots X_{i,n}Y_{i,1}Y_{i,2}\ldots Y_{i,n}0^{2n-4}$.
2. For C'_{m+i}'s, $1 \leq i \leq n$, we define a string $s_{m+i} = \{10\}^{n+i-1}00\{10\}^{2n-2-i}$ and add them to S_1.
3. We add the following four strings to S_2 for all $i \in [3n - 2]$:
   (a) $\{10\}^{i-1}01\{10\}^{3n-2-i}$, (b) $\{10\}^{i-1}10\{10\}^{3n-2-i}$,
   (c) $\{01\}^{i-1}01\{01\}^{3n-2-i}$ and (d) $\{01\}^{i-1}10\{01\}^{3n-2-i}$.

So, we get an instance (S, d) of (d, n)CLOSEST STRING by putting $S = S_1 \uplus S_2$ and $d = 3n - 2$. Here, the length n' of each of the strings is equal to $6n - 4$ and the number of strings k is equal to $m + 13n - 8$ where m and n are the number of clauses and the number of variables respectively in F.

Lemma 1. *Let s be a string of length $6n - 4$. If $d_H(s, x) \leq 3n - 2 \; \forall \; x \in S_2$ then $s[2i] = s[2i - 1] \; \forall i \in [3n - 2]$.*

Proof. For any $i \in [3n - 2]$, we look at the four strings corresponding to it in S_2. Then we look at two strings of length $6n - 6$, which are $\{10\}^{i-1}\{10\}^{3n-2-i}$ and $\{01\}^{i-1}\{01\}^{3n-2-i}$. The first is present as a subsequence in (a) and (b), and the second is present as a subsequence in (c) and (d). These subsequences occupy the same positions in the strings and are complements of each other, hence any string has to be at a distance at least $3n - 3$ from (a) and (b) or a distance at least $3n - 3$ from (c) and (d). Now, let us assume that there exists a string s such that $d_H(s, x) \leq 3n - 2 \ \forall \ x \in S_2$. As argued earlier, it is at a distance at least $3n - 3$ from (a) and (b) or at distance at least $3n - 3$ from (c) and (d). Let us look at the case when it is at a distance $3n - 3$ from (a) and (b). The strings (a) and (b) differ at just two positions, $2i - 1$ and $2i$. In these two positions, (a) has 01 and (b) has 10. So, s will have to have 00 or 11 at these two positions. Otherwise, it will differ at these two positions with one of the strings, which, in addition to already differing $3n - 3$ positions will result in a total Hamming distance of $3n - 1$, which is a contradiction. Similarly we can argue that s will have to have 00 or 11 at these positions if it is at distance at least $3n - 3$ from (c) and (d). In this way, we have shown that if $d_H(s, x) \leq 3n - 2 \ \forall x \in S_2$, then $s[2i] = s[2i - 1] \ \forall i \in [3n - 2]$. We call such a string a "doubled" string. $\square$

Lemma 2. (S, d) *is a YES instance of* (d, n)CLOSEST STRING *if and only if* F' *is satisfiable.*

Proof. ($\Leftarrow$) Let f be a satisfying assignment for F'. Now, we have to show that there exists a string s, which is at a distance at most $3n - 2$ from all the strings in S. For that, we first define a two bit string t_z for each variable, where $z = x_i$ or y_i for $i \in [n]$. We define t_z to be 00 if z is set to be 0 by f, 11 otherwise. We take $s = t_{x_1} t_{x_2} \ldots t_{x_n} t_{y_1} t_{y_2} \ldots t_{y_n} 0^{2n-4}$, which is doubled. All the strings in S_2 will differ at exactly $3n - 2$ places with a doubled string. In the set S_1, each string comes from a clause. For $1 \leq i \leq m$, each clause C_i' has exactly n variables. By the definition of s_i, it matches at the trailing $2n - 4$ zeroes with s. Now, in the remaining $4n$ positions, let us first look at the $2n$ positions corresponding to n variables which are not present in the clause. Clearly, s being a doubled string matches at n positions and differs at n positions of these $2n$ positions. Of the remaining $2n$ positions, s has to match s_i at at least two places, precisely at the variable which satisfies the clause. In the worst case, it differs at all the remaining $2n - 2$ places and we get $d_H(s, s_i) \leq 3n - 2$. For $(m + 1) \leq i \leq (m + n)$, each clause C_i' has only one variable, and the string s has to match the string s_i at the positions corresponding to that variable. In the remaining $6n - 6$ positions, it differs at exactly half the positions, i.e. at $3n - 3$ places. So, $d_H(s, s_i) = 3n - 3 \leq 3n - 2$.

($\Rightarrow$) From Lemma 1, it is clear that the target string s, which makes (S, d) a YES instance of (d, n)CLOSEST STRING, has to be doubled. To get an assignment for F', we assign a variable x_i to 0 if $s[2i - 1] = s[2i] = 0$, 1 otherwise. Similarly, we assign y_i to 0 if $s[2n + 2i - 1] = s[2n + 2i] = 0$, 1 otherwise. Since the string s is doubled, each of the variables get unique assignments.

Now, to prove that this assignment is a satisfying assignment for F', we have to prove that each clause contains at least one literal that is assigned to be true by this assignment. To see that, we look at the string s_i corresponding to the clause C_i'. If $i \leq m$, then the clause has n variables. On the $2n$ positions corresponding to the variables absent in the clause, the string s, which is doubled, differs at exactly n places. On the positions corresponding to the remaining variables (variables present in the clause C_i'), bits assigned in s_i are either 00 or 11. Since s is a doubled string, s can not differ with exactly one bit out of the two bits corresponding to each of these variables. Now, we claim that s matches at at least two places corresponding to a variable in C_i'. If not, then s differs at two positions for each of the variables present in the clause, in addition to differing at n positions for the variables not present in the clause. So, s differs s_i at at least $3n$ positions, which is a contradiction. The variable corresponding to the matching position satisfies the clause. For $(m + 1) \leq i \leq (m + n)$, each clause contains only one literal. The doubled string s differs with s_i at exactly $3n - 3$ positions, which is half of the locations filled with 10's. The only places left are where s_i has 00. If s has 11 in that place, the distance between s and s_i becomes $3n - 1$, which is not possible. Hence, s has 00 at positions corresponding to variable y_i, and the variable is assigned 0, which satisfies the clause.   □

We proved that (S, d) is a YES instance of (d, n)CLOSEST STRING if and only if F' is satisfiable. We know that F' is satisfiable if and only if F is. Hence, we have obtained an instance of (d, n)CLOSEST STRING from an instance of CNF-SAT, where the parameters n' and d are polynomial in the number of variables (n) in the original CNF-SAT instance. It is easy to see that the above procedure can be carried out in polynomial time. So essentially, we have shown a Polynomial Parameter Transformation from CNF-SAT parameterized by the number of variables to (d, n)CLOSEST STRING. We also know the following.

Theorem 3 ([11]). CNF-SAT *does not admit a polynomial kernel when parameterized by the number of variables unless* $coNP \subseteq NP/poly$.

Combining theorems 2 and 3, we get the following.

Theorem 4. (d, n)CLOSEST STRING *does not admit a polynomial kernel unless* $coNP \subseteq NP/poly$.

We observe that the kernelization lower bound for (d, n)CLOSEST STRING also works for any fixed alphabet Σ of size at least two. Also, we can give an easy PPT reduction from (d, n)CLOSEST STRING to (d_1, d_2, n)DSS and get the following.

Theorem 5. $(*)$[1] (d_1, d_2, n)DSS *does not admit a polynomial kernel unless* $coNP \subseteq NP/poly$.

4 Kernel Lower Bounds for CLOSEST SUBSTRING

In an instance of CLOSEST SUBSTRING, we call the s_i''s target strings and s the solution string. We prove that the problem does not have a polynomial kernel by giving a cross-composition from a mild restriction of the same problem. For the restricted version, we take an instance of CLOSEST SUBSTRING and assume that $l > 2d + 1$. We call this problem RESTRICTED CLOSEST SUBSTRING.

---

[1] Proofs of theorems marked with $(*)$ will be provided in the full version.

Theorem 6. (∗) RESTRICTED CLOSEST SUBSTRING *is NP-complete.*

For giving the cross-composition, we start with t instances of RESTRICTED CLOSEST SUBSTRING belonging to the same equivalence class and produce an instance of the (k, l, d)CLOSEST SUBSTRING, which is a YES instance if and only if at least one of the original t instances of RESTRICTED CLOSEST SUBSTRING was a YES instance.

For that, we first divide the input into equivalence classes. We say that two instances (S_1, l_1, d_1) and (S_2, l_2, d_2) are in the same equivalence class if and only if $|S_1| = |S_2|$, $l_1 = l_2$ and $d_1 = d_2$. Clearly, the number of such equivalence classes is at most polynomial in the size of the largest input instance. Also, given two instances as input, it can be checked in polynomial time whether they belong to the same equivalence class. Let $(S_1, l, d), (S_2, l, d), \ldots, (S_t, l, d)$ be the instances in the same equivalence class. Now, for arriving at the new instance of (k, l, d)CLOSEST SUBSTRING, we first extend the alphabet Σ by four new symbols. We take $\Sigma' = \Sigma \cup \{0, 1, *, \Box\}$ and we assume that the new symbols were not already present in Σ. We call the jth string from the ith instance $s_{i,j}$ for the sake of simplicity. Also, let us define $x = \max(\lceil \log t \rceil, \lceil \log k \rceil)$. Now, for the composition, we go through the following steps.

Step 1. For each $s_{i,j}$ we put $*B_i^{4d+2}$ between each of the consecutive symbols in that string as well as at the start and the end of the string. Here B_i is the standard binary representation of i with so many preceding 0's added that the total length is x. We call these strings $r_{i,j}$ and the substrings $*B_i^{4d+2}$ "codes". This increases the size of each string by a factor of $\mathcal{O}(dx)$.

Step 2. In this step, we first define sets $Q_j = \{r_{i,j} \mid i \in [t]\}$ for $j \in [k]$. For each Q_j, we construct the string p_j by putting the strings in Q_j in ascending order of the instance number (the first index i) separated by a delimiter. The delimiter between two strings $r_{i,j}$ and $r_{i+1,j}$ is represented by $D_{i,j}$ and is defined as follows.

$$D_{i,j} = \begin{cases} \Box * \{B_i B_j\}^{2d+1} \Box, & \text{if } i < j \\ \Box * \{B_{i+1} B_j\}^{2d+1} \Box, & \text{if } i \geq j \end{cases}$$

We call the substrings $\{B_i B_j\}^{2d+1}$ or $\{B_{i+1} B_j\}^{2d+1}$ "special codes", which have length $(4d + 2)x$ each. So, after the second step, we get k strings, which are

$$p_j = r_{1,j} D_{1,j} r_{2,j} D_{2,j} \ldots r_{i,j} D_{i,j} \ldots r_{t-1,j} D_{t-1,j} r_{t,j} \ \forall j \in [k]$$

The distance between two consecutive $*$'s or two symbols of Σ in any p_j is $(4d + 2)x + 1$. We call this quantity y. Now, to get an instance (S', l', d') of (k, l, d)CLOSEST SUBSTRING, we set $l' = l + (l+1)y$, $d' = d$ and $S' = \{p_i \mid i \in [k]\}$. Now we have to show that (S', l', d') is a YES instance if and only if there exists an i, such that (S_i, l, d) is a YES instance. For that, we first prove the following lemma.

Lemma 3. (∗) *Let s_1 and s_2 be two strings of the same length (say n) such that $d_H(s_1, s_2) \geq 2d + 1$, then there does not exist a string s of length n such that $d_H(s, s_1) \leq d$ and $d_H(s, s_2) \leq d$.*

Lemma 4. $(*)$ *If* (S', l', d') *is a YES instance, then the* $*$*'s are at the same place in all the target strings.*

Lemma 5. *If* (S', l', d') *is a YES instance, at most* $(4d + 2)x - 1$ *bits from any special code are part of any of the target strings.*

Proof. Let A be the $k \times l'$ matrix corresponding to the target strings when (S', l', d') is a YES instance. From the Lemma 3, we know that the $*$'s match, and they are equidistant. In the matrix A, they constitute of columns which have $*$'s at all the positions. This implies that all the rows of the matrix are aligned in the sense that the codes or the special codes in every row occupy the same columns. In other words, any code or special code "lies above" another code or special code. Let us take a special code $\{B_u B_v\}^{2d+1}$. It can come from delimiter $D_{u,v}$ when $u < v$, or it can come from a delimiter $D_{u-1,v}$, when $u - 1 \geq v$. In any case, we see that $u \neq v$. So, the special code $\{B_u B_v\}^{2d+1}$ differs at at least $2d+1$ positions with any code B_z^{4d+2}. Now, we look at a special code lying above another special code. Let us say they come from delimiters $D_{w,u}$ and $D_{v,z}$. The first special code will have a bit representation of u and the second will have a bit representation of z. Also, u and z can not be the same for any two delimiters coming from different strings p_i and p_j. Hence any two special codes differ at at least $2d + 1$ positions. Now, if we assume that all $(4d + 2)x$ bits from a special code are part of some target string, then this string differs with all other target strings at at least $2d + 1$ positions among the $(4d + 2)x$ positions corresponding to columns occupied by the special code. Hence, by using Lemma 3, we have that (S', l', d') a NO instance, which is a contradiction. $\square$

Lemma 6. (S', l', d') *is a YES instance of* (k, l, d)CLOSEST SUBSTRING *if and only if* (S_i, l, d) *is a YES instance of* RESTRICTED CLOSEST SUBSTRING *for some* i.

Proof. ($\Leftarrow$) Let the solution string be s and the target strings be $s'_{i,1}, s'_{i,2}, \ldots, s'_{i,k}$ which make (S_i, l, d) a YES instance. Now, for S', we transform the solution string by inserting $*B_i^{4d+2}$ between any consecutive pair of symbols as well as at start and end of the string. We call this string s''. For the substrings, we look at $r_{i,j}$ for all $j \in [k]$ and take the substrings corresponding to $s'_{i,j}$ with codes on either side. We call them $r'_{i,j}$ for all $j \in [k]$. Clearly, these $k + 1$ strings all match on the codes, i.e. at all the places which hold symbols which are not from Σ. Also, $|s''| = |r'_{i,j}| = l + (l + 1)y$. Now, if the target string s'' differs at more than d places with some substring $r'_{i,j}$, then s differs at more than d places with $s'_{i,j}$ making (S_i, l, d) a NO instance which is a contradiction. Hence (S', l', d') is a YES instance.

($\Rightarrow$) If (S', l', d') is a YES instance, then by Lemma 5, at most $(4d + 2)x - 1$ bits from any special code are part of any of the target strings. This means that the target string coming from p_j, for all $j \in [k]$, will not contain substrings of $r_{i_1,j}$ and $r_{i_2,j}$ for $i_1 \neq i_2$ because otherwise it will contain one special code fully. Now we look at target strings coming from different p_j's, say p_{j_1} and p_{j_2}. We want to show that if the target strings contain substrings of r_{i_1,j_1} and r_{i_2,j_2} for

$j_1 \neq j_2$, then $i_1 = i_2$ for all $j_1, j_2 \in [k]$. By Lemma 4, we know that the $*$'s are at the same place in all the target strings, so the codes are occupying the same positions in all target strings. If i_1 is not equal to i_2, then each of the codes in r_{i_1,j_1} will differ with any code in r_{i_2,j_2} in at least $4d+2$ positions. This will make r_{i_1,j_1} and r_{i_2,j_2} differ at at least $4d+2$ places, making (S', l', d') a NO instance. Hence, there exists an i, such that for all $j \in [k]$ the target string coming from p_j is a substrings of $r_{i,j}$, which corresponds to a unique instance, namely S_i.

Let us denote $\Sigma \cup \{\Box\}$ by Σ''. Since, we know that the strings match at the $*$'s, so they will have a symbol from Σ'' at the same positions. Since $d' = d$, we just need to show that all the target strings will contain at least l symbols from Σ, which are lying above each other. We know that two consecutive symbols of Σ'' are separated by a distance $y = (4d+2)x + 1$ in S'. Also, we have chosen $l' = l + (l+1)y$. Hence, any target string will have at least l and at most $l+1$ symbols from Σ''.

Case 1: There are l symbols from Σ'' in some target string. In this case, any codes or special codes on either side of a symbol from Σ'' must be part of the target string. We also know that all the $\Box$'s have special codes on at least one side, which can not be part of any target string. Hence, we conclude that $\Box$'s can not be part of target string in this case and hence the target string will contain at least l symbols from Σ.

Case 2: There are $l+1$ symbols from Σ'' in some target string. Any target string can contain at most one $\Box$, because otherwise it will either contain a special code fully or we have that $\exists i, j$ such that $|s_{i,j}| < l$. So, the target string contains at least l symbols from Σ even in this case.

We have proved that all the target strings contain at least l symbols from Σ and they correspond to strings from a single instance S_i. Also, since $d' = d$, these strings can differ by at most distance d and this makes (S_i, l, d) a YES instance. $\Box$

We have given a cross-composition for (k, l, d)CLOSEST SUBSTRING from RE-STRICTED CLOSEST SUBSTRING, since all the parameters in the new instance are polynomial in size of the maximum sized instance and $\log t$. We also have that RESTRICTED CLOSEST SUBSTRING is NP-complete. Hence, by application of theorem 1, we get the following.

Theorem 7. (k, l, d)CLOSEST SUBSTRING *does not admit a polynomial kernel unless coNP $\subseteq$ NP/poly.*

5   Kernels

In this section, we reproduce the result of Gramm et al. [13] which implies a kernel of size $\mathcal{O}(k^2 d \log k)$ for (d, k)CLOSEST STRING. We also give a polynomial kernel for (d_1, d_2, k_1, k_2)DSS which is based on ideas similar to those of Gramm et al. [13]. For reproducing the result, we first apply the lemma stated in (Lemma 3 of [13]), which reduces the number of symbols in the alphabet Σ to at most k, the number of strings, in an instance of CLOSEST STRING.

Then we look at the set of strings of the instance of (d, k)CLOSEST STRING as an $k \times n$ matrix. We use the term columns for the columns of this matrix. We call a column *dirty* if it has at least two different symbols from Σ. We call a column *full* if it has all symbols from Σ. We first consider the following lemma.

Lemma 7. [13] *Given a* CLOSEST STRING *instance with k strings $s_1, \ldots, s_k$ of length n and a positive integer d. If the resulting $k \times n$ matrix has more than kd dirty columns, then there is no string s with $d_H(s, s_i) \leq d \ \forall i \in [k]$.*

Proof. Let the number of dirty columns be more than kd. Let us assume that there exists a solution string s such that $d_H(s, s_i) \leq d \ \forall i \in [k]$. We first observe that the solution string has to differ with at least one of the strings in each dirty column. By pigeonhole principle, the solution strings has to differ with at least one string at more than d columns. This means that s is not a solution string, which is a contradiction. □

Theorem 8. (∗) (d, k)CLOSEST STRING *has a kernel of size $\mathcal{O}(k^2 d \log k)$.*

Now we consider (d_1, d_2, k_1, k_2)DSS. We first state a lemma similar to (Lemma 3 of [13]), which reduces the alphabet size to $k_1 + k_2$.

Lemma 8. (∗) *A DSS instance with arbitrary alphabet Σ, $|\Sigma| > k_1 + k_2$, is isomorphic to a DSS instance with alphabet Σ', $|\Sigma'| = k_1 + k_2$.*

Let $k = k_1 + k_2$. Now, we look at the $k \times n$ matrix corresponding to sets S_1 and S_2 put together. We say a column is "hit" by the solution string s if either s mismatches with at least one of the first k_1 entries in the column or it matches with at least one of the last k_2 entries in the column.

Lemma 9. (∗) *At most $k_1 d_1 + k_2 d_2$ columns are hit by a solution string s in the $k \times n$ matrix corresponding to $S = S_1 \cup S_2$.*

Next, we say a column is "inconsistent" if at least one of the following happens:

1. If the first k_1 entries are not all same.
2. If the last k_2 entries have all the symbols from Σ.
3. None of 1 and 2 happen, but the symbol occurring in the first k_1 entries is also present at at least one place in the last k_2 entries.

Lemma 10. *For an instance (S_1, S_2, d_1, d_2) of DSS, if the number of inconsistent columns is more than $k_1 d_1 + k_2 d_2$, then it is a NO instance.*

Proof. We first show that a solution string s must hit all the inconsistent columns. This fact, combined with Lemma 9 that a solution string can hit at most $k_1 d_1 + k_2 d_2$ columns gives the desired result.

Case 1. When the first k_1 entries of the inconsistent column are not all the same then s must mismatch with at least one of the entries. Hence, the column is hit.
Case 2. If last k_2 entries of the inconsistent column have all the symbols from Σ, then s must match with at least one of the entries. Hence, the column is hit.

Case 3. In this case, first k_1 entries are the same symbol (say a) and last k_2 entries contain at least one a. Then, in s, if we don't put a at the index of the column, it mismatches with first k_1 entries. If we put a, it matches with at least one of the last k_2 entries. In any case, the column is hit.   □

Theorem 9. (*) (d_1, d_2, k_1, k_2)DSS *has a kernel of size* $\mathcal{O}((k_1 + k_2)(k_1 d_1 + k_2 d_2)(\log(k_1 + k_2)))$.

6   Conclusions

In this paper we study several string problems from kernelization complexity. We were able to settle most of the problems except a few. The most important open problem that remains open from this work is CLOSEST STRING parameterized by k, the number of strings in the input.

References

1. Andoni, A., Indyk, P., Patrascu, M.: On the optimality of the dimensionality reduction method. In: FOCS, pp. 449–458 (2006)
2. Bodlaender, H.L., Downey, R.G., Fellows, M.R., Hermelin, D.: On problems without polynomial kernels. J. Comput. Syst. Sci. 75(8), 423–434 (2009)
3. Bodlaender, H.L., Jansen, B.M.P., Kratsch, S.: Cross-composition: A new technique for kernelization lower bounds. In: STACS, pp. 165–176 (2011)
4. Bodlaender, H.L., Thomassé, S., Yeo, A.: Kernel bounds for disjoint cycles and disjoint paths. Theor. Comput. Sci. 412(35), 4570–4578 (2011)
5. Boucher, C., Ma, B.: Closest string with outliers. BMC Bioinformatics 12(S-1), S55 (2011)
6. Buhler, J., Tompa, M.: Finding motifs using random projections. Journal of Computational Biology 9(2), 225–242 (2002)
7. Dell, H., van Melkebeek, D.: Satisfiability allows no nontrivial sparsification unless the polynomial-time hierarchy collapses. In: STOC, pp. 251–260 (2010)
8. Dom, M., Lokshtanov, D., Saurabh, S.: Incompressibility through colors and ids. In: Albers, S., Marchetti-Spaccamela, A., Matias, Y., Nikoletseas, S., Thomas, W. (eds.) ICALP 2009, Part I. LNCS, vol. 5555, pp. 378–389. Springer, Heidelberg (2009)
9. Fellows, M.R., Gramm, J., Niedermeier, R.: On the parameterized intractability of CLOSEST SUBSTRING and related problems. In: Alt, H., Ferreira, A. (eds.) STACS 2002. LNCS, vol. 2285, pp. 262–273. Springer, Heidelberg (2002)
10. Fellows, M.R., Gramm, J., Niedermeier, R.: On the parameterized intractability of motif search problems. Combinatorica 26(2), 141–167 (2006)
11. Fortnow, L., Santhanam, R.: Infeasibility of instance compression and succinct PCPs for NP. In: STOC, pp. 133–142 (2008)
12. Frances, M., Litman, A.: On covering problems of codes. Theory Comput. Syst. 30(2), 113–119 (1997)
13. Gramm, J., Niedermeier, R., Rossmanith, P.: Exact solutions for closest string and related problems. In: Eades, P., Takaoka, T. (eds.) ISAAC 2001. LNCS, vol. 2223, pp. 441–453. Springer, Heidelberg (2001)

14. Lanctot, J.K., Li, M., Ma, B., Wang, S., Zhang, L.: Distinguishing string selection problems. Inf. Comput. 185(1), 41–55 (2003)
15. Li, M., Ma, B., Wang, L.: Finding similar regions in many sequences. J. Comput. Syst. Sci. 65(1), 73–96 (2002)
16. Li, M., Ma, B., Wang, L.: On the closest string and substring problems. J. ACM 49(2), 157–171 (2002)
17. Ma, B.: A polynominal time approximation scheme for the closest substring problem. In: Giancarlo, R., Sankoff, D. (eds.) CPM 2000. LNCS, vol. 1848, pp. 99–107. Springer, Heidelberg (2000)
18. Marx, D.: The closest substring problem with small distances. In: FOCS, pp. 63–72 (2005)
19. Pevzner, P.A.: Computational molecular biology - an algorithmic approach. MIT Press (2000)

Complexity of Dense Bicluster Editing Problems

Peng Sun[1], Jiong Guo[3] and Jan Baumbach[2]

[1] Max Planck Institute for Informatics, Campus E1 4, 66123 Saarbrücken, Germany
psun@mpi-inf.mpg.de
[2] Institute for Mathematics and Computer Science University of Southern Denmark
Campusvej 55 5230 Odense M Denmark
[3] MMCI Cluster of Excellence, Campus E1 7, 66123 Saarbrücken, Germany
jguo@mmci-uni.saarland.de

Abstract. Given a density measure Π, an undirected graph G and a nonnegative integer k, a Π-CLUSTER EDITING problem is to decide whether G can be modified into a graph where all connected components are Π-cliques, by at most k edge modifications. Previous studies have been conducted on the complexity and fixed-parameter tractability (FPT) of Π-CLUSTER EDITING based on several different density measures. However, whether these conclusions hold on bipartite graphs is yet to be examined. In this paper, we focus on three different density measures for bipartite graphs: (1) having at most s missing edges for each vertex (s-biplex), (2) having average degree at least $|V| - s$ (average-s-biplex) and (3) having at most s missing edges within a single disjoint component (s-defective bicliques). First, the NP-completeness of the three problems is discussed and afterwards we show all these problems are fixed-parameter tractable with respect to the parameter (s, k).

Keywords: Bicluster editing, Parameterized complexity, Data reduction, NP-hardness.

1 Introduction

Graph-based data clustering methodologies have been of great importance in the scientific analyses of the real-world data, ranging from biological to social network data. In most scenarios, data entities are modeled as vertices and a certain function is defined to quantify the "relationship" between two vertices (e.g. similarities). Thresholds specified by systems or users are then used to model the edges of the graph. The clustering problem is usually defined mathematically as a "partition" of the whole vertex set into different dense subsets, so-called *clusters*, such that there are as few edges as possible between different clusters and as few missing edges as possible within clusters. The problem could also be viewed from the graph modification angle, i.e., to modify the graph by edge insertions and deletions into a Π-cluster graph, the so-called Π-CLUSTER EDITING problems. Here, a graph is a Π-cluster graph, if each of its connected components satisfies Π, where Π is a certain density measure.

Z. Cai et al. (Eds.): COCOON 2014, LNCS 8591, pp. 154–165, 2014.

The most famous problem among the Π-CLUSTER EDITING is CLUSTER EDITING, where Π is "being a clique". CLUSTER EDITING has been extensively studied and proved as NP-complete among the earliest NP-complete problems [18,3]. Successful applications of the algorithms for CLUSTER EDITING could be found in the field of computational biology [22] and machine learning [3]. In terms of parameterized complexity, CLUSTER EDITING is proved to be solved in $O(1.83^k + |E|)$ time [4] and several different data reduction schemes have been derived [5,6,8,11,1]. Furthermore, the current best approximation factor fors CLUSTER EDITING is 2.5 [20]. In computational biology, a number of applicable heuristic algorithms have also been designed [21,23].

In some real-world applications, "being a clique" is increasingly criticized as over-restrictive [17]. Thus some relaxed models might be more advantageous in a variety of application scenarios. Theoretical studies have also been conducted on relaxed versions of CLUSTER EDITING. Guo et al.[14] studied the fixed-parameter tractability of s-PLEX EDITING. In another study, Guo et al. extended their research further to several other relaxed models: s-defective cliques, average-s-plexes and μ-cliques [13], in which the NP-completeness and the fixed-parameter tractability are proved.

In some real-world scenarios which data contains information in more than one dimension, the standard clustering model is not so powerful. For instance, the microarray data analysis requires a simultaneous clustering on rows (genes) and columns (conditions) to find consistent behaviors for groups of genes under a certain number of conditions. The traditional clustering model, which clusters data of one dimension, is not feasible for such scenarios where data from different sources must be clustered together simultaneously. The concept of "biclustering" was thus introduced by Cheng and Church [7]. Biclustering allows to *simultaneously* partition both rows and columns, which is particularly useful in capturing biologically meaningful genes and conditions in one run [7]. The major reason is that the expression of gene subjects may be correlated only under some conditions while being independent under other conditions. Biclustering approaches are generally capable of discovering such local patterns and have been widely used for various types of genes expression analysis [10]. Note that biclustering problems can be easily modeled as clustering problems on bipartite graphs by forming similarities between two different sets of vertices, described as follows:

> BICLUSTER EDITING
> **Input:** A bipartite graph $G = (U, V, E)$ and an integer $k > 0$
> **Question:** Can G be converted into a *bicluster graph* where every connected component is a biclique, by at most k edge insertions and deletions?

A biclique is a bipartite graph with all possible edges. Though not so extensively studied as CLUSTER EDITING, several theoretical conclusions of bicluster editing have been published regarding parameterized tractability [12] and approximation results [2]. Applications other than microarray data analysis can be found mostly in the computational biology field, for instance, biomedical data analysis [19], drug adverse events prediction [15] and etc.. Similarly, by relaxing the criteria of "biclique" in different directions, Π-BICLUSTER EDITING can be yielded:

> **Input:** A bipartite graph $G = (U, V, E)$ and an integer $k > 0$
> **Question:** Can G be converted into a Π-*bicluster graph*, where every connected component is a Π-biclique, by at most k edge insertions and deletions?

Here in our study, we focus on the three cases of Π-BICLUSTER EDIT-ING:(1) s-biplex, (2) average-s-biplex and (3) s-defective biclique, and provide both NP-complete and FPT results.

2   Problem Definitions and Results

An undirected graph $G = (U, V, E)$, where U and V are two sets of vertices and E is the set of edges, is a *bipartite graph* if $\forall e \in E$, edge e has exactly one end vertex in U and the other end vertex in V. Let $W = U \cup V$. For an arbitrary $W' \subseteq W$, the *induced subgraph* $G[W']$ is the subgraph over the vertex set W' with the edge set $\{\{u, v\} \in E | u, v \in W'\}$. An induced subgraph $G[W'] = (U', V', E')$ is a biclique if $\forall u \in U'$ and $\forall v \in V'$, we have $\{u, v\} \in E$. The *open neighborhood* $N(v)$ of $v \in W$ is the set of vertices that are adjacent to v in G. The *degree* of a given vertex v is denoted by $d(v)$, referring to the cardinality of $N(v)$. The *closed neighborhood* of v is denoted by $N[v]$, i.e., $N[v] = N(v) \cup \{v\}$. The open and closed neighborhoods of a set of vertices $W' \subseteq W$ are defined as $N(W') = \bigcup_{u \in W'} N(u) \backslash W'$ and $N[W'] = N(W') \cup W'$, respectively. Let $W' \subseteq W$, we use $G - W'$ as the abbreviation for $G[W \backslash W']$ and for a vertex $v \in W$, let $G - v$ denote $G - \{v\}$. If $G - v$ has more connected components than G, then we call v as a *cut vertex*. Similarly, let E' be a set of edges, then $G - E'$ denotes the graph $G' = (U, V, E \backslash E')$. For a graph $G = (U, V, E)$, denote $\overline{E} = \{\{u, v\} | u \in U \land v \in V \land \{u, v\} \notin E\}$ as the set of *missing edges*. A pair of vertices $\{u, v\}$ is called a *missing edge* if $\{u, v\} \in \overline{E}$. For two sets of vertices X and Y, let $E(X, Y)$ be the set of edges between X and Y, i.e., $E(X, Y) = \{\{u, v\} \mid u \in X \land v \in Y \land \{u, v\} \in E\}$. For a vertex set X, denote $E(X)$ as the abbreviation for $E(X, X)$. For a set of vertex X' and a bipartite graph $H = (X, Y, E)$, denote the intersection between X' and H as the set of common vertices, i.e., $X' \cap H = (X' \cap X) \cup (X' \cap Y)$.

An s-biplex is a connected bipartite graph $G = (U, V, E)$ with $d(u) \geq |V| - s$ for all $u \in U$ and $d(v) \geq |U| - s$ for all $v \in V$. Note that a normal biclique is thus a 0-biplex. A bipartite graph G is called an s-*biplex cluster graph* if all its connected components are s-biplexes. Therefore, s-BIPLEX EDITING is the special case of bicluster editing with Π equal to "s-biplex". In this paper, the NP-completeness of s-BIPLEX EDITING is shown. Then the sizes of *minimal forbidden induced subgraphs* are upper-bounded by $O(s)$ and a branching strategy can be derived which indicates the FPT of s-BIPLEX EDITING.

In general graphs, average-s-plex is proposed as a "density measure", defined as the mean of the degrees of all vertices in a given graph [14]. In a bipartite graph, we define the *average degree* for two vertex sets separately: $\overline{d}_U = |E|/|U|$

and $\overline{d}_V = |E|/|V|$. A connected graph $G = (U, V, E)$ is thus an *average-s-biplex* if $\overline{d}_U \geq |V| - s$ and $\overline{d}_V \geq |U| - s$, with $1 \leq s \leq \min\{|U|, |V|\}$. This density measure can be considered as a further relaxation of s-biplex, with no requirement on the *minimum* degree. In this work, we show the NP-completeness of AVERAGE-s-BIPLEX EDITING. Afterwards, a reduction to a more general problem is conducted, followed by a polynomial-time kernelization procedure which produces a graph with at most $2k((s + 1)(4k + 6s) + 1)$ vertices. This implies FPT for AVERAGE-s-BIPLEX EDITING.

The concept of *defective clique* has been reported previously to be useful in biological network analysis [24]. NP-completeness and FPT of s-DEFECTIVE CLIQUE EDITING and DELETION are already known [13]. A connected bipartite graph $G = (U, V, E)$ is an s-defective biclique if $|E| \geq |U| \cdot |V| - s$. We prove that s-DEFECTIVE BICLUSTER EDITING is NP-complete. Then, the sizes of minimal fobidden induced subgraphs of s-defective bicluster graphs are shown to be bounded by $2s + 3$, which leads directly to the FPT of s-DEFECTIVE BICLUSTER EDITING with respect to the parameter (s, k). For more information on parameterized complexity, we refer to [9] and [16]. Due to limited space, some proofs are deferred to Appendix.

3   s-Biplexes

3.1   NP-Completeness

In this section, we show the of NP-completeness of s-BIPLEX EDITING by a reduction from 3-EXACT-3-COVER.

Theorem 1. *For every constant $s \geq 0$, s-BIPLEX EDITING is NP-complete.*

Proof. If $s{=}0$, then the problem is equivalent to BICLUSTER EDITING and thus is NP-complete. For any $s \geq 1$, we reduce the NP-complete 3-EXACT-3-COVER (3X3C), where given a collection $\mathcal{C}$ of triplets (a set of 3 elements is called a triplet) from an element set $A = \{a_1, a_2, a_3, ..., a_{3n}\}$ such that each element of A is a member of at most three triplets, one asks to find out a subcollection $\mathcal{I} \subseteq \mathcal{C}$ of size n that covers A, i.e., every element of A appears in some triplet in $\mathcal{I}$. The set $\mathcal{I}$ is called an "exact cover".

We construct an s-BIPLEX EDITING instance as follows: Let $m = (72 + s)n$. A bipartite graph $G = (U, V, E)$ is then constructed, based on the following procedure: For each element in A, one corresponding vertex is created in U, and for each triplet $S \in \mathcal{C}$, a set of m vertices is added to U. The same construction is performed to create vertices in V, that is: $U = U_1 \cup U_2$, $V = V_1 \cup V_2$, $U_1 = \{u_1, u_2, ..., u_{3n}\}$, $V_1 = \{v_1, v_2, ..., v_{3n}\}$, $U_2 = \bigcup_{S \in \mathcal{C}}\{u_{S_1}, u_{S_2}, ..., u_{S_m}\}$, $V_2 = \bigcup_{S \in \mathcal{C}}\{v_{S_1}, v_{S_2}, ..., v_{S_m}\}$.

The edge set E in G consists of five subsets: First, we connect every $u_i \in U_1$ to its corresponding $v_i \in V_1$, $1 \leq i \leq 3n$. Second, for each triplet $S \in \mathcal{C}$, let $S = \{a_x, a_y, a_z\}$, $(1 \leq x, y, z \leq 3n)$. We connect $u_i \in U_1$ and $v_j \in V_1$ for all $i, j \in \{x, y, z\}$ with $i \neq j$. Third, between U_2 and V_2, for each $S \in \mathcal{C}$, denote

$U_S^m = \{u_{S_1}, u_{S_2}, , ..., u_{S_m}\}$ and $V_S^m = \{v_{S_1}, v_{S_2}, ..., v_{S_m}\}$. We connect $u_{S_i} \in U_S^m$ to $v_{S_j} \in V_S^m$ for all $1 \leq i, j \leq m$. Finally, for each $S = \{a_x, a_y, a_z\} \in \mathcal{C}$, $(1 \leq x, y, z \leq 3n)$, every $u_i \in U_1 (i \in \{x, y, z\})$ is connected to all vertices in $V_S^m \subseteq V_2$, and every $v_i \in V_1$ $(i \in \{x, y, z\})$ is connected to all vertices in $U_S^m \subseteq U_2$. More precisely: $E = \bigcup_{i=1}^5 E_i$, $E_1 = \{\{u_i, v_i\}|\ i = 1, ..., 3n\}$, $E_2 = \{\{u_i, v_j\}|\ \exists S = \{a_x, a_y, a_z\} \in \mathcal{C} \ \wedge \ i, j \in \{x, y, z\} \ \wedge \ i \neq j\}$, $E_3 = \{\{u_{S_i}, v_{S_j}\}|\ \exists S \in \mathcal{C} \ \wedge \ u_{S_i} \in U_S^m \wedge v_{S_j} \in V_S^m\}$, $E_4 = \{\{u_i, v_{S_j}\}|\ \exists S = \{a_x, a_y, a_z\} \in \mathcal{C} \wedge i \in \{x, y, z\} \wedge v_{S_j} \in V_S^m\}$, $E_5 = \{\{v_i, u_{S_j}\}|\ \exists S = \{a_x, a_y, a_z\} \in \mathcal{C} \ \wedge \ i \in \{x, y, z\} \ \wedge \ u_{S_j} \in U_S^m\}$.

For each triplet set $S \in \mathcal{C}$, we denote: $U_S = \{u_x, u_y, u_z | \{a_x, a_y, a_z\} \in S\}$, $V_S = \{v_x, v_y, v_z | \{a_x, a_y, a_z\} \in S\}$, $W_S = U_S \cup V_S$, $U_S^m = \{u_{S_1}, ..., u_{S_m}\}$, $V_S^m = \{v_{S_1}, ..., v_{S_m}\}$, $W_S^m = U_S^m \cup V_S^m$.

Obviously, the construction can be carried out in polynomial time. Let $M = 2m(3|\mathcal{C}| - 3n)$ and $N = |E_2| - 6n$. The parameter k is equal to $M + N$. For the rest of the proof, please refer to Appendix. □

3.2   Forbidden Induced Subgraphs

In this section, we describe a set of *forbidden induced subgraphs* $\mathcal{G}_F$, that a graph G is an s-biplex cluster graph if and only if G does not contain any induced subgraphs in $\mathcal{G}_F$. If $s = 0$, then we have BICLUSTER EDITING problem and the forbidden subgrah is a path of four vertices. If $s \geq 1$, the structures of forbidden induced subgraphs are much more complex and we are faced with an exponentially increasing number of different possibilities. To solve the problem, we show in the following that the sizes of forbidden induced graphs are bounded by $O(s)$ vertices. Based on this characterization, a branching strategy solving s-BIPLEX EDITING can be established.

We start with some preliminaries. A connected induced subgraph $H = (R, T, E')$ is *minimal forbidden induced subgraph* if H is not an s-biplex but every induced proper subgraph of H is an s-biplex cluster graph. We call a vertex v in H "*forbidden*" if v is incident to more than s missing edges. A subset of vertices R' is called "*forbidden subset*" if R' contains at least one forbidden vertex. To show the upper-bound for minimal forbidden induced subgraphs, two distinct cases are studied separately: (1) subgraph H contains forbidden vertex (vertices) only in R (or T), and (2) H contains forbidden vertices in both R and T. We first prove four claims regarding the properties of a minimal forbidden induced subgraph. Next, as a summary, we show that every minimal forbidden induced subgraph of biplexes contains at most $3s + 3$ vertices in both the two cases mentioned above, for all $s \geq 1$.

Lemma 1. *Let $H = (R, T, E')$ be a minimal forbidden induced subgraph. If R is a forbidden subset, then $\min_{u \in R}\{d(u)\} = |T| - s - 1$*

Lemma 2. *Let $H = (R, T, E')$ be a minimal forbidden induced subgraph. If H has forbidden vertices both in R and T, then H can only be a path of length $2s + 3$, and only the two endpoints of the path are forbidden vertices.*

Proof. First we prove the claim that H contains no more than two forbidden vertices if H has forbidden vertices both in R and T. By contradiction, assume there are > 2 forbidden vertices in H. Since we know that in every graph, there are at least 2 non-cut vertices. Let u, v be the 2 non-cut vertices. We have 6 cases with respect to u and v:

Case i: u, v are both forbidden vertices, and $u, v \in R$. Then we can remove u without separating H. In the subgraph $H - u$, we have $d_{H-u}(v) = |T| - s - 1$ and $H - u$ is also forbidden, contradicting with minimal forbidden induced subgraph. If $u, v \in T$, same proof applies.

Case ii: u, v are both non-forbidden vertices, and $u, v \in R$. Since R, T are both forbidden subsets, there exists a forbidden vertex $w \in R$, such that $d(w) = |T| - s - 1$. The subgraph $H - u$ is also forbidden since $d_{H-u}(w) = |T| - s - 1$. If $u, v \in T$, the same proof applies.

Case iii: u is a forbidden vertex, v is a non-forbidden vertex and $u, v \in R$. Then just remove v and $H - v$ is still forbidden.

Case iv: u is a forbidden vertex in R, v is a non-forbidden vertex in T. Since R, T are both forbidden subsets, there exists a vertex $w \in T$, such that $d(w) = |R| - s - 1$. We remove v from T. Then $d_{H-v}(w) = |R| - s - 1$ and thus $H - v$ is still forbidden.

Case v: u is a non-forbidden vertex in U and v is a non-forbidden vertex in V. A proof similar to Case iv applies.

Case vi: u is a forbidden vertex in R, v is a forbidden vertex in T. Let w be a forbidden vertex in H and $w \neq u$, $w \neq v$. Without loss of generality, we assume $w \in R$. Then we can remove u from H. Then $d_{H-v}(w) = |T| - s - 1$ and thus $H - v$ is still forbidden.

To summarize the six cases, since we have two non-cut vertices and at least three forbidden vertices with at least one forbidden vertex in R and at least one in T, we can always find a forbidden vertex x and a non-cut vertex y in the same vertex set (in R or in T). Clearly, removing y does not affect the property of the forbidden vertex x and thus the subgraph $H - y$ is still forbidden. This contradicts with the assumption. Hence, if H is a minimal forbidden induced subgraph with forbidden vertices in both R and T, then H cannot contain more than two forbidden vertices.

Next, we prove that H can only be a path of $2s + 2$ vertices. Let u^* and v^* be the two forbidden vertices in H. Suppose $u^* \in R$ and $v^* \in T$. Consider a third vertex w^*, $w^* \neq u^*$ and $w^* \neq v^*$. Clearly, such vertex w^* exists. If w^* is a non-cut vertex, then consider the subgraph $H - w^*$. If $w^* \in R$, in $H - w^*$, u^* is still a forbidden vertex; if $w^* \in T$, then in $H - w^*$, v^* is still a forbidden vertex. In either case, H is not minimal. Therefore, we know that in H, all vertices other than u^* and v^* must be cut vertices. Thus in H, we have $|R| + |T| - 2$ cut vertices. Obviously, H can only be a path and u^*, v^* can only be the two endpoints of the path in R and T.   □

Lemma 3. *Let $H = (R, T, E')$ be a minimal forbidden induced subgraph with forbidden vertices only in R. Let $R_0 \subseteq R$ be the subset of all forbidden vertices and $R_1 = R \backslash R_0$. Let $T_0 = N(R_0)$ and $T_1 = T \backslash T_0$. Then we have:*

1. $\forall u \in R_1$, u is a cut vertex.
2. $\forall v \in T_0$, v is a cut vertex.
3. If $|R_0| > 1$, then $\forall u \in R_0$, u is a cut vertex.
4. If $|T_0| > 1$, then for an arbitrary vertex $v^* \in T_0$, let $\mathcal{H} = \{H_1, H_2, ..., H_l\}$ be the set of disjoint components after removing v^*. Then for each $H_i = (X_i, Y_i, E_i)$, $1 \leq i \leq l$, we have $X_i \cap R_1 \neq \emptyset$.
5. There exists at least one vertex $v \in T_1$ with $d(v) = 1$.

Lemma 4. Let $H = (R, T, E')$ be a minimal forbidden induced subgraph with forbidden vertices only in R. Let $R_0 \subseteq R$ be the subset of all forbidden vertices, $R_1 = R \backslash R_0$. Let $T_0 = N(R_0)$ and $T_1 = T \backslash T_0$. Then we have $|R| + |T| \leq 3s + 3$.

Proof. Consider an arbitrary vertex $v^* \in T_0$. By Lemma 3, v^* is a cut vertex. Let $\mathcal{H} = \{H_1, H_2, ..., H_r\}$, $r > 1$, $H_i = (X_i, Y_i, E_i)$ be the set of disjoint connected components after removing v^*. Without loss of generality, let $\{H_1, H_2, ..., H_l\}$ be the subset of $\mathcal{H}$, $l \leq r$, such that $X_i \cap R_0 \neq \emptyset$, for all $1 \leq i \leq l$. We have the following two cases:

Case i. If $l \geq 2$, we know there is at least two disjoint components that intersect with R_0. Hence consider $\forall u \in X_1$ and $\forall v \in Y_j$ $(1 < j \leq l)$, we have $\{u, v\} \notin E'$. Similarly, we have $\{u', v'\} \notin E'$, for all $u' \in X_j$ $(1 < j \leq l)$ and all $v' \in Y_1$. Thus, we have $|Y_1| \leq s + 1$ and $\sum_{j=2}^{l} |Y_j| \leq s + 1$, since otherwise we would have a $u \in R$ incident to more than $s + 1$ missing edges, contradicting with Lemma 1. Because the number of missing edges incident to any vertex in R cannot be larger than $s + 1$, we have $|Y_1| + |T_1| \leq s + 1$ and $\sum_{j=2}^{l} |Y_j| + |T_1| \leq s + 1$. Thus we have $|T| = |Y_1| + \sum_{j=2}^{l} |Y_j| + |T_1| \leq s + 1 + s + 1 = 2s + 2$. Moreover, since we know $\min_{w \in T} d(w) = 1$ and T does not contain forbidden vertex, we have $|R| \leq s + 1$. Thus the total size of the forbidden induced subgraph H is $|R| + |T| \leq 2s + 2 + s + 1 = 3s + 3$.

Case ii. if $\forall v \in T_0$, we have $l = 1$, then for all $v \in T_0$, the removal of v will not separate R_0. For an arbitrary vertex v^* in T_0, let $\mathcal{H} = \{H_1, H_2, ...H_r\}$, $H_i = (X_i, Y_i, E_i)$ be the disjoint connected components after removing v^*. Without loss of generality, let $R_0 \subseteq X_1$. Then $T_0 \subseteq Y_1$ and we can find at least one u^* with $u^* \in (N(v^*) \cap R_1)$, such that $u^* \notin N(v')$ for $v' \in T_0$, $v' \neq v^*$. Therefore, each vertex v^* in T_0 has at least one "unique" neighbor in R_1. Thus $|T_0| \leq |R_1| \leq s + 1 - |R_0| \leq s$. Moreover, we have $|T_1| \leq s + 1$, since otherwise the vertices in R_0 are incident to more than $s + 1$ missing edges. Then the total size of the forbidden induced subgraph H is $|R| + |T| \leq s + 1 + s + s + 1 \leq 3s + 2$. In summary, the claim is proved. $\square$

Combining Lemma 2 and Lemma 4, we have the following theorem:

Theorem 2. If a graph G is not an s-biplex cluster graph, then we can find a forbidden subgraph in G in polynomial time with the size bounded by $3s+3$.

Finally, we have:

Corollary 1. *S-BIPLEX CLUSTER EDITING is fixed-parameter tractable with respect to (s,k).*

4 Average-s-Plexes

In this section, we consider the AVERAGE-s-BIPLEX EDITING, proving its NP-completeness and its FPT with respect to parameters (s, k). To show its NP-hardness, a two-step reduction is demonstrated: First, we reduce a well-known NP-complete MAXIMUM BALANCED BICLIQUE (MBB) to EQUAL-SIZE BICLUSTER EDITING, afterwards, a reduction from EQUAL-SIZE BICLUSTER EDITING to AVERAGE-s-BIPLEX EDITING is conducted. The EQUAL-SIZE BICLUSTER EDITING (ESBE) is defined as follows:

> **Input**: An undirected bipartite graph $G = (U, V, E)$ and two integers $k, d \geq 0$.
>
> **Question**: Can G be transformed by editing at most k edges into d disjoint bicliques $\{C_1, C_2, C_3, ...C_d\}$, $C_i = (U_i, V_i, E_i)$, $1 \leq i \leq d$, such that $|U_i| = |U_j|$ and $|V_i| = |V_j|$ for all $1 \leq i, j \leq d$?

The edge deletion version of this problem requires only edge deletions.

Theorem 3. *EQUAL-SIZE BICLUSTER EDITING is NP-complete.*

Theorem 4. *For every constant $s \geq 1$, AVERAGE-s-BIPLEX EDITING is NP-complete.*

We present a kernalization procedure for AVERAGE-s-BIPLEX EDITING with respect to the parameter (s, k). In order to show this, first we reduce the problem into an integer-weighted version and afterwards we describe three reduction rules that can be carried out within polynomial time.

We introduce two types of weights to describe the weighted version of AVERAGE-s-BIPLEX EDITING: Vertex weights and edge weights, inspired by the idea of the reduction of the weighted version of CLUSTER EDITING, i.e. for any pair of vertices that cannot be separated by k edge modifications, we merge them into one "multi-vertex". Obviously, for all vertices merged, they end up in the same average-s-biplex in all optimal solutions.

We denote the vertex weight as $\sigma(u)$ which keeps track of the number of vertices merged into u. The vertex weight of a set of vertices S is defined as: $\sigma(S) = \sum_{v \in S} \sigma(v)$. Moreover, let $\delta(u)$ be the subset of vertices $\{u_1, u_2, ..., u_r\}$, $r \geq 1$ that merged into u, i.e., $\sigma(u) = |\delta(u)|$. The edge weight, $\omega(u, v)$, is defined between two arbitrary entities (The concept "entity" represents vertices, multi-vertices and sets of vertices), storing the number of edges between them. The degree of a vertex u is defined as: $d'(u) = \omega(u, N(u))$. Thus, for a weighted bipartite graph $G = (U, V, E)$, the average degree of the vertices in U is defined as: $\overline{d_U} = \frac{\omega(U,V)}{\sigma(U)}$.

Hence a bipartite graph $G = (U, V, E)$ is a weighted average-s-biplex, if $\overline{d_U} \geq \sigma(V) - s$ and $\overline{d_V} \geq \sigma(U) - s$. The weighted version of the problem can be defined as:

> **Input:** A graph $G = (U, V, E)$, with vertex weight $\sigma(u)$ as a function:
>
> $$\sigma(u) : \begin{cases} U \to [\,1, |U|\,] \\ V \to [\,1, |V|\,] \end{cases}$$
>
> and edge weight $\omega(u, v)$ as a function:
>
> $$\omega(u, v) : E \quad \to \quad [\,1, |U||V|\,]$$
>
> and a nonnegative integer k.
>
> **Question:** With edge modifications whose total weight is at most k, can G be edited into a weighted average-s-biplex cluster graph?

Note that if we set $\sigma(u) := 1$, $\delta(u) := \{u\}$ and for each $\{u, v\} \in E$, $\omega(u, v) := 1$, an instance of AVERAGE-s-BIPLEX EDITING can be easily reduced to an instance of WEIGHTED AVERAGE-s-BIPLEX EDITING. In this reduction, parameters k and s are not changed.

The following three reduction rules are designed for WEIGHTED AVERAGE-s-BIPLEX EDITING, which lead to a problem kernel with no more than $2k((s + 1)(4k + 6s) + 1)$ vertices.

- **Rule 1.** Remove all connected components in G that are already weighted average-s-biplexes.

- **Rule 2.** For two vertex $u, v \in U$ or $u, v \in V$, let $S(u, v) := N(u) \cap N(v)$. If $\min\{\omega(u, S(u, v)), \omega(v, S(u, v))\} > k$, then we merge u and v, by replacing u and v with a new vertex v', such that v' satisfies:
  - $\sigma(v') = \sigma(u) + \sigma(v)$
  - $\omega(v', x) = \omega(u, x) + \omega(v, x)$ for every x with $\{u, x\} \in E$, $\{v, x\} \in E$

Lemma 5. *Rule 2. is correct.*

The function of Rule 2. is to merge (or replace) the vertices that we cannot afford separating. Based on the same idea, we consider another scenario: If a vertex u has a large set of neighbors that only connects to u but no other vertex, then we cannot possibly deleted the edges between u and all its "unique" neighbors. Let $N^*(u) \subseteq N(u)$ be a set of vertices such that $\forall v \in N^*(u)$, v satisfies: (1) $N(v) = \{u\}$ and (2) $\sigma(v) = 1$. Rule 3. is then presented to reduce the size of $N^*(u)$:

- **Rule 3** For each $u \in G$, if $|N^*(u)| > k$, then we replace $N^*(u)$ with a subset of vertex containing $(k + 1)$ vertices: $\{v_0, v_1, v_2, ..., v_k\}$, such that $\omega(u, v_0) = \omega(u, N^*(u)) - k$ and $\omega(u, v_i) = 1$ for all $1 \leq i \leq k$.

Lemma 6. *Rule 3 is correct.*

Theorem 5. *(WEIGHTED) AVERAGE-s-BIPLEX EDITING is fixed-parameter tractable with respect to parameter (s, k) and admits a kernel of at most $2k((s + 1)(4k + 6s) + 1)$ vertices.*

5   Defective Bicliques

We prove now the NP-completeness of s-DEFECTIVE BICLUSTER EDITING

Theorem 6. *For every $s \geq 0$, s-DEFECTIVE BICLUSTER EDITING is NP-complete.*

Next, we show the FPT of s-DEFECTIVE BICLUSTER EDITING by proving that for every $s \geq 1$, all minimal forbidden induced subgraphs contain at most $2s + 3$ vertices and hence we are able to find a minimal forbidden subgraph in polynomial time.

Lemma 7. *For every $s \geq 1$, every minimal forbidden induced subgraph of s-defective bicluster graphs contains at most $2s + 3$ vertices. Given a graph that is not an s-defective bicluster graph, a minimal forbidden induced subgraph can be found in $O((|U| + |V|) \cdot |E|)$ time.*

Proof. Denote $H = (R, T, E')$ as a minimal forbidden induced subgraph of s-defective bicluster graph. Clearly, H is connected. Towards contradiction we assume H contains more than $2s + 3$ vertices. We distinguish 2 cases:

Case i. There exists a cut vertex in H. Let $u^* \in R$ be a cut vertex. Obviously, by $s + 4 \leq 2s + 3$ for $s \geq 1$, we can always find in H a connected subgraph H' such that H' contains u^* and other $s + 3$ vertices and u^* is a cut vertex in H'. Let $H' = (R', T', E'')$ We prove that H' is forbidden. By removing u^*, we obtain a set of disjoint connected components $\mathcal{H} = \{H_1, H_2, ..., H_l\}$, $H_i = (R_i, T_i, E_i)$. Thus, we have the number of missing edges e_m in H' is at least:

$$e_m \geq \frac{1}{2} \sum_{i=1}^{l} (|R_i|(|T'| - |T_i|) + |T_i|(|R'| - 1 - |R_i|))$$

$$= \frac{1}{2} \sum_{i=1}^{l} |R_i||T'| + \frac{1}{2} \sum_{i=1}^{l} |T_i|(|R'| - 1) - \sum_{i=1}^{l} |R_i||T_i|$$

$$= (|R'| - 1)|T'| - (|R_1||T_1| + \sum_{i=2}^{l} |R_i||T_i|)$$

$$\geq (|R'| - 1)|T'| - (|R_1||T_1| + (\sum_{i=2}^{l} |R_i|)(\sum_{i=2}^{l} |T_i|)) \; (*)$$

$$= (|R'| - 1)|T'| - (|R_1||T_1| + (|R'| - 1 - |R_1|)(|T'| - |T_1|))$$

$$= |T_1|(|R'| - 1 - |R_1|) + |R_1|(|T'| - |T_1|)$$

$$\geq |R'| + |T'| - 3 \; (**)$$

Inequality (*) holds because for any integer $a_1, b_1, a_2, b_2 > 0$, we have $a_1 \cdot b_1 + a_2 \cdot b_2 \leq (a_1 + b_1)(a_2 + b_2)$. Inequality (**) is the minimum value of the function $f(|T_1|, |R_1|) = |T_1|(|R'| - 1 - |R_1|) + |R_1|(|T'| - |T_1|)$, with $1 \leq |R_1| \leq |R'| - 1$ and $1 \leq |T_1| \leq |T'|$. Thus, we have $e_m \geq s + 1$. Since $|R'| + |T'| = s + 4$, we have H' being a forbidden subgraph, thus contradicts the assumption.

Case ii. If there is no cut vertex in H, then we know that $\forall v \in H$, v must be incident to missing edge(s), otherwise we can just remove v from H without changing the forbidden subgraph property. Let $n = |U|+|V|$, m_0 be the minimum "anti-degree" ("anti-degree" is the number of missing edges incident to a given vertex) in H and m_t be the total number of missing edges in H. Hence we have the inequalities: (1) $\frac{1}{2} \cdot n \cdot m_0 \leq m_t$ and (2) $m_t - m_0 \leq s$.

Inequality (1) holds because each vertex is incident to at least m_0 missing edges and altogether we have no more than m_t missing edges. Inequality (2) holds because H is a minimal forbidden subgraph and the removal of vertex v will decrease the number of total missing edges by at least m_0. Since H is minimal, $\forall u \in H$, $H - u$ is not forbidden and thus has no more than s missing edges. Solving the inequalities, we have: $n \leq \frac{2s+2m_0}{m_0} = \frac{2}{m_0}s + 2 \leq 2s + 3$.

Thus if $n > 2s+3$, then at least one inequality above is not satisfied and hence H is not a minimal forbidden induced subgraph. To locate a minimal forbidden induced subgraph, we first check if the given connected graph G is an s-defective bicluster. If not, we check for each $v \in G$, the subgraph $G - v$. If $G - v$ is still not an s-defective biclique, then we remove v from G. Thus to find a minimal forbidden induced subgraph takes at most $O((|U| + |V|)|E|)$ time.   □

Theorem 7. *s-DEFECTIVE   BICLUSTER   EDITING   is   fixed-parameter tractable with respect to (s,k).*

6   Outlook

We point out some further directions of this research topic. For all the three problems, further algorithmic improvements are necessary: For s-BIPLEX and s-DEFECTIVE BICLUSTER EDITING, a more elegant and efficient problem kernel is needed, and for average-s-BIPLEX EDITING, an efficient branching strategy other than brute-force is beneficial to be applied on the reduced problem kernel. Moreover, in many practical applications, for example in computational biology, high-quality heuristic algorithms should always be taken into account. Finally, it is also interesting to consider other meaningful density measurements and study their classical and parameterized complexity.

References

1. Abu-Khzam, F.N.: The multi-parameterized cluster editing problem. In: Widmayer, P., Xu, Y., Zhu, B. (eds.) COCOA 2013. LNCS, vol. 8287, pp. 284–294. Springer, Heidelberg (2013)
2. Ailon, N., Avigdor-Elgrabli, N., Liberty, E.: An improved algorithm for bipartite correlation clustering. arXiv preprint arXiv:1012.3011 (2010)
3. Bansal, N., Blum, A., Chawla, S.: Correlation clustering. Machine Learning 56(1-3), 89–113 (2004)
4. Böcker, S., Briesemeister, S., Bui, Q.B.A., Truß, A.: Going weighted: Parameterized algorithms for cluster editing. Theoretical Computer Science 410(52), 5467–5480 (2009)

5. Cao, Y., Chen, J.: Cluster editing: Kernelization based on edge cuts. In: Raman, V., Saurabh, S. (eds.) IPEC 2010. LNCS, vol. 6478, pp. 60–71. Springer, Heidelberg (2010)
6. Chen, J., Meng, J.: A 2k kernel for the cluster editing problem. Journal of Computer and System Sciences 78(1), 211–220 (2012)
7. Cheng, Y., Church, G.M.: Biclustering of expression data. In: ISMB, vol. 8, pp. 93–103 (2000)
8. Fellows, M., Langston, M., Rosamond, F., Shaw, P.: Efficient parameterized preprocessing for cluster editing. In: Csuhaj-Varjú, E., Ésik, Z. (eds.) FCT 2007. LNCS, vol. 4639, pp. 312–321. Springer, Heidelberg (2007)
9. Fellows, M.R., Downey, R.G.: Parameterized complexity (1999)
10. Gonçalves, J.P., Madeira, S.C., Oliveira, A.L.: Biggests: Integrated environment for biclustering analysis of time series gene expression data. BMC Research Notes 2(1), 124 (2009)
11. Guo, J.: A more effective linear kernelization for cluster editing. Theoretical Computer Science 410(8), 718–726 (2009)
12. Guo, J., Hüffner, F., Komusiewicz, C., Zhang, Y.: Improved algorithms for bicluster editing. In: Agrawal, M., Du, D.-Z., Duan, Z., Li, A. (eds.) TAMC 2008. LNCS, vol. 4978, pp. 445–456. Springer, Heidelberg (2008)
13. Guo, J., Kanj, I.A., Komusiewicz, C., Uhlmann, J.: Editing graphs into disjoint unions of dense clusters. Algorithmica 61(4), 949–970 (2011)
14. Guo, J., Komusiewicz, C., Niedermeier, R., Uhlmann, J.: A more relaxed model for graph-based data clustering: S-plex cluster editing. SIAM Journal on Discrete Mathematics 24(4), 1662–1683 (2010)
15. Harpaz, R., Perez, H., Chase, H.S., Rabadan, R., Hripcsak, G., Friedman, C.: Biclustering of adverse drug events in the fda's spontaneous reporting system. Clinical Pharmacology & Therapeutics 89(2), 243–250 (2010)
16. Niedermeier, R.: Invitation to fixed-parameter algorithms, vol. 3. Oxford University Press, Oxford (2006)
17. Seidman, S.B., Foster, B.L.: A graph-theoretic generalization of the clique concept. Journal of Mathematical Sociology 6(1), 139–154 (1978)
18. Shamir, R., Sharan, R., Tsur, D.: Cluster graph modification problems. Discrete Applied Mathematics 144(1), 173–182 (2004)
19. Sun, P., Guo, J., Baumbach, J.: Integrated simultaneous analysis of different biomedical data types with exact weighted bi-cluster editing. Journal of Integrative Bioinformatics 9(2), 197 (2012)
20. van Zuylen, A.: Deterministic approximation algorithms for ranking and clustering problems. Technical Report 1431. School of Operations Research and Industrial Engineering, Cornell University, Ithaca, NY (2005)
21. Wittkop, T., Baumbach, J., Lobo, F.P., Rahmann, S.: Large scale clustering of protein sequences with force-a layout based heuristic for weighted cluster editing. BMC Bioinformatics 8(1), 396 (2007)
22. Wittkop, T., Emig, D., Lange, S., Rahmann, S., Albrecht, M., Morris, J.H., Böcker, S., Stoye, J., Baumbach, J.: Partitioning biological data with transitivity clustering. Nature Methods 7(6), 419–420 (2010)
23. Wittkop, T., Emig, D., Truss, A., Albrecht, M., Böcker, S., Baumbach, J.: Comprehensive cluster analysis with transitivity clustering. Nature Protocols 6(3), 285–295 (2011)
24. Yu, H., Paccanaro, A., Trifonov, V., Gerstein, M.: Predicting interactions in protein networks by completing defective cliques. Bioinformatics 22(7), 823–829 (2006)

Parameterized Complexity
of Edge Interdiction Problems[*]

Jiong Guo and Yash Raj Shrestha

Universität des Saarlandes,
Campus E 1.7, 66123 Saarbrücken, Germany
{jguo,yashraj}@mmci.uni-saarland.de

Abstract. For an optimization problem on edge-weighted graphs, the corresponding interdiction problem can be formulated as a game consisting of two players, namely, an interdictor and an evader, who compete on an objective with opposing interests. In an edge interdiction problem, every edge of the input graph is associated with an interdiction cost. The interdictor interdicts the graph by modifying the edges in the graph and the number of such modifications is bounded by the interdictor's budget. The evader then solves the given optimization problem on the modified graph. The action of the interdictor must impede the evader as much as possible.

We study the parameterized complexity of edge interdiction problems related to minimum spanning tree, maximum matching, maximum flow and minimum maximal matching problems. These problems arise in different real world scenarios. We derive several fixed-parameter tractability and W[1]-hardness results for these interdiction problems with respect to various parameters. Hereby, we reveal close relation between edge interdiction problems and partial covering problems on bipartite graphs.

1 Introduction

For an optimization problem on graphs, the corresponding interdiction problem can be formulated as a game consisting of two players, namely, an interdictor and an evader, who compete on an objective with opposite interests. In an edge interdiction problem, every edge of the input graph has an interdiction cost associated with it and the interdictor interdicts the network by modifying edges in the graph, and the number of such modifications are constrained by the interdictor's budget. The evader then solves the given optimization problem on the modified graph. The action of the interdictor must impede the evader as much as possible.

In this paper, we focus on edge interdiction problems related to minimum spanning tree, maximum matching, maximum flow and minimum maximal matching problems. These interdiction problems arise in different real world scenarios,

---

[*] Supported by the DFG Excellence Cluster MMCI and the DFG research project DARE GU 1023/1.

Z. Cai et al. (Eds.): COCOON 2014, LNCS 8591, pp. 166–178, 2014.

e.g., detecting drug smuggling [22,20], military planning [21], analyzing power grid vulnerability [19] and hospital infection control [3].

A *spanning tree* of a connected and edge-weighted graph G is a tree composed of all the vertices and some of the edges of G. The *minimum spanning tree (MST) problem* is to find a spanning tree whose total weight is minimum. Let $\eta(G)$ be the weight of MST of G. A *matching* in a graph is a set of edges such that no two edges share an endpoint. Let $\nu(G)$ be the weight of maximum matching in G. We say a matching M *saturates* a set $U \subseteq V$ if for each vertex $u \in U$, there exists one edge in M with u as its endpoint. For $G = (V, E)$, $G - I$ is the graph resulting by removing a set of edges I from G. A set of edges M of $G = (V, E)$ is called an *edge dominating set* if every edge of $E \setminus M$ is adjacent to at least one edge in M. An *independent edge dominating set* is an edge dominating set in which no two edges are adjacent. A *minimum maximal matching* in a graph G is a maximal matching of the minimum size, denoted by $\lambda(G)$. A smallest independent edge dominating set is a minimum maximal matching [9].

We start with the definition of the edge interdiction problem related to the minimum spanning tree problem, which is also known as t-MOST VITAL EDGES IN MST (t-MVE) in literature:

Input: An edge-weighted graph $G = (V, E)$ with weight function $w : E \to \mathbb{Z}_{\geq 0}$, two positive integers t and r.
Output: A subset $I \subseteq E$ with $|I| \leq t$ such that $\eta(G - I) \geq r$.

Frederickson and Solis-Oba [10] proved that t-MVE is NP-hard even if the weights of the edges are either 0 or 1. They also gave an $\Omega(1 \setminus \log t)$-approximation algorithm for t-MVE. This problem has also been studied from the view point of exact algorithms and randomized algorithms [14,15].

The MAXIMUM MATCHING EDGE INTERDICTION (MMEI) problem, introduced by Zenklusen [25], is defined as follows:

Input: A edge-weighted graph $G = (V, E)$ with weight function $w : E \to \mathbb{Z}_{\geq 0}$, an interdiction cost function $c : E \to \mathbb{Z}_{\geq 1}$, and two positive integers b and m.
Output: A subset $I \subseteq E$ with $c(I) \leq b$ such that $\nu(G - I) \leq m$.

MMEI is NP-hard on bipartite graphs, even with unit edge weights and unit interdiction costs [25]. Zenklusen [25] introduced a constant factor approximation algorithm for MMEI on graphs with unit edge weights. Recently, Dinitz and Gupta provided a constant-factor approximation for a generalization of matching interdiction called *packing interdiction* [6]. Zenklusen [25] also showed that MMEI is solvable in pseudo-polynomial time on graphs with bounded treewidth. Pan and Schild [17] proved that weighted MMEI remains NP-hard even on planar graphs.

The s-t FLOW EDGE INTERDICTION (s-t FEI) is defined as follows [21]:

Input: A directed graph $G = (N, A)$ with distinguished vertices s and t, a positive integer capacity u_{ij} for each arc $(i, j) \in A$, an interdiction cost function $c : A \to \mathbb{Z}_{\geq 0}$, two positive integers l and f.

Output: A set of arcs A' with $c(A') \leq l$ such that the maximum s-t flow in $G - A'$ has value at most f.

The s-t FEI was shown to be strongly NP-complete on general graphs and weakly NP-complete when restricted to planar graphs [18,22]. Different algorithms for finding exact algorithms were proposed [21,22] and Burch et al. [5] gave a pseudo-approximation algorithm.

We study the *parameterized complexity* of the three edge interdiction problems defined above. Parameterized complexity is a two-dimensional framework for studying the computational complexity of hard problems [16] . A problem is *fixed parameter tractable* (FPT) with respect to parameter k (e.g. solution size) if for any instance of size n it can be solved in time $O(f(k)n^c)$ for some constant c. A core tool in the development of fixed-parameter algorithms is polynomial-time preprocessing by applying *data reduction rules*, often yielding a reduction to a *problem kernel (kernelization)*. Herein, the goal is, given a problem instance x with parameter k, to transform it in polynomial time into an new instance x' with parameter k' such that the size of x' is bounded from above by some function only depending on $k, k' \leq k$, and (x, k) is a yes-instance iff (x', k') is a yes-instance. In prameterized complexity, the principal analogue of the classical intractability class NP is called W[1].

First, we show that t-MVE is W[1]-hard with respect to r. In graphs with edge weights of only 0 or 1, t-MVE is FPT with respect to t. The reduction from CLIQUE to MMEI by Zenklusen [24] already shows that MMEI on bipartite input graphs with unit edge weights and interdiction costs is W[1]-hard with respect to b. Complementing this result, we prove that MMEI on graphs with unit edge weights and unit interdiction costs is also W[1]-hard with respect to m as parameter. In contrast to parameter b, MMEI becomes FPT with respect to m, if further restricted to bipartite input graphs. Moreover, taking both b and m as parameters leads also to FPT when restricted to instances with unit edge weights. Concerning s-t FEI, we prove that the parameterization with l is W[1]-hard, complementing the result by Wood [22], that s-t FEI is W[1]-hard with respect to f.

We observe some close relation between partial covering problems on bipartite graphs and edge interdiction problems. The goal of a partial covering problem is not to cover all elements but to minimize/maximize the number of covered elements with a specific number of sets. For instance, the PARTIAL VERTEX COVER (k-PVC) problem asks for k vertices maximizing the number of covered edges. Partial covering problems have been studied intensively not only because they generalize classical covering problems, but also because of many real life applications, see for example [2,4,1,8].

Our findings about the relation between partial covering problems and edge interdiction problems can be summarized as follows: First, we give a parameterized reduction from the W[1]-hard k-PVC problem to MMEI, leading to the W[1]-hardness of MMEI with respect to m. Then, we prove an equivalent relation between a special version of MMEI and k-PVC on bipartite graphs and thus derive the FPT result of this special case of MMEI. Moverover, we prove the W[1]-hardness of k-PVC on bipartite graphs with the number of uncovered edges as parameter by a reduction from MMEI with respect to parameter b. Further, we introduce the following edge interdiction problem called MINIMUM MAXIMAL MATCHING EDGE INTERDICTION which turns out to be equivalent to the PARTIAL EDGE DOMINATING SET problem and prove W[1]-hardness for both.

MINIMUM MAXIMAL MATCHING EDGE INTERDICTION (MMMEI)
Input: A simple graph $G = (V, E)$, and an integer interdiction budget $p \geq 0$ and an integer q.
Output: Is there a subset $I \subseteq E$ with $|I| \leq p$ such that $\lambda(G - I) \leq q$?

k-PARTIAL EDGE DOMINATING SET (k-PEDS)
Input: A graph $G = (V, E)$ and two positive integers k and x.
Output: Is there a subset $S \subseteq E$ with $|S| \leq k$ such that at least x edges are dominated by S?

Preliminary. For a vertex v, the vertices which are adjacent to v in G form the neighborhood $N(v)$ of v. For $U \subseteq V$, let $N(U)$ denote the set of all vertices not in U but adjacent to those in U. We denote the size of the neighborhood of v in G as $\deg_G(v)$. A *degree-1* vertex is a vertex with $\deg_G(v) = 1$. A *path* from vertex a to vertex b is an ordered sequence $a = v_0, v_1, \ldots, v_m = b$ of distinct vertices in which each adjacent pair (v_{j-1}, v_j) is linked by an edge. The *distance* between two vertices is the number of edges on the shortest path between them, while the distance between two edges e_1 and e_2 is the minimum of the distances of their endpoints. A subgraph $H = (V', E')$ of a graph $G = (V, E)$ is a pair $V' \subseteq V$ and $E' \subseteq E$. We say that $H = (V', E')$ is an *induced* subgraph of G if $V' \subseteq V$ and $E' = \{\{u, v\} \in E | u, v \in V'\}$ and we denote $H = G[V']$. For a set of edges S, let $V(S)$ denote the set of endpoints of S. An edge e is *dominated* by another edge e' if they share one endpoint. An edge e is covered by a vertex v if v is one of the endpoints of e. A set of edges F is called *disconnecting*, if $G - F$ has more connected components than G does. A connected graph G is k-*edge-connected* if every disconnecting edge set has at least k edges.

2 t-MOST VITAL EDGES IN MST

We consider two parameterizations for t-MVE. Firstly, we define MINIMUM k-WAY EDGE CUT which is vital in the proofs of the following two results.

MINIMUM k-WAY EDGE CUT
Input: An undirected connected graph $G = (V, E)$ with unit edge weights and non-negative integers k and s.
Question: Is there a set $S \subseteq E$ with $|S| \leq s$ such that, $G - S$ has at least k connected components?

Theorem 1. t-MOST VITAL EDGES *is W[1]-hard with respect to the weight r of the MST in $G - I$.*

Proof. We give a parameterized reduction from MINIMUM k-WAY EDGE CUT. Downey et al. [7] proved that MINIMUM k-WAY EDGE CUT is W[1]-hard with parameter k. Given an instance $G = (V, E)$ for MINIMUM k-WAY EDGE CUT, we create an instance $G' = (V', E')$ for t-MVE as follows: For each vertex $v \in V$, we create a vertex $v \in V'$. For each edge $\{u, v\} \in E$, create an edge $\{u, v\} \in E'$ with weight 0. For each pair of vertices u, v in V' such that $\{u, v\} \notin E$, we add a connection gadget M between u and v in the following way: Create a clique with $t + 1$ vertices as gadget M such that all edges in M have weight 0. Now, connect u to all vertices in M with edges of weight 1 and v to all vertices in M with edges of weight 0. Let X' be the set of all vertices in connection gadgets and Y' be the set of edges in G' with at least one endpoint in X'. Now we show that G has a k-way edge cut of size s iff at most $t = s$ edges can be deleted from G' such that MST of the remaining graph is at least $r = k - 1$.

($\Rightarrow$) Let S be the solution for MINIMUM k-WAY EDGE CUT on G. Now, $G - S$ consists of at least k connected components. We take the edges in G' corresponding to those in S as the solution S' for t-MVE. Observe that $G'[V' \setminus X'] - S'$ consists of at least k connected components. Hence, every spanning tree of $G' - S'$ must pass through at least $k - 1$ connection gadgets M. Let M' be one connection gadget through which the MST in $G' - S'$ passes. Let M' be connected to $u \in V' \setminus X'$ with edges of weight 0 and to $v \in V' \setminus X'$ with edges of weight 1. The MST can span all vertices in M' by using $t + 1$ edges between u and M' and connect u to v by taking exactly one edge between M' and v. The MST must pass through at least $k - 1$ such connection gadgets. Moreover, each connected component of $G'[V' \setminus X'] - S'$ has a MST of weight 0 and the vertices in the remaining connection gadgets can be included in the MST by taking only weight-0 edges. This gives a MST of G' of weight at least $k - 1$.

($\Leftarrow$) Let S' be the solution for t-MVE on G' and the MST of $G' - S'$ has weight at least $k - 1$. We first prove that $S' \cap Y' = \emptyset$. Let M be the connection gadget between u and v. We can observe that the minimum weight of a MST of $G[M \cup \{u, v\}]$ is 1. Since M is a clique of size $t+1$ and both u and v are connected to all vertices in M, the removal of arbitrary t edges from $G[M \cup \{u, v\}]$ cannot increase the weight of a MST of $G[M \cup \{u, v\}]$. Hence, $S' \cap Y' = \emptyset$. Now, in order to increase the weight of MST of $G' - S'$, the interdictor must force the maximum usage of weight-1 edges (which are available only in connection gadgets) of the MST. To this end, we need to maximize the number of connected components in $G' - Y'$. Hence, S' is chosen in such a way that $G' - \{Y' \cup S'\}$ has the maximum number of connected components. This corresponds to a MINIMUM k-WAY EDGE CUT in G, which completes the proof. $\square$

Theorem 2. *Given an instance with edge weights only 0 and 1, t-Most Vital Edges is fixed-parameter tractable (FPT) with respect to t.*

Proof. Kawarabayashi and Thorup [13] proved that MINIMUM k-WAY EDGE CUT is FPT with respect to s. Here, we use their algorithm as a black box. If the input graph G is d-edge-connected with $d \leq t$, we can find an edge cut S of size at most t for G. Since $G - S$ is disconnected, we take S as a solution for

t-MVE and the weight of any MST of disconnected graphs is ∞. On the other hand, if G is $(t+1)$-edge-connected, we need the following claim:

Claim. Given a $(t+1)$-edge-connected graph, there exists a solution of t-MVE that contains no weight-1 edges.

Proof. Let V_1 and V_2 be an arbitrary partition of vertices of G such that $V_1 \cap V_2 = \emptyset$ and S be a solution of t-MVE. Let T be a minimum spanning tree of $G - S$. Now, we show that if G is $(t+1)$-edge-connected, S does not contain any weight-1 edge between vertices in V_1 and V_2. Since G is $(t+1)$-edge-connected, there is at least one edge between V_1 and V_2 in $G - S$. Hence, the worst-case cost of connecting V_1 and V_2 in T is 1. So, it is never profitable to delete any edge of weight 1 between V_1 and V_2. □

By this claim, if a $(t+1)$-edge-connected graph has only weight-1 edges, then $S = \emptyset$ is the the solution . Let G be an instance of t-MVE, we run the following:

Step 1. Delete all weight-1 edges from G. Let $G - X$ be the resulting graph where X is the set of all weight-1 edges in G.

Step 2. In each connected component of $G - X$, we run the FPT-algorithm from [13] with s ranging from 1 to t. For each connected component C of $G - X$ we maintain a table A, where for each integer $1 \le i \le t$, we store the maximal number of connected components that can be achieved by deleting i edges in C. This table is of size t for each connected component of $G - X$ and can clearly be filled in FPT time with respect to t.

Step 3. For each integer $1 \le i \le t$, we sort the connected components of $G - X$ according to the decreasing order of the number of resulting components with i edge deletions as returned by Step 2. For each i we save the top t entries in this sorted list, resulting in a table B of size $t \times t$.

Step 4. Now, we enumerate all additive partitions of t. The partition function $p(t)$ gives the number of different additive partitions of t without respect to order which is clearly bounded by 2^{t-1}. Such a partition can be computed in time polynomial in t. Let, $P_1, P_2, \cdots, P_{p(t)}$ be the additive partitions of t.

Step 5. For each additive partition P_i we do the following: Assume that P_i consists of $j \le t$ integers. Now for each integer $x \in P_i$, we branch on the first j entries corresponding to x from Table B, each branch assigning exactly one entry to x, that is, one connected component for x from $G - X$. This will take $O^*(t^t)$ time for each P_i. Since there are at most n connected components in $G - X$, Step 5 runs in $O^*(t^t)$ time.

Correctness. Correctness of Step 1 follows directly from the above claim. We prove now in Step 5 it is sufficient to branch only on the first j entries corresponding to x from Table B. For each integer $x \in P_i$ exactly one connected component from $G - X$ is assigned. Now, the top candidate for x in Table B will not be assigned to x if and only if it is assigned to another integer $y \in P_i$. There can be at most $j - 1$ such integers $y \in P_i$. Hence, it is sufficient to consider only the first $j \le t$ entries corresponding to x from Table B. Steps 1 and 3 can be achieved in time polynomial in n. Steps 2, 4 and 5 are FPT with respect to t. Hence we have an overall running time exponentially depending on t.   □

3   Maximum Matching and S-T Flow Interdiction

In this section, we study the edge interdiction problems for maximum matching and s-t flow from parameterized complexity point of view. The reduction from CLIQUE to MMEI in [24] is also a parameterized one, proving that MMEI on bipartite graphs with unit edge weight and unit interdiction cost is W[1]-hard with respect to b. Now, we prove a similar result for the parameter m.

Theorem 3. *MMEI with unit edge weights and interdiction costs is W[1]-hard with respect to m.*

Proof. We give a parameterized reduction from the W[1]-hard PARTIAL VERTEX COVER (k-PVC) problem with parameter k [11]. Given an instance $G = (V, E)$ for k-PVC, we create an instance $G' = (V', E')$ for MMEI in the following: We initialize G' with G. Now, for each vertex in G', we add $|E|$ degree-1 neighbors. Let Y be the set of degree-1 neighbors added in this way. Next, we show that G has a set S of size k which covers at least x edges in G iff G' has a solution I with $b \le |E|(|V| - k) + (|E| - x)$ and $\nu(G' - I) \le m = k$.

($\Rightarrow$) Given a solution S of k-PVC on G, we construct the MMEI solution I for G' as follows: We add all edges in G' which are not incident to any vertex in S to I. Since S covers at least x edges in G, we add at most $|E| - x$ edges from E and $|E|(|V| - k)$ edges between Y and V to I. In the subgraph $G' - I$, every edge is incident to vertices in S. Hence $\nu(G' - I)$ is at most k.

($\Leftarrow$) Let I be the given solution of MMEI for G' with $b \le |E|(|V| - m) + (|E| - x)$ and $\nu(G' - I) \le m$. Since $\nu(G' - I)$ is at most m, at most m vertices in $G'[V' \setminus Y] - I$ can have degree-1 vertices attached to them. Let X denote the set of vertices in $G'[V' \setminus Y] - I$ which have degree-1 neighbors. To remove all degree-1 neighbors of the vertices in $V' \setminus X$ requires addition of $|E|(|V| - m)$ edges to I. Hence, the vertices in X must cover at least x edges in G and X is solution for k-PVC for G.   □

However, unlike for the parameter b, the parameterization of MMEI with m becomes tractable, if we restrict the input graphs to be bipartite.

Theorem 4. *MMEI is fixed-parameter tractable (FPT) with respect to m when restricted to bipartite graphs with unit edge weights and interdiction costs.*

Proof. We prove the theorem by showing that, for a bipartite graph $G = (X, Y, E)$, there is a partial vertex cover S with $|S| \le k$ covering at least x edges, if and only if there is a set $I \subseteq E$ with $|I| \le |E| - x$ and $\nu(G - I) \le m = k$. Note that k-PVC on bipartite graphs is solvable in $O^*(2k^{(2k)})$ time [1], proving the theorem.

Let S be a size-k partial vertex cover of G and I be the set of edges not incident to the vertices in S. Then, $|I| \le |E| - x$. Since, all edges in $G - I$ are incident to vertices in S and each vertex in S must have an incident edge whose other endpoint is not in S, $\nu(G - I) \leq k$. This is true because, if each vertex in S does not have an incident edge whose other endpoint is not in S, then the cover can be smaller. The reverse direction can be shown in similar way.   □

Using both b and m as parameters, we can achieve another FPT result for MMEI.

Theorem 5. *MMEI in graphs with unit edge weights is fixed-parameter tractable (FPT) with respect to both b and m.*

Proof. We show that in the instances with unit edge weights MMEI with both b and m as parameters admits a kernel. We apply the following reduction rules:

Reduction Rule 1: If a vertex v has more than b degree-1 neighbors, then keep arbitrary b of them and remove other degree-1 neighbors of v.

The correctness of Rule 1 can be shown as follows: Assume that a vertex v has more than b degree-1 neighbors. Let X be the set of all degree-1 neighbors of v. It is not possible to remove all edges between v and X with at most b edge deletions allowed. Then, one of the edges between v and X can be in the matching. Now, keeping b of them obviously does not omit any optimal solution.

Reduction Rule 2: If $U \subseteq V$ and $W \subseteq V$ with $U \cap W \neq \emptyset$ satisfy: 1) W is an independent set, 2) $N(W) = U$ and there are all possible edges between U and W, and 3) $|W| \geq \max\{|U|, b+1\}$, then keep only $\max\{|U|, b+1\}$ vertices in W and remove the rest.

We prove now the correctness of Rule 2. Let H be the bipartite graph $G[U \cup W] - E(G[U])$. Notice that we have $\nu(H) = |U|$ and there are at least $|W|$ disjoint matchings in H which saturate U. Now we show that removing any b edges from H does not decrease the cardinality of a maximum matching in G. This property is obtained by observing that since in H there are at least $b + 1$ disjoint matchings which saturate U, we have after removing up to b edges in G there is at least one matching of H which saturates U. We therefore have the desired property that H is immune to "edge removals". Hence, removing all but $\max\{b + 1, |U|\}$ vertices from $|U|$ still maintains this property.

Rule 2 runs in polynomial time, since W is clearly a module and all modules of a graph can be found in linear time [12].

Claim. MMEI admits a kernel with respect to both b and m as parameters.

Proof. Let B be the set of edges which form the solution of MMEI and let M be the maximum matching of the remaining graph. Since the costs of edges are positive integers, $|B| \leq b$ and $|M| = m$. Moreover, $|V(B)| \leq 2b$ and $|V(M)| = 2m$. As M is the maximum matching in $G - B$, each edge in $G - B$ must have at least one endpoint in $V(M)$. Hence, there exist at most b edges in G which do not have its endpoints in $V(M)$.

Now, we bound the number of edges with its endpoints in $V(M)$. There are at most $4m^2$ edges in $G[V(M)]$. Let X be the set of degree-1 neighbors of $V(M)$. Rule 1 bounds the number of degree-1 neighbors of each vertex by b; hence, there are at most $2bm$ edges between $V(M)$ and X. The number of edges between $V(B)$ and $V(M)$, is clearly bounded by $4bm$. Now, the remaining edges are between $V(M)$ and $Y := V \setminus (X \cup V(B) \cup V(M))$. We can observe that the vertices in Y have degree at least two and $N(Y) \subseteq V(M)$. For each edge $\{m_1, m_2\} \in M$, there exists at most one vertex in Y which is adjacent to both m_1 and m_2. We have m edges in M, hence there are at most m vertices in Y which are adjacent to both endpoints of an edge in M. Next, we bound the number of vertices in Y, which are adjacent to several edges in M. By Rule 2, for a size-i subset of vertices $I \subset V(M)$ such that no two vertices in I are connected by an edge in M, there can be at most $\max\{i, b+1\}$ vertices y in Y, such that $N(y) = I$. There can be at most $\sum_i 2^i \cdot \binom{m}{i}$ such subsets I in $V(M)$. Hence we have at most $l = \sum_i 2^i \cdot \binom{m}{i} \cdot \max\{i, b+1\} + m$ vertices in Y. Hence, there are at most lm edges between Y and M. In total we have $lm + 4bm + 4m^2$ edges which is a function depending only on m and b. Hence we have a kernel for MMEI with both b and m as parameters. $\square$

Wood [22] proved the NP-hardness of s-t FEI by a reduction from CLIQUE, which sets the flow amount in the resulting graph f equal to k. This implies that s-t Flow Edge Interdiction with unit edge costs and the edge capacities being 1 or 2 is W[1]-hard with respect to f. Complementing this result, we achieve the W[1]-hardness of s-t FEI with respect to parameter l.

Theorem 6. *s-t FLOW EDGE INTERDICTION in bipartite graphs with unit flow capacity is W[1]-hard with respect to l.*

Proof. We give a parameterized reduction from W[1]-hard MMEI with b as parameter. Note that this parametrization remains W[1]-hard on bipartite graphs. Let a bipartite graph $G = (X, Y, E)$ be an instance of MMEI. We create an instance G' for s-t FLOW EDGE INTERDICTION in the following way: Initialize G' with G such that each edge has unit interdiction cost and flow capacity and each arc is directed from vertex in X to the one in Y. Add two new vertices s and t to G'. Now, we add arcs with interdiction cost $l + 1$ from s to each vertex in X. Similarly, arcs with interdiction cost $l + 1$ directed from each vertex in Y to t are added. Let the set of the arcs added in this way to G be Q and each arc in Q has unit flow capacity. With this construction we can show that G has a yes answer to MMEI with b total budget and maximum matching with weight at most m allowed in the resulting graph iff G' has yes answer to the s-t FEI with total budget l and maximum flow allowed in the resulting graph at most $f = m$.

The key argument is that only the arcs in $G' - Q$ will belong to an optimal solution of s-t FEI. Hence, the amount of the s-t flow in the resulting graph is equivalent to the corresponding matching in $G' - Q$. $\square$

4   Minimum Maximal Matching Edge Interdiction

From the proof of Theorem 4, we can already observe some equivalent relation between edge interdiction problems and partial covering problems. In the following, we prove that MMMEI is W[1]-hard with respect to j by relating it to k-PEDS. Firstly, we prove the following lemma:

Lemma 1. *k-PEDS on bipartite graphs is W[1]-hard with respect to k.*

Proof. We give a parameterized reduction from W[1]-hard k-INDEPENDENT SET [16] to k-PEDS. Given a graph $G = (V, E)$ as an instance of k-INDEPENDENT SET, we create an instance $G' := (V_1, V_2, E')$ for k-PEDS in the following way: For each vertex $v \in G$, we create two vertices v_1 and v_2 and an edge $\{v_1, v_2\}$ in G'. For each edge $\{u, v\} \in G$, we create two edges $\{u_1, v_2\}$ and $\{u_2, v_1\}$ in G'. Moreover, for every vertex $v_i \in G'$, we add $n - deg_G(v)$ degree-1 neighbors where $n = |V|$. Now, we show that k edges in G' dominate at least $2kn$ edges iff there exists an independent set of size k in G.

($\Leftarrow$) Let S be an independent set of size k in G. For each vertex $v \in S$, we add the corresponding edge $\{v_1, v_2\} \in E'$ to the solution set S' for k-PEDS in G'. Given that S is an independent set, for any pair of vertices u, v in S, the corresponding edges $\{u_1, u_2\} \in E'$ and $\{v_1, v_2\} \in E'$ do not dominate any common edge. Hence, since each edge $e \in S'$ dominates exactly $2n$ edges, the set S' dominates $2kn$ edges.

($\Rightarrow$) Now, let S' be a set of k edges in G' which dominate $2kn$ edges. Each edge in G' can dominate at most $2n$ edges; hence, no two edges in S' share a dominated edge. This ensures that the shortest distance between every two edges in S' is at least two. Now, we present an algorithm to convert a given solution S' for PEDS in G' to a size-k independent set in G. For this purpose, first we define a *conflict cycle*. Let S' be a solution of PEDS in $G' = (V_1, V_2, E')$ and let $T \subseteq S'$. We say that T forms a conflict cycle C in G' if we can construct a cycle in G' containing T and a set U of $|T|$ vertices from $G' - S'$ such that in cycle C, between any two edges from T there exists exactly one vertex from U. A vertex t_j can be in U only if t_{3-j} is contained in S' for $j = \{1, 2\}$.

Assume that there exists no conflict cycle with respect to S'. Then it is easy to get the size-k independent set corresponding to S' in G. Construct a graph Y that represents the connectivity relation of edges in S': For each edge $i \in S'$, we create a vertex y_i in Y. We create an edge between two vertices y_i and y_j in Y iff their corresponding edges i and j are separated by distance exactly 2 in G'. Observe that in the absence of *conflict cycles* in G', Y is a tree. Now, we give a procedure to get S from S', given Y is a tree. We start in bottom-up fashion from leaves. Consider a leaf of Y, if $\{x_1, x_2\}$ is the edge corresponding to the leaf, then we add the corresponding vertex x to S. Let $\{x_1, y_2\}$ be the edge corresponding to the leaf and v_i be the vertex connecting the leaf to its parent in T. If $v_i \in V_1$, we add x to S, else we add y to S. Now, we remove this leave and proceed iteratively for every leaf. We can observe that, since T is a tree, no conflict will arise during this procedure. The obtained solution after all vertices in T are processed is an independent set in G. In the scenario when there exist *conflict cycles* in G', we first prove the following claim:

Claim. If there exist *conflict cycles* in G', the corresponding graph Y is bipartite.

Proof. Firstly, we observe that $T \subseteq S'$ forms a *conflict cycle*, only if there exists no edge $\{x_1, x_2\}$ in T. If $\{x_1, x_2\} \in T$, no vertex v_i with $i \in \{1, 2\}$ adjacent to x_1 or x_2 can be in U as $v_{(3-i)} \notin V(S')$. We can further observe that since G'

is bipartite, in the *conflict cycle*, the vertices will alternate between V_1 and V_2. Moreover, in the cycle, every two consecutive vertices from U will also alternate between V_1 and V_2. Now, as the edges in Y are analogous to vertices in U, each cycle in Y corresponding to a *conflict cycle* in S' is of even length.   □

For vertices in T not belong to any cycle, we can obtain the corresponding vertices in S in a bottom-up fashion recursively as for the case that T is tree. After all vertices not belonging to any cycles are dealt with, let the resulting T be T', where only even cycles remain. Since T' is bipartite, it is 2-colorable. We color T' with two colors, say black and white. If a vertex in T' corresponding to $\{x_1, y_2\}$ is black, we add x to S, else y. Since the vertices in U alternate between V_1 and V_2, this resolves all conflicts and gives a valid independent set S.   □

Theorem 7. MINIMUM MAXIMAL MATCHING EDGE INTERDICTION *on bipartite graphs is W[1]-hard with respect to parameter* p.

Proof. For a bipartite graph $G = (X, Y, E)$, we show that there is a set $S \subseteq E$ with $|S| \leq k$ that dominates at least x edges in G, iff there is a set $I \subseteq E$ with $|I| \leq |E| - x$ and $\lambda(G - I) \leq q = k$. The theorem then follows from Lemma 1. Let I be the set of edges not dominated by S, $|I| \leq |E| - x$. Clearly, S is the minimum edge dominating set of $G - I$. It is well-known that the size of the minimum edge dominating set of a graph is equal to the size of its minimum independent edge dominating set. In fact, given a minimum edge dominating set F of $G - I$ we can construct in polynomial time a minimum independent edge dominating set of $G - I$ [23]. Moreover, a minimum independent edge dominating set is also a minimum maximal matching of $G - I$. Hence, $G - I$ has a minimum maximal matching of size at most q. The reverse direction can be shown similarly.   □

Finally, we use the equivalence relation between edge interdiction problems and partial covering problems to prove the following hardness result, whose proof is in deferred to the full-version of the paper.

Corollary 1. *k-PVC on bipartite graphs is W[1]-hard with respect to the number of uncovered edges.*

5 Outlook

We proved that t-MVE on the instances of edge weights 0 or 1 is FPT with respect to t. The case with integer positive weights remains open. Another open question is the complexity of MMEI on integer edge weights with respect to b and m. Moreover, structural parameters like treewidth could be a promising alternative for parameterizing interdiction problems. Finally, the vertex interdiction problems have been studied from the viewpoints of classical complexity and approximation algorithms, but seem unexplored from the parameterized complexity perspective.

References

1. Amini, O., Fomin, F.V., Saurabh, S.: Implicit branching and parameterized partial cover problems. J. Comput. Syst. Sci. 77(6), 1159–1171 (2011)
2. Arora, S., Karakostas, G.: A $2 + \epsilon$ approximation algorithm for the k-mst problem. Math. Program. 107(3), 491–504 (2006)
3. Assimakopoulos, N.: A network interdiction model for hospital infection control. Bio. Med. 17(6), 413–422 (1987)
4. Bar-Yehuda, R.: Using homogeneous weights for approximating the partial cover problem. J. Algorithms 39(2), 137–144 (2001)
5. Burch, C., Carr, R., Krumke, S., Marathe, M., Phillips, C., Sundberg, E.: A decomposition-based pseudoapproximation algorithm for network flow inhibition. Network Interdiction and Stochastic Integer Programming 22, 51–68 (2003)
6. Dinitz, M., Gupta, A.: Packing interdiction and partial covering problems. In: Goemans, M., Correa, J. (eds.) IPCO 2013. LNCS, vol. 7801, pp. 157–168. Springer, Heidelberg (2013)
7. Downey, R.G., Estivill-Castro, V., Fellows, M.R., Prieto, E., Rosamond, F.A.: Cutting up is hard to do: The parameterized complexity of k-cut and related problems. Electr. Notes Theor. Comput. Sci. 78, 209–222 (2003)
8. Fomin, F.V., Lokshtanov, D., Raman, V., Saurabh, S.: Subexponential algorithms for partial cover problems. Inf. Process. Lett. 111(16), 814–818 (2011)
9. Forcade, R.: Smallest maximal matching in the graph of the d-dimensional cube. J. Combinatorial Theory Ser. B 14(14), 153–156 (1973)
10. Frederickson, G.N., Solis-Oba, R.: Increasing the weight of minimum spanning trees. J. Algorithms 33(2), 244–266 (1999)
11. Guo, J., Niedermeier, R., Wernicke, S.: Parameterized complexity of generalized vertex cover problems. In: Dehne, F., López-Ortiz, A., Sack, J.-R. (eds.) WADS 2005. LNCS, vol. 3608, pp. 36–48. Springer, Heidelberg (2005)
12. Hsu, W.-L., Ma, T.-H.: Substitution decomposition on chordal graphs and applications. In: Hsu, W.-L., Lee, R.C.T. (eds.) ISA 1991. LNCS, vol. 557, pp. 52–60. Springer, Heidelberg (1991)
13. Kawarabayashi, K., Thorup, M.: The minimum k-way cut of bounded size is fixed-parameter tractable. In: FOCS, pp. 160–169 (2011)
14. Liang, W.: Finding the k most vital edges with respect to minimum spanning trees for fixed k. Discrete Applied Mathematics 113(2-3), 319–327 (2001)
15. Liang, W., Shen, X.: Finding the k most vital edges in the minimum spanning tree problem. Parallel Computing 23(13), 1889–1907 (1997)
16. Niedermeier, R.: Invitation to Fixed Parameter Algorithms. Oxford Lecture Series in Mathematics and Its Applications. Oxford University Press, USA (2006)
17. Pan, F., Schild, A.: Interdiction problems on planar graphs. In: Raghavendra, P., Raskhodnikova, S., Jansen, K., Rolim, J.D.P. (eds.) APPROX/RANDOM 2013. LNCS, vol. 8096, pp. 317–331. Springer, Heidelberg (2013)
18. Phillips, C.A.: The network inhibition problem. In: Proceedings of the Twenty-fifth Annual ACM Symposium on Theory of Computing, STOC 1993, pp. 776–785. ACM, New York (1993)
19. Salmeron, J., Wood, K., Baldick, R.: Worst-case interdiction analysis of large-scale electric power grids. IEEE Transactions on Power Systems 24(1), 96–104 (2009)
20. Washburn, A., Wood, R.K.: Two-person zero-sum games for network interdiction 43(2), 243–251 (1995)

21. Wood, R.K.: Optimal interdiction policy for a flow network 18, 37–45 (1971)
22. Wood, R.K.: Deterministic network interdiction 17(2), 1–18 (1993)
23. Yannakakis, M., Gavril, F.: Edge dominating sets in graphs. SIAM J. Appl. Math. 38(3), 364–372 (1980)
24. Zenklusen, R.: Matching interdiction. CoRR, abs/0804.3583 (2008)
25. Zenklusen, R.: Matching interdiction. Discrete Applied Mathematics 158(15), 1676–1690 (2010)

Vertex Cover Gets Faster
and Harder on Low Degree Graphs

Akanksha Agrawal, Sathish Govindarajan, and Neeldhara Misra

Indian Institute of Science, Bangalore, India
{akanksha.agrawal,gsat,neeldhara}@csa.iisc.ernet.in

Abstract. The problem of finding an optimal vertex cover in a graph is a classic NP-complete problem, and is a special case of the hitting set question. On the other hand, the hitting set problem, when asked in the context of induced geometric objects, often turns out to be exactly the vertex cover problem on restricted classes of graphs. In this work we explore a particular instance of such a phenomenon. We consider the problem of hitting all axis-parallel slabs induced by a point set P, and show that it is equivalent to the problem of finding a vertex cover on a graph whose edge set is the union of two Hamiltonian Paths. We show the latter problem to be NP-complete, and also give an algorithm to find a vertex cover of size at most k, on graphs of maximum degree four, whose running time is $1.2637^k n^{O(1)}$.

1 Introduction

Let P be a set of n points in $\mathbb{R}^2$ and let $\mathcal{R}$ be the family of all distinct objects of a particular kind (disks, rectangles, triangles, ...), such that each object in $\mathcal{R}$ has a distinct tuple of points from P on its boundary. For example, $\mathcal{R}$ could be the family of $\binom{n}{2}$ axis parallel rectangles such that each rectangle has a distinct pair of points of P as its diagonal corners. $\mathcal{R}$ is called the set of all objects induced (spanned) by P.

Various questions related to geometric objects induced by a point set have been studied in the last few decades. A classical result in discrete geometry is the *First Selection Lemma* [1] which shows that there exists a point that is present in a constant fraction of triangles induced by P. Another interesting question is to compute the minimum set of points in P that "hits" all the induced objects in $\mathcal{R}$. This is a special case of the classical Hitting Set problem, which we will refer to as *Hitting Set for Induced Objects*.

For most geometric objects, it is not known if the *Hitting Set for induced objects* problem is polynomial time solvable. It is known to be polynomially solvable for skyline rectangles (3-sided rectangles) and halfspaces [2]. Recently, Rajgopal et al [10] showed that this problem is NP-complete for lines.

The problem of finding an optimal vertex cover in a graph is a classic NP-complete problem, and is a special case of the Hitting Set problem. On the other hand, the hitting set for induced objects problem often turns out to be exactly

Z. Cai et al. (Eds.): COCOON 2014, LNCS 8591, pp. 179–190, 2014.

the vertex cover problem, even on restricted classes of graphs. For example, the problem of hitting set for induced axis-parallel rectangles is equivalent to the vertex cover on the Delaunay graph of the point set with respect to axis-parallel rectangles.

We study a particular phenomenon of this type, where the hitting set question in the geometric setting boils down to a vertex cover problem on a structured graph class. We consider the problem of hitting set for induced axis-parallel slabs (rectangles whose horizontal or vertical sides are unbounded). Note that this is even more structured than general axis-parallel rectangles, and indeed, it turns out that the corresponding Delaunay graph has a very special property — its edge set is the union of two Hamiltonian paths. Since any hitting set for the class of axis-parallel slabs induced by a point set P is exactly the vertex cover of the Delaunay graph with respect to axis-parallel slabs for P, our problem reduces to solving vertex cover on the class of graphs whose edge set is simply the union of two Hamiltonian Paths.

Despite the appealing structure, we show that – surprisingly – deciding k-vertex cover on this class of graphs is NP-complete.[1] This involves a rather intricate reduction from the problem of finding a vertex cover on cubic graphs. We also appeal to the fact that the edge set of four-regular graphs can be partitioned into two two-factors, and the main challenge in the reduction involves stitching the components of the two-factors into two Hamiltonian paths while preserving the size of the vertex cover in an appropriate manner.

Having established the NP-hardness of the problem, we pursue the question of improved fixed-parameter algorithms on this special case. Vertex Cover is one of the most well-studied problems in the context of fixed-parameter algorithm design, it enjoys a long list of improvements even on special graph classes. We note that for VERTEX COVER, the goal is to find a vertex cover of size at most k in time $\mathcal{O}(c^k)$, and the "race" involves exploring algorithms that reduce the value of the best known constant c.

In particular, even for sub-cubic graphs (where the maximum degree is at most three, and the problem remains NP-complete), Xiao [12] proposed an algorithm with running time $\mathcal{O}^\star(1.1616^k)$, improving on the previous best record of $\mathcal{O}^\star(1.1864^k)$ by Razgon [11] and $\mathcal{O}^\star(1.1940^k)$ by Chen, Kanj and Xia [6]. The best-known algorithm for Vertex Cover [5] on general graphs has a running time of $\mathcal{O}^\star(1.2738^k)$ and uses polynomial-space.

Typically, these algorithms involve extensive case analysis on a cleverly designed search tree. In the second part of this work[1], we propose a branching algorithm with running time $\mathcal{O}^\star(1.2637^k)$ for graphs with maximum degree bounded by at most four. This improves the best known algorithm for this class, which surprisingly has been no better than the algorithm for general graphs. We note that this implies faster algorithms for the case of graphs that can be decomposed into the union of two Hamiltonian Paths (since they have maximum degree at

---

[1] Due to space constraints, some proofs have been omitted. We refer the reader to http://arxiv.org/abs/1404.5566 for the full version of this paper.

most four), however, whether they admit additional structure that can be exploited for even better algorithms remains an open direction.

2 Preliminaries

In this section, we state some basic definitions and introduce terminology from graph theory and algorithms. We also establish some of the notation that will be used throughout.

We denote the set of natural numbers by $\mathbb{N}$ and set of real numbers by $\mathbb{R}$. For a natural number n, we use $[n]$ to denote the set $\{1, 2, \ldots, n\}$. For a finite set A we denote by $\mathfrak{S}_A$ the set of all permutations of the elements of set A. To describe the running times of our algorithms, we will use the $\mathcal{O}^*$ notation. Given $f : \mathbb{N} \to \mathbb{N}$, we define $\mathcal{O}^*(f(n))$ to be $\mathcal{O}(f(n) \cdot p(\cdot))$, where $p(\cdot)$ is some polynomial function. That is, the $\mathcal{O}^*$ notation suppresses polynomial factors in the running-time expression.

Graphs. In the following, let $G = (V, E)$ be a graph. For any non-empty subset $W \subseteq V$, the subgraph of G induced by W is denoted by $G[W]$; its vertex set is W and its edge set consists of all those edges of E with both endpoints in W. For $W \subseteq V$, by $G \setminus W$ we denote the graph obtained by deleting the vertices in W and all edges which are incident to at least one vertex in W.

A *vertex cover* is a subset of vertices S such that $G \setminus S$ has no edges. For $v \in V$ we denote the open-neighborhood of v by $N(v) = \{u \in V | (u, v) \in E\}$, closed-neighborhood of v by $N[v] = N(v) \cup \{v\}$, second-open neighborhood by $N_2(v) = \{u \in V | \exists u' \in N(v) \text{ s.t. } (u, u') \in E\}$ second-closed neighborhood by $N_2[v] = N_2(v) \cup N[v]$.

When we are discussing a pair of vertices u, v, then the common neighborhood of u and v is the set of vertices that are adjacent to both u and v. In this context, a vertex w is called a *private neighbor* of u if (w, u) is an edge and (w, v) is not an edge. We denote the degree of a vertex $v \in V$ by $d(v)$.

A *path* in a graph is a sequence of distinct vertices $v_0, v_1, \ldots, v_k$ such that (v_i, v_{i+1}) is an edge for all $0 \leq i \leq (k-1)$. A *Hamiltonian path* of a graph G is a path featuring every vertex of G. The following class of graphs will be of special interest to us.

Definition 1 (Braid graphs). *A graph G on the vertex set $[n]$ is a braid graph if the edges of the graph can be covered by two Hamiltonian paths. In other words, there exist permutations σ, τ of the vertex set for which $E(G) = \{(\sigma(i), \sigma(i + 1)) \mid 1 \leq i \leq n-1\} \cup \{(\tau(i), \tau(i + 1)) \mid 1 \leq i \leq n-1\}$.*

Induced axis-parallel slabs: Axis-parallel slabs are a special class of axis-parallel rectangles where two horizontal or two vertical sides are unbounded. Each pair of points $p(x_1, y_1)$ and $q(x_2, y_2)$ induces two axis-parallel slabs of the form $[x_1, x_2] \times (-\infty, +\infty)$ and $(-\infty, +\infty) \times [y_1, y_2]$. Let $\mathcal{R}$ represent the family of $2\binom{n}{2}$ axis-parallel slabs induced by P.

We refer the reader to [7] for details on standard graph theoretic notation and terminology we use in the paper.

Parameterized Complexity. A parameterized problem Π is a subset of $\Gamma^* \times \mathbb{N}$, where Γ is a finite alphabet. An instance of a parameterized problem is a tuple (x, k), where x is a classical problem instance, and k is called the parameter. A central notion in parameterized complexity is *fixed-parameter tractability (FPT)* which means, for a given instance (x, k), decidability in time $f(k) \cdot p(|x|)$, where f is an arbitrary function of k and p is a polynomial in the input size.

3   Hitting Set for Induced Axis-Parallel Slabs

We show here that the problem of finding a hitting set of size at most k for the family of all axis-parallel slabs induced by a point set is equivalent to the problem of finding a vertex cover of a graph whose edges can be partitioned into two Hamiltonian Paths. In subsequent sections, we establish the NP-hardness of the latter problem, and also provide better FPT algorithms. Due the equivalence of these problems, we note that both the hardness and the algorithmic results apply to the problem of finding a hitting set for induced axis parallel slabs.

Lemma 1. *Let $G = (V, E)$ be an instance of k-vertex cover, where G is a braid graph with $V = [n]$ and associated permutations σ, τ. Then we can compute, in polynomial time, an equivalent instance of hitting set for the collection of all axis-parallel slabs induced by a point set.*

Proof (Sketch). Given an instance of vertex cover on a braid graph G, with $V(G) = [n]$ and permutations σ and τ, we create n points in $\mathbb{R}^2$ in an $(n \times n)$-grid as follows. We assume, by renaming if necessary, that σ is the identity permutation. For every $1 \leq i \leq n$, we let $p_i = (i, \tau^{-1}(i))$. Since we only need to hit empty vertical and horizontal slabs, in the induced setting this amounts to hitting all consecutive slabs in the horizontal and vertical directions. It is easy to check that a hitting set for such slabs would exactly correspond to a vertex cover of G.   □

Lemma 2. *The problem of finding a hitting set for all induced axis-parallel slabs by a point set P can be reduced to the problem of finding a vertex cover in a braid graph.*

Proof. From the given point set P, we sort the points in P according to their x-coordinates to obtain a permutation of the point set σ. Similarly, we sort with respect to y-coordinate to get a permutation τ. Note that there exists a empty axis-parallel slab between two points if and only if they are adjacent with respect to at least one of the x- or y-coordinates, These are, on the other hand, precisely the edges in the braid graph with σ and τ as the permutations, which shows the equivalence.   □

4   NP-completenes of Vertex Cover on Braids

In this section, we show that the problem of determining a vertex cover on the class of braids is hard even when the permutations of the braid are given as input.

The intuition for the hardness is the following. Consider a four-regular graph. By a theorem of Peterson, we know that the edges of such a graph can be partitioned into two sets, each of which would be a two-factor in the graph G. In other words, every four-regular graph can be thought of as a union of two collections of disjoint cycles, defined on same vertex set. It is conceivable that these cycles can be patched together into paths, leading us to a braid graph. As it turns out, for such a patching, we need to have some control over the cycles in the decomposition to begin with. So we start with an instance of Vertex Cover on a cubic 2-connected planar graphs, morph such an instance to a four-regular graph while keeping track of a special cycle decomposition, which we later exploit for the "stitching" of cycles into Hamiltonian paths.

Formally, therefore, the proof is by a reduction from Vertex Cover on a cubic 2-connected planar graph to an instance of k-vertex cover on a braid graph, noting that [8] shows the NP-hardness of Vertex Cover for cubic planar 2-connected graphs. We describe the construction in two stages, first showing the transformation to a four-regular graph and then proceeding to illustrate the transformation to a braid graph.

Due to space constraints, we only provide the highlights of the reduction. One of the main tasks is to merge the cycles in each decomposition. Let us first illustrate a gadget that combines two cycles into a longer one.[2] Note that the gadget itself must be a braid, and of course, we need to ensure equivalence.

For the purpose of this brief discussion, our starting point is a four-regular graph G. Recall that the edge set of G can be decomposed into two collections of cycles. Note that every vertex v participates in two cycles, say C_v and C_v' — these would be cycles from different collections. Now let the neighbors of v in C_v be v_1, v_2, and let the neighbors in C_v' be v_3 and v_4.

We are now ready to describe the gadget W_v. This gadget has four entry points, namely v', v'', and a, b. The gadget is shown in Figure 1(b). It is easy to check that the gadget induces a braid. Now, in G, to insert this gadget, we remove v from G, and make v_1, v_3 adjacent to a and v_2, v_4 adjacent to b. Let us denote this graph by G'. Note that there is a path from v_1 to v_2 along the cycle C_v and there is a path from v_3 to v_4 along the cycle C_v'.

For equivalence, we need to be sure that if even one of v_1, v_2, v_3, v_4 is not picked in a vertex cover of G', then we have enough room for the vertex v in the reverse direction. To this end, we show the following crucial property of the gadget W_v.

Lemma 3. *Let S' be any vertex cover of G. If one of a or b belongs to S', then $|S' \cap V(W_v)| = 10$. On the other hand, there exists a vertex cover S' of G', that contains neither a nor b, for which $|S' \cap V(W_v)| = 9$.*

Proof. The vertices $\{v', x, x'\}$, $\{v'', y, y'\}$, $\{w_1, w_2, w_3\}$, $\{w_4, w_5, w_6\}$ form triangles, and (w, a) is an edge disjoint from these triangles. Therefore, we clearly

---

[2] At this point, we are not concerned that this is leading us to, eventually, a Hamiltonian cycle rather than a path, because it is quite easy to convert the former to the latter.

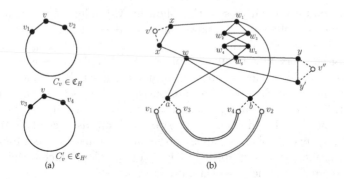

Fig. 1. Stitching together one pair of cycles with a common vertex v using W_v

require minimum of 9 vertices to cover edges of W_v alone. If we have $S' = \{x, x', y, y', w_1, w_2, w_4, w_6, w\}$ then we can cover all edges of W_v with 9 vertices. This proves the second part of the claim.

Now, let S' be a vertex cover such that $a \in S'$. Then apart from the K_3's present we have an edge (w, b), so including a we need at least 10 vertices. An analogous argument holds when b is present (edge (w, a) left). If we have $V' = \{x, x', y, y', w_1, w_2, w_4, w_6, a, b\}$ then we can cover all edges of W_v with 10 vertices. □

Corollary 1. *We have a vertex cover of size k in G if and only if G' admits a vertex cover of size $k + 9$.*

Proof. Let S be a vertex cover of G. If $v \notin S$, then $\{v_1, v_2, v_3, v_4\} \subseteq S$. Therefore, we may cover the edges of W_v using the vertices $\{x, x', y, y', w_1, w_2, w_4, w_6, w\}$ since there is no external obligation to pick either a or b, and this would be an extension of S with at most nine additional vertices. If $v \in S$, then let $S^\star := S \setminus \{v\}$. We extend $S^\star$ by the set $\{x, x', y, y', w_1, w_2, w_4, w_6, a, b\}$, which also adds up to $k+9$. In the reverse direction, given a vertex cover of size $k+9$, we know that at least nine vertices of S' are from W_v. Let $S^\dagger$ denote the remaining vertices of S'. If all of $\{v_1, v_2, v_3, v_4\} \in S'$, then note that $S^\star$ is a vertex cover of size k for G. If one of $\{v_1, v_2, v_3, v_4\} \notin S'$, then $S^\dagger \cup \{v\}$ is a vertex cover of size at most k in G. The size bound comes from Lemma 3 and the fact that either a or b belongs to S' due to the case we are considering.

We should be able to use this gadget repeatedly to stitch all cycles into two single long cycles. However, the iterative process involves several challenges. For instance, if some gadgets are already inserted, the paths along the cycles that we had before may not be so readily available. Also, while the process of breaking the cycle at a vertex v is clear, it is not obvious as to how one would mimic this construction for a neighbor of v after v has been suitably replaced. The concern here is that a straightforward application of the gadget will cause vertices from two different gadgets to become adjacent, which we would like to avoid if we are to maintain the braid structure of the gadget itself.

To address the former problem we create a slightly different gadget that creates artificial paths that can be used if the original cycle is broken by some previously inserted gadget. For the second problem, we start our reduction from cubic graphs and proceed in a manner so as to ensure that the cycle decompositions of the reduced graph are somewhat special. This allows us to choose mutually non-adjacent breakpoints v.

Finally, these gadgets must be tied together into a path, which we organize with the help of connection gadgets. The reader is referred to the full version of this work for the complete details.

Theorem 1. *The problem of finding a vertex cover of size at most k in a braid graph is* NP-complete.

5 An Improved Branching Algorithm

In this section we describe an improved FPT algorithm for the vertex cover problem on graphs with maximum degree at most four. The algorithm is essentially a search tree, and the analysis is based on the branch-and-bound technique. We use standard notation with regards to branching vectors as described in [9]. The input to the algorithm is denoted by a pair (G, k), where G is a graph, and the question is whether G admits a vertex cover of size at most k.

We work with k, the size of the vertex cover sought, as the measure — sometimes referred to as the *budget*. When we say that we *branch on a vertex v*, we mean that we recursively generate two instances, one where v belongs to the vertex cover, the other where v does not belong to the vertex cover. This is a standard method of exhaustive branching, where the measure drops, respectively, by one and $d(v)$ in the two branches (since the neighbors of v are forced to be in the vertex cover when v does not belong to the vertex cover).

Preprocessing. We begin by eliminating simplicial vertices, that is, vertices whose neighborhoods form a clique. If the graph induced by $N[v]$ is a clique, then it is easy to see that there is a minimum vertex cover containing $N(v)$ and not containing v (by a standard shifting argument). We therefore preprocess the graph in such a situation by deleting $N[v]$ and reducing the budget to $k - |N(v)|$.

Our algorithm makes extensive use of the *folding* technique, as described in past work [3, 4]. This allows us to preprocess vertices of degree two in polynomial time, while also reducing the size of the vertex cover sought by one. We briefly describe how we might handle degree-2 vertices in polynomial time. Suppose v is a degree-2 vertex in the graph G with two neighbors u and w such that u and w are not adjacent to each other. We construct a new graph G' as follows: remove the vertices v, u, and w and introduce a new vertex $v^\star$ that is adjacent to all neighbors of the vertices u and w in G (other than v). We say that the graph G' is obtained from the graph G by "folding" the vertex v, and we say that $v^\star$ is the vertex generated by folding v, or simply that $v^\star$ is the *folded vertex* (when the context is clear). It turns out that the folding operation preserves equivalence, as shown below.

Proposition 1. *[3, Lemma 2.3] Let G be a graph obtained by folding a degree-2 vertex v in a graph G, where the two neighbors of v are not adjacent to each other. Then the graph G has a vertex cover of size bounded by k if and only if the graph G' has a vertex cover of size bounded by (k − 1).*

Note that the new vertex generated by the folding operation can have more than four neighbors, especially if the vertices adjacent to the degree-2 vertex have, for example, degree four to begin with. The branching algorithm that we will propose assumes that we will always find a vertex whose degree is bounded by 3 to branch on, therefore it is important to avoid the situation where the graph obtained after folding all available degree two vertices is completely devoid of vertices of degree bounded by three (which is conceivable if all degree three vertices are adjacent to degree two vertices that in turn get affected by the folding operation). Therefore, we apply the folding operation somewhat tactfully– we apply it only when we are sure that the folded vertex has degree at most four. We call such a vertex a *foldable* vertex. Further, a vertex is said to be *easily foldable* if, after folding, it has degree at most 3. We avert the danger of leading ourselves to a four-regular graph recursively by explicitly ensuring that vertices of degree at most three are created whenever a folded vertex has degree four. Note that in the preprocessing step we will be folding only easily foldable vertices.

Typically, we ensure a reasonable drop on all branches by creating the following win-win situation: if a vertex is foldable, then we fold it, if it is not, then this is the case since there are sufficiently many neighbors in the second neighborhood of the vertex, and in many situations, this would lead to a good branching vector. Also, during the course of the branching, we appeal to a couple of simple facts about the structure of a vertex cover, which we state below.

Lemma 4. *[3, First part of Lemma 3.2] Let v be a vertex of degree 3 in a graph G. Then there is a minimum vertex cover of G that contains either all three neighbors of v or at most one neighbor of v.*

This follows from the fact that a vertex cover that contains v (where $d(v) = 3$) and two of its neighbors can be easily transformed into one, of the same size, that omits v and contains all of its neighbors.

Proposition 2. *If x, a, y, b form a cycle of length four in G (in that order), and the degree of a and b in G is two, then there exists an optimal vertex cover that does not pick a or b and contains both x and y.*

Overall Algorithm. To begin with, the branching algorithm tries to branch mainly on a vertex of degree three or two. If the input graph is four-regular, then we simply branch on an arbitrary vertex to create two instances both of which have at least one vertex of degree at most three. We note that this is an off-branching step, in the future, the algorithm maintains the invariant that at each step, the smaller graph produced has at least one vertex whose degree is at most three.

After this, we remove all the simplicial vertices and then fold all easily-foldable vertices. If a degree two vertex v with neighbors u and w is not easily-foldable, then note that there exists an optimal vertex cover that either contains v or does not contain v and includes both its neighbors. Indeed, if an optimal vertex cover S contains, say v and u, then note that $(S \setminus \{v\}) \cup \{w\}$ is a vertex cover of the same size. So we branch on the vertex v:

- when v does not belong to the vertex cover, we pick u, w in the vertex cover, leading to a drop of two in the measure,
- when v does belong to the vertex cover, we have that $N(u) \cup N(w)$ must belong to the vertex cover, and we know that $|N(u) \cup N(w) \setminus \{v\}| \geq 4$ (otherwise, v would be easily-foldable), and this leads to a drop of five in the measure.

So we either preprocess degree two vertices in polynomial time, or branch on them with a branching vector of $(2, 5)$. At the leaves of this branching tree, if we have a sub-cubic graph, then we employ the algorithm of [12]. Otherwise, we have at least one degree three vertex which is adjacent to at least one degree four vertex. We branch on these vertices next. The case analysis is based on the neighborhood of the vertex — broadly, we distinguish between when the neighborhood has at least one edge, and when it has no edges. The latter case is the most demanding in terms of a case analysis. For the rest of this section, we describe all the scenarios that arise in this context.

Degree three vertices with edges in their neighborhood. For this part of the algorithm, we can always assume that we are given a degree three vertex with a degree four neighbor. Let v be a degree three vertex, and let $N(v) := \{u, w, x\}$, where we let u denote a degree four vertex. Note that u, w, x does not form a triangle, otherwise v would be a simplicial vertex and we would have handled it earlier. So, we deal with the case when $N(v)$ is not a triangle, but has at least one edge. If (w, x) is an edge, then we branch on u:

- when u does not belong to the vertex cover, we pick four of its neighbors in the vertex cover, leading to a drop of four in the measure,
- when u does belong to the vertex cover, we delete u from the graph, and we are left with v, w, x being a triangle where v is a degree two vertex, and therefore we may pick w, x in the vertex cover — together, this leads to a drop of three in the measure.

On the other hand, if (w, x) is not an edge, then there is an edge incident to u. Suppose the edge is (u, w) (the case when the edge is (u, x) is symmetric). In this case, we branch on x exactly as above. The measure may drop by three when x does not belong to the vertex cover, if x happens to be a degree three vertex. Therefore, our worst-case branching vector in the situation when $N(v)$ is not a triangle, but has at least one edge is $(3, 3)$.

Degree three vertices whose neighborhoods are independent. Here we consider several cases. Broadly, we have two situations based on whether u, w, x have any common neighbors or not.

Before embarking on the case analysis, we describe a branching strategy for some specific situations — these mostly involve two non-adjacent vertices that have more than two neighbhors in common, with at least one of them of degree 4. This will be useful in scenarios that arise later.

We consider the case when a degree four vertex p non-adjacent to a vertex q has at least three neighbhors in commmon, say a, b, c and let x be the other neighbhor of p that may or may not be adjacent to q. Notice that there always exists an optimal vertex cover that either contains both p and q or omits both p and q. To see this, consider an optimal vertex cover S that contains p and omits q. Then, S clearly contains a, b, c. Notice now that $T := (S \setminus \{p\}) \cup \{x\}$ is also a vertex cover, and T contains neither p or q, and has the same size as S. This suggests the following branching strategy:

1. If p and q both belong to the vertex cover, then the measure clearly drops by two. We proceed by deleting p and q from G. Now note that the degree of the vertices $\{a, b, c\}$ reduces by two, and they become vertices of degree one or two (note that they cannot be isolated because we always begin by eliminating vertices of degree two by preprocessing or branching). If any one of these vertices is simplicial or foldable then we process it or fold it respectively. Otherwise, we branch on a:
   (a) when a does not belong to the vertex cover, we pick its neighbhors in the vertex cover, leading to a drop of two in the measure.
   (b) when a does belong to the vertex cover, we have that its second neighborhood must belong to the vertex cover, and this leads to a drop of six in the measure.
2. If p and q are both omitted from the vertex cover, then we pick a, b, c, x in the vertex cover and the measure drops by four.

Note that if a is foldable in $G \setminus \{p, q\}$, then we have the branch vector $(3, 4)$, otherwise, we have the branch vector $(4, 8, 4)$. We refer to the branching strategies outlined above as the **CommonNeighborBranch** strategy.

A broad overview of all the other cases is as follows.

1. **Scenario A.** There exists a vertex t that is adjacent to at least two vertices in $N(v)$. Further, t is adjacent to u and one other vertex.
   – The vertex t has degree four.
   – The vertex t has degree three, u, w, x have degree four, and $(t, x) \notin E$. We let u' and u'' denote the neighbors of u other than t and v.
     • The degree of both u' and u'' is four.
     • At least one of u' and u'' has degree three.
2. **Scenario B.** There exists a vertex t that is adjacent to at least two vertices in $N(v)$. The vertex t is not adjacent to u and is therefore adjacent to w and x.

3. **Scenario C.** The vertices u, v, w have no common neighbors other than v. We have the following cases based on degree of w, x.
   - The degree of both w and x is three.
   - The degree of both w and x is four.
   - The degree of w is four and x is three.

Theorem 2. *There is an algorithm that determines if a graph with maximum degree at most four has a vertex cover of size at most k in $\mathcal{O}^*(1.2637^k)$ worst-case running time.*

Scenario	Cases	Branch Vector	c
Scenario A	Case 1	$(2,5)$	1.2365
		$(7,4,5)$	1.2365
		$(7,9,5,5)$	1.2498
		$(2,10,6)$	1.2530
		$(7,4,10,6)$	1.2475
		$(7,9,5,10,6)$	1.2575
	Case 2 (I)	$(4,7,5)$	1.2365
		$(9,5,7,5)$	1.2498
		$(4,7,10,6)$	1.2475
		$(9,5,7,10,6)$	1.2575
	Case 2 (II)	$(4,5,6)$	1.2498
		$(4,10,6,6)$	1.2590

Scenario	Cases	Branch Vector	c
CNB		$(2,5)$	1.2365
		$(3,4)$	1.2207
		$(4,8,4)$	1.2465
Degree Two Edge in $N(v)$		$(2,6)$	1.2365
		$(3,3)$	1.2599
Scenario B		$(2,5)$	1.2365
		$(2,6,10)$	1.2530
Scenario C	Case 1	$(2,10,6)$	1.2530
	Case 2	$(8,3,8,7)$	1.2631
	Case 3	$(7,3,5)$	1.2637
		$(5,7,7,6)$	1.2519
		$(10,6,7,7,6)$	1.2592

Fig. 2. The branch vectors and the corresponding running times across various scenarios and cases. (This table is a truncated version due to lack of space.)

6   Conclusions

In this work we showed that the problem of hitting all axis-parallel slabs induced by a point set P is equivalent to the problem of finding a vertex cover on a graph whose edge set is the union of two Hamiltonian Paths. We established that this problem is NP-complete. Finally, we also gave an algorithm for Vertex Cover on graphs of maximum degree four whose running time is $\mathcal{O}^*(1.2637^k)$. It would be interesting to know if there are better algorithms for braid graphs in particular.

References

[1] Boros, E., Füredi, Z.: The number of triangles covering the center of an n-set. Geometriae Dedicata 17, 69–77 (1984)
[2] Chan, T.M., Grant, E.: Exact algorithms and APX-hardness results for geometric packing and covering problems. Comput. Geom. 47(2), 112–124 (2014)
[3] Chen, J., Kanj, I.A., Jia, W.: Vertex Cover: Further Observations and Further Improvements. In: Widmayer, P., Neyer, G., Eidenbenz, S. (eds.) WG 1999. LNCS, vol. 1665, pp. 313–324. Springer, Heidelberg (1999)
[4] Chen, J., Kanj, I.A., Xia, G.: Improved Parameterized Upper Bounds for Vertex Cover. In: Královič, R., Urzyczyn, P. (eds.) MFCS 2006. LNCS, vol. 4162, pp. 238–249. Springer, Heidelberg (2006)

[5] Chen, J., Kanj, I.A., Xia, G.: Improved upper bounds for vertex cover. Theor. Comput. Sci. 411(40-42), 3736–3756 (2010)

[6] Chen, J., Kanj, I.A., Xia, G.: Labeled Search Trees and Amortized Analysis: Improved Upper Bounds for NP-Hard Problems. Algorithmica 43(4), 245–273 (2005)

[7] Diestel, R.: Graph Theory, 3rd edn. Springer, Heidelberg (2005)

[8] Mohar, B.: Face Covers and the Genus Problem for Apex Graphs. Journal of Combinatorial Theory, Series B 82(1), 102–117 (2001)

[9] Niedermeier, R.: Invitation to Fixed Parameter Algorithms (Oxford Lecture Series in Mathematics and Its Applications). Oxford University Press, USA (2006)

[10] Rajgopal, N., Ashok, P., Govindarajan, S., Khopkar, A., Misra, N.: Hitting and Piercing Rectangles Induced by a Point Set. In: Du, D.-Z., Zhang, G. (eds.) COCOON 2013. LNCS, vol. 7936, pp. 221–232. Springer, Heidelberg (2013)

[11] Razgon, I.: Faster computation of maximum independent set and parameterized vertex cover for graphs with maximum degree 3. J. Discrete Algorithms 7(2), 191–212 (2009)

[12] Xiao, M.: A Note on Vertex Cover in Graphs with Maximum Degree 3. In: Thai, M.T., Sahni, S. (eds.) COCOON 2010. LNCS, vol. 6196, pp. 150–159. Springer, Heidelberg (2010)

A Protocol for Generating Random Elements with Their Probabilities

Thomas Holenstein and Robin Künzler

ETH Zurich, Department of Computer Science, 8092 Zurich, Switzerland
{thomas.holenstein,robink}@inf.ethz.ch

Abstract. We give an AM protocol that allows the verifier to sample elements x from a probability distribution P, which is held by the prover. If the prover is honest, the verifier outputs $(x, P(x))$ with probability close to $P(x)$.

In case the prover is dishonest, one may hope for the following guarantee: if the verifier outputs (x, p), then the probability that the verifier outputs x is close to p. Simple examples show that this cannot be achieved. Instead, we show that the following weaker condition holds (in a well defined sense) *on average*: If (x, p) is output, then p is an upper bound on the probability that x is output.

Our protocol yields a new transformation to turn interactive proofs where the verifier uses private random coins into proofs with public coins. The verifier has better running time compared to the well-known Goldwasser-Sipser transformation (STOC, 1986). For constant-round protocols, we only lose an arbitrarily small constant in soundness and completeness, while our public-coin verifier calls the private-coin verifier only once.

1 Introduction

In an interactive proof [GMR89, Bab85, BM88], an all-powerful prover tries to convince a computationally bounded verifier that some statement is true. The study of such proofs has a rich history, and has lead to numerous important and surprising results.

We are interested in interactive protocols that allow the verifier to sample elements from a probability distribution. Such protocols have proved to be very useful, and their applications include the study of private versus public coins in interactive proof systems [GS86], perfect zero knowledge [For87], basing average-case hardness or cryptographic security on worst-case hardness [FF93, BT06, AGGM06, HMX10], and many more.

We consider constant-round protocols that allow the verifier to sample an element x from a probability distribution P together with an approximation p of the probability $P(x)$. The verifier outputs pairs (x, p), and for a fixed prover we let (X, P) be the random variables corresponding to the verifier's output, P_{XP} is their joint distribution, and P_X is the marginal distribution defined as $P_X(x) = \sum_p P_{XP}(x, p)$. We would like to achieve the following properties:

Z. Cai et al. (Eds.): COCOON 2014, LNCS 8591, pp. 191–202, 2014.
© Springer International Publishing Switzerland 2014

Property 1: For every x we have $\mathsf{P}_X(x) \approx \mathsf{P}(x)$.

Property 2: If the verifier outputs (x, p), then $p \approx \mathsf{P}_X(x)$.

Recently, Haitner et al. [HMX10] gave such a protocol for sampling a distribution P on bit strings, which is given as $\mathsf{P} = f(\mathsf{P}_U)$ for an efficiently computable function f, where P_U is the uniform distribution on n-bit strings. For *any* (possibly cheating) prover their protocol achieves property 1 with equality (i.e. $\mathsf{P}_X(x) = \mathsf{P}(x)$), and property 2 for polynomially small error (i.e. $p = (1 \pm \varepsilon)\mathsf{P}(x)$ for ε polynomially small in n). The protocol extends to distributions $f(\mathsf{P}_S)$, where P_S is the uniform distribution on an efficiently decidable set $\mathcal{S} \subseteq \{0,1\}^n$.

1.1 Contributions of this Paper

We give a sampling protocol that is similar to the one of [HMX10]. However, in our protocol only the prover gets as input the distribution P. This distribution can be arbitrary, and in particular does not have to be efficiently samplable. The verifier does *not* get P as input, and in particular does not have the ability to sample from P.

We obtain the following completeness guarantee:

Completeness: If the prover is honest, then both properties 1 and 2 are satisfied.
More precisely, property 1 is satisfied with polynomially small error (i.e. $\mathsf{P}_X(x) = (1 \pm \varepsilon)\mathsf{P}(x)$ for polynomially small ε), and instead of property 2 we even guarantee that the verifier only outputs pairs $(x, \mathsf{P}(x))$.

In case the prover is not honest, since the verifier does not know P, we cannot hope to satisfy property 1. However, one could hope that property 2 is satisfied for any (possibly cheating) prover. Unfortunately, simple examples show that this cannot be achieved. Instead, we prove the following weaker guarantee:

Soundness: The following condition holds (in a well defined sense) *on average*: If (x, p) is output, then $\mathsf{P}_X(x) \leq p$.

To illustrate the usefulness of our protocol, we apply it to obtain a private-coin to public-coin transformation for interactive proofs in Section 3. Compared to the original transformation by Goldwasser and Sipser [GS86], our transformation is more efficient in terms of the verifier's running time. In particular, for constant-round protocols we only lose an arbitrarily small constant in soundness and completeness, while executing the private-coin verifier exactly once. We show that this transformation can be viewed as an interactive sampling process, where the public-coin verifier iteratively samples messages for the private-coin verifier and requests the corresponding answers from the prover.

We remark that the soundness guarantee of our protocol is weaker than the one of the [HMX10] protocol. However, as mentioned above, the guarantee of [HMX10] cannot be achieved in the more general setting we consider.

Furthermore, to achieve a private-coin to public-coin transformation, it is also possible to employ the [HMX10] sampling protocol. However, this yields a less efficient public-coin verifier, as the private coin-verifier must be executed many times when running the sampling protocol.

1.2   Related Work

Interactive protocols. Interactive proof systems were introduced by Goldwasser et al. [GMR89]. Independently, Babai and Moran [Bab85, BM88] defined the public-coin version. As mentioned above, Goldwasser and Sipser [GS86] showed that the two definitions are equivalent with respect to language recognition (for a nice exposition of the proof, we also refer to the book of Goldreich [Gol08]). The study of interactive proofs has a long and rich history, influencing the study of zero knowledge and probabilistic checkable proofs, and has lead to numerous important and surprising results such as IP = PSPACE [LFKN92, Sha92]. For historical overviews we refer for example to [Bab90, Gol08, AB09].

Interactive sampling protocols. Goldwasser and Sipser [GS86] show that private coins in interactive proofs can be made public. A constant-round set lower bound protocol is introduced which can be viewed as a sampling process: the verifier uses pairwise independent hashing to randomly select a few elements in a large set. This protocol is used in many subsequent works such as [For87, AH91, GVW02, BT06, AGGM06, HMX10]. To study the complexity of perfect zero knowledge, Fortnow [For87] introduces a constant-round protocol that allows to prove set upper bounds assuming that the verifier is given a uniform random element in the set, which is not known to the prover. The same protocol is used in a similar context by Aiello and Håstad [AH91]. As in the lower bound protocol, hashing is used to sample a small number of elements from the set. To prove that any interactive proof system can have perfect completeness, Goldreich et al. [GMS87] give a protocol that allows to sample a perfectly uniform random element from a decidable set. Their protocol requires a polynomial number of rounds (depending on the set size), and they show that no constant-round protocol can achieve this task. The upper and lower bound protocols of [GS86, For87, AH91] are used by Bogdanov and Trevisan [BT06] in their proof that the worst-case hardness of an NP-complete problem cannot be used to show the average-case hardness of an NP problem via non-adaptive reductions, unless the polynomial hierarchy collapses. The ideas used in the set lower and upper bound protocols can be employed to sample a single element from an NP set in case the verifier knows the set size. This is done in [GVW02] in the context of studying interactive proofs with bounded communication. This protocol is refined by Akavia et al. [AGGM06], where it is used to give protocols for proving the size of a set (both upper and lower bound) under the assumption that the verifier knows some approximate statistics about the size of the set. These protocols are then used to study the question whether one-way functions can be based on NP-hardness. The ideas behing the sampling protocol by Akavia et al. are refined and extended by Haitner et al. [HMX10] in order to give a protocol that allows to sample an element $x \leftarrow \mathsf{P}$ of a given distribution, which is specified as $\mathsf{P} = f(\mathsf{P}_U)$ for an efficiently computable function f, where P_U is the uniform distribution on $\{0,1\}^n$. The verifier outputs (x,p), where x is sampled from P, and $p = (1 \pm \varepsilon)\mathsf{P}(x)$ for polynomially small ε. Their protocol extends to distributions $f(\mathsf{P}_S)$, where P_S is the uniform distribution on an efficiently decidable set $S \subseteq \{0,1\}^n$. It is shown that this sampling protocol can be employed to allow the verifier to verify the shape of the distribution P, in terms of its histogram. These protocols are used to prove that a number of cryptographic primitives, such as statistically hiding commitment, cannot be based on NP-hardness via certain classes of reductions (unless the polynomial hierarchy collapses).

2   The Sampling Protocol

2.1   Informal Theorem Statement and Discussion

In our sampling protocol, the verifier will output pairs $(x,p) \in \{0,1\}^n \times (0,1]$. For a fixed prover, we let (X, P) be the random variables corresponding to the verifier's output, and we denote their joint distribution by P_{XP}. Also, P_X is defined by $\mathsf{P}_X(x) = \sum_p \mathsf{P}_{XP}(x,p)$. Informally, our sampling theorem can be stated as follows.

Theorem 1 (The Sampling Protocol, informal). *There exists a constant-round public-coin interactive protocol such that the following holds. The verifier and the prover take as input $n \in \mathbb{N}, \varepsilon, \delta \in (0,1)$, and the prover additionally gets as input a probability distribution P over $\{0,1\}^n$. The verifier runs in time* poly $\left(n \left(\frac{1}{\varepsilon} \right)^{1/\delta} \right)$ *and we have:*

Completeness: *If the prover is honest, then the verifier outputs $(x, \mathsf{P}(x))$ with probability $(1 \pm \varepsilon)\mathsf{P}(x)$.*

Soundness: *Fix any (possibly cheating) prover. Then for all $x \in \{0,1\}^n$ we have $\sum_p \frac{\mathsf{P}_{XP}(x,p)}{p} \leq 1 + \varepsilon + \delta.$*

The soundness condition may not be very intuitive at first sight. We therefore discuss it in detail below.

To keep the discussion simple, the above theorem statement is only almost true: in fact, the completeness only holds with probability $1 - \varepsilon$ over the choice of x from P, and the soundness condition only holds if we condition on good protocol executions (where an execution is bad with probability at most ε). We omit this here, and refer to the full version of this paper for the exact statement.

We will prove soundness only for deterministic provers, but the same statement holds in case the prover is probabilistic: this follows easily by conditioning on the prover's random choices, and applying the result for the deterministic prover.[1]

---

[1]   We remark that typical definitions of interactive protocols does not allow the prover to use randomness. This is because when considering decision problems, the prover

In the sampling protocol of [HMX10] (as described above) the verifier gets access to the distribution P, in the sense that a circuit computing a function f is provided as input to the protocol, where $P = f(P_U)$ and P_U is the uniform distribution. During the execution of the sampling protocol, the function f needs to be evaluated many times. In contrast, in our protocol the verifier does not get access to the distribution, and never needs to evaluate such a circuit.

We note that the verifier runs in polynomial time for any polynomially small ε and constant δ. It is an interesting open problem if it is possible to improve the protocol to allow both polynomially small δ and ε and an efficient verifier.

Discussion of the soundness condition. The main motivation for the soundness condition is that it is actually sufficient for applying the sampling protocol to obtain the private-coin to public-coin transformation in Section 3.

However, there are several remarks we would like to discuss. We give an overview of these remarks, and discuss them in detail in the full version of the paper.

- **Remark 1:** Fix any (possibly cheating) prover and recall property 2 as described in the introduction: If the verifier outputs (x, p), then $p \approx P_X(x)$. One may hope to give a protocol that satisfies this property. However, we show below that this cannot be achieved in our setting.
- **Remark 2:** It is possible to interpret the soundness condition as follows: The situation $P_X(x) \gg p$ cannot occur too often. In that sense, the protocol provides an upper bound on $P_X(x)$ "on average".
- **Remark 3:** Assume the completeness condition is satisfied. Then the soundness condition actually holds if the prover behaves like a convex combination of honest provers, i.e. it first chooses a distribution P from a set of distributions, and then behaves honestly for P.
- **Remark 4:** It is natural to ask if the converse of the statement in remark 3 is also true. That is, we ask the following. Suppose we have some protocol that satisfies both our completeness and soundness conditions. Fix any prover and consider the verifier's output distribution P_{XP}. Is there a convex combination of probability distributions, such that if the prover first chooses P and then behaves honestly for P, the verifier's output distribution equals P_{XP}? We show below that this is not the case in general.

Proving the soundness condition that any cheating prover can be seen as a convex combination of honest provers would imply that the protocol is optimal, since for any protocol a probabilistic prover actually *can* first choose some distribution and then behave honestly for it. Remark 4 implies that our soundness condition does not imply this in general.

---

can always be assumed to be deterministic. However, since our theorem does not consider a decision problem, and the sampling protocol might be used as a subprotocol of some other protocol, we do not assume that the prover is deterministic.

2.2 Technical Overview

We informally describe a simplified version of the sampling protocol of Theorem 1, and sketch how correctness and soundness can be proved. Then we discuss how to get rid of the simplifying assumptions.

Histograms. Let P be a distribution over $\{0,1\}^n$, and consider the $(1,n)$-histogram of P, which is defined as the vector $h = (h_0, \ldots, h_n)$, where $h_i := \Pr_{y \leftarrow \mathsf{P}}[y \in \mathcal{B}_i]$ for $\mathcal{B}_i := \{x : \mathsf{P}(x) \in (2^{-(i+1)}, 2^{-i}]\}$. For simplicity we assume that for all x either $\mathsf{P}(x) = 0$ or $\mathsf{P}(x) \geq 2^{-n}$, which implies that $\sum_{i \in (n)} h_i = 1$.

A simplified sampling protocol with an inefficient verifier. If efficiency were not an issue, the honest prover could just send all pairs $(x, \mathsf{P}(x))$ to the verifier, who then outputs $(x, \mathsf{P}(x))$ with probability $\mathsf{P}(x)$. It is clear that this protocol achieves even stronger completeness and soundness guarantees than stated in our theorem. We now change this protocol, still leaving the verifier inefficient. But, using hashing, the verifier of this modified protocol can later be made efficient.

To describe the intuition, we make the following simplifications: we let the verifier output probabilities of the form 2^{-j}, and when interacting with the honest prover, the verifier will output pairs (x, p) where p is a 2-approximation of $\mathsf{P}(x)$. We will also make an assumption on P, but it is the easiest to state it while describing the protocol.

The protocol. The honest prover sends the histogram h of P to the verifier. The verifier splits the interval $[0, n]$ into intervals $\mathcal{J}_j$ of length $\log_2(n)$. We denote by $\mathcal{I}_i := \mathcal{J}_{2i}$ the even intervals, and we will call the odd intervals *gaps*. For simplicity we assume $\log_2(n)$ is an even integer, and that $\sum_k \sum_{i \in \mathcal{I}_k} h_i = 1$, i.e. that P is such that h has no probability mass in the gaps. Now the verifier selects an interval $\mathcal{I}_k$ at random, where the probability of $\mathcal{I}_k$ corresponds to its weight according to h, i.e. the probability of $\mathcal{I}_k$ is $w_k := \sum_{j \in \mathcal{I}_k} h_j$. The prover sends sets $\mathcal{X}_i$ for $i \in \mathcal{I}_k$ to the verifier, where the honest prover lets $\mathcal{X}_i = \mathcal{B}_i$. The verifier checks that the $\mathcal{X}_i$ are disjoint, and that $|\mathcal{X}_i| \in 2^{\pm 1} h_i 2^i$.[2] It then randomly chooses one of the sets $\mathcal{X}_j$, where $\mathcal{X}_j$ has probability $\frac{h_j}{\sum_{i \in \mathcal{I}_k} h_i}$. Finally, the verifier chooses a uniform random element x from $\mathcal{X}_j$, and outputs $(x, 2^{-j})$.

Completeness: It is not hard to see that if the prover is honest, then for any x and j the following holds. If $x \in \mathcal{B}_j$, then $\mathsf{P}_{XP}(x, 2^{-j}) \in 2^{\pm 1}\mathsf{P}(x)$. Otherwise, $\mathsf{P}_{XP}(x, 2^{-j}) = 0$.

Soundness: We sketch a proof for the following soundness guarantee: for any (possibly dishonest) prover there is an event Bad such that $\Pr[\mathsf{Bad}] \leq 50/\sqrt{n}$, and for all x we have $\sum_p \frac{\Pr[(X,P)=(x,p)|\neg\mathsf{Bad}]}{p} \leq 2 + 50/\sqrt{n}$.

---

[2] Actually, the verifier can check the stronger condition $|\mathcal{X}_i| \in [1/2, 1]h_i 2^i$, but to simplify our statements we use $2^{\pm 1}$.

We let $k(j)$ be the function that outputs the interval of j (i.e. k such that $j \in \mathcal{I}_k$). We first observe that[3]

$$P_{XP}(x, 2^{-j}) \in 2^{\pm 1} \cdot \Pr[x \in \mathcal{X}_j | k = k(j)] \cdot 2^{-j}, \tag{1}$$

where $\Pr[x \in \mathcal{X}_j | k = k(j)]$ is the probability that the prover puts x into $\mathcal{X}_j$ given that $k(j)$ was chosen by the verifier. To see this, note that $P_{XP}(x, 2^{-j}) = \Pr[x \in \mathcal{X}_j | k = k(j)] \cdot w_k \cdot \frac{h_j}{\sum_{i \in \mathcal{I}_k} h_i} \cdot \frac{1}{|\mathcal{X}_j|}$, where w_k is the probability of choosing k, the third factor is the probability of choosing j given k was chosen, and $1/|\mathcal{X}_j|$ is the probability of choosing x, given k, j were chosen and x is in $\mathcal{X}_j$. Indeed, by the definition of w_k and using the verifier's check $|\mathcal{X}_j| \in 2^{\pm 1} h_j 2^j$, this implies (1).

To prove our soundness claims, we consider the following sets representing small, medium and large probabilities, respectively.

$$\mathcal{S}(x) = \left\{ j : 2^{-j} \leq \frac{\mathsf{P}_X(x)}{\sqrt{n}} \right\} = \{ j : j \geq \log_2(1/\mathsf{P}_X(x)) + 1/2\log_2(n) \},$$

$$\mathcal{M}(x) = \left\{ j : 2^{-j} > \frac{\mathsf{P}_X(x)}{\sqrt{n}} \wedge 2^{-j} < \sqrt{n}\mathsf{P}_X(x) \right\}$$
$$= \{ j : \log_2(1/\mathsf{P}_X(x)) - 1/2\log_2(n) < j < \log_2(1/\mathsf{P}_X(x)) + 1/2\log_2(n) \},$$

$$\mathcal{L}(x) = \{ j : 2^{-j} \geq \sqrt{n}\mathsf{P}_X(x) \} = \{ j : j \leq \log_2(1/\mathsf{P}_X(x)) - 1/2\log_2(n) \}.$$

We define the event Bad to occur if the verifier outputs a probability that is much too small, i.e. it outputs some $(x, 2^{-j})$ where $j \in \mathcal{S}(x)$. We find

$$\Pr[\mathsf{Bad}] = \sum_x \sum_{j \in \mathcal{S}(x)} P_{XP}(x, 2^{-j}) \overset{(1)}{\leq} \sum_x \sum_{j \in \mathcal{S}(x)} 2 \cdot \Pr[x \in \mathcal{X}_j | k = k(j)] \cdot 2^{-j}$$

$$\leq 2 \sum_x \sum_{j \in \mathcal{S}(x)} 2^{-j} \leq 2 \sum_x \frac{\mathsf{P}_X(x)}{\sqrt{n}} \sum_{i=0}^{\infty} \frac{1}{2^i} \leq \frac{4}{\sqrt{n}},$$

where the third inequality follows by definition of $\mathcal{S}$.

To prove the second soundness claim, we find $\sum_p \frac{\Pr[(X,P)=(x,p) \wedge \neg \mathsf{Bad}]}{p} = \sum_{j \in \mathcal{M}(x)} \frac{P_{XP}(x, 2^{-j})}{2^{-j}} + \sum_{j \in \mathcal{L}(x)} \frac{P_{XP}(x, 2^{-j})}{2^{-j}}$. Now

$$\sum_{j \in \mathcal{L}(x)} \frac{P_{XP}(x, 2^{-j})}{2^{-j}} \leq \frac{1}{\sqrt{n}\mathsf{P}_X(x)} \underbrace{\sum_{j \in \mathcal{L}(x)} P_{XP}(x, 2^{-j})}_{\leq \mathsf{P}_X(x)} \leq \frac{1}{\sqrt{n}},$$

$$\sum_{j \in \mathcal{M}(x)} \frac{P_{XP}(x, 2^{-j})}{2^{-j}} \overset{(1)}{\leq} 2 \sum_{j \in \mathcal{M}(x)} \Pr[x \in \mathcal{X}_j | k = k(j)] \leq 2,$$

---

[3] In Eqn. (1) we implicitly assume that the prover always sends disjoint sets $\mathcal{X}_j$ that are of appropriate size. Removing this assumption is a minor technicality that is dealt with in the full proof.

where the last inequality again follows since the prover must send disjoint sets $\mathcal{X}_i$, and by definition of $\mathcal{M}(x)$ we have $|\mathcal{M}(x)| < \log_2(n)$ and thus $\mathcal{M}(x) \cap \mathcal{I}_k$ is non-empty for at most one $\mathcal{I}_k$. (This is one reason for defining the intervals, gaps and the sets $\mathcal{S}, \mathcal{M}, \mathcal{L}$ as we did!) Thus we have $\sum_p \frac{\Pr[(X,P)=(x,p)|\neg\text{Bad}]}{p} = \frac{1}{\Pr[\neg\text{Bad}]} \sum_p \frac{\Pr[(X,P)=(x,p)\wedge\neg\text{Bad}]}{p} \leq (1+8/\sqrt{n})(2+1/\sqrt{n}) \leq 2 + 17/\sqrt{n} + 8/n.$

We refine this protocol to obtain our main result. In particular, we make the verifier efficient, use a more accurate histogram, handle general distributions and finally make the protocol output exact probabilities instead of 2-approximations. A technical overview of these refinements can be found in the full version of this paper.

Due to space constrains, the formal theorem statement, the protocol, and the analysis can all be found in the full version of this paper.

3   Private Coins versus Public Coins in Interactive Proofs

3.1   Theorem Statement and Discussion

We state our theorem for the case where the public-coin verifier calls the private-coin verifier exactly once. This highlights the overhead in the verifier's running time, and allows to compare our transformation to the one of [GS86] in a natural way. We remark that before applying the transformation as given in the theorem, one can repeat the private-coin protocol in parallel to amplify completeness and soundness. However, this requires that the original private-coin verifier is called several times. Our main result can be stated as follows:

Theorem 2. *For any functions* $c, s, \varepsilon, \delta : \mathbb{N} \to (0,1)$, *and* $k, t, m, \ell : \mathbb{N} \to \mathbb{N}$, *if* $1/\varepsilon, 1/\delta, k, t, m, \ell$ *are time-constructible, then*

$$\text{IP} \begin{pmatrix} \textit{rounds} & = k \\ \textit{time} & = t \\ \textit{msg size} & = m \\ \textit{coins} & = \ell \\ \textit{compl} & \geq c \\ \textit{sound} & \leq s \end{pmatrix} \subseteq \text{AM} \begin{pmatrix} \textit{rounds} & = 4k+3 \\ \textit{time} & = t + k \cdot \text{poly}\left((m+\ell) \cdot \left(\frac{1}{\varepsilon}\right)^{1/\delta}\right) \\ \textit{msg size} & = \text{poly}\left((m+\ell) \cdot \left(\frac{1}{\varepsilon}\right)^{1/\delta}\right) \\ \textit{coins} & = \text{poly}\left((m+\ell) \cdot \left(\frac{1}{\varepsilon}\right)^{1/\delta}\right) \\ \textit{compl} & \geq c - 2(k+1)\varepsilon \\ \textit{sound} & \leq (1+\varepsilon+\delta)^{k+1}s + (k+1)\varepsilon \end{pmatrix}$$

Moreover, in the protocol that achieves this transformation, the public-coin verifier calls the private-coin verifier exactly once.

We state the following corollary that shows two interesting special cases for specific parameter choices.

Corollary 1. *The following inclusions hold:*

(i) *For any functions* $t, m, \ell : \mathbb{N} \to \mathbb{N}$, *polynomial* $k(n)$ *and inverse polynomial* $\gamma(n) < 1/5$, *if* t, m, ℓ, k *and* $1/\gamma$ *are time-constructible, we have*

$$\text{IP}\begin{pmatrix} \text{rounds} & = k \\ \text{time} & = t \\ \text{msg size} = m \\ \text{coins} & = \ell \\ \text{compl} & \geq 2/3 + \gamma \\ \text{sound} & \leq 2^{-(k+5)} \end{pmatrix} \subseteq \text{AM}\begin{pmatrix} \text{rounds} & = 4k+3 \\ \text{time} & = t + \text{poly}((m+\ell)\frac{k}{\gamma}) \\ \text{msg size} = \text{poly}((m+\ell)\frac{k}{\gamma}) \\ \text{coins} & = \text{poly}((m+\ell)\frac{k}{\gamma}) \\ \text{compl} & \geq 2/3 \\ \text{sound} & \leq 1/3 \end{pmatrix}$$

(ii) For any functions $k, t, m, \ell : \mathbb{N} \to \mathbb{N}$, and $\gamma, \nu : \mathbb{N} \to (0, 1)$, $\gamma \leq \nu$, if $k, t, m, \ell, 1/\gamma, 1/\nu$ are time-constructible, then

$$\text{IP}\begin{pmatrix} \text{rounds} & = k \\ \text{time} & = t \\ \text{msg size} = m \\ \text{coins} & = \ell \\ \text{compl} & \geq 2/3 + \gamma \\ \text{sound} & \leq 1/3 - \nu \end{pmatrix} \subseteq \text{AM}\begin{pmatrix} \text{rounds} & = 4k+3 \\ \text{time} & = t + \text{poly}\left((m+\ell) \cdot (\frac{k}{\gamma})^{k/\nu}\right) \\ \text{msg size} = \text{poly}\left((m+\ell) \cdot (\frac{k}{\gamma})^{k/\nu}\right) \\ \text{coins} & = \text{poly}\left((m+\ell) \cdot (\frac{k}{\gamma})^{k/\nu}\right) \\ \text{compl} & \geq 2/3 \\ \text{sound} & \leq 1/3 \end{pmatrix}$$

Moreover, in the protocols that achieve the transformations in (i) and (ii), the public-coin verifier calls the private coin verifier exactly once.

The proof can be found in the full version of this paper. Note that part (i) implies IP(rounds $= k$) $\subseteq$ AM(rounds $= 4k + 3$) (where time $= \text{poly}(n)$, compl $\geq 2/3$, sound $\leq 1/3$), as the error probabilities of the IP protocol can be decreased by repeating it in parallel. As mentioned above, due to the repetition, the private coin verifier needs to be called multiple times in the resulting protocol.

Next, we would like to compare our result to the [GS86] transformation. For this comparison, the theorem below states what their transformation achieves *after* repeating the private-coin protocol in parallel, i.e. we again consider the setting where the private-coin verifier is called exactly once by the public-coin verifier.

Theorem 3 ([GS86]). *For any time-constructible polynomials $k(n), t(n), m(n)$, and $\ell(n)$ we have*

$$\text{IP}\begin{pmatrix} \text{rounds} & = k \\ \text{time} & = t \\ \text{msg size} = m \\ \text{coins} & = \ell \\ \text{compl} & \geq 1 - \ell^{-12k^2} \\ \text{sound} & \leq \ell^{-12k^2} \end{pmatrix} \subseteq \text{AM}\begin{pmatrix} \text{rounds} & = k+2 \\ \text{time} & = t + \text{poly}((m+\ell)k) \\ \text{msg size} = \text{poly}(m+\ell) \\ \text{coins} & = \text{poly}(m+\ell) \\ \text{compl} & \geq 2/3 \\ \text{sound} & \leq 1/3 \end{pmatrix}$$

Moreover, in the protocol that achieves this transformation, the public-coin verifier calls the private coin verifier exactly once.

By first repeating the IP protocol in parallel, this implies IP(rounds $= k$) $\subseteq$ AM(rounds $= k + 2$). Comparing this theorem to our Corollary 1 (i), we see that our result only loses a polynomially small fraction γ in completeness, and both losses in soundness and completeness are independent of ℓ. If we apply Corollary 1 (ii) for constant k, polynomial $t(n)$, any γ that is inverse polynomial in n, and any constant $\nu > 0$, we obtain a verifier that runs in polynomial time. This is stronger than [GS86] in all parameters except the number of rounds.

If there exists a constant-round sampling protocol that achieves the guarantees of our theorem, but the verifier runs in time $\text{poly}(\frac{n}{\varepsilon\delta})$, then we would get the transformation

$$\text{IP} \begin{pmatrix} \text{rounds} = k \\ \text{time} \ = t \\ \text{compl} \ \geq 2/3 + \gamma \\ \text{sound} \ \leq 1/3 - \gamma \end{pmatrix} \subseteq \text{AM} \begin{pmatrix} \text{rounds} = \Theta(k) \\ \text{time} \ = \text{poly}(\frac{k \cdot t}{\gamma}) \\ \text{compl} \ \geq 2/3 \\ \text{sound} \ \leq 1/3 \end{pmatrix}$$

for any polynomially small γ and polynomial k (where the private-coin verifier is called exactly once). It is an interesting open problem if this can be achieved.

3.2   The Protocol

Let (V, P) be the IP (i.e. private-coin) protocol for some language L, as given in the theorem. We assume that on input x, V chooses randomness $r \in \{0, 1\}^{\ell(|x|)}$. For a fixed input $x \in \{0, 1\}^n$ we let $\ell = \ell(n)$, and $M_0, \ldots, M_{k-1}$, and $A_0, \ldots, A_{k-1}$ be the random variables over the choice of r that correspond to V's messages $m_0, \ldots m_{k-1}$ and P's answers $a_0, \ldots, a_{k-1}$ in the protocol execution $(V, P)(x)$. Furthermore, we let $\Gamma_i := (M_0, A_0, \ldots, M_i, A_i)$ be the random variable over the entire communication. We now describe the protocol (V', P') that achieves the transformation described by Theorem 2. The protocol will use our sampling protocol several times, always using parameters δ and ε. $(V', P')(x)$ is defined as follows:

For $i = 0, \ldots, k - 1$ **do**
    Prover and Verifier: Use the sampling protocol to sample (m_i, p_i).
    If the prover is honest, it honestly executes the sampling protocol for the distribution P defined by

$$\mathsf{P}(m_i) := \Pr_r[M_i = m_i | \Gamma_{i-1} = (m_0, a_0, \ldots, m_{i-1}, a_{i-1})].$$

    Prover: Send a_i to the verifier.
    If the prover is honest, it sends $a_i := P(x, i, m_0, \ldots, m_i)$.[4]
Prover and Verifier: Use the sampling protocol to sample (r^*, p_k).
If the prover is honest, it honestly executes the sampling protocol for the distribution P defined by

$$\mathsf{P}(r^*) := \Pr_r[r^* = r | \Gamma_{k-1} = (m_0, a_0, \ldots, m_{k-1}, a_{k-1})].$$

Verifier: Accept if and only if all of the following conditions hold:
(a) $V(x, k, r^*, m_0, a_0, \ldots, m_{k-1}, a_{k-1}) = \text{accept}$
(b) $\prod_{i=0}^{k} p_i = \frac{1}{2^\ell}$

---

[4] Recall that $P(x, i, m_0, \ldots, m_i)$ is the prover's answer in round i given input x and the previous verifier messages m_i.

3.3 High Level Proof Sketch

We give a brief sketch of how the sampling protocol is used to prove soundness. Suppose $x \notin L$ and let P^* be any prover that satisfies the above assumption. We denote the randomness of V' by r', let M_i', P_i', A_i', R^* be the random variables over the choice of r' corresponding to m_i, p_i, a_i, r^* in the protocol $(V', P^*)(x)$, and define $\Gamma_i' := (M_0', P_0', A_0', \ldots, M_i', P_i', A_i')$. Soundness is proved by induction, and one first shows that for any $\gamma_{k-1}' = (m_0, p_0, a_0, \ldots, m_{k-1}, p_{k-1}, a_{k-1})$, letting $\gamma_{k-1} = (m_0, a_0, \ldots, m_{k-1}, a_{k-1})$, we have $\Pr_{r'}\left[(V'(r'), P^*)(x) = \text{accept} \mid \Gamma_{k-1}' = \gamma_{k-1}'\right] = \frac{1+\varepsilon+\delta}{\prod_{i=0}^{k-1} p_i} \cdot \Pr_{r \in \{0,1\}^\ell}[V(x, k, r, \gamma_{k-1}) = \text{accept}]$. Using this, we can proceed to bound the acceptance probability conditioned on any fixed $\gamma_{k-2}' = (m_0, p_0, a_0, \ldots, m_{k-2}, p_{k-2}, a_{k-2})$ as follows:

$$
\Pr_{r'}\left[(V'(r'), P^*)(x) = \text{accept} \mid \Gamma_{k-2}' = \gamma_{k-2}'\right]
$$

$$
= \sum_{m_{k-1}, p_{k-1}} \Pr_{r'}\left[(V'(r'), P^*)(x) = \text{accept}\right.
$$

$$
\left. \mid (\Gamma_{k-2}', M_{k-1}', P_{k-1}') = (\gamma_{k-2}', m_{k-1}, p_{k-1})\right] \cdot
$$

$$
\Pr_{r'}\left[(M_{k-1}', P_{k-1}') = (m_{k-1}, p_{k-1}) \mid \Gamma_{k-2}' = \gamma_{k-2}'\right]
$$

$$
= \sum_{m_{k-1}, p_{k-1}} \Pr_{r'}\left[(V'(r'), P^*)(x) = \text{accept} \mid (\Gamma_{k-1}') = (\gamma_{k-2}', m_{k-1}, p_{k-1}, a_{k-1}^*)\right]
$$

$$
\cdot \Pr_{r'}\left[(M_{k-1}', P_{k-1}') = (m_{k-1}, p_{k-1}) \mid \Gamma_{k-2}' = \gamma_{k-2}'\right]
$$

$$
\leq \sum_{m_{k-1}, p_{k-1}} \frac{1+\varepsilon+\delta}{\prod_{i=0}^{k-1} p_i} \cdot \Pr_{r}\left[V(x, k, r, \gamma_{k-2}, m_{k-1}, a_{k-1}^*) = \text{accept}\right]
$$

$$
\cdot \Pr_{r'}\left[(M_{k-1}', P_{k-1}') = (m_{k-1}, p_{k-1}) \mid \Gamma_{k-2}' = \gamma_{k-2}'\right]
$$

$$
= \frac{1+\varepsilon+\delta}{\prod_{i=0}^{k-2} p_i} \sum_{m_{k-1}} \Pr_{r}\left[V(x, k, r, \gamma_{k-2}, m_{k-1}, a_{k-1}^*) = \text{accept}\right]
$$

$$
\cdot \sum_{p_{k-1}} \frac{\Pr_{r'}\left[(M_{k-1}', P_{k-1}') = (m_{k-1}, p_{k-1}) \mid \Gamma_{k-2}' = \gamma_{k-2}'\right]}{p_{k-1}}
$$

$$
\leq \frac{(1+\varepsilon+\delta)^2}{\prod_{i=0}^{k-2} p_i} \sum_{m_{k-1}} \Pr_{r}\left[V(x, k, r, \gamma_{k-2}, m_{k-1}, a_{k-1}^*) = \text{accept}\right]
$$

The second equality above follows by the simplifying assumption that for fixed values $\gamma_{k-2}', m_{k-1}, p_{k-1}$, the prover's answer A_{k-1} is also fixed, and denoted by a_{k-1}^*. This assumption can be easily removed. The first inequality follows from the base case, and the final inequality follows from the soundness guarantee of the sampling protocol. Iterating this step yields the final result. We refer to the full version of the paper for more details.

References

[AB09]   Arora, S., Barak, B.: Computational Complexity - A Modern Approach. Cambridge University Press (2009)

[AGGM06]   Akavia, A., Goldreich, O., Goldwasser, S., Moshkovitz, D.: On basing one-way functions on NP-hardness. In: Kleinberg, J.M. (ed.) STOC, pp. 701–710. ACM (2006), See also errata on author's webpage, http://www.wisdom.weizmann.ac.il/~oded/p_aggm.html

[AH91]   Aiello, W., Håstad, J.: Statistical zero-knowledge languages can be recognized in two rounds. J. Comput. Syst. Sci. 42(3), 327–345 (1991)

[Bab85]   Babai, L.: Trading group theory for randomness. In: Sedgewick, R. (ed.) STOC, pp. 421–429. ACM (1985)

[Bab90]   Babai, L.: E-mail and the unexpected power of interaction. In: Structure in Complexity Theory Conference, pp. 30–44. IEEE Computer Society (1990)

[BM88]   Babai, L., Moran, S.: Arthur-merlin games: A randomized proof system, and a hierarchy of complexity classes. J. Comput. Syst. Sci. 36(2), 254–276 (1988)

[BT06]   Bogdanov, A., Trevisan, L.: On worst-case to average-case reductions for NP problems. SIAM J. Comput. 36(4), 1119–1159 (2006)

[FF93]   Feigenbaum, J., Fortnow, L.: Random-self-reducibility of complete sets. SIAM J. Comput. 22(5), 994–1005 (1993)

[For87]   Fortnow, L.: The complexity of perfect zero-knowledge. In: Structure in Complexity Theory Conference. IEEE Computer Society (1987)

[GMR89]   Goldwasser, S., Micali, S., Rackoff, C.: The knowledge complexity of interactive proof systems. SIAM J. Comput. 18(1), 186–208 (1989)

[GMS87]   Goldreich, O., Mansour, Y., Sipser, M.: Interactive proof systems: Provers that never fail and random selection (extended abstract). In: FOCS, pp. 449–461. IEEE Computer Society (1987)

[Gol08]   Goldreich, O.: Computational complexity - a conceptual perspective. Cambridge University Press (2008)

[GS86]   Goldwasser, S., Sipser, M.: Private coins versus public coins in interactive proof systems. In: Hartmanis, J. (ed.) STOC, pp. 59–68. ACM (1986)

[GVW02]   Goldreich, O., Vadhan, S.P., Wigderson, A.: On interactive proofs with a laconic prover. Computational Complexity 11(1-2), 1–53 (2002)

[HMX10]   Haitner, I., Mahmoody, M., Xiao, D.: A new sampling protocol and applications to basing cryptographic primitives on the hardness of NP. In: IEEE Conference on Computational Complexity, pp. 76–87. IEEE Computer Society (2010)

[LFKN92]   Lund, C., Fortnow, L., Karloff, H.J., Nisan, N.: Algebraic methods for interactive proof systems. J. ACM 39(4), 859–868 (1992)

[Sha92]   Shamir, A.: IP = PSPACE. J. ACM 39(4), 869–877 (1992)

A New View on Worst-Case to Average-Case Reductions for NP Problems

Thomas Holenstein and Robin Künzler

ETH Zurich, Department of Computer Science, 8092 Zurich, Switzerland
{thomas.holenstein,robink}@inf.ethz.ch

Abstract. We study the result by Bogdanov and Trevisan (FOCS, 2003), who show that under reasonable assumptions, there is no non-adaptive reduction that bases the average-case hardness of an NP-problem on the worst-case complexity of an NP-complete problem. We replace the hiding and the heavy samples protocol in [BT03] by employing the histogram verification protocol of Haitner, Mahmoody and Xiao (CCC, 2010), which proves to be very useful in this context. Once the histogram is verified, our hiding protocol is directly public-coin, whereas the intuition behind the original protocol inherently relies on private coins.

1 Introduction

One-way functions are functions that are easy to compute on any instance, and hard to invert on average. Assuming their existence allows the construction of a wide variety of secure cryptographic schemes. Unfortunately, it seems we are far from proving that one-way functions indeed exist, as this would imply BPP $\neq$ NP. Thus, the assumption that NP $\not\subseteq$ BPP, which states that there exists a worst-case hard problem in NP, is weaker. The following question is natural: **Question 1:** Does NP $\not\subseteq$ BPP imply that one-way functions (or other cryptographic primitives) exist? A positive answer to this question implies that the security of the aforementioned cryptographic schemes can be based solely on the worst-case assumption NP $\not\subseteq$ BPP.

Given a one-way function f and an image y, the problem of finding a preimage $x \in f^{-1}(y)$ is an NP-problem: provided a candidate solution x, one can efficiently verify it by checking if $f(x) = y$. In this sense, a one-way function provides an NP problem that is hard to solve on average, and Question 1 asks whether it can be based on worst-case hardness. Thus, the question is closely related to the study of *average-case complexity*, and in particular to the set distNP of distributional problems $(L, \mathcal{D})$, where $L \in$ NP, and $\mathcal{D}$ is an ensemble of efficiently samplable distributions over problem instances. We say that a distNP problem $(L, \mathcal{D})$ is hard if there is no efficient algorithm that solves the problem (with high probability) on instances sampled from $\mathcal{D}$. In this setting, analogously to Question 1, we ask: **Question 2:** Does NP $\not\subseteq$ BPP imply that there exists a hard problem in distNP?

A natural approach to answer Question 2 affirmatively is to give a so-called *worst-case to average-case reduction* from some NP-complete L to $(L', \mathcal{D}) \in$

Z. Cai et al. (Eds.): COCOON 2014, LNCS 8591, pp. 203–214, 2014.

distNP: such a reduction R^O is a polynomial time algorithm with black-box access to an oracle O that solves $(L', \mathcal{D})$ on average, such that $\Pr_R[R^O(x) = L(x)] \geq 2/3$. We say a reduction is *non-adaptive* if the algorithm R fixes all its queries to O in the beginning. Bogdanov and Trevisan [BT06b] (building on work by Feigenbaum and Fortnow [FF93]) show that it is unlikely that a non-adaptive worst-to-average-case reduction exists:

> **Main Result of [BT06b] (informal):** If there exists a non-adaptive worst-case to average-case reduction from an NP-complete problem to a problem in distNP, then NP $\subseteq$ coNP/poly.

The consequence NP $\subseteq$ coNP/poly implies a collapse of the polynomial hierarchy to the third level [Yap83], which is believed to be unlikely.

The work of Impagliazzo and Levin [IL90] and Ben-David et al. [BDCGL92] shows that an algorithm that solves a problem in distNP can be turned (via a non-adaptive reduction) into an algorithm that solves the search version of the same problem. Thus, as inverting a one-way function well on average corresponds to solving a distNP search problem well on average, the result of [BT06b] also implies that Question 1 cannot be answered positively by employing non-adaptive reductions, unless the polynomial hierarchy collapses.

1.1 Contributions of this Paper

The proof of the main result in [BT06b] proceeds as follows. Assuming that there exists a non-adaptive worst-case to average-case reduction R from an NP-complete language L to $(L', \mathcal{D}) \in$ distNP, it is shown that L and its complement both have a constant-round interactive proof with advice (i.e. L and $\overline{L}$ are in AM/poly. As AM/poly = NP/poly, this gives coNP $\subseteq$ NP/poly. The final AM/poly protocol consists of three sub-protocols: the heavy samples protocol, the hiding protocol, and the simulation protocol. Using the protocol to verify the histogram of a probability distribution by Haitner et al. [HMX10], we replace the heavy samples protocol and the hiding protocol. Our protocols have several advantages. The heavy samples protocol becomes quite simple, as one only needs to read a probability from the verified histogram. Furthermore, once the histogram is verified, our hiding protocol is directly public-coin, whereas the intuition behind the original hiding protocol crucially uses that the verifier can hide its randomness from the prover. Our protocol is based on a new and different intuition and achieves the same goal. Clearly, one can obtain a public-coin version of the original hiding protocol by applying the Goldwasser-Sipser transformation [GS86], but this might not provide a different intuition. Finally, our protocols show that the histogram verification protocol of [HMX10] is a very useful primitive to approximate probabilities using AM-protocols.

1.2   Related Work

Recall that our Question 2 above asked if average-case hardness can be based on the worst-case hardness of an NP-complete problem. The question if cryptographic primitives can be based on NP-hardness was stated as Question 1.

Average-case complexity. We use the definition of distNP and average-case hardness from [BT06b]. The hardness definition is essentially equivalent to Impagliazzo's notion of heuristic polynomial-time algorithms [Imp95]. We refer to the surveys of Impagliazzo [Imp95], Goldreich [Gol97], and Bogdanov and Trevisan [BT06a] on average-case complexity.

Negative results on Question 2. Feigenbaum and Fortnow [FF93] study a special case of worst-case to average-case reductions, called random self-reductions. Such a reduction is non-adaptive, and reduces L to itself, such that the queries are distributed uniformly at random (but not necessarily independently). They showed that the existence of a random self-reduction for an NP-complete problem is unlikely, as it implies coNP $\subseteq$ NP/poly and the polynomial hierarchy collapses to the third level. This result generalizes to the case of non-adaptive reductions from $L \in$ NP to $L' \in$ distNP where the queries are distributed according to a distribution P that does not depend on the input x to the reduction, but only on the length of x.

The study of random self-reductions was motivated by their use to design interactive proof systems and (program-) checkers[1]. Checkers are introduced by Blum and Blum and Kannan [Blu88, BK95]. Rubinfeld [Rub90] shows that problems that have a random self-reduction and are downward self-reducible (i.e. they can be reduced to solving the same problem on smaller instances) have a program checker. Random self-reductions can be used to prove the worst-case to average-case equivalence of certain PSPACE-complete and EXP-complete problems [STV01]. A long-standing open question is whether SAT is checkable. In this context, Mahmoody and Xiao [MX10] show that if one-way functions can be based on NP-hardness via a randomized, possibly adaptive reduction, then SAT is checkable.

In the context of program checking, Blum et al. [BLR93] introduce the notion of self-correctors. A self-corrector is simply a worst-case to average-case reduction from L to $(L', \mathcal{D})$, where $L = L'$. Clearly, a random self-reduction is also a self-corrector.

As discussed earlier, based on [FF93], Bogdanov and Trevisan [BT06b] show that the average-case hardness of a problem in distNP cannot be based on the worst-case hardness of an NP-complete problem via non-adaptive reductions (unless the polynomial hierarchy collapses). In particular, this implies that SAT does not have a non-adaptive self-corrector (unless the polynomial hierarchy

---

[1]   Checkers allow to ensure the correctness of a given program on an input-by-input basis. Formally, a checker is an efficient algorithm C that, given oracle access to a program P which is supposed to decide a language L, has the following properties for any instance x. Correctness: If P is always correct, then $C^P(x) = L(x)$ with high probability. Soundness: $C^P(x) \in \{L(x), \bot\}$ with high probability.

collapses). It is an important open question if the same or a similar result can be proved for adaptive reductions.

Watson [Wat12] shows that there exists an oracle O such that there is no worst-case to average-case reduction for NP relative to O. Impagliazzo [Imp11] then gives the following more general result: any proof that gives a positive answer to Question 2 must use non-relativizing techniques. More precisely, it is shown that there exists an oracle O such that $\mathsf{NP}^O \not\subseteq \mathsf{BPP}^O$, and there is no hard problem in distNP^O. Note that this does not rule out the existence of a worst-case to average-case reduction, as such reductions do not necessarily relativize. In particular, the result of Bogdanov and Trevisan [BT06b] also applies to reductions that are non-adaptive and do *not* relativize: there is no such reduction, unless the polynomial hierarchy collapses.

Negative results on Question 1. This question goes back to the work of Diffie and Hellman [DH76]. Even and Yacobi [EY80] give a cryptosystem that is NP-hard to break. However, their notion of security requires that the adversary can break the system in the worst-case (i.e. for every key). Their cryptosystem can in fact be broken on most keys, as shown by Lempel [Lem79]. It is now understood that breaking a cryptosystem should be hard on average, which is, for example, reflected in the definition of one-way functions.

Brassard [Bra83] shows that public-key encryption cannot be based on NP-hardness in the following sense: under certain assumptions on the scheme, if breaking the encryption can be reduced to deciding L, then $L \in \mathsf{NP} \cap \mathsf{coNP}$. In particular, if L is NP-hard this implies that $\mathsf{NP} = \mathsf{coNP}$. Goldreich and Goldwasser [GG98] show the same result under relaxed assumptions.

To give a positive answer to Question 1, one can aim for a reduction from an NP-complete problem to inverting a one-way function well on average (see for example [AGGM06] for a formal definition). As discussed earlier, the work of Impagliazzo and Levin [IL90] and Ben-David et al. [BDCGL92] allows to translate the results of [FF93] and [BT06b] to this setting. That is, there is no non-adaptive reduction from an NP-complete problem L to inverting a one-way function, unless the polynomial hierarchy collapses to the third level. Akavia et al. [AGGM06] directly use the additional structure of the one-way function to prove that the same assumption allows the stronger conclusion $\mathsf{coNP} \subseteq \mathsf{AM}$, which implies a collapse of the polynomial hierarchy to the second level.

Haitner et al. [HMX10] show that if constant-round statistically hiding commitment can be based on an NP-complete problem via $O(1)$-adaptive reductions (i.e. the reduction makes a constant number of query rounds), then $\mathsf{coNP} \subseteq \mathsf{AM}$, and the polynomial hierarchy collapses to the second level. In fact, they obtain the same conclusion for any cryptographic primitive that can be broken by a constant-depth collision finding oracle (such as variants of collision resistant hash functions and oblivious transfer). They also obtain non-trivial, but weaker consequences for $\mathrm{poly}(n)$-adaptive reductions.

Bogdanov and Lee [BL13] explore the plausibility of basing homomorphic encryption on NP-hardness. They show that if there is a (randomized, adaptive) reduction from some L to breaking a homomorphic bit encryption scheme (that supports the evaluation of any sufficiently "sensitive" collection of functions), then $L \in$ AM $\cap$ coAM. In particular, if L is NP-complete this implies a collapse of the polynomial hierarchy to the second level.

Positive results. We only know few problems in distNP that have worst-case to average-case reductions where the worst-case problem is believed to be hard. Most such problems are based on lattices, and the most important two are the short integer solution problem (SIS), and the learning with errors problem (LWE).

The SIS problem goes back to the breakthrough work of Ajtai [Ajt96]. He gives a reduction from an approximate worst-case version of the shortest vector problem to an average-case version of the same problem, and his results were subsequently improved [Mic04, MR07]. Many cryptographic primitives, such as one-way functions, collision-resistant hash functions, identification schemes, and digital signatures have been based on the SIS problem, and we refer to [BLP+13] for an overview. He gives a reduction from an approximate worst-case version of the shortest vector problem to an average-case version of the same problem, and his results were subsequently improved [Mic04, MR07]. Many cryptographic primitives, such as one-way functions, collision-resistant hash functions, identification schemes, and digital signatures have been based on the SIS problem, and we refer to [BLP+13] for an overview. He gives a reduction from an approximate worst-case version of the shortest vector problem to an average-case version of the same problem, and his results were subsequently improved [Mic04, MR07]. Many cryptographic primitives, such as one-way functions, collision-resistant hash functions, identification schemes, and digital signatures have been based on the SIS problem, and we refer to [BLP+13] for an overview.

Regev [Reg09] gives a worst- to average-case reduction for the LWE problem in the quantum setting. That is, an algorithm for solving LWE implies the existence of a quantum algorithm to solve the lattice problem. The work of Peikert [Pei09] and Lyubashevsky and Micciancio [LM09] makes progress towards getting a reduction that yields a classical worst-case algorithm. The first classical hardness reduction for LWE (with polynomial modulus) is then given by Brakerski et al. [BLP+13]. A large number of cryptographic schemes are based on LWE, and we refer to Regev's survey [Reg10], and to [BLP+13] for an overview.

Unfortunately, for all lattice-based worst-case to average-case reductions, the worst-case problem one reduces to is contained in NP $\cap$ coNP, and thus unlikely to be NP-hard. We note that several of these reductions (such as the ones of [Ajt96, Mic04, MR07]) are adaptive.

Gutfreund et al. [GSTS07] make progress towards a positive answer to Question 2: they give a worst-case to average-case reduction for NP, but sampling an input from the distribution they give requires quasi-polynomial time. Furthermore, for any fixed BPP algorithm that tries to decide SAT, they give a distribution that is hard for that specific algorithm. Note that this latter statement does not give a polynomial time samplable distribution that is hard for *any* algorithm. Unlike in [FF93, BT06b], where the reductions under consideration get black-box access to the average-case oracle, the reduction given by [GSTS07] is not black-box, i.e. it requires access to the code of an efficient average-case algorithm. Such reductions (even non-adaptive ones) are not ruled out by the results of [FF93, BT06b]. Gutfreund and Ta-Shma [GTS07] show that even though the techniques of [GSTS07] do not yield an average-case hard problem in distNP, they bypass the negative results of [BT06b]. Furthermore, under a certain derandomization assumption for BPP, they give a worst-case to average-case reduction from NP to an average-case hard problem in $\mathsf{NTIME}(n^{O(\log n)})$.

2   Technical Overview

For a formal definition of non-adaptive worst-case to average-case reductions, we refer to the full version of the paper. In the introduction we stated an informal version of the result of [BT06b]. We now state their main theorem formally. Let $\mathcal{U}$ be the set $\{\mathsf{P}_n\}_{n \in \mathbb{N}}$ where P_n is the uniform distribution on $\{0,1\}^n$.

Theorem 1 ([BT06b]). *For any L and L' and every constant c the following holds. If L is NP-hard, $L' \in$ NP, and there exists a non-adaptive $1/n^c$-worst-to-average reduction from L to $(L',\mathcal{U})$, then $\mathsf{coNP} \subseteq \mathsf{NP/poly}$.*

As discussed earlier, the conclusion implies a collapse of the polynomial hierarchy to the third level. The theorem is stated for the set of uniform distributions $\mathcal{U}$. Using the results of Ben-David et al. [BDCGL92] and Impagliazzo and Levin [IL90], the theorem can be shown to hold for any polynomial time samplable set of distributions $\mathcal{D}$. This is nicely explained in [BT06b] (Section 5). We first give an overview of the original proof, and then describe how our new protocols fit in.

2.1   The Proof of Bogdanov and Trevisan

Suppose R reduces the NP-complete language L to $(L',\mathcal{U}) \in$ distNP. The goal is to give a (constant-round) AM/poly protocol for L and its complement. As NP/poly = AM/poly, this will give the result. The idea is to simulate an execution of the reduction R on input x with the help of the prover. The verifier then uses the output of R as its guess for $L(x)$. R takes as input the instance x, randomness $r \in \{0,1\}^n$, and produces (non-adaptively) queries $y_1, \ldots, y_k \in \{0,1\}^m$ for the average-case oracle. The reduction is guaranteed to guess $L(x)$ correctly with high probability, provided the oracle answers are correct with high probability. We may assume that the queries $y_1, \ldots, y_k$ are identically (but not necessarily independently) distributed. We denote the resulting distribution of individual queries by $\mathsf{P}^{R,x}$, i.e. $\mathsf{P}^{R,x}(y) = \Pr_r[R(x,r) = y]$ (where $R(x,r)$ simply outputs the first query of the reduction on randomness r).

Handling uniform queries: the Feigenbaum-Fortnow protocol. The proof of [BT06b] relies on the following protocol by Feigenbaum and Fortnow [FF93]. The protocol assumes that the queries are uniformly distributed, i.e. $\mathsf{P}^{R,x}(y) = 2^{-m}$ for all y. The advice for the AM/poly protocol is $g_{\mathrm{UY}} = \Pr_{y \leftarrow \{0,1\}^m}[y \in L']$, i.e. the probability of a uniform sample being a yes-instance. The protocol proceeds as follows. First, the verifier chooses random strings $r_1, \ldots, r_\ell$ and sends them to the prover. The honest prover defines $(y_{i1}, \ldots, y_{ik}) := R(x, r_i)$ for all i, and indicates to the verifier which y_{ij} are in L' (we call them yes-instances), and provides the corresponding NP-witnesses. The verifier checks the witnesses, expects to see approximately a g_{UY} fraction of yes-answers, and rejects if this is not the case. The verifier then chooses a random i and outputs $R(x, r_i, y_{i1}, \ldots, y_{ik})$ as its guess for $L(x)$.

To see completeness, one uses a concentration bound to show that the fraction of yes-answers sent by the prover is approximately correct with high probability (one must be careful at this point, because the outputs of the reduction for a fixed r_i are not independent). Finally, the reduction decides $L(x)$ correctly with high probability.

To argue that the protocol is sound, we note that the prover cannot increase the number of yes-answers at all, as it must provide correct witnesses. Furthermore, the prover cannot decrease the number of yes-answers too much, as the verifier wants to see approximately a g_{UY} fraction. This gives that most answers provided by the prover are correct, and thus with high probability the reduction gets good oracle answers, in which case it outputs 0 with high probability.

We note that the Feigenbaum-Fortnow simulation protocol is public-coin.

The case of smooth distributions: the Hiding Protocol. Bogdanov and Trevisan [BT06b] generalize the above protocol so that it works for distributions that are α-smooth, i.e. where $P^{R,x}(y) \leq \alpha 2^{-m}$ for all y and some threshold parameter $\alpha = \text{poly}(n)$ (we say all samples are α-light). If the verifier knew the probability $g_Y := \Pr_{y \leftarrow P^{R,x}}[y \in L']$, it is easy to see that the Feigenbaum-Fortnow protocol (using g_Y instead of g_{UY} as above) can be used to simulate the reduction. Unfortunately, g_Y cannot be handed to the verifier as advice, as it may depend on the instance x. Thus, [BT06b] give a protocol, named the *Hiding Protocol*, that allows the verifier to obtain an approximation of g_Y, given g_{UY} as advice.

The idea of the protocol is as follows: the verifier hides a $1/\alpha$-fraction of samples from $P^{R,x}$ among uniform random samples (i.e. it permutes all samples randomly). The honest prover again indicates the yes-instances and provides witnesses for them. The verifier checks the witnesses and that the fraction of yes-answers among the uniform samples is approximately g_{UY}. If this is true, it uses the fraction of yes-answers among the samples from $P^{R,x}$ as an approximation of g_Y.

Completeness Follows Easily. The intuition to see soundness is that since the distribution is α-smooth, and as the verifier hides only a $1/\alpha$ fraction of $P^{R,x}$ samples among the uniform ones, the prover cannot distinguish them.

We note that the intuition behind this protocol crucially relies on the fact that the verifier can keep some of its random coins private: the prover is not allowed to know where the distribution samples are hidden.

General distributions and the Heavy Samples Protocol. Finally, [BT06b] remove the restriction that $P^{R,x}$ is α-smooth as follows. We say y is α-heavy if $P^{R,x}(y) \geq$

$\alpha 2^{-m}$, and let $g_H := \Pr_{y \leftarrow \mathsf{P}^{R,x}}[\mathsf{P}^{R,x}(y) \geq \alpha 2^{-m}]$ be the probability of a distribution sample being heavy, and $g_{YL} := \Pr_{y \leftarrow \mathsf{P}^{R,x}}[y \in L' \wedge \mathsf{P}^{R,x}(y) < \alpha 2^{-m}]$ the probability of a distribution sample being a yes-instance and light.

We first note that if the verifier knows (an approximation of) both g_H and g_{YL}, it can use the Feigenbaum-Fortnow approach to simulate the reduction: the verifier simply uses g_{YL} instead of g_{UY} in the protocol, and ignores the heavy samples. It can do this by having the prover indicate the α-heavy instances, and checking that their fraction is close to g_H. Using the lower bound protocol of Goldwasser and Sipser [GS86], the prover must prove that these samples are indeed heavy. Finally, for the heavy samples the verifier can simply set the oracle answers to 0: this changes the oracle answers on at most a polynomially small (i.e. a $1/\alpha$) fraction of the inputs, as by definition at most a $1/\alpha$ fraction of the y's can be α-heavy. Completeness is not hard to see, and soundness follows because a cheating prover cannot claim light samples to be heavy (by the soundness of the lower bound protocol), and thus, by the verifier's check, cannot lie much about which samples are heavy.

If the verifier knows (an approximation of) g_H, then it can use the hiding protocol to approximate g_{YL}: the verifier simply ignores the heavy samples. This is again done by having the prover additionally tell which samples are α-heavy (and prove this fact using the lower bound protocol). The verifier additionally checks that the fraction of heavy samples among the distribution samples is approximately g_H, and finally uses the fraction of light distribution samples as approximation for g_{YL}.

It only remains to approximate g_H. This is done using the *Heavy Samples Protocol* as follows: the verifier samples $y_1, \ldots, y_k$ from $\mathsf{P}^{R,x}$ by choosing random $r_1, \ldots, r_k$ and letting $y_i := R(x, r_i)$. It sends the y_i to the prover. The honest prover indicates which of them are heavy, and proves to the verifier using the lower bound protocol of [GS86] that the heavy samples are indeed heavy and using the upper bound protocol of Aiello and Håstad [AH91] that the light samples are indeed light. The verifier then uses the fraction of heavy samples as its approximation for g_H. It is intuitive that this protocol is complete and sound. The upper bound protocol requires that the verifier knows a uniform random element (which is unknown to the prover) in the set on which the upper bound is proved. In our case, the verifier indeed knows the value r_i, which satisfies this condition. We note that this protocol relies on private-coins, as the verifier must keep the r_i secret for the upper bound proofs.

2.2 Our Proof

We give two new protocols to approximate the probabilities g_H and g_{YL}, as defined in the previous section. These protocols can be used to replace the Hiding Protocol and the Heavy Samples Protocol of [BT06b], respectively. Together with the Feigenbaum-Fortnow based simulation protocol of [BT06b], this then yields a different proof of $\mathsf{coNP} \subseteq \mathsf{AM/poly}$ under the given assumptions.

Verifying histograms. We are going to employ the VerifyHist protocol by Haitner et al. [HMX10] to verify the histogram of a probability distribution. The (ε, t)-histogram $h = (h_0, \ldots, h_t)$ of a distribution P is defined by letting $h_i := \mathrm{Pr}_{y \leftarrow \mathsf{P}}[y \in \mathcal{B}_i]$, where $\mathcal{B}_i := \{x : \mathsf{P}(x) \in (2^{-(i+1)\varepsilon}, 2^{-i\varepsilon}]\}$. We will use the Verify-Hist protocol for the distribution $\mathsf{P}^{R,x}$, as defined by the reduction $R(x, \cdot)$ under consideration, i.e. $\mathsf{P}^{R,x}(y) = \mathrm{Pr}_r[R(x, r) = y]$. Intuitively, this protocol allows to prove that some given histogram h is close to the true histogram of $\mathsf{P}^{R,x}$ in terms of the 1st Wasserstein distance (also known as Earth Mover's distance). This distance between h and h' measures the minimal amount of work that is needed to push the configuration of earth given by h to get the configuration given by h': moving earth over a large distance is more expensive than moving it over a short distance. For formal definitions of histograms and the 1st Wasserstein distance we refer to the full version of the paper.

Lemma 1 (VerifyHist protocol of [HMX10], informal). *There is a constant-round public-coin protocol VerifyHist where the prover and the verifier get as input the circuit $R(x, \cdot)$ and a histogram h, and we have:* **Completeness:** *If h is the histogram of $\mathsf{P}^{R,x}$, then the verifier accepts with high probability.* **Soundness:** *If h is far from the histogram of $\mathsf{P}^{R,x}$ in the 1st Wasserstein distance, then the verifier rejects with high probability.*

The new Heavy Samples Protocol. The idea to approximate the probability g_{H} is very simple. The honest prover sends the histogram of $\mathsf{P}^{R,x}$, and the verifier uses the VerifyHist protocol to verify it. Finally, the verifier simply reads the probability g_{H} from the histogram.

There is a technical issue that comes with this approach. For example, it may be that all y's with nonzero probability have the property that $\mathsf{P}^{R,x}(y)$ is very close, but just below $\alpha 2^{-m}$. In this case, a cheating prover can send a histogram claiming that these y's have probability slightly above this threshold. This histogram has small Wasserstein distance from the true histogram, as the probability mass is moved only over a short distance. Clearly, the verifier's guess for g_{H} is very far from the true value in this case.

We note that the same issue appears in the proof of [BT06b], and we deal with it in exactly the same way as they do: we choose the threshold α randomly, such that with high probability $\mathrm{Pr}_{y \leftarrow \mathsf{P}^{R,x}}[\mathsf{P}^{R,x}(y)$ is close to $\alpha 2^{-m}]$ is small.

A public-coin Hiding Protocol for smooth distributions. We would like the verifier to only send uniform random samples to the prover (as opposed to the original hiding protocol, where a few samples from the distribution are hidden among uniform samples). We first describe the main idea in the special and simpler case where $\mathsf{P}^{R,x}$ is α-smooth. In this case, we can give the following protocol, which uses g_{UY} as advice:

The verifier sends uniform random samples $y_1, \ldots, y_k$. The prover indicates for each sample whether it is a yes-instance, and provides witnesses. Furthermore, the prover tells $\mathsf{P}^{R,x}(y_i)$ to the verifier, and proves a lower bound on this probability. The verifier checks the witnesses and if the fraction of yes-instances is approximately g_{UY}, and considers the histogram h induced by the probabilities $\mathsf{P}^{R,x}(y_i)$, and in particular checks if the probability mass of h is 1. Finally, the verifier considers the histogram h_Y induced by only considering the yes-instances, and uses the total mass in h_Y as its approximation of g_{YL}.

To see completeness, the crucial point is that the smoothness assumption implies that the verifier can get a good approximation of the true histogram.

Soundness follows because the prover cannot claim the probabilities to be too large (as otherwise the lower bound protocol rejects), and it cannot claim many probabilities to be too small, as otherwise the mass of h gets significantly smaller than 1. As it cannot lie much about yes-instances, this implies a good approximation of g_{YL}.

Dealing with general distributions. The above idea can be applied even to general distributions, assuming that the verifier knows the probability $g_{UH} := \Pr_{y \leftarrow \{0,1\}^m}[\mathsf{P}^{R,x}(y) \geq \alpha 2^{-m}]$ of a uniform random sample being heavy. The prover still provides the same information. The verifier only considers the part of the induced histogram h below the $\alpha 2^{-m}$ threshold, and checks that the mass of h below the threshold is close to $1 - g_{UH}$. As in the heavy samples protocol, we again encounter the technical issue that many y's could have probability close to the threshold, in which case the prover can cheat. But, as discussed earlier, this situation occurs with small probability over the choice of α.

Approximating the probability of a uniform sample being heavy. Thus, it remains to give a protocol to approximate g_{UH}. We do this in exactly the same way as the Heavy Samples protocol approximates g_H. That is, given the histogram that was verified using VerifyHist, the verifier simply reads the approximation of g_{UH} from the histogram. The proof that this works is rather technical, as we must show that small Wasserstein distance between the true and the claimed histogram implies a small difference of the probability g_{UH} and its approximation read from the claimed histogram. We note that we include the protocol for approximating g_{UH} directly into our Heavy Samples protocol.

References

[AGGM06] Akavia, A., Goldreich, O., Goldwasser, S., Moshkovitz, D.: On basing one-way functions on np-hardness. In: Kleinberg, J.M. (ed.) STOC, pp. 701–710. ACM (2006), See also errata on author's webpage: `http://www.wisdom.weizmann.ac.il/~oded/p_aggm.html`

[AH91] Aiello, W., Håstad, J.: Statistical zero-knowledge languages can be recognized in two rounds. J. Comput. Syst. Sci. 42(3), 327–345 (1991)

[Ajt96] Ajtai, M.: Generating hard instances of lattice problems (extended abstract). In: Miller, G.L. (ed.) STOC, pp. 99–108. ACM (1996)

[BDCGL92] Ben-David, S., Chor, B., Goldreich, O., Luby, M.: On the theory of average case complexity. J. Comput. Syst. Sci. 44(2), 193–219 (1992)

[BK95] Blum, M., Kannan, S.: Designing programs that check their work. J. ACM 42(1), 269–291 (1995)

[BL13] Bogdanov, A., Lee, C.H.: Limits of provable security for homomorphic encryption. In: Canetti, R., Garay, J.A. (eds.) CRYPTO 2013, Part I. LNCS, vol. 8042, pp. 111–128. Springer, Heidelberg (2013)

[BLP+13]   Brakerski, Z., Langlois, A., Peikert, C., Regev, O., Stehlé, D.: Classical
            hardness of learning with errors. In: Boneh, D., Roughgarden, T., Feigen-
            baum, J. (eds.) STOC, pp. 575–584. ACM (2013)
[BLR93]   Blum, M., Luby, M., Rubinfeld, R.: Self-testing/correcting with applica-
            tions to numerical problems. J. Comput. Syst. Sci. 47(3), 549–595 (1993)
[Blu88]   Blum, M.: Designing programs to check their work. Technical Report 88-
            09, ICSI (1988)
[Bra83]   Brassard, G.: Relativized cryptography. IEEE Transactions on Informa-
            tion Theory 29(6), 877–893 (1983)
[BT06a]   Bogdanov, A., Trevisan, L.: Average-case complexity. Foundations and
            Trends in Theoretical Computer Science 2(1) (2006)
[BT06b]   Bogdanov, A., Trevisan, L.: On worst-case to average-case reductions for
            NP problems. SIAM J. Comput. 36(4), 1119–1159 (2006)
[DBL10]   Proceedings of the 25th Annual IEEE Conference on Computational
            Complexity, CCC 2010, June 9-12. IEEE Computer Society (2010)
[DH76]   Diffie, W., Hellman, M.E.: New directions in cryptography. IEEE Trans-
            actions on Information Theory 22(6), 644–654 (1976)
[EY80]   Even, S., Yacobi, Y.: Cryptocomplexity and NP-completeness. In: de
            Bakker, J.W., van Leeuwen, J. (eds.) ICALP 1980. LNCS, vol. 85, pp.
            195–207. Springer, Heidelberg (1980)
[FF93]   Feigenbaum, J., Fortnow, L.: Random-self-reducibility of complete sets.
            SIAM J. Comput. 22(5), 994–1005 (1993)
[GG98]   Goldreich, O., Goldwasser, S.: On the possibility of basing cryptography
            on the assumption that P ≠ NP, Unpublished manuscript (1998)
[Gol97]   Goldreich, O.: Notes on levin's theory of average-case complexity. Elec-
            tronic Colloquium on Computational Complexity (ECCC) 4(58) (1997)
[GS86]   Goldwasser, S., Sipser, M.: Private coins versus public coins in interactive
            proof systems. In: Hartmanis, J. (ed.) STOC, pp. 59–68. ACM (1986)
[GSTS07]   Gutfreund, D., Shaltiel, R., Ta-Shma, A.: If NP languages are hard on
            the worst-case, then it is easy to find their hard instances. Computational
            Complexity 16(4), 412–441 (2007)
[GTS07]   Gutfreund, D., Ta-Shma, A.: Worst-case to average-case reductions re-
            visited. In: Charikar, M., Jansen, K., Reingold, O., Rolim, J.D.P. (eds.)
            APPROX and RANDOM 2007. LNCS, vol. 4627, pp. 569–583. Springer,
            Heidelberg (2007)
[HMX10]   Haitner, I., Mahmoody, M., Xiao, D.: A new sampling protocol and ap-
            plications to basing cryptographic primitives on the hardness of NP. In:
            IEEE Conference on Computational Complexity [DBLP10], pp. 76–87
[IL90]   Impagliazzo, R., Levin, L.A.: No better ways to generate hard NP in-
            stances than picking uniformly at random. In: FOCS, pp. 812–821. IEEE
            Computer Society (1990)
[Imp95]   Impagliazzo, R.: A personal view of average-case complexity. In: Structure
            in Complexity Theory Conference, pp. 134–147. IEEE Computer Society
            (1995)
[Imp11]   Impagliazzo, R.: Relativized separations of worst-case and average-case
            complexities for NP. In: IEEE Conference on Computational Complexity,
            pp. 104–114. IEEE Computer Society (2011)
[Lem79]   Lempel, A.: Cryptology in transition. ACM Comput. Surv. 11(4), 285–303
            (1979)

[LM09]   Lyubashevsky, V., Micciancio, D.: On bounded distance decoding, unique
            shortest vectors, and the minimum distance problem. In: Halevi, S. (ed.)
            CRYPTO 2009. LNCS, vol. 5677, pp. 577–594. Springer, Heidelberg
            (2009)

[Mic04]   Micciancio, D.: Almost perfect lattices, the covering radius problem, and
            applications to Ajtai's connection factor. SIAM J. Comput. 34(1), 118–
            169 (2004)

[MR07]   Micciancio, D., Regev, O.: Worst-case to average-case reductions based
            on gaussian measures. SIAM J. Comput. 37(1), 267–302 (2007)

[MX10]   Mahmoody, M., Xiao, D.: On the power of randomized reductions and the
            checkability of sat. In: IEEE Conference on Computational Complexity
            [DBL10], pp. 64–75

[Pei09]   Peikert, C.: Public-key cryptosystems from the worst-case shortest vec-
            tor problem: Extended abstract. In: Mitzenmacher, M. (ed.) STOC,
            pp. 333–342. ACM (2009)

[Reg09]   Regev, O.: On lattices, learning with errors, random linear codes, and
            cryptography. J. ACM 56(6) (2009)

[Reg10]   Regev, O.: The learning with errors problem (invited survey). In: IEEE
            Conference on Computational Complexity [DBL10], pp. 191–204 (2010)

[Rub90]   Rubinfeld, R.: A mathematical theory of self-checking, self-testing and
            self-correcting programs. PhD thesis. UC Berkeley (1990)

[STV01]   Sudan, M., Trevisan, L., Vadhan, S.P.: Pseudorandom generators without
            the xor lemma. J. Comput. Syst. Sci. 62(2), 236–266 (2001)

[Wat12]   Watson, T.: Relativized worlds without worst-case to average-case reduc-
            tions for NP. TOCT 4(3), 8 (2012)

[Yap83]   Yap, C.-K.: Some consequences of non-uniform conditions on uniform
            classes. Theor. Comput. Sci. 26, 287–300 (1983)

The Power of Duples (in Self-Assembly): It's Not So Hip to Be Square

Jacob Hendricks[1,*], Matthew J. Patitz[1,*], Trent A. Rogers[2,*], and Scott M. Summers[3]

[1] Department of Computer Science and Computer Engineering, University of Arkansas, Fayetteville, AR 72701, USA
jhendric@uark.edu, patitz@uark.edu
[2] Department of Mathematical Sciences, University of Arkansas, Fayetteville, AR 72701, USA
tar003@uark.edu
[3] Department of Computer Science, University of Wisconsin–Oshkosh, Oshkosh, WI 54901, USA
summerss@uwosh.edu

Abstract. In this paper we define the Dupled abstract Tile Assembly Model (DaTAM), which is a slight extension to the abstract Tile Assembly Model (aTAM) that allows for not only the standard square tiles, but also "duple" tiles which are rectangles pre-formed by the joining of two square tiles. We show that the addition of duples allows for powerful behaviors of self-assembling systems at temperature 1, meaning systems which exclude the requirement of cooperative binding by tiles (i.e., the requirement that a tile must be able to bind to at least 2 tiles in an existing assembly if it is to attach). Cooperative binding is conjectured to be required in the standard aTAM for Turing universal computation and the efficient self-assembly of shapes, but we show that in the DaTAM these behaviors can in fact be exhibited at temperature 1. We then show that the DaTAM doesn't provide asymptotic improvements over the aTAM in its ability to efficiently build thin rectangles. Finally, we present a series of results which prove that the temperature-2 aTAM and temperature-1 DaTAM have mutually exclusive powers. That is, each is able to self-assemble shapes that the other can't, and each has systems which cannot be simulated by the other. Beyond being of purely theoretical interest, these results have practical motivation as duples have already proven to be useful in laboratory implementations of DNA-based tiles.

1 Introduction

The abstract Tile Assembly Model (aTAM) [30] is a simple yet elegant mathematical model of self-assembling systems. Despite the simplicity of its formulation, theoretical results within the aTAM have provided great insights into many fundamental properties of self-assembling systems. These include results showing the power of these systems to perform computations [15, 21, 30], the ability to build shapes efficiently (in terms of the number of unique types of

---

* Supported in part by National Science Foundation Grant CCF-1117672.

Z. Cai et al. (Eds.): COCOON 2014, LNCS 8591, pp. 215–226, 2014.
© Springer International Publishing Switzerland 2014

components, i.e. tiles, needed) [1, 25, 29], limitations to what can be built and computed [15, 16], and many other important properties (see [11, 22] for more comprehensive surveys). From this broad collection of results in the aTAM, one property of systems that has been shown to yield enormous power is *cooperation*. Cooperation is the term used to specify the situation where the attachment of a new tile to a growing assembly requires it to bind to more than one tile (usually 2) already in the assembly. The requirement for cooperation is determined by a system parameter known as the *temperature*, and when the temperature is equal to 1 (a.k.a. temperature-1 systems), there is no requirement for cooperation. A long-standing conjecture is that temperature-1 systems are in fact not capable of universal computation or efficient shape building (although temperature ≥ 2 systems are) [9, 13, 18, 20]. However, in actual laboratory implementations of DNA-based tiles [2, 17, 24, 26, 32], the self-assembly performed by temperature-2 systems does not match the error-free behavior dictated by the aTAM, but instead, a frequent source of errors is the binding of tiles using only a single bond. Thus, temperature-1 behavior erroneously occurs and can't be completely prevented. This has led to the development of a number of error-correction and error-prevention techniques [5, 23, 26, 28, 31] for use in temperature-2 systems.

Despite the conjectured weakness of temperature-1 systems, an alternative approach has been to try to find ways of modifying them in the hope of developing systems which can operate at temperature-1 while exhibiting powers of temperature-2 systems but without the associated errors. Research along this path has resulted in an impressive variety of alternatives in which temperature-1 systems are capable of Turing universal computation: using 3-D tiles [9], allowing probabilistic computations with potential for error [9], including glues with repulsive forces [20], and using a model of staged assembly [3]. While these are theoretically very interesting results, the promise for use in the laboratory of each is limited by current technologies. Therefore, in this paper we introduce another technique for improving the power of temperature-1 systems, but one which makes use of building blocks which are already in use in laboratory implementations: *duples* (a.k.a. "double tiles" [2, 6, 26, 27]).

We first introduce the *Dupled abstract Tile Assembly Model* (DaTAM), which is essentially the aTAM extended to allow both square and rectangular, duple, tile types. We then show a series of results within the DaTAM which prove that at temperature 1 it is quite powerful: it is computationally universal and able to build $N \times N$ squares using $O(\log N)$ tile types. We next demonstrate that, while the addition of duples does provide significant power to temperature-1 systems, it doesn't allow for asymptotic gains over the aTAM in terms of the tile complexity required to self-assemble thin rectangles, with the lower bound for an $N \times k$ rectangle being $\Omega\left(\frac{N^{1/k}}{k}\right)$. We then provide a series of results which show that the neither the aTAM at temperature-2 nor the DaTAM at temperature-1 is strictly more powerful than the other, namely that in each there are shapes which can be self-assembled which are impossible to self-assemble in the other, and that there are also systems in each which cannot be simulated by the other. These mutually exclusive powers provide a very interesting framework for further study of the unique abilities provided by the incorporation of duples into self-assembling systems. Furthermore, as previously mentioned, the use of duples has already been proven possible in laboratory experiments, providing even further motivation for the model. Please note that the online version of this paper contains color images.

2 Preliminaries

In this version of the paper, we provide only high-level sketches of the definitions used in this paper.

2.1 Informal Description of the Dupled Abstract Tile Assembly Model

In this section, we give a very brief, informal description of the abstract Tile Assembly Model (aTAM) and the Dupled abstract Tile Assembly Model (DaTAM).

The abstract Tile Assembly Model (aTAM) was introduced by Winfree [30]. In the aTAM, the basic components are translatable but non-rotatable *tiles* which are unit squares with *glues* on their edges. Each glue consists of a string *label* value and an integer *strength* value. A *tile type* is a unique mapping of glues (including possibly the *null* glue) to 4 sides, and a tile is an instance of a tile type. Assembly begins from a specially designated *seed* which is usually a single tile but maybe be a pre-formed collection of tiles, and continues by the addition of a single tile at a time until no more tiles can attach. A tile is able to bind to an adjacent tile if the glues on their adjacent edges match in label and strength, and can attach to an assembly if the sum of the strengths of binding glues meets or exceeds a system parameter called the *temperature* (which is typically set to either 1 or 2). A *tile assembly system* (TAS) is an ordered 3-tuple (T, σ, τ) where T is the set of tile types (i.e. tile set), σ is the seed configuration, and τ is the temperature.

The Dupled abstract Tile Assembly Model (DaTAM) is an extension of the aTAM which allows for systems with square tiles as well as rectangular tiles. The rectangular tiles are 2×1 or 1×2 rectangles which can logically be thought of as two square tiles which begin pre-attached to each other along an edge, hence the name *duples*. A *dupled tile assembly system* (DTAS) is an ordered 5-tuple (T, S, D, σ, τ) where T, σ, and τ are as for a TAS, and S is the set of singleton (i.e. square) tiles which are available for assembly, and D is the set of duple tiles. The tile types which make up S and D all belong to T, with those in D each being a combination of two tile types from T.

2.2 Zig-Zag Tile Assembly Systems

Originally defined in [8], we define zig-zag tile assembly systems and compact zig-zag tile assembly systems in the same manner as [20]. In [20] they called a system $\mathcal{T} = (T, \sigma, \tau)$ a zig-zag tile assembly system provided that $\mathcal{T}$ is directed with a single assembly sequence, and for any producible assembly α of $\mathcal{T}$, α does not contain a tile with an exposed south glue. More intuitively, a zig-zag tile assembly system is a system which grows to the left or right, grows up some amount, and then continues growth again to the left or right. Moreover, we call a zig-zag tile assembly system $\mathcal{T} = (T, \sigma, \tau)$ a *compact zig-zag tile assembly system* if and only if for every tile t in any assembly α of $\mathcal{T}$, the sum of the strengths of the north and south glues of t is less than 2τ. Informally, this can be thought of as a zig-zag tile assembly system which is only able to travel upwards one tile at a time before being required to zig-zag again.

2.3   Simulation

In this section, we present a high-level sketch of what we mean when saying that one system *simulates* another (our definitions are based on those of [19]).

For one system $\mathcal{S}$ to simulate another system $\mathcal{T}$, we allow $\mathcal{S}$ to use square (or rectangular when simulating duples) blocks of tiles called *macrotiles* to represent the simulated tiles from $\mathcal{T}$. The simulator must provide a scaling factor c which specifies how large each macrotile is, and it must provide a *representation function*, which is a function mapping each macrotile assembled in $\mathcal{S}$ to a tile in $\mathcal{T}$. Since a macrotile may have to grow to some critical size (e.g. when gathering information from adjacent macrotiles about the simulated glues adjacent to its location) before being able to compute its identity (i.e. which tile from $\mathcal{T}$ it represents), it's possible for non-empty macrotile locations in $\mathcal{S}$ to map to empty locations in $\mathcal{T}$, and we call such growth *fuzz*. In standard simulation definitions (e.g. those in [10,12,14,19]), fuzz is restricted to being laterally or vertically adjacent to macrotile positions in $\mathcal{S}$ which map to non-empty tiles in $\mathcal{T}$. We follow this convention for the definition of simulation of aTAM systems by DaTAM systems. However, since duples occupy more than a unit square of space, for our definition of aTAM systems simulating DaTAM systems, we allow fuzz to extend to a Manhattan distance of 2 from a macrotile which maps to a non-empty tile in $\mathcal{T}$. As a further concession to the size of duples, for that simulation definition we also allow empty macrotile locations in $\mathcal{S}$ to map to tiles in $\mathcal{T}$, provided they are half of a duple for which the other half has sufficiently grown. Thus, while our result for aTAM systems simulating DaTAM systems (Theorem 5) shows its impossibility in general, our intent with the simulation definitions is to relax them sufficiently that, if simulation equivalent to the standard notions of simulation were possible, these definitions would allow it.

Given the notion of block representations, we say that $\mathcal{S}$ simulates $\mathcal{T}$ if and only if (1) for every producible assembly in $\mathcal{T}$, there is an equivalent producible assembly in $\mathcal{S}$ when the representation function is applied, and vice versa (thus we say the systems have *equivalent productions*), and (2) for every assembly sequence in $\mathcal{T}$, the exactly equivalent assembly sequence can be followed in $\mathcal{S}$ (modulo the application of the representation function), and vice versa (thus we say the systems have *equivalent dynamics*). Thus, equivalent production and equivalent dynamics yield a valid simulation.

3   The Dupled aTAM is Computationally Universal

In this section, we show constructively that for every compact zig-zag tile assembly system, there exists a DTAS which simulates it. It will then follow from [8] that the DaTAM can simulate an arbitrary Turing machine.

Theorem 1. *Let* $\mathcal{T} = (T, \sigma, 2)$ *be a compact zig-zag TAS and let* G_N *be the set consisting of all glues that appear on the north side of a tile in* T. *Then there exists an DTAS* $\mathcal{T}' = (T', S, D, \gamma, 1)$ *such that* S *simulates* $\mathcal{T}$ *at scale factor* $O(\log|G_N|)$ *with* $|S| + |D| = O(|T||G_N|)$.

In this version of the paper, we only provide a brief sketch of our construction. Suppose that $\mathcal{T} = (T, \sigma, 2)$ is a compact zig-zag TAS. We construct a $\tau = 1$ DTAS which simulates $\mathcal{T}$ using macrotiles. Since $\mathcal{T}$ is a compact zig-zag TAS, we need to only consider the assembly of a handful of different genres of macrotiles. In Figure 1 we see all of the genres of macrotiles up to reflection which we will need to be able to assemble in order to simulate a compact zig-zag TAS. We can separate these macrotiles into two categories: macrotiles which are simulating tile types in T that bind with strength 2 glues and macrotiles which are simulating tile types in T which require cooperation to bind.

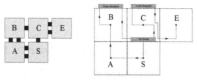

Fig. 1. (Left) A simple assembly produced by a compact zig-zag system. (Right) A system consisting of macrotiles which simulates the system on the left and demonstrates the genres of macrotiles involved in simulating compact zig-zag TASes up to rotation. The dashed boxes represent the boundaries of the macrotiles and the solid lines through the macrotiles represent single-tile wide paths which build the macrotiles.

Assembling the macrotiles which are simulating tile types in T that bind with a single strength 2 glue is straight forward. The interesting part of the construction is the assembly of macrotiles which are simulating tile types in T which require cooperation to bind. These macrotiles consist of two parts: 1) a north geometry and 2) a bit reader. The north geometry section of the macrotile encodes the information about the north glue of the tile which it is simulating. This is done by assigning each glue in T a palindromic binary string (assigning 0 to the null glue) and then encoding the glue's binary representation using the bit encoding scheme shown in Figure 2. Our use of the palindrome is just a convention so that the bits encode the same value from east to west that they do from west to east.

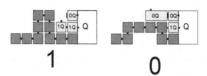

Fig. 2. A single bit example of how the assembly is able to read geometry to gain information about a north glue and still retain information about the west glue. We use the following conventions. The small black rectangles represent glues which allow singletons to bind. The longer black rectangles represent glues that can potentially bind to a duple (note that these glues are the same types of glues as the others, just drawn differently for extra clarity). The red rectangles represent glues that have mismatched.

The bit reader is able to "read" bits by means of the bit reading gadget shown in Figure 2 and works by trying to place a singleton and a duple. By way of our construction, it is the case that only one of them can be placed, and this allows the bit reader to distinguish between bits. Together, the north geometry and the bit reader of the macrotiles allow them to recreate the cooperation that takes place in $\mathcal{T}$ by passing information about the east and west glues of the simulated tiles through the glues of the tile wide paths while encoding information about the north glues as geometry. The overall growth pattern of these macrotiles follows the same assembly sequence as C in Figure 1.

Notice that the scale factor of simulation will depend on the number of bits required to represent the number of north glues in T. Also, for each tile in T we must have a tile in our simulator, say t, which has $|G_N|$ tiles associated with it so that t may grow a path and read the north geometry of the next tile. Hence, $|S| + |D| = |T||G_N|$.

Corollary 1. *For every standard Turing Machine M and input w, there exists an DTAS that simulates M on w.*

This follows directly from Lemma 7 of [8] and Theorem 1.

Corollary 2. *For every $N \in \mathbb{N}$, there exists a DTAS which assembles an $N \times N$ square with $O(\log N)$ tile complexity and constant scale factor.*

4 Self-assembly of Thin Rectangles in the DaTAM

In this section, we study the self-assembly of thin rectangles in the DaTAM. As in [7], we say that an $N \times k$ rectangle $R_{N,k} = \{0, \ldots, k-1\} \times \{0, \ldots, N-1\}$ is *thin* if $k < \frac{\log N}{\log \log N - \log \log \log N}$. We say that the temperature $\tau \in \mathbb{N}$ *tile complexity* of a shape $X \subseteq \mathbb{Z}^2$ in the DaTAM is the minimum number of unique (duple) tile types required to strictly self-assemble X, i.e., $K_{DSA}^\tau(X) = \min\{|S \cup D| \mid X$ strictly self-assembles in $\mathcal{D} = (T, S, D, \sigma, \tau)\}$. In the aTAM, the lower bound for the tile complexity of an $N \times k$ rectangle is $\Omega\left(\frac{N^{1/k}}{k}\right)$ [7]. Perhaps not too surprisingly, duple tile types do not offer any asymptotic advantage when it comes to the self-assembly of thin rectangles, i.e., we have the following lower bound for the tile complexity of thin rectangles in the DaTAM.

Theorem 2. *Let $N, k, \tau \in \mathbb{N}$. If $R_{N,k}$ is thin, then $K_{DSA}^\tau(R_{N,k}) = \Omega\left(\frac{N^{1/k}}{k}\right)$.*

The proof of Theorem 2, omitted from this version of the paper, uses a straightforward counting argument similar the proof of Theorem 3.1 of [7].

5 Mutually Exclusive Powers

In this section, we demonstrate a variety of shapes and systems in the DaTAM at $\tau = 1$ and the aTAM at $\tau = 2$ which can be self-assembled and simulated, respectively, by only one of the models.

5.1   A Shape in the DaTAM But not the aTAM

In this section, we show that there exists an infinite shape which can self-assemble in the DaTAM at $\tau = 1$ but not in the aTAM at $\tau = 2$. Figure 3 shows a high-level sketch of a portion of this shape.

Theorem 3. *There exists a shape $W \subset \mathbb{Z}^2$ such that there exists DTAS $\mathcal{D} = (T_\mathcal{D}, S, D, \sigma, 1)$ in the DaTAM which self-assembles W, but no TAS $\mathcal{T} = (T, \sigma', 2)$ in the aTAM which self-assembles W.*

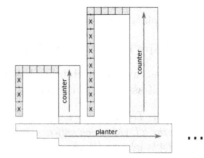

Fig. 3. A high-level sketch of a portion of the infinite shape which can self-assemble in the DaTAM at $\tau = 1$ but not in the aTAM at $\tau = 2$. (Modules not to scale.)

Here we give an intuitive overview of why the aTAM cannot simulate the shape depicted in Figure 3. First, we call the shape in Figure 3 W.

Since, by Theorem 1, DaTAM systems are capable of simulating compact zig-zag systems, W assembles in the DaTAM as follows. A horizontal counter called the **planter** begins growth from a single tile seed and continues to grow indefinitely. The topmost tiles of the **planter** expose glues that allow vertical counters to grow. Each of these vertical counters is a finite subassembly whose height is an even number of tile locations and, from left to right, each successive counters grows to a height that is greater than the previous counter. When a vertical counter finishes upward grow, a single tile wide path of 6 tiles binds to the left of the counter. The leftmost tile of this single tile wide path exposes a south glue that allows for duples to attach. Equipped with matching north and south glues, these duples form a single tile wide path of duples, called a **finger**, that grows downward toward the **planter**. Since the height of each vertical counter is even and the first duple of a **finger** is placed 1 tile location below this height, there are an odd number of tile locations for the duples of a **finger** to occupy. As a result, each finger is forced to cease growth exactly 1 tile location away from the **planter**.

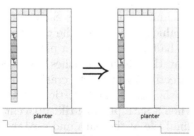

Fig. 4. Left: A **finger** containing two occurrences of a tile of type T_0. Right: A valid producible assembly that results in a shape that differs from W.

Since in an aTAM system, any tile of an assembly takes up a single location of the infinite grid-graph, it is impossible to grow the **finger** component of the shape W. This essentially follows from the fact that for a single tile wide line of length l assembled in a TAS, if the number of tiles in l is greater than the number of tile types in the TAS, then at least two tiles of l must have the same type. Therefore, by repeating the tiles between these two tiles of the same type, we can attempt to grow a line indefinitely. Hence, when a TAS attempts to grow a `finger` that is longer than the number of tile types in the TAS, we can always find an assembly sequence such that the line forming this `finger` places a tile one tile location above the tiles forming the `planter`. Figure 4 depicts this invalid assembly. Therefore, no TAS can assemble W.

5.2   A Shape in the aTAM But Not the DaTAM

In this section, we give a high-level sketch of the proof that there exists a shape which can self-assemble in the aTAM at $\tau = 2$ but not in the DaTAM at $\tau = 1$.

Theorem 4. *There exists a shape $S \subset \mathbb{Z}^2$ such that there exists a TAS $\mathcal{T} = (T, \sigma, 2)$ in the aTAM which self-assembles S, but no DTAS $\mathcal{D} = (T_\mathcal{D}, S_\mathcal{D}, D_\mathcal{D}, \sigma', 1)$ in the DaTAM which self-assembles S.*

See Figure 5 for a high-level sketch of a portion of the infinite shape, which is based on the shape used in the proof of Theorem 4.1 of [4] (which in turn is based on that of Theorem 4.1 of [15]). Essentially, $\mathcal{T}$ assembles S in the following way. Beginning from the seed, it grows a module called the `planter` eastward. The `planter` is a modified log-height counter which counts from 1 to ∞, and for each number - at a well-defined location - places a binary representation of that number on its north side. From each such location, modules called `rays` and Turing machine simulations begin. Each `ray` grows at a unique and carefully defined slope so that it can direct the growth of its adjacent Tur-

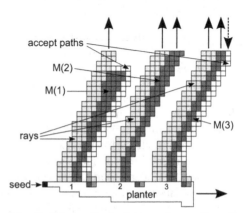

Fig. 5. Schematic depiction of a portion of the infinite shape which can self-assemble in the aTAM at $\tau = 2$ but not in the DaTAM at $\tau = 1$.

ing machine simulation in such a way the no Turing machine simulation will collide with another `ray`, but it also potentially has infinite tape space for its computation. The infinite series of Turing machine computations each run the same machine, M, on input n where n is the value presented by the `planter` at that location. If and only if each computation halts and accepts, a path of tiles grows down along the right side of the computation until it reaches a position from which it grows a vertical path of tiles directly downward to crash into the `planter` (blue in Figure 5). It's important that the height of the vertical portions (blue) of the paths increase for each. If and when a path places a tile adjacent to the `planter`, glue cooperation between the final tile of the path and a `planter` tile allow for the placement of a final tile (yellow in Figure 5). S is the infinite shape resulting from the growth of all portions.

The reason that S cannot assemble in the DaTAM at $\tau = 1$ is that glue cooperation cannot be used to place the yellow tiles, so each must be able to attach to just a tile in the blue portion of a path or the `planter` tile in a green location. It is impossible for all yellow tiles to be placed correctly because if they attach to 1) the blue portions of paths, since those get arbitrarily long, they must have repeating tile types which could be used to grow blue paths of the wrong height which allow yellow tiles to attach too far above the planter, or 2) the `planter` tiles, then the `planter` would have to be able to allow yellow tiles to attach exactly in all positions corresponding to halting and accepting computations, but the Turing machine being simulated accepts a language which is computably enumerable but not decidable, thus that is impossible. Thus, no DaTAM system can assemble S.

5.3   A DaTAM System Which Cannot Be Simulated by the aTAM

In this section, we give a single directed DaTAM system $\mathcal{D}$ at $\tau = 1$ which cannot be simulated by any aTAM system at $\tau = 2$. The fact that the aTAM at temperature 2 is incapable of simulating a single directed temperature 1 DTAS shows that the addition of duples fundamentally changes the aTAM model. The DTAS constructed in this section is similar to the system given in Section 5.1. See Figure 6 for a depiction of a producible assembly of $\mathcal{D}$. In order to show that the aTAM cannot simulate this DTAS, we use a technique used in [19]. This technique relies on the notion of a *window movie* (we modify some of the definitions given in [19]).

Theorem 5. *There exists a single directed DaTAM system $\mathcal{D} = (T_{\mathcal{D}}, S, D, \sigma, \tau)$ such that $\mathcal{D}$ cannot be simulated by any temperature 2 aTAM system.*

To prove Theorem 5, we prove that there is no TAS that can simulate the DTAS, $\mathcal{D}$, described as follows. First, the system is identical to the DTAS described in Section 5.1 with one exception. Just as with the DTAS in Section 5.1, $\mathcal{D}$ grows a `planter`, vertical counters and `fingers`. In addition to these subassemblies, $\mathcal{D}$ grows an 8 tile long, single tile wide line, l, of tiles from the base of each vertical counter. As seen in Figure 6, l, consisting of tiles $S_1, S_2, \ldots, S_8$, grows to the left of each vertical counter and

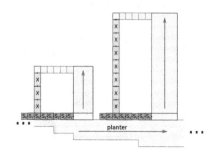

Fig. 6. A portion of a producible assembly of the temperature 1 DaTAM system which cannot be simulated in the aTAM at $\tau = 2$.

extends past the single tile wide gap between a `finger` and the `planter`. The intuitive idea behind the proof of Theorem 5 is that for any aTAM system, $\mathcal{T}$, that attempts to simulate $\mathcal{D}$, it must be able to simulate the growth of `fingers`. Therefore, for any $n > 0$, $\mathcal{T}$ must be able to grow a subassembly that simulates a `finger` consisting of n duples. This subassembly must have a constant width based on the block replacement scheme used in the simulation with block size m say, and a length of roughly $2nm$. We show that any such system $\mathcal{T}$ capable of such growth must also grow a simulated `finger` that crashes into the simulation of the `planter`. To show this, we use a window movie lemma (similar to

Lemma 3.1 in [19]). The lemma shown here holds for closed rectangular windows (in this version of the paper, we omit the details and a formal statement of the window movie lemma used here). Then, since the simulated `planter`, `finger`, and vertical counter separate the infinite grid-graph into two disjoint sets, there is no way to ensure that a subassembly representing the tile labeled S_8 of $\mathcal{T}$ grows *only* after the subassemblies representing the tiles labeled $S_1, S_2, \ldots, S_5$ grow. In other words, there is no way to ensure that $\mathcal{T}$ and $\mathcal{D}$ have equivalent dynamics, and therefore $\mathcal{T}$ does not simulate $\mathcal{D}$.

5.4   An aTAM System Which Cannot Be Simulated by the DaTAM

In Section 5.3 we showed the aTAM can't simulate every DaTAM system. Here we show the converse; the DaTAM can't simulate all aTAM systems. The particular aTAM system that we show can't be simulated by the DaTAM is the same given in [19] that is used to show that temperature 1 aTAM systems cannot simulate every temperature 2 aTAM system. Intuitively, this shows that cooperation, which is possible for temperature 2 aTAM systems, cannot be simulated using duples when temperature is restricted to 1. (See Figure 7 for the tile set.)

Theorem 6. *There exists a temperature 2 aTAM system $\mathcal{T} = (T, \sigma, 2)$ such that $\mathcal{T}$ cannot be simulated by any temperature 1 DaTAM system.*

Here we give a brief overview of the TAS, $\mathcal{T}$, that we show cannot be simulated by any DTAS and provide a sketch of the proof.

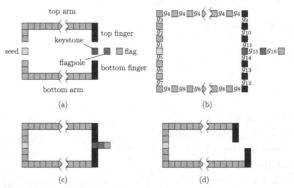

Fig. 7. (Figure taken from [19]) (a) An overview of the tile assembly system $\mathcal{T} = (T, \sigma, 2)$. $\mathcal{T}$ runs at temperature 2 and its tile set T consists of 18 tiles. (b) The glues used in the tileset T. Glues g_{11} and g_{14} are strength 1, all other glues are strength 2. Thus the keystone tile binds with two "cooperative" strength 1 glues. Growth begins from the pink seed tile σ: the top and bottom arms are one tile wide and grow to arbitrary, nondeterministically chosen, lengths. Two blue figures grow as shown. (c) If the fingers happen to meet then the keystone, flagpole and flag tiles are placed, (d) if the fingers do not meet then growth terminates at the finger "tips".

See Figure 7 for an overview of the TAS $\mathcal{T}$. The proof that there is no DTAS that simulates $\mathcal{T}$ is briefly described as follows. For any DTAS, $\mathcal{D}$, that attempts to simulate $\mathcal{T}$, it is shown that $\mathcal{D}$ is capable of an invalid assembly sequence. Intuitively, the idea is that when an arm (the bottom arm say) is sufficiently long, an assembly sequence in $\mathcal{D}$ of a subassembly α that represents this arm must contain repetition. Using a window movie lemma similar to Lemma 3.3 in [19], this repetition is removed to produce an assembly in $\mathcal{D}$ that is essentially equivalent to removing a section of tiles from α and splicing together the exposed ends along matching glues. This results in a shorter arm α' that still attempts to grow a keystone and flagpole, and hence leads to an invalid simulation of $\mathcal{T}$.

References

1. Adleman, L., Cheng, Q., Goel, A., Huang, M.D.: Running time and program size for self-assembled squares. In: Proceedings of the 33rd Annual ACM Symposium on Theory of Computing, Hersonissos, Greece, pp. 740–748 (2001)
2. Barish, R.D., Schulman, R., Rothemund, P.W.K., Winfree, E.: An information-bearing seed for nucleating algorithmic self-assembly. Proceedings of the National Academy of Sciences 106(15), 6054–6059 (2009), http://dx.doi.org/10.1073/pnas.0808736106
3. Behsaz, B., Maňuch, J., Stacho, L.: Turing universality of step-wise and stage assembly at temperature 1. In: Stefanovic, D., Turberfield, A. (eds.) DNA 2012. LNCS, vol. 7433, pp. 1–11. Springer, Heidelberg (2012), http://dx.doi.org/10.1007/978-3-642-32208-2_1
4. Bryans, N., Chiniforooshan, E., Doty, D., Kari, L., Seki, S.: The power of nondeterminism in self-assembly. Theory of Computing 9, 1–29 (2013)
5. Chen, H.-L., Kao, M.-Y.: Optimizing tile concentrations to minimize errors and time for dna tile self-assembly systems. In: Sakakibara, Y., Mi, Y. (eds.) DNA 16. LNCS, vol. 6518, pp. 13–24. Springer, Heidelberg (2011)
6. Chen, H.L., Schulman, R., Goel, A., Winfree, E.: Reducing facet nucleation during algorithmic self-assembly. Nano Letters 7(9), 2913–2919 (2007), http://dx.doi.org/10.1021/nl070793o
7. Cheng, Q., Aggarwal, G., Goldwasser, M.H., Kao, M.Y., Schweller, R.T., de Espanés, P.M.: Complexities for generalized models of self-assembly. SIAM Journal on Computing 34, 1493–1515 (2005)
8. Cook, M., Fu, Y., Schweller, R.T.: Temperature 1 self-assembly: Deterministic assembly in 3D and probabilistic assembly in 2D. In: Proceedings of the 22nd Annual ACM-SIAM Symposium on Discrete Algorithms (2011)
9. Cook, M., Fu, Y., Schweller, R.T.: Temperature 1 self-assembly: Deterministic assembly in 3D and probabilistic assembly in 2D. In: SODA 2011: Proceedings of the 22nd Annual ACM-SIAM Symposium on Discrete Algorithms. SIAM (2011)
10. Demaine, E.D., Patitz, M.J., Rogers, T.A., Schweller, R.T., Summers, S.M., Woods, D.: The two-handed assembly model is not intrinsically universal. In: Fomin, F.V., Freivalds, R., Kwiatkowska, M., Peleg, D. (eds.) ICALP 2013, Part I. LNCS, vol. 7965, pp. 400–412. Springer, Heidelberg (2013)
11. Doty, D.: Theory of algorithmic self-assembly. Commun. ACM 55(12), 78–88 (2012), http://doi.acm.org/10.1145/2380656.2380675
12. Doty, D., Lutz, J.H., Patitz, M.J., Schweller, R.T., Summers, S.M., Woods, D.: The tile assembly model is intrinsically universal. In: Proceedings of the 53rd Annual IEEE Symposium on Foundations of Computer Science, FOCS 2012, pp. 302–310 (2012)
13. Doty, D., Patitz, M.J., Summers, S.M.: Limitations of self-assembly at temperature 1. Theoretical Computer Science 412, 145–158 (2011)
14. Hendricks, J., Padilla, J.E., Patitz, M.J., Rogers, T.A.: Signal transmission across tile assemblies: 3D static tiles simulate active self-assembly by 2D signal-passing tiles. In: Soloveichik, D., Yurke, B. (eds.) DNA 2013. LNCS, vol. 8141, pp. 90–104. Springer, Heidelberg (2013), http://dx.doi.org/10.1007/978-3-319-01928-4_7

15. Lathrop, J.I., Lutz, J.H., Patitz, M.J., Summers, S.M.: Computability and complexity in self-assembly. Theory Comput. Syst. 48(3), 617–647 (2011)
16. Lathrop, J.I., Lutz, J.H., Summers, S.M.: Strict self-assembly of discrete Sierpinski triangles. Theoretical Computer Science 410, 384–405 (2009)
17. Mao, C., LaBean, T.H., Relf, J.H., Seeman, N.C.: Logical computation using algorithmic self-assembly of DNA triple-crossover molecules. Nature 407(6803), 493–496 (2000)
18. Maňuch, J., Stacho, L., Stoll, C.: Two lower bounds for self-assemblies at temperature 1. Journal of Computational Biology 17(6), 841–852 (2010)
19. Meunier, P.E., Patitz, M.J., Summers, S.M., Theyssier, G., Winslow, A., Woods, D.: Intrinsic universality in tile self-assembly requires cooperation. In: Proceedings of the ACM-SIAM Symposium on Discrete Algorithms (SODA 2014), Portland, OR, USA, January 5-7, pp. 752–771 (2014)
20. Patitz, M.J., Schweller, R.T., Summers, S.M.: Exact shapes and turing universality at temperature 1 with a single negative glue. In: Cardelli, L., Shih, W. (eds.) DNA 17. LNCS, vol. 6937, pp. 175–189. Springer, Heidelberg (2011), http://dl.acm.org/citation.cfm?id=2042033.2042050
21. Patitz, M.J., Summers, S.M.: Self-assembly of decidable sets. Natural Computing 10(2), 853–877 (2011)
22. Patitz, M.: An introduction to tile-based self-assembly and a survey of recent results. Natural Computing, 1–30 (2013), http://dx.doi.org/10.1007/s11047-013-9379-4
23. Reif, J.H., Sahu, S., Yin, P.: Compact error-resilient computational DNA tiling assemblies. In: Ferretti, C., Mauri, G., Zandron, C. (eds.) DNA 2004. LNCS, vol. 3384, pp. 293–307. Springer, Heidelberg (2005)
24. Rothemund, P.W.K., Papadakis, N., Winfree, E.: Algorithmic self-assembly of DNA sierpinski triangles. PLoS Biol. 2(12), e424 (2004), http://dx.doi.org/10.1371%2Fjournal.pbio.0020424
25. Rothemund, P.W.K., Winfree, E.: The program-size complexity of self-assembled squares (extended abstract). In: STOC 2000: Proceedings of the Thirty-second Annual ACM Symposium on Theory of Computing, pp. 459–468. ACM, Portland (2000)
26. Schulman, R., Winfree, E.: Synthesis of crystals with a programmable kinetic barrier to nucleation. Proceedings of the National Academy of Sciences 104(39), 15236–15241 (2007)
27. Schulman, R., Yurke, B., Winfree, E.: Robust self-replication of combinatorial information via crystal growth and scission. Proc. Natl. Acad. Sci. U.S.A. 109(17), 6405–6410 (2012), http://www.biomedsearch.com/nih/Robust-self-replication-combinatorial-information/22493232.html
28. Soloveichik, D., Cook, M., Winfree, E.: Combining self-healing and proofreading in self-assembly. Natural Computing 7(2), 203–218 (2008), http://dblp.uni-trier.de/db/journals/nc/nc7.html#SoloveichikCW08
29. Soloveichik, D., Winfree, E.: Complexity of self-assembled shapes. SIAM Journal on Computing 36(6), 1544–1569 (2007)
30. Winfree, E.: Algorithmic Self-Assembly of DNA. Ph.D. thesis. California Institute of Technology (June 1998)
31. Winfree, E., Bekbolatov, R.: Proofreading tile sets: Error correction for algorithmic self-assembly. In: Chen, J., Reif, J.H. (eds.) DNA 2003. LNCS, vol. 2943, pp. 126–144. Springer, Heidelberg (2004), http://dblp.uni-trier.de/db/conf/dna/dna2003.html#WinfreeB03
32. Winfree, E., Liu, F., Wenzler, L.A., Seeman, N.C.: Design and self-assembly of two-dimensional DNA crystals. Nature 394(6693), 539–544 (1998)

A Lin-Kernighan Heuristic for the DCJ Median Problem of Genomes with Unequal Contents

Zhaoming Yin[2], Jijun Tang[1,3,*], Stephen W. Schaeffer[4], and David A. Bader[2,*]

[1] School of Computer Science and Technology, Tianjin University, China
[2] School of Compuational Science and Engineering,
Georgia Institute of Technology, USA
[3] Dept. of Computer Science and Engineering, University of South Carolina, USA
[4] The Huck Institutes of Life Sciences, Pennsylvania State University, USA

Abstract. In this paper, we designed a distance metric as *DCJ-Indel-Exemplar* distance to estimate the dissimilarity between two genomes with unequal contents (with gene insertions/deletions (*Indels*) and duplications). Based on the aforementioned distance metric, we proposed the *DCJ-Indel-Exemplar* median problem, to find a median genome that minimize the *DCJ-Indel-Exemplar* distance between this genome and the given three genomes. We adapted *Lin-Kernighan* (*LK*) heuristic to calculate the median quickly by utilizing the features of adequate subgraph decomposition and search space reduction technologies. Experimental results on simulated gene order data indicate that our distance estimator can closely estimate the real number of rearrangement events; while compared with the exact solver using equal content genomes, our median solver can get very accurate results as well. More importantly, our median solver can deal with *Indels* and duplications and generates results very close to the synthetic cumulative number of evolutionary events.

Keywords: Genome Rearrangement, Double-cut and Join (*DCJ*), Lin-Kernighan Heuristic.

1 Introduction

Inferring phylogenies (evolutionary history) of a set of given species is a fundamental problem in computational biology [23]. For decades, biologists and computer scientists have studied how to infer phylogenies by the measurement of genome rearrangement events using gene order data [13]. While evolution is not an inherently parsimonious process, maximum parsimony (*MP*) phylogenetic analysis has been widely applied to the phylogeny inference to study the evolutionary patterns of genome rearrangements. Given the input of gene order data with unequal contents (with gene insertions/deletions and duplications of genes), even the computation of distance between two genomes with only duplications is **NP**-hard [7,9,10] and **APX**-hard [1,11] by various distance measurement methods. There are attempts to perform phylogenetic reconstruction from

---

* Corresponding authors.

Z. Cai et al. (Eds.): COCOON 2014, LNCS 8591, pp. 227–238, 2014.
© Springer International Publishing Switzerland 2014

genome rearrangement data with unequal gene content, which can be roughly divided into distance-based methods [28], *MP* methods [29] and adjacency-based methods [17]. However, the first two approaches are generally quite limited by methods in distance and median computation.

Various distance metrics have been proposed to calculate the dissimilarity between two genomes, such as breakpoint distance [4], signed reversal distance [2], translocation distance [15], and Double-cut-and-join (*DCJ*) distance [34], which is currently the most extensively studied. However, there are still a lot of unclear subjects in distance computation between unequal content genomes, and computational biologists tried multiple ways to surpass this limit. Traditional approaches are based on breakpoint or reversal distances, such as efforts of employing exemplar distance [21, 25] to keep only one copy of duplicated gene families, or the methods by extending polynomial time reversal distance algorithm introduced by Hannenhalli Pevzner (*HP*), to handle *Indels* as well as duplications [18]. Contemporary research focusing on unequal contents are more concerned on *DCJ* model: For genomes with *Indels* only, there are exact algorithms to compute their *DCJ* distance [6, 12]; For genomes with duplications, there are several very useful methods to approximate or compute the exact *DCJ* distance [26, 27]. However, there are few efforts to combine these methods to measure distance of genomes with gene orders that contain both *Indels* and duplications.

The median problem is defined as to find a genome that minimizes sum of distances from itself to the three input genomes [5, 19]; it's **NP**-hard under most distance metrics [3, 8, 22, 31]. Several exact algorithms have been implemented to solve the *DCJ* median problems on both circular [31, 33] and linear chromosomes [30, 32]. Some heuristics are introduced to improve the speed of median computation, such as linear programming (*LP*) [8], local search [16], evolutionary programming [14], or simply searching on one promising direction [24]. As all these algorithms are intended for solving the median problems with equal content genomes, their usage is limited in practice.

2   Background

2.1   Genome Rearrangement Events and Their Graph Representations

Genome Rearrangement Events. The content of the *DNA* molecules are often similar, but their organizations often differ dramatically. The mutation that affect the organization of genes are called genome rearrangements. Fig 1 shows examples of different rearrangement events of a single chromosome. In the examples, we use signed numbers to represent different genes and their orientation in the genome strand. Genome rearrangements events involve with multiple combinatorial optimization problems, and graph representation is a very common way to abstract these problems. In this part, we will address the foundations of using breakpoint graph to model the genome rearrangement events.

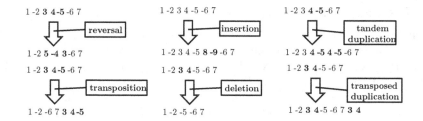

Fig. 1. Example of different rearrangement events

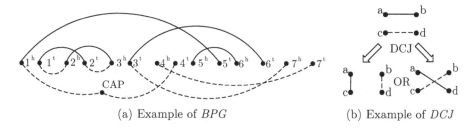

(a) Example of *BPG*   (b) Example of *DCJ*

Fig. 2. Examples of *BPG*; and *DCJ* operations

Breakpoint Graph. Given an alphabet $\mathcal{A}$, and two genomes Γ and Π are represented by two strings of signed ($+$ or $-$) symbols (representing genes) from $\mathcal{A}$. Each gene $a \in \mathcal{A}$ is represented by a pair of vertices head a_h and tail a_t, if a is positive a_h is putted in front of a_t, otherwise a_t is putted in front of a_h. For $a \in \mathcal{A}$ and $b \in \mathcal{A}$, if $a \in \Gamma$ (or Π) and $b \in \Gamma$ (or Π), and in Γ (or Π) a and b are adjacent to each other, their adjacent vertices will be connected by an edge. As for telomere genes, if they exist in a circular chromosome, two end vertices will be connected by an edge, and if they exist in a linear chromosome, two end vertices will be connected to a special vertex called *CAP* vertex. If we use one type of edges to represent adjacencies of Γ and another type of edges to represent adjacencies of Π, the resulting graph with two types of edges is called breakpoint graph (*BPG*). Fig 2(a) shows the *BPG* for gene order Γ (1,-2,3,-6,5) (solid edges) which is a genome with one circular chromosome and Π (1,2,3,7,4) (dashed edges) which is a genome with one linear chromosome.

DCJ Operation. Double-cut and join (*DCJ*) operations are able to simulate all aforementioned rearrangement events applying *BPG*. The operations cut two edges (within one genome) and rejoin them using two possible combinations of end vertices (shown in Fig 2(b)). *DCJ* distance of genomes with the same content can be easily calculated by enumerating the number of cycles/paths in the *BPG*, which is linear [34]. Comparing with the complex model based on reversal operations, *DCJ* operations are simple and powerful.

2.2 Distance Computation

In the *BPG* with two genomes Γ and Π, the vertices and the edges of a closed walk form a cycle. In Fig 2(a), the walk $(1^t, (1^t; 2^h), 2^h, (2^h; 3^h), 3^h, (3^h; 2^t), 2^t, (2^t; 1^t), 1^t)$ is a cycle. A vertex v is π-*open* (γ-*open*) if $v \notin \Gamma$ ($v \notin \Pi$). An unclosed walk in *BPG* is a path. Based on different kinds of end points of the paths, we can classify paths into different types. If the two ends of a path are *CAP* vertices, we simply denote this path as p^0. If a path is ended by one open vertex and one *CAP*, we denote it as p^π (p^γ). If a path is ended by two open vertices, it is denoted by the type of its two open vertices, for example, $p^{\pi,\gamma}$ represent a path that ends with a π-*open* vertex and a γ-*open* vertex. In Fig 2(a), the walk $(5^t, (5^t; 1^h), 1^h, (1^h; CAP), CAP)$ is a p^γ path, and the walk $(6^t, (6^t; 3^t), 3^t, (3^t; 7^h), 7^h)$ is a $p^{\gamma,\pi}$ path. A path is even (odd), if it contains even (odd) number of edges. In [12], the *DCJ* distance between two genomes with *Indels* but without duplications is calculated by equation (1). We call this distance *DCJ-Indel* distance. From this equation, we can easily get the *DCJ-Indel* distance between Γ and Π in Fig 2(a) as 4.

$$distance_{indel}(\Gamma, \Pi) = N - [c + p^{\pi,\pi} + p^{\gamma,\gamma} + \lfloor p^{\pi,\gamma} \rfloor]$$
$$+ \frac{1}{2}(p^0_{even} + min(p^\pi_{odd}, p^\pi_{even}) + min(p^\gamma_{odd}, p^\gamma_{even}) + \delta) \tag{1}$$

Where $\delta = 1$ only if $p^{\pi,\gamma}$ is odd and either $p^\pi_{odd} > p^\gamma_{even}, p^\gamma_{odd} > p^\gamma_{even}$ or $p^\pi_{odd} < p^\gamma_{even}, p^\gamma_{odd} < p^\gamma_{even}$; Otherwise, $\delta = 0$.

There are in general two approaches to cope with duplicated genes. One is by removing all but keeping one copy of duplications in gene family to generate an exemplar pair [25] and another is by relabling duplicates such that all duplicated genes will have an unique label [26, 27]. Lastly, mathematically optimized distance might not reflect the true number of biological events, distance estimation methods such as *EDE* or *IEBP* are used to rescale these computed distances [20].

2.3 Median Computation

If there are three given genomes, the graph constructed by borrowing the previous defined rule in *BPG* is called Multiple Breakpoint Graph (*MBG*). Figure 3(a) shows an example of *MBG*, With the input of three genomes: (1,2,3,4) (solid edges); (1,2,-3,4) (dashed edges) and (2,3,1,-4) (dotted edges). The *DCJ* median algorithm can be briefly described by a branch and bound (*BnB*) process [30, 31, 33] on *MBG*, which is to find a maximum matching (which is called 0-*matching*) in *MBG*. Figure 3(b) shows an example of 0-*matching* which is represented by gray edges. In [30, 31, 33], it's been proved that a type of sub-graph called adequate sub-graph (*AS*) could be used to decompose the graph with edge shrinking operations. Figure 3(c) shows an example of *AS* and edge shrinking. The *BnB* algorithm is served to solve the *DCJ* median problem with equal content genomes. Unfortunately, there is no *BnB* based algorithm that deals with unequal content cases, and we will show that it's actually hard to design such algorithm in the following section.

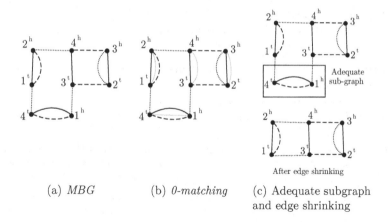

(a) *MBG*   (b) *0-matching*   (c) Adequate subgraph
                                                 and edge shrinking

Fig. 3. Examples of *MBG*; 0-matching and edge shrinking operations

3 Approaches

3.1 Applying *DCJ-Indel-Exemplar* Distance to Evaluate Dissimilarity

The *DCJ-Indel* distance can handle genomes which only have *Indels*, while the exemplar distance can only handle duplications. To process genomes with both *Indels* and duplications, a new distance metric named *DCJ-Indel-Exemplar* distance is designed by combining these two distances together. For gene families with duplicated genes, only one gene copy of a gene family in each genome is selected, and the rest of the gene copies in the gene family are deleted from both of the genomes. The resulting genomes are called 'exemplar' genomes. Of all possible selection of exemplar genomes, the one with the minimum *DCJ-Indel* distance is the *DCJ-Indel-Exemplar* distance for the original two genomes.

The *DCJ-Indel-Exemplar* distance does not reflect the true number of evolutionary events. For one thing, the number of duplications are not counted; furthermore, when there are large number of mutations, *DCJ* distance will underestimate the distance. Therefore, two steps are followed to adjust the *DCJ-Indel-Exemplar* distance. The first step is to use *EDE* [20] to rescale the distance. The second step is to add the count of duplicated genes by comparing the difference of the count of the same gene family in two genomes, if they are different, a duplication count is added. The *DCJ-Indel-Exemplar* distance after the adjustment of *EDE* distance and the addition of number of duplications is the final distance.

3.2 Adapting *Lin-Kernighan* Heuristic to Find Median

Problem Statement. Not surprisingly, finding the median genome that minimize the *DCJ-Indel-Exemplar* distance, is challenging. To begin with, given three input genomes, there are multiple choices of possible gene content selections for a median genome. Therefore, to make the problem easier, we can define a relaxed version of the median problem by providing known gene contents.

DCJ-Indel-Exemplar median

Instance. Given the gene content of a median genome, and gene orders of three modern genomes.

Question. Find an adjacency of the genes of the median genome that minimize the *DCJ-Indel-Exemplar* distance between the median genome and the three input genomes.

The *DCJ-Indel-Exemplar* median problem is not even in the class of **NP** because there is no polynomial time algorithm to verify the results. Furthermore, it's hard to design an exact *BnB* algorithm for *DCJ-Indel-Exemplar* median problem mainly because: To begin with, distance under DCJ does not hold when considering *Indels* [35]. when a *0-matching* edge is selected, edge shrinking is performed to generate the new *MBG*. The question is, when there are duplicated genes in a genome, it's possible that there are multiple edges of the same type connecting to the same vertex of a *0-matching*. This leads to ambiguity in the edge shrinking step, which makes the followed *BnB* search process very complicated and extremely hard to implement. Hence, we provided an adaption of Lin-Kernighan (*LK*) heuristic to help solving this challenging problem.

Design of *Lin-Kernighan* Heuristic. The *LK* heuristic can generally be divided into two steps: initialization of 0-*matching* for the median genome, and *LK* search to get the result.

The initialization problem can be described as: given gene contents of three genomes, find a median genome gene content that minimizes the sum of the number of *Indels* and duplications operations that transfer the median gene content to gene contents of other three genomes. In this paper, we designed a very simple rule to initialize the gene content of the median genome, which is, given the counts of one gene family of three genomes. If two or three counts are the same, we simply select this count as the number of occurence of the gene family in the median genome. If all three counts are different, we select the median count as the number of occurence of the gene family in the median genome.

After fixing the gene content for median genome, the next step is to set up the *0-matching* in the *MBG* and perform the *LK* heuristic. In this paper, we randomly set up the *0-matching*. As for the *LK* strategy, by selecting two *0-matching* edges on *MBG* of a given search node, and perform a *DCJ* operation, we can get the *MBG* of a neighbor search node. We expand the search frontier by keeping all neighboring search nodes to up until the search level $L1$. Then we only examine and add the most promising neighbors to the search list until level $L2$. The search is continued by the time when there is a neighbor solution yielding a better median score. This solution is then accepted and with it a new search is initiated from the scratch. The search will be terminated if there are no improvement on the result as the search level limit has been reached and all possible neighbors has been enumerated. If $L1 = L2 = K$, the algorithm is called K-OPT algorithm.

Adopting Adequate Subgraphs to Simplify Problem Space. There are two categories of vertices in the MBG. One connected with exactly one edge of each edge type, is called "regular" vertices; another connected with less or more than one edges of each edge type, is classified as "irregular" vertices. A subgraph in the MBG that only contains regular vertices, is defined as regular subgraph [30]. By using the adequate subgraphs [30,33], we can prove that they are still applicable for decomposing the graph in DCJ-$Indel$-$Exemplar$ median problem.

Lemma 1. *As long as the irregular vertices do not involve, regular subgraphs are applicable to decompose* MBG.

Proof. If there are d number of vertices that contain duplicated edges in MBG, then we can disambiguate the MBG by generating different subgraphs that contain only one of the duplicate edges (we call these subgraphs disambiguate MBG, d-MBG). And there are $O(\prod_{i<d} deg(i))$ number of d-MBGs. Suppose a regular adequate subgraph exists in the MBG, then it must also exist in every d-MBG. Based on the *0-matching* solution, we can transform every d-MBG into completed d-MBG (cd-MBG) by constructing the optimal completion [12] between *0-matching* and all the other 3 types of edges. After this step, the adequate subgraphs exist in every d-MBG still exist in every cd-MBG. Which means, we can use these adequate subgraphs to decompose cd-MBG for each median problem without losing accuracy.   □

Search Space Reduction Methods. The performance bottleneck with the median computation is in the exhaustive search step, because for each search level we need to consider $O(2g)^2$ possible number of edge pairs, which is $O((2g)^{2L1})$ in total. In traveling salesman problem (TSP), it's cheap to find the best neighbor, but for DCJ operations, to evaluate a neighbor, we need to compute NP-hard DCJ-$Indel$-$Exemplar$ distance, which makes this step extremely expensive to conclude. Noticing that if we search neighbors on edges that are on the same *0-i* color altered connected component (*0-i-comp*), the DCJ-$Indel$-$Exemplar$ distance for genome 0 and genome i is more likely to reduce [36]. We can sort each edge pair by how many *0-i-comp* they share. Suppose the number of *0-i-comp* that an edge pair x share is $num\_pair(x)$. When the algorithm is in the exhaustive search step ($currentLevel < L1$), we set a threshold δ and select the edge pairs that satisfy: $num\_pair(x) > \delta$ to be added into the search list. When it comes to the recursive deepening step; we select the edge pair that satisfy $\underset{x}{argmax}\ num\_pair(x)$ to be added into the search list. This strategy has two merits, 1) some of the non-promising neighbor solution is eliminated to reduce the search space. 2) the expensive evaluation step which make a function call to DCJ-$Indel$-$Exemplar$ distance is postponed to the time when a solution is retrieved from the search list.

4   Experimental Results

Distance Estimation. We simulated the data sets using genomes with 200 genes. To show how *Indels* and duplications affect the estimation of the distance, we divide the data set into multiple groups with varied *Indels* rate (γ, which varies from 5% to 10%), and duplication rate (ϕ, which varies from 5% to 10% as well). For each *Indels* or duplication event, only one gene is inserted/deleted or duplicated. We compare the change of distance estimation with the change of mutation rate (θ, which varies from 10% to 100%, we used reversal operation to simulate the mutation mainly because *DCJ* distance and reversal distance are quite similar when using genome data of same contents), with each specific setting of γ and ϕ. With two genomes (one is called target and the other is called subject) we conduct experiments on two sets of data. One set of data that set target genome as identity genome (for example $(1, 2, 3, ..., i, j, ..., n)$), and the subject genome is evolved from the identity genome with full ratio of θ, γ, ϕ, we call this set *'identity'*. Another set of data assigns half ratio of θ, γ, ϕ to both of target and subject genomes to let them evolve from identity genome, we call this set *'dual'*.

The result for *DCJ-Indel-Exemplar* distance and *DCJ-Indel-Exemplar* distance corrected by *EDE* are shown in Fig 4. As for the impact of different evolution operation rates, the main factor that affects the accuracy of distance estimation is the change of rate γ and ϕ. This is mainly because an *Indel* after a duplication can cancel the count of both *Indel* and duplication and makes the distance underestimated. As for the effect of two different data sets, it seems that the *'dual'* set underestimates the result more than *'identity'* set, which is mainly because both of two genomes will delete a common set of genes, which makes the actual size of alphabet $\mathcal{A}$ shrunk.

Median Computation. We simulate the median data of three genomes using the same simulation strategy as in the distance simulation. In our experiments, each genome is "evolved" from a seed genome, which is identity, and they all have the same evolution rate (θ, γ and ϕ). We compare the result of using *LK* algorithm with $L1 = 2$ and $L2 = 3$, and the *K-OPT* algorithm of $K = 2$. We use the search space reduction methods and set $\delta = 2$ and $\delta = 3$ respectively.

To test the accuracy of our *LK* and *K-OPT* methods, we first set both γ and ϕ to 0 and increased the mutation rate θ from 10% to 100%, so that each of the three genomes has the same gene content. We run the exact *DCJ* median solver (we use the one in [36]) to compare the exact result with our heuristic. In Fig 5(a), it shows the accuracy of our heuristic compared with the exact result. It is shown that when $\theta \leq 60\%$, all results of the *LK* and *K-OPT* methods are quite close to the exact solver. For parameter of $\delta = 2$, both *LK* and *K-OPT* methods can generate exact results for most of the cases.

As for the median results for unequal contents, we set both γ and ϕ to 5% and increase the mutation (inversion) rate θ from 10% to 60%. We compare our result with the accumulated distance of three genomes to their simulation seed. Although it can not show the accuracy of our method (since we do not have an

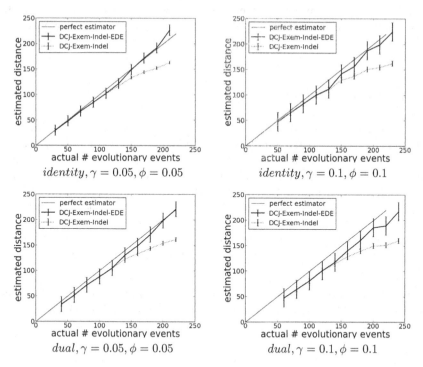

Fig. 4. Distance computation results, the x-axis represents the actual number of *DCJ* operations and the y-axis represent the computed distance for the methods using *DCJ-Indel-Exemplar* distance, *DCJ-Indel-Exemplar* distance rectified by *EDE*, and the true estimator. γ is the rate of *Indels* and ϕ is the rate of duplications. The results are grouped by two sets of data, which are *identity* and *dual*.

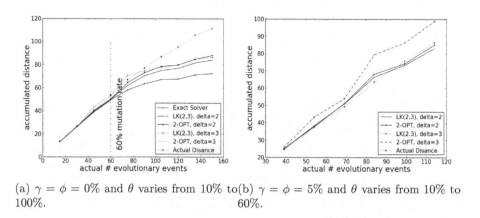

(a) $\gamma = \phi = 0\%$ and θ varies from 10% to 100%.

(b) $\gamma = \phi = 5\%$ and θ varies from 10% to 60%.

Fig. 5. Experimental results for median computation

exact solver), it can be used as an indicator of how close of our method was to the real evolution. Fig 5(b) shows the median results for unequal gene contents. It indicates that when $\delta = 3$, both *LK* and *K-OPT* algorithms get results quite close to the real evolutionary distance.

5   Conclusion

In this paper, we proposed a new way to compute the distance and median between genomes with unequal contents (with *Indels* and duplications). Nevertheless, there are still a lot of aspects to be improved. For example, we need to design a scheme to better estimate the gene contents. A way to deal with ambiguation when shrinking an edge is needed; therefore, a branch and bound algorithm could be designed to infer the exact median genome. Last but not least, since the *LK* algorithm can only process hundreds of genes, algorithm engineering and high performance computing methods are required to provide a way helping us to design faster algorithms to deal with high resolution data.

Acknowledgements. This Research was sponsored in part by the NSF OCI-0904461 (Bader), OCI-0904179, IIS-1161586 (Tang) and OCI- 0904166 (Schaeffer).

References

1. Angibaud, S., Fertin, G., Rusu, I., Thévenin, A., Vialette, S.: On the approximability of comparing genomes with duplicates. J. Graph Algorithms Appl. 13(1), 19–53 (2009)
2. Bader, D.A., Moret, B.M.E., Yan, M.: A linear-time algorithm for computing inversion distance between signed permutations with an experimental study. Journal of Computational Biology 8, 483–491 (2001)
3. Bergeron, A., Mixtacki, J., Stoye, J.: On sorting by translocations. Journal of Computational Biology, 615–629 (2005)
4. Blin, G., Chauve, C., Fertin, G.: The breakpoint distance for signed sequences. In: Proc. CompBioNets 2004. Text in Algorithms, vol. 3, pp. 3–16. King's College, London (2004)
5. Bourque, G., Pevzner, P.A.: Genome-Scale Evolution: Reconstructing Gene Orders in the Ancestral Species. Genome Res. 12(1), 26–36 (2002)
6. Braga, M.D.V., Willing, E., Stoye, J.: Genomic distance with DCJ and indels. In: Moulton, V., Singh, M. (eds.) WABI 2010. LNCS, vol. 6293, pp. 90–101. Springer, Heidelberg (2010)
7. Bryant, D.: The complexity of calculating exemplar distances. In: Sankoff, D., Nadeau, J. (eds.) Comparative Genomics. Kluwer (2001)
8. Caprara, A.: The Reversal Median Problem. INFORMS Journal on Computing 15(1), 93–113 (2003)
9. Chauve, C., Fertin, G., Rizzi, R., Vialette, S.: Genomes containing duplicates are hard to compare. In: Alexandrov, V.N., van Albada, G.D., Sloot, P.M.A., Dongarra, J. (eds.) ICCS 2006. Part II. LNCS, vol. 3992, pp. 783–790. Springer, Heidelberg (2006)

10. Chen, X., Zheng, J., Fu, Z., Nan, P., Zhong, Y., Lonardi, S., Jiang, T.: Assignment of orthologous genes via genome rearrangement. IEEE/ACM Trans. Comput. Biology Bioinform. 2(4), 302–315 (2005)
11. Chen, Z., Fu, B., Zhu, B.: Erratum: The approximability of the exemplar breakpoint distance problem. In: Snoeyink, J., Lu, P., Su, K., Wang, L. (eds.) FAW-AAIM 2012. LNCS, vol. 7285, p. 368. Springer, Heidelberg (2012)
12. Compeau, P.E.C.: A simplified view of dcj-indel distance. In: Raphael, B., Tang, J. (eds.) WABI 2012. LNCS (LNBI), vol. 7534, pp. 365–377. Springer, Heidelberg (2012)
13. Fertin, G., Labarre, A., Rusu, I., Tannier, E., Vialette, S.: Combinatorics of Genome Rearrangements, 1st edn. The MIT Press (2009)
14. Gao, N., Yang, N., Tang, J.: Ancestral genome inference using a genetic algorithm approach. PLoS One 8(5) (2013)
15. Hannenhalli, S.: Polynomial-time algorithm for computing translocation distance between genomes. Discrete Applied Mathematics 71(1-3), 137–151 (1996)
16. Lenne, R., Solnon, C., Stützle, T., Tannier, E., Birattari, M.: Reactive Stochastic Local Search Algorithms for the Genomic Median Problem. In: van Hemert, J., Cotta, C. (eds.) EvoCOP 2008. LNCS, vol. 4972, pp. 266–276. Springer, Heidelberg (2008)
17. Lin, Y., Hu, F., Tang, J., Moret, B.M.: Maximum likelihood phylogenetic reconstruction from high-resolution whole-genome data and a tree of 68 eukaryotes. In: Proc. 18th Pacific Symp. on Biocomputing, PSB 2013, pp. 285–296. IEEE Computer Society, Washington, DC (2013)
18. Marron, M., Swenson, K.M., Moret, B.M.E.: Genomic distances under deletions and insertions. In: Warnow, T., Zhu, B. (eds.) COCOON 2003. LNCS, vol. 2697, pp. 537–547. Springer, Heidelberg (2003)
19. Moret, B.M.E., Tang, J., San Wang, L., Warnow, Y.: Steps toward accurate reconstructions of phylogenies from gene-order data. J. Comput. Syst. Sci 65, 508–525 (2002)
20. Moret, B.M.E., Wang, L.S., Warnow, T., Wyman, S.K.: New approaches for reconstructing phylogenies from gene order data. In: ISMB (Supplement of Bioinformatics), pp. 165–173 (2001)
21. Nguyen, C.T., Tay, Y.C., Zhang, L.: Divide-and-conquer approach for the exemplar breakpoint distance. Bioinformatics 21(10), 2171–2176 (2005)
22. Pe'er, I., Shamir, R.: The median problems for breakpoints are np-complete. Elec. Colloq. on Comput. Complexity 71 (1998)
23. Pevzner, P.A.: Computational Molecular Biology: An Algorithmic Approach, 1st edn. Computational Molecular Biology. A Bradford Book (August 2000)
24. Rajan, V., Xu, A.W., Lin, Y., Swenson, K.M., Moret, B.M.E.: Heuristics for the inversion median problem. BMC Bioinformatics 11(S-1), 30 (2010)
25. Sankoff, D.: Genome rearrangement with gene families. Bioinformatics 15(11), 909–917 (1999)
26. Shao, M., Lin, Y.: Approximating the edit distance for genomes with duplicate genes under dcj, insertion and deletion. BMC Bioinformatics 13(S-19), S13 (2012)
27. Shao, M., Lin, Y., Moret, B.: An exact algorithm to compute the DCJ distance for genomes with duplicate genes. In: Sharan, R. (ed.) RECOMB 2014. LNCS (LNBI), vol. 8394, pp. 280–292. Springer, Heidelberg (2014)
28. Swenson, K.M., Marron, M., Earnest-DeYoung, J.V., Moret, B.M.E.: Approximating the true evolutionary distance between two genomes. In: Demetrescu, C., Sedgewick, R., Tamassia, R. (eds.) ALENEX/ANALCO, pp. 121–129. SIAM (2005)

29. Tang, J., Moret, B.M.E.: Phylogenetic reconstruction from gene-rearrangement data with unequal gene content. In: Dehne, F., Sack, J.-R., Smid, M. (eds.) WADS 2003. LNCS, vol. 2748, pp. 37–46. Springer, Heidelberg (2003)

30. Xu, A.W.: DCJ median problems on linear multichromosomal genomes: Graph representation and fast exact solutions. In: Ciccarelli, F.D., Miklós, I. (eds.) RECOMB-CG 2009. LNCS (LNBI), vol. 5817, pp. 70–83. Springer, Heidelberg (2009)

31. Xu, A.W.: A fast and exact algorithm for the median of three problem: A graph decomposition approach. Journal of Computational Biology 16(10), 1369–1381 (2009)

32. Xu, A.W., Moret, B.M.E.: Gasts: Parsimony scoring under rearrangements. In: Przytycka, T.M., Sagot, M.-F. (eds.) WABI 2011. LNCS (LNBI), vol. 6833, pp. 351–363. Springer, Heidelberg (2011)

33. Xu, A.W., Sankoff, D.: Decompositions of multiple breakpoint graphs and rapid exact solutions to the median problem. In: Crandall, K.A., Lagergren, J. (eds.) WABI 2008. LNCS (LNBI), vol. 5251, pp. 25–37. Springer, Heidelberg (2008)

34. Yancopoulos, S., Attie, O., Friedberg, R.: Efficient sorting of genomic permutations by translocation, inversion and block interchange. Bioinformatics 21(16), 3340–3346 (2005)

35. Yancopoulos, S., Friedberg, R.: Sorting genomes with insertions, deletions and duplications by DCJ. In: Nelson, C.E., Vialette, S. (eds.) RECOMB-CG 2008. LNCS (LNBI), vol. 5267, pp. 170–183. Springer, Heidelberg (2008)

36. Yin, Z., Tang, J., Schaeffer, S.W., Bader, D.A.: Streaming breakpoint graph analytics for accelerating and parallelizing the computation of dcj median of three genomes. In: ICCS, pp. 561–570 (2013)

Diffuse Reflection Radius in a Simple Polygon[*]

Eli Fox-Epstein[1], Csaba D. Tóth[2], and Andrew Winslow[3]

[1] Department of Computer Science, Brown University, Providence, RI 02912, USA
`ef@cs.brown.edu`
[2] Department of Mathematics, California State University,
Northridge, Los Angeles, CA 91330, USA
`cdtoth@acm.org`
[3] Department of Computer Science, Tufts University, Medford, MA 02155, USA
`awinslow@cs.tufts.edu`

Abstract. Light reflecting diffusely off of a surface leaves in all directions. It is shown that every simple polygon with n vertices can be illuminated from a single point light source s after at most $\lfloor (n-2)/4 \rfloor$ *diffuse reflections*, and this bound is the best possible. A point s with this property can be computed in $O(n \log n)$ time.

1 Introduction

When light diffusely reflects off of a surface, it scatters in all directions. This is in contrast to specular reflection, where the angle of incidence equals the angle of reflection. We are interested in the minimum number of diffuse reflections needed to illuminate all points in the interior of a simple polygon P with n vertices from a single light source s in the interior of P. A *diffuse reflection path* is a polygonal path γ contained in P such that every interior vertex of γ lies in the relative interior of some edge of P, and the relative interior of every edge of γ is in the interior of P (see Fig. 1 for an example). Our main result is the following.

Theorem 1. *For every simple polygon P with $n \geq 3$ vertices, there is a point $s \in \text{int}(P)$ such that for all $t \in \text{int}(P)$, there is an s-to-t diffuse reflection path with at most $\lfloor (n-2)/4 \rfloor$ internal vertices. This upper bound is the best possible. A point $s \in \text{int}(P)$ with this property can be computed in $O(n \log n)$ time.*

Our main result is, in fact, a tight bound on the diffuse reflection radius (defined below) for simple polygons. Denote by $V_k(s) \subseteq P$ the part of the polygon illuminated by a light source s after at most k diffuse reflections. Formally, $V_k(s)$ is the set of points $t \in P$ such that there is a diffuse reflection path from s to t with at most k interior vertices. Hence $V_0(s)$ is the visibility polygon of point s within the polygon P. The *diffuse reflection depth* of a point $s \in \text{int}(P)$ is the minimum $r \geq 0$ such that $\text{int}(P) \subseteq V_r(s)$. The *diffuse reflection radius* $R(P)$ of a simple polygon P is the minimum diffuse reflection depth over all

---

[*] A full version of this paper can be found at http://arxiv.org/abs/1402.5303

Z. Cai et al. (Eds.): COCOON 2014, LNCS 8591, pp. 239–250, 2014.
© Springer International Publishing Switzerland 2014

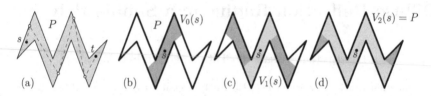

Fig. 1. (a) A diffuse reflection path between s to t in a simple polygon P. (b)–(d) The regions of a polygon illuminated by a light source s after 0, 1, and 2 diffuse reflections. The diffuse reflection radius of a zig-zag polygon with n vertices is $\lfloor (n-2)/4 \rfloor$.

points $s \in \mathrm{int}(P)$. The set of points $s \in \mathrm{int}(P)$ that attain this minimum is the *diffuse reflection center* of P. With this terminology, Theorem 1 implies that $R(P) \le \lfloor (n-2)/4 \rfloor$ for every simple polygon P with $n \ge 3$ vertices. A family of zig-zag polygons (see such polygon in Fig. 1) shows that this bound is the best possible for all $n \ge 3$. The *diffuse reflection diameter* $D(P)$ of P is the *maximum* diffuse reflection depth over all $s \in \mathrm{int}(P)$. Barequet et al. [6] recently proved, confirming a conjecture by Aanjaneya et al. [1], that $D(P) \le \lfloor n/2 \rfloor - 1$ for all simple polygons with n vertices, and this bound is the best possible.

Proof Technique. The regions $V_k(s)$ are notoriously difficult to handle. Brahma et al. [7] constructed examples where $V_2(s)$ is not simply connected, and where $V_3(s)$ has $\Omega(n)$ holes. In general, the maximum complexity of $V_k(s)$ is known to be $\Omega(n^2)$ and $O(n^9)$ [2]. Rather than consider $V_k(s)$, we use the simply connected regions $R_k(s) \subseteq V_k(s)$ defined by Barequet et al. [6] (reviewed in Section 2.1) and prove that $\mathrm{int}(P) \subseteq R_{\lfloor (n-2)/4 \rfloor}(s)$ for some point $s \in \mathrm{int}(P)$.

In Section 2, we establish a simple sufficient condition (Lemma 1) for a point s to determine if $\mathrm{int}(P) \subseteq R_{\lfloor (n-2)/4 \rfloor}(s)$. We use a generalization of the kernel of a simple polygon (Section 3.1) and the weak visibility polygon for a line segment (Section 3.2) to prove that there exists a point satisfying these considitions, with the exception of two extremal cases that are resolved directly (Section 2.2). The existential proof is turned into an efficient algorithm by computing the generalized kernel in $O(n \log n)$ time, and maintaining the visibility of a point moving along a line segment with a persistent data structure undergoing $O(n)$ updates in $O(\log n)$ time each.

Motivation and Related Work. The diffuse reflection path is a special case of a *link path*, which has been studied extensively due to its applications in motion planning, robotics, and curve compression [13,17]. The *link distance* between two points, s and t, in a simple polygon P is the minimum number of edges in a polygonal path between s and t that lies entirely in P. In a polygon P with n vertices, the link distance between two points can be computed in $O(n)$ time [20]. The *link diameter* of P, the maximum link distance between two points in P, can be computed in $O(n \log n)$ time [21]. The *link depth* of a point s is the smallest number d such that all other points in P are within link distance d of s. The *link radius* is the minimum over all link depths, and the *link center* is the set of points with minimum link depth. It is known that the link center is a convex region, and can be computed in $O(n \log n)$ time [12].

The *geodesic center* of a simple polygon is a point inside the polygon which minimizes the maximum internal (geodesic) distance to any point in the polygon. Pollack et al. [18] show how to compute the geodesic center of a simple polygon with n vertices in $O(n \log n)$ time. Hershberger and Suri [15] give an $O(n)$ time algorithm for computing the *geodesic diameter*. Bae et al. [5] show that the geodesic diameter and center under the L_1 metric can be computed in $O(n)$ time in every simple polygon with n vertices.

Note that the link distance, geodesic distance and the L_1-geodesic distance are all metrics, while the minimum number of reflections on a diffuse reflection path between two points is *not* a metric (the triangle inequality fails). This partly explains the difficulty of handling diffuse reflections.

In contrast to link paths, the currently known algorithm for computing a minimum diffuse reflection path (one with the minimum number of reflections) between two points in a simple polygon with n vertices takes $O(n^9)$ time [2,13]; and no polynomial time algorithm is known for computing the diffuse reflection diameter or radius of a polygon.

2 Preliminaries

For a planar set $U \subseteq \mathbb{R}^2$, we denote the interior by $\text{int}(U)$, the boundary by ∂U, and the closure by $\text{cl}(U)$. Let P be a simply connected closed polygonal domain (for short, *simple polygon*) with n vertices. A *chord* of P is a closed line segment ab such that $a, b \in \partial P$, and the relative interior of ab is in $\text{int}(P)$.

We assume that the vertices of P are in general position, and we only consider light sources $s \in \text{int}(P)$ that do not lie on any line spanned by two vertices of P. Recall that $V_0(s)$ is the visibility polygon of the point $s \in P$ with respect to P. The *pockets* of $V_0(s)$ are the connected components of $P \setminus \text{cl}(V_0(s))$. See Fig. 2(a) for examples. The common boundary of $V_0(s)$ and a pocket is a chord ab of P (called a *window*) such that a is a reflex vertex of P that lies in the relative interior of segment sb. We say that a pocket with a window ab is *induced by* the reflex vertex a. Note that every reflex vertex induces at most one pocket of $V_0(s)$. We define the *size* of a pocket as the number of vertices of P on the boundary of the pocket. Since the pockets of $V_0(s)$ are pairwise disjoint, the sum of the sizes of the pockets is at most n, the number of vertices of P.

A pocket is a *left* (resp., *right*) pocket if it lies on the left (resp., right) side of the directed line $\overrightarrow{ab}$. Two pockets of $V_0(s)$ are *dependent* if some chord of P crosses the window of both pockets; otherwise they are *independent*. One pocket is called independent if it is independent of all other pockets.

Proposition 1. *All left (resp., right) pockets of $V_0(s)$ are pairwise independent.*

The main result of this section is a sufficient condition (Lemma 1) for a point $s \in \text{int}(P)$ to fully illuminate $\text{int}(P)$ within $\lfloor (n-2)/4 \rfloor$ diffuse reflections. A proof of the lemma is offered in the full version of the paper. It relies on the following subsection, techniques developed in [6], and the bound $D(P) \leq \lfloor n/2 \rfloor - 1$ on the diffuse reflection diameter.

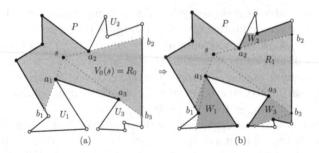

Fig. 2. (a) A polygon P where $V_0(s)$ has three pockets U_1, U_2 and U_3, of size 4, 3, and 5, respectively. The left pockets are U_1 and U_2, the only right pocket is U_3. Pocket U_1 is independent of both U_2 and U_3; but U_2 and U_3 are dependent. (b) The construction of region R_1 from $R_0 = V_0(s)$ in [6]. Pocket U_1 is saturated, and pockets U_2 and U_3 are unsaturated.

Lemma 1. *We have* $\mathrm{int}(P) \subseteq V_{\lfloor (n-2)/4 \rfloor}(s)$ *for a point* $s \in \mathrm{int}(P)$ *if the pockets of* $V_0(s)$ *satisfy these conditions:*

C_1 *every pocket has size at most* $\lfloor n/2 \rfloor - 1$; *and*
C_2 *the sum of the sizes of any two dependent pockets is at most* $\lfloor n/2 \rfloor - 1$.

2.1   Review of Regions R_k

We briefly review the necessary tools from [6]. Let $s \in \mathrm{int}(P)$ be a point in general position. Recall that $V_k(s)$, the set of points reachable from s with at most k diffuse reflections, is not necessarily simply connected when $k \geq 1$ [7]. Instead of tackling $V_k(s)$ directly, Barequet et al. [6] recursively define simply connected subsets $R_k = R_k(s) \subseteq V_k(s)$ for all $k \in \mathbb{N}_0$, starting with $R_0 = V_0(s)$. We review how R_{k+1} is constructed from R_k. Each region R_k is bounded by chords of P and segments along the boundary ∂P. The connected components of $P \setminus \mathrm{cl}(R_k)$ are the *pockets* of R_k. Each pocket U_{ab} of R_k is bounded by a chord ab such that a is a reflex vertex of P, b is an interior point of an edge of P, and the two edges of P incident to a are on the same side of the line ab (these properties are maintained in the recursive definition). A pocket U_{ab} of R_k is *saturated* if every chord of P that crosses ab has one endpoint in R_k and the other endpoint in U_{ab}. Otherwise, U_{ab} is *unsaturated*. Recall that for a point $s' \in P$, $V_0(s')$ is the set of points in P visible from s'; and for a line segment $pq \subseteq P$, $V_0(pq)$ is the set of points in P visible from any point in pq.

The regions R_k are defined as follows (refer to Fig. 2(b)). Let $R_0 = V_0(s)$. If $\mathrm{int}(P) \subseteq R_k$, then let $R_{k+1} = \mathrm{cl}(R_k) = P$. If $\mathrm{int}(P) \not\subseteq R_k$, then R_k has at least one pocket. For each pocket U_{ab}, define a set $W_{ab} \subseteq U_{ab}$: If ab is saturated, then let $W_{ab} = V_0(ab) \cap U_{ab}$. If ab is unsaturated, then let $p_{ab} \in R_k \cap \partial P$ be a point infinitely close to b such that no line determined by two vertices of P separates b and p_{ab}; and then let $W_{ab} = V_0(p_{ab}) \cap U_{ab}$. Let R_{k+1} be the union of $\mathrm{cl}(R_k)$ and the sets W_{ab} for all pockets U_{ab} of R_k. Barequet et al. [6] prove that $R_k \subseteq V_k(s)$ for all $k \in \mathbb{N}_0$.

We say that a region R_k weakly covers an edge of P if the boundary ∂R_k intersects the relative interior of that edge. On the boundary of every pocket U_{ab} of R_k, there is an edge of P that R_k does not weakly cover, namely, the edge of P incident to a. We call this edge the *lead edge* of U_{ab}. The following observation follows from the way the regions R_k are constructed in [6].

Proposition 2 ([6]). *For every pocket U of region R_k, $k \in \mathbb{N}_0$, the lead edge of U is weakly covered by region R_{k+1} and is not weakly covered by R_k.*

Proposition 3. *If a pocket U_{ab} of $V_0(s)$ has size m, then R_k weakly covers at least $\min(k+1, m)$ edges of P on the boundary of U.*

The following lemma is a direct consequence of Proposition 3. It will be used for unsaturated pockets of $V_0(s)$.

Lemma 2. *If U is a size-m pocket of $V_0(s)$, then $\mathrm{int}(U) \subseteq R_{m-1}$.*

For saturated pockets, the diameter bound allows a significantly better result.

Lemma 3. *If U is a size-m saturated pocket of R_k, then $\mathrm{int}(U) \subseteq R_{k+\lfloor m/2 \rfloor}$.*

Lemmas 2 and 3 combined yield the following for dependent pockets of $V_0(s)$.

Lemma 4. *Let U be a pocket of $V_0(s)$ of size m. If each pocket dependent on U has size at most $m' < m$, then $\mathrm{int}(U) \subseteq R_{\lfloor (m+m')/2 \rfloor}$.*

2.2   Double Violators

Recall that the sum of sizes of the pockets of $V_0(s)$ is at most n, the number of vertices of P. Therefore, it is possible that several pockets or dependent pairs of pockets violate conditions $\mathbf{C}_1$ or $\mathbf{C}_2$ in Lemma 1. We say that a point $s \in \mathrm{int}(P)$ is a *double violator* if $V_0(s)$ has either (i) two disjoint pairs of dependent pockets, each pair with total size at least $\lfloor n/2 \rfloor$, or (ii) a pair of dependent pockets of total size at least $\lfloor n/2 \rfloor$ and an independent pocket of size at least $\lfloor n/2 \rfloor$. (We do not worry about the possibility of two independent pockets, each of size at least $\lfloor n/2 \rfloor$.) In this section, we show that if there is a double violator $s \in \mathrm{int}(P)$, then there is a point $s' \in \mathrm{int}(P)$ (possibly $s' = s$) for which $\mathrm{int}(P) \subseteq V_{\lfloor (n-2)/4 \rfloor}(s')$, and such an s' can be found in $O(n)$ time.

The key technical tool is the following variant of Lemma 4 for a pair of dependent pockets that are adjacent to a common edge (i.e., *share* an edge).

Lemma 5. *Let U_{ab} and $U_{a'b'}$ be two dependent pockets of $V_0(s)$ such that neither is dependent on any other pocket, and points b and b' lie in the same edge of P. Let the size of U_{ab} be m and $U_{a'b'}$ be m'. Then $R_{\lfloor (m+m'-1)/2 \rfloor}$ contains the interior of both U_{ab} and $U_{a'b'}$.*

Lemma 6. *Suppose that $V_0(s)$ has two disjoint pairs of dependent pockets, each pair with total size $\lfloor n/2 \rfloor$. Then there is a point $s' \in \mathrm{int}(P)$ such that $\mathrm{int}(P) \subseteq V_{\lfloor (n-2)/4 \rfloor}(s')$, and s' can be computed in $O(n)$ time.*

Lemma 7. *Suppose that $V_0(s)$ has a pair of dependent pockets of total size $\lceil n/2 \rceil$ and an independent pocket of size $\lfloor n/2 \rfloor$. Then there is a point $s' \in \mathrm{int}(P)$ with $\mathrm{int}(P) \subseteq V_{\lfloor (n-2)/4 \rfloor}(s')$, and s' can be computed in $O(n)$ time.*

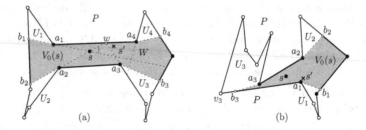

Fig. 3. Two instances of a double violator point s. (a) A polygon P with $n = 13$ vertices where $V_0(s)$ has four pockets: two pairs of dependent pockets, the sum of sizes of each pair is $\lfloor n/2 \rfloor = 6$. (b) A polygon P with $n = 13$ vertices where $V_0(s)$ has three pockets: two dependent pockets of total size $\lfloor n/2 \rfloor = 6$ and an independent pocket of size $\lfloor n/2 \rfloor = 6$.

3   Finding a Witness Point

In Section 3.1, we show that in every simple polygon P, there is a point $s \in \text{int}(P)$ that satisfies condition $\mathbf{C}_1$ In Section 3.2, we pick a point $s \in \text{int}(P)$ that satisfies condition $\mathbf{C}_1$, and move it continuously until either (i) it satisfies both conditions $\mathbf{C}_1$ and $\mathbf{C}_2$, or (ii) it becomes a double violator. In both cases, we find a witness point for Theorem 1 (by Lemmas 1, 6, and 7).

3.1   Generalized Kernel

Let P be a simple polygon with n vertices. Recall that the set of points from which the entire polygon P is visible is the *kernel*, denoted $K(P)$, which is the intersection of all halfplanes bounded by a supporting line of an edge of P and facing towards the interior of P. Lee and Preparata [16] designed an optimal $O(n)$ time algorithm for computing the kernel of a simple polygon with n vertices. We now define a generalization of the kernel. For an integer $q \in \mathbb{N}_0$, let $K_q(P)$ denote the set of points $s \in P$ such that every pocket of $V_0(s)$ has size at most q. Clearly, $K(P) = K_0(P) = K_1(P)$, and $K_q(P) \subseteq K_{q+1}(P)$ for all $q \in \mathbb{N}_0$. The set of points that satisfy condition $\mathbf{C}_1$ is $K_{\lfloor n/2 \rfloor}(P)$. For every reflex vertex v, we define two polygons $L_q(v) \subseteq P$ and $M_q(v) \subseteq P$: Let $L_q(v)$ (resp. $M_q(v)$) be the set of points $s \in P$ such that v does not induce a left (resp., right) pocket of size more than q in $V_0(s)$. We have

$$K_q(P) = \bigcap_{v \text{ reflex}} (L_q(v) \cap M_q(v)).$$

We show how to compute the polygons $L_q(v)$ and $M_q(v)$. Refer to Fig. 4. Denote the vertices of P by $(v_0, v_1, \ldots, v_{n-1})$, and use arithmetic modulo n on the indices. For a reflex vertex v_i, let $v_i a_i$ be the first edge of the shortest (geodesic) path from v_i to v_{i-q} in P. If the chord $v_i a_i$ and $v_i v_{i+1}$ meet at a reflex angle, then $v_i a_i$ is on the boundary of the *smallest* left pocket of size at

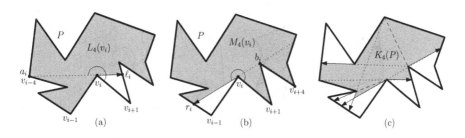

Fig. 4. (a) Polygon $L_4(v_i)$. (b) Polygon $M_4(v_i)$. (c) Polygon $K_4(P)$.

least q induced by v_i (for any source $s \in P$). In this case, the ray $\overrightarrow{a_i v_i}$ enters the interior of P, and we denote by ℓ_i the first point hit on ∂P. The polygon $L_q(v_i)$ is the part of P lying on the left of the chord $\overrightarrow{v_i \ell_i}$. However, if the chord $v_i a_i$ and $v_i v_{i+1}$ meet at convex angle, then every left pocket induced by v_i has size less than q, and we have $L_q(v_i) = P$. Similarly, let $v_i b_i$ be the first edge of the shortest path from v_i to v_{i+q}. Vertex v_i can induce a right pocket of size more than q only if $b_i v_i$ and $v_i v_{i-1}$ make a reflex angle. In this case, $v_i b_i$ is the boundary of the *largest* right pocket of size at most q induced by v_i, the ray $\overrightarrow{b_i v_i}$ enters the interior of P, and hits ∂P at a point m_i, and $M_q(v_i)$ is the part of P lying on the right of the chord $\overrightarrow{v_i m_i}$. if $b_i v_i$ and $v_i v_{i-1}$ meet at a convex angle, then $M_q(v_i) = P$.

Note that every set $L_q(v_i)$ (resp., $M_q(v_i)$) is *P-convex* (a.k.a. *geodesically convex*), that is, $L_i(v_i)$ contains the shortest path between any two points in $L_q(v_i)$ with respect to P [5,11,22]. Since the intersection of P-convex polygons is P-convex, $K_q(P)$ is also P-convex for every $q \in \mathbb{N}_0$. There exists a point $s \in \text{int}(P)$ satisfying condition $\mathbf{C}_1$ iff $K_{\lfloor n/2 \rfloor}(P)$ is nonempty. We prove $K_{\lfloor n/2 \rfloor}(P) \neq \emptyset$ using a Helly-type result by Breen [8].

Theorem 2 ([8]). *Let $\mathcal{P}$ be a family of simple polygons in the plane. If every three (not necessarily distinct) members of $\mathcal{P}$ have a simply connected union and every two members of $\mathcal{P}$ have a nonempty intersection, then $\bigcap \{P : P \in \mathcal{P}\} \neq \emptyset$.*

Lemma 8. *For every simple polygon P with $n \geq 3$ vertices, $K_{\lfloor n/2 \rfloor}(P) \neq \emptyset$.*

Proof. We apply Theorem 2 for the polygons $L_{\lfloor n/2 \rfloor}(v_i)$ and $M_{\lfloor n/2 \rfloor}(v_i)$ for all reflex vertices v_i of P. By definition, $L_{\lfloor n/2 \rfloor}(v_i)$ (resp., $M_{\lfloor n/2 \rfloor}(v_i)$) is incident to at least $\lfloor n/2 \rfloor + 1$ vertices of P, namely $v_{i-\lfloor n/2 \rfloor}, \ldots, v_i$ (resp., $v_i, \ldots, v_{i+\lfloor n/2 \rfloor}$). Hence the intersection of any two sets is incident to at least at most $2(\lfloor n/2 \rfloor + 1) - n > n$ vertices of P. It remains to show that the union of any three of them is simply connected.

Suppose, to the contrary, that there are three sets whose union has a hole. Since each set is bounded by a chord of P, the hole must be a triangle bounded by the three chords on the boundary of the three polygons. Each chord is incident to a reflex vertex of P and is collinear with *another* chord of P that weakly

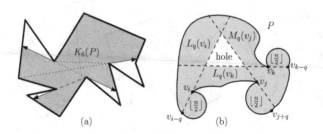

Fig. 5. (a) A simple polygon P with $n = 13$ vertices, and the generalized kernel $K_{\lfloor n/2 \rfloor}(P) = K_6(P)$. (b) A schematic picture of a triangular hole in the union of three polygons in P.

separates the vertices $\{v_i, v_{i+1}, \ldots, v_{i+\lfloor n/2 \rfloor}\}$ or $\{v_i, v_{i-1}, \ldots, v_{i-\lfloor n/2 \rfloor}\}$ from the hole. Figure 5(b) shows a schematic image. The three chords together weakly separate disjoint sets of vertices of total size at least $3\lfloor n/2 \rfloor + 3 > n$ from the hole, contradicting the fact that P has n vertices altogether. □

Lemma 9. *For every $q \in \mathbb{N}_0$, $K_q(P)$ can be computed in $O(n \log n)$ time.*

3.2   Finding a Witness

In this section, we present an algorithm that, given a simple polygon P with n vertices in general position, finds a witness $s \in \text{int}(P)$ such that $\text{int}(P) \subseteq V_{\lfloor (n-2)/4 \rfloor}(s)$.

Let s_0 be an arbitrary point in $\text{int}(K_{\lfloor n/2 \rfloor}(P))$. By Lemma 8, such a point exists. We can compute the visibility polygon $V_0(s_0)$ and its pockets in $O(n)$ time [14]. The definition of $K_{\lfloor n/2 \rfloor}(P)$ ensures that s_0 satisfies condition $\mathbf{C_1}$ of Lemma 1. If it also satisfies $\mathbf{C_2}$, then $s = s_0$ is a desired witness.

Assume that s_0 does not satisfy $\mathbf{C_2}$, that is, $V_0(s_0)$ has two dependent pockets of total size at least $\lfloor n/2 \rfloor$, say a left pocket U_{ab} and (by Proposition 1) a right pocket $U_{a'b'}$. We may assume that U_{ab} is at least as large as $U_{a'b'}$, by applying a reflection if necessary, and so the size of U_{ab} is at least $\lfloor n/4 \rfloor$. Refer to Fig. 6(a). Let $c \in \partial P$ be a point sufficiently close to b such that segment bc is disjoint from all lines spanned by the vertices of P, segment s_0c is disjoint from the intersection of any two lines spanned by the vertices of P, and $s_0c \subseteq P$. In Lemma 10 (below), we find a point on segment s_0c that is a witness, or double violator, or improves a parameter (spread) that we introduce now.

For a pair of dependent pockets, a left pocket U_{ab} and (by Proposition 1) a right pocket $U_{a'b'}$, let $spread(a, a')$ be the number of vertices on ∂P clockwise from a to a' (inclusive). Note that the spread is always at least the sum of the sizes of the two dependent pockets, as all vertices incident to the two pockets are counted. For a pair of pockets of total size at least $\lfloor n/2 \rfloor$, we have $\lfloor n/2 \rfloor \leq spread(a, a') \leq n$.

The visibility polygons of two points are combinatorially equivalent if there is a bijection between their pockets such that corresponding pockets are incident to the same sets of vertices of P. The combinatorial changes incurred by a moving point s have been thoroughly analysed in [3,4,10]. The set of points $s \in P$ that induces combinatorially equivalent visibility polygons $V_0(s)$ is a cell in the *visibility decomposition* $VD(P)$ of polygon P. It is known that each cell is convex and there are $O(n^3)$ cells, but a line segment in P intersects only $O(n)$ cells [4,9]. A combinatorial change in $V_0(s)$ occurs if s crosses a *critical line* spanned by two vertices of P, and the circular order of the rays from s to the two vertices is reversed. The possible changes are: a pocket of size 2 appears or disappears; (2) the size of a pocket increases or decreases by one; (3) two pockets merge into one pocket or a pocket splits into two pockets. Importantly, the combinatorics of $V_0(s)$ does not include the dependence between pockets: Proposition 1 will prove critical for tracking when two dependent pockets become independent.

Proposition 4. *Let $s_1 s_2$ be a line segment in* $\text{int}(P)$. *Then*

(i) *Every left (resp., right) pocket of $V_0(s_2)$ induced by a vertex on the left (right) of $\overrightarrow{s_1 s_2}$ is contained in a left (right) pocket of $V_0(s_1)$.*

(ii) *Let U_{left} and U_{right} be independent pockets of $V_0(s_1)$. Then every two pockets of $V_0(s_2)$ contained in U_{left} and U_{right}, respectively, are independent.*

Lemma 10. *There is a point $s \in s_0 c$ such that one of the following holds.*

- *s satisfies both $\mathbf{C}_1$ and $\mathbf{C}_2$;*
- *s is a double violator;*
- *s satisfies $\mathbf{C}_1$ but violates $\mathbf{C}_2$ due to two pockets of spread $\leq \text{spread}(a, a') - \lfloor n/4 \rfloor$.*

Proof. We move a point $s \in s_0 c$ from s_0 to c and trace the combinatorial changes of the pockets of $V_0(s)$, and their dependencies. Initially, when $s = s_0$, all pockets have size at most $\lfloor n/2 \rfloor - 1$; and there are two dependent pockets, a left pocket U_{ab} on the left of $\overrightarrow{s_0 c}$ and, by Proposition 1, a right pocket $U_{a'b'}$ on the right of $\overrightarrow{s_0 c}$, of total size at least $\lfloor n/2 \rfloor$. When $s = c$, every left pocket of $V_0(s)$ on the left of $\overrightarrow{s_0 c}$ is independent of any right pocket on the right of $\overrightarrow{s_0 c}$.

Consequently, when s moves from s_0 to c, there is a critical change from $s = s_1$ to $s = s_2$ such that $V_0(s_1)$ still has two dependent pockets of size at least $\lfloor n/2 \rfloor$ where the left (resp., right) pocket is on the left (right) of $\overrightarrow{s_0 c}$; but $V_0(s_2)$ has no two such pockets. (See Fig. 6 for examples.) Let U_{left} and U_{right} denote the two violator pockets of $V_0(s_1)$. The critical point is either a combinatorial change (i.e., the size of one of these pockets drops), or the two pockets become independent. By Proposition 4, we have $U_{\text{left}} \subseteq U_{ab}$ and $U_{\text{right}} \subseteq P \setminus U_{ab}$, and the spread of U_{left} and U_{right} is at most $\text{spread}(a, a')$. We show that one of the statements in Lemma 10 holds for s_1 or s_2.

If s_2 satisfies both $\mathbf{C}_1$ and $\mathbf{C}_2$, the our proof is complete (Fig. 6(a-b)). If s_2 violates $\mathbf{C}_1$, i.e., $V_0(s_2)$ has a pocket of size $\geq \lfloor n/2 \rfloor$, then $V_0(s_1)$ also has a combinatorially equivalent pocket independent of U_{left} and U_{right}, and so s_1 is a double violator. Finally, if s_2 violates $\mathbf{C}_2$, i.e., $V_0(s_2)$ has two dependent pockets of total size $\lfloor n/2 \rfloor$, then the left pocket of this pair is not contained in U_{ab}.

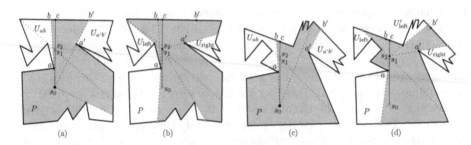

Fig. 6. (a) A polygon with $n = 21$ vertices where s_0 violates $\mathbf{C}_2$ a pair of dependent pockets U_{ab} and $U_{a'b'}$. (b) Point $s_2 \in s_0 c$ satisfies both $\mathbf{C}_1$ and $\mathbf{C}_2$. (c) A polygon with $n = 21$ vertices where s_0 violates $\mathbf{C}_2$ with a pair of pockets U_{ab} and $U_{a'b'}$ of spread 19. (d) Point s_2 also violates $\mathbf{C}_2$ with a pair of pockets of spread 13.

We have two subcases to consider: (i) If the right pocket of this new pair is contained in U_{right}, then their spread is at most $\text{spread}(a, a') - \lfloor n/4 \rfloor$ (Fig. 6(c-d)). (ii) If the right pocket of the new pair is disjoint from U_{right}, then $V_0(s_1)$ also has a combinatorially equivalent pair of pockets, which is different from U_{left} and U_{right}, and so s_1 is a double violator. ☐

Lemma 11. *A point $s \in s_0 c$ described in Lemma 10 can be found in $O(n \log n)$ time.*

Proof. It is enough to show that the critical positions, s_1 and s_2, in the proof of Lemma 10 can be computed in $O(n \log n)$ time. We use the data structure of Chen and Daescu [9], which is constructed by decomposing $s_0 c$ into a set of $O(n)$ intervals with combinatorially distinct region and recording the changing region in a persistent search tree.

However, the data structure of [9] only stores the visible region, not whether the region induces dependent pockets. The main technical difficulty is that $\Omega(n^2)$ dependent pairs might become independent as s moves along $s_0 s$ (even if we consider only pairs of total size at least $\lfloor n/2 \rfloor$), in contrast to only $O(n)$ combinatorial changes of the visibility region. We reduce the number of relevant events by focusing on only the "large" pockets (pockets of size at least $\lfloor n/4 \rfloor$), and maintaining at most one pair that violates $\mathbf{C}_2$ for each large pocket. (In a dependent pair of size $\geq \lfloor n/2 \rfloor$, one of the pockets has size $\geq \lfloor n/4 \rfloor$.)

We augment the data structure of [9] as follows. We maintain the list of all left (resp., right) pockets of $V_0(s)$ lying on the left (right) of $\overrightarrow{s_0 c}$, sorted in counterclockwise order along ∂P. We also maintain the set of *large* pockets of size at least $\lfloor n/4 \rfloor$ from these two lists. There are at most 4 large pockets for any $s \in s_0 c$. For a large pocket $U_{\alpha\beta}$ of $s \in s_0 c$, we maintain one possible other pocket $U_{\alpha'\beta'}$ of $V_0(s)$ such that they together violate $\mathbf{C}_2$. If there are several such pockets $U_{\alpha'\beta'}$, we maintain only the one where α' (the reflex vertex that induces $U_{\alpha'\beta'}$) is farthest from c along ∂P. Thus, we maintain a set $\mathcal{U}(s)$ of at most 4 pairs $(U_{\alpha\beta}, U_{\alpha'\beta'})$. Finally, for each of pair $(U_{\alpha\beta}, U_{\alpha'\beta'}) \in \mathcal{U}$, we maintain the positions $s' = sc \cap \alpha\alpha'$ where the pair $(U_{\alpha\beta}, U_{\alpha'\beta'})$ becomes independent assuming that neither $U_{\alpha\beta}$ nor $U_{\alpha'\beta'}$ goes through combinatorially before s reaches s'. We use [9], combined with these supplemental structures, to find critical points $s_1, s_2 \in s_0c$ such that $\mathcal{U}(s_1) \neq \emptyset$ but $\mathcal{U}(s_2) = \emptyset$.

We still need to show that $\mathcal{U}(s)$ can be maintained in $O(n \log n)$ time as s moves from s_0 to c. A pair $(U_{\alpha\beta}, U_{\alpha'\beta'})$ has to be updated if $U_{\alpha\beta}$ or $U_{\alpha'\beta'}$ undergoes a combinatorial change, or if they become independent (i.e., $s \in \alpha\alpha'$). Each large pocket undergoes $O(n)$ combinatorial changes affect them by Proposition 4, and there are $O(n)$ reflex vertices along ∂P between a and a' (these are the candidates for α'). No update is necessary when β or β' changes but $U_{\alpha\beta}$ remains large and the total size of the pair is at least $\lfloor n/2 \rfloor$. If the size of $U_{\alpha\beta}$ drops below $\lfloor n/4 \rfloor$, we can permanently eliminate the pair from $\mathcal{U}$. In all other cases, we search for a new vertex α', by testing the reflex vertices that induce pockets from the current α' towards c along ∂P until we either find a new pocket $U_{\alpha'\beta'}$ or determine that $U_{\alpha\beta}$ is not dependent of any other pocket with joint size $\geq \lfloor n/2 \rfloor$. We can test dependence between $U_{\alpha\beta}$ and a candidate for $U_{\alpha'\beta'}$ in $O(\log n)$ time (test $\alpha\alpha' \subseteq P$ by a ray shooting query). Each update of $(U_{\alpha\beta}, U_{\alpha'\beta'})$ decreases the size of the large pocket $U_{\alpha\beta}$ or moves the vertex α' closer to c. Therefore, we need to test dependence between only $O(n)$ candidate pairs of pockets. Overall, the updates to $\mathcal{U}(s)$ take $O(n \log n)$ time.   □

We are now ready to prove Theorem 1.

Proof (of Theorem 1). Let P be a simple polygon with $n \geq 3$ vertices. Compute the generalized kernel $K_{\lfloor n/2 \rfloor}(P)$, and pick an arbitrary point $s_0 \in \text{int}(K_{\lfloor n/2 \rfloor}(P))$, which satisfies $\mathbf{C}_1$. If s_0 also satisfies $\mathbf{C}_2$, then $\text{int}(P) \subseteq V_{\lfloor (n-2)/4 \rfloor}(s_0)$ by Lemma 1. Otherwise, there is a pair of dependent pockets, U_{ab} and $U_{a'b'}$, of total size at least $\lfloor n/2 \rfloor$ and $\lfloor n/2 \rfloor \leq \text{spread}(a, a') \leq n$. Invoke Lemma 10 up to four times to find a point $s \in \text{int}(P)$ that either satisfies both $\mathbf{C}_1$ and $\mathbf{C}_2$, or is a double violator. If s satisfies $\mathbf{C}_1$ and $\mathbf{C}_2$ then Lemma 1 completes the proof. If s is a double violator, apply Lemma 6 or Lemma 7 as appropriate to complete the proof. The overall running time of the algorithm is $O(n \log n)$ from the combination of Lemmas 6, 7, 9, and 11.

For every $k \geq 1$, the diffuse reflection diameter of the zig-zag polygon (cf. Fig. 1) with $n = 4k + 2$ vertices is $k = \lfloor (n-2)/4 \rfloor$. Adding up to 3 dummy vertices on the boundary of a zig-zag polygon gives n-vertex polygons P_n with $R(P_n) = \lfloor (n-2)/4 \rfloor$ for all $n \geq 6$. Finally, every simple polygon with $3 \leq n \leq 5$ vertices is star-shaped, and so its diffuse reflection radius is $0 = \lfloor (n-2)/4 \rfloor$. □

References

1. Aanjaneya, M., Bishnu, A., Pal, S.P.: Directly visible pairs and illumination by reflections in orthogonal polygons. In: Abstracts of 24th European Workshop on Comput. Geom., pp. 241–244 (2008)
2. Aronov, B., Davis, A.R., Iacono, J., Yu, A.S.C.: The complexity of diffuse reflections in a simple polygon. In: Correa, J.R., Hevia, A., Kiwi, M. (eds.) LATIN 2006. LNCS, vol. 3887, pp. 93–104. Springer, Heidelberg (2006)

3. Aronov, B., Guibas, L.J., Teichmann, M., Zhang, L.: Visibility queries and maintenance in simple polygons. Discrete Comput. Geom. 27(4), 461–483 (2002)
4. Bose, P., Lubiw, A., Munro, J.I.: Efficient visibility queries in simple polygons. Comput. Geometry Theory Appl. 23(3), 313–335 (2002)
5. Bae, S.W., Korman, M., Okamoto, Y., Wang, H.: Computing the L_1 Geodesic diameter and center of a simple polygon in linear time. In: Pardo, A., Viola, A. (eds.) LATIN 2014. LNCS, vol. 8392, pp. 120–131. Springer, Heidelberg (2014)
6. Barequet, G., Cannon, S.M., Fox-Epstein, E., Hescott, B., Souvaine, D.L., Tóth, C.D., Winslow, A.: Diffuse reflections in simple polygons. Electronic Notes in Discrete Math. 44(5), 345–350 (2013)
7. Brahma, S., Pal, S.P., Sarkar, D.: A linear worst-case lower bound on the number of holes in regions visible due to multiple diffuse reflections. J. of Geometry 81(1-2), 5–14 (2004)
8. Breen, M.: A Helly-type theorem for simple polygons. Geometriae Dedicata 60(3), 283–288 (1996)
9. Chen, D.Z., Daescu, O.: Maintaining visibility of a polygon with a moving point of view. Information Processing Letters 65(5), 269–275 (1998)
10. Chen, D.Z., Wang, H.: Weak visibility queries of line segments in simple polygons. In: Chao, K.-M., Hsu, T.-s., Lee, D.-T. (eds.) ISAAC 2012. LNCS, vol. 7676, pp. 609–618. Springer, Heidelberg (2012)
11. Demaine, E.D., Erickson, J., Hurtado, F., Iacono, J., Langerman, S., Meijer, H., Overmars, M., Whitesides, S.: Separating point sets in polygonal environments. Intl. J. Comput. Geom. Appl. 15(4), 403–419 (2005)
12. Djidjev, H.N., Lingas, A., Sack, J.-R.: An $O(n \log n)$ algorithm for computing the link center of a simple polygon. Discrete. Comput. Geom. 8, 131–152 (1992)
13. Ghosh, S.K.: Visibility algorithms in the plane, ch. 7. Cambridge Univ. Press (2007)
14. Guibas, L., Hershberger, J., Leven, D., Sharir, M., Tarjan, R.E.: Linear-time algorithms for visibility and shortest path problems inside triangulated simple polygons. Algorithmica 2(1-4), 209–233 (1987)
15. Hershberger, J., Suri, S.: Matrix searching with the shortest path metric. SIAM J. Comput. 26(6), 1612–1634 (1997)
16. Lee, D.T., Preparata, F.: An optimal algorithm for finding the kernel of a polygon. J. ACM 26, 415–421 (1979)
17. Maheshwari, A., Sack, J., Djidjev, H.N.: Link distance problems. In: Handbook of Computational Geometry, ch.12. Elsevier (2000)
18. Pollack, R., Sharir, M., Rote, G.: Computing the geodesic center of a simple polygon. Discrete Comput. Geom. 4, 611–626 (1989)
19. Schuierer, S.: Computing the L_1-diameter and center of a simple rectilinear polygon. In: Proc. Intl. Conf. on Computing and Information (ICCI), pp. 214–229 (1994)
20. Suri, S.: A linear time algorithm for minimum link paths inside a simple polygon. Comput. Vision. Graph. Image Process. 35, 99–110 (1986)
21. Suri, S.: On some link distance problems in a simple polygon. IEEE Trans. Robot. Autom. 6, 108–113 (1990)
22. Toussaint, G.T.: An optimal algorithm for computing the relative convex hull of a set of points in a polygon. In: Signal Processing III: Theories and Applications (EURASIP 1986), Part 2, pp. 853–856 (1986)

On Edge-Unfolding One-Layer Lattice Polyhedra with Cubic Holes

Meng-Huan Liou* Sheung-Hung Poon*,** and Yu-Jie Wei*

Department of Computer Science,
Institute of Information Systems and Applications,
National Tsing Hua University,
No. 101, Sec. 2, Kuang Fu Rd., Hsinchu, Taiwan, R.O.C..
{roy3190,kolz1234}@gmail.com, spoon@cs.nthu.edu.tw

Abstract. An *edge-unfolding* of a polyhedron is a cutting of the polyhedron's surface along its edges so that its surface can be flattened into a single connected flat patch on the plane without any self-overlapping. A *one-layer lattice polyhedron* is a polyhedron of height one, whose surface faces are grid squares. We consider the edge-unfolding problem on several classes of one-layer lattice polyhedra with cubic holes. We propose linear-time algorithms for one-layer lattice polyhedra with rectangular external boundary and cubic holes, one-layer lattice polyhedra with cubic holes strictly enclosed by an orthogonally convex polygon, and one-layer lattice polyhedra with sparse cubic holes, respectively. The algorithms use two different novel techniques to cut the edges of cubic holes of the given polyhedron so that no self-overlapping can occur in the flattened patch. Our algorithms are the *first algorithms* especially designed to edge-unfold a *polyhedron of genus greater than zero* to a single connected flattened patch. We leave open the question whether any of these edge-cutting methods can be extended to edge-unfold general one-layer lattice polyhedra with cubic holes.

1 Introduction

Folding and unfolding problems are classical problems in geometry and mathematics. They were studied as early as the year 1525 by Dürer [8], and have been studied extensively in discrete and computational geometry in recent years. This research area finds applications in various applied fields. For instance, one industrial application of unfolding methods is the automated process planning for a robotic sheet metal bending operations [14,9].

A *polyhedron* is a closed surface, each of whose edges is adjacent to two polygonal plane faces. In general, an *unfolding* of a polyhedron is a cutting along the polyhedron's surface so that its surface can be flattened into a single connected flat patch on the plane without any self-overlapping. In particular, an *edge-unfolding* of a polyhedron is a cutting of the polyhedron's surface along its

---

* Supported in part by grant NSC 100-2628-E-007-020-MY3 in Taiwan, R.O.C.
** Corresponding author.

Z. Cai et al. (Eds.): COCOON 2014, LNCS 8591, pp. 251–262, 2014.
© Springer International Publishing Switzerland 2014

edges. A polyhedron is *orthogonal* if all of its faces meet at right angles, and all of its edges are parallel to coordinate axes. Furthermore, a *lattice polyhedron* is an orthogonal polyhedron whose faces are unit grid squares.

In computer graphics, there are heuristic algorithms to edge-unfold a 3-dimensional object mesh into several non-self-overlapping flattened patches [12,13], which can then be used to reconstruct the original object model. Takahashi et al. [13] recently designed an algorithm based on the genetic algorithm approach.

In computational geometry, there are two long-standing unsolved open problems, which are the questions of whether a convex polyhedron can be edge-unfolded [11], and whether a nonconvex polyhedron can be unfolded [3]. Here we focus on the survey of the edge-unfolding results and some related unfolding results for orthogonal polyhedra. O'Rourke [10] proposed an edge-unfolding algorithm for orthogonal terrains. Biedl et al. [4] proposed an edge-unfolding algorithm for lattice orthotubes and an unfolding algorithm for orthostacks. Damian and Meijer [7] later gives an edge-unfolding algorithm for lattice orthostacks with orthogonally convex slabs. However, the question whether a general lattice orthostack can be edge-unfolded remains open. Furthermore, Damian et al. [5] showed that well-separated lattice orthotrees can be edge-folded. However, the question whether a general lattice orthotree can be edge-unfolded again remains open. Moreover, Damian et al. [6] present an unfolding algorithm for Manhattan towers, but the question of whether a general lattice Manhattan tower can be edge-unfolded is still open. Recently, Abel and Demaine [1] showed that the decision question whether an orthogonal polyhedron can be edge-unfolded is strongly NP-complete. The orthogonal polyhedra concerned in the above related work are all of genus zero.

In this paper, we study the edge-unfolding problem on several classes of lattice polyhedra of genus greater than zero. To the best of our knowledge, our algorithms are the *first algorithms* especially designed for edge-unfolding *polyhedra of non-zero genus*.

2   Definitions

In this section, we define some terminologies used in this paper. A *one-layer lattice polyhedron* is a polyhedron of height one, whose surface faces are unit grid squares. See Fig. 1 for an example. In this paper, we consider the problem of edge-unfolding one-layer lattice polyhedra with cubic holes, say P, into a single connected flat patch Q on the xy-plane without any self-overlapping.

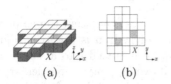

Fig. 1. (a) Side view of P; (b) Top view of P

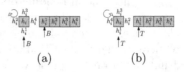

Fig. 2. Unfold hole faces of h_i: (a) in clockwise orientation; (b) in counterclockwise orientation

The faces on the upper horizontal surface T of P are called its *top faces*, and those on the bottom horizontal surface B of P its *bottom faces*. The outer cyclic strip of height one around P is called the *external boundary* (or simply *boundary*) of P, and is denoted by X. See Fig. 1(a) for an example of the boundary X of P. Note that the top view of the boundary X of P is a simple orthogonal polygon— see Fig. 1(b). A continuous segment from face X_1 to face X_2 along boundary X is denoted as the *boundary strip* $[X_1, X_2]$. A *column* is a vertical square prism space with unit square base along the whole y-axis, which includes its top faces, its bottom faces, boundary faces in the column, the holes in this column and their corresponding hole faces. Let $C = \{c_1, c_2, ..., c_w\}$ be the set of all columns ordered from left to right, where w is the total number of columns.

In the following terminology, we assume that each column c_j of P is a connected component. Column c_j contains the *front* and *rear boundary faces*, denoted by X_j^f and X_j^r, respectively. A *vertical strip* (resp. *horizontal strip*) of faces is a sequence of consecutive faces along the y-axis (resp. x-axis). A strip is called *extremal* if one of its two end edges is attached to a boundary face, and the other end edge is attached to a hole face or a boundary face. The *length* of a strip is the number of faces in the strip. A *k-strip* is a strip of length k. In column c_j, X_j^f is attached to two extremal vertical strips, the *top front strip* s_j^{fT} on T and the *bottom front strip* s_j^{fT} on B; and X_j^r is also attached to two extremal vertical strips, the *top rear strip* s_j^{fT} on T and the *bottom rear strip* s_j^{fT} on B.

Let h_i and h_{i+1} be two consecutive holes along a column, and let the *unfolding connection path* π be a path going from h_i to h_{i+1}. Then the two vertical strips between h_i and h_{i+1} are called a *double bridge*, denoted by e_i. e_i consists of its top bridge e_i^T and bottom bridge e_i^B, which are opposite to each other in P. The bridge which is used by the unfolding connection path π from h_i to h_{i+1} is called the *connecting bridge*; and the other bridge is called the *redundant bridge*.

In a flattened patch, the space along a column below or above a face f is called a *half column*; the space along a row on the left or right of a face f is called a *half row*. A half row or column is said to be *free* if it does not contain any other face in the current flattened patch.

We define the *vertical distance* $d_V^U(s_1, s_2)$ (resp. *horizontal distance* $d_H^U(s_1, s_2)$) as the y- (resp. x-) coordinate difference between the center point of two entities s_1 and s_2, where s_1 and s_2 belong to object U.

3 One-Layer Lattice Polyhedra with Cubic Holes and Special Boundary

3.1 One-Layer Lattice Polyhedra with Rectangular Boundary and Cubic Holes

First in this subsection, we consider the edge-unfolding of a one-layer lattice polyhedron P, which possesses an external boundary of rectangular shape and internal cubic holes. See Fig. 3(a) for an example. In the following, we propose a linear-time algorithm to edge-unfold such a polyhedron.

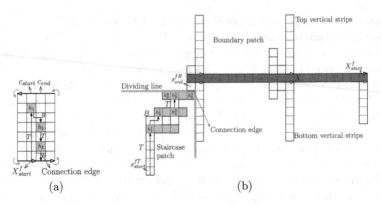

Fig. 3. (a) One-layer lattice polyhedron P with rectangular boundary and cubic holes; (b) The flattened patch Q of (a) contains two pieces, separated by a dividing line

Our algorithm consists of two execution phases. The first phase is called the *hole unfolding phase*, in which the holes are unfolded incrementally from left to right and consecutive holes are connected by strip paths on the top or bottom surface of P so that the unfolded hole strips and the connecting strip paths all run either rightward or upward consistently in the flattened patch Q. The flattened patch obtained from the hole unfolding phase is called the *staircase patch*. The second phase is called the *boundary unfolding phase*, in which the remaining front and rear vertical strips are attached to their corresponding boundary faces X_j^f and X_j^r respectively and the external boundary X is straightened as a horizontal strip on the plane. The flattened patch obtained from the boundary unfolding phase is called the *boundary patch*. The staircase patch and the boundary patch are connected together to form the final flattened patch Q. See Fig.3(b) for an example. The details of the two phases are described as follows.

Hole Unfolding Phase. First we describe the hole unfolding phase. Let c_{start} be the column where the first hole h_1 is located, and c_{end} the column where the last hole h_m is located. The hole unfolding procedure starts by first placing the strip front top s_{start}^{fT} at c_{start} to run in upward direction in the flattened patch Q. From there, we start to unfold first hole h_1, and then make a connection path to reach the second hole h_2. During this process, we force the unfolded 4-strip for h_1 and the path strip connecting h_1 and h_2 to only run in the right or upward direction so that no overlapping can occur. Repeatedly using the same step, we proceed to unfold hole h_2, and other holes up to last hole h_m, which locates in column c_{end}. This is the overall hole unfolding procedure, whose details are further described as follows.

For the purpose that the unfolded strips of h_i, connection path π and h_{i+1} always extend either rightward or upward consistently in the flattened patch Q, the running direction of the straight connection paths between the holes in the same column has to go consistently in one direction, which is either upward or downward, and the relationship of running directions of columns are described as follows. At the starting column c_{start}, since the starting connection path runs from the front boundary face to hole h_1 in the upward direction, the running direction of column c_{start} is upward. Then whenever the unfolding process reaches a new column with some hole, the running direction of this new column switches to the direction opposite to the running direction of the former column with some hole. This traversal order of the holes results in an ordering for all holes, $h_1, h_2, \ldots, h_m$, where h_m is the rightmost hole. Such specific running directions of the columns are essential for the purpose of correctly unfolding the holes and the connection paths into strips in Q running upward or rightward consistently.

In the hole unfolding process, we need to unfold the holes and need to select an appropriate connection path for each pair of consecutive holes. The four faces of a hole h_i can be unfolded in two *hole unfolding orientations*, clockwise or counterclockwise. A face of h_i connected by incoming unfolding path belongs to T or B is called h_i^1, and the following unfolded faces h_i^2, h_i^3, and h_i^4 are labeled sequentially according to the hole unfolding orientation—see Fig. 2. Such an operation is called a *basic hole unfolding operation*. On the other hand, the connection path between two consecutive holes h_i and h_{i+1} can be very different under specific conditions. The connection path can be straight or non-straight. If h_i and h_{i+1} lie in the same column, then we call the unfolding step between h_i and h_{i+1} as a *straight path unfolding operation*. Without loss of generality we assume h_{i+1} is above h_i and h_i is unfolded as a horizontal 4-strip running rightward. The case that h_{i+1} is below h_i can be handled similarly. If the incoming connection path of h_i lies on T, then we need to select e_i^B as the connecting bridge from h_i to h_{i+1}—see Fig. 4.

If the incoming connection path of h_i lies on B, then we need to select e_i^T as the connecting bridge from h_i to h_{i+1}. Moreover, in order to force the unfolded strips of both h_i and h_{i+1} to extend rightward, we have to set the hole unfolding orientation of h_{i+1} to be the reverse of the hole unfolding orientation of h_i. By applying basic hole unfolding operation to both h_i and h_{i+1}, it is clear that the connecting bridge between h_i and h_{i+1} connects face h_i^3 to face h_{i+1}^1 and runs upward. Since the half column below h_{i+1}^3 is free—see Fig. 4(b), the redundant bridge between h_i and h_{i+1} can be

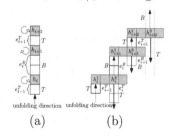

Fig. 4. (a) The holes in the same column; (b) The unfolding result of (a) where e_{i-1}^T runs in the increasing y-direction

safely attached to the bottom edge of h_{i+1}^3. This last operation is called a *redundant bridge placement operation*. In this manner, the corresponding unfolded strips can run either upward or rightward consistently. Otherwise if h_i and h_{i+1} lie in different columns, then we call the unfolding step between h_i and h_{i+1} as a *non-straight path unfolding operation*.

We use the non-straight path unfolding operation to select an appropriate specific connection path between the highest (resp. lowest) hole h_i on c_j and the highest (resp. lowest) hole h_{i+1} on c_{j+p}, where $p \geq 1$. There are two main

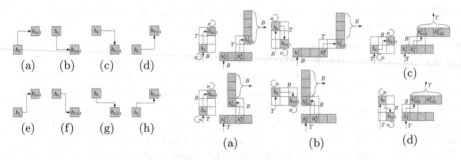

Fig. 5. (a) L_1-path; (b) L_2-path; (c) L_3-path; (d) L_4-path; (e) Z_1-path; (f) Z_2-path; (g) Z_3-path; (h) Z_4-path

Fig. 6. (a) L_1-path unfolding operation; (b) L_3-path unfolding operation; (c) Z_1-path unfolding operation; (d) Z_3-path unfolding operation

classes of non-straight paths, L-paths and Z-paths, and each class of paths has four subclasses—see Fig. 5. For L-paths, depending on different cases, the hole unfolding orientations of h_i and h_{i+1} may be the same or different. For Z-paths, no matter in which case, the hole unfolding orientations of h_i and h_{i+1} are always different. The restrictions are in order to force the flattened patch to extend rightward or upward. The selection of a suitable connection path is affected by three entities: (i) the incoming unfolding direction of h_i (upward or downward); (ii) h_i is higher or lower than h_{i+1}; and (iii) the connecting bridge of e_i belongs to T or B (on top or bottom surface). For any set of values of these three entities, we show that there exists one suitable connection path to use so that the corresponding unfolded patch extends in upward or rightward direction. However, due to lack of space, the details for selecting suitable connection paths for all different situations are omitted. Now, we suppose that we have selected the suitable connection path for any possible situation. Then we proceed to consider how to prevent the unfolding result of any non-straight path unfolding operation from overlapping any redundant bridge. The case analysis is as follows.

First, we consider that h_i and h_{i+1} are connected by an L-path. Fig. 6 illustrates the unfolding results of different kinds of connection L-paths shown in Fig. 5. Fig. 6(a) shows the two possible unfolding results of the L_1-path shown in Fig. 5(a). The L_2-path in Fig. 5(b) has the two similar unfolding results as the L_1-path. Fig. 6(b) shows the two possible unfolding results of the L_3-path shown in Fig. 5(c). The L_4-path in Fig. 5(d) has the two similar unfolding structures as the L_3-path. If we use an L-path unfolding operation, and h_i is unfolded as a horizontal (resp. vertical) 4-strip, then h_{i+1} is unfolded as a vertical (resp. horizontal) 4-strip. Suppose that h_i is unfolded as a horizontal 4-strip and h_{i+1} as a vertical 4-strip. The half row on the left of h_{i+1}^3 is free; so the redundant bridge originally opposite to e_{i+1} can be attached to the left of h_{i+1}^3. Hence, the unfolding patch of h_i, h_{i+1}, and their connection L-path cannot have any overlap with any redundant bridge appeared up to h_{i+1} is unfolded.

Next, we consider that h_i and h_{i+1} are connected by a Z-path. Fig. 6(c) shows the possible unfolding result of the Z_1-path shown in Fig. 5(e). The Z_2-path in Fig. 5(f) has the similar unfolding result as the Z_1-path. Fig. 6(d) shows the possible unfolding result of the Z_3-path shown in Fig. 5(g). The Z_4-path in Fig. 5(h) has the similar unfolding result as the Z_3-path. If we use a Z-path unfolding operation, h_i and h_{i+1} are both unfolded as horizontal or vertical 4-strips. Suppose that h_i and h_{i+1} are both unfolded as horizontal 4-strips. For a Z_1- or Z_2-path unfolding operation, the corresponding path starts at h_i^2 or h_i^4; so the half column below h_{i+1}^3 is free since $d_H^Q(h_i^4, h_{i+1}^3) \geq 1$. Thus the unfolding patch of h_i, h_{i+1}, and their connection Z_1- or Z_2-path cannot have any overlap with any redundant bridge appeared up to h_{i+1} is unfolded.

We proceed to consider a Z_3- and Z_4-path unfolding operation. For these two operations, the corresponding path starts at either h_i^1 or h_i^3. If it starts at h_i^3, the half column below h_{i+1}^3 is free since $d_H^Q(h_i^4, h_{i+1}^3) \geq 1$. Otherwise, if it starts at h_i^1, the half column below h_{i+1}^3 is free when $d_H^P(h_i, h_{i+1}) \geq 2$. The only possible trouble case is that the Z_3 or Z_4-path starts at h_i^1 and $d_H^P(h_i, h_{i+1}) = 1$. In such situation, h_{i+1}^3 is vertically above h_i^4 in Q. If h_{i+2} is not in the same col-

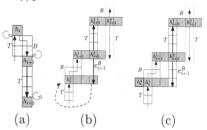

(a)   (b)   (c)

Fig. 7. (a) Problematic case; (b) Redundant bridge e_{i+1}^B overlaps with h_i^4; (c) Hole face displacement operation is performed, and the overlap is avoided

umn as h_{i+1}, then h_{i+2} is connected from h_{i+1} via a non-straight path, and thus there is no redundant bridge between holes h_{i+1} and h_{i+2} and thus the unfolded strip of h_i needs no further modification. But if h_{i+2} is in the same column as h_{i+1}, the redundant bridge originally opposite to e_{i+1} may penetrate through h_i^4 when the redundant bridge placement operation is applied—see Fig. 7(b). Thus we have to apply the *hole face displacement operation*, which we will described below. In such an operation, h_i^4 is moved and attached to the left of h_i^1 so that the half column below h_{i+1}^3 becomes free. See Fig. 7 for an example. Thus we can directly attach the redundant bridge originally opposite to e_{i+1} to the bottom edge of h_{i+1}^3—see Fig. 7(c). In all, by applying the hole face displacement operation on the unfolded 4-strip of h_i in the unfolding patch of h_i, h_{i+1}, and their connection Z_3- or Z_4-path has no overlap with any redundant bridge appeared up to h_{i+1} is unfolded.

Using the straight path unfolding operation and the non-straight path unfolding operation together, we unfold the holes $h_1, h_2, \ldots$, and h_m. When we have done the unfolding of the last hole h_m, the hole unfolding phase completes, and the flattened staircase patch is thus obtained.

Boundary Unfolding Phase. After the hole unfolding phase is executed, we obtain the staircase patch, which contains all the double bridges and the faces of all the holes in P. The remaining faces constitute the external boundary X and the extremal vertical strips. In the boundary unfolding phase, we continue to unfold these remaining faces on P to become the flattened boundary patch. We then describe the boundary unfolding phase as follows. If the incoming connection path of h_m in the hole unfolding phase lies on T (resp. B), then we take extremal strip s_{end}^B on B (resp. s_{end}^T on T) in column c_{end} to reach the boundary face X_{end}. Then we take a ride on the external boundary cycle X, and straighten out the whole boundary X such that the remaining extremal vertical strips remain attached to their corresponding front or rear boundary faces. In order to force the unfolded strip of boundary X extend upward or rightward in the flattened patch, the unfolding orientation of X is determined by the following rule: if the initial extremal strip used is s_{end}^B, then we unfold X in counterclockwise orientation; otherwise if the initial extremal strip used is s_{end}^T, then we unfold X in clockwise orientation. It is clear that the flattened boundary patch obtained from the above procedure does not have any self-overlapping.

Now we will stitch the boundary patch together with the staircase patch obtained from the hole unfolding phase. At the beginning of this phase, we starts from an edge on h_m and go straight to reach the external boundary via an extremal strip s_{end} in column c_{end}. The starting edge is the *connection edge* e between the boundary patch and the staircase patch. We have two cases to consider depending on which face of h_m the connection edge e belongs to. We first consider the case that e belongs to h_m^4. This case happens when the incoming direction of the connection path reaching h_m is rightward. In this situation, since the overall shape of boundary patch is like a strip backbone with all extremal strips sticking out and face h_m^4 is just one more grid face extending from one extremal strip on one side of boundary face X_{end}. Thus there is no overlap between the staircase and boundary patches. Next we consider the other case that e belongs to h_m^3. This case happens when the incoming direction of the connection path reaching h_m is upward or downward. Let X_2 be the boundary face next to X_{end} in the boundary patch, and let s_2 be the extremal strip attached to X_2 adjacent to s_{end}. Then it is easy to see that if the extremal strip s_2 is longer than s_{end}, then s_2 penetrates through the hole face h_m^4 in the staircase patch. To present such an overlap situation, we perform the hole face displacement operation of the unfolded strip of h_m so that h_m^4 is moved to the other end of the strip next to the hole face h_m^1. See Fig. 3(b) for an illustration of the result. Thus such an overlap situation is prevented. Finally, we obtain a connected flattened patch Q, which does not contain any self-overlap. This completes the whole unfolding algorithm. Hence, we obtain the following theorem.

Theorem 1. *There is an $O(n)$-time algorithm to edge-unfold one-layer lattice polyhedra with rectangular boundary and cubic holes.*

3.2   One-Layer Lattice Polyhedra with Cubic Holes Strictly Enclosed by an Orthogonally Convex Polygon

The algorithm in the previous subsection can be extended to handle more general one-layer lattice polyhedra with cubic holes, as stated in the following theorem.

Theorem 2. *There is an $O(n)$-time algorithm to edge-unfold a one-layer lattice polyhedron with cubic holes strictly enclosed by an orthogonally convex polygon.*

4   One-Layer Lattice Polyhedra with Sparse Cubic Holes

4.1   One Column with One Cubic Hole

In this subsection, we first consider one-layer lattice polyhedra such that there is at most one connected component in each column and there is at most one hole in such a connected component. Then we present the algorithm for general one-layer lattice polyhedra with sparse cubic holes in Section 4.2.

Suppose that we are going to unfold column c_j. Let h_j be the only hole in column c_j. Hole h_j consists of four faces parallel to the z-axis, its front face h_j^f, rear face h_j^r, left face h_j^L and right face h_j^R. Column c_j contains two boundary faces, its front boundary face X_j^f and rear boundary face X_j^r. If column c_j contains a hole, X_j^f is attached to two extremal vertical strips, the *front top strip* s_j^{fT} on T and the *front bottom strip* s_j^{fB} on B, which have equal length; and X_j^r is also attached to two extremal vertical strips, the *rear top strip* s_j^{rT} on T and the *rear bottom strip* s_j^{rB} on B, which have equal length. If any c_j does not contain a hole, then we treat the top and bottom vertical strips of column c_j as s_j^{fT} and s_j^{fB} respectively, which are both attached to X_j^f. In this manner, it seems that there is a virtual hole at the position of the rear end X_j^r of column c_j. By treating virtual holes as real holes, we suppose that each column contains a hole in the following description.

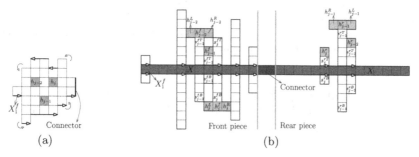

Fig. 8. (a) A special case; (b) The unfolding result of (a)

Our algorithm runs as follows. First, we take boundary face X_1^f as the first face, and then straighten out the whole boundary X in counterclockwise order as a horizontal strip. All the extremal vertical strips are flattened and remain attached to their corresponding front or rear boundary faces. Thus the extremal vertical strip lies vertically on either side of the flattened boundary X. For each column c_j, s_j^{fT} and s_j^{rT} lie vertically above X, and s_j^{fB} and s_j^{rB} lie below X—see Fig. 8(b). Then we unfold the holes in the columns from left to right in the order of column sequence $c_1, c_2, \ldots, c_w$ using the algorithmic procedure described below. More precisely, for each column c_j, h_j^f is attached to s_j^{fT} or s_j^{fB}, and h_j^r is attached to s_j^{rT} or s_j^{rB} depending on different cases. Moreover, h_j^L (resp. h_j^R) will be attached to h_j^f or h_j^r depending on different cases. Finally, we obtain the flattened patch Q.

The output patch Q contains two pieces as follows—see Fig. 8(b). The front strips and the part of corresponding boundary attaching to them are unfolded to become *front piece*. The rear strips and the part of corresponding boundary attaching to them are unfolded to become *rear piece*. In Q, front piece and rear piece are connected by a boundary segment, which is the boundary strip on X attached to the right wall of the rightmost column. When we unfold each column from left to right in P, the front piece extends rightward in Q and the rear piece extends leftward.

Before the introduction of the detailed procedure to edge-unfold the hole h_j in column c_j, we suppose that by applying the following procedure, columns c_k for $1 \leq k \leq j-1$ have been edge-unfolded into current flattened patch Q_{i-1} which does not have any self-overlapping. Moreover, after applying the following procedure for columns $c_1, \ldots, c_{j-1}$, we obtain that each h_k^L may be attached to the left of h_k^f or the right of h_k^r in Q_{i-1}, and each h_k^R may be attached to the right of h_k^f or the left of h_k^r in Q_{i-1}; h_k^f is attached to just above s_j^{fT} or just below s_j^{fB}, and h_j^r is attached to just above s_j^{rT} or just below s_j^{rB}. After we unfold column c_j, we obtain a new flattened patch Q_i. In our algorithmic procedure to unfold h_j in column c_j, we maintain the current unfolded patch Q_i to satisfy two properties: (i) Q_i has no self-overlapping; (ii) h_j^R does not overlap with $h_{j+1}^f \cup s_{j+1}^{fT}$ and $h_{j+1}^r \cup s_{j+1}^{rT}$.

Now, we assume that Q_{i-1} satisfies properties (i) and (ii). Under this precondition, we can proceed to describe the procedure to edge-unfold the hole h_j in column c_j so that after the procedure, there is no self-overlap introduced and the above two properties are still satisfied. The procedure consists of two main steps, which will be described below in details.

Column Unfolding Step. First, we consider how to unfold c_j so that it cannot overlap with the unfolded patch of adjacent column c_{j-1}. We have three main cases. Initially, we set $Q = Q_{i-1}$ and attach h_j^f to s_j^{fT} and h_j^r to s_j^{rT}, then we determine whether h_j^L (resp. h_j^R) is to be attached to h_j^f or h_j^r.

Case 1: $d_V^P(X_{j-1}^f, X_j^f) \geq 1$, i.e., the boundary strip $[X_{j-1}^f, X_j^f]$ is not flat. We first place h_j^L to the left of h_j^f in Q. If h_{j-1}^R is attached on the left of h_{j-1}^r in Q, then by the precondition there is no self-overlap in Q. Now as $d_H^Q(X_{j-1}^f, X_j^f) \geq 2$, we can safely attach h_j^L to the left of h_j^f in Q without causing any self-overlap. If h_{j-1}^R is attached on the right of h_{j-1}^f in Q_{i-1}, we have two cases. If $d_V^P(X_{j-1}^f, X_j^f) = 1$, i.e., $d_H^Q(X_{j-1}^f, X_j^f) = 2$, then $d_V^Q(h_{j-1}^R, h_j^L) \geq 1$ since $d_V^P(h_{j-1}, h_j) \geq 2$. If $d_V^P(X_{j-1}^f, X_j^f) \geq 2$, i.e., $d_H^Q(X_{j-1}^f, X_j^f) \geq 3$, then $d_H^Q(h_{j-1}^R, h_j^L) \geq 1$ since $d_H^Q(s_{j-1}^{fT}, s_j^{fT}) \geq 3$. In these conditions, h_j^L cannot overlap with h_{j-1}^R according to the above arguments, and h_j^L also cannot overlap with $s_{j-1}^{fT} \cup h_{j-1}^f$ because $d_H^Q(s_{j-1}^{fT}, s_j^{fT}) \geq 3$ when $d_V^P(X_{j-1}^f, X_j^f) \geq 1$. According to the precondition, the rear piece also has no overlapping. Thus again we can attach h_j^L directly to the left of h_j^f in Q.

Case 2: $d_V^P(X_{j-1}^r, X_j^r) \geq 1$, i.e., the boundary strip $[X_{j-1}^r, X_j^r]$ is not flat. This is symmetrical to Case 1 since it is the same as Case 1 when we look from the rear. Thus we can directly attach h_j^L to the right of h_j^r in Q.

Case 3: $d_V^P(X_{j-1}^f, X_j^f) = 0$ and $d_V^P(X_{j-1}^r, X_j^r) = 0$, i.e., the boundary strip $[X_{j-1}^f, X_j^f]$ and $[X_{j-1}^r, X_j^r]$ are both flat. Then $h_j^f \cup s_j^{fT}$ is adjacent to $h_{j-1}^f \cup s_{j-1}^{fT}$, and $h_j^r \cup s_j^{rT}$ is adjacent to $h_{j-1}^r \cup s_{j-1}^{rT}$ in Q. We need to avoid h_j^L from overlapping with $h_{j-1}^f \cup s_{j-1}^{fT}$ or $h_{j-1}^r \cup s_{j-1}^{rT}$. As columns c_{j-1} and c_j are adjacent, we have that h_{j-1} and h_j lie in different rows. Thus we have two subcases depending on which of the two holes is higher.

Subcase 3-1: h_{j-1} is lower than h_j. Then the length of s_{j-1}^{fT} is shorter than s_j^{fT}. Thus h_j^f is higher than h_{j-1}^f in Q. We then attach h_j^L to the left of h_j^f, and h_j^L does not overlap with $s_{j-1}^{fT} \cup h_{j-1}^f$ in Q.

Subcase 3-2: h_{j-1} is higher than h_j. Then the length of s_{j-1}^{fT} is longer than s_j^{fT}. Thus, h_j^L cannot be attached to the left of h_j^f. Now suppose that we view the scene from the rear end. Then the situation is similar to Subcase 3-1. The strip s_{j-1}^{rT} is shorter than s_j^{rT}. Thus h_j^r is higher than h_{j-1}^r in Q. We then attach h_j^L to the right of h_j^r, and h_j^L does not cause any overlap with $s_{j-1}^{rT} \cup h_{j-1}^r$ in Q.

For the case of attaching h_j^R to h_j^f or h_j^r in order to maintain the property (ii) for Q_i, i.e., h_j^R does not overlap $h_{j+1}^f \cup s_{j+1}^{fT}$ and $h_{j+1}^r \cup s_{j+1}^{rT}$, we perform the similar procedure as we have just done for attaching h_j^L to h_j^f or h_j^r in order that h_j^L does not overlap with $h_{j-1}^f \cup s_{j-1}^{fT}$ and $h_{j-1}^r \cup s_{j-1}^{rT}$. This completes the description of the column unfolding step for column c_j.

Overlap Repairing Step. The unfolding of c_j in the previous step does not cause any overlapping with the unfolded patch of c_{j-1}. However, it may cause some overlapping with the unfolded patch of c_{j-2}. In fact, h_{j-2}^R and h_j^L may possibly overlap each other when $d_H^Q(h_{j-2}^f, h_j^f)$ and $d_H^Q(h_{j-2}^r, h_j^r)$ equals 2 and h_j^L is unfolded under Case 3 in the column unfolding step. Hence, we have to proceed to the second step, overlap repairing step, to modify the unfolding result of column c_j so that such overlap is avoided. Note that the unfolded patch Q_{i-1} obtained in the previous round is not modified during the execution of our repairing step. However due to lack of space, the details of the overlap repairing step are omitted.

By repeatedly applying the above two steps to unfold all the holes in the columns of polyhedron P from left to right, we finally obtain a flattened patch Q without self-overlapping. Hence, we obtain the following theorem.

Theorem 3. *There is an $O(n)$-time algorithm to edge-unfold a one-layer lattice polyhedron with at most one hole in each column.*

4.2   Extending to Polyhedra with Sparse Cubic Holes

The algorithm in the previous subsection can be extended to handle *one-layer lattice polyhedra with sparse cubic holes*, which are polyhedra such that each connected component in a column contains at most one hole. The result is stated in the following theorem.

Theorem 4. *There is an $O(n)$-time algorithm to edge-unfold one-layer lattice polyhedra with sparse cubic holes.*

5   Conclusions

We design three linear time edge-unfolding algorithms to unfold three special classes of one-layer lattice polyhedra with cubic holes. However, the edge-unfolding problem on general one-layer lattice polyhedra with arbitrary boundary and holes is still open. For such general polyhedra, even the problem of grid-unfolding which allows cutting along not only the original edges but also the edges of the refined grid inside any lattice polyhedron face is also open.

References

1. Abel, Z., Demaine, E.D.: Edge-Unfolding Orthogonal Polyhedra is Strongly NP-Complete. In: Proceedings of the 23rd Canadian Conference on Computational Geometry (2011)
2. Aloupis, G., et al.: Common unfoldings of polyominoes and polycubes. In: Akiyama, J., Bo, J., Kano, M., Tan, X. (eds.) CGGA 2010. LNCS, vol. 7033, pp. 44–54. Springer, Heidelberg (2011)
3. Bern, M., Demaine, E.D., Eppstein, D., Kuo, E., Mantler, A., Snoeyink, J.: Ununfoldable polyhedra with convex faces. Comput. Geom. Theory Appl. 24(2), 51–62 (2003)
4. Biedl, T., Demaine, E., Demaine, M., Lubiw, A., Overmars, M., OŘourke, J., Robbins, S., Whitesides, S.: Unfolding Some Classes of Orthogonal Polyhedra. In: Proceedings of the 10th Canadian Conference on Computational Geometry (1998)
5. Damian, M., Flatland, R., Meijer, H., OŘourke, J.: Unfolding well-separated orthotrees. In: 15th Annu. Fall Workshop Comput. Geom., pp. 23–25 (2005)
6. Damian, M., Flatland, R.Y., OŘourke, J.: Unfolding manhattan towers. Comput. Geom. 40(2), 102–114 (2008)
7. Damian, M., Meijer, H.: Grid Edge-unfolding Orthostacks with Orthogonally Convex Slabs. In: Proc. of the 14th Workshop on Computational Geometry, pp. 25–26 (2004)
8. Dürer, A.: Unterweysung der Messung mit dem Zirkel und Richtscheyt, in Linien Ebnen und gantzen Corporen, 1525. Reprinted 2002, Verlag Alfons Uhl, Nördlingen; translated as The Painter's Manual. Abaris Books, New York (1977)
9. Gupta, S.K., Bourne, D.A., Kim, K.H., Krishnan, S.S.: Automated process planning for sheet metal bending operations. Journal of Manufacturing Systems 17, 338–360 (1998)
10. OŘourke, J.: Unfolding Orthogonal Terrains, Smith Technical Report 084, arXiv:0707.0610v4 [cs.CG] (July 2007)
11. Shephard, G.C.: Convex polytopes with convex nets. In: Mathematical Proceedings of the Cambridge Philosophical Society, pp. 389–403 (1975)
12. Straub, R., Prautzsch, H.: Creating optimized cut-out sheets for paper models from meshes. In: SIAM Conf. Geometric Design and Computing (2005)
13. Takahashi, S., Wu, H.-Y., Saw, S.H., Lin, C.-C., Yen, H.-C.: Optimized topological surgery for unfolding 3D meshes. Computer Graphics Forum, 2077–2086 (2011)
14. Wang, C.-H.: Manufacturability-driven decomposition of sheet metal products (1997)

Directed Steiner Tree with Branching Constraint

Dimitri Watel[1,3], Marc-Antoine Weisser[1], Cédric Bentz[2],
and Dominique Barth[3]

[1] SUPELEC System Sciences, Computer Science DPT., 91192 Gif Sur Yvette, France
{dimitri.watel,marc-antoine.weisser}@supelec.fr
[2] CEDRIC-CNAM 292 rue Saint-Martin 75141 Paris Cédex 03, France
cedric.bentz@cnam.fr
[3] University of Versailles, 45 avenue des Etats-Unis, 78035, Versailles, France
dominique.barth@prism.uvsq.fr

Abstract. Given a directed weighted graph G, a root r and k terminals, the k-*Directed Steiner Tree* problem is to find a minimum cost tree rooted at r and spanning all terminals. If this problem has several applications in multicast routing in packet switching networks, the modeling is not adapted anymore in networks based upon the circuit switching principle in which some nodes, called *non diffusing nodes*, are not able to duplicate packets. We define a more general problem, named Directed Steiner Tree with Limited number of Diffusing nodes (DSTLD), able to model the multicast in a network containing at most d diffusing nodes. We show that DSTLD is XP with respect to d, and use this result to build a $\lceil \frac{k-1}{d} \rceil$-approximation XP in d for DST. Finally, we prove that, under the assumption that NP $\not\subseteq$ DTIME$[n^{O(\log \log n)}]$, there is no polynomial approximation algorithm for DSTLD with ratio $1 + (\frac{1}{e} - \varepsilon) \cdot \frac{k}{d-1}$ for every constant $\varepsilon > 0$.

Keywords: Directed Steiner Tree, Approximation, Diffusing node.

1 Introduction

Givent a directed weighted graph, the Directed Steiner Tree problem (DST) asks for a minimum cost tree rooted at a specific node r and spanning a specific set X of k nodes called *terminals*. This problem is known to have applications essentially in multicast routing where one wants to minimize the bandwidth consumption [1–4]. DST is used, instead of the undirected version [5], when the model of a symmetric network is not sufficient.

Contrary to the undirected problem which can be polynomially approximated within a constant ratio [6, 7], it was proved in [5] that DST is a generalization of the Set Cover problem and therefore is inapproximable within a $O(\log(k))$ ratio unless $NP \subseteq DTIME[n^{O(\log \log n)}]$, where k is the number of terminals [8]. It was later proved under the same assumption that for $\varepsilon > 0$ there is no $O(\log^{2-\varepsilon}(k))$ approximation algorithm [9]. The best known approximation ratio for DST is $O(k^\varepsilon)$ for any $\varepsilon > 0$ [10]. This approximation uses the following result: a tree of fixed height l with minimum cost is a $O(k^{\frac{1}{l}})$-approximation [11, 12]. Note

Z. Cai et al. (Eds.): COCOON 2014, LNCS 8591, pp. 263–275, 2014.
© Springer International Publishing Switzerland 2014

that this last approximation is neither polynomial nor XP in the parameter l. The two Steiner problems are FPT in k as it exists an exact algorithm in time $O(3^k n + 2^k(k + \log(n))n + n^2)$ and in space $O(2^k n)$ [13, 14], but are W[2]-hard with respect to the parameter "Optimal solution cost" [15, 16].

Nevertheless, the DST model assumes that when a *branching node* (with at least 2 successors) of the tree receives a data, it can transmit it to its multiple successors. This is the case in classical packet switching networks. However, previous works emphasize the fact that in optical networks, this assumption does not hold anymore as most of the nodes, called *non diffusing nodes*, can not copy any data but only send it to one of its children. As a consequence, a non diffusing branching node has to receive p copies of the data to transmit it to p children. Fortunately, some routers, called *diffusing nodes*, can duplicate data and thus need to receive it only once. The DST model assumes every node is diffusing. This new constraint was first introduced in undirected wavelength division multiplexing networks [17]. The optimal placement of diffusing nodes considering given multicast trees was then studied in [18]. Polynomial and approximability results can be found considering the undirected case where the diffusing nodes are already placed [19–21] or where a multicast tree is given [22].

We previously studied a restricted model in the directed and undirected cases where no arc could be used twice to send the same data: every branching node of the solution must be a diffusing node [23]. This problem is equivalent to finding a minimum cost Steiner Tree with Limited number of Branching nodes [23, 24]. When reducing the number of authorized branching nodes, the Directed Steiner problem become much harder to solve and approximate, even if the number of authorized branching nodes and terminals are fixed. This is due to instances where it is NP-Complete to build a feasible solution.

If no node is diffusing, including the root, the solution can not contain any branching node. This problem is equivalent to the Steiner Cycle problem [25, 26] where one wants to find a cycle of minimum cost containing the root and the terminals. As for DST with a limited number of branching nodes, it may be hard to find a feasible solution for instances of this problem.

In DST, every node is diffusing and thus there is no need to use the same arc more than once. As DST solutions are trees with at most k leaves, note that there is no need to select more than $k - 1$ diffusing nodes: the at most $k - 1$ branching nodes of the tree.

This new constraint defines a new cost for feasible solutions, previously introduced in [21, 22]: the *load* of an arc in a solution is the number of data transiting through it to transfer information from the root to terminals, and the cost of that arc is equal to its weight multiplied by its load. In the undirected model defined in [21, 22], the feasible solutions returned were trees, because it was supposed that the best solution for the network was a tree. Such a model allows to define recursively the load as 1 for an arc entering a diffusing node or a terminal, and the sum of the load of its outgoing arcs for any other node. If this definition can be generalized to directed acyclic graphs, it can not be applied to general digraphs.

Our Results. We define the Directed Steiner Tree problem with Limited number of Diffusing nodes (DSTLD), improving the models defined in [19–22], and compensating the difficulties encountered in [23–26]. In DSTLD, we do not impose the optimal solution to be a tree. This is a more accurate model for asymmetric network applications with diffusing routers. We prove this problem is XP but W[2]-Hard in the number of authorized diffusing routers d, and NP-Complete even if $k - d$ is fixed.

We finally prove two approximation results. 1) As explained, computing the tree of best cost with a fixed height l is a $O(k^{\frac{1}{l}})$-approximation [11, 12] but this approximation is neither polynomial nor XP in the parameter l. We are here not interested in the height of the tree, but in the number of diffusing nodes it contains. We claim that one can transform any instance of DST to get an instance of DSTLD with $d < k$ authorized diffusing routers, solve that instance to get a solution T_d, and build from T_d a $\lceil \frac{k-1}{d} \rceil$-approximation for the DST instance. As DSTLD is XP in d, this approximation for DST is also XP in d. 2) There is a strong inapproximability result for DSTLD. For every $\varepsilon > 0$ it cannot be approximated within $1 + (1 - \varepsilon)\frac{k}{e \cdot d}$ (where e denotes the Euler constant) unless $NP \subseteq DTIME[n^{O(\log \log n)}]$.

The next section gives an example where returning a graph containing a cycle is a better answer than returning a tree. Then, Section 3 details some notations useful for this paper and Section 4 defines the new model. Finally, Sections 5 and 6 study the parameterized complexity of this problem and its approximability.

2   The Optimal Solution Is Not Always a Tree

The undirected model developed in [21, 22] assumes that optimal routings are always trees. This property cannot be claimed for directed cases. Figure 1 illustrates a case where returning a graph containing a cycle is a better answer.

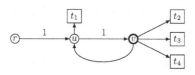

Fig. 1. In this example, the maximal number of allowed diffusing node is 1. Every non specified weight is 0. v is here chosen as a diffusing node.

We assume only one node can be selected as a diffusing node, including r. If r is selected, then it sends the data once per terminal. The arc (r, u) is used 4 times, and (u, v) 3 times, for a total cost of 7. If u is diffusing, then the root sends the data to u, and u dispatches it to each terminal, for a total cost of 4. Finally, if v is selected, the data is transmitted only once over the path $\{(r, u), (u, v)\}$, the node v is able to send one data to t_1 using the $\{(v, u), (u, t_1)\}$ path. The number of data per arc is one for every arc, and the total cost is 2. This optimal routing is not a tree but it is always possible to describe the routing as a tree. For example, this solution can be described as $\{(r, v), (v, t_1), (v, t_2), (v, t_3), (v, t_4)\}$, a tree where each arc describes a shortest path in the original graph.

3   Notations

We define as n the number of nodes, m the number of arcs, k the number of terminals and d the maximum number of diffusing nodes in the returned solution.

Let u and v be two nodes in a directed graph G weighted over its arcs with function ω. We define $P(u, v)$ as the shortest path linking u and v and $\omega^\triangleright(u, v)$ as the weight of this path. If $P(u, v)$ does not exist, then $\omega^\triangleright(u, v) = +\infty$. If multiple shortest paths exist, one is arbitrary chosen as $P(u, v)$. If G' is a subgraph of G, then $\omega(G') = \sum_{a \in G'} \omega(a)$, and similarly $\omega^\triangleright(G') = \sum_{(u,v) \in G'} \omega^\triangleright(u, v)$.

We assume the root has not any predecessor and only one successor. We also assume the terminals to be leaves. If not, we first preprocess the graph to ensure those properties. For instance, if the root r has a predecessor or two successors, we simply replace it by a copy r' and add an arc (r', r) with weight 0.

Definition 1. *Let $\mathcal{I} = (G = (V, A), r, X, \omega)$ be an instance of the DST problem. Then the shortest paths instance $\mathcal{I}^\triangleright = (G^\triangleright = (V, A^\triangleright), r, X, \omega^\triangleright)$ defines the instance where $G^\triangleright$ is a complete graph weighted by the lengths of the shortest paths in G.*

$\omega^\triangleright$ satisfies the triangle inequality. As we assume that the root has no predecessor and that the terminals are all leaves, then each arc entering r and each arc leaving a terminal has infinite weight in the shortest paths instance.

Definition 2. *A branching node is a node with at least two successors.*

4   The Directed Steiner Tree with Limited Number of Diffusing Nodes

In this section, we define our model and explain its relation with the optical network problem with diffusing nodes.

Problem 1. Given a DST instance $\mathcal{I} = (G, r, X, \omega)$ and an integer $d \in [0; k - 1]$, the Directed Steiner Tree problem with Limited number of Diffusing nodes (DSTLD) consists, in the shortest paths instance $\mathcal{I}^\triangleright$, in the search of a tree T rooted at r, spanning X, with at most d branching nodes minimizing the cost $\omega^\triangleright(T) = \sum_{a \in T} \omega^\triangleright(a)$.

4.1 Link with DST

When $d = k - 1$, the instance of DSTLD, and the instance of DST are equivalent:

- From a feasible solution T for DST, one can build a feasible solution $T^\triangleright$ for DSTLD by replacing each arc (u, v) of T by the equivalent arc (u, v) in $G^\triangleright$. As $G^\triangleright$ is weighted with shortest paths, $\omega^\triangleright(T^\triangleright) \leq \omega(T)$
- Conversely, from a feasible solution $T^\triangleright$ for DSTLD, one can build a feasible solution T for DST by returning $\bigcup_{(u,v)\in T^\triangleright} P(u, v)$. The two solutions satisfy, $\omega(T) \leq \omega^\triangleright(T^\triangleright)$.

As a consequence, any approximation algorithm for DSTLD with a ratio $\alpha(k)$ implies an approximation algorithm within $\alpha(k)$ for DST. As, in this particular case, $d = k - 1$, this results holds for an $\alpha(d)$-approximation for DSTLD.

Theorem 1. *DSTLD is NP-Complete and any approximation algorithm for DSTLD with ratio $\alpha(d)$ or $\alpha(k)$ implies an approximation algorithm with ratio $\alpha(k)$ for DST.*

4.2 Application of DSTLD to Multicast in an Optical Network

From a feasible solution of DSTLD, we can determine the diffusing nodes of the network and the load inside each arc necessary to transmit the data from the root to all terminals. Let T be a feasible solution of an instance of DSTLD.

- Each branching node in T is a *diffusing node* in G.
- The *load* $l(a)$ of an arc a in G is the number of time it appears in path $P(u, v)$ for every $(u, v) \in T$.

We now suppose the data is transmitted from the root to each terminal using T. It is copied at each diffusing node in the original graph. As a consequence, the data goes through each path $P(u, v)$ for $(u, v) \in T$ exactly once, thus through each arc a number of times equal to its load. Note that some diffusing nodes in G can be non branching nodes, and some branching nodes in G can be non diffusing. The cost of T is, as defined in the previous model of [21, 22], the sum, over all the arcs, of the weights of each arcs multiplied by its load.

Lemma 1. $\omega^\triangleright(T) = \sum\limits_{a\in A} l(a)\omega(a)$

Remark 1. The DSTLD model does not allow a non diffusing root to send the data more than once although the root has no such limitation in a real network. This permits us to simplify the proofs of this paper by reducing the number of different types of node: diffusing or not.

Remark 2. Unlike in [23–26], we can decide in polynomial time whether or not an instance contains a feasible solution. Indeed, if the original graph contains a path from the root to every terminals, we can return the star (with one branching node) in $\mathcal{I}^\triangleright$ centered at the root and containing every terminal. And if some terminal is unreachable, then the instance has no solution of finite cost.

4.3 Application of DSTLD to Multicast in an Optical Network Where the Diffusing Nodes are Already Chosen

DSTLD assumes every node in the network is able to diffuse or not, but at most d branching nodes will actually diffuse in order to limit, for example, energy consumption or signal loss quality. We can assume, as in [19–21] on the contrary that the possible diffusing nodes D are already placed in the graph, for instance because the technology for diffusing routers is specific. We set d to $|D|$.

Note that any data sent from the root or a diffusing node either joins with a shortest path another diffusing node or a terminal. Consequently, we can search, as it is done in [19, 20], a solution in the shortest paths instance $\mathcal{I}_r^\triangleright$ restricted to $\{r\} \cup X \cup D$, instead of the complete instance $\mathcal{I}^\triangleright$.

Because $d = |D|$, no tree in $\mathcal{I}_r^\triangleright$ rooted at r spanning X can have more than d branching nodes. As a consequence, an optimal multicast is then described by an optimal directed steiner tree T^* in $\mathcal{I}_r^\triangleright$. Moreover, any α-approximation algorithm for DST gives an α-approximation algorithm for this problem.

5 Parameterized Complexity

In this section, we establish three parameterized complexity results over DSTLD. The two first ones claim that DSTLD parameterized with d belongs to the class XP although it is W[2]-Hard. The last one studies DSTLD parameterized by $k - d$ and shows it is an NP-Complete problem.

In order to show the first result, we firstly prove that any tree with internal non branching nodes can be reduced to a smaller tree.

Lemma 2. *Let T be a feasible solution of an instance $\mathcal{I}$ of DSTLD. A directed tree with cost at most that of T, rooted at r, containing only all the terminals X and the branching nodes B of T can be obtained in polynomial time from T.*

Proof. We delete all the cycles and leaves not in $X \cup B$. We get a directed tree rooted at r with lower cost, with less branching nodes, where all the leaves are in $X \cup B$. Let E be the set $\{r\} \cup X \cup B$. We now replace each path with endpoints in E and internal nodes not in E by a single arc of $G^\triangleright$. As the weights satisfy the triangular inequality, the cost of the tree does not increase. The obtained tree is a feasible solution and contains only r, X and B.   □

By Lemma 2, one of the optimal solutions is such a tree. As any terminal in the original graph is a leaf, it is also a leaf in any optimal solution. Thus, we can reduce the search space to the trees with k leaves and at most d internal nodes.

Theorem 2. *The DSTLD problem, when parameterized by d, is XP.*

Proof. We prove there is an exact polynomial algorithm for DSTLD when d is fixed. Let $\mathcal{I}$ be an instance of DSTLD. Let $\kappa = (u_1, u_2, \ldots, u_j)$ be j distinct nodes of V, with $j \leq d$. We now search a minimum cost feasible solution T_κ containing every nodes of κ and where each branching node is in κ. Obviously, if κ is exactly the set of branching nodes of an optimal solution T^* of $\mathcal{I}$, then T_κ is also an optimal solution: by iterating over all possible sets κ, we return an optimal solution.

By Lemma 2, if T_κ has a finite cost, it can be searched among all the trees rooted at r containing only κ and X. We now point out that T_κ is a spanning tree of the shortest paths graph $G^\triangleright$ rooted at r restricted to r, κ and X. We then search for a minimum spanning tree in that graph. Every branching node of T is in κ. Indeed, every terminal is a leaf in G, then, unless the cost of T_κ is infinite, a minimum spanning tree does not use any terminal as a branching node. Similarly, the root is not used as a branching node.

A minimum directed spanning tree can be find in time $O((1 + k + d)^2)$ [27], then this algorithm runs in time $O(n^d(1 + k + d)^2)$. □

Theorem 3. *The DSTLD problem, when parameterized by d, is W[2]-Hard.*

Sketch of proof. An FPT reduction from the problem Set Cover parameterized with the cost of an optimal solution can be obtained by adapting the classical reduction from Set Cover to DST. A similar proof is given in [23]. □

Theorem 4. *Even if $k - d$ is a fixed parameter, the DSTLD problem is NP-Complete and any $\alpha(d)$-approximation for DSTLD implies an $\alpha(k')$-approximation for DST where k' is the number of terminals in an instance of DST.*

Sketch of proof. This theorem extends Theorem 1. Let p be a fixed integer. From an instance $\mathcal{I}$ for DST with k' terminals, one can build an instance $\mathcal{I}_d$ for DSTLD with k terminals satisfying $k - d = p$ by adding to $\mathcal{I}$ a star of weight 0 centered at r and with p terminals. Thus, $k' = k - p = d$. Obviously, from any feasible solution for $\mathcal{I}_d$, one can build a solution with same cost for $\mathcal{I}$ in polynomial time and vice versa. Consequently, any $\alpha(d)$-approximation for $\mathcal{I}_d$ is an $\alpha(k')$-approximation for $\mathcal{I}$. □

6 Approximability

In this section we are interested in approximation results over DSTLD. The first subsection establishes an approximation ratio between an optimum solution of a DSTLD instance and an optimum solution of the associated DST instance. The second subsection shows an inapproximability result for DSTLD.

6.1 How DST Can be Approximated by DSTLD

It was previously proved that restricting the search space to trees with fixed height l gives a $O(k^{\frac{1}{l}})$-approximation for DST [11, 12]. However, this approximation is neither polynomial nor XP in l. We are interested here in reducing the number of authorized diffusing nodes instead of the height of the tree.

Any DST instance $\mathcal{I} = (G, r, X, \omega)$ can be transformed into a DSTLD instance $\mathcal{I}_d$ by adding a parameter $d \leq k - 1$. We now prove that computing an optimal solution for $\mathcal{I}_d$ gives an approximated solution for $\mathcal{I}$. As DSTLD is XP with respect to the parameter d, this is possible to compute this solution for small values of d. We now assume that $k > 1$.

Let T^* be an optimal solution for $\mathcal{I}$ and T_d^* be an optimal solution for $\mathcal{I}_d$.

Lemma 3. $\frac{\omega^\triangleright(T_d^*)}{\omega(T^*)} \leq \lceil \frac{k-1}{d} \rceil$.

Proof. We will transform T^* into a feasible solution T_d of $\mathcal{I}_d$ by replacing d subtrees of T^* by stars. We first build the equivalent tree T^* in $G^\triangleright$: each arc (u, v) of T^* is replaced by the arc (u, v) in $G^\triangleright$. As (u, v) is weighted by the length $\omega^\triangleright(u, v)$ of shortest path between u and v, its cost does not increase.

We define for a node u of T^* the subtree rooted at u by $T^*(u)$ and the terminals it reaches by $X(u)$. Let v be a node such that:

$$\begin{cases} |X(v)| \geq 1 + \lceil \frac{k-1}{d} \rceil \\ |X(w)| < 1 + \lceil \frac{k-1}{d} \rceil \text{ for each successor } w \text{ of } v. \end{cases}$$

If no node can satisfy those properties, then $|X(u)| < 1 + \lceil \frac{k-1}{d} \rceil$ for every node u. In that case, we choose v as the first branching node reached by r in T (or r itself if it is a branching node).

$S(v)$ is the star containing the arc (v, t) for all $t \in X(v)$. The cost $\omega^\triangleright(S(v))$ of $S(v)$ is at most $\sum_{t \in X(v)} \sum_{a \in P_{T^*}(v,t)} \omega^\triangleright(a)$ where $P_{T^*}(v, t)$ is the path from v to t in T^*. Moreover, for an arc a in $T^*(v)$, the number of distinct paths $P_{T^*}(v, t)$ containing a cannot be more than $\lceil \frac{k-1}{d} \rceil$. Indeed, each successor of v reaches at most $\lceil \frac{k-1}{d} \rceil$ terminals.

$$\omega^\triangleright(S(v)) \leq \lceil \frac{k-1}{d} \rceil \cdot \omega^\triangleright(T^*(v)) \tag{1}$$

We temporary replace in T^* the subtree $T^*(v)$ by a terminal $t^{(1)}$. The obtained tree $T^{(1)}$ contains at most $k - |X(v)| + 1 \leq k - \lceil \frac{k-1}{d} \rceil \leq k - \frac{k-1}{d}$ terminals. We repeat this operation and build the trees $T^{*(2)}$, $T^{*(3)}$, ... until it remains only one terminal. As each operation removes at least $\frac{k-1}{d}$ terminals, this operation is repeated at most d times.

We now expand all the stars in reverse order. The resulting graph is a tree T_d containing all the terminals in X and at most d branching nodes (the root of each star). By equation (1), $\omega^\triangleright(T_d) \leq \lceil \frac{k-1}{d} \rceil \omega^\triangleright(T^*) \leq \lceil \frac{k-1}{d} \rceil \omega(T^*)$. □

Remark 3. There is no $\lceil \frac{d_2}{d_1} \rceil$ approximation ratio between two optimal solutions of $\mathcal{I}_{d_1}$ and $\mathcal{I}_{d_2}$, because the Steiner tree can have less than $k-1$ branching nodes. The only known result is $\omega(T^*) = \omega^\triangleright(T_{k-1}^*) \leq \cdots \leq \omega^\triangleright(T_2^*) \leq \omega^\triangleright(T_1^*)$.

Theorem 5. *For any α-approximation algorithm for DSTLD, one can build an approximation algorithm of ratio $\alpha \lceil \frac{k-1}{d} \rceil$ for DST.*

Proof. If we compute our α-approximation algorithm over $\mathcal{I}_d$, we get a tree T_d satisfying $\omega^{\triangleright}(T_d) \leq \alpha \cdot \omega^{\triangleright}(T_d^*)$. The tree $T = \bigcup_{(u,v) \in T_d} P(u,v)$ is a feasible solution for $\mathcal{I}$ and costs at most $\omega^{\triangleright}(T_d)$.

By Lemma 3, $\omega(T) \leq \alpha \lceil \frac{k-1}{d} \rceil \cdot \omega(T^*)$ ◻

With this technique, one can either choose a fixed parameter d to get a $\lceil \frac{k-1}{d} \rceil$ polynomial approximation for DST (to be more exact, this approximation is XP with respect to d), or choose a variable parameter d, for instance $d = \log(k)$, and compute an $\alpha(d, k)$ polynomial approximation for DSTLD to get an $\alpha(d, k) \lceil \frac{k-1}{d} \rceil$ polynomial approximation for DST. Unfortunately, this second approximation seems pointless as the next part proves a strong inapproximability result for DSTLD.

6.2   A Strong Inapproximability Result for DSTLD

In this section, we build a reduction from the Maximum Coverage problem in order to prove, under the assumption P $\neq$ NP there is no polynomial approximation for DSTLD of ratio better than $1 + \frac{k}{e \cdot (d-1)}$ where e is the Euler constant.

Problem 2. Given a *universe* X of *elements*, a *cover* $\mathcal{S}$ of X, and an integer $d \in \mathbb{N}^*$, the maximum Coverage problem (max-SC) asks for a cover $C \subset \mathcal{S}$ with at most d sets maximizing the number of covered elements.

Theorem 6. *[8]: Unless NP $\subseteq$ DTIME$[n^{O(\log \log n)}]$, for every constant $\varepsilon > 0$, there is no polynomial approximation algorithm with ratio $1 - \frac{1}{e} + \varepsilon$ for max-SC, even restricted to instances where every optimal solution covers all the universe with exactly d sets.*

We now describe the reduction. Let $d \geq 2$ and let $\mathcal{I} = (X, \mathcal{S}, d - 1)$ be a max-SC instance where every optimal solution covers all the elements with exactly $d - 1$ sets. Let k be the number of elements.

We remove from $\mathcal{I}$ each set included in another set. Any feasible solution with strictly less than $d - 1$ sets can be completed to get a strictly better solution. We will now only consider inclusion-wise maximal feasible solutions.

We will define an instance $\mathcal{I}' = (G = (V, A), r, X', \omega, d)$ of DSTLD. The nodes V consist of the root r, one node for each set of $\mathcal{S}$ and one terminal for each element of X. Each set and element is identified with its corresponding node in V. Let $B > 0$ be a constant we will fix later. The root is linked to each set with an arc of weight B. Each set s is linked to the elements it covers. Finally, the root is linked to each terminal with an arc of weight $B + 1$. Note that to obtain $G^{\triangleright}$, we complete G with infinite cost arcs. As a result, any feasible solution of finite cost can be described by a tree using only arcs of G. An example is given in Figure 2.

We now define how to build a feasible solution of $\mathcal{I}$ from a feasible solution of $\mathcal{I}'$ and vice versa.

The main idea of the reduction is that the sets in an optimal solution for $\mathcal{I}$ are the diffusing nodes in an optimal solution for $\mathcal{I}'$. In order to prove the

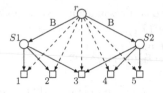

Fig. 2. An example of reduction from the instance $\mathcal{I}$ of max-SC with $X = \{1, 2, 3, 4, 5\}$ and $\mathcal{S} = \{S1 = \{1, 2, 3\}, S2 = \{3, 4, 5\}\}$. Each dashed arc is weighted with $B + 1$.

inapproximation result, we will set the value of B to a sufficiently big value. This will ensure that a terminal not covered by a set costs a lot more than the other terminals.

Build a solution $\mathcal{T}(C)$ for $\mathcal{I}'$ from a solution C for $\mathcal{I}$.

Let C be an inclusion-wise maximal feasible solution of $\mathcal{I}$. Then we define $\mathcal{T}(C)$ as a tree of $\mathcal{I}'$ containing the root, each set in C and all the terminals. For each element t not covered by C, we add the arc (r, t) to $\mathcal{T}(C)$, and for each other element t, we add the arc (s, t) where s is chosen among the sets in C covering t. As C is inclusion-wise maximal, it contains exactly $d - 1$ sets. Then the number of branching nodes in $\mathcal{T}(C)$ is at most d, and $\mathcal{T}(C)$ is a feasible solution. The cost of $\mathcal{T}(C)$ is

$$\omega^{\triangleright}(\mathcal{T}(C)) = B(d - 1 + k - |X(C)|) + k \qquad (2)$$

where $X(C)$ are the terminals covered by C. As an optimal solution C^* covers all the terminals with exactly $d - 1$ sets, then $\omega^{\triangleright}(\mathcal{T}(C^*)) = B(d - 1) + k$.

Build a solution $\mathcal{C}(T)$ for $\mathcal{I}$ from a solution T for $\mathcal{I}'$.

Let T be a feasible solution of $\mathcal{I}'$ with finite cost. We define $\mathcal{C}(T)$ as a feasible solution of $\mathcal{I}$ with all the sets corresponding to branching nodes of T. If $\mathcal{C}(T)$ is not inclusion-wise maximal, we add new sets covering non covered elements until it reaches $d - 1$ sets.

As $d \geq 2$, any feasible solution T of $\mathcal{I}'$ with finite cost contains at most $d - 1$ branching nodes corresponding to sets. Indeed, a tree with d such sets would not use r as a branching node, and some sets would be linked together with arcs of infinite weight. Then $\mathcal{C}(T)$ contains at most $d - 1$ sets.

Lemma 4. *Let T be a tree feasible solution of $\mathcal{I}'$ with finite cost then we have* $\omega^{\triangleright}(\mathcal{T}(\mathcal{C}(T))) \leq \omega^{\triangleright}(T)$.

Proof. Let $S = \{s_1, s_2, \ldots, s_p\}$ be the sets branching nodes of T, and let X_S be the terminals covered by those nodes in T. The cost of T is then $\omega^{\triangleright}(T) = B(p + k - |X_S|) + k$. By equation (2), $\omega^{\triangleright}(\mathcal{T}(\mathcal{C}(T))) = B(d - 1 + k - |X(\mathcal{C}(T))|) + k$. We recall $B > 0$. We want to prove that $\omega^{\triangleright}(T) - \omega^{\triangleright}(\mathcal{T}(\mathcal{C}(T))) \geq 0$, equivalent to the following equations:

$$B \cdot (p - |X_S| - (d - 1) + |X(\mathcal{C}(T))|) \geq 0$$
$$|X(\mathcal{C}(T))| - |X_S| \geq (d - 1) - p$$

Each set in $\mathcal{C}(T)$ not in S was added to get an inclusion-wise maximal solution. Each time such a set is added in $\mathcal{C}(T)$, it covers at least one element that was not yet covered because all optimal solutions contain $d-1$ sets. Then the number of terminals in $X(\mathcal{C}(T))$ not in X_S is at least $(d-1)-p$.   □

Lemma 5. *If C^* is an optimal solution of $\mathcal{I}$, then $\mathcal{T}(C^*)$ is optimal for $\mathcal{I}'$.*

Proof. Assume it exists a feasible solution T for $\mathcal{I}'$ of cost strictly better than $\mathcal{T}(C^*)$. By Lemma 4, $\omega^{\triangleright}(\mathcal{T}(\mathcal{C}(T))) \leq \omega^{\triangleright}(T) < \omega^{\triangleright}(\mathcal{T}(C^*))$. We recall $B > 0$. By Equation (2): $B(d-1+k-|X(\mathcal{C}(T))|)+k < B(d-1+k-|X(C^*)|)+k$ and then $|X(\mathcal{C}(T))| > |X(C^*)|$. This is a contradiction, hence the lemma is proved.   □

Theorem 7. *Under the assumption that $NP \not\subseteq DTIME[n^{O(\log \log n)}]$, for every constant $\varepsilon > 0$, there is no polynomial-time approximation algorithm for DSTLD with ratio $1 + (\frac{1}{e} - \varepsilon) \cdot \frac{k}{d-1}$.*

Proof. Let $\varepsilon > 0$. We assume there is an approximation algorithm with ratio $\rho = 1 + (\frac{1}{e} - \varepsilon) \cdot \frac{k}{d-1}$ for DSTLD. Then, by Lemma 5, the algorithm returns a feasible solution T with cost at most $\rho \cdot \omega^{\triangleright}(\mathcal{T}(C^*))$ where C^* is an optimal solution for $\mathcal{I}$.

We recall that, by Lemma 4, $\omega^{\triangleright}(\mathcal{T}(\mathcal{C}(T))) \leq \omega^{\triangleright}(T)$. Moreover, any optimal solution of $\mathcal{I}$ spans all elements in X: $|X(C^*)| = k$.

$$\frac{\omega^{\triangleright}(\mathcal{T}(\mathcal{C}(T)))}{\omega^{\triangleright}(\mathcal{T}(C^*))} \leq \frac{\omega^{\triangleright}(T)}{\omega^{\triangleright}(\mathcal{T}(C^*))} \leq \rho$$

$$\Leftrightarrow \quad \frac{B(d-1+k-|X(\mathcal{C}(T))|)+k}{B \cdot (d-1)+k} \leq \rho$$

$$\Leftrightarrow \quad d-1+k-|X(\mathcal{C}(T))| + \frac{k}{B} \leq (d-1+\frac{k}{B}) \cdot \rho$$

$$\Leftrightarrow \quad d-1+k-|X(\mathcal{C}(T))| + \frac{k}{B} \leq (d-1) \cdot (1+(\frac{1}{e}-\varepsilon) \cdot \frac{k}{d-1})+\frac{k}{B} \cdot \rho$$

$$\Leftrightarrow \quad k \cdot (1 - \frac{1}{e} + \varepsilon + \frac{1-\rho}{B}) \leq |X(\mathcal{C}(T))|$$

$$\Leftrightarrow \quad |X(C^*)| \cdot (1 - \frac{1}{e} + \varepsilon + \frac{1-\rho}{B}) \leq |X(\mathcal{C}(T))|$$

We now fix $B = 2 \cdot \frac{\rho-1}{\varepsilon} > 0$. Then $1 - \frac{1}{e} + \frac{\varepsilon}{2} \leq \frac{X(\mathcal{C}(T))}{X(C^*)}$, which implies a contradiction with Theorem 6, thus this theorem is proved.   □

7   Conclusion and Perspectives

We proposed a generalization of the Directed Steiner Tree problem in order to model multicast in network containing at most d diffusing nodes. We have proved that this problem is XP in d by providing an algorithm to solve it. We have used this algorithm to design an approximation which runs in time XP in d. However, we have also shown a strong inapproximability result for DSTLD.

Even so, the only polynomial approximation algorithm considered for DSTLD has a ratio depending only on k. It would be interesting to build new approximation algorithms for BO with better ratios, depending on k and d.

References

1. Cheng, X., Du, D.Z.: Steiner trees in industry, vol. 11. Kluwer (2001)
2. Voß, S.: Steiner tree problems in telecommunications. In: Handbook of Optimization in Telecommunications, pp. 459–492 (January 2006)
3. Rugeli, J., Novak, R.: Steiner tree algorithms for multicast protocols (1995)
4. Novak, R.: A note on distributed multicast routing in point-to-point networks. Computers & Operations Research, 1149–1164 (October 2001)
5. Karp, R.: Reducibility among combinatorial problems. Springer (1972)
6. Kou, L., Markowsky, G., Berman, L.: A fast algorithm for Steiner trees. Acta Informatica, 141–145 (1981)
7. Robins, G., Zelikovsky, A.: Improved Steiner tree approximation in graphs. In: Proc. SODA, pp. 770–779 (2000)
8. Feige, U.: A threshold of ln n for approximating set cover. JACM, 634–652 (1998)
9. Halperin, E., Krauthgamer, R.: Polylogarithmic inapproximability. In: Proc. STOC, pp. 585–594. ACM (2003)
10. Charikar, M., Chekuri, C., Cheung, T., Dai, Z.: Approximation algorithms for directed Steiner problems. In: Proc. SODA, pp. 192–200 (1998)
11. Zelikovsky, A.: A series of approximation algorithms for the acyclic directed Steiner tree problem. Algorithmica, 99–110 (1997)
12. Helvig, C., Robins, G., Zelikovsky, A.: An improved approximation scheme for the group Steiner problem. Networks (2001)
13. Dreyfus, S.E., Wagner, R.A.: The steiner problem in graphs. Networks 1(3), 195–207 (1971)
14. Ding, B., Yu, J.X., Wang, S., Qin, L.: Finding top-k min-cost connected trees in databases. In: ICDE (2007)
15. Downey, R.G., Fellows, M.R.: Parameterized complexity. Monographs in computer science edn. Springer (1999)
16. Jones, M., Lokshtanov, D., Ramanujan, M.S., Saurabh, S., Suchý, O.: Parameterized complexity of directed steiner tree on sparse graphs. In: Bodlaender, H.L., Italiano, G.F. (eds.) ESA 2013. LNCS, vol. 8125, pp. 671–682. Springer, Heidelberg (2013)
17. Malli, R., Zhang, X., Qiao, C.: Benefit of Multicasting in All-Optical Networks. In: SPIE Proc. Conf. All-Optical Networking (1998)
18. Lin, H.-c., Wang, S.-W.: Splitter Placement in All-Optical WDM Networks. In: Global Telecommunications Conference (2005)
19. Du, H., Jia, X., Wang, F., Thai, M.Y., Li, Y.: A Note on Optical Network with Nonsplitting Nodes. JCO (2005)
20. Guo, L., Wu, W., Wang, F., Thai, M.: Approximation for Minimum Multicast Route in Optical Network with Nonsplitting Nodes. JCO (2005)
21. Reinhard, V., Tomasik, J., Barth, D., Weisser, M.-A.: Bandwidth Optimization for Multicast Transmissions in Virtual Circuit Networks. In: Fratta, L., Schulzrinne, H., Takahashi, Y., Spaniol, O. (eds.) NETWORKING 2009. LNCS, vol. 5550, pp. 859–870. Springer, Heidelberg (2009)
22. Reinhard, V., Cohen, J., Tomasik, J., Barth, D., Weisser, M.A.: Optimal configuration of an optical network providing predefined multicast transmissions. Comput. Netw. 56(8), 2097–2106 (2012)
23. Watel, D., Weisser, M.-A., Bentz, C., Barth, D.: Steiner Problems with Limited Number of Branching Nodes. In: Moscibroda, T., Rescigno, A.A. (eds.) SIROCCO 2013. LNCS, vol. 8179, pp. 310–321. Springer, Heidelberg (2013)

24. Gargano, L., Hell, P., Stacho, L., Vaccaro, U.: Spanning trees with bounded number of branch vertices. In: Widmayer, P., Triguero, F., Morales, R., Hennessy, M., Eidenbenz, S., Conejo, R. (eds.) ICALP 2002. LNCS, vol. 2380, pp. 355–365. Springer, Heidelberg (2002)
25. Salazar-González, J.J.: The Steiner cycle polytope. European Journal of Operational Research 147(3), 671–679 (2003)
26. Steinová, M.: Approximability of the Minimum Steiner Cycle Problem (2010)
27. Tarjan, R.: Finding optimum branchings. Networks 7(1), 25–35 (1977)

On the Parameterized Complexity
of Labelled Correlation Clustering Problem*

Xianmin Liu, Jianzhong Li**, and Hong Gao

Harbin Institute of Technology, Harbin, Heilongjiang, China
xianmliu@gmail.com, {lijzh,honggao}@hit.edu.cn

Abstract. The Labelled Correlation Clustering problem, a variant of Correlation Clustering problem, is defined and studied in this paper. Since the
problem is NP-*complete*, we consider the parameterized complexities.
Three different parameterizations are considered, and the corresponding
parameterized complexities are studied.

1 Introduction

A general *graph*[1] G is composed of a node set V_G and an edge set $E_G \subseteq V_G \times V_G$,
denoted by $G = (V_G, E_G)$. A graph G is *clustered* if every connected component
of G is a clique. An *edge labelled* graph, *el*-graph for short, can be denoted by
$\widetilde{G} = (V_{\widetilde{G}}, E_{\widetilde{G}}, f_{\widetilde{G}})$ where $V_{\widetilde{G}}$ and $E_{\widetilde{G}}$ define a general graph and $f_{\widetilde{G}}$ is a mapping
$E_{\widetilde{G}} \mapsto \{0, 1\}$. Graph G *disagrees* with a *el*-graph $\widetilde{G}$, if there is some edge $e \in E_{\widetilde{G}}$
such that $f_{\widetilde{G}}(e) = 1 \wedge e \notin E_G$ or $f_{\widetilde{G}}(e) = 0 \wedge e \in E_G$, and such edges are called
disagreed edges between G and $\widetilde{G}$. G *agrees* with $\widetilde{G}$, if there are no disagreed
edges between them. Given an *el*-graph set $\widetilde{\mathcal{G}} = \{\widetilde{G}_1, \ldots, \widetilde{G}_m\}$ and a graph G,
let DISAGREE be a function such that DISAGREE$(G, \widetilde{\mathcal{G}})$ is the number of graphs
in $\widetilde{\mathcal{G}}$ with which G disagrees.

In the classic Correlation Clustering problem (CC for short) [1], given a *el*-
graph $\widetilde{G}$, the goal is to find a *clustered* graph G such that the size of disagreed
edges between G and $\widetilde{G}$ is minimum. It has many applications, such as entity
identification [2], coreference resolution [3] and so on. The CC problem is NP-
hard, and there have been lots of works focusing on it, for example [1,4,5,6].

In this paper, a variant of Correlation Clustering is studied, which is called
Labelled Correlation Clustering (LCC for short) and can be defined as follows. The
input of a LCC instance is a *el*-graph set $\widetilde{\mathcal{G}} = \{\widetilde{G}_1, \ldots, \widetilde{G}_m\}$, the goal is to find
a *clustered* graph G such that the size DISAGREE$(G, \widetilde{\mathcal{G}})$ is minimum.

* This work was supported in part by the National Grand Fundamental Research
973 Program of China under grant 2012CB316200, the Major Program of National
Natural Science Foundation of China under grant 61190115, the Key Program of
the National Natural Science Foundation of China under grant 61033015, 60933001,
and the Shandong Provincial Natural Science Foundation of China under grant no.
ZR2013FQ028.
** Corresponding author.
[1] We only consider *undirected graph* here.

Z. Cai et al. (Eds.): COCOON 2014, LNCS 8591, pp. 276–287, 2014.

This problem formulation can be motivated by the following example in entity identification problem. Suppose there are a set of records, each of them contains the values of several attributes about one person. For example, the record {*name*="Bob", *age*="12", ...} represents one person named Bob is 12-year-old. There may be several records describing the same person, and the entity identification problem is to find a clustering way for the records such that each cluster exactly contains all records of one person. To solve entity identification problem directly is difficult, previous works usually focus on *entity matching* problem. Given two records, an entity matching algorithm will determine whether they represent the same person. Treating each record as a node in graph and using the edge between nodes to represent the two corresponding records belong to the same person, the output of an entity matching algorithm can be a general graph, while the output of entity identification problem is required to be a *clustered* graph. To fill the gap between the output of entity matching and entity identification, intuitively, the CC problem is to transform the entity matching result to the entity identification result while minimizing the difference between them. Given the same records, different matching algorithms may output different results, combining the results of multiple matching algorithms brings opportunities to improve the accuracy of identification result[7]. Given the results of several matching algorithms, the LCC problem is to generate the identification result such that it is compatible to as many matching algorithms as possible.

Obviously, the LCC problem is NP-*hard*, because, treating each labelled edge in CC as a *el*-graph, CC is indeed a special case of LCC. Therefore, we analyze the LCC problem from the point of view of parameterized complexity [8], to study whether there are efficient algorithms when some parameters of the input are small. According to [9], a parameterized problem is a set $L \subseteq \Sigma^* \times \mathbb{N}$, where Σ is a fixed alphabet and $\mathbb{N}$ is the positive integer set. For $(x, k) \in \Sigma^* \times \mathbb{N}$, x is the input and k is the parameter. A parameterized problem P is *fixed-parameter tractable* if there is a computable function $f : \mathbb{N} \to \mathbb{N}$, a constant $c \in \mathbb{N}$, and an algorithm that, given a pair $(x, k) \in \Sigma^* \times \mathbb{N}$, decides if $(x, k) \in P$ in at most $f(k) \cdot |x|^c$ steps. The class of all the fixed-parameter tractable problems is FPT. Beyond FPT, a hierarchy of parameterized complexity classes FPT $\subseteq$ W[1] $\subseteq$ W[2] $\subseteq \cdots \subseteq$ W[P] have been defined by [10,11,12], which play a central role in identifying parameterized intractable problems. For example, the standard parameterization version of classic CLIQUE problem is W[1]-*complete*, and the standard parameterization version of DOMINATING SET problem is W[2]-*complete*, which implies that DOMINATING SET is intuitively harder than CLIQUE. To study the LCC problem from the perspective of parameterized complexity, we consider different parameterization methods for LCC and study which class of *W*-hierarchy each parameterized LCC belongs to.

In this paper, the parameterized versions of the LCC problem, denoted by *p*-LCC, are defined on the following three parameters.

Problem: p-LCC

> *Instance*: A el-graph set $\widetilde{\mathcal{G}} = \{\widetilde{G}_1, \ldots, \widetilde{G}_m\}$ and a positive integer k.
>
> *Question*: Is there a clustered graph G such that $\mathrm{DISAGREE}(G, \widetilde{\mathcal{G}})$ is not larger than k?
>
> *Parameter 1*: $m = |\widetilde{\mathcal{G}}|$, the size of the el-graph set.
>
> *Parameter 2*: $n = |\bigcup\{V_{\widetilde{G}_i}\}|$, the size of node set of graphs in $\widetilde{\mathcal{G}}$.
>
> *Parameter 3*: k.

The three parameterized problems are denoted by p-LCCm, p-LCCn, and p-LCCk respectively.

Our Results. Because the LCC problem is NP-*hard*, a natural question is whether it is fixed-parameter tractable. That is, when some parameter of the instances of LCC is small, whether or not there are efficient algorithms. Three parameterization methods are considered. When the LCC problem is parameterized with m and n, we give the positive answers. Specifically, the corresponding problems p-LCCm and p-LCCn are shown to be fixed-parameter tractable. When it is parameterized with k, negative answer is given. By means of checking *bad* circles which will be introduced later, it is shown that the p-LCCk problem is W[t]-*hard* for any $t > 0$ and in W[P]. It means that, unless for any $t > 0$ we have FPT $=$ W[t], the p-LCCk problem is not fixed-parameter tractable.

1.1   Related Work

To the best of our knowledge, there are no previous works focusing on the Labelled Correlation Clustering problem. The most related one is the classic Correlation Clustering problem. Treating each edge e labelled 1 (resp. 0) as a requirement for the existence (resp. inexistence) of e, the similarity of that two problems is to seek to find a clustered graph G, which represents a solution of clustering nodes in G, to satisfy the requirements as many as possible. Additionally, treating a el-graph $\widetilde{G}$ as the requirements of one user, the difference is that the CC problem seeks a clustered G satisfying as many requirements as possible of one user, while the LCC problem considers the requirements of several users and seeks a G satisfying as many users as possible. The main results of the previous works on CC problem can be summarized as follows. For the case that $\widetilde{G}$ is defined over a complete graph, CC is proved to be NP-*complete* in [1], and a series of approximation algorithms with best constant factor 3 are designed by [1,4,5]. For the case that $\widetilde{G}$ is defined over a general graph, the CC problem is proved to be APX-*hard* by [6], and two $O(\log n)$ approximation algorithms are designed independently by [5] and [6]. Weighted variants of CC are also considered by [1,4,6].

The LCC problem is also motivated by several recent works on the minimum label graph problems, whose definition is given in [13]. In that problem, the edges in a graph are associated with labels, and the goal is to find the minimum label set consisting of edges satisfying some property. Such problems include

---

Algorithm p-LCCm-Solver

Input: A set of *el*-graphs $\widetilde{\mathcal{G}} = \{\widetilde{G}_1, \ldots, \widetilde{G}_m\}$,
    and a positive integer k.
Output: **true** or **false**.
1. **if** $k \geq m$ **then**
2.   **return true;**
3. **for** each *el*-graph set $\widetilde{\mathcal{G}}' \subseteq \widetilde{\mathcal{G}}$ satisfying $|\widetilde{\mathcal{G}}'| = k$ **do**
4.   Initialize an empty graph G;
5.   **for** each $\widetilde{G}_i \in \widetilde{\mathcal{G}} \setminus \widetilde{\mathcal{G}}'$ **do**
6.   Add all edges labelled 1 of $\widetilde{G}_i$ to G;
7.   **if** there is an edge (u, v) labelled 0 in $\widetilde{\mathcal{G}} \setminus \widetilde{\mathcal{G}}'$
        s.t. u and v are connected in G **then**
8.   **continue;**
9.   **else**
10.   **return true;**
11. **return false;**

Fig. 1. Algorithm p-LCCm-Solver

the Minimum Label Spanning Tree problem [14,15], the Minimum Label Path problem [16,17], and the Minimum Label Cut problem [18,19].

Parameterized complexity is a method of identifying 'easy' fragments of NP-*hard* problems. Traditionally, if multiple paramters are involved in the input of one problem, they are treated equally when considering the complexity. However, in real applications, if some parameter is always very small, there may be efficient algorithms. In parameterized complexity, it is studied whether or not there are algorithms with time cost $O(f(k)p(n))$, where f is a computable function, p is a polynomial function, k is the small parameter, and n is input length. The parameterized complexity is introduced by Downey and Fellows in a series of works [10,11,12]. Some recent results and new perspectives can be found in the monograph [9].

2   The p-LCCm and p-LCCn Problem

Before further discussions, a useful concept is introduced. An *el*-graph set $\widetilde{\mathcal{G}}$ is *consistent*, iff there exists a *clustered* graph G such that G agrees with every *el*-graph in $\widetilde{\mathcal{G}}$, otherwise, it is *inconsistent*.

The p-LCCm problem is shown to be fixed-parameter tractable by giving the FPT-algorithm p-LCCm-Solver in Fig. 1. The idea of p-LCCm-Solver can be summarized as follows. First, if k is not smaller than m, any clustered graph G satisfies $\text{DISAGREE}(G, \widetilde{\mathcal{G}}) \leq k$, therefore, **true** is returned (line 1-2). Then, if there exists some subset $\widetilde{\mathcal{G}}'$ of $\widetilde{\mathcal{G}}$ with size k such that $\widetilde{\mathcal{G}} \setminus \widetilde{\mathcal{G}}'$ is consistent, **true** is returned, otherwise, **false** is returned (line 3-11).

Theorem 1. *The* p-*LCCm* *problem is* fixed-parameter tractable.

Proof. First, the correctness of p-LCCm-Solver can be implied immediately, if it can be proved that, for each $\widetilde{\mathcal{G}}'$, the iteration (line 4-10) determines whether $\widetilde{\mathcal{G}} \backslash \widetilde{\mathcal{G}}'$ is consistent correctly. We show that by following two claims. (1) If **true** is returned by line 10, a graph G' built based on G by transforming each connected component to a clique will satisfy $\text{DISAGREE}(G', \widetilde{\mathcal{G}}) \leq k$, and $\widetilde{\mathcal{G}} \backslash \widetilde{\mathcal{G}}'$ is consistent. (2) If $\widetilde{\mathcal{G}} \backslash \widetilde{\mathcal{G}}'$ is consistent, let G' be a clustered graph agreeing all ones in $\widetilde{\mathcal{G}} \backslash \widetilde{\mathcal{G}}'$. Then, in $\widetilde{\mathcal{G}} \backslash \widetilde{\mathcal{G}}'$, all edges labelled 1 belong to G', and all edges labelled 0 do not belong to G'. Therefore, G is a subgraph of G', and it is obvious that the condition in line 7 is not satisfied and **true** will be returned.

Second, the cost of p-LCCm-Solver can be analyzed as follows. Obviously, there are at most m^k iterations of line 3. The cost of line 4-10 can be bounded by some polynomial function p. Therefore, the total cost of p-LCCm-Solver can be bounded by $m^k \cdot p(|x|) \leq m^m \cdot p(|x|)$, where $|x|$ is the length of input.

Finally, p-LCCm-Solver is an FPT algorithm for p-LCCm, and the problem p-LCCm is in FPT.

The p-LCCn problem can be verified to be *fixed-parameter tractable* by following simple solution. The idea is simple and obviously correct. For each possible graph G defined over the node set $V = \bigcup\{V_{\widetilde{G}_i}\}$, it is checked whether the conditions (a) G is *clustered* and (b) $\text{DISAGREE}(G, \widetilde{\mathcal{G}}) \leq k$ are satisfied. It is also easy to check that the time cost of the above solution can be bounded by $O(2^{n^2} \cdot p(|x|))$, where p is a polynomial function. Therefore, we have the following theorem.

Theorem 2. *The problem* p-LCCn *is* fixed-parameter tractable.

3   The p-LCCk Problem

In this section, first, a concept of *bad* circles is introduced, then, based on the technique of checking bad circles, the p-LCCk is shown to be W[t]-*hard* for any $t > 0$ and belong to W[P].

3.1   Bad Circle

Before showing the proof, an important observation is introduced first. Given a *el*-graph set $\widetilde{\mathcal{G}}$, $\widetilde{\mathcal{G}}$ is *canonical*, if there does not exist an edge e such that $f_{\widetilde{G}_i}(e) = 1$ and $f_{\widetilde{G}_j}(e) = 0$ for some $\widetilde{G}_i, \widetilde{G}_j \in \widetilde{\mathcal{G}}$, that is there are no edges labelled differently in $\widetilde{\mathcal{G}}$. Then, for a *canonical el*-graph set $\widetilde{\mathcal{G}} = \{\widetilde{G}_i\}$, a new *el*-graph $\widetilde{G} = \text{MERGE}(\widetilde{\mathcal{G}})$ can be defined by letting $V_{\widetilde{G}} = \bigcup\{V_{\widetilde{G}_i}\}$, $E_{\widetilde{G}} = \bigcup\{E_{\widetilde{G}_i}\}$, and $f_{\widetilde{G}}(e) = f_{\widetilde{G}_i}(e)$ for each $e \in E_{\widetilde{G}_i}$. Because there are no edges having different labels in $\widetilde{\mathcal{G}}$, obviously, $\text{MERGE}(\widetilde{\mathcal{G}})$ is well-defined. Here, it should be noticed that because the graph G required is *clustered*, a canonical set can still be inconsistent. In a *el*-graph $\widetilde{G}$, a circle is *bad*, if it contains exactly one edge labelled 0. Then, we have the following observation.

Proposition 1. *A canonical el-graph set $\widetilde{\mathcal{G}}$ is consistent, if and only if there are no* bad *circles in* $\text{MERGE}(\widetilde{\mathcal{G}})$.

3.2   Parameterized Complexity Results

For the lower bound, it is shown that, for any $t > 0$, p-LCCk is W[t]-*hard*, and for the upper bound, it is shown that p-LCCk is in W[P]. Therefore, unless W[t] equals to FPT, there are no algorithms in time $O(f(k) \cdot n^c)$ solving problem p-LCCk, where $f(k)$ is arbitrary computable function and c is some positive constant.

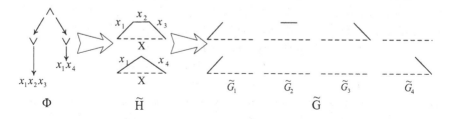

Fig. 2. Example for the reduction in Lemma 2

First, a problem utilized in the proof, Weighted Monotone t-Normalized Satisfiability, is introduced. Let X be the propositional variables set, and $\Gamma_{t,1}^+$ be the class of propositional formulas from [9].

- $\Gamma_{0,1}^+ := \{x | x \in X\}$,
- $\Delta_{0,1}^+ := \Gamma_{0,1}^+$,
- $\Gamma_{t+1,1}^+ := \{\bigwedge_{i \in I} \delta_i \mid I$ finite nonempty index set,
  and $\delta_i \in \Delta_{t,1}$ for all $i \in I\}$,
- $\Delta_{t+1,1}^+ := \{\bigvee_{i \in I} \gamma_i \mid I$ finite nonempty index set,
  and $\gamma_i \in \Gamma_{t,1}$ for all $i \in I\}$.

A truth assignment τ for X is a mapping from X to $\{1,0\}$, and the *weight* of τ is the number of variables mapped to 1 by τ. Then, the Weighted Monotone t-Normalized Satisfiability problem, p-WSAT$(\Gamma_{t,1}^+)$ for short, can be defined as follows.

Problem: p-WSAT$(\Gamma_{t,1}^+)$

Instance: A $\Gamma_{t,1}^+$ formula Φ over variable set X, and a positive integer k.
Question: Does Φ have a satisfying truth assignment τ of weight k?
Parameter: k.

The problem on $\Gamma_{2,1}^+$ is also denoted by p-WCNF$^+$, since $\Gamma_{2,1}^+$ is indeed a conjunctive normal form (CNF) formula.

Lemma 1. *(From [9]). For every $t > 1$, if t is even, p-WSAT$(\Gamma_{t,1}^+)$ is W[t]-complete under FPT-reduction. Specially, p-WCNF$^+$ is W[2]-complete.*

To show the lower bound of p-LCCk, an inductive proof is introduced here. First, basically, the p-LCCk problem is shown to be W[2]-*hard* by a reduction from p-WCNF$^+$ in Lemma 2. Then, inductively, for *even t*, the p-LCCk problem is shown to be W[t]-*hard* by a reduction from p-WSAT($\Gamma_{t,1}^+$) in Lemma 3. Finally, based on those two lemmas, Theorem 3 for the lower bound is given.

Lemma 2. *The* p-LCC*k problem is* W[2]-hard.

Proof. It is proved by making an FPT reduction from p-WCNF$^+$ to p-LCCk.

Given an instance (Φ, k) of problem p-WCNF$^+$, an instance $(\widetilde{\mathcal{G}}, k')$ of p-LCCk problem will be constructed. The idea of the construction is to first build a el-graph $\widetilde{H}$ and a function $h : E_{\widetilde{H}} \mapsto 2^X$ and then build $\widetilde{\mathcal{G}}$ according to $\widetilde{H}$ and h. Intuitively, $\widetilde{H}$ is the graph MERGE($\widetilde{\mathcal{G}}$) (it will be seen that $\widetilde{\mathcal{G}}$ is canonical), and h indicates which graphs in $\widetilde{\mathcal{G}}$ each edge of $\widetilde{H}$ comes from.

1) Let $\Phi = C_1 \wedge \cdots \wedge C_I$, each C_i ($i \in [1, I]$) is in the form $l_{i1} \vee l_{i2} \vee \cdots \vee l_{iJ_i}$ where each l_{ij} ($j \in [1, J_i]$) is a variable in X.

2) $\widetilde{H}$ and h can be built by following steps. For each C_i, a set of nodes $V_{C_i} = \{v_{i1}, \ldots, v_{i(J_i+1)}\}$ is added to $\widetilde{H}$ first. Then, for each $j \in [1, J_i]$, the edge $e_{ij} = (v_{ij}, v_{i(j+1)})$ is labelled 1 and added to $\widetilde{H}$, and let $h(e_{ij}) = \{l_{ij}\}$. Finally, the edge $e_{i(J_i+1)} = (v_{i(J_i+1)}, v_{i1})$ is labelled 0 and added to $\widetilde{H}$, and let $h(e_{i(J_i+1)}) = X$.

3) Then, based on $\widetilde{H}$ and h, $\widetilde{\mathcal{G}}$ can be built as follows. First, for each $x_i \in X$, a el-graph $\widetilde{G}_i = \{V_{\widetilde{H}}, \emptyset, \emptyset\}$ is initialized and added to $\widetilde{\mathcal{G}}$. Then, for each edge $e \in E_{\widetilde{H}}$ and each $x_i \in X$, if $x_i \in h(e)$, e is added to $\widetilde{G}_i$ and let $f_{\widetilde{G}_i}(e) = f_{\widetilde{H}}(e)$.

4) Finally, let $k' = k$.

A reduction example is given in Fig. 2, where the solid edges are labelled 1 and dash edges are labelled 0, and, in $\widetilde{H}$, the values of h function are marked on the corresponding edges.

According to [9], to show the correctness of the reduction, we must prove that (1) there is an assignment τ weighted k satisfying Φ, iff there is a clustered graph G satisfying DISAGREE($G, \widetilde{\mathcal{G}}$) $\leq k'$, (2) the reduction is computable by an FPT algorithm, and (3) there is a computable function $g : \mathbb{N} \to \mathbb{N}$ such that $k' \leq g(k)$. Obviously, the reduction satisfies the last two conditions. In the following, we will prove the first condition.

First, assume that there is an assignment τ weighted k satisfying Φ. According to the reduction, each $x_i \in X$ has a corresponding graph $\widetilde{G}_i$ in $\widetilde{\mathcal{G}}$. Let $\widetilde{\mathcal{G}}'$ be the graph set $\{\widetilde{G}_i | \tau(x_i) = 1 \text{ and } \widetilde{G}_i \in \widetilde{\mathcal{G}}\}$. According to the definition of the reduction, obviously, there are exactly I *bad* circles in $\widetilde{H}$, which are the sources of inconsistencies of $\widetilde{\mathcal{G}}$. Consider the graph MERGE($\widetilde{\mathcal{G}} \setminus \widetilde{\mathcal{G}}'$). For arbitrary C_i in Φ, since τ satisfies C_i, there is one literal l_{ij} of C_i is set to 1 by τ, and, thus, it is known $\widetilde{G}_{l_{ij}} \in \widetilde{\mathcal{G}}'$ and the edge $e_{ij} = (v_{ij}, v_{i(j+1)})$ is in $\widetilde{\mathcal{G}}'$. According to an observation that, during the reduction, each edge labelled 1 in $\widetilde{H}$ is distributed to only one graph of $\widetilde{\mathcal{G}}$, we know that e_{ij} is not in $\widetilde{\mathcal{G}} \setminus \widetilde{\mathcal{G}}'$. Therefore, there are no *bad* circles in MERGE($\widetilde{\mathcal{G}} \setminus \widetilde{\mathcal{G}}'$). Since $\widetilde{\mathcal{G}}$ is canonical, by Proposition 1, the graph set $\widetilde{\mathcal{G}} \setminus \widetilde{\mathcal{G}}'$ is consistent, which implies that there exists a clustered graph G such that DISAGREE($G, \widetilde{\mathcal{G}}$) $\leq |\widetilde{\mathcal{G}}'| = k = k'$.

Second, assume that there is a graph G such that $\mathrm{DISAGREE}(G,\widetilde{\mathcal{G}}) \le k'$. An assignment τ can be built as follows. For each $\widetilde{G}_i \in \widetilde{\mathcal{G}}$, if G agrees with $\widetilde{G}_i$, let $x_i = 0$, otherwise, let $x_i = 1$. Then, consider τ for each clause C_i of Φ. Let the induced subgraph of V_{C_i} on $\widetilde{H}$ be $\widetilde{H}_{C_i}$, according to the reduction, obviously, $\widetilde{H}_{C_i}$ is a *bad* circle. Since G is clustered, there is at least one edge e in $\widetilde{H}_{C_i}$ on which G does not agree with $\widetilde{H}_{C_i}$.

(a) If e is $e_{i(J_i+1)}$, because $h(e) = X$, G must disagree with every $\widetilde{G}_i$ in $\widetilde{\mathcal{G}}$. Then, we have $k = k' \ge |X|$, and we can extend τ trivially to τ' by setting all variables of X to be 1. Obviously, τ' satisfy Φ.

(b) If e is e_{ij} for some $j \in [1, J_i]$, because $h(e) = \{l_{ij}\}$, G must disagree with $\widetilde{G}_{l_{ij}}$. According to definition of τ, we have $\tau(l_{ij}) = 1$, and τ satisfies C_i.

Therefore, τ satisfies Φ. Finally, if the weight w_τ of τ is smaller than k, we can extend τ to τ'' by setting arbitrary additional $k - w_\tau$ variables in X to be 1. Because all literals in Φ are positive, τ'' still satisfy Φ.

Finally, in conclusion, an FPT reduction from $p\text{-}\mathrm{WCNF}^+$ to $p\text{-}\mathsf{LCC}k$ has been given, and $p\text{-}\mathsf{LCC}k$ is W[2]-*hard*.

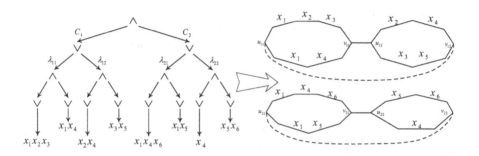

Fig. 3. Example for the reduction in Lemma 3

Lemma 3. *The* p-LCC*k problem is* W[t]-hard *for any even* $t > 0$.

Proof. Based on the result in Lemma 2, for any *even* $t > 0$, we give an FPT reduction from $p\text{-}\mathrm{WSAT}(\Gamma_{t,1}^+)$ to $p\text{-}\mathsf{LCC}k$. Then, according to Lemma 1, the lemma can be proved.

Given an instance (Φ, k) of $p\text{-}\mathrm{WSAT}(\Gamma_{t,1}^+)$, an instance $(\widetilde{\mathcal{G}}, k')$ of problem $p\text{-}\mathsf{LCC}k$ will be constructed. Similar with the proof of Lemma 2, we also construct graph $\widetilde{H}$ and function h first, and then build $\widetilde{\mathcal{G}}$ based on them.

1) Given Φ in $\Gamma_{t,1}^+$, $\widetilde{H}$ and h will be built by an induction on t. It is satisfied that, in each induction step, each connected component of the outputted *el*-graph $\widetilde{H}$ contains exactly one edge labelled 0.

   - *Basically*, if $t = 2$, $\widetilde{H}$ and h can be built as the reduction in Lemma 2. Additionally, it is obvious that each connected component of $\widetilde{H}$ is a *bad* circle which contains exactly one edge labelled 0.

- *Inductively*, it is assumed that, for each even $t \leq 2k$ ($k > 0$), if Φ is in $\Gamma_{t,1}^+$, $\widetilde{H}$ and h can be well defined such that each connected component of $\widetilde{H}$ contains only one edge labelled 0. Then, for $t = 2k + 2$, given the $\Gamma_{t,1}^+$ formula

$$\Phi = \bigwedge_{i \in I} (\bigvee_{j \in J_i} \lambda_{ij})$$

where each $\lambda_{ij} \in \Gamma_{t-2,1}^+$, the construction can be defined as follows.

Let the graph built for λ_{ij} is $\widetilde{H}_{ij}$. According to the assumption, it is known that $\widetilde{H}_{ij}$ consists of several connected components each of which has exactly one edge labelled 0. Assume there are L edges labelled 0 in $\widetilde{H}_{ij}$, which are denoted by $E_{ij}^0 = \{(u_1, v_1) \ldots (u_L, v_L)\}$. First, delete the edges of E_{ij}^0 from $\widetilde{H}_{ij}$. Then, two nodes u_{ij} and v_{ij} are added, and, for each node u_y (resp. v_y) for $y \in [1, L]$, an edge $e_u = (u_y, u_{ij})$ (resp. $e_v = (v_y, v_{ij})$) is added and let $h(e_u) = X$ (resp. $h(e_v) = X$).

Then, for each $i \in I$, $\widetilde{H}_i$ is built by linking the graphs $\widetilde{H}_{ij}$ for all $j \in [1, J_i]$ together as follows. (a) For each $j \in [1, J_i - 1]$, an edge $(v_{ij}, u_{i(j+1)})$ labelled 1 is added, and its h value is set to be X. (b) An edge (v_{iJ_i}, u_{i1}) labelled 0 is added, and the h value is also set to be X.

Finally, $\widetilde{H}$ can be obtained by taking a union of $\{\widetilde{H}_i | i \in I\}$. Obviously, $\widetilde{H}$ contains I connected components, and each component has only one edge labelled 0.

2) Then, based on $\widetilde{H}$ and h, the graph set $\widetilde{\mathcal{G}}$ containing $|X|$ el-graphs can be built in the same way of Lemma 2.

3) Finally, let $k' = k$.

An example for this reduction is shown in Fig. 3, where solid edges are labelled 1 and dash edges are labelled 0, and if h value of some edge is not X, it is marked.

Similar with the proof in Lemma 2, to show the correctness, we only need to prove that there is an assignment τ weighted k satisfying Φ iff there is a clustered graph G satisfying DISAGREE$(G, \widetilde{\mathcal{G}}) \leq k'$.

First, assume that there is an assignment τ weighted k satisfying Φ. Without loss of generality, it is assumed that $k < |X|$ in the follow, because if $k \geq |X|$ arbitrary clustered graph G will satisfy the conditions. Let $\widetilde{\mathcal{G}}'$ be the set $\{\widetilde{G}_i | \tau(x_i) = 1 \text{ and } \widetilde{G}_i \in \widetilde{\mathcal{G}}\}$ again. It will be proved that the set $\widetilde{\mathcal{G}} \backslash \widetilde{\mathcal{G}}'$ is consistent by an induction on t. (a) Basically, for $t = 2$, the proof in Lemma 2 can be applied here. (b) Inductively, it is assumed that, for formula Φ in $\Gamma_{t,1}^+$ where t is even and $t \leq 2k$ ($k > 0$), if τ satisfies Φ, the set $\widetilde{\mathcal{G}} \setminus \widetilde{\mathcal{G}}'$ will be consistent. Then, for $t = 2k + 2$, given the formula $\Phi \in \Gamma_{t,1}^+$, let $\widetilde{\mathcal{G}}$ be the el-graph set produced by the reduction. Consider the graph $\widetilde{H}$ which $\widetilde{\mathcal{G}}$ is built on. According to the reduction, there are I connected components in $\widetilde{H}$, and each of them has only one edge labelled 0. For clause C_i ($i \in I$), its corresponding component is $\widetilde{H}_i$, all bad circles in $\widetilde{H}_i$ follows the following form.

$$v_{iJ_i} - u_{i1} \sim v_{i1} - u_{i2} \sim v_{i2} - u_{i3} \cdots\cdots v_{i(J_i - 1)} - u_{iJ_i} \sim v_{iJ_i}$$

Here, '$-$' means an edge, and '$\sim$' is a path. Then, because τ satisfies Φ, for each $i \in I$, the formula $\bigvee_{j \in J_i} \lambda_{ij}$ must be satisfied by τ. Therefore, there must exist $j \in J_i$ such that λ_{ij} is satisfied by τ. Obviously, λ_{ij} is in $\Gamma_{2k,1}^+$. Let $\widetilde{\mathcal{G}}_{ij}$ be the graph produced by the reduction for λ_{ij}, and $\widetilde{\mathcal{G}}'_{ij}$ be the set $\{\widetilde{G}_z | \tau(x_z) = 1 \text{ and } \widetilde{G}_z \in \widetilde{\mathcal{G}}_{ij}\}$. According to the assumption for the induction, obviously, $\widetilde{\mathcal{G}}_{ij} \setminus \widetilde{\mathcal{G}}'_{ij}$ is consistent. Then, for each edge (u, v) labelled 0 in $\widetilde{\mathcal{G}}_{ij}$, we have (u, v) exists in $\text{MERGE}(\widetilde{\mathcal{G}}_{ij} \setminus \widetilde{\mathcal{G}}'_{ij})$, because (u, v) is labelled X in $\widetilde{\mathcal{G}}_{ij}$ and $|\widetilde{\mathcal{G}}'_{ij})| \leq k < |X|$. Therefore, there is no path from u to v such that all edges on the path are labelled 1 in $\text{MERGE}(\widetilde{\mathcal{G}}_{ij} \setminus \widetilde{\mathcal{G}}'_{ij})$. Then, it can be implied that in $\text{MERGE}(\widetilde{\mathcal{G}} \setminus \widetilde{\mathcal{G}}')$ there is no path with all edges labelled 1 from u_{ij} to v_{ij}. Therefore, in $\text{MERGE}(\widetilde{\mathcal{G}} \setminus \widetilde{\mathcal{G}}')$, the bad circles in $\widetilde{H}_i$ are broken, and $\widetilde{\mathcal{G}} \setminus \widetilde{\mathcal{G}}'$ is consistent. Finally, there must be a graph G such that $\text{DISAGREE}(G, \widetilde{\mathcal{G}}) \leq |\widetilde{\mathcal{G}}'| = k = k'$.

Second, assume that there is a clustered graph G satisfying $\text{DISAGREE}(G, \widetilde{\mathcal{G}}) \leq k'$. If $k' \geq |X|$, a trivial assignment setting all variables to 1 will satisfies Φ. In the following, we only consider the case that $k' < |X|$. An assignment τ can be built by setting $x_i = 0$ if and only if G agrees with $\widetilde{G}_i$. We prove that τ satisfies Φ also by an induction on t. (a) Basically, for $t = 2$, the proof in Lemma 2 can be applied again. (b) Inductively, assume that if the formula Φ is in $\Gamma_{t,1}^+$ for even t satisfying $t \leq 2k$ ($k > 0$), if G satisfies $\text{DISAGREE}(G, \widetilde{\mathcal{G}}) \leq k'$, the corresponding τ satisfies Φ. Then, for $t = 2k + 2$, consider each C_i in Φ for $i \in I$. For $\widetilde{H}_i$, we have $\text{DISAGREE}(G, \widetilde{H}_i) \leq k' < |X|$ since $\text{DISAGREE}(G, \widetilde{\mathcal{G}}) \leq k'$. Therefore, the edges $\{(v_{ij}, u_{i(j+1)}) | j \in [1, J_i - 1]\}$ are in G and the edge (v_{iJ_i}, u_{i1}) is not in G, otherwise, we will have the contradiction that $\text{DISAGREE}(G, \widetilde{\mathcal{G}}) \geq |X| > k'$. Then, it is known that there must exist a $j \in [1, J_i]$ such that the edges in $E_{ij}^0 = \{(u_1, v_1) \ldots (u_L, v_L)\}$ which are labelled 0 in $\widetilde{H}_{ij}$ do not appear in G. Otherwise, we will have the contradiction that there is a bad circle in G. Then, we have $\text{DISAGREE}(G, \widetilde{\mathcal{G}}_{ij}) \leq k'$, because all edges labelled 1 in $\widetilde{\mathcal{G}}_{ij}$ belong to $\widetilde{\mathcal{G}}$ too. According to the assumption, it is known that τ satisfies λ_{ij}. Obviously, for each $i \in I$, there exists such a j. Therefore, τ satisfies Φ.

Finally, in conclusion, an FPT reduction from $p\text{-WSAT}(\Gamma_{t,1}^+)$ to $p\text{-LCC}k$ has been given, and $p\text{-LCC}k$ is W[t]-*hard* for any $t > 0$.

Theorem 3. *The p-LCCk problem is W[t]-hard for any $t > 0$.*

Proof. It should be noticed that, different from the $p\text{-WSAT}(\Gamma_{t,1}^+)$ problem, the definition of the $p\text{-LCC}k$ problem does not depend on t, therefore, the W[t]-*hard* results of $p\text{-LCC}k$ for *even* t implies its W[t]-*hard* results for all $t > 0$. Therefore, the theorem can be obtained by Lemma 2 and 3 easily.

The upper bound of $p\text{-LCC}k$ is shown in the following theorem.

Theorem 4. *The problem p-LCCk is in W[P].*

Proof. According to [20], we prove that the problem can be solved by a k-restricted nondeterministic Turing machine, which, for each input x and parameter k, makes decisions by at most $f(k) \cdot p(|x|)$ steps and $h(k) \cdot \log |x|$ nondeterministic steps where $f, h : \mathbb{N} \to \mathbb{N}$ are two computable functions and p is a polynomial function.

Then, p-LCCk problem can be determined by the following nondeterministic algorithm.

- First, guess a subset $\widetilde{\mathcal{G}}'$ of $\widetilde{\mathcal{G}}$ such that $|\widetilde{\mathcal{G}}'| = k$.
- Then, return **true** if and only if $\widetilde{\mathcal{G}} \setminus \widetilde{\mathcal{G}}'$ is consistent.

To represent $\widetilde{\mathcal{G}}'$, at most $k \cdot \log |\widetilde{\mathcal{G}}| \leq k \cdot \log |x|$ bits are needed. Therefore, the number of nondeterministic steps can be bounded by $k \cdot \log |x|$. To determine whether $\widetilde{\mathcal{G}} \setminus \widetilde{\mathcal{G}}'$ is consistent, we can use the method in the Algorithm p-LCCm-Solver in Fig. 1 (line 4-10) which can be implemented in polynomial time cost.

Obviously, the algorithm above can be computed by a k-restricted nondeterministic Turing machine, and the p-LCCk problem is in W[P].

4   Conclusion

In this paper, the paremeterized complexity of LCC is studied. For parameters m (the size of graph set) and n (the size of node set), the corresponding parameterized LCC problems are shown to be fixed-parameter tractable. For parameter k, the parameterized LCC problem is in W[P] and W[t]-*hard* for any $t > 0$.

References

1. Bansal, N., Blum, A., Chawla, S.: Correlation clustering. Machine Learning 56(1-3), 89–113 (2004)
2. Chaudhuri, S., Ganti, V., Motwani, R.: Robust identification of fuzzy duplicates. In: ICDE, pp. 865–876 (2005)
3. Soon, W.M., Ng, H.T., Lim, D.C.Y.: A machine learning approach to coreference resolution of noun phrases. Computational Linguistics 27(4), 521–544 (2001)
4. Ailon, N., Charikar, M., Newman, A.: Aggregating inconsistent information: Ranking and clustering. Journal of ACM 55(5), 23:1–23:27 (2008)
5. Charikar, M., Guruswami, V., Wirth, A.: Clustering with qualitative information. Journal of Computer and System Sciences 71(3), 360–383 (2005)
6. Demaine, E.D., Emanuel, D., Fiat, A., Immorlica, N.: Correlation clustering in general weighted graphs. Theoretical Computer Science 361(2), 172–187 (2006)
7. Chen, Z., Kalashnikov, D.V., Mehrotra, S.: Exploiting context analysis for combining multiple entity resolution systems. In: SIGMOD, pp. 207–218 (2009)
8. Downey, R.G., Fellows, M.R.: Parameterized Complexity. Monographs in Computer Science. Springer (1999)
9. Flum, J., Grohe, M.: Parameterized Complexity Theory. Springer (2006)
10. Downey, R.G., Fellows, M.R.: Fixed-parameter tractability and completeness I: Basic results. SIAM Journal of Computing 24(4), 873–921 (1995)
11. Downey, R.G., Fellows, M.R.: Fixed-parameter tractability and completeness II: On completeness for W[1]. Theoretical Computer Science 141(1&2), 109–131 (1995)

12. Downey, R., Fellows, M.: Fixed-parameter tractability and completeness III: Some structural aspects of the W hierarchy. In: Complexity Theory: Current Research – Proceedings of the 1992 Dagstuhl Workshop on Structural Complexity, pp. 191–225 (1993)
13. Fellows, M.R., Guo, J., Kanj, I.: The parameterized complexity of some minimum label problems. Journal of Computer and System Sciences 76, 727–740 (2010)
14. BrüGgemann, T., Monnot, J., Woeginger, G.J.: Local search for the minimum label spanning tree problem with bounded color classes. Operations Research Letters 31(3), 195–201 (2003)
15. Hassin, R., Monnot, J., Segev, D.: Approximation algorithms and hardness results for labeled connectivity problems. Journal of Combinatorial Optimization 14, 437–453 (2007)
16. Broersma, H.J., Li, X., Woeginger, G., Zhang, S.G.: Paths and cycles in colored graphs. Australasian J. Combin. 31, 299–311 (2005)
17. Carr, R.D., Doddi, S., Konjevod, G., Marathe, M.: On the red-blue set cover problem. In: SODA, pp. 345–353 (2000)
18. Jha, S., Sheyner, O., Wing, J.: Two formal analys s of attack graphs. In: Proceedings of the 15th IEEE Workshop on Computer Security Foundations, pp. 49–63 (2002)
19. Zhang, P., Cai, J.-Y., Tang, L.-Q., Zhao, W.-B.: Approximation and hardness results for label cut and related problems. Journal of Combinatorial Optimization 21, 192–208 (2011)
20. Grohe, M.: Parameterized complexity for the database theorist. ACM SIGMOD Record 31(4), 86–96 (2002)

Shortest Color-Spanning Intervals

Minghui Jiang and Haitao Wang*

Department of Computer Science, Utah State University, Logan, UT 84322, USA
mjiang@cc.usu.edu, haitao.wang@usu.edu

Abstract. Given a set of n points on a line, where each point has one of k colors, and given an integer $s_i \geq 1$ for each color i, $1 \leq i \leq k$, the problem SHORTEST COLOR-SPANNING t INTERVALS (SCSI-t) aims at finding t intervals to cover at least s_i points of each color i, such that the maximum length of the intervals is minimized. Chen and Misiolek introduced the problem SCSI-1, and presented an algorithm running in $O(n)$ time if the input points are sorted. Khanteimouri et al. gave an $O(n^2 \log n)$ time algorithm for the special case of SCSI-2 with $s_i = 1$ for all colors i. In this paper, we present an improved algorithm with running time of $O(n^2)$ for SCSI-2 with arbitrary $s_i \geq 1$. We also obtain some interesting results for the general problem SCSI-t. From the negative direction, we show that approximating SCSI-t within any ratio is NP-hard when t is part of the input, is W[2]-hard when t is the parameter, and is W[1]-hard with both t and k as parameters. Moreover, the NP-hardness and the W[2]-hardness with parameter t hold even if $s_i = 1$ for all i. From the positive direction, we show that SCSI-t with $s_i = 1$ for all i is fixed-parameter tractable with k as the parameter, and admits an exact algorithm running in $O(2^k n \cdot \max\{k, \log n\})$ time.

1 Introduction

Given a set of n points on a line, where each point has one of k colors, and given an integer $s_i \geq 1$ for each color i, $1 \leq i \leq k$, the problem SHORTEST COLOR-SPANNING t INTERVALS (SCSI-t) aims at finding t intervals to cover at least s_i points of each color i, such that the maximum length of the intervals is minimized.

Chen and Misiolek [3] introduced the problem SCSI-1, and presented an algorithm running in $O(n)$ time if the input points are sorted. Khanteimouri et al. [13] gave an $O(n^2 \log n)$ time algorithm for the special case of SCSI-2 with $s_i = 1$ for all colors i. Our first result in this paper is an improved algorithm for SCSI-2 with arbitrary $s_i \geq 1$:

Theorem 1. *SCSI-2 admits an exact algorithm running in $O(n^2)$ time.*

The problems SCSI-1 and SCSI-2 naturally generalize to SCSI-t for $t \geq 1$. Our next theorem shows that SCSI-t is intractable in a very strong sense:

---

* H. Wang's research was supported in part by NSF under Grant CCF-1317143.

Z. Cai et al. (Eds.): COCOON 2014, LNCS 8591, pp. 288–299, 2014.

Theorem 2. *Approximating SCSI-t within any ratio is NP-hard when t is part of the input, is W[2]-hard when t is the parameter, and is W[1]-hard with both t and k as parameters. Moreover, the NP-hardness and the W[2]-hardness with parameter t hold even if $s_i = 1$ for all i.*

Optimization problems that are hard to approximate within any ratio are no longer a novelty. A recent example is the exemplar distance problem in comparative genomics; see [11] and the references therein. The study of intractability combining both parameterized complexity and approximation hardness is not new either; see e.g. [15]. But to our best knowledge, SCSI-t is the first natural problem that is known to be intractable in the special way that obtaining any approximation is W[2]-hard.

In contrast to the very negative result in Theorem 2, our following theorem shows that the special case of SCSI-t with $s_i = 1$ for all i is fixed-parameter tractable when the parameter is the number k of colors:

Theorem 3. *The special case of SCSI-t with $s_i = 1$ for all i admits an exact algorithm running in $O(2^k n \cdot \max\{k, \log n\})$ time.*

In particular, we can solve SCSI-t with $s_i = 1$ for all i in $O(n \log n)$ time if k is a constant, and in $O(n^2 \log n)$ time if $k \leq \log n$. Thus the problem SCSI-t may still be manageable in practice.

1.1   Related Work

Instead of finding t intervals to cover at least $s_i \geq 1$ points of each color i as in SCSI-t, another generalization of the problem SCSI-1 aims at finding one geometric object to cover at least $s_i \geq 1$ points of each color i in the plane rather than on a line. This planar problem is typically studied with $s_i = 1$ for all colors i. Abellanas et al. [1] proposed an $O(n(n - k) \log^2 k)$ time algorithm for computing the smallest (by perimeter or area) axis-parallel rectangle that contains at least one point of each color. Das et al. [6] gave an improved algorithm with $O(n(n - k) \log k)$ time for this problem, and moreover gave an $O(n^3 \log k)$ time algorithm for computing the smallest color-spanning rectangle of arbitrary orientation. Khanteimouri et al. [14] gave an $O(n \log^2 n)$ time algorithm for computing the smallest color-spanning axis-parallel square. Algorithms for computing the smallest color-spanning strips were also given in [1,6]. Recently, Barba et al. [2] considered the related problem of computing a region (e.g., rectangle, square, or disc) that contains *exactly* s_i points of each color i.

Given a set of colored points, a *color-spanning set* is a subset of the input points including at least one point of each color. The various color-spanning problems can be viewed as finding a color-spanning set such that certain geometric property of the set is optimized. In this framework, Fleischer and Xu [9,10] gave polynomial time algorithms for finding a minimum-diameter color-spanning set under the L_1 or L_∞ metric, and proved that the problem is NP-hard for all L_p with $1 < p < \infty$. Ju et al. [12] gave an efficient algorithm for computing a color-spanning set with the maximum diameter, and proved that several other problems are NP-hard, e.g., finding the color-spanning set with the largest closest pair. Fan et al. [7] studied the problem of finding a color-spanning set with the minimum connection radius in the corresponding disk intersection graph.

2 An $O(n^2)$-Time Exact Algorithm for SCSI-2

In this section we prove Theorem 1. We present an $O(n^2)$ time algorithm for solving the problem SCSI-2, which improves the $O(n^2 \log n)$ time algorithm in [13].

Let $P = \{p_1, p_2, \ldots, p_n\}$ be a set of n points given on a line L, say, the x-axis, sorted from left to right. Each point p_i has one of k colors. A line segment on L is also called an *interval* of L. We say an interval of L *covers* a point if the point is on the interval. The problem SCSI-2 is to find two intervals on L to cover at least s_i points of each color i with $1 \le i \le k$ such that the maximum length of the intervals is minimized. In the following, we assume that for any i, the number of points of color i in P is at least s_i, since otherwise there would be no solution for the problem.

If two intervals of L together cover at least s_i points of each color i in P, then we say the two intervals form a *feasible solution* for SCSI-2. For any interval I, let $d(I)$ denote the length of I. An interval I_1 is said to be *longer* than another interval I_2 if and only if $d(I_1) \ge d(I_2)$. We first prove the following lemma:

Lemma 1. *There must exist an optimal solution for the problem SCSI-2 that consists of two intervals such that the longer interval have both left and right endpoints in P.*

Proof. Consider any optimal solution for SCSI-2 that consists of two intervals I_1 and I_2. If both the left and right endpoints of both I_1 and I_2 are in P, then we are done with the proof. Otherwise, without loss of generality, assume the left endpoint of I_1 is not at any point of P. Then, we can shrink I_1 by moving its left endpoint rightwards for an infinitesimal distance such that the new interval I_1' covers the same subset of points of P as I_1 does (e.g., see Fig. 1). Clearly, I_1' and I_2 together still form a feasible solution.

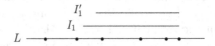

Fig. 1. Illustrating an example for the proof of Lemma 1: the left endpoint of interval I_1 is not at any point of P. We can obtain another interval I_1' by moving the left endpoint of I_1 rightwards for an infinitesimal distance such that I_1' and I_1 cover the same subset of points of P.

If some endpoints of I_1' and I_2 are not in P, then we use the same technique as above to shrink them. Eventually, we can obtain two intervals I_1'' and I_2'' whose endpoints are all in P and they form a feasible solution. Since $d(I_1'') \le d(I_1)$, $d(I_2'') \le d(I_2)$, and I_1 and I_2 form an optimal solution, the two new intervals I_1'' and I_2'' must also form an optimal solution. The lemma thus follows.   □

Definition 1. *For any interval I on L, I is a* candidate interval *if there is another interval I' such that*

1. $d(I') \leq d(I)$,
2. *I and I' form a feasible solution for SCSI-2.*

Given any interval I, we can determine whether I is a candidate interval in $O(n)$ time in the following way. First, we discard all points from P whose colors have already been covered by I. Let s'_i be the number of points covered by I for each color i. Second, we find a shortest interval I' that covers at lease $s_i - s'_i$ points of each color i among the remaining points of P. Finally, I is a candidate interval if and only if $d(I') \leq d(I)$. The first step and computing s'_i for all i can be easily done in $O(n)$ time. The second step is essentially the problem SCSI-1, which can be done in $O(n)$ time [3].

Let $\mathcal{I}$ denote the set of all intervals each of which has its two endpoints in P. Clearly, $|\mathcal{I}| = \Theta(n^2)$. Lemma 1 implies that we can solve the problem SCSI-2 by finding the shortest candidate interval in $\mathcal{I}$. Let I^* denote the shortest candidate interval in $\mathcal{I}$. Since $|\mathcal{I}| = \Theta(n^2)$, if we check every interval of $\mathcal{I}$, then we can solve the problem SCSI-2 in $O(n^3)$ time. In the following we present an $O(n^2)$ time algorithm for finding I^*.

For any $1 \leq i \leq j \leq n$, denote by I_{ij} the interval with left endpoint at p_i and right endpoint at p_j. For any $1 \leq i \leq n$, let $\mathcal{I}_i = \{I_{ij} \mid i \leq j \leq n\}$. Therefore, the sets $\mathcal{I}_i$ for $i = 1, \ldots, n$ form a partition of $\mathcal{I}$. For each $1 \leq i \leq n$, let I_i^* be the shortest candidate interval in $\mathcal{I}_i$. Hence, I^* is the shortest interval among the intervals $I_1^*, I_2^*, \ldots, I_n^*$. In the sequel we compute these intervals $I_1^*, I_2^*, \ldots, I_n^*$ in $O(n^2)$ time.

For each $1 \leq i \leq n$, let $h(i)$ be the smallest index j such that $I_i^* = I_{ij}$. Note that since $d(I_{ij})$ is monotonically increasing as j increases from i to n, $h(i)$ is also the smallest index j such that I_{ij} is a candidate interval. The following lemma, which is crucial to our algorithm, shows a monotonicity property of the indices $h(i)$:

Lemma 2. *It holds that $h(1) \leq h(2) \leq \cdots \leq h(n)$.*

Proof. Consider any i with $1 \leq i \leq n - 1$. Below we show that $h(i) \leq h(i+1)$, which will prove the lemma. Let $j = h(i)$. Note that $i \leq j$ holds. If $i = j$, then $h(i+1) \geq h(i)$ simply follows since $h(i+1) \geq i+1$. In the following, we assume $i + 1 \leq j$.

To prove $h(i + 1) \geq j$, it is sufficient to show that $I_{i+1,m}$ for any $m < j$ is not a candidate interval. Assume to the contrary that $I_{i+1,m}$ is a candidate interval for some m with $m < j$, which implies that $i + 2 \leq j$ because $i + 1 \leq m$. According to the definition of the candidate interval, there is another interval I such that $d(I) \leq d(I_{i+1,m})$ and I and $I_{i+1,m}$ together form a feasible solution.

Consider the interval I_{im}, which covers one more point that $I_{i+1,m}$ does, i.e., the point p_i. Therefore, I and I_{im} also form a feasible solution. Since $d(I) \leq d(I_{i+1,m})$ and $d(I_{i+1,m}) \leq d(I_{im})$, we have $d(I) \leq d(I_{im})$. This implies that the interval I_{im} is also a candidate interval. But since $m < j$, it contradicts with the definition of $j = h(i)$ (i.e., $h(i)$ is the smallest index such that $I_{i,h(i)}$ is a candidate interval).

The lemma thus follows. $\square$

Based on the preceding lemma, our algorithm works as follows.

For convenience of discussion, we let $h(0) = 1$. For any $0 \leq i \leq n-1$, suppose $h(i)$ has already been computed. According to Lemma 2, to compute $h(i+1)$, we only need to check the intervals $I_{i+1,j}$ in $\mathcal{I}_{i+1}$ for $j = h(i), h(i) + 1, \cdots$, to find the smallest index j such that $I_{i+1,j}$ is a candidate interval, and the above index j is $h(i+1)$ and the above candidate interval is I_{i+1}^*. Therefore, we can find the n intervals I_i^* for $i = 1, 2, \ldots, n$ by checking only $O(n)$ intervals of $\mathcal{I}$ in total and for each such interval, we can check whether it is a candidate interval in $O(n)$ time, as discussed earlier. Hence, we can find all intervals I_i^* for $i = 1, 2, \ldots, n$ in a total of $O(n^2)$ time. Consequently, the shortest candidate interval I^* of $\mathcal{I}$ and an optimal solution for SCSI-2 can be computed in additional $O(n)$ time. Theorem 1 is thus proved.

3   Intractability of SCSI-t

In this section we prove Theorem 2.

3.1   W[2]-hardness with Parameter t

We show that approximating SCSI-t within any ratio is NP-hard when t is part of the input, and moreover is W[2]-hard when t is the parameter, even for the special case of SCSI-t with $s_i = 1$ for all i. This is achieved by a polynomial FPT reduction from the NP-hard and W[2]-hard problem COLORFUL RED-BLUE DOMINATING SET [5].

Given a bipartite graph $G = (R \cup B, E)$ where each vertex in R has one of $\hat{\kappa}$ colors, COLORFUL RED-BLUE DOMINATING SET is the problem of deciding whether there exists a *colorful dominating set* D of $\hat{\kappa}$ vertices in R, including exactly one vertex of each color, such that each vertex in B is adjacent to at least one vertex in D. Put $\hat{n}_r = |R|$, $\hat{n}_b = |B|$, and $\hat{m} = |E|$. Our reduction constructs a colored point set of $n = \hat{n}_r + \hat{m}$ points with $k = \hat{\kappa} + \hat{n}_b$ colors, including one *r-color* i for each color i of the vertices in R, and one *b-color* v for each vertex v in B.

Place n points in $\hat{n}_r$ clusters, one cluster for each vertex in R. For each vertex u of color i in R, the cluster for u contains one point of r-color i, and contains one point of b-color v for each vertex v in B that is adjacent to u. Arrange the clusters of points such that the maximum distance between two points in each cluster is 1, and that the minimum distance between two points from different clusters is $\gamma > 1$. Set $t = \hat{\kappa}$. Refer to Figure 2 for an example.

The reduction can be easily made polynomial; it is also an FPT reduction with parameter t since t is a function of $\hat{\kappa}$ only. The following two lemmas imply that it is both NP-hard and W[2]-hard (with parameter t) to approximate SCSI-t for the special case when $s_i = 1$ for all i within γ for any approximation ratio $\gamma > 1$:

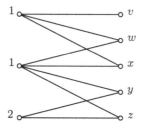

Fig. 2. A graph with $\hat{\kappa} = 2$, $\hat{n}_r = 3$, $\hat{n}_b = 5$, and $\hat{m} = 9$. The corresponding three clusters are $\{1, v, w, x\}$, $\{1, w, x, y, z\}$, and $\{2, y, z\}$. The two vertices for $\{1, v, w, x\}$ and $\{2, y, z\}$ are a colorful dominating set.

Lemma 3. *There is a colorful dominating set of size $\hat{\kappa}$ in the graph only if there is a color-spanning set of t intervals for the colored points with maximum length 1.*

Proof. Given a colorful dominating set in the graph, the $\hat{\kappa}$ clusters corresponding to the $\hat{\kappa}$ vertices in the dominating set clearly span all colors of the points, and each cluster can be covered with an interval of length 1.   □

Lemma 4. *There is a colorful dominating set of size $\hat{\kappa}$ in the graph if there is a color-spanning set of t intervals for the colored points with maximum length less than γ.*

Proof. Given a color-spanning set of $t = \hat{\kappa}$ intervals with maximum length less than γ, each interval can cover points from at most one cluster due to the restriction on the maximum interval length. Thus to span all $\hat{\kappa}$ r-colors of the points, each of the $\hat{\kappa}$ intervals must cover a point of a distinct r-color in a distinct cluster. Since all $\hat{n}_b$ b-colors of the points are spanned by the $\hat{\kappa}$ intervals, the $\hat{\kappa}$ vertices in R corresponding to the $\hat{\kappa}$ clusters must dominate all $\hat{n}_b$ vertices in B.   □

Note that the approximation lower bound γ can be arbitrarily large: it can even be a function of the instance size for the SCSI-t problem; indeed with suitable (say, binary) encoding of the interval coordinates as rational numbers, the lower bound can be exponential in the size of the problem instance. Moreover, if the colored points are allowed to coincide, or equivalently, if each point can have more than one color, then each cluster in our reduction can be compressed into a single point, and consequently the problem cannot be approximated at all.

3.2 W[1]-hardness with Parameters t and k

We show that approximating SCSI-t within any ratio is W[1]-hard with parameters t and k by an FPT reduction from the W[1]-hard problem COLORED CLIQUE [8]. Given a graph $G = (V, E)$ where each vertex has one of $\hat{\kappa}$ colors, COLORED CLIQUE is the problem of deciding whether there exists in G a *colored*

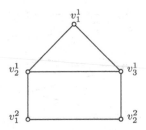

Fig. 3. A graph with $\hat{\kappa} = 3$, $\hat{n} = 5$, and $\hat{m} = 6$. The five vertices v_1^1, v_2^1, v_3^1, v_1^2, v_2^2 correspond to the five vertex clusters (here ji_1^p and ji_2^p mean p points of colors ji_1 and ji_2, respectively) $\{1, 21_1^1, 21_2^4, 31_1^1, 31_2^4\}$, $\{2, 12_1^1, 12_2^4, 32_1^4, 32_2^4\}$, $\{3, 13_1^1, 13_2^4, 23_1^1, 23_2^4\}$, $\{1, 21_1^2, 21_2^3, 31_1^2, 31_2^3\}$, $\{2, 12_1^2, 12_2^3, 32_1^2, 32_2^3\}$. The six edges $\{v_1^1, v_2^1\}$, $\{v_1^1, v_3^1\}$, $\{v_2^1, v_3^1\}$, $\{v_2^1, v_1^2\}$, $\{v_3^1, v_2^2\}$, $\{v_1^2, v_2^2\}$ correspond to the six edge clusters $\{12, 21_1^4, 21_2^1, 12_1^4, 12_2^1\}$, $\{13, 31_1^4, 31_2^1, 13_1^4, 13_2^1\}$, $\{23, 32_1^4, 32_2^1, 23_1^4, 23_2^1\}$, $\{12, 21_1^3, 21_2^2, 12_1^4, 12_2^1\}$, $\{23, 32_1^3, 32_2^2, 23_1^4, 23_2^1\}$, $\{12, 21_1^3, 21_2^2, 12_1^3, 12_2^2\}$, The colored clique consisting of the three vertices v_1^1, v_2^1, v_3^1 and the three edges $\{v_1^1, v_2^1\}$, $\{v_1^1, v_3^1\}$, $\{v_2^1, v_3^1\}$ correspond to three vertex clusters and three edge clusters whose union is $\{1, 2, 3, 12, 13, 23, 12_1^5, 12_2^5, 21_1^5, 21_2^5, 13_1^5, 13_2^5, 31_1^5, 31_2^5, 23_1^5, 23_2^5, 32_1^5, 32_2^5\}$.

clique of $\hat{\kappa}$ pairwise-adjacent vertices including exactly one vertex of each color. Put $\hat{n} = |V|$ and $\hat{m} = |E|$. For each i, $1 \leq i \leq \hat{\kappa}$, let $\hat{n}_i$ denote the number of vertices of color i, and let $v_i^1, \ldots, v_i^{\hat{n}_i}$ denote these vertices.

Our reduction constructs a colored point set with $k = \hat{\kappa} + \binom{\hat{\kappa}}{2} + 4 \cdot \binom{\hat{\kappa}}{2}$ colors of three types (i.e., *vertex type, edge type, and consistency type*):

- one *vertex color* i for each color i of the vertices in V;
- one *edge color* ij with $i < j$ for each unordered pair of colors $\{i, j\}$ of the edges in E;
- four *consistency colors* ji_1, ji_2, ij_1, ij_2 for each unordered pair of colors $\{i, j\}$ of the edges in E.

Place $n = \hat{n} + \hat{m} + (\hat{\kappa} - 1)\hat{n} \cdot \hat{n} + 2\hat{n} \cdot \hat{m}$ points in $\hat{n} + \hat{m}$ clusters, including exactly one *vertex cluster* for each vertex in V, and exactly one *edge cluster* for each edge in E:

- For each vertex of color i in V, put a point of vertex color i in the corresponding vertex cluster.
- For each edge of color pair $\{i, j\}$ in E, put a point of edge color ij in the corresponding edge cluster.
- For each vertex v_i^p in V, and for each color $j \neq i$, put p points of consistency color ji_1 and $\hat{n} - p$ points of consistency color ji_2 in the vertex cluster for v_i^p.
- For each edge $e = \{v_i^p, v_j^q\}$ in E, put $\hat{n} - p$ points of consistency color ji_1, p points of consistency color ji_2, $\hat{n} - q$ points of consistency color ij_1, and q points of consistency color ij_2 in the edge cluster for e.

Arrange the clusters of points such that the maximum distance between two points in each cluster is 1, and that the minimum distance between two points from different clusters is $\gamma > 1$. Set $t = \hat{\kappa} + \binom{\hat{\kappa}}{2}$, then set the required number of points to be covered to 1 for each vertex color and for each edge color, and to $\hat{n}$ for each consistency color. Refer to Figure 3 for an example.

The reduction is FPT with both t and k as parameters since they are functions of $\hat{\kappa}$ only. The following two lemmas, analogous to Lemmas 3 and 4, imply that it is W[1]-hard to approximate SCSI-t within γ for any approximation ratio $\gamma > 1$:

Lemma 5. *There is a colored clique of size $\hat{\kappa}$ in the graph only if there is a color-spanning set of t intervals for the colored points with maximum length 1.*

Proof. Given a colored clique in the graph, it is straightforward to verify that the $t = \hat{\kappa} + \binom{\hat{\kappa}}{2}$ clusters, including the $\hat{\kappa}$ vertex clusters for the $\hat{\kappa}$ vertices and the $\binom{\hat{\kappa}}{2}$ edge clusters for the $\binom{\hat{\kappa}}{2}$ edges in the clique, span all colors of the points, and moreover each cluster can be covered with an interval of length 1. □

Lemma 6. *There is a colored clique of size $\hat{\kappa}$ in the graph if there is a color-spanning set of t intervals for the colored points with maximum length less than γ.*

Proof. Given a color-spanning set of $t = \hat{\kappa} + \binom{\hat{\kappa}}{2}$ intervals with maximum length less than γ, each interval can cover points from at most one cluster due to the restriction on the maximum interval length. Thus to span the $\hat{\kappa}$ vertex colors and the $\binom{\hat{\kappa}}{2}$ edge colors of the points, each of the $\hat{\kappa} + \binom{\hat{\kappa}}{2}$ intervals must cover a point of a distinct vertex color or a distinct edge color in a distinct cluster.

To show that the set V' of $\hat{\kappa}$ vertices and the set E' of $\binom{\hat{\kappa}}{2}$ edges corresponding to these $\hat{\kappa} + \binom{\hat{\kappa}}{2}$ clusters form a colored clique of size $\hat{\kappa}$, it remains to prove that they are consistent, that is, for each edge $e = \{v_i^p, v_j^q\}$ in E', the two vertices v_i^p and v_j^q must be in V'. But this property is clearly ensured by our requirement that at least $\hat{n}$ points of each of the four consistency colors ji_1, ji_2, ij_1, ij_2 must be covered. □

4   An FPT Algorithm for SCSI-t with $s_i = 1$ for All i

In this section we prove Theorem 3. We show that the special case of SCSI-t with $s_i = 1$ for all i is fixed-parameter tractable when the parameter is the number k of colors, and admits an exact algorithm running in $O(2^k n \cdot \max\{k, \log n\})$ time. In the following of this section, unless otherwise stated, SCSI-t refers to the special case when $s_i = 1$ for all i.

Let $P = \{p_1, p_2, \ldots, p_n\}$ be a set of n points on a line, say, the x-axis, sorted from left to right. Each point p_i has one of k colors.

We first have the following lemma, which generalizes Lemma 1:

Lemma 7. *There must exist an optimal solution for the problem SCSI-t that consists of t intervals such that a longest interval have both left and right endpoints in P.*

For any two indices i and j with $1 \leq i \leq j \leq n$, let d_{ij} be the distance between the points p_i and p_j. Let $\mathcal{D}$ be the set of the distances d_{ij} for all $1 \leq i \leq j \leq n$. Let λ^* be the maximum length of the intervals in any optimal solution of the problem SCSI-t. By Lemma 7, $\lambda^* \in \mathcal{D}$ holds.

To solve SCSI-t, we will first give a *decision algorithm* to determine whether $d \geq \lambda^*$ for any given distance d, without knowing λ^*. With the decision algorithm, we can find λ^* by doing binary search after sorting the values of $\mathcal{D}$. However, since $|\mathcal{D}| = \Theta(n^2)$, to avoid the quadratic time, we represent $\mathcal{D}$ implicitly in such a way that λ^* can be found by calling the decision algorithm $O(\log n)$ times, and we achieve this by using a technique called *binary search on sorted arrays* [4].

We define the *decision problem* of SCSI-t as follows: Given a distance value d, determine whether $d \geq \lambda^*$, which is equivalent to determining whether there exists a color-spanning set of t intervals with uniform length d. In the sequel, we present an algorithm for solving the decision problem, and we refer to the algorithm as the *decision algorithm*.

Our decision algorithm consists of two main steps. The second step is a dynamic programming procedure, and the first step can be considered as a preprocessing for the second step. Let d be any given distance value.

In the first step, for each i with $1 \leq i \leq n$, we compute the following information: (1) an index $g(i)$, which is smallest index j such that the points $p_j, p_{j+1}, \ldots, p_i$ are covered by I_i^r, where I_i^r is the interval of length d with right endpoint at p_i (e.g., see Fig. 4), and (2) a color set C_i that consists of the colors of the points covered by the interval I_i^r.

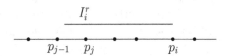

Fig. 4. Illustrating the definition of the index $g(i)$. I_i^r is an interval of length d with right endpoint at p_i and its left endpoint is between p_{j-1} and p_j. The index $g(i)$ is j in this example.

The indices $g(i)$ for all $i = 1, 2, \ldots, n$ can be easily computed in $O(n)$ time by scanning all points from right to left. Below we focus on computing the color sets C_i for all $i = 1, 2, \ldots, n$.

The first issue is how to maintain the color set C_i. For simplicity, we use a k-sized array $C_i[1, 2, \ldots, k]$ to represent C_i in the way that $C_i[j] = 1$ if and only if the j-th color is in C_i. To compute C_i, we sweep the points of P from right to left by using an interval I of length d. During the sweeping, for each color we maintain the number of points of that color covered by the current interval I. Suppose we have just computed the set C_i and the right endpoint of I is now at p_i. To compute C_{i-1}, we first copy all values of the array C_i to C_{i-1}. Then, we keep sweeping I leftwards. Since p_i will not be in I, we decrease the number for the color of p_i by 1 and if this number becomes zero, we set the value for that color in C_{i-1} to zero. During the sweeping, if the left endpoint point of I encounters a point p of P, then we increase the number for the color of p by 1 and if this number was originally zero (and now becomes 1), then we set the value for that color in C_{i-1} to 1. We keep sweeping I until the right endpoint of I encounters a point, which is p_{i-1}. At this moment, we have obtained C_{i-1}. In this way, we can compute all color sets C_i for $i = 1, 2, \ldots, n$ in $O(nk)$ time and space.

Note that one may wonder that there may be other better implementations for the first step. We should point out that here we only want to give an algorithm that is simple because the running time and the space of the entire decision algorithm is dominated by the second step anyway.

Hence, in the first step, we compute the index $g(i)$ and the color set C_i for all $i = 1, 2, \ldots, n$, which takes $O(nk)$ time and space.

The second step of our decision algorithm is a dynamic programming procedure. Let C be the set of all k colors. For each subset S of C, for any $1 \leq i \leq n$, denote by $N[S, i]$ the minimum number of intervals of length d for covering at least one point of each color in S among the points $p_1, p_2, \ldots, p_i$. Note that $d \geq \lambda^*$ if and only if $N[C, n] \leq t$. We have the following recurrence

$$N[S, i] = \min\{N[S, i - 1], N[S \setminus C_i, g(i) - 1] + 1\}$$

with the base case $N[\emptyset, i] = 0$. The recurrence distinguishes two cases: either the point p_i is not covered, or it is the last point covered by (without loss of generality it is the right endpoint of) an interval of length d, which covers points from $p_{g(i)}$ to p_i with color set C_i, according to the definitions of $g(i)$ and C_i.

By representing the color set S as an array of size k as in the first step, the running time of the dynamic programming procedure is $O(2^k k n)$: the table $N[S, i]$ has $2^k n$ entries; each entry takes $O(k)$ time to compute since $g(i)$ and C_i have been computed in the first step and the set minus operation $S \setminus C_i$ takes $O(k)$ time. In addition, with standard techniques, the actual intervals in the corresponding solution can be computed in the same amount of time asymptotically.

In summary, we can solve the decision problem in $O(2^k k n)$ time.

By exploring the power of RAM, we can further improve the running time of our above decision algorithm to $O(2^k n \lceil k/\log n \rceil)$. Specifically, suppose we work in a standard word RAM model with word size $w \geq \log n$, such that each bitwise logical operation on computer words can be done in constant time. Then we represent a color set by a sequence of k bits, which can be stored in $\lceil k/\log n \rceil$ words. In this way, the first step of the algorithm can be done in $O(n \lceil k/\log n \rceil)$ time because each update on any color set takes $O(\lceil k/\log n \rceil)$ time (and there are $O(n)$ updates in total). For the second step, we claim that each set minus operation on $S \setminus C_i$ can be done in $O(\lceil k/\log n \rceil)$ time. Indeed, suppose we want to do a set minus operation $A \setminus B$ on two computer words A and B (e.g., if $A = 10011$ and $B = 01001$, then $A \setminus B = 10010$); one can easily verify that we can obtain $A \setminus B$ by first doing a bitwise complement operation on B and then doing a bitwise AND operation on A and the new B. Since each bitwise operation takes constant time, we can obtain $A \setminus B$ in constant time. Hence, since the set minus operation $S \setminus C_i$ in our problem is on $\lceil k/\log n \rceil$ words, we can perform each such operation in $O(\lceil k/\log n \rceil)$ time. Thus, the second step of our decision algorithm takes $O(2^k n \lceil k/\log n \rceil)$ time.

Therefore, we obtain the following lemma:

Lemma 8. *For any given distance d, we can determine whether $d \geq \lambda^*$ in $O(2^k n \lceil k/\log n \rceil)$ time.*

In the following, we solve the problem SCSI-t by using the above decision algorithm. As mentioned above, a straightforward approach is to compute the set $\mathcal{D}$ and sort all values in $\mathcal{D}$; the value λ^* can be computed by using binary search and our decision algorithm. The total running time is $O(n^2 \log n + 2^k n \lceil k/\log n \rceil \log n)$. Below, we present a faster algorithm by using the technique *binary search on sorted arrays* [4], without computing the set $\mathcal{D}$ explicitly.

The technique binary search on sorted arrays is used for solving the following problem. Suppose there is a "black-box" decision procedure Π available, such that given any value a, Π can report whether a is a *feasible value* (with respect to a certain problem) in $O(T)$ time, and further, if a is a feasible value, then any value larger than a is a feasible value. Given M arrays A_i, $1 \leq i \leq M$, each containing $O(N)$ values in sorted order, the goal is to find the smallest feasible value δ in $\cup_{i=1}^{M} A_i$. An algorithm is given in [4] that can compute the value δ in $O((M + T) \log(MN))$ time. The algorithm is similar in spirit to the linear time selection algorithm; refer to Lemma 13 in [4] for the details.

In our problem, we consider our decision algorithm as the decision procedure Π, with $T = O(2^k n \lceil k/\log n \rceil)$. A value d is considered as a *feasible value* if and only if $d \geq \lambda^*$. Hence, if d is a feasible value, any value larger than d is also feasible. Note that λ^* is the smallest feasible value in $\mathcal{D}$. For each $1 \leq i \leq n$, we define the set D_i as $\{d_{ij} \mid i \leq j \leq n\}$. Hence, the sets D_i for all $i = 1, 2, \ldots, n$ form a partition of $\mathcal{D}$. Note that each set D_i can be considered as a sorted array of size $O(n)$ since for any two indices $j_1 \leq j_2$, it holds that $d_{i,j_1} \leq d_{i,j_2}$; further, given any index j, the value d_{ij} can be obtained in constant time. Our goal is to compute λ^*, which is the smallest feasible value in $\mathcal{D} = \cup_{i=1}^{n} D_i$. Hence, by using the algorithm in [4], we can compute λ^* in $O(n \log n + 2^k n \lceil k/\log n \rceil \log n) = O(2^k n \lceil k/\log n \rceil \log n)$ time. After having λ^*, an actual solution set of color-spanning intervals can be found by applying the decision algorithm on $d = \lambda^*$, in additional $O(2^k n \lceil k/\log n \rceil)$ time. Hence, the overall running time for solving SCSI-t is $O(2^k n \lceil k/\log n \rceil \log n)$. Note that if $k \geq \log n$, then $O(2^k n \lceil k/\log n \rceil \log n)$ is $O(2^k k n)$; otherwise, it is $O(2^k n \log n)$. Therefore, the overall running time is $O(2^k n \cdot \max\{k, \log n\})$. Theorem 3 is thus proved.

Note that in a pointer machine model, the decision algorithm runs in $O(2^k k n)$ time, and therefore, by using the above technique of binary search on sorted arrays, we can solve the problem SCSI-t in $O(2^k k n \log n)$ time.

References

1. Abellanas, M., Hurtado, F., Icking, C., Klein, R., Langetepe, E., Ma, L., Palop, B., Sacristán, V.: Smallest color-spanning objects. In: Meyer auf der Heide, F. (ed.) ESA 2001. LNCS, vol. 2161, pp. 278–289. Springer, Heidelberg (2001)

2. Barba, L., Durocher, S., Fraser, R., Hurtado, F., Mehrabi, S., Mondal, D., Morrison, J., Skala, M., Wahid, M.A.: On k-enclosing objects in a coloured point set. In: Proceedings of the 25th Canadian Conference on Computational Geometry (CCCG 2013), pp. 229–234 (2013)

3. Chen, D.Z., Misiolek, E.: Algorithms for interval structures with applications. Theoretical Computer Science 508, 41–53 (2013)

4. Chen, D.Z., Wang, C., Wang, H.: Representing a functional curve by curves with fewer peaks. Discrete and Computational Geometry 46, 334–360 (2011)

5. Cygan, M., Philip, G., Pilipczuk, M., Pilipczuk, M., Wojtaszczyk, J.O.: Dominating set is fixed parameter tractable in claw-free graphs. Theoretical Computer Science 412, 6982–7000 (2011)

6. Das, S., Goswami, P.P., Nandy, S.C.: Smallest color-spanning objects revisited. International Journal of Computational Geometry and Applications 19, 457–478 (2009)

7. Fan, C., Luo, J., Zhu, B.: Tight approximation bounds for connectivity with a color-spanning set. In: Cai, L., Cheng, S.-W., Lam, T.-W. (eds.) ISAAC 2013. LNCS, vol. 8283, pp. 590–600. Springer, Heidelberg (2013)

8. Fellows, M.R., Hermelin, D., Rosamond, F., Vialette, S.: On the parameterized complexity of multiple-interval graph problems. Theoretical Computer Science 410, 53–61 (2009)

9. Fleischer, R., Xu, X.: Computing minimum diameter color-spanning sets. In: Lee, D.-T., Chen, D.Z., Ying, S. (eds.) FAW 2010. LNCS, vol. 6213, pp. 285–292. Springer, Heidelberg (2010)

10. Fleischer, R., Xu, X.: Computing minimum diameter color-spanning sets is hard. Information Processing Letters 111, 1054–1056 (2011)

11. Jiang, M.: The zero exemplar distance problem. Journal of Computational Biology 18, 1077–1086 (2011)

12. Ju, W., Fan, C., Luo, J., Zhu, B., Daescu, O.: On some geometric problems of color-spanning sets. Journal of Combinatorial Optimization 26, 266–283 (2013)

13. Khanteimouri, P., Mohades, A., Abam, M.A., Kazemi, M.R.: Spanning colored points with intervals. In: Proceedings of the 25th Canadian Conference on Computational Geometry (CCCG 2013), pp. 265–270 (2013)

14. Khanteimouri, P., Mohades, A., Abam, M.A., Kazemi, M.R.: Computing the smallest color-spanning axis-parallel square. In: Cai, L., Cheng, S.-W., Lam, T.-W. (eds.) ISAAC 2013. LNCS, vol. 8283, pp. 634–643. Springer, Heidelberg (2013)

15. Marx, D.: Parameterized complexity and approximation algorithms. The Computer Journal 51, 60–78 (2008)

Fixed Parameter Tractable Algorithms in Combinatorial Topology

Benjamin A. Burton* and William Pettersson

The University of Queensland, Brisbane, Australia
bab@maths.uq.edu.au, william@ewpettersson.se

Abstract. To enumerate 3-manifold triangulations with a given property, one typically begins with a set of potential face pairing graphs (also known as dual 1-skeletons), and then attempts to flesh each graph out into full triangulations using an exponential-time enumeration. However, asymptotically most graphs do not result in *any* 3-manifold triangulation, which leads to significant "wasted time" in topological enumeration algorithms. Here we give a new algorithm to determine whether a given face pairing graph supports any 3-manifold triangulation, and show this to be fixed parameter tractable in the treewidth of the graph.

We extend this result to a "meta-theorem" by defining a broad class of properties of triangulations, each with a corresponding fixed parameter tractable existence algorithm. We explicitly implement this algorithm in the most generic setting, and we identify heuristics that in practice are seen to mitigate the large constants that so often occur in parameterised complexity, highlighting the practicality of our techniques.

1 Introduction

In combinatorial topology, a triangulated 3-manifold involves abstract tetrahedra whose faces are identified or "glued" in pairs. Many research questions involve looking for a triangulated manifold which fits certain requirements, or is pathologically bad for certain algorithms, or breaks some conjecture. One invaluable tool for such tasks is an exhaustive *census* of triangulated 3-manifolds.

The first of these was the census of cusped hyperbolic 3-manifold triangulations on ≤ 5 tetrahedra by Hildebrand and Weeks [18] in 1989, later extended to ≤ 9 tetrahedra [8,12,27]. Another much-used example is the census of closed orientable prime minimal triangulations of ≤ 6 tetrahedra by Matveev [24], later extended to ≤ 12 tetrahedra [22,23].

In all of these prior works, the authors enumerate all triangulated manifolds on n tetrahedra by first enumerating all 4-regular multigraphs on n nodes (very fast), and then for each graph G essentially modelling every possible triangulation with G as its dual graph (very slow). If any such triangulation built from G is the triangulation of a 3-manifold, we say that G is *admissible*. If G admits a 3-manifold triangulation with some particular property p, we say that G is *p-admissible*.

* Supported by the Australian Research Council (DP1094516, DP110101104).

Z. Cai et al. (Eds.): COCOON 2014, LNCS 8591, pp. 300–311, 2014.
© Springer International Publishing Switzerland 2014

Using state-of-the-art public software [9], generating such a census on 12 tetra-hedra takes 1967 CPU-days, of which over 1588 CPU-days is spent analysing non-admissible graphs. Indeed, for a typical census on ≤ 10 tetrahedra, less than 1% of 4-regular graphs are admissible [7]. Moreover, Dunfield and Thurston [17] show that the probability of a random 4-regular graph being admissible tends toward zero as the size of the graph increases. Clearly an efficient method of determining whether a given graph is admissible could have significant effect on the (often enormous) running time required to generating such a census.

We use parameterized complexity [16] to address this issue. A problem is *fixed parameter tractable* if, when some parameter of the input is fixed, the problem can be solved in polynomial time in the input size. In Theorem 14 we show that to test whether a graph G is admissible is fixed parameter tractable, where the parameter is the treewidth of G. Specifically, if the treewidth is fixed at $\leq k$ and G has size n, we can determine whether G is admissible in $O(n \cdot f(k))$ time.

Courcelle showed [14,13] that for graphs of bounded treewidth, an entire class of problems have fixed parameter tractable algorithms. However, employing this result for our problem of testing admissibility looks to be highly non-trivial. In particular, it is not clear how the topological constraints of our problem can be expressed in monadic second-order logic, as Courcelle's theorem requires. Even if Courcelle's theorem could be used, our results here provide significantly better constants than a direct application of Courcelle's theorem would.

Following the example of Courcelle's theorem, however, we generalise our result to a larger class of problems (Theorem 18). Specifically, we introduce the concept of a *simple property*, and give a fixed parameter tractable algorithm which, for any simple property p, determines whether a graph admits a triangulated 3-manifold with property p (again the parameter is treewidth).

We show that these results are practical through an explicit implementation, and identify some simple heuristics which improve the running time and memory requirements. To finish the paper, we identify a clear potential for how these ideas can be extended to the more difficult enumeration problem, in those cases where a graph *is* admissible and a complete list of triangulations is required.

Parameterised complexity is very new to the field of 3-manifold topology [10,11], and this paper marks the first exploration of parameterised complexity in 3-manifold enumeration problems. Given that 3-manifold algorithms are often extremely slow and complex, our work here highlights a growing potential for parameterised complexity to offer practical alternative algorithms in this field.

2   Background

To avoid ambiguity with the words "vertex" and "edge", we use the terms *node* and *arc* instead for graphs, and *vertex* and *edge* in the context of triangulations.

Many NP-hard problems on graphs are fixed parameter tractable in the *treewidth* of the graph (e.g., [1,2,4,5,13]). Introduced by Robertson and Seymour [26], the treewidth measures precisely how "tree-like" a graph is:

Definition 1 (Tree decomposition and treewidth). *Given a graph G, a tree decomposition of G is a tree H with the following additional properties:*

- *Each node of H, also called a bag, is associated with a set of nodes of G;*
- *For every arc a of G, some bag of H contains both endpoints of a;*
- *For any node v of G, the subforest in H of bags containing v is connected.*

If the largest bag of H contains k nodes of G, we say that the tree decomposition has width $k + 1$. The treewidth *of G, denoted $\mathrm{tw}(G)$, is the minimum width of any tree decomposition of G.*

Bodlaender [4] gave a linear time algorithm for determining if a graph has treewidth $\leq k$ for fixed k, and for finding such a tree decomposition, and Kloks [21] demonstrated algorithms for finding "nice" tree decompositions.

A closed 3-manifold is essentially a topological space in which every point has some small neighbourhood homeomorphic to $\mathbb{R}^3$. We first define *general triangulations*, and then give conditions under which they represent 3-manifolds.

Definition 2 (General triangulation). *A general triangulation is a set of abstract tetrahedra $\{\Delta_1, \Delta_2, \ldots, \Delta_n\}$ and a set of face identifications or "gluings" $\{\pi_1, \pi_2, \ldots, \pi_m\}$, such that each π_i is an affine identification between two distinct faces of tetrahedra, and each face is a part of at most one such identification.*

Note that this is more general than a simplicial complex (e.g., we allow an identification between two distinct faces of the same tetrahedron), and it need not represent a 3-manifold. Any face which is not identified to another face is called a *boundary face* of the triangulation. If a triangulation has no such boundary faces, we say it is *closed*. We also note that there are six ways to identify two faces, given by the six symmetries of a regular triangle.

We can partially represent a triangulation by its face pairing graph, which describes *which* faces are identified together, but not *how* they are identified.

Definition 3 (Face pairing graph). *The face pairing graph of a triangulation $\mathcal{T}$ is the multigraph $\Gamma(\mathcal{T})$ constructed as follows. Start with an empty graph G, and insert one node for every tetrahedron in $\mathcal{T}$. For every face identification between two tetrahedra T_i and T_j, insert the arc $\{i, j\}$ into the graph G.*

Note that a face pairing graph will have parallel arcs if there are two distinct face identifications between T_i and T_j, and loops if two faces of the same tetrahedron are identified together. $\mathcal{T}$ is connected if and only if $\Gamma(\mathcal{T})$ is connected.

Some edges of tetrahedra will be identified together as a result of these face identifications (and likewise for vertices). Some edges may be identified directly via a single face identification, while others may be identified indirectly through a series of face identifications.

We assign an arbitrary orientation to each edge of each tetrahedron. Given two tetrahedron edges e and e' that are identified together via the face identifications, we write $e \simeq e'$ if the orientations agree, and $e \simeq \overline{e'}$ if the orientations are

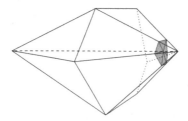

Fig. 1. A triangulation of a 3-ball with 6 tetrahedra meeting along an internal edge

reversed. In settings where we are not interested in orientation, we write $e \sim e'$ if the two edges are identified (i.e. one of $e \simeq e'$ or $e \simeq \overline{e'}$ holds).

This leads to the natural notation $[e] = \{e' : e \sim e'\}$ as an equivalence class of identified edges (ignoring orientation). We refer to $[e]$ as an *edge of the triangulation*. Likewise, we use the notation $v \sim v'$ for vertices of tetrahedra that are identified together via the face identifications, and we call an equivalence class $[v]$ of identified vertices a *vertex of the triangulation*.

A *boundary edge / vertex* of a triangulation is an edge / vertex of the triangulation whose equivalence class contains some edge / vertex of a boundary face.

The *link* of a vertex $[v]$ is the (2-dimensional) frontier of a small regular neighbourhood of $[v]$. Figure 1 shows the link of the top vertex shaded in grey; in this figure, the link is homeomorphic (topologically equivalent) to a disc. The link is a 2-dimensional triangulation (in the example it has six triangles), and we use the term *arc* to denote an edge in this triangulation. In this paper, whether "arc" refers to a graph or a vertex link is always clear from context.

Definition 4 (Closed 3-manifold triangulation). *A closed 3-manifold triangulation $\mathcal{T}$ is a general triangulation for which (i) $\mathcal{T}$ is connected; (ii) for any vertex v in $\mathcal{T}$, the link of v is homeomorphic to a 2-sphere; and (ii) no edge e in $\mathcal{T}$ is identified with itself in reverse (i.e. $e \not\simeq \overline{e}$).*

These properties are necessary and sufficient for the underlying topological space to be a 3-manifold. We say that a graph G is *admissible* if it is the face pairing graph for any closed 3-manifold triangulation $\mathcal{T}$.

Definition 5 (Partial-3-manifold triangulation). *A partial-3-manifold triangulation $\mathcal{T}$ is a general triangulation for which (i) for any vertex v in $\mathcal{T}$, the link of v is homeomorphic to a 2-sphere with zero or more punctures; and (ii) no edge e in $\mathcal{T}$ is identified with itself in reverse (i.e. $e \not\simeq \overline{e}$).*

These are in essence "partially constructed" 3-manifold triangulations; the algorithms of Section 4.1 build these up into full 3-manifold triangulations. Note that the underlying space of $\mathcal{T}$ might not even be a 3-manifold with boundary: there may be "pinched vertices" whose links have many punctures.

We can make some simple observations: (i) the boundary vertices of a partial 3-manifold triangulation are precisely those whose links have at least one puncture; (ii) a connected partial-3-manifold triangulation with no boundary faces is a closed 3-manifold triangulation, and vice-versa; (iii) a partial-3-manifold triangulation with a face identification removed, or an entire tetrahedron removed, is still a partial-3-manifold triangulation.

3 Configurations

The algorithms in Section 4.1 build up 3-manifold triangulations one tetrahedron at a time. As we add tetrahedra, we must track what happens on the boundary of the triangulation, but we can forget about the parts of the triangulation not on the boundary—this is key to showing fixed parameter tractability. In this section we define and analyse edge and vertex configurations of general triangulations, which encode exactly those details on the boundary that we must retain.

Definition 6 (Edge configuration). *The edge configuration of a triangulation $\mathcal{T}$ is a set C_e of triples detailing how the edges of the boundary faces are identified together. Each triple is of the form $((f, e), (f', e'), o)$, where: f and f' are boundary faces; e and e' are tetrahedron edges that lie in f and f' respectively; e and e' are identified in $\mathcal{T}$; and o is a boolean "orientation indicator" that is true if $e \simeq e'$ and false if $e \simeq \overline{e'}$.*

This mostly encodes the 2-dimensional triangulation of the boundary, though additional information describing "pinched vertices" is still required.

Example 7 (2-tetrahedra pinched pyramid). In all examples, we use the notation $t_i : a$ to denote vertex a of tetrahedron t_i, and $t_i : abc$ to denote face abc of tetrahedron t_i. Face identifications are denoted as $t_i : abc \leftrightarrow t_j : def$, which means that face abc of t_i is mapped to face def of t_j such that $a \leftrightarrow d$, $b \leftrightarrow e$ and $c \leftrightarrow f$.

Take two tetrahedra t_0 and t_1, each with vertices labelled $0, 1, 2, 3$, and apply the face identifications $t_0 : 012 \leftrightarrow t_1 : 012$ and $t_0 : 023 \leftrightarrow t_1 : 321$.

The resulting triangulation is a square based pyramid with one pair of opposing faces identified (see Figure 2(a)). The final space resembles a hockey puck with a pinch in the centre, as seen in Figure 2(b). Note that the vertex at top of the pyramid, which becomes the pinched centre of the puck, has a link homeomorphic to a 2-sphere with two punctures. Therefore, although this is a partial 3-manifold triangulation, the underlying space is not a 3-manifold.

The edge configuration of this triangulation is:

$$\{((t_0 : 013, 03), (t_1 : 013, 13), f), \quad ((t_0 : 013, 01), (t_1 : 013, 01), t),$$
$$((t_0 : 013, 13), (t_0 : 123, 13), t), \quad ((t_0 : 123, 12), (t_0 : 123, 23), f),$$
$$((t_1 : 013, 03), (t_1 : 023, 03), t), \quad ((t_1 : 023, 02), (t_1 : 023, 23), f)\};$$

here t and f represent *true* and *false* respectively.

Definition 8 (Vertex configuration). *The vertex configuration C_v of a triangulation $\mathcal{T}$ is a partitioning of those tetrahedron vertices that belong to boundary faces, where vertices v and v' are in the same partition if and only if $v \sim v'$.*

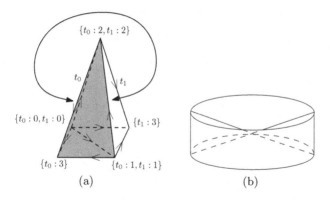

(a)   (b)

Fig. 2. The triangulation from Example 7. The grey shaded tetrahedron is t_0. Edges are marked with their orientations, and the double-ended arrow indicates the identification of two opposing faces of the pyramid. The resulting space resembles a hockey puck with the centre pinched into a point. This pinch is the vertex $\{t_0\!:\!2, t_1\!:\!2\}$.

In partial-3-manifold triangulations, vertex links may have multiple punctures; the vertex configuration then allows us to deduce which punctures belong to the same link. In essence, the vertex configuration describes how the triangulation is "pinched" inside the manifold at vertices whose links have too many punctures.

For instance, the vertex configuration of Example 7 is given by

$$\{\{t_0\!:\!0, t_1\!:\!0, t_1\!:\!3\}, \quad \{t_0\!:\!1, t_0\!:\!3, t_1\!:\!1\}, \quad \{t_0\!:\!2, t_1\!:\!2\}\}.$$

The partition $\{t_0 : 2, t_1 : 2\}$ represents the pinch at the center of the "hockey puck".

Definition 9 (Boundary configuration). *The boundary configuration C of a triangulation $\mathcal{T}$ is the pair (C_e, C_v) where C_e is the edge configuration and C_v is the vertex configuration.*

Lemma 10. *For b boundary faces, there are $\frac{(3b)!}{(3b/2)!}$ possible edge configurations.*

Proof Note that b must be even; let $b = 2m$. Each boundary face has three edges, so there are $6m$ possible pairs (f, e) where e is an edge on a boundary face f. Each such pair must be identified with exactly one other pair, with either $e \simeq e'$ or $e \simeq \overline{e'}$, and so the number of possible edge configurations is

$$2 \cdot (6m - 1) \cdot 2 \cdot (6m - 3) \cdot \ldots \cdot 2 \cdot 3 \cdot 2 \cdot 1 = \frac{(6m)!}{(3m)!} = \frac{(3b)!}{(3b/2)!}. \qquad \square$$

Lemma 11. *For b boundary faces, the number of possible boundary configurations is bounded from above by*

$$\frac{(3b)!}{(3b/2)!} \cdot \left(\frac{2.376b}{\ln(3b + 1)} \right)^{3b}.$$

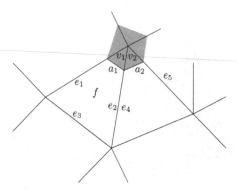

Fig. 3. Part of the boundary of a triangulation. The link of the top vertex is shaded grey; this link does not contain the vertex, but instead cradles the vertex from below.

Proof   There are $3b$ tetrahedron vertices on boundary faces, and so the number of possible vertex configurations is the Bell number B_{3b}. The result now follows from Lemma 10 and the following inequality of Berend [3]:

$$B_{3b} = \frac{1}{e} \sum_{i=0}^{\infty} \frac{i^{3b}}{i!} < \left(\frac{2.376b}{\ln(3b+1)} \right)^{3b}. \qquad \square$$

Corollary 12. *The number of possible boundary configurations for a triangulation on n tetrahedra with b boundary faces depends on b, but not on n.*

The boundary configuration can be used to partially reconstruct the links of vertices on the boundary of the triangulation. In particular:

- The edge configuration allows us to follow the arcs around each puncture of a vertex link—in Figure 3 for instance, we can follow the sequence of arcs $a_1, a_2, \dots$ that surround the puncture in the link of the top vertex.
- The vertex configuration tells us whether two sequences of arcs describe punctures in the *same* vertex link, versus *different* vertex links.

In this way, we can reconstruct all information about punctures in the vertex links, even though we cannot access the full (2-dimensional) triangulations of the links themselves. As the next result shows, this means that the boundary configuration retains all data required to build up a partial-3-manifold triangulation, without knowledge of the full triangulation of the underlying space.

Lemma 13. *Let $\mathcal{T}$ be a partial 3-manifold triangulation with b boundary faces, and let $\mathcal{T}'$ be formed by introducing a new identification between two boundary faces of $\mathcal{T}$. Given the boundary configuration of $\mathcal{T}$ and the new face identification, we can test whether $\mathcal{T}'$ is also a partial-3-manifold triangulation in $O(b)$ time.*

A full proof appears in the full version of this paper. The basic idea is to check whether the conditions in Definition 5 are preserved. The edge configuration allows us to easily test for edges identified together in reverse, and the partial reconstruction of the vertex links (as described above) allows us to test whether all vertex links are still 2-spheres with zero or more punctures.

4   Algorithms and Simple Properties

Recall that the motivating problem for our work was to quickly detect whether a given graph admits a closed 3-manifold triangulation. To this end we show:

Theorem 14. *Given a connected 4-regular multigraph G, the problem of determining whether there exists a closed 3-manifold triangulation $\mathcal{T}$ such that $\Gamma(\mathcal{T}) = G$ is fixed parameter tractable in the treewidth of G.*

This is a special case of our more general Theorem 18, and so we do not prove it in detail here. The basic idea is as follows.

We say that a boundary configuration C is *viable* for a graph G if there exists some partial-3-manifold triangulation $\mathcal{T}$ with $\Gamma(\mathcal{T}) = G$ and with C as its boundary configuration. Our algorithm starts with an empty triangulation, and then introduces tetrahedra and face identifications in a way that essentially works from the leaves up to the root of the tree decomposition of G. For each subtree in the tree decomposition we compute which configurations are viable for the corresponding subgraph of G, and then propagate these configurations further up the tree. The running time at each node depends only on the number of boundary faces, which is bounded in terms of the bag size and thereby tw(G).

4.1   A Generalisation to Simple Properties

Here we generalise Theorem 14 to many other settings. For this we define a *simple property* of a partial 3-manifold triangulation (see below).

We extend boundary configurations to include an extra piece of data ϕ based on the partial triangulation that helps test our property. For instance, if p is the simple property that the triangulation contains ≤ 3 internal vertices, then ϕ might encode the number of internal vertices thus far in the partial 3-manifold triangulation (here ϕ takes one of the values $0, 1, 2, 3, \mathsf{too\_many}$).

As before: for a simple property p, we say that a boundary configuration C is *p-viable* for a graph G if there exists some partial-3-manifold triangulation $\mathcal{T}$ with property p, with $\Gamma(\mathcal{T}) = G$ and with C as its boundary configuration.

Shortly we solve the problem of testing whether a graph G admits any closed 3-manifold triangulation with property p, for any simple property p. The basic idea is as before: for each subtree of our tree decomposition of G, we compute all viable configurations and propagate this information up the tree.

Definition 15 (Simple property). *A boolean property p of a partial-3-manifold triangulation is called* simple *if all of the following hold. Here all configurations have $\leq b$ boundary faces, and f, g, h are some computable functions.*

1. *The extra data ϕ in the boundary configuration satisfies $\phi \in P$ for some universal set P with $|P| \leq f(b)$.*

2. *We can determine whether a triangulation satisfies p based only on its boundary configuration (including the extra data ϕ).*

3. *Given any viable configuration and a new face identification π between two of its boundary faces, we can in $O(g(b))$ time test whether introducing this identification yields another viable configuration and, if so, calculate the corresponding value of ϕ.*

4. *Given viable configurations for two disjoint triangulations, we can in $O(h(b))$ time test whether the configuration for their union is also viable and, if so, calculate the corresponding value of ϕ.*

The four conditions above can be respectively interpreted as meaning:

1. the upper bound on the number of viable configurations (including the data ϕ) still depends on b but not the number of tetrahedra;
2. we can still test property p without examining the full triangulation;
3. new face identifications can still be checked for p-viability in $O(g(b))$ time;
4. configurations for disjoint triangulations can be combined in $O(h(b))$ time.

Example 16. Let p be the property that a triangulation contains at most x internal vertices (i.e., vertices with links homeomorphic to a 2-sphere), for some fixed integer x. Then p is simple.

Here we define $\phi \in P = \{0, 1, \ldots, x, \texttt{too\_many}\}$ to be the number of vertices in our partial 3-manifold triangulation with 2-sphere links. This clearly satisfies conditions 1 and 2. For condition 3: when identifying two faces together, a new vertex acquires a 2-sphere link if and only if the identification closes off all punctures in the link (which we can test from the edge and vertex configurations). Condition 4 is easily satisfied by summing ϕ over the disjoint configurations.

The case when $x = 1$ is highly relevant: much theoretical and computational work has gone into 1-vertex triangulations of 3-manifolds [19,23], and these are of particular use when searching for 0-efficient triangulations [20].

We can now state the main result of this paper:

Problem 17 p-ADMISSIBILITY(G) *Let p be a simple property. Given a connected 4-regular multigraph G, determine whether there exists a closed 3-manifold triangulation $\mathcal{T}$ with property p such that $\Gamma(\mathcal{T}) = G$.*

Theorem 18. *Let p be a simple property. Given a connected 4-regular multigraph G on n nodes with treewidth $\leq k$, and a corresponding tree decomposition with $O(n)$ nodes where each bag has at most two children, we can solve p-ADMISSIBILITY(G) in $O(n \cdot f(k))$ time for some computable function f.*

Our requirement for such a tree decomposition is not restrictive: Bodlaender [4] gives a fixed parameter tractable algorithm to find a tree decomposition of width $\leq k$ for fixed k, and Kloks [21] gives an $O(n)$ time algorithm to transform this into a tree decomposition where each bag has at most two children. The "two children" constraint can be relaxed; we use it here to keep the proof simple.

A full proof appears in the full version of this paper; the main ideas are as follows. For each bag ν of the tree decomposition we define a corresponding subgraph G_ν of G, which contains precisely those nodes of G that do *not* appear in bags outside the subtree rooted at ν. As before we use a dynamic programming approach, working from the leaves of the tree decomposition up to the root: for each ν we construct all viable configurations for G_ν, by combining the viable configurations at the child nodes of ν and analysing any new face identifications that might appear. We bound the running time at each ν by a function of the bag size, using the properties of Definition 15 and the observation that any partial triangulation admitted by G_ν must have $\leq 4(\mathrm{tw}(G) + 1)$ boundary faces.

Once we reach the root node of the tree decomposition, the final list contains a p-viable configuration if and only if G admits a closed 3-manifold triangulation with property p.

5   Implementation and Experimentation

The algorithm was implemented Java, using the treewidth library from [15]. Although our theoretical bound on the number of configurations is extremely large (Lemma 11), we store all configurations using hash maps to exploit situations where in practice the number of viable configurations is much smaller. As seen below, we find that such a discrepancy does indeed arise (and significantly so).

We also introduce another modification that yields significant speed improvements in practice. The algorithm builds up a complete list of all viable configurations at each bag ν of the tree decomposition. However, for an affirmative answer to the problem, only a small subset of these may be required. We take advantage of this as follows.

For any bag ν with no children, configurations are computed as normal. Once a viable configuration is found, it is immediately propagated up the tree in a depth-first manner. This means that, rather than calculating every possible viable configuration for every subgraph G_ν, the improved algorithm can identify a full triangulation with property p quickly and allow early termination.

We implemented the program with p defined to be *one-vertex and possibly minimal*, using criteria on the degrees of edges from [6]. This allowed us to compare both correctness and timing with the existing software *Regina* [9]. We ran our algorithm on all 4-regular graphs on 4, 5 or 6 nodes to verify correctness. We see that the average time to process a graph increases with treewidth, as expected. We also see that the number of viable configurations is indeed significantly lower than the upper bound of Lemma 11, as we had hoped.

Regina significantly outperforms our algorithm on all of these graphs, though these are small problems for which asymptotic behaviour plays a less important role. What matters more is performance on larger graphs, where existing software begins to break down.

We therefore ran a sample of 12-node graphs through our algorithm, selected randomly from graphs which cause significant slowdown in existing software. This "biased" sampling was deliberate—our aim is not for our algorithm to always outperform existing software, but instead to seek new ways of solving those difficult cases that existing software cannot handle. Here we do find success: our algorithm was at times 600% faster at identifying non-admissible graphs than *Regina,* though this improvement was not consistent across all trials. More detailed experiments will appear in the full version of this paper.

In summary: for larger problems, our proof-of-concept code already exhibits far superior performance for some cases that *Regina* struggles with. With more careful optimisation (e.g., for dealing with combinatorial isomorphism), we believe that this algorithm would be an important tool that complements existing software for topological enumeration.

The full source code for the implementation of this algorithm is available at http://www.github.com/WPettersson/AdmissibleFPG.

6   Applications and Extensions

We first note that our meta-theorem is useful: here we list several simple properties p that are important in practice, with a brief motivation for each.

1. *One-vertex triangulations* are crucial for computation: they typically use very few tetrahedra, and have desirable combinatorial properties. This is especially evident with 0-efficient triangulations [20].
2. Likewise, *minimal triangulations* (which use the fewest possible tetrahedra) are important for both combinatorics and computation [6,7]. Although minimality is not a simple property, it has many simple necessary conditions, which are used in practical enumeration software [7,23].
3. *Ideal triangulation of hyperbolic manifolds* play a key role in 3-manifold topology. An extension of Theorem 18 allows us to support several necessary conditions for hyperbolicity, which again are used in real software [12,18].

Finally: a major limitation of all existing 3-manifold enumeration algorithms is that they cannot "piggyback" on prior results for fewer tetrahedra, a technique that has been remarkably successful in other areas such as graph enumeration [25]. This is not a simple oversight: it is well known that we cannot build all "larger" 3-manifold triangulations from smaller 3-manifold triangulations. The techniques presented here, however, may allow us to overcome this issue—we can modify the algorithm of Theorem 18 to store entire families of triangulations at each bag of the tree decomposition. We would lose fixed parameter tractability, but for the first time we would be able to cache and reuse partial results across different graphs and even different numbers of tetrahedra, offering a real potential to extend census data well beyond its current limitations.

References

1. Arnborg, S.: Efficient algorithms for combinatorial problems on graphs with bounded decomposability—a survey. BIT 25(1), 2–23 (1985)
2. Arnborg, S., Lagergren, J., Seese, D.: Easy problems for tree-decomposable graphs. J. Algorithms 12(2), 308–340 (1991)
3. Berend, D., Tassa, T.: Improved bounds on Bell numbers and on moments of sums of random variables. Probab. Math. Statist. 30(2), 185–205 (2010)

4. Bodlaender, H.L.: A linear-time algorithm for finding tree-decompositions of small treewidth. SIAM J. Comput. 25(6), 1305–1317 (1996)
5. Bodlaender, H.L., Kloks, T.: Efficient and constructive algorithms for the pathwidth and treewidth of graphs. J. Algorithms 21(2), 358–402 (1996)
6. Burton, B.A.: Face pairing graphs and 3-manifold enumeration. J. Knot Theory Ramifications 13(8), 1057–1101 (2004)
7. Burton, B.A.: Enumeration of non-orientable 3-manifolds using face-pairing graphs and union-find. Discrete & Computational Geometry 38(3), 527–571 (2007)
8. Burton, B.A.: The cusped hyperbolic census is complete (preprint, 2014), http://arxiv.org/abs/1405.2695
9. Burton, B.A., Budney, R., Pettersson, W.: Regina: Software for 3-manifold topology and normal surface theory (1999-2013), http://regina.sourceforge.net
10. Burton, B.A., Lewiner, T., Paixão, J., Spreer, J.: Parameterized complexity of discrete Morse theory. In: SCG 2013: Proceedings of the 29th Annual Symposium on Computational Geometry, pp. 127–136. ACM (2013)
11. Burton, B.A., Spreer, J.: The complexity of detecting taut angle structures on triangulations. In: SODA 2013: Proceedings of the Twenty-Fourth Annual ACM-SIAM Symposium on Discrete Algorithms, pp. 168–183. SIAM (July 2012)
12. Callahan, P.J., Hildebrand, M.V., Weeks, J.R.: A census of cusped hyperbolic 3-manifolds. Math. Comp. 68(225), 321–332 (1999), with microfiche supplement
13. Courcelle, B., Makowsky, J.A., Rotics, U.: On the fixed parameter complexity of graph enumeration problems definable in monadic second-order logic. Discrete Appl. Math. 108(1-2), 23–52 (2001), International Workshop on Graph-Theoretic Concepts in Computer Science, Smolenice Castle (1998)
14. Courcelle, B.: The monadic second-order logic of graphs. I. Recognizable sets of finite graphs. Inform. and Comput. 85(1), 12–75 (1990)
15. van Dijk, T., van den Heuvel, J.P., Slob, W.: Computing treewidth with LibTW (2006)
16. Downey, R.G., Fellows, M.R.: Parameterized complexity. Monographs in Computer Science. Springer, New York (1999)
17. Dunfield, N.M., Thurston, W.P.: Finite covers of random 3-manifolds. Invent. Math. 166(3), 457–521 (2006)
18. Hildebrand, M., Weeks, J.: A computer generated census of cusped hyperbolic 3-manifolds. In: Computers and Mathematics, Cambridge, MA, pp. 53–59. Springer, New York (1989)
19. Jaco, W., Letscher, D., Rubinstein, J.H.: Algorithms for essential surfaces in 3-manifolds. In: Topology and geometry: commemorating SISTAG. Contemp. Math., vol. 314, pp. 107–124. Amer. Math. Soc., Providence (2002)
20. Jaco, W., Rubinstein, J.H.: 0-efficient triangulations of 3-manifolds. J. Differential Geom. 65(1), 61–168 (2003)
21. Kloks, T. (ed.): Treewidth. LNCS, vol. 842. Springer, Heidelberg (1994)
22. Martelli, B., Petronio, C.: Three-manifolds having complexity at most 9. Experimental Mathematics 10(2), 207–236 (2001)
23. Matveev, S.: Algorithmic topology and classification of 3-manifolds, 2nd edn. Algorithms and Computation in Mathematics, vol. 9. Springer, Berlin (2007)
24. Matveev, S.V.: Computer recognition of three-manifolds. Experiment. Math. 7(2), 153–161 (1998)
25. McKay, B.D.: Isomorph-free exhaustive generation. J. Algorithms 26(2), 306–324 (1998)
26. Robertson, N., Seymour, P.D.: Graph minors. II. Algorithmic aspects of tree-width. J. Algorithms 7(3), 309–322 (1986)
27. Thistlethwaite, M.: Cusped hyperbolic manifolds with 8 tetrahedra (October 2010), http://www.math.utk.edu/~morwen/8tet/

Improved Approximation Algorithms
for a Bilevel Knapsack Problem

Xian Qiu[1] and Walter Kern[2]

[1] Faculty of Computer Science, Zhejiang University, China
xianqiu@zju.edu.cn
[2] Department of Applied Mathematics, University of Twente, The Netherlands
w.kern@utwente.nl

Abstract. We study the Stackelberg/bilevel knapsack problem as proposed by Chen and Zhang [4]: Consider two agents, a leader and a follower. Each has his own knapsack. (Knapsack capacities are possibly different). As usual, there is a set of items $i = 1, ..., n$ of given weight w_i and profits p_i. It is allowed to pack item i into both knapsacks, but in this case the corresponding profit for each player becomes $p_i + a_i$, where a_i is a given (positive or negative) number. The objective is to find a packing for the leader such that the total profit of the two knapsacks is maximized, assuming that the follower acts selfishly. We present tight approximation algorithms for all settings considered in [4].

1 Introduction

The standard knapsack problem is one of the most fundamental and well-studied problems in combinatorial optimization: There is a knapsack of prescribed capacity W and n items with given size w_i and profit p_i. The task is to select a set of items of total size at most W and maximum total profit. A first bilevel variant (in the form of a *Stackelberg game*) was introduced by Dempe and Richter [7]: There are two decision makers (players) – a *leader* and a *follower* – as well as a (universal) knapsack with flexible capacity and a set of items with given sizes as above, yet item profits may vary w.r.t the leader and the follower, respectively. The leader first determines the capacity of the knapsack, and afterwards the follower, assumed to be selfish, packs items to the knapsack, maximizing his own profit. The (leader's bilevel) problem is to compute the knapsack capacity such that the leader's profit – defined by a linear function of the knapsack capacity plus his total profit of packed items – is maximized.

Several other bilevel variants of knapsack have been proposed as well. For example, Mansi *et al.* [13] study a setting in which both the leader and the follower pack items into a knapsack (of fixed capacity). DeNegre [8] investigates a bilevel version where both players own a private knapsack each and pack items from a common item set. Again, the leader acts first, selecting a set of items for his own knapsack, then the follower packs items from the remaining item set into his own knapsack, seeking to maximize his total profit. The objective of

Z. Cai et al. (Eds.): COCOON 2014, LNCS 8591, pp. 312–323, 2014.
© Springer International Publishing Switzerland 2014

the (hostile) leader is to choose his set of items such that the follower's profit is minimized.

In this paper we consider yet another variant of the bilevel knapsack problem, due to Chen and Zhang [4]. In this setting, again, each player has his own knapsack of fixed capacity W_1 and W_2, respectively. Items $1, \ldots, n$ have fixed weights w_i and profits p_i. The characteristic feature of the model in [4] is that items may be *double-packed*, *i.e.* packed by both players. In case item i is packed only by one player, it accounts for a profit of p_i, as usual, however, if i is packed by both players, its profit (for both players) is modified to $p_i + a_i$ for given *profit modifier* $a_i \in \mathbb{R}$. Again, the setting is that of a Stackelberg game, and the objective is to exhibit an optimal packing for the leader, *i.e.*, one that maximizes the total profit assuming that the second player (the follower) acts selfishly (disregarding the impact any double packing may have on the items packed by the leader). As a motivating example, Chen and Zhang mention the case of two investors, say, the government and a company with budgets W_1 and W_2, respectively. Items correspond to potential projects of cost w_i and reward p_i, resp. $p_i + a_i$ with $a_i > 0$ if both players invest in project i . Depending on the application, the numbers a_i may be positive or negative ("double booking"). In case all a_i are positive, Chen and Zhang [4] call it the *beneficial model* and if all a_i are negative, it is referred to as the *competitive model*.

Bilevel optimization is often computationally difficult and likely to extend beyond NP. In the last decades, bilevel and multilevel optimization have received much attention in the literature (*cf.* books by Migdalas, Pardalos & Värbrand [14] and Dempe [6], a survey by Colson *et al.* [5]). Dempe and Richter [7] introduced a mixed integer bilevel program for their problem variant and proposed an algorithm based on branch and bound. Afterwards, a dynamic programming algorithm for this problem was given by Brotcorne *et al.* [1]. Recently, Caprara *et al.* [2] proved that the first three problem variants mentioned above are Σ_2^P-hard (probably the fourth one is as well), *i.e.*, there is no way of formulating them as single-level integer programs of polynomial size unless *the polynomial hierarchy collapses* (*cf.* [2] for more details). In particular, they showed that the first two variants (*cf.* Dempe and Richter [7], Mansi *et al.* [13]) do not possess a polynomial approximation algorithm with finite worst case guarantee unless $P = NP$ and proposed a polynomial time approximation scheme for the third variant (*cf.* DeNegre [8]), which is known as the first approximation scheme for a Σ_2^P-hard problem. For other variants and related problems, *cf.* [12,18,17,9,13].

Regarding the problem to be considered in this paper, Chen and Zhang [4] proposed a $(2 + \epsilon)$-approximation algorithm for the competitive model ($a_i \leq 0$), and, for the beneficial model ($a_i \geq 0$), a $(1 + \sqrt{2} + \epsilon)$-approximation for the case $W_1 > W_2$ and a $(2 + \epsilon)$-approximation for the case $W_1 \leq W_2$.

In this paper, we present better approximation algorithms for the beneficial model as well as the competitive model and show that the approximation ratios are tight in each case, *i.e.*, the approximation ratios can be made arbitrarily close to the known lower bounds (*cf.* Fig. 1). The main ingredients of our approach are: An ϵ-approximation of the maximum profit problem in case both players

cases		approx. ratios [4]	lower bounds
$a_i \leq 0$		$2 + \epsilon$	1.5
$a_i \geq 0$	$W_1 > W_2$	$1 + \sqrt{2} + \epsilon$	1.5
	$W_1 < W_2$	$2 + \epsilon$	2

Fig. 1. Known lower bounds

cooperate – which may be of independent interest, *cf.* (P3) in section 2 – and a factor revealing LP for estimating the quality of our approximation algorithms (*cf.* Jain *et al.* [11]).

The rest of the paper is organized as follows: In the section below, we formally introduce the bilevel problem (*cf.* (P1) in Section 2) and its "cooperative" counterpart (*cf.* (P3)). In Section 3, we describe a polynomial time approximation scheme (PTAS) for the cooperative problem version (P3). In Section 4, we present new approximation algorithms and analyze their approximation ratios. Finally, in Section 5, we mention some open problems.

2   Bilevel Knapsack with Independent Knapsacks

Let W_1, W_2 be capacities of the knapsacks owned by player 1 (leader) and player 2 (follower), respectively. Let $A = \{1, 2, \ldots, n\}$ be a set of items of weight w_i, profit p_i and "double packing modifier" a_i for all $i \in A$. Let $x_i, y_i \in \{0, 1\}$ indicate whether item i is packed by player 1 and player 2, respectively.

Recall that the profit of item i is modified to $p_i + a_i$ if i is packed by both players. Thus the leader's problem can be formulated as a bilevel integer program as follows:

$$\max_x \sum_{i=1}^n p_i(x_i + y_i) + 2 \sum_{i=1}^n a_i x_i y_i \tag{P1}$$

$$\text{s.t.} \sum_{i=1}^n w_i x_i \leq W_1,$$

$$x_i \in \{0, 1\}, \quad i = 1, 2, \ldots, n,$$

y is an optimal solution of (P2) *below.*

$$\max_y \sum_{i=1}^n p_i y_i + \sum_{i=1}^n a_i x_i y_i \tag{P2}$$

$$\text{s.t.} \sum_{i=1}^n w_i y_i \leq W_2,$$

$$y_i \in \{0, 1\}, \quad i = 1, 2, \ldots, n.$$

Note that for fixed x there may exist multiple corresponding optimal solutions y for player 2. In our analysis, we always assume a worst case scenario, *i.e.* we focus on a "pessimistic" version of the above bilevel problem, where player 2 chooses an optimal solution y of (P2) minimizing the objective of (P1). We call the outcome of this pessimistic version the *competitive optimum* and – as in the paper by Chen and Zhang [4] – compare it to the so-called *cooperative* optimum, *i.e.*, the maximum total profit the two players could achieve. The latter can be expressed by a (single level) integer program

$$\max \sum_{i=1}^{n} p_i(x_i + y_i) + 2\sum_{i=1}^{n} a_i x_i y_i \tag{P3}$$

$$\text{s.t.} \sum_{i=1}^{n} w_i x_i \leq W_1,$$

$$\sum_{i=1}^{n} w_i y_i \leq W_2,$$

$$x_i, y_i \in \{0, 1\}.$$

Example. Consider two knapsacks of capacity $W_1 = 1$ (for the leader) and capacity $W_2 = 2$ (for the follower). The item set contains two items of weights $w_1 = 1$, $w_2 = 2$, profits $p_1 = 2$, $p_2 = 1$ and modifiers $a_1 = -1$ (a_2 is arbitrary).

Observe that player 1 only has two options: Packing item 1 or packing nothing. If player 1 packs item 1, then player 2 gets a maximal profit 1 by either packing item 1 or item 2, resulting in a total profit 2 or 3 respectively. If player 1 packs nothing, then player 2 packs item 1, resulting in a total profit 2. Thus, the cooperative optimum may be as large as 1.5 times the competitive optimum. We aim at showing that this example is a worst case example in the sense that if all modifiers a_i are negative, then the ratio between the cooperative and competitive optimum is bounded by 1.5 and that solutions x for the leader's problem ensuring a ratio of $1.5 + \epsilon$ can be found in polynomial time. Similarly, we present tight bounds for the case of non-negative a_i and, eventually, the mixed case with both positive and negative double packing modifiers.

As a first step, we compute approximately optimal cooperative solutions.

3   Polynomial Time Approximation Scheme for (P3)

It is well known that knapsack can be solved by a fully polynomial time approximation scheme (FPTAS) (*cf.* [10,16]). As (P3) with $a_i \to -\infty$ becomes a multiple knapsack problem with two knapsacks, we can not expect an FPTAS for (P3) unless $P = NP$ (*cf.* [3]). In the following, we seek for a polynomial time approximation scheme (PTAS) for (P3).

We start with some notations. For any set S of items, let $w(S)$ and $p(S)$ denote the total weight and the total profit of items in S, respectively, *i.e.*, $w(S) = \sum_{i \in S} w_i$ and $p(S) = \sum_{i \in S} p_i$. Since the profits of items may be modified due to double-packing, $p(S)$ may also denote the modified total profit of items in S if no misunderstanding is possible.

As it turns out, *negative items*, i.e., those with $a_i < 0$, and *non-negative items* (those with $a_i \geq 0$) can be treated independently. Therefore, we simplify matters by first assuming that all items are negative (non-negative items will be dealt afterwards). For technical reasons (to be explained in the proof) we slightly generalize our problem, assuming that certain items, say, items in the set $N_1 \subseteq N$, are not allowed to be double-packed. We let $n_1 = |N_1|$ denote the number of items that are prescribed to be single, and n_2 is the number of items in $N_2 = N \backslash N_1$ that may be double-packed. Thus $n_1 + n_2 = n$. Our proof for the approximation ratio will be by induction on $n_1 + 2n_2$.

We describe an $O(1 - 1/k)$-approximation algorithm – again denoted by ALG – proceeding in a similar way to that of Sahni [15]. One difference is that for a double-packable item i we distinguish between its *primary* copy i with an associated profit p_i and its *secondary* copy i with profit $p_i + 2a_i < p_i$. Let $S^* = S_1^* \cup S_2^* = \text{supp } x^* \cup \text{supp } y^*$ be an optimal solution of (P3). Observe that – as we distinguish primary and secondary copies – the set S^* may be understood as a *set* rather than a multiset. In phase I, ALG seeks to "guess" the k most profitable items from the optimum solution $S_1^* \cup S_2^*$ to include them in the initial packing. In case an item i is double-packed in S^*, and (the primary copy of) i belongs to the k most profitable items, we want ALG to include also the secondary copy into the initial packing. For this reason, we let ALG start from all initial packings with up to k primary items plus some of their secondary copies and let $S = S_1 \cup S_2$ be this set of items, with S_1, S_2 packed on knapsacks 1 and 2, respectively.

The second phase, again, considers the remaining items in order of non-increasing profit rates. As in the single knapsack case, it proceeds in a true online manner as explained below. In particular, whenever ALG checks an item i, it immediately decides upon packing or not packing i, but does not yet decide whether i should be double-packed. Note that, as we stick to the case of negative items here, we have $p_i + 2a_i < p_i$, so that primary items come with higher profit rates and are checked for inclusion before their corresponding secondary copies arrive.

Summarizing, in phase II, ALG considers the (copies of) items in $\{1, \ldots, n\} \backslash S$ in order of non-increasing profit rates

$$r_{i_1} \geq r_{i_2} \geq \ldots \geq r_{i_m} \quad (m = n_1 + 2n_2 - |S|)$$

defined as explained above.

Whenever ALG checks a primary item $i = i_t$, the item is packed wherever it fits. In case it does not fit anywhere, the item is skipped. Whenever ALG checks a secondary item $i = i_t$, it is perfectly clear on which knapsack i should be packed and ALG seeks to accommodate item i there, say, on knapsack 1, by "switching" single items from knapsack 1 to knapsack 2 if necessary. More precisely, ALG considers all single (primary) items from $\{1, ..., n\} \backslash S$ currently packed on knapsack 1 in some order and switches them onto knapsack 2 whenever they fit there, until either item i can eventually be accommodated on knapsack 1 or no further single item can be switched to knapsack 2 while item i still cannot be added to knapsack 1. In the latter case, item $i = i_t$ is skipped. The order in which items are considered for switching from knapsack 1 to knapsack 2 is not relevant, but it is convenient to use a "last in first out" order as switching rule.

Note that any packed item remains packed, only the assignment to a particular knapsack may be revised (possibly even several times), due to switching.

Lemma 1. *ALG as described above yields a $(1 - 2/k)$-approximation for (P3) (assuming that all items are negative).*

It is an easy exercise to bound the running time: There are $O(n^{k+1})$ sets S of size $|S| \leq k$ to choose for the primary items. Given S, we are left to fix for each $i \in S$ whether to double-pack i or not, and, in the latter case, where to pack it (knapsack 1 or 2). Thus in total the number of "guesses" is bounded by $O(n^{k+1}3^k)$.

Next let us turn to the beneficial model, which turns out to be easier: Assume that all items in N are non-negative, i.e., $a_i \geq 0$. Our approximation algorithm – denoted by ALG – in this case, will again have a "guessing" phase I, followed by a phase II, where items are processed in order of non-increasing profit rates. Note that this time, however, double-packing yields higher profit rates than single packing, so ALG will first decide about double-packing item i (at profit rate $r_i = (p_i + a_1)/w_i$) and then – in case of non-acceptance consider single-packing i at a later stage. Correspondingly, in the beneficial model, it does not make sense to distinguish between a primary and secondary copy of item i (as both arrive at the same time). Thus, in what follows we will interpret an optimal solution of (P3) as a multiset $S^* = S_1^* \cup S_2^*$, where $D^* = S_1^* \cap S_2^*$ is the set of double-packed items.

In phase I, ALG guesses a set $D \subseteq \{1, ..., n\}$ of size $|D| \leq k$ (as a candidate for a set of most profitable double-packed items in an optimal solution) as well as a set $S \subseteq \{1, ..., n\}\backslash D$, $|S| \leq k$ of most profitable single-packed items. In addition, it guesses a number $l \leq n$ indicating the total number of items that should be double-packed by ALG (on top of the already chosen set D). In phase II, ALG processes the remaining items in order of non-increasing profit rates as usual. More precisely, let

$$r_{i_1} = \frac{p_{i_1} + a_{i_1}}{w_{i_1}} \geq ... \geq r_{i_l} = \frac{p_{i_l} + a_{i_l}}{w_{i_l}}$$

be the l highest double-packing profit rates in $N_2\backslash D$. Then ALG double-packs as much as possible of $i_1, ..., i_l$ (on top of D) and then continues (after packing S_1 and S_2) with single packing of the remaining items (i.e., items in $(N_1 \cup N_2)\backslash(D \cup S \cup \{i_1, ..., i_l\})$) in order of non-increasing profit rates.

There are $O(n^{2(k+1)})$ possible choices for D and S, and $O(2^k)$ bipartitions of S. Together with the different choices for l, there are $O(n^{2k+3}2^k)$ different choices of parameters.

Lemma 2. *For suitable choice of parameters D, S_1, S_2 and l, ALG yields an approximation ratio $(1 - 4/k)$.*

Remark 1. We like to point out that negative items cannot be treated in such an easy way: Consider, for example a list of $2n$ items with $p_i = 10$, $a_i = -5$ plus $2n$ items with $p_i = 5, a_i = -\epsilon$ and $w_i = 1$ for all $4n$ items. An optimum packing for two knapsacks of capacity $3n$ each would single-pack all items with $p_i = 10$ and double-pack all remaining items. Yet, a simple greedy heuristic like the "positive item variant" of ALG would always start double-packing the "high profit" items – at least we cannot prevent it from doing so by prescribing the number l of items to be double-packed.

We are left to combine the two algorithms for negative and non-negative items in order to deal with "mixed" sets of items. The only way we found works in a sense by "brute force": We guess the amount W_1^+ and W_2^+ of capacity an optimum solution uses on knapsack 1 and 2 for non-negative items respectively and then split the problem accordingly into one with negative and one with non-negative items. In our case, it is sufficient to approximate W_1^+ and W_2^+ up to a factor $1/k$, so we actually guess $m_i := \lceil \log_{1+1/k} W_i \rceil$ and run the algorithm with $\tilde{W}_i^+ = (1 + 1/k)^{m_i}$ instead of W_i^+. The total number of guesses we need is $O(k \log W_i)$ for $i = 1, 2$, thus $O(k^2 (\log W_1)(\log W_2))$ in total.

This introduces another possible loss of order $1/k$ on the total profit gained with positive items, so that after all, the combined algorithm ALG will yield a total profit P with $P \geq (1 - 1/k)(1 - 4/k)P^*$, where P^* is the optimum profit. Thus we have shown

Theorem 1. *For fixed k, there exists a polynomial time algorithm ALG that approximates (P3) up to a factor $(1 - 5/k)$ in time $O(n^{2k+5} 2^k \log W_1 \log W_2)$.*

4   Approximation Algorithms for the Leader

Let S_1 and S_2 be optimal solutions for the standard single knapsack instances with knapsack capacities W_1 and W_2, resp., and profits p_i for all items. In other words, S_1 and S_2 denote the support of optimal solutions of the following problem with $W = W_1$ and $W = W_2$, resp.:

$$\max \sum_{i=1}^{n} p_i x_i \tag{P4}$$

$$\text{s.t.} \sum_{i=1}^{n} w_i x_i \leq W,$$

$$x_i \in \{0, 1\}, \quad i = 1, 2, \ldots, n.$$

Let (S_1^*, S_2^*) be an optimal solution of the cooperative relaxation (P3). We first note that the value OPT of (P3) satisfies

$$OPT = p(S_1^*) + p(S_2^*) + 2h^*, \quad \text{where } h^* := \sum_{i \in S_1^* \cap S_2^*} a_i. \tag{4.1}$$

Furthermore, we trivially have

$$p(S_1^*) \leq p(S_1) \quad \text{and} \quad p(S_2^*) \leq p(S_2). \tag{4.2}$$

As (S_1^*, S_2^*) can be found by a straightforward dynamic programming and can be approximated arbitrarily closely by a PTAS (*cf.* Section 3), we first present an algorithm by assuming that S_1^*, S_2^*, S_1 and S_2) can be found exactly. It turns out that we lose a factor of ϵ if these solutions are approximated correspondingly by PTAS. We (first) treat the beneficial and the competitive model separately.

4.1   The Beneficial Model: $a_i \geq 0$

We present a very simple algorithm and show that the approximation ratios are tight in both cases ($W_1 \geq W_2$ and $W_1 < W_2$).

Algorithm 1: Pack one of S_1^* and S_1 (whichever results in a maximum total profit [1]).

Assume first that player 1 packs S_1^* and player 2 packs some set $\hat{S}$. Let $\hat{h} = \sum_{i \in S_1^* \cap \hat{S}_2} a_i$. Then $p(\hat{S}_2) + \hat{h} \geq p(S_2^*) + h^*$. Denote by ALG the value obtained by the algorithm. Thus

$$ALG = p(S_1^*) + p(\hat{S}_2) + 2\hat{h}$$
$$\geq p(S_1^*) + p(S_2^*) + h^* + \hat{h}.$$

Since $p(\hat{S}_2) \leq p(S_2)$, we have

$$\hat{h} \geq p(S_2^*) - p(\hat{S}_2) + h^* \geq p(S_2^*) - p(S_2) + h^*,$$

implying

$$ALG \geq p(S_1^*) + 2p(S_2^*) - p(S_2) + 2h^*. \tag{4.3}$$

Now assume that player 1 packs S_1. Then player 2 will get at least $p(S_2)$ by packing S_2, implying

$$ALG \geq p(S_1) + p(S_2).$$

Besides, we observe the following "knapsack constraints":

$$p(S_1) \geq p(S_2) \text{ if } W_1 \geq W_2, \tag{4.4}$$
$$p(S_1) \leq p(S_2) \text{ if } W_1 \leq W_2. \tag{4.5}$$

Let $\alpha = ALG/OPT$. We derive the following linear program for estimating the approximation ratio:

$$\text{minimize } \alpha \tag{4.6}$$

---

[1] *cf.* Remark 2 at the end of Section 4.2 explaining how this decision can be taken.

$$\text{subject to } p(S_1^*) + p(S_2^*) + 2h^* = 1,$$
$$\alpha \geq p(S_1^*) + 2p(S_2^*) - p(S_2) + 2h^*,$$
$$\alpha \geq p(S_1) + p(S_2),$$
$$p(S_1) \geq p(S_1^*),$$
$$p(S_2) \geq p(S_2^*),$$
$$p(S_1), p(S_2), p(S_1^*), p(S_2^*), h^* \geq 0$$

and the knapsack constraints hold.

The minimum value equals 2/3 if $W_1 \geq W_2$ and 1/2 if $W_1 < W_2$, proving that OPT/ALG is at most 3/2 resp. 2, matching the lower bounds (*cf.* [4]).

4.2 Th Competitive Model: $a_i \leq 0$

In principle, we could also apply Algorithm 1 to this case: If player 1 packs S_1^*, similar to the above argument, we have

$$ALG \geq p(S_1^*) + 2p(S_2^*) - p(S_2) + 2h^*.$$

In case of packing S_1, we want to show that $ALG \geq p(S_1) + p(S_2) + 2h$, where $h = \sum_{i \in S_1 \cap S_2} a_i$. This is clearly true if player 2 packs S_2. Otherwise, player 2 packs a set, say, $\hat{S}_2$, with $\hat{h} = \sum_{i \in S_1 \cap \hat{S}_2} a_i$, such that $p(\hat{S}_2) + \hat{h} \geq p(S_2) + h$. As $p(\hat{S}_2) \leq p(S_2)$, this implies $\hat{h} \geq h$. Thus,

$$ALG \geq p(S_1) + p(\hat{S}_2) + 2\hat{h}$$
$$\geq p(S_1) + p(S_2) + h + \hat{h}$$
$$\geq p(S_1) + p(S_2) + 2h,$$

as claimed.

Now we obtain the following linear program bounding the approximation ratio of Algorithm 1.

$$\min \alpha \tag{4.7}$$
$$\text{s.t. } p(S_1^*) + p(S_2^*) + 2h^* = 1,$$
$$\alpha \geq p(S_1^*) + 2p(S_2^*) - p(S_2) + 2h^*,$$
$$\alpha \geq p(S_1) + p(S_2) + 2h,$$
$$p(S_1) \geq p(S_1^*),$$
$$p(S_2) \geq p(S_2^*),$$
$$p(S_1), p(S_2), p(S_1^*), p(S_2^*) \geq 0, \ h, h^* \leq 0$$

and the knapsack constraints hold.

Letting $p(S_1^*) = p(S_1) = p(S_2) = 1$, $h = -1$ and $h^* = p(S_2^*) = 0$, we can easily see that the optimal objective value is 0. We observe that in this worst-case instance the penalty h is too large: A better choice for player 1 is to pack nothing. Then player 2 must pack S_2, implying $ALG \geq p(S_2)$. Adding this simple constraint to the above program yields an optimal objective value 0.5.

Thus, excluding items with large penalty from S_1 can improve the performance of ALG, however, it does not yield a tight approximation yet. The idea for further improvement is not only to avoid packing items with large penalties but also to pack items with *small* penalties. We define the following sets:

$$S_2^+ = \left\{ i \in S_2 \mid |a_i| > \frac{p_i}{2} \right\}, \quad S_2^- = S_2 \backslash S_2^+, \tag{4.8}$$

where S_2^+, S_2^- are sets of items having large penalties and small penalties respectively. Our algorithm can now be described as follows.

Algorithm 2: Pack one of $S_1^*, S_1 \backslash S_2^+$ and $\emptyset$ (whichever results in a maximum total profit).

To analyze its performance, we distinguish three cases:

Case 1. Player 1 packs S_1^*. This is already considered in the above analysis and yields $ALG \geq p(S_1^*) + 2p(S_2^*) - p(S_2) + 2h^*$.

Case 2. Player 1 packs $S_1 \backslash S_2^+$. We prove that $ALG \geq p(S_1) + p(S_2^-) + 2h^-$, where $h^- = \sum_{i \in S_2^-} a_i$. Clearly, this is true when player 2 packs S_2. If player 2 packs some other set, say $\hat{S}_2 \neq S_2$, with $\hat{h} = \sum_{i \in (S_1 \backslash S_2^+) \cap \hat{S}_2} a_i$, then $p(\hat{S}_2) + \hat{h} \geq p(S_2) + h^-$, implying $\hat{h} \geq h^-$ (recall that $p(\hat{S}_2) \leq p(S_2)$). Hence,

$$\begin{aligned} ALG &\geq p(S_1) - p(S_2^+) + p(\hat{S}_2) + 2\hat{h} \\ &\geq p(S_1) - p(S_2^+) + p(S_2) + h^- + \hat{h} \\ &\geq p(S_1) - p(S_2^+) + p(S_2) + 2h^- \\ &= p(S_1) + p(S_2^-) + 2h^-. \end{aligned}$$

Case 3. Player 1 packs nothing. Then player 2 can guarantee a total profit $p(S_2)$ by packing S_2. Thus, $ALG \geq p(S_2)$.

Furthermore, in addition to the knapsack constraints (4.4), (4.5) we have the following constraints:

$$p(S_2) = p(S_2^+) + p(S_2^-) \quad \text{and} \quad h^- \geq -\frac{p(S_2^-)}{2}.$$

This finally yields the following linear program with an optimal value of $2/3$ (for both $W_1 \geq W_2$ and $W_1 < W_2$).

$$\begin{aligned} \text{minimize } &\alpha \tag{4.9}\\ \text{subject to } &p(S_1^*) + p(S_2^*) + 2h^* = 1, \\ &\alpha \geq p(S_1^*) + 2p(S_2^*) - p(S_2) + 2h^*, \\ &\alpha \geq p(S_1) + p(S_2^-) + 2h^-, \\ &\alpha \geq p(S_2), \\ &p(S_2) = p(S_2^+) + p(S_2^-), \end{aligned}$$

$$h^- \geq -\frac{p(S_2^-)}{2},$$
$$p(S_1) \geq p(S_1^*),$$
$$p(S_2) \geq p(S_2^*),$$
$$p(S_1), p(S_2), p(S_2^+), p(S_2^-), p(S_1^*), p(S_2^*) \geq 0, \ h^*, h^- \leq 0$$

and the knapsack constraints hold.

Hence $OPT/ALG \leq 1.5$, which is tight (*cf.* the example in Section 2).

Remark 2. Since player 2 acts after player 1, the reader may wonder how to find the maximal total profit over all cases (in Algorithm 1 and 2). This can be done by checking the inequalities "$\alpha \geq ...$" in (4.6) and (4.9) respectively: If any of these inequalities gets tight, the algorithm may pick the associated set (for player 1).

4.3   The Mixed Case

Finally, let us turn to the case where the item set contains both positive and negative items. This case is easily reduced to the two cases (beneficial and competitive model resp.) considered above: All we have to do is to split the item set A into the set A^+ and A^- of non-negative and negative items, resp. and to guess – again, up to a certain factor, say, $1/k$ – the associated parts of knapsacks 1 and 2 that are filled with non-negative and negative items, resp., in an optimal solution $S_1^* \cup S_2^*$ of (P3). Solving the "pure" (beneficial resp. competitive) cases for the approximately correct choice of corresponding knapsack capacities will then give solutions for the leader's problem that differ from the cooperative value by at most a factor of $(2 + \epsilon)$ at the expense of an additional $O((\log W_1)(\log W_2))$ factor in the running time.

5   Remarks

We have presented a unified approach for computing solutions to the leader's problem with approximately optimal ratio, if compared to the outcome of the cooperative version of the problem. Without further going into details, we mention that in the lower bound examples (*cf.* the example in section 2 and the examples in [4], the maximum cooperative value equals the maximum value of (the optimistic version of) the bilevel problem. Thus our results also provide tight bounds for the ratio between the optimistic and pessimistic version of the bilevel problem itself.

A natural question to ask is about side payments: Player 1 can certainly enforce the optimistic value of the bilevel problem with arbitrarily small side payments. Thus, it seems natural to also investigate approximation algorithms in the optimistic setting.

References

1. Brotcorne, L., Hanafi, S., Mansi, R.: A dynamic programming algorithm for the bilevel knapsack problem. Operation Research Letters 37(3), 215–218 (2009)
2. Caprara, A., Carvalho, M., Lodi, A., Woeginger, G.J.: A complexity and approximability study of the bilevel knapsack problem. In: Goemans, M., Correa, J. (eds.) IPCO 2013. LNCS, vol. 7801, pp. 98–109. Springer, Heidelberg (2013)
3. Chekuri, C., Khanna, S.: A ptas for the multiple knapsack problem. In: Proceedings of SODA 2000, pp. 213–222. Society for Industrial and Applied Mathematics, Philadelphia (2000)
4. Chen, L., Zhang, G.: Approximation algorithms for a bi-level knapsack problem. In: Wang, W., Zhu, X., Du, D.-Z. (eds.) COCOA 2011. LNCS, vol. 6831, pp. 399–410. Springer, Heidelberg (2011)
5. Colson, B., Marcotte, P., Savard, G.: Bilevel programming: A survey. 4OR: A Quarterly Journal of Operations Research 3(2), 87–107 (2005)
6. Dempe, S.: Foundations of Bilevel Programming. Kluwer Academic Publishers, Dordrecht (2002)
7. Dempe, S., Richter, K.: Bilevel programming with knapsack constraints. Central European Journal of Operation Research 8, 93–107 (2000)
8. DeNegre, S.: Interdiction and discrete bilevel linear programming. PhD thesis, Bethlehem, PA, USA (2011), AAI3456385.
9. Deng, X.: Complexity issues in bilevel linear programming. In: Multilevel Optimization: Algorithms and Applications, pp. 149–164. Kluwer Academic Publishers, Dordrecht (1998)
10. Ibarra, O.H., Kim, C.E.: Fast approximation algorithms for the knapsack and sum of subset problems. Journal of the ACM 22(4), 463–468 (1975)
11. Jain, K., Mahdian, M., Saberi, A.: A new greedy approach for facility location problems. In: Proceedings of STOC 2002, pp. 731–740 (2002)
12. Loridan, P., Morgan, J.: Weak via strong stackelberg problem: New results. Journal of Global Optimization 8(3), 263–287 (1996)
13. Mansi, R., Alves, C., de Carvalho, J., Hanafi, S.: An exact algorithm for bilevel 0-1 knapsack problems. Mathematical Problems in Engineering, Article ID 504713 (2012)
14. Migdalas, A., Pardalos, P.M., Värbrand, P.: Multilevel Optimization: Algorithms and Applications. Kluwer Academic Publishers, Dordrecht (1998)
15. Sahni, S.: Approximate algorithms for the 0/1 knapsack problem. Journal of the ACM 22(1), 115–124 (1975)
16. Vazirani, V.V.: Approximation algorithms. Springer-Verlag New York, Inc., New York (2001)
17. Wang, Z., Xing, W.: Two-person knapsack game. Journal of Industrial and Management Optimization 6(4), 847–860 (2010)
18. Wang, Z., Xing, W., Fang, S.-C.: Two-group knapsack game. Theoretical Computer Science 411(7-9), 1094–1103 (2010)

A Complex Semidefinite Programming Rounding Approximation Algorithm for the Balanced Max-3-Uncut Problem

Chenchen Wu[1], Dachuan Xu[1,*], Donglei Du[2], and Wen-qing Xu[1,3]

[1] College of Science, Tianjin University of Technology, Tianjin 300384, China
xudc@bjut.edu.cn
[2] Faculty of Business Administration, University of New Brunswick,
Fredericton, NB E3B 5A3, Canada
[3] Department of Mathematics and Statistics, California State University,
Long Beach, CA 90840, USA

Abstract. In this paper, we consider the balanced Max-3-Uncut problem which has several applications in the design of VLSI circuits. We propose a complex discrete linear program for the balanced Max-3-Uncut problem. Applying the complex semidefinite programming rounding technique, we present a 0.3456-approximation algorithm by further integrating a greedy swapping process after the rounding step. One ingredient in our analysis different from previous work for the traditional Max-3-Cut is the introduction and analysis of a bivariate function rather than a univariate function.

Keywords: Complex semidefinite programming, Approximation algorithm, Rounding, Balanced Max-3-Uncut.

1 Introduction

Graph partition problems have been investigated extensively in combinatorial optimization. In the Max-k-Cut problem (cf. [7]) on a weighted graph, we want to find a partition of the vertex set into k subsets such that the total weight of the edges from different subsets is maximized. If the subsets are required to have equal cardinality, the problem is then called the Max-k-Section (cf. [1]). When $k = 2$, the above two problems become the well-known Max-Cut and Max-Bisection problems respectively.

The capacitated Max-k-Uncut problem (a.k.a. the k-general partition problem [17]) is to partition the vertices of the graph into k subsets of prescribed sizes such that the total weight of the edges from the same subset is maximized. When we impose the equal cardinality constraint, the capacitated Max-k-Uncut problem becomes the balanced Max-k-Uncut problem (a.k.a. the k-partition problem [6], or when $k = 2$, the Max-$\frac{n}{2}$-Uncut problem). These graph partition problems are all NP-hard, and have applications in the design of VLSI circuits (cf.

---

* Corresponding author.

Z. Cai et al. (Eds.): COCOON 2014, LNCS 8591, pp. 324–335, 2014.
© Springer International Publishing Switzerland 2014

[6],[9],[17]). In this paper, we are particularly interested in the balanced Max-3-Uncut
problem.

Semidefinite programming (SDP) relaxation is a powerful tool in designing
approximation algorithm for graph partition problems. In their influential work,
Goemans and Williamson [9] propose a 0.87856-approximation algorithm for
the Max-Cut problem by integrating SDP relaxation and hyperplane rounding.
Their techniques are later applied to many graph partition problems (cf. [4, 5],
[11–13],[19–21]). Using SDP rounding together with greedy swap, Frieze and Jerrum [7] propose a 0.6514-approximation algorithm for the Max-Bisection problem, followed by further improvements to 0.699 [20], 0.7016 [11], and 0.7028 [5],
respectively. Raghavendra and Tan [16] propose a semidefinite programming hierarchies (SDPH) rounding 0.85-approximation algorithm for the Max-Bisection
problem, which is further improved to the currently best ratio 0.8776 by Austrin
et al. [2]. For the balanced Max-2-Uncut problem, we only mention two results:
one is the 0.6436-approximation algorithm by the standard SDP rounding [11];
and the other is the 0.8776-approximation algorithm by the SDPH rounding
[18]. Each of them gives the currently best known approximation ratios by using
SDP/SDPH rounding techniques respectively. We note that SDPH, which has
many applications in the area of complexity and approximation algorithm, is
first introduced by Lasserre [14].

In a subsequent work, Goemans and Williamson [10] use the three cubic roots
$(1, \omega, \omega^2)$ of unity (i.e., each of the roots satisfies the equation $z^3 = 1$) to
represent ternary decision variables, leading to a complex discrete program for
the Max-3-Cut problem. They extend their earlier SDP relaxation and hyperplane rounding technique [9] to the complex case with complex semidefinite
programming (CSDP) relaxation and complex hyperplane rounding [10], resuling in an approaximation algorithm with ratio $\left(\frac{7}{12} + \frac{3}{4\pi^2} \arccos^2(-1/4) - \epsilon\right) \approx$
$(0.8360 - \epsilon)$, for any given $\epsilon > 0$. This result improves an earlier 0.800217-approximation algorithm due to Frieze and Jerrum [7]. Further work along
this line include a $\frac{\pi}{4}$-approximation algorithm for a class of discrete complex
quadratic optimization problems when the coefficient matrix of the objective
is a positive semidefinite Hermitian matrix [22], and a 0.6733-approximation
algorithm for the Max-3-Section problem [15].

In this paper, we formulate the balanced Max-3-Uncut problem as a complex
discrete program and provide an approximation algorithm based on the technique of CSDP relaxation and rounding. After the random rounding, the obtained solution is infeasible. We then modify the solution by adopting a greedy
adjustment of the size of the subsets to obtain a 0.3456-approximation algorithm. This approximation ratio improves previous ratio 1/3 by Choudhury et
al. [3] who propose a local search $(1/(d(k-1) + 1))$-approximation algorithm
for the capacitated Max-k-Uncut problem, where d is the ratio of the largest
cardinality to the smallest cardinality in the partition. One ingredient in our
analysis different from previous work for the Max-3-Cut is the introduction and
analysis of a bivariate function rather than a univariate function.

In the remaining of this paper, we use $W(A) = \sum_{i \in A} w_i$ to denote the weight summation of the set A. Moreover, we use $W(A, B)$ to denote $W(A) + W(B)$ for any two sets A and B, and this notation can be analogously defined for more than two sets. The inner product of any two complex vectors $a, b \in \mathbb{C}^n$ is defined as $a \cdot b = b^*a$, where b^* is the conjugation transpose of b.

The organization of this paper is as follows. In Section 2, we present a complex discrete program along with its CSDP relaxation for the balanced Max-3-Uncut problem. We offer a CSDP rounding approximation algorithm in Section 3, followed by the analysis in Section 4. Finally, some discussions are given in Section 5.

2   Formulation

In this section, we first introduce a complex discrete program for the balanced Max-3-Uncut problem. Then, we relax this program to obtain a CSDP relaxation, whose optimal solution will be rounded to obtain a discrete (complex) solution. For any given complex number z, we adopt the standard notations $\text{Re}(z)$ and $\text{Im}(z)$ to represent the real and imaginary parts of z, respectively.

Formally, in the balanced Max-3-Uncut problem, we are given a graph $G = (V, E)$ with vertex set V whose size $n := |V|$ is a multiple of 3 and edge set E. There is also a weight function $w : E \to \mathbb{R}_+$ defined on E. The goal is to partition V into three subsets S_1, S_2, and S_3 with equal cardinality such that the total weight of the edges from the same subsets is maximized; that is,

$$\max_{\substack{(S_1, S_2, S_3) \in \mathcal{P}(V) \\ |S_1| = |S_2| = |S_3|}} \sum_{i,j \in S_1} w_{ij} + \sum_{i,j \in S_2} w_{ij} + \sum_{i,j \in S_3} w_{ij},$$

where

$$\mathcal{P}(V) := \{(S_1, S_2, S_3) : S_1 \cup S_2 \cup S_3 = V, \text{ and } S_k \cap S_l = \emptyset \text{ for all } k \neq l\}$$

and we define $w_{ij} = 0$ if $(i, j) \notin E$ for completeness. We remark that if $|V|$ is not divisible by 3, one can add isolated vertices if necessary to guarantee this divisibility.

The balanced Max-3-Uncut problem can be formulated as the following.

$$\max \frac{1}{3} \sum_{i<j} w_{ij}(1 + 2\text{Re}(y_i \cdot y_j))$$

$$\text{s. t. } \sum_{i \in V} y_i = 0, \tag{1}$$

$$y_i \in \{1, \omega, \omega^2\}, \quad \forall i \in V.$$

Recall that $(1, \omega, \omega^2)$ are the three distinct complex cubic roots of unity, where

$$\omega = e^{i\frac{2\pi}{3}} = -\frac{1}{2} + i\frac{\sqrt{3}}{2}, \qquad \omega^2 = e^{i\frac{4\pi}{3}} = -\frac{1}{2} - i\frac{\sqrt{3}}{2}.$$

In the above program, the variable y_i represents the subset to which vertex i is assigned: If $y_i = 1$, vertex i is assigned to S_1; if $y_i = \omega$, vertex i is assigned to S_2; and finally if $y_i = \omega^2$, vertex i is assigned to S_3.

Lemma 1. *The program (1) is a valid representation of the balanced Max-3-Uncut problem.*

Proof. We prove this lemma by the following two steps.

First, we consider two cases by showing that the weight of (i, j) is counted only when vertices i and j are in the same subset.

Case 1. Vertices i and j are in the same subset. We obtain the desired result from the following three subcases.

$$\frac{1}{3}(1 + 2\operatorname{Re}(y_i \cdot y_j)) = \frac{1}{3}(1 + 2) = 1, \text{ when } y_i = y_j = 1;$$

$$\frac{1}{3}(1 + 2\operatorname{Re}(y_i \cdot y_j)) = \frac{1}{3}(1 + 2\operatorname{Re}(\omega \cdot \omega)) = 1, \text{ when } y_i = y_j = \omega;$$

$$\frac{1}{3}(1 + 2\operatorname{Re}(y_i \cdot y_j)) = \frac{1}{3}(1 + 2\operatorname{Re}(\omega^2 \cdot \omega^2)) = 1, \text{ when } y_i = y_j = \omega^2.$$

Case 2. Vertices i and j are in different subsets. Due to symmetry of i and j, we consider the following three subcases.

$$\frac{1}{3}(1 + 2\operatorname{Re}(y_i \cdot y_j)) = \frac{1}{3}(1 + 2\operatorname{Re}(1 \cdot \omega)) = 0, \text{ when } y_i = 1, y_j = \omega;$$

$$\frac{1}{3}(1 + 2\operatorname{Re}(y_i \cdot y_j)) = \frac{1}{3}(1 + 2\operatorname{Re}(1 \cdot \omega^2)) = 0, \text{ when } y_i = 1, y_j = \omega^2;$$

$$\frac{1}{3}(1 + 2\operatorname{Re}(y_i \cdot y_j)) = \frac{1}{3}(1 + 2\operatorname{Re}(\omega \cdot \omega^2)) = 0, \text{ when } y_i = \omega, y_j = \omega^2.$$

Second, we will show that the first constraint of (1) is equivalent to

$$|\{i : y_i = 1\}| = |\{j : y_j = \omega\}| = |\{k : y_k = \omega^2\}| \tag{2}$$

under the condition $y_i \in \{1, \omega, \omega^2\}$ for all $i \in V$. Although this fact is already used in [15], we give a formal proof for completeness. Note that ω and ω^2 are conjugate to each other. On one hand, suppose that $\{y_i\}$ satisfies (2). We have

$$\sum_{j:y_j=\omega} y_j + \sum_{k:y_k=\omega^2} y_k = -|\{j : y_j = \omega\}| = -|\{k : y_k = \omega^2\}|.$$

The above equalities indicate

$$\sum_{i\in V} y_i = \sum_{i:y_i=1} y_i + \sum_{j:y_j=\omega} y_j + \sum_{k:y_k=\omega^2} y_k$$
$$= |\{i : y_i = 1\}| - |\{j : y_j = \omega\}|$$
$$= 0.$$

On the other hand, assume that $\{y_i\}$ satisfies the first constraint of (1). Since the summation of y_i is 0 which is a real number, we must have $|\{j : y_j = \omega\}| = |\{k : y_k = \omega^2\}|$. Then,

$$\sum_{j:y_j=\omega} y_j + \sum_{k:y_k=\omega^2} y_k = -|\{j : y_j = \omega\}| = -|\{k : y_k = \omega^2\}|.$$

Again from the first constraint of (1), we have

$$0 = \sum_{i:y_i=1} y_i + \sum_{j:y_j=\omega} y_j + \sum_{k:y_k=\omega^2} y_k = |\{i : y_i = 1\}| - |\{j : y_j = \omega\}|.$$

The above two equalities indicate that $\{y_i\}$ satisfies (2).   □

Since $y_i \in \{1, \omega, \omega^2\}$ for all $i \in V$, we must have

$$y_i \cdot y_j + y_j \cdot y_i \geq -1, \qquad \forall i, j \in V,$$
$$\omega \cdot (y_i \cdot y_j) + \omega^2 \cdot (y_j \cdot y_i) \geq -1, \qquad \forall i, j \in V,$$
$$\omega^2 \cdot (y_i \cdot y_j) + \omega \cdot (y_j \cdot y_i) \geq -1, \qquad \forall i, j \in V,$$

which can be rewritten as

$$\mathrm{Re}(y_i \cdot y_j) \geq -\frac{1}{2}, \qquad \forall i, j \in V,$$

$$\mathrm{Re}(\omega \cdot (y_i \cdot y_j)) \geq -\frac{1}{2}, \qquad \forall i, j \in V,$$

$$\mathrm{Re}(\omega^2 \cdot (y_i \cdot y_j)) \geq -\frac{1}{2}, \qquad \forall i, j \in V.$$

By adding the above extra inequalities into (1) (cf. [10]), we get the CSDP relaxation as follows.

$$\max \frac{1}{3} \sum_{i<j} w_{ij}(1 + 2\mathrm{Re}(v_i \cdot v_j))$$

$$\text{s. t. } \mathrm{Re}(v_i \cdot v_j) \geq -\frac{1}{2}, \qquad \forall i, j \in V,$$

$$\mathrm{Re}(\omega \cdot (v_i \cdot v_j)) \geq -\frac{1}{2}, \qquad \forall i, j \in V, \qquad (3)$$

$$\mathrm{Re}(\omega^2 \cdot (v_i \cdot v_j)) \geq -\frac{1}{2}, \qquad \forall i, j \in V,$$

$$\sum_{i,j} v_i \cdot v_j = 0,$$

$$\| v_i \| = 1, \qquad \forall i \in V,$$

$$v_i \in \mathbb{C}^n, \qquad \forall i \in V.$$

The above CSDP is polynomially solvable as it is equivalent to an SDP with double size comparing with (3) (cf. [10]).

3   Algorithm

Based on the CSDP relaxation (3), we propose the following algorithm for the balanced Max-3-Uncut problem.

Algorithm 1

Step 1. *Solve the CSDP (3) to obtain an optimal solution* $\{v_i\}$, *leading to a complex semidefinite matrix* $V := (v_i \cdot v_j)$.

Step 2. *For a given parameter* $\theta \in [0, 1]$, *choose a random vector* $\xi \sim N(0, \theta V + (1 - \theta)I)$, *where* I *is the* $n \times n$ *identity matrix.*

Step 3. *Define*

$$
\hat{y}_i = \begin{cases} 1, & \mathrm{Arg}(\xi_i) \in [0, \frac{2}{3}\pi); \\ \omega, & \mathrm{Arg}(\xi_i) \in [\frac{2}{3}\pi, \frac{4}{3}\pi); \\ \omega^2, & \mathrm{Arg}(\xi_i) \in [\frac{4}{3}\pi, 2\pi). \end{cases}
$$

Let $S_1 := \{i : \hat{y}_i = 1\}$, $S_2 := \{i : \hat{y}_i = \omega\}$, *and* $S_3 := \{i : \hat{y}_i = \omega^2\}$.

Step 4. *Assume, without loss of generality,* $|S_1| \geq |S_2| \geq |S_3|$. *In this step, we will perform a size-adjustment operation to equalize the cardinality of the three sets dependening on two cases. Initialize* $\hat{S}_\ell = S_\ell$ *(*$\ell = 1, 2, 3$*). Denote the final partition with equal cardinality as* $\tilde{S}_1$, $\tilde{S}_2$, *and* $\tilde{S}_3$.

Case 4.1. *If* $|S_1| \geq |S_2| \geq \frac{n}{3} \geq |S_3|$, *then iteratively, perform the following operations (i)-(ii) until* $|\hat{S}_\ell| = \frac{n}{3}$ *for each* $\ell = 1, 2$:

(i) *Sort the vertices in* $\hat{S}_\ell$ *such that* $\delta(i_1) \geq \ldots \geq \delta(i_{|\hat{S}_\ell|})$ *where* $\delta(i) = \sum_{i' \in \hat{S}_\ell} w_{i'i}$ *(*$i \in \hat{S}_\ell$*).*

(ii) *Move the point* $i_{|\hat{S}_\ell|}$ *from* $\hat{S}_\ell$ *to* $\hat{S}_3$; *namely,* $\hat{S}_\ell = \hat{S}_\ell \backslash \{i_{|\hat{S}_1|}\}$, *and* $\hat{S}_3 = \hat{S}_3 \cup \left\{i_{|\hat{S}_\ell|}\right\}$.

Case 4.2. *If* $|S_1| \geq \frac{n}{3} \geq |S_2| \geq |S_3|$, *then iteratively, perform the following operations (i)-(ii) until* $|\hat{S}_\ell| = \frac{n}{3}$ *for each* $\ell = 2, 3$:

(i) *Sort the vertices in* $\hat{S}_1$ *such that* $\delta(i_1) \geq \ldots \geq \delta(i_{|\hat{S}_1|})$ *where* $\delta(i) = \sum_{i' \in \hat{S}_1} w_{i'i}$ *(*$i \in \hat{S}_1$*).*

(ii) *Move the point* $i_{|\hat{S}_1|}$ *from* $\hat{S}_1$ *to* $\hat{S}_\ell$; *namely,* $\hat{S}_1 = \hat{S}_1 \backslash \{i_{|\hat{S}_1|}\}$, *and* $\hat{S}_\ell = \hat{S}_\ell \cup \left\{i_{|\hat{S}_1|}\right\}$.

4   Analysis

To establish the approximation ratio of Algorithm 1 in Theorem 9, we need several lemmas (Lemmas 2-8) whose proofs are omitted in this conference version.

First, based on Step 2 of Algorithm 1, we consider two cases (Lemmas 2-3) to estimate the relationship of the weights before and after the size-adjustment.

Lemma 2. *In Case 4.1. at Step 4 of Algorithm 1, we have*

$$
\frac{W\left(\tilde{S}_1, \tilde{S}_2, \tilde{S}_3\right)}{W(S_1, S_2, S_3)} \geq \frac{1}{81 x_1^2 x_2^2},
$$

where $x = (x_1, x_2, x_3) = \left(\frac{|S_1|}{n}, \frac{|S_2|}{n}, \frac{|S_3|}{n} \right)$.

□

Similar to the analysis in Lemma 2, we have the following result for Case 4.2 in Algorithm 1.

Lemma 3. *In Case 4.2. at Step 4 of Algorithm 1, we have*

$$\frac{W\left(\tilde{S}_1, \tilde{S}_2, \tilde{S}_3\right)}{W(S_1, S_2, S_3)} \geq \frac{1}{9x_1^2},$$

where $x = (x_1, x_2, x_3) = \left(\frac{|S_1|}{n}, \frac{|S_2|}{n}, \frac{|S_3|}{n} \right)$.

□

Second, we estimate the quality for (S_1, S_2, S_3) (Lemmas 4-5). Denote W^* as the optimal value of (3). The following lemma is given in [10] (see also [22]), which estimates the real part of the expected value of $\hat{y}_i \cdot \hat{y}_j$ for each i, j.

Lemma 4. *([10, 22]) The real part of the expected value of $\hat{y}_i \cdot \hat{y}_j$ is*

$$\frac{9}{8\pi^2} \left[\arccos^2\left(-\mathrm{Re}(\theta v_i \cdot v_j)\right) - \frac{1}{2} \arccos^2(-\mathrm{Re}(\omega \cdot (\theta v_i \cdot v_j))) \right.$$
$$\left. - \frac{1}{2} \arccos^2(-\mathrm{Re}(\omega^2 \cdot (\theta v_i \cdot v_j))) \right].$$

□

By Lemma 4, we can obtain the following lemma.

Lemma 5. *For a given $\theta \in [0, 1]$, the ratio of the expected weight of (S_1, S_2, S_3) and W^* is no less than $\alpha(\theta)$, where $\alpha(\theta)$ is*

$$\min \ g(\theta, z_1, z_2)$$
$$\text{s. t. } -\frac{1}{2} \leq z_1 \leq 1,$$
$$-\frac{1}{2} \leq -\frac{1}{2}z_1 + \frac{\sqrt{3}}{2}z_2 \leq 1,$$
$$-\frac{1}{2} \leq -\frac{1}{2}z_1 - \frac{\sqrt{3}}{2}z_2 \leq 1,$$
$$z_1^2 + z_2^2 \leq 1.$$

In the above,

$$g(\theta, z_1, z_2) := \frac{1}{1 + 2z_1} \left\{ 1 + \frac{9}{4\pi^2} \left[\arccos^2(-\theta z_1) - \frac{1}{2} \arccos^2\left(\frac{1}{2}\theta z_1 - \frac{\sqrt{3}}{2}\theta z_2 \right) \right. \right.$$
$$\left. \left. - \frac{1}{2} \arccos^2\left(\frac{1}{2}\theta z_1 + \frac{\sqrt{3}}{2}\theta z_2 \right) \right] \right\}.$$

□

Third, we will estimate the violation of the equal cardinality constraint (Lemma 6). We need some notations first. Introduce a new random variable C such that

$$C := |S_1||S_2| + |S_1||S_3| + |S_2||S_3|.$$

From $|S_1| + |S_2| + |S_3| = n$ and the AM-GM inequality, we have that $C \leq \frac{n^2}{3}$. Let $e = (1, 1, \ldots, 1)^T$ be the all-one vector. Noting that $\hat{y}$ is an discrete (not necessarily feasible) solution before the adjustment, we obtain

$$e^T \hat{y} = |S_1| + \omega|S_2| + \omega^2|S_3|.$$

Then, we have

$$
\begin{aligned}
&|S_1||S_2| + |S_1||S_3| + |S_2||S_3| \\
&= |S_1|^2 + |S_2|^2 + |S_3|^2 - \left(e^T \hat{y}\right)^* \left(e^T \hat{y}\right) \\
&= \left(|S_1| + |S_2| + |S_3|\right)^2 - \left(e^T \hat{y}\right)^* \left(e^T \hat{y}\right) - 2\left(|S_1||S_2| + |S_1||S_3| + |S_2||S_3|\right) \\
&= n^2 - \left(e^T \hat{y}\right)^* \left(e^T \hat{y}\right) - 2\left(|S_1||S_2| + |S_1||S_3| + |S_2||S_3|\right),
\end{aligned}
$$

which implies

$$C = |S_1||S_2| + |S_1||S_3| + |S_2||S_3| = \frac{1}{3}n^2 - \frac{1}{3}\left(e^T \hat{y}\right)^* \left(e^T \hat{y}\right).$$

Lemma 6 below is given in [15].

Lemma 6. ([15]) *Define*

- $f(x) := \frac{9}{8\pi^2} \left(\arccos^2(-x) - \arccos^2\left(\frac{1}{2}x\right)\right).$
- $c(\theta) := \min\limits_{-\frac{1}{2} \leq x \leq 1} \frac{f(\theta) - f(\theta x)}{1 - x}.$
- $\beta(\theta) := \left(1 - \frac{1}{n}\right)(1 - f(\theta) + c(\theta)).$
- $C^* := \frac{n^2}{3}.$

For a given $\theta \in (0, 1)$, we can estimate the ratio between C and C^* as follows:

$$E\left[\frac{C}{C^*}\right] \geq \beta(\theta).$$

$\square$

Fourth, in Lemmas 7-8 below, we solve the optimization problem $\min_{x \in \Delta} R(x; \theta, \gamma)$, where

- $\mu(x) := \frac{C}{C^*} = 3\left(x_1 x_2 + (x_1 + x_2)(1 - x_1 - x_2)\right).$
- $r(x) := \begin{cases} \dfrac{1}{81x_1^2 x_2^2}, & \text{if } x_2 \geq \dfrac{1}{3}; \\ \dfrac{1}{9x_1^2}, & \text{if } x_2 < \dfrac{1}{3}. \end{cases}$

$$- \Delta := \left\{ x = (x_1, x_2, x_3) \left| \begin{array}{l} \sum_{i=1}^{3} x_i = 1 \\ x_1 \geq x_2 \geq x_3 \geq 0 \end{array} \right. \right\}.$$

$- R(x; \theta, \gamma) := r(x)(\alpha(\theta) + \gamma\beta(\theta) - \gamma\mu(x)), \ \gamma > 0.$

Lemma 7. *When* $\gamma \in \left[\frac{8\alpha(\theta)}{9-8\beta(\theta)}, \frac{\alpha(\theta)}{1-\beta(\theta)} \right]$, *the minimum of* $R(x; \theta, \gamma)$ *for all* $x \in \{x \in \Delta | x_2 \geq \frac{1}{3}\}$ *is*

$$R_1(\theta, \gamma) := \frac{27}{256} \gamma^4 \left(\frac{1 + \sqrt{1 - \frac{8(\alpha(\theta)+\gamma\beta(\theta))}{9\gamma}}}{\alpha(\theta) + \gamma\beta(\theta)} \right)^3 \left(1 - 3\sqrt{1 - \frac{8(\alpha(\theta) + \gamma\beta(\theta))}{9\gamma}} \right).$$

□

Lemma 8. *When* $\gamma \in \left[\frac{8\alpha(\theta)}{9-8\beta(\theta)}, \frac{\alpha(\theta)}{1-\beta(\theta)} \right]$, *the minimum of* $R(x; \theta, \gamma)$ *for all* $x \in \{x \in \Delta | x_2 \leq \frac{1}{3}\}$ *is*

$$R_2(\theta, \gamma) := \gamma \frac{\alpha(\theta) + \gamma\beta(\theta) - \gamma}{4(\alpha(\theta) + \gamma\beta(\theta)) - 3\gamma}.$$

□

Finally, we are ready to present the main result in this section.

Theorem 9. *The approximation ratio of Algorithm 1 is*

$$\max_{\theta \in (0,1)} \max_{\gamma \in \left[\frac{8\alpha(\theta)}{9-8\beta(\theta)}, \frac{\alpha(\theta)}{1-\beta(\theta)} \right]} \min \{R_1(\theta, \gamma), R_2(\theta, \gamma)\}.$$

Proof. Define

$$z(\gamma) := \frac{W(S_1, S_2, S_3)}{W^*} + \gamma \frac{C}{C^*}.$$

From Lemmas 5-6, we have

$$E\left[\frac{W(S_1, S_2, S_3)}{W^*} \right] \geq \alpha(\theta),$$

and

$$E\left[\frac{C}{C^*} \right] \geq \beta(\theta).$$

The above two inequalities indicate

$$E[z(\gamma)] = E\left[\frac{W(S_1, S_2, S_3)}{W^*} + \gamma \frac{C}{C^*} \right] \geq \alpha(\theta) + \gamma\beta(\theta), \ \forall \gamma > 0.$$

For any $\epsilon > 0$, we can run Algorithm 1 independently $T = O\left(\frac{1}{\epsilon} \log\left(\frac{1}{\epsilon}\right)\right)$ times to output the maximum value of $z(\gamma)$ by $z_T(\gamma)$. We may assume that $z_T(\gamma) \geq \alpha(\theta) + \gamma\beta(\theta)$ almost surely (cf. [12]). Then, we have

$$\frac{W(S_1, S_2, S_3)}{W^*} \geq \alpha(\theta) + \gamma\beta(\theta) - \gamma \frac{C}{C^*}.$$

Recall the definitions of $\mu(x)$, $r(x)$, and $R(x; \theta, \gamma)$. From the last inequality and Lemmas 2-3, we have

$$\frac{W(\tilde{S}_1, \tilde{S}_2, \tilde{S}_3)}{W^*} \geq r(x)\frac{W(S_1, S_2, S_3)}{W^*}$$
$$\geq r(x)(\alpha(\theta) + \gamma\beta(\theta) - \gamma\mu(x))$$
$$= R(x; \theta, \gamma).$$

Thus, the approximation ratio of Algorithm 1 is

$$\max_{\gamma, \theta} \min_{x \in \Delta} R(x; \theta, \gamma).$$

Now Lemmas 7-8 imply the desired result.   □

Setting $\theta := 0.3115$ and $\gamma = 12.1855$, then, we obtain the approximation ratio of Algorithm 1 is 0.3456. In this case, $\alpha(\theta) = 0.4521$ and $\beta(\theta) = 0.9952$.

5 Discussions

In this paper, we propose a CSDP rounding approximation algorithm for the balanced Max-3-Uncut problem with the approximation ratio 0.3456. There are several directions for future research.

- It will be very interesting to develop a complex SDPH and its corresponding rounding technique, extending the real case in [14]. We believe that this technique should significantly improve the approximation ratios for the Max-3-Section and balanced Max-3-Uncut problems.
- We believe that our technique for the balanced Max-3-Uncut problem can be extended to the general capacitated Max-3-Uncut problem by extending the greedy adjustment and the corresponding analysis.
- It is also of interest to consider other NP-hard problems using the CSDP rounding and investigate the inapproximability of the CSDP relaxation.

Acknowledgments. The authors would like to thank the three anonymous reviewers, Yishui Wang, and Peng Zhang for their helpful comments on an earlier version of this paper. This work was partially done while the first author was a visiting doctorate student at the Department of Applied Mathematics, Beijing University of Technology and supported in part by NSF of China (No. 11071268). The research of the second author is supported by NSF of China (No. 11371001) and China Scholarship Council. The third author's research is supported by the Natural Sciences and Engineering Research Council of Canada (NSERC) grant 283106.

References

1. Andersson, G.: An Approximation Algorithm for Max p-Section. In: Meinel, C., Tison, S. (eds.) STACS 1999. LNCS, vol. 1563, pp. 237–247. Springer, Heidelberg (1999)
2. Austrin, P., Benabbas, S., Georgiou, K.: Better Balance by Being Biased: A 0.8776-Approximation for Max Bisection. In: 24th Annual ACM-SIAM Symposium on Discrete Algorithms, pp. 277–294. SIAM Press, New Orleans (2013)
3. Choudhury, S., Gaur, D., Krishnamurti, R.: An Approximation Algorithm for Max k-Uncut with Capacity Constraints. Optim. 61, 143–150 (2012)
4. Doids, Y., Guruswami, V., Khanna, S.: The 2-Catalog Segmentation Problem. In: 17th Annual ACM-SIAM Symposium on Discrete Algorithms, pp. 897–898. SIAM Press, Baltimore (1999)
5. Feige, U., Langberg, M.: The RPR^2 Rounding Technique for Semidefinite Programs. J. Algorithms 60, 1–23 (2006)
6. Feo, T., Goldschmidt, O., Khellaf, M.: One-Half Approximation Algorithms for the k-Partition Problem. Oper. Res. 40, S170–S173 (1992)
7. Frieze, A.M., Jerrum, M.: Improved Approximation Algorithms for MAX k-CUT and MAX BISECTION. Algorithmica 18, 67–81 (1997)
8. Goldschmidt, O., Hochbaum, D.S.: A Polynomail Algorithm for the k-Cut Problem for Fixed k. Math. Oper. Res. 19, 24–37 (1994)
9. Goemans, M.X., Williamson, D.P.: Improved Approximation Algorithms for Maximum Cut and Satisfiability Problems Using Semidefinite Programming. J. ACM 42, 1115–1145 (1995)
10. Goemans, M.X., Williamson, D.P.: Approximation Algorithms for MAX-3-CUT and Other Problems via Complex Semidefinite Programming. J. Comput. Syst. Sci. 68, 442–470 (2004)
11. Halperin, E., Zwick, U.: A Unified Framework for Obtaining Improved Approximation Algorithms for Maximum Graph Bisection Problems. Random Structures & Algorithms 20, 382–402 (2002)
12. Han, Q., Ye, Y., Zhang, J.: An Improved Rounding Method and Semidefinite Programming Relaxation for Graph Partition. Math. Program. Ser. B 92, 509–535 (2002)
13. Kleinberg, J., Papadimitriou, C., Raghavan, P.: Segmentation Problems. In: 30th Annual ACM Symposium on the Theory of Computing, pp. 473–482. ACM Press, Dallas (1998)
14. Lasserre, J.B.: An Explicit Equivalent Positive Semidefinite Program for Nonlinear 0-1 Programs. SIAM J. Optim. 12, 756–769 (2002)
15. Ling, A.-F.: Approximation Algorithms for Max 3-Section Using Complex Semidefinite Programming Relaxation. In: Du, D.-Z., Hu, X., Pardalos, P.M. (eds.) COCOA 2009. LNCS, vol. 5573, pp. 219–230. Springer, Heidelberg (2009)
16. Raghavendra, P., Tan, N.: Approximating CSPs with Global Cardinality Constraints Using SDP Hierarchies. In: 23rd Annual ACM-SIAM Symposium on Discrete Algorithms, pp. 373–387. SIAM Press, Kyoto (2012)
17. Sahni, S., Gonzalez, T.: P-Complete Approximation Problems. J. ACM 23, 555–565 (1976)
18. Wu, C., Du, D., Xu, D.: An Improved Semidefinite Programming Hierarchies Rounding Approximation Algorithm for Maximum Graph Bisection Problems. J. Comb. Optim. (2013), doi 10.1007/s10878-013-9673-1

19. Xu, D., Han, J., Huang, Z., Zhang, L.: Improved Approximation Algorithms for MAX $n/2$-DIRECTED-BISECTION and MAX $n/2$-DENSE-SUBGRAPH. J. Glob. Optim. 27, 399–410 (2003)

20. Ye, Y.: A.699-Approximation Algorithm for Max-Bisection. Math. Program. 90, 101-111 (2001)

21. Zwick, U.: Outward Rotations: A Tool for Rounding Solutions of Semidefinite Programming Relaxations, with Applications to MAX CUT and Other Problems. In: 31st Annual ACM Symposium on Theory of Computing, pp. 679–687. ACM Press, Atlanta (1999)

22. Zhang, S., Huang, Y.: Complex Quadratic Optimization and Semidefinite Programming. SIAM J. Optim. 16, 871–890 (2006)

Primal-Dual Approximation Algorithms for Submodular Vertex Cover Problems with Linear/Submodular Penalties

Dachuan Xu[1,*], Fengmin Wang[1], Donglei Du[2], and Chenchen Wu[1]

[1] College of Science, Tianjin University of Technology, Tianjin 300384, China
xudc@bjut.edu.cn
[2] Faculty of Business Administration, University of New Brunswick,
Fredericton, NB E3B 5A3, Canada

Abstract. In this paper, we introduce two variants of the submodular vertex cover problem, namely, the submodular vertex cover problems with linear and submodular penalties, for which we present two primal-dual approximation algorithms with approximation ratios of 2 and 4, respectively. Implementing the primal-dual framework directly on the dual programs of the linear program relaxations for these two variants cannot guarantee the dual ascending process terminates in polynomial time. To overcome this difficulty, we relax the two dual programs to slightly weaker versions which lead to two primal-dual approximation algorithms with the aforeclaimed approximation ratios.

Keywords: Primal-dual, Approximation algorithm, Submodular function, Vertex cover problem.

1 Introduction

1.1 Vertex Cover and Submodular Function

The vertex cover problem is a fundamental and widely investigated problem in combinatorial optimization (cf. [17]). It is well-known that the vertex cover problem is NP-hard (cf. [23]), and cannot be solved in polynomial time unless $P = NP$. The vertex cover problem is defined on an undirected graph $G = (V, E)$ with vertex set V and edge set E. Each vertex $i \in V$ has an associated nonnegative cost $c(i)$. A vertex subset $S \subseteq V$ is called a vertex cover in G if every edge in E is incident to vertex in S —that is, the vertex subset covers all the edges. The objective is to find a vertex cover with the minimum cost.

A submodular function f is defined on a collection of subsets, and satisfies $f(X \cup Y) + f(X \cap Y) \leq f(X) + f(Y)$, for any two subsets X and Y. Submodular functions arise naturally in the fields of operations research, computer science, and economics (cf. [11]) due to the decreasing marginal return property. There have been extensive work on submodular function optimization (cf.

---

* Corresponding author.

Z. Cai et al. (Eds.): COCOON 2014, LNCS 8591, pp. 336–345, 2014.

[10, 11],[13],[19],[26]). Lovász [25] establishes a direct connection between submodularity and convexity: the submodularity of a set function can be characterized by the convexity of a continuous function obtained by extending the set function in an appropriate manner.

Many combinatorial optimization problems have their submodular counterparts. The classical vertex cover problem was extended to the submodular case by Iwata and Nagano [20], where, given a nonnegative submodular function $C : 2^V \to R_+$, the objective is to find a vertex cover $S \subseteq V$ that minimizes the cost $C(S)$. The notion of penalty has been considered earlier in the context of the Steiner tree, TSP, and Facility location (see [6–8],[14],[24] and references therein). In this paper, we relax further the requirement that a vertex cover has to cover all the edges by penalizing the uncovered edges, resulting in the submodular vertex cover problem with penalties (SVCWP). We consider two variants depending on whether the penalty cost function is linear or submodular, namely the submodular vertex cover problem with linear penalties (SVCLP) and the submodular vertex cover problem with submodular penalties (SVCSP).

1.2 Related Work

Over the last decades, the vertex cover problem has received a particular attention. There are a number of approximation algorithms that have been proposed for this problem and many variants exist with various performance guarantees (cf. [17]).

Several techniques have been developed in designing approximation algorithms for the vertex cover problem, including LP-rounding, primal-dual, and greedy techniques. Hochbaum [16] give an LP-rounding 2-approximation algorithm which is the first constant approximation algorithm for this problem. Bar-Yehuda and Even [2] propose a primal-dual 2-approximation algorithm. Depending on the number of the vertices or the maximum degree of the graph, there are some approximation algorithms that achieve $2 - o(1)$ ratio (cf. [3],[15],[21]). On the other hand, Knote and Segev [22] prove that the lower bound is $2 - \epsilon$ for any $\epsilon > 0$ under the unique game conjecture.

Various generalizations and variants of the vertex cover problem have been studied (cf. [1],[12]). We only review some results related to the SVCWP. Hochbaum [18] introduces the generalized vertex cover problem and presents an LP-rounding 2-approximation algorithm. This problem (a.k.a. the prize-collecting vertex cover problem [4]) is essentially the vertex cover problem with linear penalties in our terminology. Using primal-dual technique, Bar-Yehuda and Rawitz [5] propose a 2-approximation algorithm. Taking the maximum degree d of the given graph into consideration, Bar-Yehuda et al. [4] give a local-ratio $(2 - 2/d)$-approximation algorithm. Iwata and Nagano [20] introduce the submodular vertex cover problem and propose a convex programming rounding 2-approximation algorithm.

1.3 Our Contribution

The main contributions of this paper are summarized as follows.

- We firstly introduce the SVCWP which generalizes the classic vertex cover problem, submodular vertex cover problem [20], and generalized vertex cover problem [18].
- We present two primal-dual approximation algorithms with 2 and 4 ratios for the SVCLP and SVCSP respectively. Implementing the primal-dual framework directly on the dual programs of the linear program relaxations for the SVCLP and SVCSP cannot control the dual ascending process in polynomial time. To overcome this difficulty, we relax these two dual programs to slightly weaker versions which lead to two primal-dual approximation algorithms with the claimed approximation ratios.

1.4 Organization

The paper is organized as follows. We offer two primal-dual approximation algorithms with ratios of 2 and 4 for the SVCLP and SVCSP in Sections 2 and 3 respectively. We conclude the paper in Section 4. All proofs are deferred to the Appendix.

2 Primal-dual Approximation Algorithm for the SVCLP

2.1 Formulation

In the SVCLP, we are given an undirected graph $G = (V, E)$ with vertex set V and edge set E, and a nonnegative submodular function $C : 2^V \to R_+$ with $C(\emptyset) = 0$. Each subset $S \subseteq V$ has a covering cost $C(S)$ and each edge $e \in E$ has a penalty cost p_e. The objective is to select a vertex subset to cover some edges and penalize the uncovered edges such that the total cost including covering and penalty is minimized. Let us denote $e := (i, j)$ for each edge $e \in E$ such that i and j are the two adjacent vertices of edge e. When there is no confusion, we abuse $e = (i, j)$ to denote a vertex subset consisting of the two adjacent vertices i and j.

To formulate the SVCLP as an integer linear program, we introduce two types of binary variables: X_S for each subset $S \subseteq V$ and z_e for each edge $e \in E$. For an arbitrarily subset $S \subseteq V$, X_S indicates whether S is selected to cover some edges. For an arbitrarily $e \in E$, z_e indicates whether e is penalized. Then, we have the following integer linear program for the SVCLP.

$$OPT_{SVCLP} := \min \sum_{S \subseteq V} C(S) X_S + \sum_{e \in E} p_e z_e$$

$$\text{s. t.} \sum_{S \subseteq V : S \cap e \neq \emptyset} X_S + z_e \geq 1, \qquad \forall e = (i, j) \in E, \qquad (1)$$

$$X_S, z_e \in \{0, 1\}, \qquad \forall S \subseteq V, e \in E.$$

The first constraint of (1) guarantees that each edge $e \in E$ is either covered by some vertex subset S or penalized. Due to the submodularity of $C(\cdot)$, there must exist exactly one $S \subseteq V$ such that $X_S = 1$ in the optimal solution of (1). Relaxing the integer constraints, we obtain

$$\min \quad \sum_{S \subseteq V} C(S) X_S + \sum_{e \in E} p_e z_e$$

$$\text{s. t.} \quad \sum_{S \subseteq V : S \cap e \neq \emptyset} X_S + z_e \geq 1, \qquad \forall e = (i,j) \in E, \tag{2}$$

$$X_S, z_e \geq 0, \qquad \forall S \subseteq V, e \in E.$$

The dual program of (2) is

$$\max \quad \sum_{e \in E} y_e$$

$$\text{s. t.} \quad \sum_{e \in E : e \cap S \neq \emptyset} y_e \leq C(S), \qquad \forall S \subseteq V, \tag{3}$$

$$y_e \leq p_e, \qquad \forall e \in E,$$

$$y_e \geq 0, \qquad \forall e \in E.$$

If we adopt the standard primal-dual framework to ascend the dual variables, we will deal with the ratio between two submodular functions which cannot be computed in polynomial time in general. Instead of (3), we propose the following weak dual program.

$$\max \quad \sum_{e \in E} y_e$$

$$\text{s. t.} \quad \sum_{e \in E : |e \cap S| = 1} \frac{1}{2} y_e + \sum_{e \in E : e \subseteq S} y_e \leq C(S), \quad \forall S \subseteq V, \tag{4}$$

$$y_e \leq p_e, \quad \forall e \in E,$$

$$y_e \geq 0, \qquad \forall e \in E.$$

We present the relationship among (2)-(4) in the following lemma.

Lemma 1. *Given any feasible solution $\{y\}$ for (4), $\{\frac{1}{2}y\}$ is feasible for (3). Furthermore, $\frac{1}{2}\sum_{e \in E} y_e$ is a lower bound for the primal linear program (2).*

2.2 Algorithm

Based on the weak dual program (4), we give a primal-dual algorithm for the SVCLP as follows.

Algorithm 1

STEP 0. *Introduce a notion of time t, and start the algorithm at time $t = 0$. Let $\tilde{S}$ denote the vertex subset which is used to cover edges, along with its edge set $\tilde{E}(\tilde{S})$ covered by $\tilde{S}$. Let $\tilde{Q}$ denote the temporarily penalized edge subset. Moreover, denote F as the edge set in which the corresponding dual variables stop increasing, that is $F = \tilde{E}(\tilde{S}) \cup \tilde{Q}$. We say that an edge is frozen if it belongs to F; otherwise the edge is unfrozen. Initially, all dual variables y_e are set to 0 and all edges are unfrozen. Initialize $\tilde{S} := \emptyset$, $\tilde{E}(\tilde{S}) := \emptyset$, $\tilde{Q} := \emptyset$, and $F := \emptyset$.*

STEP 1. *Increase the dual variables y_e's for all unfrozen edges $e \in E$ uniformly at unit rate with time t. One of the following events may occur:*

Event 1. *There exists a vertex subset $S \subseteq V$ such that*

$$\left(\sum_{e \in E \setminus F : |e \cap S| = 1} \frac{1}{2} t + \sum_{e \in E \setminus F : e \subseteq S} t \right)$$
$$+ \left(\sum_{e \in F : |e \cap S| = 1} \frac{1}{2} y_e + \sum_{e \in F : e \subseteq S} y_e \right)$$
$$= C(S).$$

In this case, freeze those unfrozen edges in S by setting $F := F \cup \{e \in E \setminus F : e \cap S \neq \emptyset\}$. Update $\tilde{S} := \tilde{S} \cup S$, and $\tilde{E}(\tilde{S}) := \tilde{E}(\tilde{S}) \cup \{e \in E : e \cap S \neq \emptyset\}$.

Event 2. *There exists an edge $e \in E \setminus F$ such that $t = p_e$. Freeze and temporarily penalize this edge by setting $F := F \cup \{e\}$ and $\tilde{Q} := \tilde{Q} \cup \{e\}$.*

If several events occur simultaneously, the algorithm executes them in an arbitrary order. Repeat the above process until all edges are frozen.

STEP 2. *Let $\hat{S}$ and $\hat{Q}$ denote the final vertex subset used to cover edges and the final penalized edge set respectively. Set $\hat{S} := \tilde{S}$ and $\hat{Q} := \tilde{Q} \setminus \tilde{E}(\tilde{S})$.*

2.3   Analysis

Lemma 2. *Algorithm 1 can be implemented in polynomial time.*

Lemma 3. *Consider any given time t during the implementation of Algorithm 1. Let $y_e(t)$ be the dual value of edge e at time t which will increase with time t until edge e is frozen. Then for the set $\tilde{S}$, we always have*

$$\sum_{e \in E : |e \cap \tilde{S}| = 1} \frac{1}{2} y_e(t) + \sum_{e \in E : e \subseteq \tilde{S}} y_e(t) = C(\tilde{S}).$$

Theorem 4. *Algorithm 1 is a primal-dual 2-approximation algorithm for the SVCLP.*

We remark that the framework of our analysis for the SVCLP (also for the SVCSP in Section 3) follows the work of Du et al. [8] for the facility location problem with submodular penalties. But we need to pay more attention to the vertex cover structure. Particularly, Event 1 of Algorithm is quite different comparing with the situation for the facility location problem.

3   Primal-dual Approximation Algorithm for the SVCSP

3.1   Formulation

Given a nonnegative monotonically increasing submodular function $P(\cdot) : 2^E \to R_+$ with $P(\emptyset) = 0$. The SVCSP is the same as SVCLP except that the linear penalty p_e for each edge $e \in E$ is replaced with a submodular penalty function $P(Q)$ for each edge subset $Q \subseteq E$.

Let us introduce two types of binary variables X_S for each vertex subset $S \subseteq V$ and Z_Q for each edge subset $Q \subseteq E$. For an arbitrarily subset $S \subseteq V$, X_S indicates whether S is selected to cover some edges. For an arbitrarily subset $Q \subseteq E$, Z_Q indicates whether Q is selected to be penalized. Then, we have the following integer linear program for the SVCSP.

$$OPT_{SVCSP} := \min \sum_{S \subseteq V} C(S)X_S + \sum_{Q \subseteq E} P(Q)Z_Q$$

$$\text{s. t.} \sum_{S \subseteq V : S \cap e \neq \emptyset} X_S + \sum_{Q \subseteq E : e \in Q} Z_Q \geq 1, \quad \forall e = (i, j) \in E, \quad (5)$$

$$X_S, Z_Q \in \{0, 1\}, \quad \forall S \subseteq V, Q \subseteq E.$$

The first constraint of (5) guarantees that each edge $e \in E$ is either covered by some vertex subset S or penalized in some edge subset Q. Due to the submodularity of $C(\cdot)$ and $P(\cdot)$, there must exist exactly a pair of subsets $S \subseteq V$ and $Q \subseteq E$ such that $X_S = 1$ and $Z_Q = 1$ in the optimal solution of (5). Relaxing the integer constraints, we obtain

$$\min \sum_{S \subseteq V} C(S)X_S + \sum_{Q \subseteq E} P(Q)Z_Q$$

$$\text{s. t.} \sum_{S \subseteq V : S \cap e \neq \emptyset} X_S + \sum_{Q \subseteq E : e \in Q} Z_Q \geq 1, \quad \forall e = (i, j) \in E, \quad (6)$$

$$X_S, Z_Q \geq 0, \quad \forall S \subseteq V, Q \subseteq E.$$

The dual program of (6) is

$$\max \sum_{e \in E} y_e$$

$$\text{s. t.} \sum_{e \in E : e \cap S \neq \emptyset} y_e \leq C(S), \quad \forall S \subseteq V, \quad (7)$$

$$\sum_{e \in Q} y_e \leq P(Q), \quad \forall Q \subseteq E,$$

$$y_e \geq 0, \quad \forall e \in E.$$

We propose the following weak dual program to approximate (7).

$$\max \sum_{e \in E} y_e$$

$$\text{s. t.} \sum_{e \in E : |e \cap S| = 1} \frac{1}{2} y_e + \sum_{e \in E : e \subseteq S} y_e \leq C(S), \quad \forall S \subseteq V, \tag{8}$$

$$\sum_{e \in Q} y_e \leq P(Q), \quad \forall Q \subseteq E,$$

$$y_e \geq 0, \quad \forall e \in E.$$

Similar to Lemma 1, we have the following lemma to describe the relationship among the last three linear programs.

Lemma 5. *Given any feasible solution y for (8), $\frac{1}{2}y$ is feasible for (7). Furthermore, $\frac{1}{2}\sum_{e \in E} y_e$ is a lower bound for the primal linear program (6).*

3.2 Algorithm

Based on the weak dual program (8), we give a primal-dual algorithm for the SVCSP as follows. Since Algorithm 2 is almost the same as Algorithm 1, we only address Event 2 which is the only difference between the two.

Algorithm 2

Event 2. *There exists an edge set $Q \subseteq E$ such that*

$$\sum_{e \in Q \setminus F} t + \sum_{e \in Q \cap F} y_e = P(Q).$$

Freeze those unfrozen edges in Q and temporarily penalize all edges in Q by setting $F := F \cup Q$ and $\tilde{Q} := \tilde{Q} \cup Q$.

We remark that some edges covered by $\tilde{S}$ may be included in $\tilde{Q}$ in Event 2. Moreover, some temporarily penalized edges may be covered by $\tilde{S}$ in Event 1. These two situations may occur if $\tilde{Q} \cap E(\tilde{S}) \neq \emptyset$ (see Figure 1).

3.3 Analysis

Lemma 6. *Algorithm 2 can be implemented in polynomial time.*

Lemma 7. *Consider any given time t during the implementation of Algorithm 2. Let $y_e(t)$ be the dual value of edge e at time t which will increase with time t until edge e is frozen. Then for the set $\tilde{S}$ and $\tilde{Q}$, we always have*

$$\sum_{e \in E : |e \cap \tilde{S}| = 1} \frac{1}{2} y_e(t) + \sum_{e \in E : e \subseteq \tilde{S}} y_e(t) = C(\tilde{S}),$$

and

$$\sum_{e \in \tilde{Q}} y_e(t) = P(\tilde{Q}).$$

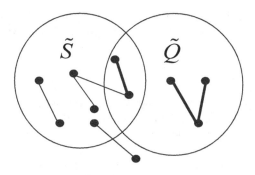

Fig. 1.

Theorem 8. *Algorithm 2 is a primal-dual 4-approximation algorithm for the SVCSP.*

4 Discussion

In this paper, we introduce the SVCWP with two variants, namely, the SVCLP and SVCSP, for which we present two primal-dual approximation algorithms.

We point out that the approximation ratio 4 for the SVCSP can be improved. Based on the double Lovász extensions for $C(\cdot)$ and $P(\cdot)$, one can obtain an integer convex programming for the SVCSP. The corresponding convex programming relaxation can be solved in polynomial time by the ellipsoid method. Recall that there is a convex programming rounding 2-approximation algorithm for the submodular vertex cover problem [20]. Applying the general rounding framework for covering problems with submodular penalties [24], we can obtain a convex programming rounding (non-combinatorial) approximation algorithm with the ratio $(1 - e^{-1/2})^{-1} \leq 2.542$ for the SVCSP. However, our primal-dual 4-approximation algorithm for the SVCSP is combinatorial and easily adaptable to other problems.

There are several directions for future research. First, what is the approximability of the SVCSP? Since the best constant ratio we can hope for the vertex cover problem is 2 under the unique game conjecture, the proposed primal-dual 2-approximation algorithm for the SVCLP is unlikely to be improved. It will be interesting to improve the 2.542 approximation ratio for the SVCSP. Second, what is the approximability of the problem if the edge penalties are arbitrary functions rather than submodular functions? Third, is it possible to extend our approach to study other variants of the vertex cover problem explained in the introduction section.

Acknowledgments. The authors would like to thank the four anonymous reviewers for their helpful comments on an earlier version of this paper. The first author is supported by NSF of China (No. 11371001) and China Scholarship

Council. The third author's research is supported by the Natural Sciences and Engineering Research Council of Canada (NSERC) grant 283106.

References

1. Bshouty, N., Burroughs, L.: Massaging a Linear Programming Solution to Give a 2-Approximation for a Generalization of the Vertex Cover Problem. In: Meinel, C., Morvan, M. (eds.) STACS 1998. LNCS, vol. 1373, pp. 298–308. Springer, Heidelberg (1998)
2. Bar-Yehuda, R., Even, S.: A Linear-Time Approximation Algorithm for the Weighted Vertex Cover. J. Algorithms 2, 198–203 (1981)
3. Bar-Yehuda, R., Even, S.: A Local-Ratio Theorem for Approximating the Weighted Vertex Cover Problem. Ann. Discrete Math. 25, 27–46 (1985)
4. Bar-Yehuda, R., Hermelin, D., Rawitz, D.: An Extension of the Nemhauser-Trotter Theorem to Generalized Vertex Cover with Applications. SIAM J. Discrete Math. 24, 287–300 (2010)
5. Bar-Yehuda, R., Rawitz, D.: On the Equivalence between the Primal-Dual Schema and the Local Technique. SIAM J. Discrete Math. 19, 762–797 (2005)
6. Bienstock, D., Goemans, M., Simchi-Levi, D., Williamson, D.: A Note on the Prize Collecting Traveling Salesman Problem. Math. Program. 59, 413–420 (1993)
7. Charikar, M., Khuller, S., Mount, D., Narasimhan, G.: Algorithms for Facility Location Problems with Outliers. In: 12th Annual ACM-SIAM Symposium on Discrete Algorithms, pp. 642–651. SIAM Press, Washington SC (2001)
8. Du, D., Lu, R., Xu, D.: A Primal-Dual Approximation Algorithm for the Facility Location Problem with Submodular Penalties. Algorithmica 63, 191–200 (2012)
9. Edmonds, J.: Submodular Functions, Matroids, and Certain Polyhedra. In: Guy, R., Hanam, H., Sauer, N., Schonheim, J. (eds.) Combinatorial Structures and Their Applications (Proc. 1969 Calgary Conference), pp. 69–87. Gordon and Breach, New York (1970)
10. Fleischer, L., Iwata, S.: A Push-Relabel Framework for Submodular Function Minimization and Applications to Parametric Optimization. Discrete Appl. Math. 131, 311–322 (2003)
11. Fujishige, S.: Submodular Functions and Optimization, 2nd edn. Elsevier, Amsterdam (2005)
12. Guha, S., Hassin, R., Khuller, S., Or, E.: Capacitated Vertex Covering. J. Algorithms 48, 257–270 (2003)
13. Grötschel, M., Lovász, L., Schrijver, A.: Geometric Algorithms and Combinatorial Optimization. Springer, Berlin (1988)
14. Goemans, M.X., Williamson, D.P.: A General Approximation Technique for Constrained Forest Problems. SIAM J. Comput. 24, 296–317 (1995)
15. Halperin, E.: Improved Approximation Algorithms for the Vertex Cover Problem in Graphs and Hypergraphs. SIAM J. Comput. 31, 1608–1623 (2002)
16. Hochbaum, D.S.: Approximation Algorithms for the Set Covering and Vertex Cover Problems. SIAM J. Comput. 11, 555–556 (1982)
17. Hochbaum, D.S.: Approximation Algorithms for NP-hard Problems. PWS Publishing Company, Boston (1997)
18. Hochbaum, D.S.: Solving Integer Programs over Monotone Inequalities in Three Variables: a Framework of Half Integrality and Good Approximations. Eur. J. Oper. Res. 140, 291–321 (2002)

19. Iwata, S., Fleischer, L., Fujishige, S.: A Combinatorial Strongly Polynomial Algorithm for Minimizing Submodular Functions. J. ACM 48, 761–777 (2001)
20. Iwata, S., Nagano, K.: Submodular Function Minimization under Covering Constraints. In: 50th Annual IEEE Symposium on Foundations of Computer Science, pp. 671–680. IEEE Press, Atlanta (2009)
21. Karakostas, G.: A Better Approximation Ratio for the Vertex Cover Problem. ACM Trans. on Algorithms 5, Article No. 41 (2009)
22. Khot, S., Regev, O.: Vertex Cover Might Be Hard to Approxmate to with $2 - \epsilon$. J. Comput. Syst. Sci. 74, 335–349 (2008)
23. Karp, R.M.: Reducibility among Combinatorial Problems. In: Miller, R.E., Thatcher, J.W., Bohlinger, J.D. (eds.) Complexity of Computer Computations, pp. 85–103. Springer, US (1972)
24. Li, Y., Du, D., Xiu, N., Xu, D.: Improved Approximation Algorithms for the Facility Location Problems with Linear/submodular Penalty. In: Du, D.-Z., Zhang, G. (eds.) COCOON 2013. LNCS, vol. 7936, pp. 292–303. Springer, Heidelberg (2013)
25. Lovász, L.: Submodular Functions and Convexity. In: Bachem, A., Grtschel, M., Korte, B. (eds.) Mathematical Programming The State of the Art, pp. 235–257. Springer, Heidelberg (1983)
26. Schrijver, A.: A Combinatorial Algorithm Minimizing Submodular Functions in Strongly Polynomial Time. J. Comb. Theory B 80, 346–355 (2000)

A New Approximation Algorithm
for the Unbalanced Min s-t Cut Problem

Peng Zhang*

School of Computer Science and Technology,
Shandong University, Jinan 250101, China
algzhang@sdu.edu.cn

Abstract. Let k be an input parameter. An s-t cut is of k-size if its s-side has size at most k. The Min k-Size s-t Cut problem asks to find a k-size s-t cut with the minimum capacity. Being the unbalanced version of the famous Min s-t Cut problem, this problem is fundamental and has extensive applications, especially in community identification in social and information networks. In this paper, we give a new $\frac{k+1}{k+1-k^*}$-approximation algorithm for the Min k-Size s-t Cut problem, where k^* is the size of s-side of an optimal solution.

1 Introduction

Let $G = (V, E)$ be an undirected graph with nonnegative edge capacities $\{c_e\}$. Given a partition (A, B) of the vertex set V (that is, $B = \overline{A}$ and $A, B \neq \emptyset$), a *cut* (A, B) of graph G is the set of edges having one endpoint in A and the other endpoint in B. The *capacity* $\delta(A, B)$ of a cut (A, B) is the sum of capacities of edges in the cut, i.e, $\delta(A, B) = \sum_{e \in (A,B)} c_e$. If every edge in the graph has unit capacity, the capacity of a cut is simply the number of edges in the cut.

Let k be an input parameter. Given two distinguished vertices s and t of graph G, known as the source and the sink respectively, an s-t cut (S, T) is a cut separating s and t. A *k-size s-t cut* of G is an s-t cut whose s-side has size at most k. Given an s-t cut (S, T), we always assume that $s \in S$. In this paper, we study the Min k-Size s-t Cut problem. It is the *unbalanced* version of the Min s-t Cut problem, in the sense that a k-size s-t cut has its s-side size-bounded.

Definition 1. The Min k-Size s-t Cut Problem.
  (Instance) *An undirected graph $G = (V, E)$ with nonnegative edge capacities $\{c_e\}$, a source-sink pair (s, t), and a positive integer k.*
  (Goal) *Find a k-size s-t cut with the minimum capacity.*

Motivations. From the viewpoint of combinatorial optimization, the Min k-Size s-t Cut problem is just a natural generalization of the classic Min s-t Cut

---

* This work was done when the author was visiting University of California, Riverside, USA. Supported by the State Scholarship Fund of China, Natural Science Foundation of Shandong Province (ZR2012Z002 and ZR2011FM021), and the Independent Innovation Foundation of Shandong University (2012TS072).

Z. Cai et al. (Eds.): COCOON 2014, LNCS 8591, pp. 346–356, 2014.

problem. While Min s-t Cut was known to be polynomial-time solvable several decades ago, it is surprising that only recently starts the algorithmic research of the unbalanced Min s-t Cut problem (i.e., Min k-Size s-t Cut). This may be partially because new applications desiring unbalanced cuts arise only recently.

In applications, the unbalanced cut problems are closely related to the clustering and classification problems, especially in community identification in social and information networks [1,2]. A community in a social network is thought of as a set of vertices such that there are many connections between its members. Sometimes we want to make clear which community a given node belongs to. Identifying community containing a given node with a given size just leads to the Min k-Size s-t Cut problem.

1.1 Related Work and Our Results

Li and Zhang [2] introduced the Min k-Size s-t Cut problem. They proved the problem is NP-hard and gave an $O(\log n)$-approximation algorithm for the problem, where n is the number of vertices in the input graph. The strategy is via a series of reductions finally to the Min Ek-Size Cut problem (find a minimum capacity cut with the small side having exactly k vertices), which is approximately solved by Räcke's elegant tree decomposition method [3].

Hayrapetyan et al. [4] study the Min-Size Bounded-Capacity Cut problem (MinSBCC, for short), which asks to find an s-t cut in a graph such that the capacity of the cut does not exceed a given budget $B \geq 0$ and the size of the s-side of the cut is minimized. For this problem, the authors [4] gave a *bi-criteria* $(\frac{1}{\lambda}, \frac{1}{1-\lambda})$-approximation algorithm for any $0 < \lambda < 1$; that is, the algorithm outputs a cut with its s-side having size at most $\frac{1}{1-\lambda}$ times that of an optimal cut (i.e., $\frac{1}{1-\lambda}$ is the approximation ratio), and with its capacity being at most $\frac{1}{\lambda}B$ (i.e., $\frac{1}{\lambda}$ is the violation of the budget). The strategy is based on the parametric push-relabel max flow algorithm due to Gallo et al. [5].

The Min k-Size s-t Cut problem and the MinSBCC problem are closely related in the sense that for the cut to be found, the former applies budget k on the size of the s-side and minimizes the capacity, while the latter applies budget B on the cut capacity and minimizes the size of its s-side. Motivated by the strategy of parametric flow used in [4], we design a new *true* $\frac{k+1}{k+1-k^*}$-approximation algorithm for the Min k-Size s-t Cut problem, where k^* is the size of s-side of an optimal solution. In the worst case, k^* could be equal to k and the approximation ratio $\frac{k+1}{k+1-k^*}$ could degenerate into $O(k)$. We get this result by deep analysis of the structure of a series of s-t cuts, and thus extend the parametric flow technique to the Min k-Size s-t Cut problem. While the algorithm itself has nothing to do with linear programming, its analysis is based on a natural linear programming relaxation to the Min k-Size s-t Cut problem. We also prove that the integrality gap of this relaxation is at least $k/2$.

Feige et al. [6] studied the problem of cutting *exactly* k vertices from a graph that separates a given source-sink pair s and t. Let us denote this problem by Min Ek-size s-t Cut. By extending Karger's random edge-contraction technique [7] for the Global Min Cut problem, the authors [6] gave a randomized $O(k/\log n)$-approximation algorithm for the Min Ek-size s-t Cut problem, where n is the number of vertices in the input graph. The approximation ratio is guaranteed with high probability. However, it is not known how to derandomize this algorithm (as well as Karger's algorithm for Global Min Cut [7]). Although not mentioned by the authors, the work of Feige et al. [6] implies that the Min k-Size s-t Cut problem studied in this paper can be approximated within $O(k/\log n)$ with high probability by reducing it to Min Ek-Size s-t Cut. In contrast, our approximation ratio for the Min k-Size s-t Cut problem, in the worst case being $O(k)$, is guaranteed deterministically.

1.2 More Related Work

The Min k-Size Cut problem asks to find a k-size cut with minimum capacity, where a k-size cut means that the smaller side of the cut has size at most k. Based on the approximation results of Min k-Size s-t Cut, one may easily try to approximate Min k-Size Cut by trying Min k-Size s-t Cut for all possible source-sink pairs. Surprisingly, Armon and Zwick [8] showed that the Min k-Size Cut problem can be optimally solved in $O(n^6 \log n)$ time, where n is the number of vertices in the input graph. Their strategy is a series of reductions finally to the enumeration of all the 2-approximate min cuts, where a 2-approximate min cut is a cut whose capacity is at most two times that of the min cut. By the work of Nagamochi et al. [9], the enumeration can be done in $O(n^6)$ time. The optimal solution to Min k-Size Cut can be found in this enumeration.

By extending Räcke's tree decomposition method [3], Li and Zhang [2] gave $O(\log n)$-approximation algorithms for the Min Ek-Size Cut problem and the Min Ek-Size s-t Cut problem (find a minimum capacity s-t cut with the s-side having exactly k vertices), where n is the number of vertices in the input graph.

Fomin et al. [10] considered the parameterized complexity for several unbalanced edge-cut and vertex-cut problems. Specifically, they showed that the Min k-Size s-t Cut problem is fixed-parameter tractable when parameterized by k and the cut capacity. Chuzhoy et al. [11] gave some approximation algorithms for the k-Route Cut problem, which is shown by the authors to be a generalization of the MinSBCC problem and the Min k-Size s-t Cut problem.

Svitkina and Tardos [12] studied the Max-Size Bounded-Capacity Cut problem (MaxSBCC, for short), which is the maximization version of the MinSBCC problem. The MaxSBCC problem asks to find an s-t cut in a graph such that the capacity of the cut does not exceed a given budget and the size of the s-side of the cut is maximized. For both MinSBCC and MaxSBCC, only bi-criteria approximation algorithms are currently known [4,12].

2 Preliminaries

Notations. For convenience, when we talk about an s-t cut (S, T), we always mean that $s \in S$ and $t \in T$. Since once one side of a cut is given, the other side is accordingly determined, a cut (S,T) is also abbreviated to just S. That is, when we talk about cut S, we mean the cut (S,T) where $T = \overline{S}$.

Let $(A_1, A_2, \cdots)$ be a sequence (of numbers, sets, etc.). We use (A_i) as its succinct expression.

As usual, given a graph G, let n denote the number of its vertices, and m denote the number of its edges. Given an instance of some optimization problem, let OPT denote the optimal value of the instance.

The Parametric Flow Technique. Hayrapetyan et al. [4] defined the following parametric network G^α, which is an instance of a class of more general parametric networks defined by Gallo et al. [5]. Given graph G and sink t, add an edge of capacity $\alpha \geq 0$ from every vertex $v \neq t$ to the sink t, introducing parallel edges if necessary. Denote by G^α the resulting parametric network.

Noticing that in G^α the capacity of an s-t cut $(S, \overline{S})$ is $\alpha|S| + \delta_G(S)$, it is easy to see that the larger α, the smaller s-side of a min s-t cut of G^α. Let S_0 be a min s-t cut of G^α when $\alpha = 0$. As α increases, the s-side of the min s-t cut of G^α will contain fewer and fewer vertices, until eventually the last cut, denoted by S_ℓ for some ℓ, contains only the source vertex s. It is well known [13,14,5] that there are at most $n - 1$ distinct cuts $S_0, S_1, \cdots, S_\ell$ produced in the above procedure. By extending the famous push-relabel max flow algorithm of Goldberg and Tarjan [15], Gallo et al. [5, Section 3.3] gave an elaborated and efficient algorithm which finds in $O(nm \log(n^2/m))$ time all these cuts and their corresponding α values, denoted by $\alpha_0, \alpha_1, \cdots, \alpha_\ell$.

Let $k_i = |S_i|$ for $0 \leq i \leq \ell$. By the above description, Properties 1 and 2 are obvious.

Property 1. $\alpha_0 < \alpha_1 < \cdots < \alpha_\ell$.

Property 2. $k_0 > k_1 > \cdots > k_\ell$.

In fact, [5] shows that the cuts $\{S_i\}$ form a nested family $S_0 \supset S_1 \supset \cdots \supset S_\ell$.

Let δ_i be the capacity of cut $(S_i, \overline{S_i})$ with respect to graph G, that is, $\delta_i = \delta_G(S_i)$. Hayrapetyan et al. [4] showed that Property 3 holds. Its proof is given here for completeness.

Property 3. $\delta_0 < \delta_1 < \cdots < \delta_\ell$.

Proof. Suppose not and for some pair i, j that $i < j$ we have $\delta_i \geq \delta_j$. By Property 2, $|S_i| > |S_j|$. So we have $\alpha|S_i| + \delta_G(S_i) > \alpha|S_j| + \delta_G(S_j)$ for any $\alpha > 0$. This means S_j can not be after S_i, contradicting the fact $i < j$.   □

3   Approximation Algorithm

In this section we present the approximation algorithm for the Min k-Size s-t Cut problem. Before doing this, we first state some lemmas which help to understand the design of the algorithm.

Let S^* be a min k-size s-t cut, and $k^* = |S^*|$.

Lemma 1 (Assumption). *By the guess skill we can assume that the value of k^* is known.*

Proof. The guess skill is a folklore in the design of approximation algorithms. It actually means that we try the algorithm (for the Min k-Size s-t Cut problem) for each possible value of k^* and output the best solution ever found.   □

Lemma 2 (Assumption). *In general we can assume that $k_0 > k \geq k^*$.*

Proof. Obviously we always have $k \geq k^*$. Consider the case $k \geq k_0$. Since S_0 (with $|S_0| = k_0$) is a min s-t cut of G, S_0 is also a min k-size s-t cut of G. So, in this case we simply output S_0 and we are done. This means that we need only to focus on the case $k_0 > k$, giving the assumption.   □

Lemma 3. *If k^* is in the sequence (k_i), then (i) the first entry in sequence (k_i) that is at most k, k_j to say, is just k^*, and (ii) the corresponding cut S_j is a min k-size s-t cut.*

Proof. (i) Prove by contradiction and suppose $k_j \neq k^*$. Since $k^* \leq k$ and k^* is in (k_i), k^* must be after k_j in the sequence (i.e., $k_j > k^*$). Since S_j is a k-size s-t cut, we have $\delta_G(S^*) \leq \delta_G(S_j)$. Therefore,

$$\alpha|S^*| + \delta_G(S^*) < \alpha|S_j| + \delta_G(S_j)$$

for any $\alpha > 0$. This means S^* should be in the sequence and at least before S_j, contradicting the fact that $k_j > k^*$.

(ii) Since S_j is a min s-t cut of graph G^{α_j}, and S^* is an s-t cut of G^{α_j}, obviously we have $\alpha_j|S_j| + \delta_G(S_j) \leq \alpha_j|S^*| + \delta_G(S^*)$. Since $k_j = k^*$, we get $\delta_G(S_j) \leq \delta_G(S^*)$.

On the other hand, S^* is a min k-size s-t cut of graph G, and S_j is a k-size s-t cut of G. So $\delta_G(S^*) \leq \delta_G(S_j)$. As a consequence, we have $\delta_G(S_j) = \delta_G(S^*)$, that is, S_j is also a min k-size s-t cut of G.   □

Lemma 4. *If k^* is not in the sequence (k_i), then all the entries in $[k, k - 1, \cdots, k^*]$ are not in (k_i).*

Proof. Take any $k' \in [k, k - 1, \cdots, k^* + 1]$. Suppose k' is in (k_i). Let S' be the cut corresponding to k', and α' be the corresponding α value.

By Lemma 2, $k_0 > k'$ and hence $\alpha' > 0$. Since S' is a min s-t cut of graph $G^{\alpha'}$, and S^* is an s-t cut of $G^{\alpha'}$, we have $\alpha'|S'| + \delta_G(S') \leq \alpha'|S^*| + \delta_G(S^*)$. Since $k' > k^*$, that is, $|S'| > |S^*|$, we have $\delta_G(S') < \delta_G(S^*)$, contradicting the fact that S^* is a min k-size s-t cut of G.   □

Based on the observations in Lemmas 1-4, we design Algorithm $\mathcal{A}$ for the Min k-Size s-t Cut problem.

Algorithm $\mathcal{A}$ for Min k-Size s-t Cut
Input: Instance $\mathcal{I} = (G, c, k, s, t)$.
Output: A k-size s-t cut (S, T).

1 By calling the push-relabel parametric flow algorithm, compute a sequence of cuts $S_0, S_1, \cdots, S_\ell$. For each $0 \le i \le \ell$, $k_i \leftarrow |S_i|$.
2 If $k \ge k_0$ then return S_0 and stop.
3 Guess the size of the smaller side of an optimal cut $(S^*, \overline{S^*})$. Denote it by k^*.
4 Find k^* in $\{k_0, k_1, \cdots, k_\ell\}$.
5 If found, say $k_j = k^*$, then return S_j.
6 Otherwise find a j such that $k_j > k \ge k^* > k_{j+1}$, return S_{j+1}.

Algorithm $\mathcal{A}$ is a purely combinatorial and efficient algorithm. The running time of step 1 dominates the time of other steps. So, the total running time of Algorithm $\mathcal{A}$ is $O(nm \log(n^2/m))$.

4 Analysis

While Algorithm $\mathcal{A}$ itself has nothing to do with linear programming, its analysis is based on the linear programming relaxation (LP_1) for the Min k-Size s-t Cut problem, and its Lagrangian relaxation (LP_2) (which is just the linear programming relaxation for the Min s-t Cut problem on graph $G^{\alpha'}$).

The following linear program (LP_1) is an LP-relaxation for the Min k-Size s-t Cut problem. To see this, just consider its integer version, that is, the integer program obtained by replacing constraint (3) with $x_v, y_e \in \{0, 1\}, \forall v, \forall e$. In the program, for a vertex v, the value of x_v being 1 means vertex v is in the s-side and the value being 0 means the t-side. For an edge e, the value of y_e being 1 means edge e is cut and the value being 0 means not. In order to minimize the objective function, a cut has to assign 1 to all vertices in its s-side and 0 to all vertices in its t-side. Constraint (2) requires the number of vertices in the s-side of a cut is at most k.

$$\min \quad \sum_{e \in E} c_e y_e \qquad\qquad\qquad (\mathrm{LP}_1)$$

$$\text{s.t.} \quad x_s = 1,$$

$$x_t = 0,$$

$$y_e \ge |x_u - x_v|, \quad \forall e = (u, v) \in E, \qquad (1)$$

$$\sum_{v \in V} x_v \le k, \qquad\qquad\qquad (2)$$

$$x_v, y_e \ge 0, \qquad \forall v \in V, \forall e \in E. \qquad (3)$$

Linear program (LP_1) has a parameter k in its constraint (2). We use $(\mathrm{LP}_1(k))$ to denote the linear program with a specified parameter k. By the way, (LP_1) is really a linear program since the inequality in constraint (1) can be replaced by $y_e \ge x_u - x_v$ and $y_e \ge x_v - x_u$.

By applying the Lagrangian relaxation technique on (LP_1), we can move to the objective function the left hand side of constraint (2) with a multiplier α', resulting in the linear program (LP_2).

$$\min \quad \alpha' \sum_{v \in V(G)} x_v + \sum_{e \in E(G)} c_e y_e \qquad (LP_2)$$

$$\text{s.t.} \quad x_s = 1,$$
$$x_t = 0,$$
$$y_e \geq |x_u - x_v|, \qquad \forall e = (u,v) \in E(G),$$
$$x_v, y_e \geq 0, \qquad \forall v \in V(G), \forall e \in E(G).$$

Hayrapetyan et al. [4] proved Lemma 5 by rounding a fractional optimal solution to (LP_2) to an integer solution without increasing the objective function value. Here, we give a new, simple proof for Lemma 5 using the total unimodularity of a matrix.

Lemma 5. *Linear program (LP_2) has integer optimal solutions.*

Proof. Consider the parametric network $G^{\alpha'}$. Define $c_e = \alpha'$ for each edge $e = (v,t)$ $(v \neq t)$ added to graph G when we construct $G^{\alpha'}$. Then the objective function of (LP_2) can be written as $\sum_{e \in E(G^{\alpha'})} c_e y_e$, if we define $y_e = x_v$ for each edge $e = (v,t) \in E(G^{\alpha'}) \setminus E(G)$. In this way, we get the linear program (LP_3) shown below. For an optimal solution to (LP_3), we actually have $y_e = |x_v - x_t| = x_v$ for each edge $e = (v,t) \in E(G^{\alpha'}) \setminus E(G)$. Therefore, (LP_2) is equivalent to (LP_3) in the sense that they have the same optimal solutions.

$$\min \quad \sum_{e \in E(G^{\alpha'})} c_e y_e \qquad (LP_3)$$

$$\text{s.t.} \quad x_s = 1,$$
$$x_t = 0,$$
$$y_e \geq |x_u - x_v|, \quad \forall e = (u,v) \in E(G^{\alpha'}),$$
$$x_v, y_e \geq 0, \qquad \forall v \in V(G^{\alpha'}), \forall e \in E(G^{\alpha'}).$$

Linear program (LP_3) is nothing but the LP-relaxation of the famous Min s-t Cut problem. It is well-known that the constraint matrix of (LP_3) is totally unimodular [16, Theorem 13.3]. Given that the right hand sides of the constraints in (LP_3) are all integers, all *basic feasible solutions* to (LP_3) are integral [16, Theorem 13.2], implying that (LP_3), and hence (LP_2), has integer optimal solutions. □

Now we are ready to give the analysis for Algorithm $\mathcal{A}$.

Lemma 6. *Let j be the index found in step 6 of Algorithm $\mathcal{A}$. Suppose $|S_j| \geq \frac{1}{1-\lambda} k^*$ for some $\lambda > 0$. Then $\delta_G(S_{j+1}) \leq \frac{1}{\lambda} \delta_G(S^*)$.*

Proof. Since the cuts S_j and S_{j+1} are neighbors, there exists a value α' ($\alpha' = (\delta_{j+1} - \delta_j)/(k_j - k_{j+1})$), such that both S_j and S_{j+1} are min s-t cuts of graph $G^{\alpha'}$.

By Lemma 5, (LP$_2$) has integer optimal solutions. By the formulation of (LP$_2$), an integer optimal solution corresponds to a min s-t cut of $G^{\alpha'}$. Therefore, S_j and S_{j+1} leads to two integer optimal solutions to (LP$_2$), denoted by (x^-, y^-) and (x^+, y^+), respectively. Therefore, the linear combination

$$(1 - \gamma)(x^-, y^-) + \gamma(x^+, y^+) = (x^*, y^*) \tag{4}$$

for some $\gamma \in (0, 1)$ is a (fractional) optimal solution to (LP$_2$), where we choose γ such that

$$(1 - \gamma) \sum_v x_v^- + \gamma \sum_v x_v^+ = k^*. \tag{5}$$

Since $\sum_v x_v^- = k_j > k^*$, we know that such γ exists.

Claim. (x^*, y^*) is a fractional optimal solution to (LP$_1(k^*)$).

Proof. Suppose not and (x', y') is an optimal solution. So, we have $\sum_e c_e y_e' < \sum_e c_e y_e^*$. Since $\sum_v x_v' \le k^*$ and $\sum_v x_v^* = k^*$, we have

$$\alpha' \sum_v x_v' + \sum_e c_e y_e' < \alpha' \sum_v x_v^* + \sum_e c_e y_e^*.$$

This means (x', y') is a solution to (LP$_2$) better than (x^*, y^*), contradicting the optimality of (x^*, y^*).   □

By the above claim and the fact that the cut S^* is a feasible solution to (LP$_1(k^*)$), we get

$$\sum_{e \in E(G)} c_e y_e^* \le \delta_G(S^*). \tag{6}$$

By equation (5), $\sum_v x_v^- \le \frac{k^*}{1-\gamma}$. By the condition given in the lemma, $\sum_v x_v^- = |S_j| \ge \frac{k^*}{1-\lambda}$. So, we have

$$\frac{1}{1 - \lambda} \le \frac{1}{1 - \gamma} \Longleftrightarrow \lambda \le \gamma. \tag{7}$$

By the linear combination (4), we have

$$y_e^* \ge \gamma \cdot y_e^+ \tag{8}$$

for any $e \in E(G^{\alpha'})$. So, finally we have

$$\delta_G(S_{j+1}) = \sum_{e \in E(G)} c_e y_e^+ \underset{(8)}{\le} \frac{1}{\gamma} \sum_{e \in E(G)} c_e y_e^* \underset{(7)}{\le} \frac{1}{\lambda} \sum_{e \in E(G)} c_e y_e^* \underset{(6)}{\le} \frac{1}{\lambda} \delta_G(S^*),$$

proving the lemma.   □

Theorem 1. *Algorithm $\mathcal{A}$ is a $\frac{k+1}{k+1-k^*}$-approximation algorithm for the Min k-Size s-t Cut problem, where k^* is the size of the s-side of an optimal solution.*

Proof. By Lemma 2 and Lemma 3, if Algorithm $\mathcal{A}$ terminates in step 2 or step 5, the algorithm returns an optimal solution.

By Lemma 4, if k^* is not in the sequence (k_i), the algorithm must find a j in step 6 such that $|S_j| > k \geq k^* > |S_{j+1}|$. In this case, S_{j+1} is a feasible solution. So, if we set $\lambda = 1 - \frac{k^*}{k+1}$, then we have $|S_j| \geq k + 1 = \frac{1}{1-\lambda}k^*$. Therefore, by Lemma 6, we have

$$\delta_G(S_{j+1}) \leq \frac{1}{\lambda}\delta_G(S^*) = \frac{1}{\lambda}\text{OPT}.$$

The approximation ratio is

$$\frac{1}{\lambda} = \frac{1}{1 - \frac{k^*}{k+1}} = \frac{k+1}{k+1-k^*}.$$

Finally, the algorithm obviously runs in polynomial time. The theorem follows. □

5   Integrality Gap

Given a minimization problem and its linear programming relaxation, recall that the integrality gap of the linear program is the supremum of the ratio OPT/OPT_f over all instances of the problem, where OPT_f is the fractional optimal value of the linear program, and OPT, as before, is the optimal value of the problem instance, that is, the optimal value of the corresponding integer program.

We can show that the integrality gap of linear program (LP_1) for the Min k-Size s-t Cut problem has integrality gap at least $k/2$. This means that any approach (e.g., LP-rounding) of designing approximation algorithm for the Min k-Size s-t Cut problem by using OPT_f of (LP_1) as the lower bound of OPT, can not beat our approximation ratio in Theorem 1 up to a constant factor.

Theorem 2. *The integrality gap of (LP_1) is at least $k/2$.*

Proof. Consider the instance of Min k-Size s-t Cut shown in Figure 1. The vertices $\{s, v_1, \cdots, v_k\}$ constitute a $(k+1)$-clique. Besides this, there is an edge between v_j and the sink t, where v_j is an arbitrary vertex in $\{v_1, \cdots, v_k\}$. Each edge in the graph has unit capacity.

Since any feasible cut can contain at most k vertices in its s-side and the instance contain a $(k+1)$-clique, $S_1^* = \{s\}$ and $S_2^* = \{s, v_1, \cdots, v_k\} \setminus \{v_j\}$ are the only two optimal solutions both with OPT $= k$.

Then consider a fractional solution to (LP_1). Set $x_s = 1$, $x_t = 0$, and $x_i = \frac{k-1}{k}$ for every $1 \leq i \leq k$. For each edge e of the k edges ending at s, set $y_e = 1 - \frac{k-1}{k} = \frac{1}{k}$; for edge $e = (v_j, t)$, set $y_e = \frac{k-1}{k}$; and for all the other edges e set $y_e = 0$. Since $\sum_v x_v = 1 + k \cdot \frac{k-1}{k} = k$, and $y_e \geq |x_u - x_v|$ for every edge e, the solution

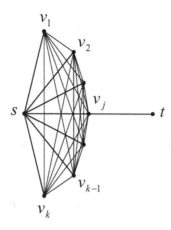

Fig. 1. An example to show the integrality gap of (LP_1)

(x, y) defined above is a feasible solution to (LP_1) on the instance in Figure 1. The objective value of this solution is

$$\sum_{e \text{ ends at } s} y_e + y_{(v_j, t)} = k \cdot \frac{1}{k} + \frac{k-1}{k} = 2 - \frac{1}{k}.$$

Summing up, the integrality gap of (LP_1) is $\geq \frac{\text{OPT}}{\text{OPT}_f} \geq \frac{k}{2 - 1/k} > \frac{k}{2}$. $\qquad\Box$

6 Conclusions

In this paper, we present a new $\frac{k+1}{k+1-k^*}$-approximation algorithm for the Min k-Size s-t Cut problem, where k^* is the size of s-side of an optimal solution. In the worst case, the approximation ratio degenerates into $O(k)$. An immediate research problem is how to improve the ratio $O(k)$. As we have already given an instance to show the integrality gap of (LP_1) is at least $k/2$, it is hopeless to do this improvement by using the pure linear programming technique based on (LP_1). On the other hand, there is yet no approximation hardness result known for the Min k-Size s-t Cut problem. To prove its approximation hardness also remains a good open problem.

Acknowledgements. The author is grateful to one of the anonymous reviewers for his/her suggestions which make the results in the paper more clear.

References

1. Leskovec, J., Lang, K., Dasgupta, A., Mahoney, M.: Statistical properties of community structure in large social and information networks. In: Proceedings of the 17th International World Wide Web Conference (WWW), pp. 695–704 (2008)

2. Li, A., Zhang, P.: Unbalanced graph partitioning. In: Cheong, O., Chwa, K.-Y., Park, K. (eds.) ISAAC 2010, Part I. LNCS, vol. 6506, pp. 218–229. Springer, Heidelberg (2010)
3. Räcke, H.: Optimal hierarchical decompositions for congestion minimization in networks. In: Proceedings of the 40th Annual ACM Symposium on Theory of Computing (STOC), pp. 255–264 (2008)
4. Hayrapetyan, A., Kempe, D., Pál, M., Svitkina, Z.: Unbalanced graph cuts. In: Brodal, G.S., Leonardi, S. (eds.) ESA 2005. LNCS, vol. 3669, pp. 191–202. Springer, Heidelberg (2005)
5. Gallo, G., Grigoriadis, M.D., Tarjan, R.E.: A fast parametric maximum flow algorithm and applications. SIAM Journal on Computing 18(1), 30–55 (1989)
6. Feige, U., Krauthgamer, R., Nissim, K.: On cutting a few vertices from a graph. Discrete Applied Mathematics 127, 643–649 (2003)
7. Karger, D., Stein, C.: A new approach to the minimum cut problem. Journal of the ACM 43(4), 601–640 (1996)
8. Armon, A., Zwick, U.: Multicriteria global minimum cuts. Algorithmica 46(1), 15–26 (2006)
9. Nagamochi, H., Nishimura, K., Ibaraki, T.: Computing all small cuts in an undirected network. SIAM Journal on Discrete Mathematics 10(3), 469–481 (1997)
10. Fomin, F., Golovach, P., Korhonen, J.: On the parameterized complexity of cutting a few vertices from a graph. CoRR abs/1304.6189 (2013)
11. Chuzhoy, J., Makarychev, Y., Vijayaraghavan, A., Zhou, Y.: Approximation algorithms and hardness of the k-route cut problem. In: Proceedings of the 23rd Annual ACM-SIAM Symposium on Discrete Algorithms (SODA), pp. 780–799 (2012)
12. Svitkina, Z., Tardos, É.: Min-max multiway cut. In: Jansen, K., Khanna, S., Rolim, J.D.P., Ron, D. (eds.) APPROX and RANDOM 2004. LNCS, vol. 3122, pp. 207–218. Springer, Heidelberg (2004)
13. Eisner, M.J., Severance, D.G.: Mathematical techniques for efficient record segmentation in large shared databases. Journal of the ACM 23, 619–635 (1976)
14. Stone, H.S.: Critical load factors in two-processor distributed systems. IEEE Transactions on Software Engineering 4, 254–258 (1978)
15. Goldberg, A.V., Tarjan, R.E.: A new approach to the maximum-flow problem. Journal of the ACM 35(4), 921–940 (1988)
16. Papadimitriou, C.H., Steiglitz, K.: Combinatorial Optimizatoin: Algorithms and Complexity. Dover Publications, Inc., Mineola (1998)

Approximability of the Minimum Weighted Doubly Resolving Set Problem⋆

Xujin Chen and Changjun Wang

Institute of Applied Mathematics, AMSS, Chinese Academy of Sciences
Beijing 100190, China
{xchen,wcj}@amss.ac.cn

Abstract. Locating source of diffusion in networks is crucial for controlling and preventing epidemic risks. It has been studied under various probabilistic models. In this paper, we study source location from a deterministic point of view by modeling it as the minimum weighted doubly resolving set problem, which is a strengthening of the well-known metric dimension problem.

Let G be a vertex weighted undirected graph on n vertices. A vertex subset S of G is a doubly resolving set (DRS) of G if for every pair of vertices u, v in G, there exist $x, y \in S$ such that the difference of distances (in terms of number of edges) between u and x, y is not equal to the difference of distances between v and x, y. The minimum weighted DRS problem consists of finding a DRS in G with minimum total weight. We establish $\Theta(\ln n)$ approximability of the minimum DRS problem on general graphs for both weighted and unweighted versions. This is the first work providing explicit approximation lower and upper bounds for minimum (weighted) DRS problem, which are nearly tight. Moreover, we design first known strongly polynomial time algorithms for the minimum weighted DRS problem on general wheels and trees with additional constant $k \geq 0$ edges.

Keywords: Source location, Doubly resolving set, Approximation algorithms, Polynomial-time solvability, Metric dimension.

1 Introduction

Locating the source of a diffusion in complex networks is an intriguing challenge, and finds diverse applications in controlling and preventing network epidemic risks [17]. In particular, it is often financially and technically impossible to observe the state of all vertices in a large-scale network, and, on the other hand, it is desirable to find the location of the source (who initiates the diffusion) from measurements collected by sparsely placed observers [16]. Placing an observer at vertex v incurs a cost, and the observer with a clock can record the time at

⋆ Research supported in part by NNSF of China under Grant No. 11222109, 11021161 and 10928102, by 973 Project of China under Grant No. 2011CB80800, and by CAS Program for Cross & Cooperative Team of Science & Technology Innovation.

Z. Cai et al. (Eds.): COCOON 2014, LNCS 8591, pp. 357–368, 2014.

which the state of v is changed (e.g., knowing a rumor, being infected or contaminated). Typically, the time when the single source originates an information is unknown [16]. The observers can only report the times they receive the information, but the senders of the information [9]. The information is diffused from the source to any vertex through shortest paths in the network, i.e., as soon as a vertex receives the information, it sends the information to all its neighbors simultaneously, which takes one time unit. Our goal is to select a subset S of vertices with minimum total cost such that the source can be uniquely located by the "infected" times of vertices in S. This problem is equivalent to finding a minimum weighted *doubly resolving set* in networks defined as follows.

DRS model. Networks are modeled as undirected connected graphs without parallel edges nor loops. Let $G = (V, E)$ be a graph on $n \geq 2$ vertices, and each vertex $v \in V$ has a nonnegative *weight* $w(v)$, representing its cost. For any $S \subseteq V$, the *weight* of S is defined to be $w(S) := \sum_{v \in S} w(v)$. For any $u, v \in V$, we use $d_G(u, v)$ to denote the *distance* between u and v in G, i.e., the number of edges in a shortest path between u and v. Let u, v, x, y be four distinct vertices of G. Following Cáceres et al. [2], we say that $\{u, v\}$ *doubly resolves* $\{x, y\}$, or $\{u, v\}$ *doubly resolves* x and y, if

$$d_G(u, x) - d_G(u, y) \neq d_G(v, x) - d_G(v, y).$$

Clearly $\{u, v\}$ doubly resolves $\{x, y\}$ if and only if $\{x, y\}$ doubly resolves $\{u, v\}$. For any subsets S, T of vertices, S *doubly resolves* T if every pair of vertices in T is doubly resolved by some pair of vertices in S. In particular, S is called a *doubly resolving set* (DRS) of G if S doubly resolves V. Trivially, V is a DRS of G. The *minimum weighted doubly resolving set* (MWDRS) problem is to find a DRS of G that has a minimum weight (i.e. a minimum weighted DRS of G). In the special case where all vertex weights are equal to 1, the problem is referred to as the *minimum doubly resolving set* (MDRS) problem [15], and it concerns with the minimum cardinality $\mathbf{dr}(G)$ of DRS of G.

Consider arbitrary $S \subseteq V$. It is easy to see that S fails to locate the diffusion source in G at some case if and only if there exist distinct vertices $u, v \in V$ such that S cannot distinguish between the case of u being the source and that of v being the source, i.e., $d_G(u, x) - d_G(u, y) = d_G(v, x) - d_G(v, y)$ for any $x, y \in S$; equivalently, S is not a DRS of G (see [5]). Hence, the MWDRS problem models exactly the problem of finding cost-effective observer placements for locating source, as mentioned in our opening paragraph.

Related work. Epidemic diffusion and information cascade in networks has been extensively studied for decades in efforts to understand the diffusion dynamics and its dependence on various factors. However, the inverse problem of inferring the source of diffusion based on limited observations is far less studied, and was first tackled by Shah and Zaman [17] for identifying the source of rumor, where the rumor flows on edges according to independent exponentially distributed random times. A maximum likelihood (ML) estimator was proposed for maximizing the correct localizing probability, and the notion of rumor-centrality was developed for approximately tracing back the source from the configuration of infected vertices at a given moment. The accuracy of estimations heavily depended on the structural properties of the networks. Along a different line, Pinto et al. [16] proposed other ML estimators that perform source detection via sparsely distributed observers who measure from which neighbors and at what time they received the information. The ML estimators were shown to be optimal for trees, and suboptimal for general networks under the assumption that the propagation delays associated with edges are i.i.d. random variables with known Gaussian distribution. In contrast to previous probabilistic model for estimating the location of the source, we study the problem from a combinatorial optimization's point of view; our goal is to find an observer set of minimum cost that guarantees deterministic determination of the accurate location of the source, i.e., to find a minimum weighted DRS.

The double resolvability is a strengthening of the well-studied resolvability, where a vertex x resolves two vertices u, v if and only if $d_G(u, x) \neq d_G(v, x)$. A subset S of V is a *resolving set* (RS) of G if every pair of vertices is resolved by some vertex of S. The minimum cardinality of a RS of G is known as the *metric dimension* $\mathtt{md}(G)$ of G, which has been extensively studied due to its theoretical importance and diverse applications (see e.g., [2,4,8,10] and references therein). Most literature on finding minimum resolving sets, known as the *metric dimension problem*, considered the unweighted case. The unweighted problem is NP-hard even for planar graphs, split graphs, bipartite graphs and bounded degree graphs [7,8,10]. On general graphs, Hauptmann et al. [10] showed that the unweighted problem is not approximable within $(1 - \varepsilon) \ln n$ for any $\varepsilon > 0$, unless $NP \subset DTIME(n^{\log \log n})$; moreover, the authors [10] gave a $(1 + o(1)) \ln n$-approximation algorithm based on approximability results of the test set problem in bioinformatics [1]. A lot of research efforts have been devoted to obtaining the exact values or upper bounds of the metric dimensions of special graphic classes [2]. Recently, Epstein et al [8] studied the weighted version of the problem, and developed polynomial time exact algorithms for finding a minimum weighted RS, when the underlying graph G is a cograph, a *k-edge-augmented tree* (a tree with additional k edges) for constant $k \geq 0$, or a (un)complete wheel.

Compared with nearly four decade research and vast literatures on resolving sets (metric dimension), the study on DRS has a relatively short history and its results have been very limited. The concept of DRS was introduced in 2007 by Cáceres et al. [2], who proved that the minimum RS of the Cartesian product of graphs is tied in a strong sense to minimum DRS of the graphs: the metric dimension of the Cartesian product of graphs G_1 and G_2 is upper bounded by $\mathtt{md}(G_1) + \mathtt{dr}(G_2)$. When restricted to the same graph, it is easy to see that a DRS must be a RS, but the reverse is not necessarily true. Thus $\mathtt{md}(G) \leq \mathtt{dr}(G)$. The ratio $\mathtt{dr}(G)/\mathtt{md}(G)$ can be arbitrarily large. This can be seen from the tree graph G depicted in Fig. 1. On the one hand, it is easily checked that $\{r_1, r_2\}$ is a RS of G, giving $\mathtt{dr}(G) \leq 2$. On the other hand, $\mathtt{md}(G) = n/2$ since $\{s_1, s_2, \ldots, s_h\}$ is the unique minimum DRS of G, as proved later in Lemma 4 of this paper.

Fig. 1. The graph tree G with $\mathtt{dr}(G) = n/2$ and $\mathtt{md}(G) = 2$

In view of the large gap, algorithmic study on DRS deserves good efforts, and it is interesting to explore the algorithmic relation between the minimum (weighted) DRS problem and its resolving set counterpart.

Previous research on DRS considered only the unweighted case. As far as general graphs are concerned, the MDRS problem has been proved to be NP-hard [14], and solved experimentally by metaheuristic approaches that use binary encoding and standard genetic operators [14] and that use variable neighborhood search [15]. To date, no efficient general-purpose algorithms with theoretically provable performance guarantees have been developed for the MDRS problem, let alone the MWDRS problem. Despite the NP-hardness, the approximability status of either problem has been unknown in literature. For special graphs, it is known that every RS of Hamming graph is also a DRS [12]. Recently, Čangalović et al. showed that $\mathtt{dr}(G) \in \{3, 4\}$ when G is a prism graph [3] or belongs to one of two classes of convex polytopes [13].

Our contributions. As far as we know, our opening example of cost-effective source location is the first real-world application of DRS explicitly addressed. Motivated by the application, we study and provide a thorough treatment of the MWDRS problem in terms of algorithmic approximability. Broadly speaking, we show that the MDRS and MWDRS problems have similar approximability to their resolving set counterparts.

Based on the construction of Hauptmann et al. [10], we prove that there is an approximation preserving reduction from the minimum dominating set problem to the MDRS problem, showing the MDRS problem does not admit $(1 - \varepsilon) \ln n$-approximation algorithm for any $\varepsilon > 0$ unless $NP \subset DTIME(n^{\log \log n})$. The strong inapproximability improves the NP-completeness established in [14]. Besides, we develop a $(\ln n + \ln \log_2 n + 1)$-approximation algorithm for solving the MWDRS problem in $O(n^4)$ time, based on a modified version of the approximation algorithms used in [1,10]. To the best of our knowledge, this paper is the first work providing explicit approximation lower and upper bounds for the MDRS and MWDRS problems, which are nearly tight (for large n). A byproduct of our algorithm gives the first logarithmic approximation for the weighted metric dimension problem on general graphs.

Despite many significant technical differences between handling DRS and RS, we establish the polynomial time solvability of the MWDRS problem for all these graph classes, with one exception of cographs, where the weighted metric dimension problem is known to admit efficient exact algorithms [8]. Our results are first known strong polynomial time algorithms for the MDRS problem on

k-edge-augmented trees and general wheels, including paths, trees and cycles. Using the fact that every minimum weighted DRS is *minimal* (with respect to the inclusion relation), our algorithms make use of the graphic properties to cleverly "enumerate" minimal doubly resolving sets that are potentially minimum weighted, and select the best one among them.

The paper is organized as follows: The inapproximability is proved in Section 2, The approximation algorithm for general graphs and exact algorithms for special graphs are presented in Sections 3 and 4, respectively. Future research directions are discussed in Section 5. The omitted details can be found in the full version [5].

2   Approximation Lower Bound

In this section, we establish a logarithmic lower bound for approximation the MDRS problem under the assumption that $NP \not\subset DTIME(n^{\log\log n})$. Hauptmann et al. [10] constructed a reduction from the *minimum dominating set* (MDS) problem to the metric dimension problem. Although their proof does not work for DRS, we show that their construction actually provides an approximation preserving reduction from the MDS problem to the MDRS problem.

A vertex subset S of graph G is a *dominating set* of G if every vertex outside S has a neighbor in S. The MDS problem is to find a dominating set of G that has the minimum cardinality $\mathbf{ds}(G)$. Unless $NP \subset DTIME(n^{\log\log n})$, the MDS problem cannot be approximated within $(1 - \varepsilon) \ln n$ for any $\varepsilon > 0$ [6].

Lemma 1. *There exists a polynomial time transformation that transfers graph* $G = (V, E)$ *to graph* $G' = (V', E')$ *such that* $\mathbf{dr}(G') \leq \mathbf{ds}(G) + \lceil \log_2 n \rceil + 3$. $\square$

Let graphs G and G' be as in Lemma 1. It has been shown that, given any RS (in particular DRS) S of G', a dominating set of G with cardinality at most $|S|$ can be found in polynomial time [10]. This, in combination with Lemma 1 and the logarithmic inapproximability of the MDS problem [6], gives the following lower bound for approximating minimum DRS.

Theorem 1. *Unless* $NP \subset DTIME(n^{\log(\log n)})$, *the* MDRS *problem cannot be approximated in polynomial time within a factor of* $(1-\epsilon) \ln n$, *for any* $\epsilon > 0$. $\square$

3   Approximation Algorithm

In this section, we present an $O(n^4)$ time approximation algorithm for the MW-DRS problem in general graphs that achieves approximation ratio $(1 + o(1)) \ln n$, nearly matching the lower bound $\ln n$ established in Theorem 1.

Our algorithm uses similar idea to that of Hauptmann et al. [10] for approximating minimum resolving sets in the metric dimension (MD) problem. The MD problem is a direct "projection" of the *unweighted* test set problem studied by Berman et al. [1] in the sense that a vertex in the MD problem can be seen as a "test" in the test set problem, which allows Hauptmann et al. to apply Berman-DasGupta-Kao algorithm [1] directly. However, in the DRS problem, one cannot simply view two vertices as a "test", because such a "test" would fail the algorithm in some situation. Besides, the algorithm deals with only unweighted cases. Thus we need conduct certain transformation that transforms the DRS problem to a series of weighted test set problems. Furthermore, we need modify Berman-DasGupta-Kao algorithm to solve these *weighted* problems within logarithmic approximation ratios.

Transformation. For any $x \in V$, let $U_x = \{\{x, v\} : v \in V \setminus \{x\}\}$. As seen later, each element of U_x can be viewed as a *test* or a certain combination of *tests* in the test set problem studied in [1]. From this point of view, we call each element of U_x a *super test*, and consider the *minimum weighted super test set* (MWSTS) problem on (V, U_x) as follows: For each super test $T = \{x, v\} \in U_x$, let its weight be $w(T) = w(v)$, The problem is to find a set of super tests $\mathcal{T} \subseteq U_x$ such that each pair of vertices in G is doubly resolved by some super test in $\mathcal{T}$ and the weight $w(\mathcal{T}) = \sum_{T \in \mathcal{T}} w(T)$ of $\mathcal{T}$ is minimized. The following lemma establishes the relation between the MWDRS problem and the MWSTS problem.

Lemma 2. *Let S be a DRS of G and $s \in S$. Then every pair of vertices in G is doubly resolved by at least one element of $\{\{s, v\} : v \in S \setminus \{s\}\}$.*

Proof. Let u, v be any two distinct vertices of G. There exist $s_1, s_2 \in S$ such that $d_G(u, s_1) - d_G(v, s_1) \neq d_G(u, s_2) - d_G(v, s_2)$. It follows that either $d_G(u, s_1) - d_G(v, s_1) \neq d_G(u, s) - d_G(v, s)$ or $d_G(u, s) - d_G(v, s) \neq d_G(u, s_2) - d_G(v, s_2)$, saying that u and v are doubly resolved by either $\{s, s_1\}$ or $\{s, s_2\}$. □

Since V is a DRS of G, Lemma 2 implies that U_x doubly resolves V. More importantly, Lemma 2 provides the following immediate corollary that is crucial to our algorithm design.

Corollary 1. *Let S^* be a minimum weighted DRS of G and $\alpha \in S^*$. Then the minimum weight of a solution to the MWSTS problem on (V, U_α) is at most $w(S^*) - w(\alpha)$.* □

Approximation. To solve the MWSTS problem, we adapt Berman-DasGupta-Kao algorithm [1] to augment a set $\mathcal{T}$ ($\subseteq U_x$) of super tests to be a feasible solution step by step. We define *equivalence relation* $\equiv^{\mathcal{T}}$ on V by: two vertices $u, v \in V$ are equivalent under $\equiv^{\mathcal{T}}$ if and only if $\{u, v\}$ is not doubly resolved by any test of $\mathcal{T}$. Clearly, the number of equivalence classes is non-decreasing with the size of $\mathcal{T}$. Let $E_1, \ldots, E_k$ be the equivalence classes of $\equiv^{\mathcal{T}}$. The value $H_{\mathcal{T}} := \log_2(\prod_{i=1}^{k} |E_i|!)$ is called the *entropy* of $\mathcal{T}$. Note that

$$H_{\mathcal{T}} = 0 \Leftrightarrow \text{every equivalent class of } \equiv^{\mathcal{T}} \text{ is a singleton}$$
$$\Leftrightarrow \cup_{T \in \mathcal{T}} T \text{ is a DRS of } G. \tag{3.1}$$

Hence our task is reduced to finding a set $\mathcal{T}$ of super tests with zero entropy $H_{\mathcal{T}}$ and weight $w(\mathcal{T})$ as small as possible.

For any super test $T \in U_x$, an equivalence class of $\equiv^{\mathcal{T}}$ is either an equivalence class of $\equiv^{\mathcal{T} \cup T}$ or it is partitioned into several (possibly more than two) equivalence classes of $\equiv^{\mathcal{T} \cup T}$. (If T partitions each equivalent class into at most two equivalent classes, then T works as a test in the test set problem.) Therefore $H_{\mathcal{T}} \geq H_{\mathcal{T} \cup T}$, and $IC(T, \mathcal{T}) := H_{\mathcal{T}} - H_{\mathcal{T} \cup T} \geq 0$ equals the decreasing amount of the entropy when adding T to $\mathcal{T}$. It is clear that

$$IC(T, \emptyset) \leq \log_2 n! - log_2 1 < n \log_2 n. \tag{3.2}$$

We now give a $(1+o(1)) \ln n$-approximation algorithm for the MWSTS problem on (V, U_x). The algorithm adopts the greedy heuristic to decrease the entropy of the current set of super tests at a minimum cost (weight).

Algorithm 1. Finding minimum weighted set $\mathcal{T}$ of super sets.

---

1. $\mathcal{T} \leftarrow \emptyset$
2. **while** $H_{\mathcal{T}} \neq 0$ **do**
3.   Select a super test $T \in U_x - \mathcal{T}$ that *maximizes* $\frac{IC(T,\mathcal{T})}{w(T)}$
4. $\mathcal{T} \leftarrow \mathcal{T} \cup T$
5. **end-while**

The major difference between Algorithm 1 and the algorithms in [1,10] is the criterion used in Step 3 for selecting T. It generalizes the previous unweighted setting. The following lemma extends the result on test set [1] to super test set.

Lemma 3. $IC(T, \mathcal{T}_0) \geq IC(T, \mathcal{T}_1)$ *for any sets* $\mathcal{T}_0$ *and* $\mathcal{T}_1$ *of super tests with* $\mathcal{T}_0 \subseteq \mathcal{T}_1$.   □

Using (3.2) and Lemma 3, the proof of performance ratio (cf. [5]) goes almost verbatim as the argument of Berman et al. [1].

Theorem 2. *Algorithm 1 is an* $O(n^3)$ *time algorithm for the* MWSTS *problem on* (V, U_x) *with approximation ratio* $\ln \left(\max_{T \in U_x} IC(T, \emptyset) \right) + 1 \leq \ln n + \ln \log_2 n + 1.$□

Suppose that given the MWSTS problem on (V, U_x), Algorithm 1 outputs a super test set $\mathcal{T}_x$. By (3.1), Running Algorithm 1 for n times, we obtain n doubly resolving sets S_x, $x \in V$ of G, from which we select the one, say S_v, that has the minimum weight, i.e. $w(S_v) = \min\{w(S_x) : x \in V\}$.

Theorem 3. *The* MWDRS *problem can be approximated in* $O(n^4)$ *time within a ratio* $\ln \left(\max_{u,v \in V} IC(\{u, v\}, \emptyset) \right) + 1 \leq \ln n + \ln \log_2 n + 1 = (1 + o(1)) \ln n.$

Proof. Let S^* be an optimal solution to the MWDRS problem. It suffices to show $w(S_v)/w(S^*) \leq (1 + o(1)) \ln n$. Take $\alpha \in S^*$, and let $\mathcal{T}_\alpha^*$ be an optimal solution to the MWSTS problem on (V, U_α). It follows from the choice of S_v, Theorem 2 and Corollary 1 that $w(S_v) \leq w(S_\alpha) = w(\alpha) + w(\mathcal{T}_\alpha) \leq w(\alpha) + (\ln n + \ln \log_2 n + 1)w(\mathcal{T}_\alpha^*) < (\ln n + \ln \log_2 n + 1)w(S^*)$.   □

Our algorithm and analysis show that the algorithm of [1] can be extended to solve the weighted test set problem, where each test has a nonnegative weight, by changing the selection criterion to be maximizing $IC(T, \mathcal{T})$ divided by the weight of T. A similar extension applied to the algorithm of Hauptmann et al. [10] gives a $(1+o(1)) \ln n$-approximate solution to the weighted metric dimension problem.

4   Exact Algorithms

Let $k \geq 0$ be a constant. A connected graph is called a k-*edge-augmented tree* if the removal of at most k edges from the graph leaves a spanning tree. Trees and cycles are 0-edge- and 1-edge-augmented trees, respectively. We design efficient algorithms for solving the MWDRS problem exactly on k-edge-augmented trees. Our algorithms run in linear time for $k = 0, 1$, and in $O(n^{12k})$ time for $k \geq 2$.

A graph is called a general wheel if it is formed from a cycle by adding a vertex and joining it to some (not necessarily all) vertices on the cycle. We solve the MWDRS problem on general wheels in cubic time by dynamic programming.

4.1 k-edge-augmented Trees

Let $G = (V, E)$ be a k-edge-augmented tree, and let L be the set of leaves (degree one vertices) in G. For simplicity, we often use $d(u, v)$ instead of $d_G(u, v)$ to denote the distance between vertices $u, v \in V$ in the underlying graph G of the MWDRS problem.

Trees: The Case of $k = 0$. When $k = 0$, graph $G = (V, E)$ is a tree. There is a fundamental difference between DRS and RS of G in terms of minimal sets. In general, G may have multiple minimal RSs and even multiple minimum weighted RSs. Nevertheless, in any case G has only one minimal DRS, which consists of all its leaves. In particular, we have $\mathbf{dr}(G) = |L|$.

Lemma 4. L *is the unique minimal DRS of G.*

Proof. For any two vertices $u, v \in V$, there exist leaves $l_1, l_2 \in L$ such that the path between l_1 and l_2 goes through u and v. It is easy to see that $d(u, l_1) - d(u, l_2) \neq d(v, l_1) - d(v, l_2)$. So L is a DRS. On the other hand, consider any leaf $l \in L$ and its neighbor $p \in V$. Since $d(l, v) - d(p, v) = 1$ for any $v \in V - \{l\}$, we see that each DRS of G contains l, and thus L. The conclusion follows.   □

Cycles: A Special Case of $k = 1$. Let $G = v_1 v_2 \cdots v_n v_1$ be a cycle, where $V = \{v_1, v_2, \ldots, v_n\}$. Suppose without loss of generality that $w(v_p) = \min_{i=1}^{n} w(v_i)$, where $p := \lceil n/2 \rceil$. It was known that any pair of vertices whose distance is not exactly $n/2$ is a minimal RS of G, and vice versa [8]. As the next lemma shows, the characterization of DRS turns out to be more complex. Each nonempty subset S of V cuts G into a set $\mathcal{P}_S$ of edge-disjoint paths such that they are internally disjoint from S and their union is G.

Lemma 5. *Given a cycle $G = (V, E)$, let S be a nonempty subset of V. Then S is a DRS of G if and only if no path in $\mathcal{P}_S$ has length longer than $\lceil n/2 \rceil$ and at least one path in $\mathcal{P}_S$ has length shorter than $n/2$.*   □

An instant corollary reads: The size of a minimal DRS of cycle G is 2 or 3 when n is odd, and is 3 when n is even; In particular, $\mathbf{dr}(G) = 2$ when n is odd, and $\mathbf{dr}(G) = 3$ when n is even. These properties together with the next one lead to our algorithm for solving the MWDRS problem on cycles.

Corollary 2. *If some minimum weighted DRS has cardinality 3, then there exists a minimum weighted DRS of G that contains vertex v_p.*   □

Algorithm 2. Finding minimum weighted DRS S in cycle G.

---

1. $\omega \leftarrow w(v_1)$, $i[1] \leftarrow 1$, $j \leftarrow 1$, $W \leftarrow w(V)$
2. **for** $h = 1$ **to** p **do**
3.   **if** $w(v_h) < \omega$ **then** $j \leftarrow j + 1$, $i[j] \leftarrow h$, $\omega \leftarrow w(v_h)$
4. **end-for**
5. **if** $j > 1$ **then** $k \leftarrow j$ **else** $k \leftarrow 2$, $i[k] \leftarrow p$
6. **if** n is odd **then** $S \leftarrow \arg\min_{i=1}^{n} w(\{v_i, v_{i+p-1}\})$, $W \leftarrow w(S)$
7. **for** $j = 1$ **to** $k - 1$ **do**
8.   let u_j be a vertex in $\{v_h : i[j]+p \le h \le i[j+1]+p\}$ with $w(u_j) = \min_{h=i[j]+p}^{i[j+1]+p} w(v_h)$
9.   **if** $w(v_p) + w(v_{i[j]}) + w(u_j) < W$ **then** $S \leftarrow \{v_p, v_{i[j]}, u_j\}$, $W \leftarrow w(S)$
10. **end-for**

Note that $V_j := \{v_h : i[j] + p \le h \le i[j+1] + p\}$, $j = 1, \ldots, k-1$ induce $k-1$ internally disjoint paths in G. It is thus clear that Algorithm 2 runs in $O(n)$ time. The vertices indices $1 = i[1] < i[2] < \cdots < i[k] = p$ found by the algorithm satisfy $w(v_{i[j]}) = \min_{h=i[j]}^{i[j+1]-1} w(v_h) = \min_{h=1}^{i[j+1]-1} w(v_h)$ for every $j = 1, \ldots, k-1$ and $w(v_{i[k]}) = w(v_p) = \min_{h=1}^{p} w(v_h)$. Moreover, either $w(v_1) = w(v_p)$ and $k = 2$, or $w(v_{i[j]}) > w(v_{i[j+1]})$ for every $j = 1, \ldots, k-1$. These facts together with the properties mentioned above verify the correctness of the algorithm.

Theorem 4. *Algorithm 2 finds in $O(n)$ time a minimum weighted DRS of cycle G.*   □

The Case of General k. Our approach resembles at a high level the one used by Epstein et al. [8]. However, double resolvablity imposes more strict restrictions, and requires extra care to overcome technical difficulties. Let $G_b = (V_b, E_b)$ be the graph obtained from $G = (V, E)$ by repeatedly deleting leaves. We call G_b the *base graph* of G. We reduce the MWDRS problem on G to the MWDRS problem on G_b (see Lemma 6). The latter problem can be solved in polynomial time by exhaustive enumeration, since, as proved in the sequel, every minimal DRS of G_b has cardinality at most $12(k - 1)$ for $k \ge 2$

Clearly, G_b is connected and has minimum degree at least 2. A vertex in V_b is called a *root* if in G it is adjacent to some vertex in $V \setminus V_b$. Let R denote the set of roots. Clearly, $R \cap L = \emptyset$. In G_b, we change the weights of all roots to zero, while the weights of other vertices remain the same as in G.

Lemma 6. *Suppose that S_b is a minimum weighted DRS of G_b. Then $(S_b \setminus R) \cup L$ is a minimum weighted DRS of G.*   □

Therefore, for solving the MWDRS problem on a weighted k-edge-augmented tree G, we only need to find a minimum weighted DRS of base graph G_b with the weights of all roots modified to be 0.

For 1-edge-augmented tree G, its base graph G_b is a cycle, whose minimum weighted DRS can be found in $O(n)$ time (recall Algorithm 2). Combining this with Lemmas 4 and 6, we have the following linear time solvability.

Theorem 5. *There is an $O(n)$ time exact algorithm for solving the* MWDRS *problem on k-edge-augmented trees, for $k = 0, 1$, including trees and cycles.*

In the remaining discussion for k-edge-augmented tree, we assume $k \geq 2$. A vertex is called a *branching vertex* of a graph if it has degree at least 3 in the graph. Recall that every vertex of the base graph $G_b = (V_b, E_b)$ has degree at least 2. It can be shown that (see [5])

- In $O(|E_b|) = O(n^2)$ time, G_b can be decomposed into at most $3k - 3$ edge-disjoint paths whose ends are branching vertices of G_b and internal vertices have degree 2 in G_b.
- For any minimal DRS set S of G_b, and any path P in the above path decomposition of G_b, at most four vertices of S are contained in P.

It follows that every minimal DRS of G_b contains at most $12(k - 1)$ vertices. Our algorithm for finding the minimum weighted DRS of G_b examines all possible subsets of V_b with cardinality at most $12(k - 1)$ by taking at most four vertices from each path in the path decomposition of G_b; among these sets, the algorithm selects a DRS of G_b with minimum weight. Have a table that stores the distances between each pair of vertices in G_b, it takes $O(n^2)$ time to test the double resolvability of a set. Recalling Lemma 6, we obtain the following strong polynomial time solvability for the general k-edge-augmented trees.

Theorem 6. *The* MWDRS *problem on k-edge-augmented trees can be solved in $O(n^{12k})$ time.*   □

4.2   Wheels

A general *wheel* $G = (V, E)$ on n (≥ 6) vertices $v_1, v_2, \ldots, v_n$ is formed by the *hub* vertex v_n and a cycle $C = (V_c, E_c)$ over the vertices $v_1, v_2, \ldots, v_{n-1}$, called *rim* vertices, where the hub is adjacent to some (not necessarily all) rim vertices. We develop dynamic programming algorithm to solve the MWDRS problem on general wheels in $O(n^3)$ time.

We start with complete wheels whose DRS has a very nice characterization that is related to the consecutive one property. A general wheel is *complete* if its hub is adjacent to every rim vertex.

The distance in G between any two vertices in G is either 1 or 2.   (4.1)

Lemma 7. *Given a complete wheel $G = (V, E)$, let S be a proper nonempty subset of V. Then S is a doubly resolving set if and only if $S \cap V(C)$ is a dominating set of C and any pair of rim vertices outside S has at least two neighbors in $S \cap V(C)$.*

Proof. If $S \cap V(C)$ is not a dominating set of C, then there exists a vertex $v_i \in V(C) - S$ such that v_i is not adjacent to any vertex of $S \cap V(C)$. In this case, S cannot doubly resolve $\{v_n, v_i\}$ because for any $s_1, s_2 \in S$, $d(s_1, v_n) - d(s_1, v_i) = -1 = d(s_2, v_n) - d(s_2, v_i)$.

If $S \cap V(C)$ is a dominating set of C but there exist two cycle vertices $v_i, v_j \in V(C) - S$ such that v_i, v_j are uniquely dominated by the same cycle vertex $v \in S$, then for any two vertices $s_1, s_2 \in S$, $d(s_1, v_i) - d(s_1, v_j) = 0 = d(s_2, v_i) - d(s_2, v_j)$, saying that S is not a doubly resolving set.

Suppose that S satisfies the condition stated in the lemma. We prove that S can resolve every pair of vertices x, y in G. When one of x and y, say x, is a rim vertex in S, since $n \geq 6$, there exists another rim vertex $z \in S - \{x\}$ that is not adjacent to x. It follows from (4.1) that $\{x, y\}$ is resolved by $\{x, z\}$ as $d(x, x) - d(x, z) = -2 < -1 \leq d(y, x) - d(y, z)$. When both x and y are rim vertices outside S, there are two rim vertices x' and y' in S dominating x and y, respectively. It follows that $\{x, y\}$ is resolved by $\{x', y'\}$. When one of x and y, say x is the hub, we only need consider the case of y is a rim vertex outside S. Take rim vertices z, z' from S such that z dominates y and z' does not dominate y. It follows that $\{x, y\}$ is resolved by $\{z, z'\}$ as $d(x, z) - d(x, z') = 0 < -1 = d(y, z) - d(y, z')$. $\qquad\square$

The characterization in Lemma 7 can be rephrased as follows: A subset $S \subseteq V$ is a DRS of G if and only if every set of three consecutive vertices on C contains at least one vertex of S, and every set of five consecutive vertices on C contains at least two vertices of S. This enables us to formulate the MWDRS problem on a complete wheel as an integer programming with consecutive 1's and circular 1's constraints, which can be solved in $O(n^3 \log^2 n)$ time by Hochbaum and Levin's algorithm [11]. (To the best of our knowledge, there is no such a concise way to formulate the metric dimension problem on complete wheels as an integer programming with consecutive one matrix.) Moreover, it is not hard to see from the characterization that linear time efficiency can be achieved by dynamic programming approach. Furthermore, we elaborate on the idea to solve the MWDRS problem on more complex general wheels.

Theorem 7. *The* MWDRS *problem on complete wheels can be solved in* $O(n)$ *time. The* MWDRS *problem on general wheels can be solved in* $O(n^3)$ *time.* $\qquad\square$

5 Conclusion

In this paper, we have established $\Theta(\ln n)$ approximability of the MDRS and MWDRS problems on general graphs. There is still a gap of $1 + \ln \log_2 n$ hidden in the big theta (see Theorems 1 and 2). It deserves good research efforts to obtain even tighter upper bounds for the approximability. The k-edge-augmented trees, general wheels and cographs are known graph classes on which the weighted metric dimension problem is polynomial time solvable. In this paper, we have extended the polynomial time solvability to the MWDRS problem for the first two graph classes. It would be interesting to see whether the problem on cographs and other graphs also admits efficient algorithms.

References

1. Berman, P., DasGupta, B., Kao, M.Y.: Tight approximability results for test set problems in bioinformatics. J. Comput. Syst. Sci. 71(2), 145–162 (2005)
2. Cáceres, J., Hernando, C., Mora, M., Pelayo, I.M., Puertas, M.L., Seara, C., Wood, D.R.: On the metric dimension of cartesian products of graphs. SIAM J. Discrete Math. 21(2), 423–441 (2007)
3. Čangalović, M., Kratica, J., Kovačević-Vujčić, V., Stojanović, M.: Minimal doubly resolving sets of prism graphs. Optimization (ahead-of-print), 1–7 (2013)
4. Chartrand, G., Zhang, P.: The theory and applications of resolvability in graphs: A survey. Congressus Numerantium 160, 47–68 (2003)
5. Chen, X., Hu, X., Wang, C.: Approximability of the minimum weighted doubly resolving set problem. CoRR abs/1404.4676 (2014)
6. Chlebík, M., Chlebíková, J.: Approximation hardness of dominating set problems in bounded degree graphs. Inf. Comput. 206(11), 1264–1275 (2008)
7. Díaz, J., Pottonen, O., Serna, M., van Leeuwen, E.J.: On the complexity of metric dimension. In: Epstein, L., Ferragina, P. (eds.) ESA 2012. LNCS, vol. 7501, pp. 419–430. Springer, Heidelberg (2012)
8. Epstein, L., Levin, A., Woeginger, G.J.: The (weighted) metric dimension of graphs: hard and easy cases. In: Golumbic, M.C., Stern, M., Levy, A., Morgenstern, G. (eds.) WG 2012. LNCS, vol. 7551, pp. 114–125. Springer, Heidelberg (2012)
9. Gomez Rodriguez, M., Leskovec, J., Krause, A.: Inferring networks of diffusion and influence. In: Proc. of the 16th ACM SIGKDD International Conference on Knowledge Discovery and Data Mining, KDD 2010, pp. 1019–1028 (2010)
10. Hauptmann, M., Schmied, R., Viehmann, C.: Approximation complexity of metric dimension problem. J. Discrete Algorithms 14, 214–222 (2012)
11. Hochbaum, D.S., Levin, A.: Optimizing over consecutive 1's and circular 1's constraints. SIAM J. Optimization 17(2), 311–330 (2006)
12. Kratica, J., Kovačević-Vujčić, V., Čangalović, M., Stojanović, M.: Minimal doubly resolving sets and the strong metric dimension of hamming graphs. Appl. Anal. Discret. Math. 6(1), 63–71 (2012)
13. Kratica, J., Kovačević-Vujčić, V., Čangalović, M., Stojanović, M.: Minimal doubly resolving sets and the strong metric dimension of some convex polytopes. Appl. Math. Comput. 218(19), 9790–9801 (2012)
14. Kratica, J., Čangalović, M., Kovačević-Vujčić, V.: Computing minimal doubly resolving sets of graphs. Comput. Oper. Res. 36(7), 2149–2159 (2009)
15. Mladenović, N., Kratica, J., Kovačević-Vujčić, V., Čangalović, M.: Variable neighborhood search for metric dimension and minimal doubly resolving set problems. Eur. J. Oper. Res. 220(2), 328–337 (2012)
16. Pinto, P.C., Thiran, P., Vetterli, M.: Locating the source of diffusion in large-scale networks. Phys. Rev. Lett. 109, 068702 (2012)
17. Shah, D., Zaman, T.: Rumors in a network: Who's the culprit? IEEE Trans. Information Theory 57(8), 5163–5181 (2011)

Approximating High-Dimensional Range Queries with kNN Indexing Techniques

Michael A. Schuh[1], Tim Wylie[2], Chang Liu[1], and Rafal A. Angryk[3]

[1] Dept. of Computer Science, Montana State University, Bozeman, MT, 59717 USA
[2] Dept. of Computer Science, University of Alberta, Edmonton, AB, T6G2E8 Canada
[3] Dept. of Computer Science, Georgia State University, Atlanta, GA, 30302 USA
michael.schuh@cs.montana.edu

Abstract. While k-nearest neighbor queries are becoming increasingly common due to mobile and geospatial applications, orthogonal range queries in high-dimensional data are extremely important in scientific and web-based applications. For efficient querying, data is typically stored in an index optimized for either kNN or range queries. This can be problematic when data is optimized for kNN retrieval and a user needs a range query or vice versa. Here, we address the issue of using a kNN-based index for range queries, as well as outline the general computational geometry problem of adapting these systems to range queries. We refer to these methods as space-based decompositions and provide a straightforward heuristic for this problem. Using iDistance as our applied kNN indexing technique, we also develop an optimal (data-based) algorithm designed specifically for its indexing scheme. We compare this method to the suggested naïve approach using real world datasets and results show that our data-based algorithm consistently performs better.

1 Introduction

Modern society has grown dependent upon large-scale data mining applications. This reliance is only increasing as our ability to record and store vast quantities of rich data improves due to new technologies and new applications. Often this data is high-dimensional in nature, which can be challenging for efficient use and data mining. For a user interacting with the data, two types of queries are often necessary: orthogonal range and k-nearest neighbor (kNN).

The use of nearest neighbor queries is essential in many types of applications such as geospatial, consumer, and mobile. These queries allow a user to search for the nearest stores of interest to their location, or to find a song, image, etc. with similar qualities to something they already know. Thus, many modern systems focus on this query methodology – especially within mobile and web-based applications. Orthogonal range queries differ in that they require a min and max range to be selected for every dimension of the query. This is often used more in scientific, user-defined, exploratory, and web search applications. Allowing the user to specify each dimensional range is important for capturing critical values or removing specificity entirely, *i.e.*, a wildcard search with some

Z. Cai et al. (Eds.): COCOON 2014, LNCS 8591, pp. 369–380, 2014.
© Springer International Publishing Switzerland 2014

set of unrestricted dimensions. This is also useful for data with non-numeric attributes where distance is meaningless.

Indexing methods allow large volumes of data to be stored in a way that is optimized to decrease retrieval time for their intended application. However, there may be times when another type of application (or query) is desired, but the data is stored in an inconvenient (and unoptimized) manner for this request. Here, we look at the situation where data is optimized for kNN retrieval, but range queries are also necessary. This assumes that we will not modify the current index, and our methods must use the inherent functionality of the existing indexing and retrieval mechanisms. We will focus on the state-of-the-art iDistance kNN index [1, 2], which has been used in a number of demanding applications, including: large-scale image retrieval [3], video indexing [4], mobile computing [5], peer-to-peer systems [6], and video surveillance retrieval [7].

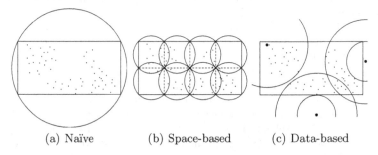

(a) Naïve   (b) Space-based   (c) Data-based

Fig. 1. Three different methods of retrieving range queries using iDistance. In (a) we naïvely encircle the range query. For (b) we cover the query space with overlapping smaller query spheres. In (c) we calculate the specific intersected iDistance partitions.

This work explores three methods for satisfying range queries within a kNN indexing technique, shown as $2D$ examples in Figure 1. The first is the naïve method, suggested by many authors including those of iDistance [1], but this has drawbacks. In higher dimensions, only a few poorly chosen dimensional ranges (or wildcards) can lead to the pessimal case as seen in Figure 3(d). Another possibility comes from the field of computational geometry. Since kNN queries are hyperspheres, we attempt to use a number of them to cover the hyperrectangle space of the range query. Unfortunately, this method breaks down in higher dimensions because it is an NP-hard problem to optimally cover the space. We provide a heuristic to accomplish this based on the ranges needed in each dimension. We assume that overlap between the query spheres is acceptable, and we are able to skirt the theoretical optimization problems based on minimum overlap, minimum number of queries, etc. These are rich and important areas of research [8], but are not critical for this work and are therefore kept brief.

These first two methods can be classified as space-based decompositions because we seek to use one geometric object to cover the same space as another. While these methods will work with any kNN system, the last method we develop is a novel data-based approach specific to iDistance. We refer to this as a data-based method because it uses the underlying indexing data structure rather than solely the coverage space of the range query. It is designed for the iDistance index, and therefore other indexing schemes would require their own specific data-based range query algorithm.

Results highlight retrieval performance of range queries over three high-dimensional real-world datasets with existing kNN indices. We find great variability in results over the various datasets which implies a certain degree of difficulty in performing useful range queries in high dimensional spaces, as well as a general lack of ability to anticipate results in practice. Regardless, we show the data-based method is the preferred choice for facilitating range queries as it out-performs the naïve method in almost all situations.

This paper is organized as follows. Section 2 will briefly overview background and related work. Sections 3 and 4 will then cover the space-based methods, followed by the data-based method in Section 5. Then we present empirical evaluations and discussion in Section 6, and the paper concludes in Section 7.

2   Preliminaries

2.1   Indexing

For our study we focus on iDistance, which is a modern kNN-based indexing method for high-dimensional data first proposed in 2001 [1]. It uses a filter-and-refine strategy based on a lossy transformation into an efficient one-dimensional B$^+$-tree for indexing and retrieval. iDistance is based on a Voronoi tessellation of the space [9], but for speed it approximates each of these areas with a hypersphere. These partitions $\mathfrak{P} = \{P_1, \ldots, P_M\}$ are stored by their central reference point, $\mathfrak{O} = \{O_1, \ldots, O_M\}$. The number of partitions, M, can be arbitrarily set, as can the location of the reference points (and thereby resultant partitions). The data points are assigned to the closest partition by reference point distance. The radius of each partition is stored as *distmax*, which is the distance of the farthest point in that partition. Figure 2 shows an example of these elements as well as a query sphere and the areas of each partition to search.

Prior studies have shown that poor placement or an insufficient (or abundant) number of partitions can greatly reduce the effectiveness of the algorithm [2, 10–12]. The general practice is to use a clustering algorithm, such as k-means which has similar spherical characteristics, to derive cluster centers to be used as reference points. It is also good practice to use a number of clusters directly related to the size and dimensionality (D) of the dataset to be indexed, with $2D$ partitions serving as a general rule of thumb [10].

For efficiency, the points are indexed in a one-dimensional B$^+$-tree [13], which allows iDistance to be incorporated into modern databases. This lossy transformation separates partitions by a spacing constant c, which can be safely set to $\sqrt{D}$ since no two data points can be farther apart than this value in a normalized dataspace. Then, for any point p assigned to a partition P_i, it is mapped in the B$^+$-tree to the value $y_p = c \cdot i + dist(O_i, p)$. This mapping can be seen in Figure 2. Note that concentric rings of data points share the same index value.

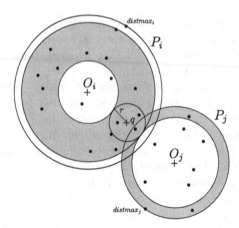

Fig. 2. A basic example of iDistance indexing partitions with a sample query q of radius r. Partitions P_i, P_j with reference points O_i, O_j, respectively.

A kNN query using iDistance involves three basic steps. Note that the query is given an initial radius to create a query hypersphere in the dataspace. First, the filter step finds all intersecting partitions and retrieves the points in the B^+-tree corresponding to this overlap. Since several points may have the same value in the B^+-tree, iDistance then refines the results to ensure they are within the query hypersphere. Lastly, if k results were not yet found, the algorithm iterates by increasing the query radius and searching the partitions again. Only when the farthest item in the full set of k items is closer than the current query radius can the search be successfully terminated with the guaranteed exact results.

2.2 Geometric Queries

Formally, we will assume that any query is in a D dimensional space such that the set of dimensions is $\mathfrak{D} = \{d_1, \ldots, d_D\}$. A point kNN query is defined as a query point q with a radius r, which returns the k nearest elements within the ball $B(q, r)$ based on a specified distance metric $dist()$.

An orthogonal (or axis-aligned) range query in a D dimensional space is the set $\mathfrak{R} = \{R_{d_1}, \ldots R_{d_D}\}$. Each range R_{d_i} is a tuple of the minimum and maximum values that define the range over that dimension, *i.e.*, $R_{d_i} = \langle v_{min}, v_{max} \rangle$, so the range of dimension i is $[v_{min}, v_{max}]$. For convenience, we will also use the notation $\min(R_{d_i})$ for v_{min} in dimension i, and similarly for v_{max}. We will assume these definitions in the subsequent sections.

Orthogonal range queries, also known as window queries, are a well-studied problem that has spawned a lot of research for new data structures and algorithms for efficient retrieval [8]. Some of the more notable methods include kd-trees, multi-layered range trees, and R-trees [14]. Optimal retrieval time is provable with a tree construction, but requires greater storage than $O(N)$ [15]. Many of these methods achieve better time efficiency through additional space usage [8]. In large high-dimensional datasets, many of these methods can be too space inefficient to be practical. We focus on iDistance since it is a modern high-dimensional method, which is currently used widely in practice.

3   A Naïve Approach

The basic idea of the naïve method is to encompass the desired range query within a query sphere, retrieve all points inside this sphere, and then prune the results outside of the original ranges. Figure 1(a) shows an example of this. What is immediately obvious is that the number of bad results returned grows quickly as the difference between the lengths of the hyperrectangle edges grows. This is especially problematic for wildcard-capable queries or several dimensions with wide ranges which can expand the query sphere to nearly the size of the entire dataspace, as shown in Figure 3(d).

The naïve method is a straight-forward approach that easily adapts a range query to a *k*NN system. The query point is the average of all the ranges, and the radius is from the center to a corner of the hyperrectangle. The value of each dimension i for the center point c of the circumscribed sphere is calculated by Equation 1, and the radius from the center to the corner by Equation 2.

$$c_i = \frac{\min(R_{d_i}) + \max(R_{d_i})}{2} \quad (1) \qquad r^2 = \sum_{i=1}^{D}(\max(R_{d_i}) - c_i)^2 \quad (2)$$

In the above equations, c_i is the coordinate value of the query center point in the ith dimension ($0 \le c_i \le 1$ in a normalized space). After we have calculated the query sphere, the *k*NN algorithm works the same except we do not bound our return set to only the k closest items and instead return everything found within the query sphere. An additional refinement step is then required to prune away all points returned from the *k*NN query that are outside of the original range query. Although this method is cited as an easy extension [1], its drawbacks often make it an impractical choice.

4   Space-Based Decomposition

Encircling a hyperrectangle can be inefficient with respect to the amount of area included outside the hyperrectangle. The optimal shape, resulting in minimal area included outside the hyperrectangle, is when the sides are equal and we have a hypercube. Our goal is to give a heuristic to decompose a range query into a set of smaller range queries that are closer to hypercubes, and thereby individually more efficient. Then each of these queries can be encircled and searched with any *k*NN system like the naïve method. An example is shown in Figure 1(b). This decomposition will result in less area searched overall, although the complexity will be higher due to the algorithm overhead and combination of multiple partially overlapping (and unbounded) *k*NN queries.

Sphere packing is a well-studied problem, but many problems related to packing and covering with overlapping spheres are still open, such as the optimal

1-density above two dimensions [16]. The 1-density, δ_D^1 in D dimensions, is the volume covered by a single sphere in a space covered with unit spheres as shown in Figure 3(a). In $2D$, the optimal is $opt(\delta_2^1) = (3\sqrt{(3)} - \pi)/\pi \approx 0.6539$, and it is believed that in $3D$ the optimal is $opt(\delta_3^1) \approx 0.315$, which was determined by empirical methods [16]. It is also known that as D increases, the optimal 1-density decreases, and the limit is zero as the dimensionality goes to infinity. This becomes important in our application. We briefly outline the space-based decomposition problem in relation to range queries and give our heuristic which may be a polynomial-time approximation scheme (PTAS) while $hal_I(\delta_D^1) > 0$, where $hal_I(\delta_D^1)$ gives the 1-density of a single sphere I from our heuristic algorithm (hal) in D dimensions.

Definition 1. *Space-based Decomposition:*
Instance: *Given an orthogonal range query* $\mathfrak{R} = \{R_{d_1}, \ldots R_{d_D}\}$ *in D dimensions where* $R_{d_i} = \langle v_{min}, v_{max}\rangle$ *is the minimum and maximum value allowed in the ith dimension, and a* $K \in \mathbb{R}^+$.
Problem: *Find a set of spheres* $\mathfrak{S} = \{s_1, \ldots, s_S\}$ *of equal radius such that* $V_{\mathfrak{S}} = \cup_{i=1}^{|\mathfrak{S}|} Vol(s_i)$ *covers* $V_{\mathfrak{R}} = \cup_{i=1}^{|\mathfrak{R}|} Vol(R_{d_i})$, *where* $V_{\mathfrak{S}} - V_{\mathfrak{R}} \leq K$ *and the 1-density* $hal_s(\delta_D^1) > 0 \,\forall s \in \mathfrak{S}$.

Essentially, we want to cover the range hyperrectangle with hyperspheres while guaranteeing that each sphere covers some volume (Vol) which no other sphere covers, and that the total volume queried outside the hyperrectangle is less than K. Our method attempts to partition each dimension such that the resulting subrectangles are close to hypercubes. This guarantees that for each sphere, we can bound the 1-density based on knowing the number of neighboring spheres and the amount of intersection. We note that all our spheres have the same radius, which is necessary in order to avoid a bin-packing problem. Thus, the spheres can be regarded as unit-spheres.

A D-cube has $2D$ sides, and thus $2D$ D-spheres which may intersect the hypersphere covering the hyperrectangle. If we let each side be represented by x, then we can calculate the 1-density for each sphere. The diameter of every sphere is $d = \sqrt{Dx^2}$. From this we can calculate the intersecting volume of the $2D$-spheres and subtract it from the volume of the hypercube $V_c = x^D$.

The overlap of a single sphere from one side is $\frac{1}{2D}(V_s - V_c)$ where V_s is the volume of a single sphere. Thus, the 1-density of our heuristic algorithm is $hal(\delta_2^1) = 2V_c - V_s + DO_s$ where O_s is the overlap volume of a sphere onto a neighboring sphere, which is zero in $2D$. The 1-density shrinks drastically in higher dimensions. For $2D$ the value is $2x^2 - \frac{x^2\pi}{2}$ and if $x = 1$ then $hal(\delta_2^1) = 2 - \pi/2 \approx .4292$ and is shown in Figure 3(a). Our heuristic is less than the optimal $opt(\delta_2^1) \approx 0.6539$, which means we have more overlap in our query spheres. The optimal value requires a difficult partitioning of the space in $2D$, and the optimal partitioning is unknown above $2D$.

The heuristic algorithm tries to find the most economical way to split each dimension such that the resulting hyperrectangles are as close to hypercubes as possible. Based on an $\varepsilon > 0$, we can find the optimal number of divisions. We find the appropriate number of divisions, X_i, for a given dimensional range, R_i where $1 \le i \le D$, with Equation 3. We can see in Figures 3(b) and 3(c) how this number of sections created in a given dimension grows exponentially. For the sake of discussion, we ignore the real world implications of retrieving and combining multiple overlapping queries in an indexing system such as iDistance.

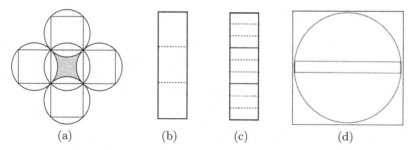

(a)   (b)   (c)   (d)

Fig. 3. (a) A $2D$ example of the 1-density in our heuristic. The shaded region is the 1-density of the circle around the center cube. (b) A range query with a base-3 split once into three sections and then (c) split once again creating nine total sections. (d) A normalized dataspace with a range query resulting in a poor naïve solution that queries most of the dataspace even though only a small area is necessary.

In order to find the best practical ratio with respect to ε, we look at several logarithmic bases b. This means $b^{X_i^b}$ is the number of resulting sections in a dimension i. Further, we will find different values for X_i depending on which base we use, and we therefore add a superscript to denote the base. We limit our algorithm to the first few unique bases to ensure that there are only a constant number of calculations each time while still satisfying ε. Given the splits, we want to choose the ratio that is closest to one for each dimension. The ratio for base b in dimension i is given in Equation 4. Note that we can use a different ε for each dimension to achieve the best ratio nearest to one.

$$X_i^b = \left\lfloor \log_b \left(\frac{R_i}{\varepsilon} \right) \right\rfloor, b = 2, 3, 5, \dots \quad (3) \qquad r_{ib} = \frac{R_i}{\varepsilon \cdot (b^{X_i^b})} \quad (4)$$

The ε term allows us to increase the number of spheres used and as we make them smaller we decrease the volume covered outside the range query. We can make this volume as small as desirable, but our running time is dependent on how many spheres there are. Thus, this may be a PTAS that is still exponential in ε. For a more practical heuristic we limit $b = \{2, 3, 5\}$ and set $\varepsilon = \min(\mathfrak{R})$. The value $\min(\mathfrak{R})$ is the smallest length of any side of the hyperrectangle, and thus a good basis for finding the closest dimensions to approximate a hypercube.

Although this method is straightforward and constant to calculate, the number of query spheres grows exponentially with respect to the number of dimensional splits. This makes the general algorithm infeasible in any high-dimensional space when there is a lot of variation between the query ranges in each dimension. One could consider bounding the algorithm to only a few splits or dimensions, but this is non-trivial and changes the desired results. Further, we know that as the dimensionality increases, the overlap of the spheres also increases, which means it becomes less effective in higher dimensions because we are repeatedly re-searching parts of the dataspace. To be efficient, the indexing method would need to track searched space across all spheres. This requires non-trivial modifications to the indexing system, which is beyond our scope of purpose.

5   Data-Based Method

Sections 3 and 4 focused on methods for ensuring that the same range query volume in the dataspace is covered. We now want to ignore the underlying space of the range query, and simply focus on the data that exists within this space. As you can see in Figure 1(c), given clustered data in the space, we only need to search where data exists. Since iDistance is an exact retrieval algorithm that indexes all data based on spherical partitions, we only need to identify which partitions the range query intersects, and to what extent they intersect.

Since our range queries are orthogonal, the closest point p in the axis aligned range $\mathfrak{R}$ to an outside reference point O_i (of partition P_i) is given on a per dimension (d) comparison. That is, we can calculate the closest possible point to O_i that is still in the range by looking at each dimension individually. We drop the i subscript here for clarity since we are using the dimension as the subscript when calculating the location of the point in relation to a single partition.

$$p_d(R_d, O_d) = \begin{cases} \min(R_d), & \text{if } O_d < \min(R_d) \\ \max(R_d), & \text{if } O_d > \max(R_d) \\ O_d, & \text{if } \min(R_d) \leq O_d \leq \max(R_d) \end{cases} \tag{5}$$

The closest point in $\mathfrak{R}$ is also the closest point to the reference point O_i in partition P_i that any data may be within the range. We can then take the distance $m = dist(O_i, p)$ and search the interval of $[distmax - m, distmax]$ for that partition in the B$^+$-tree – partition boundaries often further limit this interval. If m is greater than the $distmax$, then the range query does not intersect that partition. If it does, then we immediately know the exact amount and a single query will retrieve all the results.

There are a few special cases, such as if the partition completely encapsulates the range. In this case we can default to the naïve method to search (which would be equivalent here), but to be consistent with our algorithm we calculate the farthest point and the closest point within our range to the reference point. Equation 5 is the closest, but the farthest point can also easily be deduced in a similar manner.

To enable this data-based range query feature, we had to modify one critical function from the original iDistance implementation. By modifying the *SearchO* function previously presented in [2], we use iDistance as a state-of-the-art kNN indexing algorithm to optimally facilitate and retrieve high-dimensional range queries. Presented in Algorithm 1, *RangeSearchO* is now used instead of *SearchO* for all range queries. This procedure assumes a global variable S which contains the retrieved set of candidates from the filter step. Note this does not change any of the underlying indexing mechanisms or data structures.

Algorithm 1. Find all points within a specified range

Input: $\Re$ is the set of ranges in each dimension.

```
 1: procedure RangeSearchO (ℜ)
 2:     for each Pᵢ ∈ 𝔓 do                                          ▷ Filter Step
 3:         closest-point ← ClosestPoint(ℜ, Oᵢ)                     ▷ Equation 5
 4:         m ← dist(Oᵢ, closest-point)
 5:         if m ≤ distmaxᵢ then
 6:             farthest-point ← FarthestPoint(ℜ, Oᵢ)
 7:             if distmaxᵢ < farthest-point then
 8:                 farthest-point ← distmaxᵢ
 9:             SearchInward((c · i) + farthest-point, m)
10:     for each p ∈ S do                                          ▷ Refine Step
11:         if p ∉ ℜ then
12:             S ← S\p
```

---

6 Empirical Evaluation

We now test our three methods of kNN-based range query retrieval through general performance comparisons on real world data. Using best practices, we setup an iDistance index on each normalized dataset using k-means to derive $2D$ cluster centers for reference points. Results are presented for three commonly used public datasets of varying dimensionality and size. Synthetic datasets and other types (and amounts) of reference points were also researched here, but due to limited space and similar results, we exclude these from discussion.

To explicitly investigate the functionality of the methods, we performed range queries with a varying number of wide dimensions, defined as the entire $[0, 1]$ range. We randomly select 100 data points as queries to ensure that our query point distribution follows the underlying dataset distribution. For each experiment we show the number of candidate points returned during the filter step and the final result set size as a percentage of the dataset, as well as the total time taken (in milliseconds) to perform the query and return the final exact results. For space, we omit results for B^+-tree nodes accessed, as they directly mimic candidates accessed for all tests performed, which is typical behavoir. We also omit results for the space-based decomposition method because while reasonably effective in low dimensions, or with only a few wide dimensions, it (as expected) quickly becomes inefficient and impractical to run in higher dimensions.

Sequential scan (SS) is often used as a benchmark comparison for worst-case performance because it checks every data point in a brute force fashion. Note that all data fits in main memory, so all experiments are compared without depending on the behaviors of specific hardware-based disk-caching routines. In real-life however, disk-based I/O bottlenecks are a common concern for inefficient retrieval methods. Therefore, in these experiments it is more important that the index properly filters out data points (fewer candidates) so that they do not have to be accessed whatsoever.

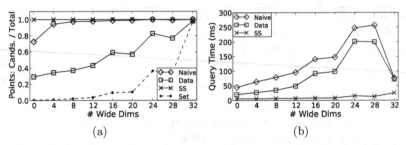

Fig. 4. Range queries with increasing number of wide dimensions for the Color dataset

We begin with the popular University of California Irvine (UCI) Corel Image Features dataset[1] (referred to herein as Color). This is the smallest of the three, containing 68,040 points in 32 dimensions derived from color histograms. Results are shown in Figure 4 for the naïve and data-based (Data) methods compared to SS. We also include the return set size (Set) in 4(a), which indicates the percentage of the dataset returned by the range query. Notice that even the initial hypercube is retrieved much more efficiently by the data-based method, whereby the naïve method starts poorly and immediately degrades to searching almost the entire dataset. Once that happens, sequential scan is a better choice because it does not have the algorithmic overhead, which is evident by the query time for each method. We also see that less time is required when the entire space is forced to be searched upfront (all wide dims), rather than methodically determining to search most of the space anyway (e.g., 28 dims). The most important factor here is the filter capability, and we see the data-based method gracefully degrades to the naïve (and SS) performance as the number of wide dimensions increases.

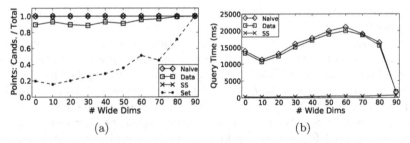

Fig. 5. Range queries with increasing number of wide dimensions for the Music dataset

Figure 5 presents results from the UCI YearPredictionMSD dataset[2] (Music), which has 90 dimensions and 515,345 data points representing musical characteristics of songs. Here we see the data-based method only performs slightly better than the naïve method. Notice in 5(a) that the initial query already returns about 20% of the dataset, which could be affecting relative performance comparisons. For each dataset we pick the query points and wide dimensions in the same random manner, so it seems to imply a difference in dataset characteristics.

---

[1] http://archive.ics.uci.edu/ml/datasets/Corel+Image+Features
[2] http://archive.ics.uci.edu/ml/datasets/YearPredictionMSD

The ANN_SIFT1M dataset[3] (Sift) contains one million points and was specifically created for nearest neighbor retrieval on 128-dimensional SIFT feature vectors extracted from images. The results, presented in Figure 6, are quite different than the Music dataset, rejecting any purely "high-dimensional" performance factors in favor of more rich factors tied closely to the dataset characteristics. Here we see exceptional performance for the data-based method with both statistics. However, it is likely in part due to the extremely small result set size, which on average is only the data point the query is centered on. To our surprise, this holds true even with almost all wide dimensions in the query. This explains why the time taken increases with all wide dimensions (the opposite of other datasets), because it finally incurs a higher cost during the refinement step. By maintaining relatively high performance in retrieving a small set from a large and high dimensional dataset, this suggests a promising use case for application of retrieving highly selective high-dimensional range queries.

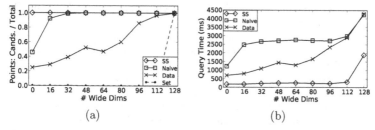

(a)   (b)

Fig. 6. Range queries with increasing number of wide dimensions for the Sift dataset

We find varying results over all three datasets, but it is clear that the data-based method is superior to the trivial naïve approach. We believe further investigation and analysis of these datasets could shed light on the results we found, which will be true of any real world dataset. The Color dataset provides reasonable results one would expect to see, while the Music and Sift datasets show varying possible outcomes. The Music dataset seems to contain relatively congested areas in the dataspace given the poor filtering ability on all queries, while the Sift dataset appears extremely sparse, allowing the query search to avoid most of the dataset and return very few results.

7   Conclusions

This paper presented solutions for approximating orthogonal range (window) queries in kNN indexing systems. We discussed the general computational geometry problem surrounding this adaptation process, and introduced a heuristic guided space-based decomposition algorithm. Using iDistance for the high-dimensional kNN indexing algorithm, we developed an optimal data-based range query retrieval algorithm that requires no modifications to the underlying index or data structures and no additional storage costs. While providing an ideal graceful degradation towards the naïve solution in worst-case scenarios, results

---

[3] http://corpus-texmex.irisa.fr/

show our novel data-based method generally performs superior and provides exceptional filtering of the dataset even when many wide dimensions are specified, satisfying any practical wildcard searching.

Future work surrounds the nature and usage of range queries in practice. For example, preliminary results in selectivity-based range queries yield drastically different results for different real world datasets, much like the return set sizes presented here. Other considerations include weighted range queries (hyperellipses), exploratory (or fuzzy boundary) range queries, and rank (distance-based) results, all of which could be performed from an existing kNN index.

References

1. Yu, C., Ooi, B.C., Tan, K.-L., Jagadish, H.V.: Indexing the Distance: An Efficient Method to KNN Processing. In: Proc. of the 27th VLDB Conf., pp. 421–430 (2001)
2. Jagadish, H.V., Ooi, B.C., Tan, K.L., Yu, C., Zhang, R.: iDistance: An adaptive B+-tree based indexing method for nearest neighbor search. ACM Trans. Database Syst. 30, 364–397 (2005)
3. Zhang, J., Zhou, X., Wang, W., Shi, B., Pei, J.: Using high dimensional indexes to support relevance feedback based interactive images retrieval. In: Proc. of the 32nd VLDB Conf., pp. 1211–1214 (2006)
4. Shen, H.T.: Towards effective indexing for very large video sequence database. In: SIGMOD Conference, pp. 730–741 (2005)
5. Ilarri, S., Mena, E., Illarramendi, A.: Location-dependent queries in mobile contexts: Distributed processing using mobile agents. IEEE Trans. on Mobile Computing 5(8), 1029–1043 (2006)
6. Doulkeridis, C., Vlachou, A., Kotidis, Y., Vazirgiannis, M.: Peer-to-peer similarity search in metric spaces. In: Proc. of the 33rd VLDB Conf., pp. 986–997 (2007)
7. Qu, L., Chen, Y., Yang, X.: iDistance based interactive visual surveillance retrieval algorithm. In: Intelligent Computation Technology and Automation (ICICTA), vol. 1, pp. 71–75. IEEE (October 2008)
8. de Berg, M., Cheong, O., van Kreveld, M., Overmars, M.: Computational Geometry: Algorithms and Applications, 3rd edn. Springer (April 2008)
9. Aurenhammer, F.: Voronoi diagrams – a survey of a fundamental geometric data structure. ACM Comput. Surv. 23, 345–405 (1991)
10. Schuh, M.A., Wylie, T., Banda, J.M., Angryk, R.A.: A comprehensive study of iDistance partitioning strategies for kNN queries and high-dimensional data indexing. In: Gottlob, G., Grasso, G., Olteanu, D., Schallhart, C. (eds.) BNCOD 2013. LNCS, vol. 7968, pp. 238–252. Springer, Heidelberg (2013)
11. Schuh, M.A., Wylie, T., Angryk, R.A.: Improving the performance of high-dimensional kNN retrieval through localized dataspace segmentation and hybrid indexing. In: Catania, B., Guerrini, G., Pokorný, J. (eds.) ADBIS 2013. LNCS, vol. 8133, pp. 344–357. Springer, Heidelberg (2013)
12. Schuh, M.A., Wylie, T., Angryk, R.A.: Mitigating the curse of dimensionality for exact knn retrieval. In: Proc. of the 27th FLAIRS Conf. AAAI (2014)
13. Bayer, R., McCreight, E.M.: Organization and maintenance of large ordered indices. Acta Informatica 1, 173–189 (1972)
14. Guttman, A.: R-trees: a dynamic index structure for spatial searching. In: Proc. of the ACM SIGMOD Int. Conf. on Management of Data, pp. 47–57 (1984)
15. Chazelle, B.: Lower bounds for orthogonal range searching: I. the reporting case. Journal of the ACM 37(2), 200–212 (1990)
16. Zhu, B.: On the 1-density of unit ball covering. CoRR abs/0711.2092 (2007)

Approximation Algorithms for Maximum Agreement Forest on Multiple Trees[*]

Feng Shi[1], Jianer Chen[1,2], Qilong Feng[1], and Jianxin Wang[1]

[1] School of Information Science and Engineering, Central South University, China
[2] Department of Computer Science and Engineering, Texas A&M University, USA

Abstract. Given a collection of phylogenetic trees with identical leaf label-set, the Maximum Agreement Forest problem (MAF) asks for a largest common subforest of these input trees. The MAF problem on two binary phylogenetic trees has been studied extensively in the literature. In this paper, we will be focused on the MAF problem on multiple (i.e., two or more) binary phylogenetic trees and present two polynomial-time approximation algorithms, one for the MAF problem on multiple rooted trees, and the other for the MAF problem on multiple unrooted trees. The ratio of our algorithm for the MAF problem on multiple rooted trees is 3, which is an improvement over the previously best ratio 8 for the problem. Our 4-approximation algorithm for the MAF problem on multiple unrooted trees is the first approximation algorithm for the problem.

1 Introduction

Phylogenetic trees have been widely used in the study of evolutionary biology to represent the tree-like evolution of a collection of species. However, different data sets and different building methods may lead to different phylogenetic trees. In order to facilitate the comparison of these different trees, several distance metrics have been proposed, such as Robinson-Foulds [1], NNI [2], TBR and SPR [3, 4].

A graph theoretical model, the *maximum agreement forest* (MAF) of two phylogenetic trees, has been formulated for the TBR distance and the SPR distance [5] for phylogenetic trees. Define the *order* of a forest to be the number of connected components in the forest.[1] Allen and Steel [6] proved that the TBR distance between two unrooted binary phylogenetic trees is equal to the order of their MAF minus 1, and Bordewich and Semple [7] proved that the rSPR distance between two rooted binary phylogenetic trees is equal to the order of their rooted version of MAF minus 1. In terms of computational complexity, it is known that computing the order of an MAF is NP-hard and MAX SNP-hard for two unrooted binary phylogenetic trees [5], as well as for two rooted binary phylogenetic trees [7].

---

[*] This work is supported by the National Natural Science Foundation of China under Grants (61103033, 61173051, 61232001).

[1] The definitions for the study of maximum agreement forests have been kind of confusing. If *size* denotes the number of edges in a forest, then for a forest, the size is equal to the number of vertices minus the order. In particular, when the number of vertices is fixed, a forest of a large size means a small order of the forest.

Z. Cai et al. (Eds.): COCOON 2014, LNCS 8591, pp. 381–392, 2014.
© Springer International Publishing Switzerland 2014

Approximation algorithms have been studied for the MAF problem, mainly on two trees. For the MAF problem on two rooted binary trees, approximation algorithms have been studied extensively [5, 8–11]. The best known approximation algorithm for the MAF problem on two rooted binary trees is a linear-time 3-approximation algorithm, which is developed by Whidden et al. [12, 13]. For the MAF problem on two unrooted binary trees, the best known approximation algorithm is also due to Whidden et al. [12, 13], who presented a linear-time approximation algorithm of ratio 3.

The MAF problem on multiple phylogenetic trees has not been studied as extensively as that on two trees. To our best knowledge, there is currently no known approximation algorithm for the MAF problem on multiple unrooted binary phylogenetic trees. The only approximation algorithm for the problem on multiple phylogenetic trees is a 8-approximation algorithm developed by Chataigner [15], which is for the problem on two or more rooted binary trees.

Note that it makes perfect sense to investigate the MAF problem on more than two trees: we may construct two or more different phylogenetic trees for the same collection of species according to different data sets and different building methods. However, it seems much more difficult to construct an MAF for more than two trees than that for two trees. For example, while there have been several polynomial-time approximation algorithms of ratio 3 for the MAF problem on two rooted binary trees [10–13], the best polynomial-time approximation algorithm [15] for the MAF problem on more than two rooted binary trees has a ratio 8. Also, to our best knowledge, there are currently no known approximation algorithms for the MAF problem on multiple unrooted binary trees.

In the current paper, we will be focused on the approximation algorithms for the MAF problem on multiple (i.e., two or more) binary phylogenetic trees, for both the version of rooted trees and the version of unrooted trees. Our algorithms are based on careful analysis of the graph structures that takes advantage of special relations among leaves in the trees. Our main contributions include a 3-approximation algorithm for the MAF problem on multiple rooted binary trees, which is an improvement over the previously best 8-approximation algorithm for the problem, and its ratio matches the best known approximation ratio for the problem on two rooted binary trees. We also present a 4-approximation algorithm for the MAF problem on multiple unrooted binary trees, which is the first approximation algorithm for the problem.

2   Definitions and the Problem Formulations

A tree is a *single-vertex tree* if it consists of a single vertex, which is the leaf of the tree. A tree is *binary* if either it is a single-vertex tree or each of its vertices has degree either 1 or 3. The degree-1 vertices are *leaves* and the degree-3 vertices are *non-leaves* of the tree. There are two versions in our discussion, one is on unrooted trees and the other is on rooted trees. In the following, we first give the terminologies on the unrooted version, then remark on the differences for the rooted version. Let X be a fixed *label-set*.

Unrooted X-trees and X-forests

A binary tree is *unrooted* if no root is specified in the tree – in this case no ancestor-descendant relation is defined in the tree. For the label-set X, an unrooted *binary phylogenetic X-tree*, or simply an unrooted X-*tree*, is an unrooted binary tree whose leaves are labeled bijectively by the label-set X (all non-leaves are unlabeled). An unrooted X-tree will also be called an (unrooted) leaf-labeled tree if the label-set X is irrelevant. A *subforest* of an unrooted X-tree T is a subgraph of T, and a *subtree* of T is a connected subgraph of T. An unrooted X-*forest* F is a subforest of an unrooted X-tree T that contains all leaves of T such that each connected component of F contains at least one leaf in T. Thus, an unrooted X-forest F is a collection of leaf-labeled trees whose label-sets are disjoint such that the union of the label-sets is equal to X. Define the *order* of the X-forest F, denoted $\mathrm{Ord}(F)$, to be the number of connected components in F. For a subset X' of the label-set X, if all leaves with labels in X' are in the same connected component of an unrooted X-forest F, then denote the *subtree induced* by X' in F by $F[X']$, which is the minimal subtree of F that contains all leaves with labels in X'.

A subtree T' of an unrooted X-tree may contain unlabeled vertices of degree less than 3. In this case we apply the *forced contraction* operation on T', which replaces each degree-2 vertex v and its incident edges with a single edge connecting the two neighbors of v, and removes each unlabeled vertex that has degree smaller than 2. Note that the forced contraction does not change the order of an X-forest. It has been well-known that the forced contraction operation does not affect the construction of an MAF for X-trees. Therefore, we will assume that the forced contraction is applied immediately whenever it is applicable. An X-forest F is *strongly reduced* if the forced contraction can not apply to F. Thus, the X-forests in our discussion are always assumed to be strongly reduced. With this assumption, a unlabeled vertex in an unrooted X-forest has degree 3.

Two leaf-labeled forests F_1 and F_2 are isomorphic if there is a graph isomorphism between F_1 and F_2 in which each leaf of F_1 is mapped to a leaf of F_2 with the same label. We will simply say that a leaf-labeled forest F' is a subforest of another leaf-labeled forest F if, up to the forced contraction, F' is isomorphic to a subforest of F.

Rooted X-trees and X-forests

A binary tree is *rooted* if a particular leaf is designated as the root (so it is *both* a root and a leaf), which specifies a unique ancestor-descendant relation in the tree. A rooted X-*tree* is a rooted binary tree whose leaves are labeled bijectively by the label-set X. The root of a rooted X-tree will always be labeled by a special label ρ in X. A subtree T' of a rooted X-tree T is a connected subgraph of T which contains at least one leaf in T. In order to preserve the ancestor-descendant relation in T, we should define the root of the subtree T'. If T' contains the leaf labeled ρ, certainly, it is the root of the subtree; otherwise, the node in T' that is in T the least common ancestor of all the labeled leaves in T' is defined to be the root of T'. A *subforest* of a rooted X-tree T is defined to be a subgraph of T. A (rooted) X-*forest* F is a subforest of a rooted X-tree

T that contains a collection of subtrees whose label-sets are disjoint such that the union of the label-sets is equal to X. Thus, one of the subtrees in a rooted X-forest F must have the leaf labeled ρ as its root.

We also assume that the forced contraction is applied immediately whenever it is applicable. However, if the root r of a subtree T' is of degree 2, then the operation will *not* be applied on r, in order to preserve the ancestor-descendant relation in T'. Thus, all unlabeled vertices in T' that are not the root of T' have degree 3.

Agreement Forests

The following terminologies are used for both rooted and unrooted versions.

Let E_0 be a subset of edges in an X-forest F, and let F' be the forest that obtained by removing the edges in E_0 from F, without the forced contraction applied. If each connected component of F' contains at least one leaf in F, then E_0 is an *essential edge-set* (abbr. EES) of F, and we denote the forest that obtained by applying the forced contraction on F' by $F \setminus E_0$. Obviously, the order of $F \setminus E_0$ is equal to $\text{Ord}(F) + |E_0|$. For any X-forest F' that is a subforest of another X-forest F, it is easy to see that there is an EES E' of $\text{Ord}(F') - \text{Ord}(F)$ edges in F such that $F' = F \setminus E'$.

An X-forest F is an *agreement forest* for a collection $\{F_1, F_2, \ldots, F_m\}$ of X-forests if F is a subforest of F_i, for all i. A *maximum agreement forest* (abbr. MAF) F^* for $\{F_1, F_2, \ldots, F_m\}$ is an agreement forest for $\{F_1, F_2, \ldots, F_m\}$ with a minimum $\text{Ord}(F^*)$ over all agreement forests for $\{F_1, F_2, \ldots, F_m\}$.

Both the rooted version and the unrooted version of the MAF problem on multiple X-forests studied in the current paper, are formally given as follows.

ROOTED MAXIMUM AGREEMENT FOREST (rooted-MAF)

Input: A set $\{F_1, \ldots, F_m\}$ of rooted X-forests

Output: a maximum agreement forest F^* for $\{F_1, \ldots, F_m\}$

UNROOTED MAXIMUM AGREEMENT FOREST (unrooted-MAF)

Input: A set $\{F_1, \ldots, F_m\}$ of unrooted X-forests

Output: a maximum agreement forest F^* for $\{F_1, \ldots, F_m\}$

3   Edge-Removal Meta-Step

The MAF problem (either rooted or unrooted) on the instance $(F_1, F_2, \ldots, F_m)$ looks for an MAF for the instance. Define the order of the MAF to be the *optimal order* for the instance, denoted $\text{Opt}(F_1, F_2, \ldots, F_m)$.

Our approximation algorithm for MAF consists of a sequence of "meta-steps". An *edge-removal meta-step* (or simply meta-step) of an algorithm is a collection of consecutive computational steps in the algorithm that on an instance $(F_1, F_2, \ldots, F_m)$ of MAF problem (either rooted or unrooted) removes an EES of F_i, $1 \leq i \leq m$ (and applies the forced contraction).

The performance of our approximation algorithm for MAF heavily depends on the quality of the meta-steps we employ in the algorithm. For this, we introduce the following concept that measures the quality of a meta-step, where $r \geq 1$ is an arbitrary real number.

Definition 1. *Let $(F_1, F_2, \ldots, F_m)$ be an instance of* MAF *problem (either rooted or unrooted), and let M be an edge-removal meta-step that removes an EES E_M of F_i, $1 \leq i \leq m$. Meta-Step M keeps ratio r if $(c' - c) \leq \frac{(r-1)}{r}|E_M|$, where c and c' are $Opt(F_1, \ldots, F_i, \ldots, F_m)$ and $Opt(F_1, \ldots, F_i \setminus E_M, \ldots, F_m)$, respectively.*

Remark 1. By definition, if an edge-removal meta-step does not change the optimal order for the instance, then it keeps ratio r for any $r \geq 1$. Define an edge-removal meta-step is *safe* if it does not change the optimal order for the instance, thus it keeps ratio r for any $r \geq 1$.

If any meta-step in the approximation algorithm keeps ratio not greater than a constant t, then we can get that the order of the agreement forest which the algorithm get for $(F_1, \ldots, F_i, \ldots, F_m)$ is at most t times $Opt(F_1, \ldots, F_i, \ldots, F_m)$, thus, the ratio of the algorithm is at most t. Therefore, performing low-ratio meta-steps in the algorithm is critical to maintain the performance guarantee of the algorithm.

The following lemma will play an important role in our discussion.

Lemma 1. *Let $(F_1, F_2, \ldots, F_m)$ be an instance of* MAF *problem (either rooted or unrooted), and let $\{e\}$ be an EES of F_i, $1 \leq i \leq m$. $Opt(F_1, \ldots, F_i \setminus \{e\}, \ldots, F_m)$ is at most one more than $Opt(F_1, \ldots, F_i, \ldots, F_m)$.*

4   MAF for Multiple X-forests

Fix a label-set X. Because of the bijection between the leaves in an X-forest F and the labels in the label-set X, sometimes we will use, without confusion, a label in X to refer to the corresponding leaf in F, or vice versa.

Let $F_1, F_2, \ldots, F_m$ be m X-forests, either all are rooted or all are unrooted. Let F^* be a fixed MAF for the instance $(F_1, F_2, \ldots, F_m)$. Note that F^* must be an agreement forest for F_1 and F_2. Therefore, in the section, we first discuss how to construct an agreement forest for F_1 and F_2. The discussion is divided into the case for the rooted version and the case for the unrooted version.

4.1   Agreement Forest for Two Rooted X-forests

In this case, both F_1 and F_2 are rooted X-forests. Two vertices v_1 and v_2 in a rooted X-forest are *siblings* if they have the common parent. By the definition, v_1 and v_2 can not be the leaf ρ.

During the construction of an agreement forest for F_1 and F_2, not only the relation between two leaves but also the relation between two non-leaf vertices in the X-forests will be analyzed. For convenience, we will *mark* each non-leaf vertex in the X-forest with a unique *symbol* which is just used for identifying the unique vertex in the X-forest such that we can use a symbol to refer to the corresponding non-leaf vertex in the forest, or vice versa. For a non-leaf vertex in the X-forest that marked with a symbol, it is still unlabeled and the forced contraction can be applied on this vertex when it is applicable. For each vertex

v in the X-forest that either marked with a symbol or labeled by a label, we maintain a label set $L(v)$ for it. If v is a leaf in the X-forest, then $L(v) = \{v\}$.

At the beginning of the construction of an agreement forest for F_1 and F_2, we maintain a set N_s. For each element s in N_s, it is either a label or a symbol that the structure of $F_1[L(s)]$ is isomorphic to that of $F_2[L(s)]$. Initially, let $N_s = X$.

Then, we proceed the construction by repeatedly removing edges from F_1 and F_2 until there do not exist two elements of N_s that are siblings in F_2. Therefore, in the following, we will assume that there exist two elements a and b of N_s that are siblings in F_2, and consider all possible cases for a and b in F_1.

Case 1. Elements a and b are also siblings in F_1.

In this case, the structures of $F_1[L(a) \cup L(b)]$ and $F_2[L(a) \cup L(b)]$ are isomorphic. Thus, the structures in both forests could remain unchanged when we construct an agreement forest for F_1 and F_2, and we can treat the structure as a single vertex that marked with symbol $\underline{ab}$ in both F_1 and F_2.

Mark Meta-Step. Mark the parent of a and b with symbol $\underline{ab}$ in both forests and let $L(\underline{ab}) = L(a) \cup L(b)$. And replace N_s with $(N_s \setminus \{a, b\}) \cup \{\underline{ab}\}$.

This meta-step does not change the structures of F_1 and F_2, and it is safe.

Case 2. Element a or b is the root of a connected component in F_1.

W.l.o.g., we assume that a is the root of a connected component in F_1. Thus, $L(a)$ contains all labels which are in the same connected component with a in F_1. Since the MAF F^* for $(F_1, F_2, \ldots, F_m)$ is a subforest of F_1, any label of $L(a)$ can not be in the same connected component with any label of $X \setminus L(a)$ in F^*. Let e_a be the edge between a and a's parent in F_2.

Meta-Step 2. Remove EES $\{e_a\}$ of F_2.

Lemma 2. *Let F_1 and F be two rooted X-forests that F is a subforest of F_1. If there exists an element v in F_1 that any label of $L(v)$ is not in the same connected component with any label of $X \setminus L(v)$ in F, then F is also a subforest of $F_1 \setminus \{e\}$, where e is the edge (if exists) between v and v's parent in F_1.*

By Lemma 2, F^* is also a subforest of $F_2 \setminus \{e_a\}$. Therefore, F^* is also an MAF for $(F_1, F_2 \setminus \{e_a\}, \ldots, F_m)$, Meta-Step 2 is safe.

Case 3. Elements a and b are in different connected components in F_1.

In this case, any label of $L(a)$ can not be in the same connected component with any label of $L(b)$ in F^*. Since a and b are siblings in F_2, thus, either any label of $L(a)$ is not in the same connected component with any label of $X \setminus L(a)$ in F^* or any label of $L(b)$ is not in the same connected component with any label of $X \setminus L(b)$ in F^*. Let e_a be the edge between a and a's parent in F_1, and let e_b be the edge between b and b's parent in F_1.

Meta-Step 3. Remove EES $\{e_a, e_b\}$ of F_1.

Lemma 3. *Meta-Step 3 keeps ratio 2.*

Case 4. Elements a and b are in the same connected component in F_1.

Let $P = \{a, c_1, c_2, \ldots, c_r, b\}$ be the path in F_1 that connects a and b, in which c_h is the least common ancestor of a and b, $1 \le h \le r$, $r \ge 2$. See Figure 1(a) for an illustration. Since a and b are siblings in F_2, there are only three situations for $L(a)$ and $L(b)$ in F^* in this case: (1) any label of $L(a)$ is not in the same connected component with any label of $X \setminus L(a)$ in F^*; (2) any label of $L(b)$ is not in the same connected component with any label of $X \setminus L(b)$ in F^*; (3) there exist some labels of $L(a)$ that are in the same connected component with some labels of $L(b)$ in F^*.

The analysis for Situation (3) is given as follows. Suppose that $u \in L(a)$ and $v \in L(b)$ are in the same connected component in F^*. Since the path that connects a and b in F_2 contains only one internal vertex (the common parent of a and b), but the path P contains at least two internal vertices, thus, in order to construct an agreement forest for F_1 and F_2 by removing edges from F_1 and F_2, only one internal vertex in P can be kept, and the other $r - 1$ internal vertices in P should be removed by the forced contraction (in this situation, all edges on P cannot be removed, these internal vertices only can be removed by the forced contraction). Because the subtrees in a rooted X-forest preserve the ancestor-descendant relation, so the least common ancestor of u and v in F^* corresponds to the common parent of a and b in F_2, and the least common ancestor c_h of a and b in F_1. Thus, the internal vertex c_h should be kept, all the other internal vertices in P should be removed. That is, all the edges that incident to a vertex c_i in P but not on P, $1 \le i \le r$, $i \ne h$, should be removed. If c_1 is not the least common ancestor of a and b, then let e be the edge that incident to c_1 but not on P; otherwise, let e be the edge that incident to c_r but not on P. Let e_a be the edge between a and c_1 in F_1, and let e_b be the edge between b and c_r in F_1.

Meta-Step 4. If vertex c_h is the root of the connected component and path P contains only two internal vertices, $r = 2$, then remove EES $\{e_a, e\}$ of F_1; otherwise, remove EES $\{e_a, e_b, e\}$ of F_1.

Note that if vertex c_h is the root of the connected component and P contains only two internal vertices, $r = 2$, then $\{e_a, e_b, e\}$ is not an EES of F_1 and $F_1 \setminus \{e_a, e\} = F_1 \setminus \{e_b, e\}$. In this special case, we just remove EES $\{e_a, e\}$ of F_1.

Lemma 4. *Meta-Step 4 keeps ratio 3.*

For F_1 and F_2, if we iteratively apply the above process, then the process will be ended up with that any two elements of N_s are not siblings in F_2. When this occurs, we apply the following two meta-steps if possible.

If there exists an element v of N_s that is the root of a connected component in F_2 (F_1), but is not the root of a connected component in F_1 (F_2). Let e be the edge between v and v's parent in F_1 (F_2). Then, we apply **Meta-Step 5** (**Meta-Step 6**) which removes the EES $\{e\}$ of F_1 (F_2). By Lemma 2, Meta-Step 5 (Meta-Step 6) is safe.

Lemma 5. *Let F_1 and F_2 be two rooted X-forests. If there do not exist two elements of N_s that are siblings in F_2, then after series of Meta-Step 5 operations and Meta-Step 6 operations in linear time, F_1 is a subforest of F_2.*

Unmark Meta-Step. Remove the symbols of non-leaf vertices in F_1 and F_2.

The construction of an agreement forest for F_1 and F_2 is finished, F_1 itself is an agreement forest for F_1 and F_2. Therefore, the symbols of non-leaf vertices in F_1 and F_2 can be removed (note that the non-leaf vertices are not removed).

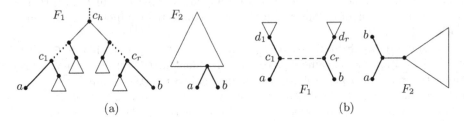

(a)   (b)

Fig. 1. The path connecting a and b in F_1 when F_1 is (a) rooted; (b) unrooted

4.2   Agreement Forest for Two Unrooted X-forests

The analysis for the construction of an agreement forest for two unrooted X-forests proceeds in a similar manner. However, since an unrooted tree enforces no ancestor-descendant relation in the tree, subtrees in the tree have no requirement of preserving such a relation. This fact induces certain subtle differences.

In this case, F_1 and F_2 are two unrooted X-forests. Two vertices v_1 and v_2 in an unrooted X-forest are *siblings* if either they are connected by an edge (*edge-siblings*), or they are adjacent to the same non-leaf vertex (*vertex-siblings*), which will be called the "parent" of v_1 and v_2.

We also maintain a set N_s for the construction of an agreement forest for F_1 and F_2, and initially let $N_s = X$. We proceed by repeatedly removing edges in F_1 and F_2 until certain condition is met. The certain condition for the unrooted version is that for any two elements of N_s, if they are siblings in F_2, then they are edge-siblings in both F_1 and F_2. Therefore, in the following, we will assume that there exist two elements a and b of N_s that are siblings (edge-siblings or vertex-siblings) in F_2, but are not edge-siblings in both F_1 and F_2. And we will consider all possible cases for a and b in F_1.

Case 1. Elements a and b are also siblings in F_1.

Since the definition of siblings in unrooted X-forests is subtle different from that in rooted X-forests. There are three subcases.

Subcase 1.1. a and b are vertex-siblings in both F_1 and F_2.

Mark Meta-Step. Mark the parent of a and b with symbol $\underline{ab}$ in both forests and let $L(\underline{ab}) = L(a) \cup L(b)$. And replace N_s with $(N_s \setminus \{a, b\}) \cup \{\underline{ab}\}$.

Subcase 1.2. a and b are vertex-siblings in F_1, but are edge-siblings in F_2.

In this subcase, $L(a) \cup L(b)$ contains all labels which are in the same connected component with a and b in F_2. Therefore, any label of $L(a) \cup L(b)$ can not be in the same connected component with any label of $X \setminus (L(a) \cup L(b))$ in F^*.

Lemma 6. *Let F_1 and F be two unrooted X-forests that F is a subforest of F_1. If there exist two elements a and b in F_1 that are vertex-siblings in F_1 and any label of $L(a) \cup L(b)$ is not in the same connected component with any label of $X \setminus (L(a) \cup L(b))$ in F, then F is also a subforest of $F_1 \setminus \{e\}$, where e is the edge that incident to the parent of a and b but not incident to a and b in F_1.*

Let p be the parent of a and b in F_1, and let v be the neighbor of p different from a and b. Let e_v be the edge between p and v.

Meta-Step 1.2. Remove EES $\{e_v\}$ of F_1.

Note that if v is a non-leaf vertex that marked with a symbol of N_s, once the edge e_v is removed, v would be a vertex of degree 2 which will be removed by the forced contraction, and the two children c_1 and c_2 of v in F_1 will be edge-siblings. Therefore, there will exist a symbol v of N_s that there does not exist a vertex in F_1 that marked with the symbol v, but there exists a vertex in F_2 that marked with the symbol v. In order to keep the bijection between the elements in N_s and the vertices in X-forests, we will replace N_s with $(N_s \setminus \{v\}) \cup \{c_1, c_2\}$, and remove the symbol of the vertex in F_2 which is marked with v. Therefore, there will exist two elements c_1 and c_2 of N_s that are edge-siblings in F_1, but are vertex-siblings in F_2, which satisfies the condition of Subcase 1.3. Summarizing the above analysis, we apply the following step:

Symbol-Removal Meta-Step. If there exists a symbol v of N_s that there only exists one vertex that marked with v in F_1 and F_2, then we replace N_s with $(N_s \setminus \{v\}) \cup \{c_1, c_2\}$, where c_1 and c_2 are the two children of v in the forest, and remove the symbol v of the vertex in the forest.

Note again that we will apply the Symbol-Removal Meta-Step after an edge-removal meta-step to keep the bijection between the elements in N_s and the vertices in X-forests.

Subcase 1.3. a and b are edge-siblings in F_1, but are vertex-siblings in F_2.

Let p be the parent of a and b in F_2, and let v be the neighbor of p different from a and b. Let e_v be the edge between p and v.

Meta-Step 1.3. Remove EES $\{e_v\}$ of F_2.

By Lemma 6, both Meta-Step 1.2 and Meta-Step 1.3 are safe.

Case 2. Element a or b is a single-vertex tree in F_1.

W.l.o.g., we assume that a is a single-vertex tree in F_1. Obviously, the label a is also a single-vertex tree in F^*. Let e_a be the edge incident to a in F_2.

Meta-Step 2. Remove EES $\{e_a\}$ of F_2.

Obviously, Meta-Step 2 is safe.

Case 3. Elements a and b are in different connected components in F_1.

In this case, again either any label of $L(a)$ can not be in the same connected component with any label of $X \setminus L(a)$ in F^* or any label of $L(b)$ can not be in the same connected component with any label of $X \setminus L(b)$ in F^*. Let e_a be the edge that incident to a and on the path from a to an arbitrary label of $X \setminus L(a)$

in F_1, and let e_b be the edge that incident to b and on the path from b to an arbitrary label of $X \setminus L(b)$ in F_1.

Meta-Step 3. Remove EES $\{e_a, e_b\}$ of F_1.

Lemma 7. *Let F_1 and F be two unrooted X-forests that F is a subforest of F_1. If there exists an element v in F_1 that any label of $L(v)$ is not in the same connected component with any label of $X \setminus L(v)$ in F, then F is also a subforest of $F_1 \setminus \{e\}$, where e is the edge (if exists) that incident to v and on the path from v to an arbitrary label of $X \setminus L(v)$ in F_1.*

Lemma 8. *Meta-Step 3 keeps ratio 2.*

Case 4. Elements a and b are in the same connected component in F_1.

Let $P = \{a, c_1, c_2, \ldots, c_r, b\}$ be the path in F_1 that connects a and b, where $r \geq 2$. See Figure 1(b) for an illustration. There are also three possible situations for $L(a)$ and $L(b)$ in F^* in this case, which are the same as that for Case 4 of the rooted version.

The analysis for Situation (3) is given as follows. Because the path connects a and b in F_2 contains at most one internal vertex, but the path P contains at least two internal vertices, thus, in order to construct an agreement forest for F_1 and F_2 by removing edges from F_1 and F_2, at most one internal vertex on P can be kept, and all the other internal vertices should be removed by forced contraction. However, since the subtrees in an unrooted X-forest do not preserve any ancestor-descendant relation, we cannot decide which one of these internal vertices in P should be kept. On the other hand, since $r \geq 2$, we know that at least one of c_1 and c_r should be removed. That is, at least one of e_{c_1} and e_{c_r}, which are not on the path P but are incident to c_1 and c_r, respectively, should be removed.

Let e_a be the edge that incident to a and on the path from a to an arbitrary label of $X \setminus L(a)$ in F_1, and let e_b be the edge that incident to b and on the path from b to an arbitrary label of $X \setminus L(b)$ in F_1.

Meta-Step 4. If P contains only two internal vertices, $r = 2$, then remove EES $\{e_a, e_{c_1}, e_{c_r}\}$ of F_1; otherwise, remove EES $\{e_a, e_b, e_{c_1}, e_{c_r}\}$ of F_1.

Lemma 9. *Meta-Step 4 keeps ratio 4.*

For F_1 and F_2, if we iteratively apply the above process, then the process will be ended up with that for any two elements of N_s, if they are siblings in F_2, then they are edge-siblings in both forests. When this occurs, we apply the following meta-step if possible.

If there is an element v of N_s that is a single-vertex tree in F_2, but is not a single-vertex tree in F_1. Let e_v be the edge incident to v in F_1. Then, we apply **Meta-Step 5** which removes the EES $\{e_v\}$ of F_1. This meta-step is safe.

Lemma 10. *Let F_1 and F_2 be two unrooted X-forests. If for any two elements of N_s that are siblings in F_2, they are edge-siblings in both F_1 and F_2, then after series of Meta-Step 5 operations in linear time, F_1 is a subforest of F_2.*

Unmark Meta-Step. Remove the symbols of non-leaf vertices in F_1 and F_2.

5 Approximation Algorithm for the MAF Problem

At first, we give the outline of our approximation algorithm for the MAF problem (either rooted or unrooted). Let $F_1, F_2, \ldots, F_m$ be m X-forests, either all are rooted or all are unrooted.

Main-Algorithm:
> **for** $i = 2$ to m
>> **do** meta-steps on F_1 and F_i until F_1 is a subforest of F_i.
>
> return F_1.

Theorem 1. *The Main-Algorithm correctly returns an agreement forest for the instance $(F_1, F_2, \ldots, F_m)$ of MAF problem (either rooted or unrooted).*

Approximation Algorithm for unrooted-MAF

Now we are ready for presenting the approximation algorithm for the unrooted-MAF problem. The algorithm is a combination of the analysis given in Subsection 4.2 and the Main-Algorithm, which is given in Figure 2.

Algorithm Unrooted Apx-MAF$(F_1, F_2, \ldots, F_m)$
Input: a collection $\{F_1, F_2, \ldots, F_m\}$ of unrooted X-trees, $m \geq 1$
Output: an agreement forest F for $\{F_1, F_2, \ldots, F_m\}$

1. **if** $(m = 1)$ **then** return F_1;
2. **for** i from 2 to m
2.1.   let $N_s = X$;
2.2.   **while** there exist two elements a and b of N_s that are siblings in F_i,
          but are not edge-siblings in both F_1 and F_i.
2.2.1.   **switch**
                case 1.1:   apply Mark Meta-Step;
                case 1.2:   apply Meta-Step 1.2;
                case 1.3:   apply Meta-Step 1.3;
                case 2:   apply Meta-Step 2;
                case 3:   apply Meta-Step 3;
                case 4:   apply Meta-Step 4;
2.2.2.   apply Symbol-Removal Meta-Step if possible;
2.3.   apply Meta-Step 5 if possible;
2.4.   apply Unmark Meta-Step;
3. return F_1;

Fig. 2. An approximation algorithm for the unrooted-MAF problem

Theorem 2. *Algorithm Unrooted Apx-MAF is a 4-approximation algorithm for the unrooted-MAF problem that runs in time $O(n \log n)$, where n is the number of vertices in the input instance.*

The approximation algorithm for rooted-MAF proceeds in a similar way, based on the corresponding analysis given in Subsection 4.1. Due to the space limit, we only present its main result below. The entire discussion for this problem will be given in the complete version.

Theorem 3. *Algorithm Rooted Apx-MAF is a 3-approximation algorithm for the rooted-*MAF *problem that runs in time* $O(n \log n)$, *where* n *is the number of vertices in the input instance.*

6   Conclusion

In this paper, we presented two approximation algorithms for the MAF problem on multiple binary phylogenetic trees: one for rooted trees with ratio 3, which is an improvement over the previously best 8-approximation algorithm for the problem; and the other one for unrooted trees with ratio 4, which is the first constant ratio approximation algorithm for the problem.

References

1. Robinson, D., Foulds, L.: Comparison of phylogenetic trees. Mathematical Biosciences 53(1-2), 131–147 (1981)
2. Li, M., Tromp, J., Zhang, L.: On the nearest neighbour interchange distance between evolutionary trees. Journal on Theoretical Biology 182(4), 463–467 (1996)
3. Hodson, F., Kendall, D., Tauta, P. (eds.): The recovery of trees from measures of dissimilarity. Mathematics in the Archaeological and Historical Sciences, pp. 387–395. Edinburgh University Press, Edinburgh (1971)
4. Swofford, D., Olsen, G., Waddell, P., Hillis, D.: Phylogenetic inference. In: Molecular Systematics, 2nd edn., pp. 407–513. Sinauer Associates (1996)
5. Hein, J., Jiang, T., Wang, L., Zhang, K.: On the complexity of comparing evolutionary trees. Discrete Applied Mathematics 71, 153–169 (1996)
6. Allen, B., Steel, M.: Subtree transfer operations and their induced metrics on evolutionary trees. Annals of Combinatorics 5(1), 1–15 (2001)
7. Bordewich, M., Semple, C.: On the computational complexity of the rooted subtree prune and regraft distance. Annals of Combinatorics 8(4), 409–423 (2005)
8. Rodrigues, E.M., Sagot, M.-F., Wakabayashi, Y.: Some approximation results for the maximum agreement forest problem. In: Goemans, M.X., Jansen, K., Rolim, J.D.P., Trevisan, L. (eds.) RANDOM 2001 and APPROX 2001. LNCS, vol. 2129, pp. 159–169. Springer, Heidelberg (2001)
9. Bonet, M., John, R., Mahindru, R., Amenta, N.: Approximating subtree distances between phylogenies. J. Comput. Biol. 13(8), 1419–1434 (2006)
10. Bordewich, M., McCartin, C., Semple, C.: A 3-approximation algorithm for the subtree distance between phylogenies. J. Discrete Algorithms 6(3), 458–471 (2008)
11. Rodrigues, E., Sagot, M., Wakabayashi, Y.: The maximum agreement forest problem: approximation algorithms and computational experiments. Theoretical Computer Science 374(1-3), 91–110 (2007)
12. Whidden, C., Zeh, N.: A unifying view on approximation and FPT of agreement forests. In: Salzberg, S.L., Warnow, T. (eds.) WABI 2009. LNCS, vol. 5724, pp. 390–402. Springer, Heidelberg (2009)
13. Whidden, C., Beiko, R., Zeh, N.: Fixed-parameter and approximation algorithms for maximum agreement forests. CoRR. abs/1108.2664 (2011)
14. Chen, J., Fan, J.-H., Sze, S.-H.: Parameterized and Approximation Algorithms for the MAF Problem in Multifurcating Trees. In: Brandstädt, A., Jansen, K., Reischuk, R. (eds.) WG 2013. LNCS, vol. 8165, pp. 152–164. Springer, Heidelberg (2013)
15. Chataigner, F.: Approximating the maximum agreement forest on k trees. Information Processing Letters 93, 239–244 (2005)

Optimal Inspection Points for Malicious Attack Detection in Smart Grids

Subhankar Mishra[1], Thang N. Dinh[2], My T. Thai[1], and Incheol Shin[3],[*]

[1] Dept. of Comp. and Info. Sci. and Eng.,
University of Florida, Gainesville, Florida 32611, USA
{mishra,mythai}@cise.ufl.edu
[2] Dept. of Comp. Sci.,
Virginia Comm. University, Richmond, VA 23284, USA
tndinh@vcu.edu
[3] Info. Security Dept.,
Mokpo National University Muan, Rep. of Korea
ishin@mokpo.ac.kr

Abstract. In this paper, we study the Optimal Inspection Points (OIP) problem, which asks us to find a subset of vertices in a given network to perform the Deep Packet Inspection so as to maximize the number of scanned packets while satisfying the delay constraints. This problem finds many applications for malicious attack detection, especially those where packet scanning is a must. Accordingly, we prove OIP is NP-complete and provide an FPTAS in the case of single path routing. For the multiple path routings, we design an FPTAS when the routing graph takes a form of series-parallel graphs, which is commonly used to model electric networks.

Keywords: Malicious Attacks Detection, Smart Grids, Optimization.

1 Introduction

A key concern for the computer dependent systems is the threat from the malicious attacks which execute almost perfectly legitimate operations to compromise the whole system security. For example, in case of Distributed Denial of Service attacks in the Internet, the intrusion detection system needs to monitor the entire network traffic [1–3]. Another notable example is the Smart Grid [4], where many new classes of cyber attacks have emerged [5–8].

A common type of attack in Smart Grid is to alter the network dynamics by valid yet malicious commands [4]. To guard against this type of attack, *Deep Packet Inspection* (DPI) is essential to search for malicious packets. However, DPI leads to significant delays in the throughput hence increasing the latency for packets to arrive at the central monitoring node. Control messages not satisfying time constraints are discarded, which includes the risk of dropping important

---

[*] The first two authors contribute equally to this work.

Z. Cai et al. (Eds.): COCOON 2014, LNCS 8591, pp. 393–404, 2014.

control messages leading to serious physical and/or financial damage, therefore inspection cannot be performed at all points and on all packets.

Based on the above motivation, we introduce and study a new optimization problem, namely *Optimal Inspection Points* (OIP). Given a network represented by a graph $G = (V, E)$, the goal is to find a subset $D \subset V$ which represents the optimal inspection points, such that the number of scanned packets at the center node is maximized without violating the latency constraint. Clearly this problem helps to inspect the packets as much as possible to search for malicious ones while ensuring all packets arrive on time.

The routing schemes in different networks together with the strict latency constraints make this problem challenging and interesting. The time constraint in IEC 61850 [9–12], for example, could be as low as 3ms for the critical fault isolation and protection control messages [4]. Also the number of the scanned packets, which in turn increases the probability of catching a malicious packet, has to be as high as possible. Therefore, it would be nice if we can devise a Fully Polynomial Time Approximation Scheme (FPTAS) [14] for the OIP. Indeed, we have developed such a solution for the single path routing scenario. As for the multiple path routing, we devised an another FPTAS to OIP when the network can be transformed to a series-parallel graph.

The remainder of this paper is organized as follows. Section 2 presents the network model and our problem definition. The complexity and FPTAS are discussed in Section 3 and 4, respectively. We introduce the FPTAS for multiple-path routing in series-parallel graphs in Section 5 and provide more discussion with different scanning scenarios in Section 6.

2   Model and Problem Definitions

We use the Smart Grid as an example to illustrate the network model for our problem. A smart grid is modeled as a directed graph $G = (V, E)$ where the vertices in $V = \{r\} \cup O \cup S$ represent the set of nodes in the grid and E represents the set of communication links among the nodes.

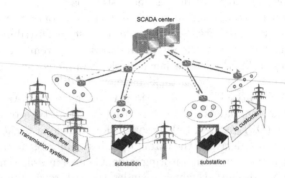

Fig. 1. A Smart Grid Structure

The set of vertices V includes the following:

- The center node r which represents the Supervisory Control And Data Acquisition (SCADA) center. All the state estimations and corresponding actions based on the message received from S are done by r. All the request packets in the smart grid communication network are routed towards r.
- S is the set of the nodes that can act as a source of malicious packets and hence can be under the control of attackers. These nodes are the Intelligent Electronic Devices (IEDs) or the Remote Terminal Units (RTUs).
- $O = V \setminus \{S \cup r\}$ is the set of intermediate nodes where DPI can be performed. If a node in O does not have DPI scanner, then equivalently the capacity of the scanner at that node is 0. We assume that there is no queueing effect and packets arrive continuously that scanners do not have to wait for packets unless the scanner capacity exceeds the amount of arrived packets.

For each $u \in V$, let $N^-(u)$ and $N^+(u)$ represent the set of incoming and outgoing neighbors of u respectively. Also let the flow $f(u,v)$ represent the network traffic from $u \to v$ (measured as the number of packets going from u to v within a time unit). Note that $f(,)$ contains the information about the routing and data forwarding in the smart grid network as follows. At a node $u \in V \setminus \{r\}$, a packet can be forwarded to any of its neighbor $v \in N^+(u)$ unless $f(u,v) = 0$. Also the probability that a packet is forwarded from u to v is proportional to $f(u,v)$, i.e., the probability is given by $\frac{f(u,v)}{f(u)}$, where

$$f(v) = \begin{cases} \sum_{u \in N^+ v} f(v,u) = \sum_{w \in N^- v} f(w,v) & v \in O \\ \sum_{u \in N^+ v} f(v,u) & v \in S \\ \sum_{u \in N^- v} f(u,v) & v = r. \end{cases} \tag{1}$$

For single path routing protocols, the out degree of every vertex in G is at most one. Thus G is a directed tree rooted at r. For multiple path routing protocols, a vertex in G may have multiple out going edges. In that case, we restrict our attention to the case when G is acyclic, i.e., there will be no routing loop problem in G.

We now formally define the following optimization problem:

Definition 1 (Optimal Inspection Points (OIP) problem). *Given a smart grid represented by a graph $G = (V, E)$, the center node r, the set of terminal nodes S, the set of intermediate nodes O, the capacity m_u of scanner at $u \in O$, the average traffic flow $f(u,v)$ for $(u,v) \in E$, the delay δ_u caused by DPI at $u \in O$, and the maximum delay δ_{max} for a packet. Find a subset $D \subset V$ to place scanners so that the total delay of any packet arriving at r, on any path is at most δ_{max} and the number of inspected packets is maximized.*

3 Complexity

In this section we show that OIP is NP-complete and the NP-hardness even holds for the simple path network.

Theorem 1. *The Optimal Inspection Points problem is NP-complete.*

Proof. We prove the NP-completeness of the problem even when the graph is a simple path. The decision version of OIP is defined as follows.

Decision version of OIP. Given an *acyclic* graph $G = (V, E)$, capacities m_u for $u \in V$, flow values $f(u, v), (u, v) \in E$, and maximum latency δ_{max} of a packet, is there a subset $D \subset V$ such that for any path $\mathcal{P}_v$ starting from a terminal node $v \in S$ to r,

$$\sum_{u \in D \cap \mathcal{P}_v} \delta_u \leq \delta_{max},$$

and the number of scanned packets is at least P for some $P \geq 0$?

Given a set of inspection points D, it is easy to verify in polynomial time if the total inspection time is less than the maximum delay allowed in the given system and the number of packets scanned is at least P. Hence, OIP is in NP.

To show the NP-hardness, we reduce from the $0 - 1$ Knapsack problem which is defined as follows. Given an instance of $0 - 1$ Knapsack problem with n items $a_1, a_2, \ldots, a_n$ where a_i has value v_i and weight w_i, and the bag can carry a maximum weight W. The decision version of the $0 - 1$ Knapsack problem asks if we can select a subset of items with total weight at most W and total value at least B, for some $B \geq 0$.

Construction. We reduce the Knapsack instance to the following instance of OIP. Construct a graph $G = (V = S \cup O \cup \{r\}, E)$ where $S = \{u_0\}, O = \{u_1, u_2, \ldots, u_{n-1}\}, r = \{u_n\}$. There is an edge (u_i, u_{i+1}) for all $i = 1 \ldots n$ (see Fig. 2). The scanner at u_i has capacity v_i and a scanning time $\delta_i = w_i$ for $i = 1 \ldots n$. The traffic flow $f(u_i, u_{i+1}) = \infty$ for $i = 0 \ldots n - 1$. Set $\delta_{max} = W$ and $P = B$.

Fig. 2. Reduction from $0 - 1$ Knapsack

($\rightarrow$)Suppose we have a solution $K \subset \{a_1, a_2, ..., a_n\}$ for the 0-1 Knapsack problem. Now with our construction we see that, K corresponds to a subset D of vertices O. Since, $\delta_{max} = W$, $\sum_{a_i \in K} w_i \leq W$ which implies it also satisfies the delay constraint in OIP and the number of scanned packets is at least P.

($\leftarrow$) Let say $D \subset O$ is a solution for OIP. The above solution satisfies the delay constraint δ_{max} which satisfies $\delta_{max} = W$, based on our construction, Hence $D \subset O$ is also a solution which satisfies the weight constraint and total value $B = \sum_{u_i \in D} v_i$. $\square$

4   One Time Scan in Single Path Routing

4.1   IP Formulation

In this section, we discuss the formulation for the given problem assuming single-path-routing protocols, e.g., packets are routed following the shortest path. First, we define the binary variable x_v for each vertex $v \in V$ as follows:

$$x_v = \begin{cases} 1 & \text{if } v \text{ is selected as an inspection point} \\ 0 & \text{otherwise} \end{cases} \qquad (2)$$

For each node $u \in S$, let P_u denote the set of nodes on the unique path from u to r. The delay constraint is given by:

$$\sum_{v \in P_u} \delta_v \cdot x_v \le \delta_{max}, u \in S \qquad (3)$$

Our objective is to maximize the total number of scanned packets. The number of packets scanned at $v \in O$ is $\min\{m_v, \#\text{unscanned packets that arrived at } v\}$. Let y_u denote the number of scanned packets going out from u, which include the packets scanned before arriving to u and also the packets scanned at u. We have

$$y_v = \min\{f(v), x_v m_v + \sum_{u \in N^-(v)} y_u\}.$$

The problem can be formulated as

$$\text{maximize} \quad y_r$$

$$\text{s.t.} \quad \sum_{v \in P_u} \delta_v \cdot x_v \le \delta_{max} \qquad\qquad u \in S$$

$$y_v \le f(v) \qquad\qquad v \notin S$$

$$y_v \le x_v m_v + \sum_{u \in N^-(v)} y_u \qquad\qquad v \notin S$$

$$y_u = 0 \qquad\qquad u \in S$$

$$x_v \in \{0, 1\} \qquad\qquad v \in V$$

4.2   FPTAS for Single-path Routing One-time Scanning

We give an $(1 - \epsilon)$-approximation algorithm that has an $O(\epsilon^{-2} n^5)$ time complexity for $\epsilon > 0$.

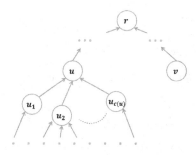

Fig. 3. In single-path-routing, the graph is a directed tree, rooted at r

Our algorithm consists of two phases: 1) First, we standardize the capacities of the deep packet inspectors as well as the flow values so that all values are bounded by a polynomial in n and $\frac{1}{\epsilon}$; 2) Second, we use the dynamic programming to find the solution in polynomial time. The algorithm is summarized in Algorithm 1. In the first phase, the scanners' capacities m_u are standardized as shown in step 2 in Algorithm 1; then both m_u and the flow values $f(u)$ are scaled down by a factor M as defined in step 3 and rounded down. This preprocessing step ensures that all m'_u are integers between 0 and $\lceil \frac{n}{\epsilon} \rceil$.

Dynamic Programming. In the second phase, we use dynamic programming to find an optimal solution for the OIP problem instance (G, m', f'). In the case of single-path-routing, there is at most one path from each node to root node r. For simplicity, we remove nodes that have no paths to r. The remaining graph is a directed tree rooted at r as shown in Fig. 3.

Now, given the tree $T = (V, E)$ with $|V| = n$, and the SCADA center $r \in V$ serving as the root. In order to describe the dynamic algorithm, we use the following the notations:

- T^u: The subtree rooted at u in T with the set of vertices V^u and the set of edges E^u is denoted by $T^u = (V^u, E^u)$.
- $c(u)$: is the in-degree of u. Note that when referring to T^u, we disregard the edge that connect u to its parent (i.e., only consider the in-degree); thus the degree of u within T^u is one less than that in T unless $u = r$.
- $u_1, u_2, \ldots, u_{c(u)}$: represent the children of u.
- d_u: the latency is defined as $d_u = \max_{v \in N^-_{(u)}} \{d_v\} + x_u \delta_u$.
- y_u: is the number of scanned packets outgoing from u and is given $y_u = \min\{f'(u), x_u m'_u + \sum_{v \in N^-(u)} y_v\}$, where x_u indicates whether the scanner at u is switched on ($x_u = 1$) or not($x_u = 0$).

We define the following recursion functions :

- $T^u(p)$:   The minimum value of latency d_u among all possible ways to deploy inspectors in the T^u rooted at u so that number of packets scanned y_u is at least p i.e. $y_u \geq p$.
- $T^u_i(p)$: The minimum value of latency d_u among the maximum value of latency among $\{d_{u_1}, d_{u_2}, \ldots, d_{u_i}\}$ among all possible ways of deploying inspectors in the subtrees $\{T^{u_1}, T^{u_2}, \ldots, T^{u_i}\}$ correspondingly, so that number of packets scanned y_u is at least p, i.e., $y_u \geq p$., where $i = 1 \ldots c(u)$.

The core of the dynamic algorithm is to compute $T^u_{c(u)}(p)$ and $T^u(p)$ through the following recursions.

$$T^u(p) = \min\{\delta_u + \max\{T^u_{c(u)}(p - m'_u)\}, T^u_{c(u)}(p)\}, \forall p = 1 \ldots \lceil n^2/\epsilon \rceil \quad (4)$$

$$T^u_i(p) = \min_{q=0..p} \{\max\{T^u_{i-1}(p - q) + T^u(q)\}\}, \forall p = 1 \ldots \lceil n^2/\epsilon \rceil, i = 1 \ldots c(u) \quad (5)$$

The basis cases are as follows.

$$T^u(p) = \begin{cases} 0 & p \leq 0 \\ \infty & p > \lceil n^2/\epsilon \rceil \end{cases}, \quad T^u_i(p) = \begin{cases} 0 & p \leq 0 \\ \infty & p > \lceil n^2/\epsilon \rceil \end{cases} \quad (6)$$

$$T_i^u(p) = \begin{cases} 0 & \text{if } u \in S \\ d_u & \text{if } u_1, u_2, \ldots, u_{c(u)} \in S \end{cases} \qquad (7)$$

Finally, the maximum objective for the OIP instance (G, m', f') is given at the root r by $\max\{p \mid T^r(p) \le \delta_{max}\}$.

Algorithm 1. FPTAS for Single-path-routing OIP

Phase 1: Preprosessing

1. Remove all nodes that have no paths to r.

2. For all $u \in V$, if $\delta_u > \delta_{max}$, set $m_u \leftarrow 0$; else set $m_u = \min\{m_u, f(u)\}$.

3. Given $\epsilon > 0$, let $K = \frac{\epsilon M}{n}$, where $M = \max_{u \in V}\{m_u\}$.

4. Let $f'(u) = \lfloor \frac{f(u)}{K} \rfloor$ and $m'_u = \lfloor \frac{m_u}{K} \rfloor$

Phase 2: Dynamic programming algorithm

5. Compute $T^u(p)$ and $T_i^u(p)$ using the recursions in Eqs. 4 and 5.

6. Find an optimal solution, say S', by tracing from $\max\{p \mid T^r(p) \le \delta_{max}\}$

7. Return S'.

Lemma 1. *Algorithm 1 finds an optimal solution for the single-path-routing OIP instance (G, m', f') in an $O(\epsilon^{-2} n^5)$ time.*

Proof. The correctness of the dynamic programming algorithm comes from the sub-optimal structure of the problem.

As for the running time, the major portion of running time is to compute $T_i^u(p)$. Since we have at most $n - 1$ possible pairs of u and i (the total number of children), and $q \le p \le \lceil n^2/\epsilon \rceil$. The running time to compute $T_i^u(p)$ is $O(n \times \lceil n^2/\epsilon \rceil \times \lceil n^2/\epsilon \rceil) = O(\epsilon^{-2} n^5)$. $\qquad \square$

Theorem 2. *For any $\epsilon > 0$, there is an $(1 - \epsilon)$-approximation algorithm for the OIP problem, single-path-routing with a time complexity $O(\frac{1}{\epsilon^2} n^5)$.*

Proof. Let $S^* \subseteq V$ be an optimal solution of the OIP instance (G, m, f) with the objective value $OPT = y_r(S^*)$ (the total number of scanned packets at any nodes).

Given $1 > \epsilon > 0$, we apply Algorithm 1 to find an optimal solution $S' \subseteq V$ for the instance (G, m', f') with an objective value $OPT' = y'_r(S')$. By Lemma 1, this takes a (polynomial) running time $O(\epsilon^{-2} n^5)$.

First, S' is also a feasible solution for the instance (G, m, f), since it satisfies the condition that the latency at r is at most δ_{max}. Let $y_r(S')$ be the objective value associated with S' w.r.t. the instance (G, m, f). We will show that

$$y_r(S') \ge (1 - \epsilon)OPT.$$

The dynamic programming must return a solution at least as good as S^* (for the OIP instance (G, m', f')). Thus $y'_r(S^*)$, the objective value associated with S^* w.r.t. the instance (G, m', f'), is at most OPT'. Due to the rounding down, Mm'_u can be smaller than m_u, but by not more than K. Hence,

$$y_r(S^*) - My'_r(S^*) \leq nK.$$

Therefore

$$y_r(S') \geq Ky'_r(S') \geq Ky'_r(S^*) \geq y_r(S^*) - nK = OPT - \epsilon M.$$

Since we filtered out nodes u with $\delta_u > \delta_{max}$, we have $OPT \geq M$. Therefore,

$$y_r(S') \geq OPT - \epsilon M \geq (1 - \epsilon)OPT.$$

Thus the objective of S' is within a factor $1 - \epsilon$ of OPT, i.e., Algorithm 1 is a $(1 - \epsilon)$ approximation algorithm for the single-path-routing OIP problem. $\square$

5 One Time Scan in Multiple Paths Routing

In this section, we study the OIP problem in which packets can be routed using different paths to the SCADA center. We present an IP formulation for the problem in Section 5.1. We study the special case when the network has form of a series-parallel graph, which is often used to model electric networks. Accordingly, we present the definition of series-parallel graphs (SP-graphs) in Section 5.2 and describe an FPTAS for Multi-path routing OIP in SP-graphs in Section 5.3.

5.1 IP Formulation

We present the formulation for the given problem when multi-path routing protocols are in use. Let $x_u = 1$ if node u is selected as an inspection point and 0, otherwise. Also, let l_u denote the latency, the maximum possible delay of a packet, at node u. Then l_u is given by

$$l_u = \max_{v \in N^-(u)} \{l_v\} + x_u \delta_u \quad \forall u \notin S \tag{8}$$

Thus the total delay constraint is $l_r \leq \delta_{max}$.

Since the probability that a packet is forwarded from u to v is proportional to $f(u, v)$, the number of scanned packets going out from v is given by

$$y_v = \min\{f(v), x_v m_v + \sum_{u \in N^- v} y_u \frac{f(u, v)}{f(u)}\}.$$

The OIP problem with multi-path routing can be formulated as follows.

$$
\begin{aligned}
\text{maximize} \quad & y_r \\
\text{s.t.} \quad & l_r \leq \delta_{max} \\
& l_v \geq l_u + x_v \delta_v && v \notin S, u \in N^-(v) \\
& y_v \leq f(v) && v \notin S \\
& y_v \leq x_v m_v + \sum_{u \in N^- v} y_u \frac{f(u,v)}{f(u)} && v \notin S \\
& y_u = 0, l_u = 0 && u \in S \\
& x_v \in \{0,1\} && v \in V
\end{aligned}
$$

It is common that the network traffic is much higher than the capacities of scanners, thus if we choose to activate the scanner at u, we can scan exactly m_u additional packets. Thus the above formulation can be simplified to

$$
\begin{aligned}
\text{maximize} \quad & x_u m_u \\
\text{s.t.} \quad & l_r \leq \delta_{max} \\
& l_u = 0 && \forall u \in S \\
& l_v \geq l_u + x_v m_v && \forall v \notin S, u \in N^-(v) \\
& x_v \in \{0,1\} && \forall v \notin S.
\end{aligned}
$$

5.2 Series-Parallel Graphs (SP-graphs)

Given two graphs G_1 and G_2 together with pairs of source-sink nodes (s_1, t_1) in G_1 and (s_2, t_2) in G_2, a *series composition* creates a new graph by merging the sink t_1 and the source s_2 and (s_1, t_2) becomes the source-sink of the composed graph. A *parallel composition* creates a new graph by merging two sources s_1 and s_2 into the new source, and two sinks t_1 and t_2 into the new sink node. A series-parallel graph (SP-graph) is a graph that is constructed by a sequence of series and parallel compositions starting from a set of single-edge graphs, i.e. cliques of size two.

An SP-graph G with a source-sink pair (s,t) can be decomposed into several single-edge base graphs. The decomposition is specified by a binary decomposition tree $T(G)$ whose nodes represent subgraphs of G. Each non-leaf node of the tree has two child subgraphs and an associated operation (either series or parallel). The parent subgraph can be constructed by applying the operation on two children subgraphs. Construction of $T(G')$ can be done in a linear time complexity [13].

5.3 FPTAS for Multi-path Routing One-time Scanning in GSP

We present an FPTAS for the OIP problem when the graph has form of an SP-graph. Specifically, let G' be an augment of G by adding a source node s and connecting s to all nodes in S. We set $f(s,u) = f(u)$ for each $u \in S$ and $m_s = 0$

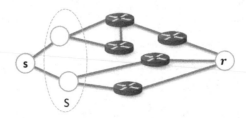

Fig. 4. A series-parallel network. Bold nodes are intermediate nodes in O.

(Fig. 4). The OIP problems G' with source-sink pair (s, r) is then equivalent to the OIP problem on G.

We present an FPTAS for the OIP problem under the following assumptions.

- Multi-path routing is possible and the augmented graph G' is an SP-graph.
- Single scanning mechanism is employed.
- The scanner capacities are relatively small to the number of arriving packets so that even when all scanners are in use, there are still unscanned packets at every node.

The last assumptions simplify the calculation of the number of scanned packets to the total capacities of deployed scanners.

We summarize the FPTAS in Algorithm 2. We begin by constructing the decomposition tree $T(G')$ [13]. Also, the scanners' capacities are scaled and rounded to the nearest integers in $[0...\lceil \frac{n}{\epsilon} \rceil]$. Then we solve the problem in each subgraph of G' starting from the leaf nodes of $T(G')$ up to the root via a set of series and parallel merging operations.

For any subgraph $\bar{G}$ with a source-sink pair $(\bar{s}, \bar{t})$, we define a function $h_{\bar{G}}(p)$ as the *minimum latency* among all possible way to deploy scanners on $\bar{G}$ so that the *total capacities of deployed scanners is at least p*. We enforce that no scanner is deployed at $\bar{s}$ to avoid duplicate scanner deployment for the series operation.

The Basis. When $\bar{G}$ is a base graph with a single edge $(\bar{s}, \bar{t})$, we have

$$h_{\bar{G}}(p) = \begin{cases} 0 & p = 0 \\ \delta_{\bar{t}} & p = 1...m_{\bar{t}} \\ \infty & p > m_{\bar{t}} \end{cases}$$

The Series Operation. Assume that the subgraph $\bar{G}$ with source-sink pair $(\bar{s}, \bar{t})$ is obtained by applying *series* operation on the subgraph G_1 with the source-sink pair (s_1, t_1) and the subgraph G_2 with the source-sink pair (s_2, t_2). For a series operation, both the latency and the number of scanned packets in $\bar{G}$ are equal to those of G_1 and G_2. Thus

$$h_{\bar{G}}(p) = \min_{p'=0...p} \{h_{G_1}(p') + h_{G_2}(p - p')\}, 0 \le p \le \sum_{u \in \bar{G}} m'_u$$

Parallel Operation. For a *parallel* operation, the latency in $\bar{G}$ is equal to the maximum of the latency in G_1 and G_2 and the number of scanned packets in $\bar{G}$ is the sum of those in G_1 and G_2. Hence

$$h_{\bar{G}}(p) = \min_{p'=0...p} \max\{h_{G_1}(p'), h_{G_2}(p-p')\}, 0 \leq p \leq \sum_{u \in \bar{G}} m_u'$$

Algorithm 2. FPTAS for Multi-path-routing OIP in SP-graphs

1. Construct $G' = (V', E')$ by adding a node s to G and setting $f(s, u) = f(u), u \in S, m_s = 0$.

2. Scale m_u: Let $K = \frac{\epsilon M}{n}$, where $M = \max_{u \in V}\{m_u\}$. Set $m_u' = \lfloor \frac{m_u}{K} \rfloor$.

3. Construct the decomposition tree $T(G')$.

4. Starting from leaf-node in $T(G')$ up to the root, compute the $h_{\bar{S}}(p)$ for each subgraph $\bar{G}$ in $T(G')$ using the formulations for series and parallel operations.

5. At the root node of $T(G')$, choose a solution with the maximum p value that $h_{G'}(p) \leq \delta_{max}$.

Time Complexity Analysis. There are no more than n operations (either series or parallel). Since $\sum_{m_u'} \leq n \times \lceil n/\epsilon \rceil$, we need to compute $h_{\bar{G}}(p)$ for at most $O(n^2/\epsilon)$ different values of p, which, in turn, requires an $O(n^2/\epsilon)$ time. Therefore, the total time complexity is $O(n^5/\epsilon^2)$.

Theorem 3. *For $\epsilon > 0$, Algorithm 2 is a $(1-\epsilon)$-approximation algorithm for the multi-path routing OIP problem when the augmented graph G' is an SP-graph.*

Proof. Let $S^* \subseteq V$ be an optimal solution of the OIP instance (G, m, f) with the objective value $OPT = \sum_{u \in S^*} m_u$ and S' be the optimal solution of the OIP instance (G, m', f') found in Algorithm 2.

Since S^* satisfies the latency constraint, it is also a feasible solution for the instance (G, m', f'). From the optimality of S', we have

$$\sum_{u \in S^*} m_u' \leq \sum_{v \in S'} m_v'.$$

Let $OPT' = \sum_{v \in S'} m_v$, we will show that $OPT' \geq (1 - \epsilon)OPT$. Thus the objective value given by S' is at least a $(1 - \epsilon)$ times the optimal objective value. We have

$$OPT' = \sum_{v \in S'} m_v \geq \sum_{v \in S'} Km_v' \geq K \sum_{u \in S^*} m_u'$$

$$\geq \sum_{v \in S'}(m_v - K) \geq OPT - nM \geq (1 - \epsilon)OPT$$

Thus Algorithm 2 gives an $(1 - \epsilon)$ approximation algorithm for the OIP problem when G' is an SP-graph. $\square$

6   Discussion

In this paper we assume that each packet will not be scanned multiple times. This can be implemented by altering the packet header to add one flag to check whether the packet has been scanned. This approach requires updating either the hardware/firmware components at the network core.

We can also relax this scanning requirement and do not check for multiple scanning of packets. This approach provides greater compatibility for legacy devices with a cost in efficacy (due to redundant scanning). Using this approach we can also formulate two new optimization problems depending on whether single path or multiple path routing is in use. These alternative formulations are more difficult than their one time scanning counterparts and are subjects of our further studies.

Acknowledgment. This work is partially supported by DTRA YIP #HDTRA1-09-1-0061.

References

1. Mirkovic, J., Reiher, P.: A Taxonomy of DDoS Attack and DDoS Defense Mechanisms. In: ACM SIGCOMM (2004)
2. Peng, T., Leckie, C., Ramamohanarao, K.: Survey of Network-Based Defense Mechanisms Countering the DoS And DDoS Problems. ACM Comput. Surv. 39(1), Article 3 (2007)
3. Kim, Y., Lau, W.C., Chuah, M.C., Chao, H.J.: PacketScore: A Statistics-Based Packet Filtering Scheme against Distributed Denial-of-Service Attacks. IEEE Trans. Dependable Secur. Comput. 3(2), 141–155 (2006)
4. Wanga, W., Lu, Z.: Cyber Security in the Smart Grid: Survey and Challenges. Computer Networks (57), 1344–1371 (2013)
5. Jeffrey, L.H., James, H.G., Sandip, C.P.: Cyber security enhancements for SCADA and DCS systems, Technical Report TR-ISRL-07-02, University of Louisville, 1–27 (2007)
6. Hamieh, A., Ben-Othman, J.: Detection of jamming attacks in wireless ad hoc networks using error distribution. In: Proc. of IEEE ICC 2009 (2009)
7. Choi, D., Kim, H., Won, D., Kim, S.: Advanced key-management architecture for secure SCADA communications. IEEE Trans. Power Delivery 24, 1154–1163 (2009)
8. Choi, D., Lee, S., Won, D., Kim, S.: Efficient secure group communications for SCADA. IEEE Trans. Power Delivery 25, 714–722 (2010)
9. Mackiewicz, R.E.: Overview of IEC 61850 and Benefits. In: Power Systems Conference and Exposition, PSCE 2006, pp. 623–630. IEEE PES (2006)
10. Mohagheghi, S., Stoupis, J., Wang, Z.: Communication protocols and networks for power systems - current status and future trends. In: Proc. of Power Systems Conference and Exposition, PES 2009 (2009)
11. IEC Standard, IEC 62351: Data and communication security
12. Brunner, C.: IEC 61850 for power system communication. In: Transmission and Distribution Conference and Exposition, pp. 1–6. IEEE/PES (2008)
13. Takamizawa, K., Nishizeki, T., Saito, N.: Linear-time Computability of Combinatorial Problems on Series-parallel Graphs. Journal of ACM, 623–641 (1982)
14. Vazirani, V.V.: Approximation Algorithms. Springer-Verlag New York, Inc., New York (2001)

Reconfiguration of Dominating Sets

Akira Suzuki[1,*], Amer E. Mouawad[2,**], and Naomi Nishimura[2,**]

[1] Graduate School of Information Sciences, Tohoku University
Aoba-yama 6-6-05, Aoba-ku, Sendai, 980-8579, Japan
a.suzuki@ecei.tohoku.ac.jp
[2] David R. Cheriton School of Computer Science
University of Waterloo, Waterloo, Ontario, Canada
{aabdomou,nishi}@uwaterloo.ca

Abstract. We explore a reconfiguration version of the dominating set problem, where a dominating set in a graph G is a set S of vertices such that each vertex is either in S or has a neighbour in S. In a reconfiguration problem, the goal is to determine whether there exists a sequence of feasible solutions connecting given feasible solutions s and t such that each pair of consecutive solutions is adjacent according to a specified adjacency relation. Two dominating sets are adjacent if one can be formed from the other by the addition or deletion of a single vertex.

For various values of k, we consider properties of $D_k(G)$, the graph consisting of a vertex for each dominating set of size at most k and edges specified by the adjacency relation. Addressing an open question posed by Haas and Seyffarth, we demonstrate that $D_{\Gamma(G)+1}(G)$ is not necessarily connected, for $\Gamma(G)$ the maximum cardinality of a minimal dominating set in G. The result holds even when graphs are constrained to be planar, of bounded tree-width, or b-partite for $b \geq 3$. Moreover, we construct an infinite family of graphs such that $D_{\gamma(G)+1}(G)$ has exponential diameter, for $\gamma(G)$ the minimum size of a dominating set. On the positive side, we show that $D_{n-\mu}(G)$ is connected and of linear diameter for any graph G on n vertices with a matching of size at least $\mu + 1$.

1 Introduction

The *reconfiguration version* of a problem determines whether it is possible to transform one feasible solution s into a *target* feasible solution t in a step-by-step manner (a *reconfiguration*) such that each intermediate solution is also feasible. The study of such problems has received considerable attention in recent literature [8,9,13,15,16] and is interesting for a variety of reasons. From an algorithmic standpoint, reconfiguration models dynamic situations in which we seek to transform a solution into a more desirable one, maintaining feasibility during the process. Reconfiguration also models questions of evolution; it can represent

* Research supported by JSPS Grant-in-Aid for Scientific Research, Grant Numbers 24.3660 and 26730001.
** Research supported by the Natural Science and Engineering Research Council of Canada.

Z. Cai et al. (Eds.): COCOON 2014, LNCS 8591, pp. 405–416, 2014.
© Springer International Publishing Switzerland 2014

the evolution of a genotype where only individual mutations are allowed and all genotypes must satisfy a certain fitness threshold, i.e. be feasible. Moreover, the study of reconfiguration yields insights into the structure of the solution space of the underlying problem, crucial for the design of efficient algorithms. In fact, one of the initial motivations behind such questions was to study the performance of heuristics [9] and random sampling methods [4], where connectivity and other properties of the solution space play a crucial role. Even though reconfiguration gained popularity in the last decade or so, the notion of exploring the solution space of a given problem has been previously considered in numerous settings. One such example is the work of Mayr and Plaxton [18], where the authors consider the problem of transforming one minimum spanning tree of a weighted graph into another by a sequence of edge swaps.

Some of the problems for which the reconfiguration version has been studied include vertex colouring [1,3,4,6,5], list edge-colouring [14], list L(2,1)-labeling [15], block puzzles [11], independent set [11,13], clique, set cover, integer programming, matching, spanning tree, matroid bases [13], satisfiability [9], shortest path [2,16], subset sum [12], dominating set [10,19], odd cycle transversal, feedback vertex set, and hitting set [19]. For most **NP**-complete problems, the reconfiguration version has been shown to be **PSPACE**-complete [13,14,17], while for some problems in **P**, the reconfiguration question could be either in **P** [13] or **PSPACE**-complete [2].

The problem of transforming input s into input t can be viewed as the problem of determining if there is a path from s to t in a graph representing feasible solutions. Such a path is called a *reconfiguration sequence*. For the problem of dominating set, the *k-dominating graph*, defined formally in Section 2, consists of a node for each feasible solution and an edge for each pair of solutions that differ by a single vertex. Finding an s-t path in this graph (not included in the input) has been shown to be **W**[2]-hard [19], and hence not likely to yield even a fixed-parameter tractable algorithm [7].

Although having received less attention than the s-t path problem, other characteristics of the solution graph have been studied. Determining the diameter of the reconfiguration graph will result in an upper bound on the length of any reconfiguration sequence. For a problem such as colouring, one can determine the *mixing number*, the minimum number of colours needed to ensure that the entire graph is connected; such a number has been obtained for the problem of list edge-colouring on trees [14].

In previous work on reconfiguration of dominating sets, Haas and Seyffarth [10] considered the connectivity of the graph of solutions of size at most k, for various values of k relative to n, the number of vertices in the input graph G. They demonstrated that the graph is connected when $k = n-1$ and G has at least two non-adjacent edges, or when k is one greater than the maximum cardinality of a minimal dominating set and G is non-trivially bipartite or chordal. They left as an open question, answered negatively here, whether the latter results could be extended to all graphs.

In this paper we extend previous work by showing in Section 3 that the solution graph is connected and of linear diameter for $k = n - \mu$ for any input graph with a matching of size least $\mu + 1$, for any nonnegative integer μ. In Section 4, we give a series of counterexamples demonstrating that $D_{\Gamma(G)+1}(G)$ is not guaranteed to be connected for planar graphs, graphs of bounded treewidth, or b-partite graphs for $b \geq 3$. In Section 5, we pose and answer a question about the diameter of $D_{\gamma(G)+1}(G)$ by showing that there is an infinite family of graphs of exponential diameter.

2   Preliminaries

We assume that each G is a simple, undirected graph on n vertices with vertex set $V(G)$ and edge set $E(G)$. The *diameter* of G is the maximum over all pairs of vertices u and v in $V(G)$ of the length of the shortest path between u and v.

A set $S \subseteq V(G)$ is a *dominating set* of G if and only if every vertex in $V(G) \backslash S$ is adjacent to a vertex in S. The minimum cardinality of any dominating set of G is denoted by $\gamma(G)$. Similarly, $\Gamma(G)$ is the maximum cardinality of any minimal dominating set in G.

For a vertex $u \in V(G)$ and a dominating set S of G, we say u is *dominated* by $v \in S$ if $u \notin S$ and u is adjacent to v. For a vertex v in a dominating set S, a *private neighbour* of v is a vertex dominated by v and not dominated by any other vertex in S; the *private neighbourhood of* v is the set of private neighbours of v. A vertex v in a dominating set S is *deletable* if $S \setminus \{v\}$ is also a dominating set of G.

Fact 1. *A vertex v is deletable if and only if v has at least one neighbour in S and v has no private neighbour.*

Given a graph G and a positive integer k, we consider the k-*dominating graph* of G, $D_k(G)$, such that each vertex in $V(D_k(G))$ corresponds to a dominating set of G of cardinality at most k. Two vertices are adjacent in $D_k(G)$ if and only if the corresponding dominating sets differ by either the addition or the deletion of a single vertex; each such operation is a *reconfiguration step*. Formally, if A and B are dominating sets of G of cardinality at most k, then there exists an edge between A and B if and only if there exists a vertex $u \in V(G)$ such that $(A \setminus B) \cup (B \setminus A) = \{u\}$. We refer to vertices in G using lower case letters (e.g. u, v) and to the vertices in $D_k(G)$, and by extension their associated dominating sets, using upper case letters (e.g. A, B). We write $A \leftrightarrow B$ if there exists a path in $D_k(G)$ joining A and B. The following fact is a consequence of our ability to add vertices as needed to form B from A.

Fact 2. *For dominating sets A and B, if $A \subseteq B$, then $A \leftrightarrow B$ and $B \leftrightarrow A$.*

3   Graphs with a Matching of Size $\mu + 1$

Theorem 1. *For any nonnegative integer μ, if G has a matching of size at least $\mu + 1$, then $D_{n-\mu}(G)$ is connected for $n = |V(G)|$.*

Proof. For G a graph with matching $M = \{\{u_i, w_i\} \mid 0 \le i \le \mu\}$, we define $U = \{u_i \mid 0 \le i \le \mu\}$, $W = \{w_i \mid 0 \le i \le \mu\}$, and the set of *outsiders* $R = V(G) \setminus (U \cup W)$.

Using any dominating set S of G, we classify edges in M as follows: edge $\{u_i, w_i\}$, $0 \le i \le \mu$, is *clean* if neither u_i nor w_i is in S, *u-odd* if $u_i \in S$ but $w_i \notin S$, *w-odd* if $w_i \in S$ but $u_i \notin S$, *odd* if $\{u_i, w_i\}$ is *u-odd* or *w-odd*, and *even* if $\{u_i, w_i\} \subseteq S$. We use clean(S) and odd(S), respectively, to denote the numbers of clean and odd edges for S. Similarly, we let *u-odd*(S) and *w-odd*(S) denote the numbers of *u-odd* and *w-odd* edges for S. In the example graph shown in Figure 1, $\mu + 1 = 7$ and $R = \emptyset$. There is a single clean edge, namely $\{u_1, w_1\}$, three *w-odd* edges, two *u-odd* edges, and a single even edge.

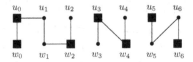

Fig. 1. Vertices in S are marked with squares

It suffices to show that for S an arbitrary dominating set of G such that $|S| \le n - \mu$, $S \leftrightarrow N$ for $N = V(G) \setminus W$; N is clearly a dominating set as each vertex $w_i \in W = V(G) \setminus N$ is dominated by u_i. By Fact 2, for S' a dominating set of G such that $S' \supseteq S$ and $|S'| = n - \mu$, since S' is a superset of S, then $S \leftrightarrow S'$. The reconfiguration from S' to N can be broken into three stages. In the first stage, for a dominating set S_0 with no clean edges, we show $S' \leftrightarrow S_0$ by repeatedly decrementing the number of clean edges (u_i or w_i is added to the dominating set for some $0 \le i \le \mu$). In the second stage, for T_μ with μ *u-odd* edges and one even edge, we show $S_0 \leftrightarrow T_\mu$ by repeatedly incrementing the number of *u-odd* edges. Finally, we observe that deleting the single remaining element in $T_\mu \cap W$ yields $T_\mu \leftrightarrow N$.

In stage 1, for $x = \text{clean}(S')$, we show that $S' = S_x \leftrightarrow S_{x-1} \leftrightarrow S_{x-2} \leftrightarrow \ldots \leftrightarrow S_0$ where for each $0 \le j \le x$, S_j is a dominating set of G such that $|S_j| = n - \mu$ and clean(S_j) = j. To show that $S_a \leftrightarrow S_{a-1}$ for arbitrary $1 \le a \le x$, we prove that there is a deletable vertex in some even edge and hence a vertex in a clean edge can be added in the next reconfiguration step. For $b = \text{odd}(S_a)$, the set E of vertices in even edges is of size $2((\mu+1) - a - b)$. Since each vertex in E has a neighbour in S_a, if at least one vertex in E does not have a private neighbour, then E contains a deletable vertex (Fact 1).

The μ vertices in $V(G) \setminus S_a$ are the only possible candidates to be private neighbours. Of these, the b vertices of $V(G) \setminus S_a$ in odd edges cannot be private neighbours of vertices in E, as each is the neighbour of a vertex in $S_a \setminus E$ (the other endpoint of the edge). The number of remaining candidates, $\mu - b$, is smaller than the number of vertices in E; $\mu \ge 2a + b$ as the vertices of $V(G) \setminus S_a$ must contain both endpoints of any clean edge and one endpoint for any odd edge.

Hence, there exists at least one deletable vertex in E. When we delete such a vertex and add an arbitrary endpoint of a clean edge, the clean edge becomes an odd edge and the number of clean edges decreases. We can therefore reconfigure from S_a to the desired dominating set, and by applying the same argument a times, to S_0.

In the second stage we show that for $y = u\text{-odd}(S_0)$, $S_0 = T_y \leftrightarrow T_{y+1} \leftrightarrow T_{y+2} \leftrightarrow \ldots \leftrightarrow T_\mu$ where for each $y \le j \le \mu$, T_j is a dominating set of G such that $|T_j| = n - \mu$, $\text{clean}(T_j) = 0$, and $u\text{-odd}(T_j) = j$. To show that $T_c \leftrightarrow T_{c+1}$ for arbitrary $y \le c \le \mu - 1$, we use a counting argument to find a vertex in an even edge that is in W and deletable; in one reconfiguration step the vertex is deleted, increasing the number of u-odd edges, and in the next reconfiguration step an arbitrary vertex in R or in a w-odd edge is added to the dominating set. We let $d = w\text{-odd}(T_c)$ (i.e. the number of w-odd edges for T_c) and observe that since there are c u-odd edges, d w-odd edges, and no clean edges, there exist $(\mu + 1) - c - d$ even edges. We define E_w to be the set of vertices in W that are in the even edges, and observe that each has a neighbour in T_c; a vertex in E_w will be deletable if it does not have a private neighbour.

Of the μ vertices in $V(G) \backslash T_c$, only those in R are candidates to be private neighbours of vertices in E_w, as each vertex in an odd edge has a neighbour in T_c. As there are c u-odd edges and d w-odd edges, the total number of vertices in $R \cap V(G) \backslash T_c$ is $\mu - c - d$. Since this is smaller than the number of vertices in E_w, at least one vertex in E_w must be deletable. When we delete such a vertex from T_c and in the next step add an arbitrary vertex from the outsiders or w-odd edges, the even edge becomes a u-odd edge and the number of u-odd edges increases. Note that we can always find such a vertex since there are $\mu - c - d$ outsiders, d w-odd edges, and $c \le \mu - 1$. Hence, we can reconfigure from T_c to T_{c+1}, and by $\mu - c$ repetitions, to T_μ. $\qquad\square$

Corollary 1 results from the length of the reconfiguration sequence formed in Theorem 1; reconfiguring to S' can be achieved in at most $n - \mu$ steps, and stages 1 and 2 require at most 2μ steps each, as $\mu \in O(n)$ is at most the numbers of clean and u-odd edges. Theorem 2 shows that Theorem 1 is tight.

Corollary 1. *The diameter of $D_{n-\mu}(G)$ is in $O(n)$ for G a graph with a matching of size $\mu + 1$.*

Theorem 2. *For any nonnegative integer μ, there exists a graph G_μ with a matching of size μ such that $D_{n-\mu}(G_\mu)$ is not connected.*

Proof. Let G_μ be a path on $n = 2\mu$ vertices. Clearly, G_μ has μ disjoint edges, $n - \mu = 2\mu - \mu = \mu$, and $D_{n-\mu}(G_\mu) = D_\mu(G_\mu)$. We let S be a dominating set of G_μ such that $|S| \ge \mu + 1$. At least one vertex in S must have all its neighbors in S and is therefore deletable. It follows that $\Gamma(G_\mu) = \mu$ and $D_{n-\mu}(G_\mu) = D_\mu(G_\mu) = D_{\Gamma(G_\mu)}(G_\mu)$ which is not connected by the result of Haas and Seyffarth [10, Lemma 3]. $\qquad\square$

4 $D_{\Gamma(G)+1}(G)$ May Not Be Connected

In this section we demonstrate that $D_{\Gamma(G)+1}(G)$ is not connected for an infinite family of graphs $G_{(d,b)}$ for all positive integers $b \geq 3$ and $d \geq 2$, where graph $G_{(d,b)}$ is constructed from $d+1$ cliques of size b. We demonstrate using the graph $G_{(4,3)}$ as shown in part (a) of Figure 2, consisting of fifteen vertices partitioned into five cliques of size 3: the *outer clique* C_0, consisting of the top, left, and right *outer vertices* o_1, o_2, and o_3, and the four *inner cliques* C_1 through C_4, ordered from left to right. We use $v_{(i,1)}$, $v_{(i,2)}$, and $v_{(i,3)}$ to denote the top, left, and right vertices in clique C_i, $1 \leq i \leq 4$. More generally, a graph $G_{(d,b)}$ has $d+1$ b-cliques C_i for $0 \leq i \leq d$. The clique C_0 consists of outer vertices o_j for $1 \leq j \leq b$, and for each inner clique C_i, $1 \leq i \leq d$ and each $1 \leq j \leq b$, there exists an edge $\{o_j, v_{(i,j)}\}$.

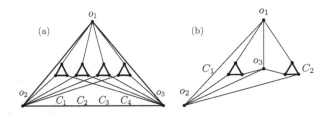

Fig. 2. Counterexamples for (a) general and (b) planar graphs

For any $1 \leq j \leq b$ a dominating set does not contain o_j, then the vertices $v_{(i,j)}$ of the inner cliques must be dominated by vertices in the inner cliques (hence Fact 3). In addition, the outer vertex o_j can be dominated only by another outer vertex or some vertex $v_{(i,j)}$, $1 \leq i \leq d$ (hence Fact 4).

Fact 3. *Any dominating set that does not contain all of the outer vertices must contain at least one vertex from each of the inner cliques.*

Fact 4. *Any dominating set that does not contain any outer vertex must contain at least one vertex of the form $v_{(\cdot,j)}$ for each $1 \leq j \leq b$.*

Lemma 1. *For each graph $G_{(d,b)}$ as defined above, $\Gamma(G_{(d,b)}) = d + b - 2$.*

Proof. We first demonstrate that there is a minimal dominating set of size $d + b - 2$, consisting of $\{v_{(1,j)} \mid 2 \leq j \leq b\} \cup \{v_{(i,1)} \mid 2 \leq i \leq d\}$. The first set dominates $b - 1$ of the outer vertices and the first inner clique and the second set dominates o_1 and the rest of the inner cliques. The dominating set is minimal, as the removal of any vertex $v_{(1,j)}$, $2 \leq j \leq b$, would leave vertex o_j with no neighbour in the dominating set and the removal of any $v_{(i,1)}$, $2 \leq i \leq d$, would leave $\{v_{(i,j)} \mid 1 \leq j \leq b\}$ with no neighbour in the dominating set.

By Fact 3, any dominating set that does not contain all outer vertices must contain at least one vertex in each of the d inner cliques. Since the outer vertices form a minimal dominating set, any other minimal dominating set must contain at least one vertex from each of the inner cliques.

We now consider any dominating set S of size at least $d + b - 1$ containing one vertex for each inner clique and show that it is not minimal. If S contains at least one outer vertex, we can find a smaller dominating set by removing all but the outer vertex and one vertex for each inner clique, yielding a total of $d + 1 < d + b - 1$ vertices (since $b \geq 3$). Now suppose that S consists entirely of inner vertices. By Fact 4, S contains at least one vertex of the form $v_{(\cdot,j)}$ for each $1 \leq j \leq b$. Moreover, for at least one value $1 \leq j' \leq b$, there exists more than one vertex of the form $v_{(\cdot,j')}$ as $d + b - 1 > b$. This allows us to choose b vertices of the form $v_{(\cdot,j)}$ for each $1 \leq j \leq b$ that dominate at least two inner cliques as well as all outer vertices. By selecting one member of S from each of the remaining $d - 2$ inner cliques, we form a dominating set of size $d + b - 2 < d + b - 1$, proving that S is not minimal. □

Theorem 3. *There exists an infinite family of graphs such that for each G in the family, $D_{\Gamma(G)+1}(G)$ is not connected.*

Proof. For any positive integers $b \geq 3$ and $d \geq 2$, we show that there is no path between dominating sets A to B in $D_{d+b-1}(G_{(d,b)})$, where A consists of the vertices in the outer clique and B consists of $\{v_{(i,\ell)} \mid 1 \leq i \leq d, 1 \leq \ell \leq b, i \equiv \ell \pmod{b}\}$.

By Fact 3, before we can delete any of the vertices in A, we need to add one vertex from each of the inner cliques, resulting in a dominating set of size $d + b = \Gamma(G_{(d,b)}) + 2$. As there is no such vertex in our graph, there is no way to connect A and B. □

Each graph $G_{(d,b)}$ constructed for Theorem 3 is a b-partite graph; we can partition the vertices into b independent sets, where the jth set, $1 \leq j \leq b$ is defined as $\{v_{(i,j)} \mid 1 \leq i \leq d\} \cup \{o_i \mid 1 \leq i \leq d, i \equiv j + 1 \pmod{b}\}$. Moreover, we can form a tree decomposition of width $2b - 1$ of $G_{(d,b)}$, for all positive integers $b \geq 3$ and $d \geq b$, by creating bags with the vertices of the inner cliques and adding all outer vertices to each bag.

Corollary 2. *For every positive integer $b \geq 3$, there exists an infinite family of graphs of tree-width $2b - 1$ such that for each G in the family, $D_{\Gamma(G)+1}(G)$ is not connected, and an infinite family of b-partite graphs such that for each G in the family, $D_{\Gamma(G)+1}(G)$ is not connected.*

Theorem 3 does not preclude the possibility that when restricted to planar graphs or any other graph class that excludes $G_{(d,b)}$, $D_{\Gamma(G)+1}(G)$ is connected. However, the next corollary follows directly from the fact that $G_{(2,3)}$ is planar (part (b) of Figure 2).

Corollary 3. *There exists a planar graph G for which $D_{\Gamma(G)+1}(G)$ is not connected.*

5   On the Diameter of $D_k(G)$

In this section, we obtain a lower bound on the diameter of the k-dominating graph of a family of graphs G_n. We describe G_n in terms of several component subgraphs, each playing a role in forcing the reconfiguration of dominating sets.

A *linkage gadget* (part (a), Figure 3) consists of five vertices, the *external vertices* (or endpoints) e_1 and e_2, and the *internal vertices* i_1, i_2, and i_3. The external vertices are adjacent to each internal vertex as well as to each other; the following results from the internal vertices having degree two:

Fact 5. *In a linkage gadget, the minimum dominating sets of size one are $\{e_1\}$ and $\{e_2\}$. Any dominating set containing an internal vertex must contain at least two vertices. Any dominating set in a graph containing m vertex-disjoint linkage gadgets with all internal vertices having degree exactly two must contain at least one vertex in each linkage gadget.*

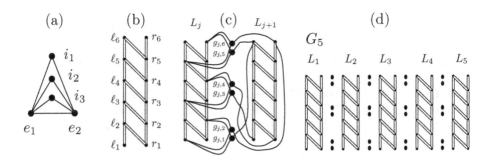

Fig. 3. Parts of the construction

A *ladder* (part (b) of Figure 3, linkages shown as double edges) is a graph consisting of twelve *ladder vertices* paired into six *rungs*, where rung i consists of the vertices ℓ_i and r_i for $1 \le i \le 6$, as well as the 45 internal vertices of fifteen linkage gadgets. Each linkage gadget is associated with a pair of ladder vertices, where the ladder vertices are the external vertices in the linkage gadget. The fifteen pairs are as follows: ten *vertical pairs* $\{\ell_i, \ell_{i+1}\}$ and $\{r_i, r_{i+1}\}$ for $1 \le i \le 5$, and five *cross pairs* $\{\ell_{i+1}, r_i\}$ for $1 \le i \le 5$. For convenience, we refer to vertices ℓ_i, $1 \le i \le 6$ and the associated linkage gadgets as the *left side of the ladder* and to vertices r_i, $1 \le i \le 6$ and the associated linkage gadgets as the *right side of the ladder*, or collectively as the *sides of the ladder*.

The graph G_n consists of n ladders L_1 through L_n and $n-1$ sets of *gluing vertices*, where each set consists of three *clusters* of two vertices each. For $\ell_{j,i}$ and $r_{j,i}$, $1 \le i \le 6$, the ladder vertices of ladder L_j, and $g_{j,1}$ through $g_{j,6}$ the gluing vertices that join ladders L_j and L_{j+1}, we have the following connections for $1 \le j \le n-1$:

- Edges connecting the *bottom cluster* to the bottom two rungs of ladder L_j and the top rung of ladder L_{j+1}: $\{\ell_{j,1}, g_{j,1}\}$, $\{\ell_{j,1}, g_{j,2}\}$, $\{r_{j,2}, g_{j,1}\}$, $\{r_{j,2}, g_{j,2}\}$, $\{\ell_{j+1,6}, g_{j,1}\}$, $\{r_{j+1,6}, g_{j,2}\}$.
- Edges connecting the *middle cluster* to the middle two rungs of ladder L_j and the bottom rung of ladder L_{j+1}: $\{\ell_{j,3}, g_{j,3}\}$, $\{\ell_{j,3}, g_{j,4}\}$, $\{r_{j,4}, g_{j,3}\}$, $\{r_{j,4}, g_{j,4}\}$, $\{\ell_{j+1,1}, g_{j,3}\}$, $\{r_{j+1,1}, g_{j,4}\}$.
- Edges connecting the *top cluster* to the top two rungs of ladder L_j and the top rung of ladder L_{j+1}: $\{\ell_{j,5}, g_{j,5}\}$, $\{\ell_{j,5}, g_{j,6}\}$, $\{r_{j,6}, g_{j,5}\}$, $\{r_{j,6}, g_{j,6}\}$, $\{\ell_{j+1,6}, g_{j,5}\}$, $\{r_{j+1,6}, g_{j,6}\}$.

Figure 3 parts (c) and (d) show details of the construction of G_n; they depict, respectively, two consecutive ladders and G_5, both with linkages represented as double edges. When clear from context, we sometimes use single subscripts instead of double subscripts to refer to the vertices of a single ladder.

We let $\mathcal{D} = \{\{\ell_{(j,2i-1)}, \ell_{(j,2i)}\}, \{r_{(j,2i-1)}, r_{(j,2i)}\} \mid 1 \leq i \leq 3, 1 \leq j \leq n\}$ denote a set of $6n$ pairs in G_n; the corresponding linkage gadgets are vertex-disjoint. Then Fact 5 implies the following:

Fact 6. *Any dominating set S of G_n must contain at least one vertex of each of the linkage gadgets for vertical pairs in the set $\mathcal{D}$ and hence is of size at least $6n$; if S contains an internal vertex, then $|S| > 6n$.*

Choosing an arbitrary external vertex for each vertical pair does not guarantee that all vertices on the side of a ladder are dominated; for example, the set $\{\ell_i \mid i \in \{1,4,5\}\}$ does not dominate the internal vertices in the vertical pair $\{\ell_2, \ell_3\}$. Choices that do not leave such gaps form the set $\mathcal{C} = \{\mathcal{C}_i \mid 1 \leq i \leq 4\}$ where $\mathcal{C}_1 = \{1,3,5\}$, $\mathcal{C}_2 = \{2,3,5\}$, $\mathcal{C}_3 = \{2,4,5\}$, and $\mathcal{C}_4 = \{2,4,6\}$.

Fact 7. *In any dominating set S of size $6n$ and in any ladder L in G_n, the restriction of S to L must be of the form S_i for some $1 \leq i \leq 7$, as illustrated in Figure 4.*

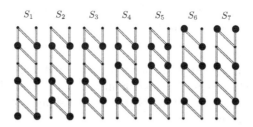

Fig. 4. Minimum dominating sets for G_1

Proof. Fact 6 implies that the only choices for the left (right) vertices are $\{\ell_i \mid i \in \mathcal{C}_j\}$ ($\{r_i \mid i \in \mathcal{C}_j\}$) for $1 \leq j \leq 4$. The sets S_i, $1 \leq i \leq 7$, are the only combinations of these choices that dominate all the internal vertices in the cross pairs. □

We say that ladder L_j *is in state* S_i if the restriction of the dominating set to L_j is of the form S_i, for $1 \leq j \leq n$ and $1 \leq i \leq 7$.

The exponential lower bound in Theorem 4 is based on counting how many times each ladder is modified from S_1 to S_7 or vice versa; we say ladder L_j undergoes a *switch* for each such modification. We first focus on a single ladder.

Fact 8. *For S a dominating set of G_1, a vertex $v \in S$ is deletable if and only if either v is the internal vertex of a linkage gadget one of whose external vertices is in S, or for every linkage gadget containing v as an external vertex, either the other external vertex is also in S or all internal vertices are in S.*

Lemma 2. *In $D_{\gamma(G_1)+1}(G_1)$ there is a single reconfiguration sequence between S_1 and S_7, of length 12.*

Proof. We define P to be the path in the graph corresponding to the reconfiguration sequence $S_1 \leftrightarrow S_1 \cup \{\ell_2\} \leftrightarrow S_2 \leftrightarrow S_2 \cup \{r_2\} \leftrightarrow S_3 \leftrightarrow S_3 \cup \{\ell_4\} \leftrightarrow S_4 \leftrightarrow S_4 \cup \{r_4\} \leftrightarrow S_5 \leftrightarrow S_5 \cup \{\ell_6\} \leftrightarrow S_6 \leftrightarrow S_6 \cup \{r_6\} \leftrightarrow S_7$ and demonstrate that there is no shorter path between S_1 and S_7.

By Facts 7 and 6, G_1 has exactly seven dominating sets of size six, and any dominating set S of size seven contains two vertices from one vertical pair d in $\mathcal{D}$ and one from each of the remaining five. The neighbours of S in $D_{\gamma(G_1)+1}(G_1)$ are the vertices corresponding to the sets S_i, $1 \leq i \leq 7$, obtained by deleting a single vertex of S. The number of neighbours is thus at most two, depending on which, if any, vertices in d are deletable.

If at least one of the vertices of S in d is an internal vertex, then at most one vertex satisfies the first condition in Fact 8. Thus, for S to have two neighbours, there must be a ladder vertex that satisfies the second condition of Fact 8, which by inspection of Figure 4 can be seen to be false.

If instead d contains two ladder vertices, in order for S to have two neighbours, the four ladder vertices on the side containing d must correspond to the union of two of the sets in $\mathcal{C}$. There are only three such unions, $\mathcal{C}_1 \cup \mathcal{C}_2$, $\mathcal{C}_2 \cup \mathcal{C}_3$, and $\mathcal{C}_3 \cup \mathcal{C}_4$, which implies that the only pairs with common neighbours are $\{S_i, S_{i+1}\}$ for $1 \leq i \leq 6$, as needed to complete the proof.   □

For $n > 2$, we cannot reconfigure ladders independently from each other, as we need to ensure that all gluing vertices are dominated. For consecutive ladders L_j and L_{j+1}, any cluster that is not dominated by L_j must be dominated by L_{j+1}; the bottom, middle, and top clusters are not dominated by any vertex in S_2, S_4, and S_6, respectively.

Fact 9. *In any dominating set S of G_n, for any $1 \leq j < n$, if L_j is in state S_2, then L_{j+1} is in state S_7; if L_j is in state S_4, then L_{j+1} is in state S_1; and if L_j is in state S_6, then L_{j+1} is in state S_7.*

Lemma 3. *For any reconfiguration sequence in which L_j and L_{j+1} are initially both in state S_1, if L_j undergoes p switches then L_{j+1} must undergo at least $2p + 1$ switches.*

Proof. We use a simple counting argument. When $p = 1$, the result follows immediately from Fact 9 since L_j can only reach state S_7 if L_{j+1} is reconfigured from S_1 to S_7 to S_1 and finally back to S_7. After the first switch of L_j, both ladders are in state S_7.

For any subsequent switch of L_j, L_j starts in state S_7 because for L_j to reach S_1 from S_2 or to reach S_7 from S_6, by Fact 9 L_{j+1} must have been in S_7. Since by definition L_j starts in S_1 or S_7, to enable L_j to undergo a switch, L_{j+1} will have to undergo at least two switches, namely S_7 to S_1 and back to S_7. $\square$

Theorem 4. *For S a dominating set of G_n such that every ladder of G_n is in state S_1 and T a dominating set of G_n such that every ladder of G_n is in state S_7, the length of any reconfiguration sequence between S and T is at least $12(2^{n+1} - n - 2)$.*

Proof. We first observe that Lemma 2 implies that the switch of any ladder requires at least twelve reconfiguration steps; since the vertex associated with a dominating set containing a gluing vertex will have degree at most one in the k-dominating graph, there are no shortcuts formed.

To reconfigure from S to T, ladder L_1 must undergo at least one switch. By Lemma 3, ladder L_2 will undergo at least $3 = 2^2 - 1$ switches, hence $2^j - 1$ switches for ladder L_j, $1 \leq j \leq n$. Since each switch requires twelve steps, the total number of steps is thus at least $12 \sum_{i=1}^{n} (2^i - 1) = 12(2^{n+1} - n - 2)$. $\square$

Corollary 4. *There exists an infinite family of graphs such that for each graph G_n in the family, $D_{\gamma(G_n)+1}(G_n)$ has diameter $\Omega(2^n)$.*

6   Conclusions and Future Work

In answering Haas and Seyffarth's question concerning the connectivity of $D_k(G)$ for general graphs and $k = \Gamma(G) + 1$, we have demonstrated infinite families of planar, bounded treewidth, and b-partite graphs for which the k-dominating graph is not connected. It remains to be seen whether k-dominating graphs are connected for graphs more general than non-trivially bipartite graphs or chordal graphs, and whether $D_{\Gamma(G)+2}(G)$ is connected for all graphs. It would also be useful to know if there is a value of k for which $D_k(G)$ is guaranteed not to have exponential diameter. Interestingly, for our connectivity and diameter examples, incrementing the size of the sets by one is sufficient to break the proofs.

References

1. Bonamy, M., Bousquet, N.: Recoloring bounded treewidth graphs. In: Proc. of the 7th Latin-American Algorithms, Graphs, and Optimization Symp. (2013)
2. Bonsma, P.: The complexity of rerouting shortest paths. In: Rovan, B., Sassone, V., Widmayer, P. (eds.) MFCS 2012. LNCS, vol. 7464, pp. 222–233. Springer, Heidelberg (2012)

3. Bonsma, P., Cereceda, L.: Finding paths between graph colourings: PSPACE-completeness and superpolynomial distances. Theor. Comput. Sci. 410(50), 5215–5226 (2009)
4. Cereceda, L., van den Heuvel, J., Johnson, M.: Connectedness of the graph of vertex-colourings. Discrete Math 308(56), 913–919 (2008)
5. Cereceda, L., van den Heuvel, J., Johnson, M.: Finding paths between 3-colorings. J. of Graph Theory 67(1), 69–82 (2011)
6. Cereceda, L., van den Heuvel, J., Johnson, M.: Mixing 3-colourings in bipartite graphs. European J. of Combinatorics 30(7), 1593–1606 (2009)
7. Downey, R.G., Fellows, M.R.: Parameterized complexity. Springer, New York (1997)
8. Fricke, G., Hedetniemi, S.M., Hedetniemi, S.T., Hutson, K.R.: γ-Graphs of Graphs. Discussiones Mathematicae Graph Theory 31(3), 517–531 (2011)
9. Gopalan, P., Kolaitis, P.G., Maneva, E.N., Papadimitriou, C.H.: The connectivity of boolean satisfiability: computational and structural dichotomies. SIAM J. on Computing 38(6), 2330–2355 (2009)
10. Haas, R., Seyffarth, K.: The k-Dominating Graph. Graphs and Combinatorics (March 2013) (online publication)
11. Hearn, R.A., Demaine, E.D.: PSPACE-completeness of sliding-block puzzles and other problems through the nondeterministic constraint logic model of computation. Theor. Comput. Sci. 343(1-2), 72–96 (2005)
12. Ito, T., Demaine, E.D.: Approximability of the subset sum reconfiguration problem. In: Ogihara, M., Tarui, J. (eds.) TAMC 2011. LNCS, vol. 6648, pp. 58–69. Springer, Heidelberg (2011)
13. Ito, T., Demaine, E.D., Harvey, N.J.A., Papadimitriou, C.H., Sideri, M., Uehara, R., Uno, Y.: On the complexity of reconfiguration problems. Theor. Comput. Sci. 412(12-14), 1054–1065 (2011)
14. Ito, T., Kamiński, M., Demaine, E.D.: Reconfiguration of list edge-colorings in a graph. Discrete Applied Math. 160(15), 2199–2207 (2012)
15. Ito, T., Kawamura, K., Ono, H., Zhou, X.: Reconfiguration of list L(2,1)-labelings in a graph. In: Chao, K.-M., Hsu, T.-S., Lee, D.-T. (eds.) ISAAC 2012. LNCS, vol. 7676, pp. 34–43. Springer, Heidelberg (2012)
16. Kamiński, M., Medvedev, P., Milanič, M.: Shortest paths between shortest paths. Theor. Comput. Sci. 412(39), 5205–5210 (2011)
17. Kamiński, M., Medvedev, P., Milanič, M.: Complexity of independent set reconfigurability problems. Theor. Comput. Sci. 439, 9–15 (2012)
18. Mayr, E.W., Plaxton, C.G.: On the spanning trees of weighted graphs. Combinatorica 12(4), 433–447 (1992)
19. Mouawad, A.E., Nishimura, N., Raman, V., Simjour, N., Suzuki, A.: On the parameterized complexity of reconfiguration problems. In: Gutin, G., Szeider, S. (eds.) IPEC 2013. LNCS, vol. 8246, pp. 281–294. Springer, Heidelberg (2013)

Back-Up 2-Center
on a Path/Tree/Cycle/Unicycle

Binay Bhattacharya[1], Minati De[2], Tsunehiko Kameda[1], Sasanka Roy[3],
Vladyslav Sokol[1], and Zhao Song[4]

[1] School of Computing Science, Simon Fraser Univ., Canada
{binay,tiko,vlads}@sfu.ca
[2] Advanced Computing and Microelectronics Unit, Indian Statistical Inst.,
Kolkata, India
minati.isi@gmail.com
[3] Chennai Mathematical Inst., Chennai, India
sasanka@cmi.ac.in
[4] Dept. of Computer Science, Univ. of Texas at Austin, USA
zhaos@utexas.edu

Abstract. This paper addresses the problem of locating two facilities
in vertex-weighted path/tree/cycle/unicyclic networks, where each facility can fail with a given probability [16]. It is assumed that the two
facilities do not fail simultaneously, and when a facility fails, the other
facility is required to service all the vertices. We show that the weighted
back-up 2-center on a path (resp. tree, cycle, unicyclic) network can be
computed in $O(n)$ (resp. $O(n \log n)$, $O(n^2)$, $O(n^2 \log n)$) time, where n
is the number of vertices, and the centers need not be at vertices. This is
the first sub-quadratic time result for vertex-weighted tree networks. For
vertex-unweighted trees, it runs in $O(n)$ time, matching the best known
result [19]. The algorithm remains linear when there are only a constant
number of distinct weights.

1 Introduction

The p-center and p-median problems in networks are the two most important
problems in facility location optimization. Both problems are known to be NP-
hard for general networks [9, 10]. Therefore, many researchers have focussed on
simple networks such as trees, cactus, etc. [2, 3, 12, 14, 17].

Any decision-making environments can be classified into three categories: (i)
certainty, (ii) risk, and (iii) uncertainty [13]. Environment (i) is deterministic and
doesn't randomly change over time, whereas in Environments (ii) and (iii), the
change over time is random. The problems under risk and uncertainty situations
have been modeled as stochastic optimization problems and robust optimization
problems, respectively [11, 15]. For Environment (iii), Snyder and Daskin [16]
proposed another reliability model, in which facilities are located to minimize
the cost, while also taking into account the expected cost after the failure of facilities. Based on the reliability model, Wang et al. [18, 19] introduced the backup

Z. Cai et al. (Eds.): COCOON 2014, LNCS 8591, pp. 417–428, 2014.
© Springer International Publishing Switzerland 2014

2-center problem, whereby they locate two facilities *at vertices* such that the expected weighted distance from any vertex to the closest functioning facility is minimized. They assume that the two facilities fail independently with constant probability ρ, but they don't fail at the same time. They solve the back-up 2-center problem on a tree with equal vertex weight and the two centers restricted to vertices in $O(n)$ time. They also proposed a naïve $O(n^3)$ time algorithm for the backup 2-center problem on vertex-weighted tree networks, and posed an open problem for improving the complexity. They then considered the back-up 2-median problems in weighted tree networks and proposed an $O(n \log n)$ time algorithm for the problem. Very recently, Hong and Kang [8] considered the back-up 2-center problem on interval graphs. In this paper, based on the reliability model, we consider the back-up 2-center problem in vertex-weighted tree, cycle, and unicyclic networks, such that the center locations are not restricted to vertices.

This paper is organized as follows. In Sec. 2, we define the terms that are used throughout the paper, and review relevant facts. Secs. 3 and 4 show that a backup 2-center on a path and a tree can be found in $O(n)$ and $O(n \log n)$ time, respectively. The backup 2-center problem in a cycle network turns out to be more difficult, because a path from a vertex to a center is not unique. In Sec. 5, we present a somewhat complicated algorithm to solve this problem in $O(n^2)$ time. A *unicyclic* network, containing just one cycle is the most general network that we discuss in this paper. Drawing on the results in the preceding sections, Sec. 6 presents an algorithm that finds a backup 2-center in a unicycle in $O(n^2 \log n)$ time. Sec.7 concludes the paper. Due to lack of space, most proofs are omitted from this extended abstract.

2   Preliminaries

2.1   Definitions

Let $G = (V, E)$ denote a network with vertex set V and edge set E, where each vertex $v \in V$ has a weight w_v (≥ 0) and each edge in E has a non-negative length. Let $X = \{x_1, x_2\}$ be a pair of points on G, not necessarily vertices. We define the distance $d(v, X)$ between a vertex v and X by

$$d(v, X) \triangleq \min_{1 \leq i \leq 2} \{d(v, x_i)\},$$

where $d(v, x_i)$ is the length of a shortest path in G between vertex v and point x_i. Let us introduce the *cost* of X by

$$\phi(X, V) \triangleq \max_{v \in V} \{d(v, X)w_v\}. \tag{1}$$

We say that a set X of points *r-covers* V (and G) if $r \geq \phi(X, V)$. The (classical) *weighted 1-center*, or just *1-center* for short, c of G is defined as a point that r-covers V with the minimum r. The (classical) *weighted 2-center* C of G, or just *2-center* for short, is defined as a pair of points that r-covers V with the minimum r. When X contains just one point x, then we write $\phi(X, V) = \phi(x, V)$. We define the *cost function* of a single point x with respect to a single vertex v by

$$f_v(x) \triangleq \phi(x, \{v\}) = d(v, x)w_v. \tag{2}$$

Let c be a 1-center of network $G = (V, E)$, and let $\{c_1, c_2\}$ be any 2-center of G. Given a pair of points $\{x_1, x_2\}$, we partition V into $V_1(x_1, x_2)$ and $V_2(x_1, x_2)$ such that

$$\forall v \in V_1(x_1, x_2) : d(v, x_1) \leq d(v, x_2); \ \forall v \in V_2(x_1, x_2) : d(v, x_1) > d(v, x_2). \tag{3}$$

Without loss of generality, we may assume that the subnetwork G_1 (resp. G_2) defined by $V_1(x_1, x_2)$ (resp. $V_2(x_1, x_2)$) is connected. The objective function of a back-up 2-center is

$$\begin{aligned}\psi_\rho(x_1, x_2) \triangleq &(1 - \rho)\max\{\phi(x_1, V_1(x_1, x_2)), \phi(x_2, V_2(x_1, x_2))\} \\ &+ \rho(\phi(x_1, V) + \phi(x_2, V)),\end{aligned} \tag{4}$$

where ρ ($0 \leq \rho < 1$) is given [19]. The motivation for this objective function is given in [19], and is related to the expected value of the maximum cost, when each center fails independently with probability ρ. We want to minimize $\psi_\rho(x_1, x_2)$ by varying $x_1 \in G_1$ and $x_2 \in G_2$. The pair $(x_1 = c_1^\rho, x_2 = c_2^\rho)$ that minimizes $\psi_\rho(x_1, x_2)$ is called the *back-up 2-center*. For a (classical) 1-center c, any vertex $v \in V$ satisfying $\phi(c, V) = d(v, x)w_v$ is called a *critical vertex* for c, and is denoted by $\gamma(c)$.[1] When $\rho = 0$, (4) becomes

$$\psi_0(x_1, x_2) = \max\{\phi(x_1, V_1(x_1, x_2)), \phi(x_2, V_2(x_1, x_2))\},$$

which is obviously minimized by $x_1 = c_1$ and $x_2 = c_2$, where $\{c_1, c_2\}$ is a 2-center of G. Given a subnetwork $G' = (V', E')$ of network G, we call $\phi(x, V')$ for $x \notin G'$ the *upper envelope* (of cost functions) for G'. A network is said to be *trivial* if $\max\{\phi(c_1, V_1(x_1, x_2)), \phi(c_2, V_2(x_1, x_2))\} = \phi(c, V)$. In this paper, we discuss only nontrivial networks.

2.2   Basic Facts

Let c be a 1-center of $G = (V, E)$. The *weighted radius* of G is given by $r_G \triangleq \max_{v \in V} d(v, c)w_v$.

Fact 1 *Let c be a 1-center of G. Then there are two* critical vertices *for c in G, u and u', $u \neq u'$, satisfying $d(u, c)w_u = d(u', c)w_{u'} = r_G$, such that c lies on the path between u and u'.*   □

Lemma 1. [1] *The 2-center of a tree can be computed in $O(n)$ time.*   □

---

[1] Whether the argument of $\gamma(\cdot)$ is a 1-center of one of the centers of a 2-center will be clear from the context.

2.3   Observations

Lemma 2. [19] *Let $\{c_1, c_2\}$ be any 2-center of a network G. Then there is a 1-center c that lies on a path between c_1 and c_2.* □

Lemma 3. *Let $\{c_1, c_2\}$ be any 2-center of G. Then there is a backup 2-center $\{c_1^\rho, c_2^\rho\}$ such that c_1^ρ (resp. c_2^ρ) lies on a path between c_1 (resp. c_2) and a 1-center c.* □

3   Path Network

3.1   Preparation

We consider a path $P = (V, E)$, whose vertices are named $v_1, v_2, \ldots, v_n$ from left to right. The cost function of a vertex v_i, $f_{v_i}(x) = \phi(x, \{v_i\})$, consists of two linear segments. We first consider the right half of each such V-shaped function, and construct their upper envelope $E_L(x)$, starting from v_1.[2] It is clear that $E_L(x)$ can be constructed in $O(n)$ time, by processing the cost functions in the left-to-right order. Similarly, we consider the left halves of the V-shaped functions, whose upper envelope is denoted by $E_R(x)$. It is easy to see that $E_L(x)$ (resp. $E_R(x)$) is monotonically increasing (resp. decreasing) with respect to point x that moves from v_1 to v_n. We clearly have

$$\phi(x, V) = E(x) \triangleq \max\{E_L(x), E_R(x)\}, \tag{5}$$

i.e., $\phi(x, V)$ can be obtained as the upper envelope of $E_L(x)$ and $E_R(x)$, and it is a convex function. See Fig. 1. We can find the lowest point of $E(x)$, which is the 1-center c, very easily by computing the intersection of $E_L(x)$ and $E_R(x)$. See Fig. 1.

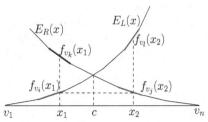

Fig. 1. $f_{v_i}(x_1) = f_{v_j}(x_2)$

As long as $\phi(x_1, V_1) > \phi(x_2, V_2)$, moving x_2 towards c to x_2' does not affect $\max\{\phi(x_1, V_1), \phi(x_2, V_2)\}$, but reduces $\phi(x_2, V)$. The reduction in $\phi(x_2, V)$ is $\Delta = \phi(x_2, V) - \phi(x_2', V)$, and the saving is $\psi_\rho(x_1, x_2) - \psi_\rho(x_1, x_2') = \rho\Delta$. This is a positive gain as long as $\rho > 0$. We thus have

Lemma 4. *If $\rho > 0$, then there is a back-up 2-center $\{c_1^\rho, c_2^\rho\}$ on a path such that $\phi(c_1^\rho, V_1) = \phi(c_2^\rho, V_2)$ holds.* □

---

[2] The subscript "L" means "due to the vertices on the Left."

3.2 Path Algorithm

Based on Lemmas 3 and 4, we now design an algorithm for computing the optimal back-up 2-center for a path. Given x_1, suppose that $E_L(x_1) = f_{v_i}(x_1)$ lies on a segment on $E_L(x)$. We find x_2 and v_j such that $f_{v_j}(x_2) = E_R(x_2) = f_{v_i}(x_1)$. In Fig. 1, $f_{v_i}(x)$ and $f_{v_j}(x)$ are indicated by red and purple line segments, respectively. We do this because Lemma 4 asserts that if $\{x_1, x_2\}$ is to be an optimal solution, then $f_{v_j}(x_2) = f_{v_i}(x_1)$ holds. Suppose that $E_R(x_1)$ (resp. $E_L(x_2)$) is on the blue (resp. green) line segment, which is $f_{v_k}(x)$ (resp. $f_{v_l}(x)$). Then (4) becomes

$$\psi_\rho(x_1, x_2) = (1 - \rho)f_{v_i}(x_1) + \rho(f_{v_k}(x_1) + f_{v_l}(x_2)). \tag{6}$$

We know that $f_{v_i}(x)$ is a linear function of the form $y = a_i z + b_i$, where z is the distance from v_1. Therefore, (6) is a piece-wise linear function in z, and it is minimized at an end of a linear segment, i.e., a bending point on $E_L(x)$ or $E_R(x)$. Let $\{c_1, c_2\}$ be a 2-center of the path. We project these bending points onto the horizontal axis, and name those lying between c_1 and c as $z_1 (= c_1), z_2, \ldots, z_m (= c)$ from left to right.

Algorithm 1 BU2Center-Path

1. *Compute $E_L(x)$, $E_R(x)$, a center c, and a 2-center (c_1, c_2).*
2. *For $k = 1, 2, \ldots, m$, carry out Steps 3 to 5.*
3. *Let $x_1 = z_k$ and find v_i such that $f_{v_i}(x_1) = E_L(x_1)$.*
4. *Determine v_j and x_2 such that $f_{v_j}(x_2) = E_R(x_2) = f_{v_i}(x_1)$.*
5. *Evaluate (6).*
6. *Find the minimum value among those computed for all k. The corresponding $\{x_1, x_2\}$ gives a candidate for a back-up 2-center $\{c_1^\rho, c_2^\rho\}$.* □

As we move x_1 to the right (i.e., as distance z increases), the corresponding x_2 moves monotonically to the left. Steps 3 and 4 can be carried out in amortized constant time per z_k. We repeat **BU2Center-Path** on the given path with the left and right ends reversed, and pick the less costly one between the two candidates for $\{c_1^\rho, c_2^\rho\}$ generated in Step 6. Each step can be carried out in linear time.

Theorem 1. *The back-up 2-center of a path can be found in $O(n)$ time.*

4 Tree Network

4.1 Observation

We assume that a given tree T is balanced and binary, so that any path from a leaf vertex to the root has length $O(\log n)$. If not, we can perform *spine tree decomposition* [3–5] to convert it into a structure that has the properties of a balanced binary tree. Similarly to Lemma 4 for a path, we also have

Lemma 5. *If $\rho > 0$, then there is a back-up 2-center $\{c_1^\rho, c_2^\rho\}$ in a tree such that $\phi(c_1^\rho, V_1) = \phi(c_2^\rho, V_2)$ holds.* □

4.2   Tree Algorithm

Based on Lemma 3, we design an algorithm for a tree. Let $\langle u_1, u_2, \ldots, u_t \rangle$ be the sequence of vertices on the path $\pi(c_1, c_2)$, where $\{c_1, c_2\}$ is a 2-center of T. We consider $\pi(c_1, c_2)$ to be a horizontal path from c_1 to c_2. For $x \in \pi(c_1, c_2)$ we define two upper envelopes. $E_L(x)$ (resp. $E_R(x)$) is the upper envelope of the cost functions of the vertices of T that lie to the left (resp. right) of x. See Fig. 2. Jumps in $E_L(x)$ and $E_R(x)$ may occur at vertices from which subtrees hang.

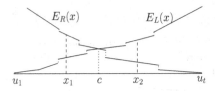

Fig. 2. $E_L(x)$ and $E_R(x)$

Algorithm 2 BU2Center-Tree.

1. *Compute a 2-center $\{c_1, c_2\}$ of T. Let $\pi(c_1, c_2) = \langle u_1(=c_1), u_2, \ldots, u_t(=c_2) \rangle$.*
2. *Compute the two upper envelopes $E_L(x)$ and $E_R(x)$.*
3. *Apply Algorithm BU2Center-Path to $\pi(c_1, c_2)$.*   □

Step 1 takes $O(n)$ time [1] by Lemma 1. We can compute $E_L(x)$ and $E_R(x)$ in Step 2 in $O(n \log n)$ time [7].

Theorem 2. *The back-up 2-center of a tree can be found in $O(n \log n)$ time.* □

As mentioned earlier, Wang et al. [19] solved the back-up 2-center problem on a tree with equal vertex weight and the two centers restricted to vertices in $O(n)$ time. Without the latter constraint, Algorithm BU2Center-Tree also runs in linear time, since $E_L(x)$ and $E_R(x)$ each consist of just one line segment. If the weight of each vertex is from a set of a fixed number of weights, then the back-up 2-center can be found in $O(n)$ time, since $E_L(x)$ and $E_R(x)$ each consist of a constant number of line segments.

5   Cycle Network

5.1   Observation

We consider a cycle $C = (V, E)$ with circumference l_C. For points $a, b \in C$ let $C(a, b)$ denote the clockwise section of C from a to b, and let $d(a, b)$ denote the length of the shortest path between a to b either clockwise or counterclockwise. Let $\{c_1^\ell, c_2^\ell\}$ be a back-up 2-center of C. See Fig. 3(a). The point $\alpha(p)$ that is at distance $l_C/2$ from $p \in C$ is called the *antipode* of p. We construct an augmented cycle $C'(V', E')$ from $C(V, E)$ by adding a vertex at the antipode $\alpha(v)$ of each

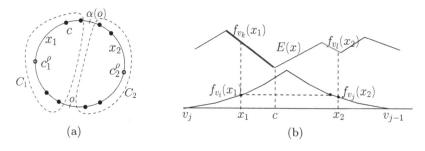

Fig. 3. (a) Critical vertex $\gamma(c_1^\rho)$ (resp. $\gamma(c_2^\rho)$) for c_1^ρ (resp. c_2^ρ); (b) $f_{v_i}(x_1) > f_{v_j}(x_2)$

$v \in V$, and assign weight 0 to it. We also add the center c of C and $\alpha(c)$. Let $V' = \{v_1, v_2, \ldots, v_n\}$. Thus each edge and its antipodal edge determine a partition of V' into V_1 and V_2.

Recall from Sec. 2.1 that, given a pair of points $\{x_1, x_2\}$, where $x_1 \neq x_2$, we partitioned V into $V_1(x_1, x_2)$ and $V_2(x_1, x_2)$ based on (3). Similarly, we can partition V' into $V_1(x_1, x_2)$ and $V_2(x_1, x_2)$ by removing an edge and its antipodal edge. We make the following obvious observation in terms of Fig. 3(a):

Proposition 1. *Let o and $\alpha(o)$ be the mid points between c_1^ρ and c_2^ρ such that $o \in C(c_2^\rho, c_1^\rho)$ and $\alpha(o) \in C(c_1^\rho, c_2^\rho)$. For any point $x \in C'(o, \alpha(o))$, such that $x \neq o, \alpha(o)$, we have $d(x, c_1^\rho) < d(x, c_2^\rho)$. Similarly, for any point $x \in C'(\alpha(o), o)$, such that $x \neq o, \alpha(o)$, we have $d(x, c_1^\rho) > d(x, c_2^\rho)$.* □

For each $v \in V'$, let $f_v^{cw}(x) = f_v(x) = \phi(x, \{v\})$ for $x \in C'(v, \alpha(v))$ and $f_v^{cw}(x) = 0$ for $x \in C'(\alpha(v), v)$. Similarly, let $f_v^{ccw}(x) = f_v(x) = \phi(x, \{v\})$ for $x \in C'(\alpha(v), v)$ and $f_v^{ccw}(x) = 0$ for $x \in C'(v, \alpha(v))$. We also define upper envelopes for these functions, $E^{cw}(x) \triangleq \max\{f_{v_i}^{cw}(x) \mid i = 1, 2, \ldots, n\}$ and $E^{ccw}(x) \triangleq \max\{f_{v_i}^{ccw}(x) \mid i = 1, 2, \ldots, n\}$. We can compute $E^{cw}(x)$, $E^{ccw}(x)$, and $E(x) \triangleq \max\{E^{cw}(x), E^{ccw}(x)\} = \phi(x, V')$ in $O(n \log n)$ time [7]. Analogously to Fig. 1, here we have Fig. 3(b), when C' was converted into a path by removing the edge (v_{j-1}, v_j).

5.2 Cycle Algorithm

Removing an edge $(v_{j-1}, v_j) \in E'$ and its antipodal edge $(v_{k-1}, v_k) \in E'$ from C' partitions V' into V_1 and V_2. We assume that both V_1 and V_2 contain at least one vertex of V. Let c_1' (resp. c_2') be a 1-center of V_1 (resp. V_2). We thus have $\phi(x_1, V_1) \geq \phi(c_1', V_1)$ for $x_1 \in \pi(v_j, v_{k-1})$ and $\phi(x_2, V_2) \geq \phi(c_2', V_2)$ for $x_2 \in \pi(v_k, v_{j-1})$. Then by Proposition 1, one such partition will correspond to the partition by c_1^ρ and c_2^ρ, shown in Fig. 3(a). Removing just edge $(v_{j-1}, v_j) \in E'$ transforms C' into a path, and the left (resp. right) half of this path consists of vertices in V_1 (resp. V_2). With respect to this path, let $E_L(x)$ (resp. $E_R(x)$) denote the upper envelopes of V_1 (resp. V_2) defined by $E_L(x) \triangleq \max\{f_v^{cw}(x) \mid v \in V_1\}$ and $E_R(x) \triangleq \max\{f_v^{ccw}(x) \mid v \in V_2\}$, which can be computed in $O(n)$ time.

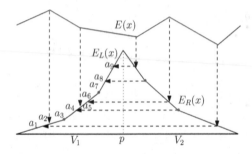

Fig. 4. Illustration for Steps 1 to 3 of Procedure `Map-Bending-Pts`

See Fig. 4. Suppose we have precomputed the upper envelope $E(x) = \phi(x, V')$ for C'. Note that $E(x)$ is different from $\max\{E_L(x), E_R(x)\}$ here, unlike (5). Let us map the bending points of $E_R(x)$ (resp. $E_L(x)$) and $E(x)$ onto $E_L(x)$ (resp. $E_R(x)$) as follows.

Procedure 1 `Map-Bending-Pts`

1. *Map the bending points of $E(x)$ downwards onto $E_L(x)$ and $E_R(x)$, as shown by the vertical dashed arrows in Fig. 4.*
2. *Map the points on $E_R(x)$ that are either a bending point of $E_R(x)$ or a projected point resulting from Step 1 horizontally to $E_L(x)$.*
3. *The points on $E_L(x)$ that are either a bending point of $E_L(x)$ or a projected point resulting from Step 1 or 2, are named $a_1, a_2, \ldots, a_t$, clockwise from v_1.*
4. *Map the points on $E_L(x)$ that are either a bending point of $E_L(x)$ or a projected point resulting from Step 1 horizontally to $E_R(x)$.*
5. *The points on $E_R(x)$ that are either a bending point of $E_R(x)$ or a projected point resulting from Step 1 or 4, are named $b_1, b_2, \ldots, b_t$, counterclockwise from v_n.*   □

Lemma 6. *Procedure `Map-Bending-Pts` runs in $O(n)$ time.*   □

We now mark the point corresponding to c'_1 (resp. c'_2) on $E_L(x)$ (resp. $E_R(x)$). We insert c'_1 (resp. c'_2) there, map it to $E_R(x)$ (resp. $E_L(x)$), and rename the resulting sequence $\langle a_1, a_2, \ldots, a_{t'} \rangle$ (resp. $\langle b_1, b_2, \ldots, b_{t'} \rangle$), where $t' = t + 2$. There is a linear segment between any adjacent pair of a_i's, and b_i's. Let $p(a_i) \in C'$ (resp. $p(b_i) \in C'$) be the point corresponding to a_i (resp. b_i). Consider a pair of points $\{x_1, x_2\}$ such that $x_1 \in C'(p(a_i), p(a_{i+1}))$ and $x_2 \in C'(p(b_i), c'_2)$. For any such x_1 and x_2, we have $\phi(x_1, V_1) \geq \phi(x_2, V_2)$, and therefore,

$$\psi_\rho(x_1, x_2) = (1 - \rho)\max\{\phi(x_1, V_1), \phi(x_2, V_2)\} + \rho(\phi(x_1, V') + \phi(x_2, V')) \quad (7)$$
$$= (1 - \rho)\phi(x_1, V_1) + \rho(\phi(x_1, V') + \phi(x_2, V')). \quad (8)$$

The part $(1 - \rho)\phi(x_1, V_1) + \rho\phi(x_1, V')$ in (8) takes its minimum value at either $x_1 = p(a_i)$ or $x_1 = p(a_{i+1})$. The minimum value of the last term $\rho\phi(x_2, V')$ for $x_2 \in C'(p(b_i), c'_2)$ can be obtained easily from $E(x)$. Similarly, if $x_1 \in$

$C'(c_1', p(a_i))$ and $x_2 \in C'(p(b_{i+1}), p(b_i))$, then we have $\phi(x_1, V_1) \leq \phi(x_2, V_2)$. In this case, we can minimize

$$\psi_\rho(x_1, x_2) = (1 - \rho)\phi(x_2, V_2) + \rho(\phi(x_1, V') + \phi(x_2, V')) \qquad (9)$$

for each i in amortized constant time. If $x_1 \in C'(p(a_i), p(a_{i+1}))$ and $x_2 \in C'(p(b_{i+1}), p(b_i))$, then (7) involves only linear functions in x_1 and x_2, and the pair $\{x_1, x_2\}$ minimizing (7) can be computed in constant time. The pair $\{x_1, x_2\}$ that corresponds to the minimum of all the minimum values we have computed for a particular edge $(v_{j-1}, v_j) \in E'$ that was removed gives a candidate for a back-up 2-center. One such edge should lead to the partition, $\{V_1(c_1^\rho, c_2^\rho), V_2(c_1^\rho, c_2^\rho)\}$ (see (3)). Let $e_1, e_2, \ldots, e_n$ be the edges of C'.

Algorithm 3 BU2Center-Cycle(C')

1. *Compute $E(x)$.*
2. *For $j = 1, 2, \ldots, n$, remove edge e_j and carry out Steps 3 and 4.*
3. *Compute $E_L(x)$ and $E_R(x)$ for the resulting path (as in Fig. 1).*
4. *Invoke Procedure* Map-Bending-Pts. *For $i = 1, 2, \ldots, t' - 1$, find three pairs $\{x_1, x_2\}$, if any, such that*
   (a) *$\{x_1, x_2\}$ minimizes (8) within the constraints $x_1 \in C'(p(a_i), p(a_{i+1}))$ and $x_2 \in C'(p(b_i), c_2')$.[3]*
   (b) *$\{x_1, x_2\}$ minimizes (9) within the constraints $x_1 \in C'(c_1', p(a_i))$ and $x_2 \in C'(p(b_{i+1}), p(b_i))$.[4]*
   (c) *$\{x_1, x_2\}$ minimizes (7) within the constraints $x_1 \in C'(p(a_i), p(a_{i+1}))$ and $x_2 \in C'(p(b_{i+1}), p(b_i))$.*
5. *From among the pairs found in Step 5, pick the one with minimum cost as a back-up 2-center $\{c_1^\rho, c_2^\rho\}$.* □

The most time-consuming step is Step 4, and we argued after Lemma 6 that it runs in $O(n)$ time per removed edge e_j.

Theorem 3. *Algorithm* BU2Center-Cycle *computes a back-up 2-center of a cycle network in $O(n^2)$ time.* □

6   Unicyclic Network

A *unicyclic* network, $G = (V, E)$, contains just one cycle C. We can assume without loss of generality that the degree of each cycle vertex is at most 3. Otherwise, we can insert dummy vertices of weight 0 and dummy edges of length 0. A tree that hangs from vertex $u \in C$, excluding u and the edge to u, is called a *graft*,[5] and is denoted by $\Gamma(u)$. See Fig. 5. We first find a 1-center c of G, using the $O(n \log n)$ time algorithm of Ben-Moshe et al. [2]. We can also find a

---

[3] $\{x_1, x_2\}$ doesn't exist if $p(b_i)$ lies to the right of c_2'.
[4] $\{x_1, x_2\}$ doesn't exist if c_1' lies to the left of $p(a_i)$.
[5] The definition of graft in [6] includes u.

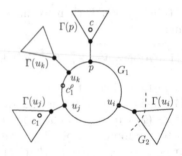

Fig. 5. Unicyclic network;

(classical) 2-center $\{c_1, c_2\}$ in $O(n \log n)$ time [1, 2]. Let $p \in C$ be the vertex such that either $\Gamma(p)$ or the edge connecting $\Gamma(p)$ and p (including p) contains c. We call $\Gamma(p)$ the *parent graft* of C. Let $p \ (= u_0), u_1, u_2, \ldots, u_{g-1}$ be the g vertices of C, clockwise along C. By our assumption, for each k, u_k is connected to at most one graft.

We assume that each graft is a balanced binary tree. If not, we can perform spine tree decomposition [3–5]. We preprocess G to construct the *cycle envelope tree* [5] for C. Using it, we can thus compute the upper envelope for all the grafts at their roots in $O(n \log n)$ time during preprocessing. Lemma 3 implies

Lemma 7. *Suppose that* $c_1 \in \Gamma(u_j)$. *Then* $c_1^\rho \in \Gamma(u_j) \cup \Gamma(p) \cup C$.   □

Table 1. Six possible cases

		(i) $c_1^\rho, c_2^\rho \in C$	(ii) $c_1^\rho \in \Gamma(u_j); c_2^\rho \in \Gamma(u_k)$	(iii) $c_1^\rho \in \Gamma(u_j); c_2^\rho \in C$
I	$c_1, c_2 \in C$	A	–	–
II	$c_1 \in \Gamma(u_j); c_2 \in \Gamma(u_k)$	B	D	E
III	$c_1 \in \Gamma(u_j); c_2 \in C$	C	–	F

The three possibilities regarding the locations of c_1 and c_2 are shown as rows I, II, and III in Table 1.[6] In the table, $c_1 \in \Gamma(u_j)$, for example, means that c_1 is in $\Gamma(u_j)$ or on the edge connecting $\Gamma(u_j)$ and u_j. The columns of Table 1 represent the possible locations of c_1^ρ and c_2^ρ. An entry marked by "–" indicates that the corresponding row-column combination, such as I-(ii), cannot occur, according to Lemma 3. Thus there are six cases to consider.

Cases A, B, C: These cases are similar to a cycle, since $c_1^\rho, c_2^\rho \in C$. Note, however, that in the case of a cycle, $\phi(x_1, V_1)$ and $\phi(x_2, V_2)$ are both continuous, as shown in Fig. 4, whereas here they are monotone but may not be continuous. As

---

[6] The case where $c_1 \in C$ and $c_2 \in \Gamma(u_k)$ is not shown, because it becomes case III, if we interchange c_1 and c_2.

in BU2Center-Cycle, we remove one edge at a time to to convert C into a path. Since grafts hang from this path, it take $O(n \log n)$ time [7] to compute $E_L(x)$ and $E_R(x)$. Therefore, (the modified) BU2Center-Cycle runs in $O(n^2 \log n)$ time in these cases.

Case D: The path $\pi(c_1, c_2)$ obviously goes through u_j and u_k, and we have $c_1^\rho \in \pi(c_1, u_j)$ and $c_2^\rho \in \pi(c_2, u_k)$. Thus this case is similar to a tree. We construct the upper envelope $E_L(x_1)$ (resp. $E_R(x_2)$) for $x_1 \in \pi(c_1, u_j)$ (resp. $x_2 \in \pi(c_2, u_k)$). To construct the upper envelope $E(x_1)$ for the entire network, we remove the edge on which $\alpha(u_j)$ lies. $E(x_1)$ is the upper envelope for the resulting tree, and is monotone for $x_1 \in \pi(c_1, u_j)$. $E(x_2)$ can be constructed similarly for $x_2 \in \pi(c_2, u_k)$ by removing the edge on which $\alpha(u_k)$ lies. It is monotone for $x_2 \in \pi(c_2, u_k)$. Thus the time required is the same as BU2Center-Tree, which is $O(n \log n)$. (Theorem 2)

Cases E and F: In these cases, $E_L(x_1)$ and $E(x_1)$ are the same as in Case D. The 1-center c_1^ρ may cover a part of C, so that c_2^ρ needs to cover the rest of C. If we remove an edge $e \in C$, a tree results. The back-up 2-center of this tree is of course different from the back-up 2-center of the unicyclic network.

Let us look at it backward as follows. Consider a back-up 2-center $\{c_1^\rho, c_2^\rho\}$ such that $c_1^\rho \in \Gamma(u_j)$ and $c_2^\rho \in C$. It partitions G into two subgraphs G_1 and G_2. If a part of C belongs to G_1, then two edges, e and e', separate G_1 from G_2. If we remove just e from G, then a tree results, and $\{c_1^\rho, c_2^\rho\}$ is a back-up 2-center of the tree, provided we use $\phi(x_1, V) = E(x_1)$ and $\phi(x_2, V) = E(x_2)$ in (4), where $E(x)$ is the upper envelope for $x \in C$, not the upper envelope of the tree. Therefore, there must be the "right" edge whose removal leads to the back-up 2-center of G. If we test all edges, then we will hit this right edge. The true back-up 2-center of G will be given by the one with the minimum cost.

If C is totally contained in G_2, then we test each edge on $\pi(c_1, u_j)$ at a time to partition G into G_1 and G_2. We can compute a backup 2-center $\{c_1^\rho, c_2^\rho\}$ in $O(n \log n)$ time per removed edge. Among all the candidates for $\{c_1^\rho, c_2^\rho\}$ computed so far, we pick the one with the least cost. (Details are omitted due to space limitations.) All this takes $O(n^2 \log n)$ time.

Theorem 4. *The back-up 2-center of a unicyclic network can be computed in $O(n^2 \log n)$ time.* □

7 Conclusion

We have shown that the back-up 2-center on a path (resp. tree, cycle, unicyclic) network can be computed in $O(n)$ (resp. $O(n \log n)$, $O(n^2)$, $O(n^2 \log n)$) time. Our model assumes that the two centers fail with the same probability. But it should be easy to extend it to the case where the probabilities are different. Another possible extension is to the back-up p-center problem, where p (> 2), about which nothing appears to be known. Snyder and Daskin [16] discussed the construction of the objective function for the p-median problem. Similar formulation should be possible for the p-center problem as well. New characterizations are needed for the back-up p-center problem.

References

1. Ben-Moshe, B., Bhattacharya, B., Shi, Q.: An optimal algorithm for the continuous/Discrete weighted 2-center problem in trees. In: Correa, J.R., Hevia, A., Kiwi, M. (eds.) LATIN 2006. LNCS, vol. 3887, pp. 166–177. Springer, Heidelberg (2006)
2. Ben-Moshe, B., Bhattacharya, B., Shi, Q., Tamir, A.: Efficient algorithms for center problems in cactus networks. Theoretical Computer Science 378(3), 237–252 (2007)
3. Benkoczi, R.: Cardinality constrained facility location problems in trees. Ph.D. thesis, School of Computing Science. Simon Fraser University, Canada (2004)
4. Benkoczi, R., Bhattacharya, B., Chrobak, M., Larmore, L.L., Rytter, W.: Faster algorithms for k-medians in trees. In: Rovan, B., Vojtáš, P. (eds.) MFCS 2003. LNCS, vol. 2747, pp. 218–227. Springer, Heidelberg (2003)
5. Bhattacharya, B., Kameda, T., Song, Z.: Minmax regret 1-center on a path/cycle/tree. In: Proc. 6th Int'l Conf. on Advanced Engineering Computing and Applications in Sciences (ADVCOMP), pp. 108–113 (2012), http://www.thinkmind.org/index.php?view=article&articleid=advcomp_2012_5_20_20093
6. Burkard, R., Krarup, J.: A linear algorithm for the pos/neg-weighted 1-median problem on a cactus. Computing 60, 193–215 (1998)
7. Hershberger, J.: Finding the upper envelope of n line segments in $O(n \log n)$ time. Information Processing Letters 33(4), 169–174 (1989)
8. Hong, Y., Kang, L.: Backup 2-center on interval graphs. Theoretical Computer Science 445, 25–35 (2012)
9. Kariv, O., Hakimi, S.: An algorithmic approach to network location problems, part 1: The p-centers. SIAM J. Appl. Math. 37, 513–538 (1979)
10. Kariv, O., Hakimi, S.: An algorithmic approach to network location problems, part 2: The p-median. SIAM J. Appl. Math. 37, 539–560 (1979)
11. Kouvelis, P., Yu, G.: Robust Discrete Optimization and its Applications. Kluwer Academic Publishers, London (1997)
12. Megiddo, N.: Linear-time algorithms for linear-programming in R^3 and related problems. SIAM J. Computing 12, 759–776 (1983)
13. Rosenhead, J., Elton, M., Gupta, S.K.: Robustness and optimality as criteria for strategic decisions. Operational Research Quarterly 23, 413–431 (1972)
14. Shi, Q.: Efficient algorithms for network center/covering location optimization problems. Ph.D. thesis, School of Computing Science, Simon Fraser University, Canada (2008)
15. Snyder, L.: Facility location under uncertainty: a review. IIE Transactions 38, 537–554 (2006)
16. Snyder, L., Daskin, M.S.: Reliability models for facility locations: The expected failure cost case. Transportation Science 39, 400–416 (2005)
17. Tamir, A.: An $O(pn^2)$ algorithm for the p-median and the related problems in tree graphs. Operations Research Letters 19, 59–64 (1996)
18. Wang, H.L., Wu, B.Y., Chao, K.M.: A linear time algorithm for the backup 2-center problem on a tree. In: 24th Workshop on Combinatorial Mathematics and Computation Theory, pp. 189–195 (2007)
19. Wang, H.L., Wu, B.Y., Chao, K.M.: The backup 2-center and backup 2-median problems on trees. Networks 53(1), 39–49 (2009)

Quantum Algorithms for Finding Constant-Sized Sub-hypergraphs

François Le Gall[1], Harumichi Nishimura[2], and Seiichiro Tani[3]

[1] Graduate School of Information Science and Technology,
The University of Tokyo, Japan
[2] Graduate School of Informatics, Nagoya University, Japan
[3] NTT Communication Science Laboratories, NTT Corporation, Japan

Abstract. We develop a general framework to construct quantum algorithms that detect if a 3-uniform hypergraph given as input contains a sub-hypergraph isomorphic to a prespecified constant-sized hypergraph. This framework is based on the concept of nested quantum walks recently proposed by Jeffery, Kothari and Magniez, and extends the methodology designed by Lee, Magniez and Santha for similar problems over graphs. As applications, we obtain a quantum algorithm for finding a 4-clique in a 3-uniform hypergraph on n vertices with query complexity $O(n^{1.883})$, and a quantum algorithm for determining if a ternary operator over a set of size n is associative with query complexity $O(n^{2.113})$.

1   Introduction

Quantum query complexity is a model of quantum computation, in which the cost of computing a function is measured by the number of queries that are made to the input given as a black-box. In this model, it was exhibited in the early stage of quantum computing research that there exist quantum algorithms superior to the classical counterparts, such as Deutsch and Jozsa's algorithm, Simon and Shor's period finding algorithms, and Grover's search algorithm. Extensive studies following them have invented a lot of powerful upper bound (i.e., algorithmic) techniques such as variations/generalizations of Grover's search algorithm or quantum walks. Although these techniques give tight bounds for many problems, there are still quite a few cases for which no tight bounds are known. Intensively studied problems among them are the k-distinctness problem [1,4,5] and the triangle finding problem [3,7,9,13,15].

A recent breakthrough is the concept of the learning graph introduced by Belovs [3], who used it to improve the long-standing $O(n^{13/10})$ upper bound [15] of the triangle finding problem to $O(n^{35/27})$. His idea was generalized by Lee, Magniez and Santha [12] and Zhu [20] to obtain a quantum algorithm that finds a constant-sized subgraph with complexity $o(n^{2-2/k})$, improving the previous best bound $O(n^{2-2/k})$ [15], where k is the size of the subgraph. Subsequently, Lee, Magniez and Santha [13] constructed a triangle finding algorithm with quantum query complexity $O(n^{9/7})$. This bound was later shown by Belovs

Z. Cai et al. (Eds.): COCOON 2014, LNCS 8591, pp. 429–440, 2014.
© Springer International Publishing Switzerland 2014

and Rosmanis [6] to be the best possible bound attained by the family of quantum algorithms whose complexities depend only on the index set of 1-certificates. Ref. [13] also gave a framework of quantum algorithms for finding a constant-sized subgraph, based on which they showed that associativity testing (testing if a binary operator over a domain of size n is associative) has quantum query complexity $O(n^{10/7})$.

Recently, Jeffery, Kothari and Magniez [9] cast the idea of the above triangle finding algorithms into the framework of quantum walks (called nested quantum walks) by recursively performing the quantum walk algorithm given by Magniez, Nayak, Roland and Santha [14]. Indeed, they presented two quantum-walk-based triangle finding algorithms of complexities $\tilde{O}(n^{35/27})$ and $\tilde{O}(n^{9/7})$, respectively. The nested quantum walk framework was further employed in [5] (but in a different way from [9]) to obtain $\tilde{O}(n^{5/7})$ complexity for the 3-distinctness problem. This achieves the best known upper bound (up to poly-logarithmic factors), which was first obtained with the learning-graph-based approach [4].

The triangle finding problem also plays a central role in several areas beside query complexity, and it has been recently discovered that faster algorithms for (weighted versions of) triangle finding would lead to faster algorithms for matrix multiplication [10,17], the 3SUM problem [16], and for Max-2SAT [18,19]. In particular, Max-2SAT over n variables is reducible to finding a triangle with maximum weight over $O(2^{n/3})$ vertices; in this context, although the final goal is a *time-efficient* classical or quantum algorithm that finds a triangle with maximum weight, studying triangle finding in the query complexity model is a first step toward this goal.

Our Results. Along this line of research, this paper studies the problem of finding a 4-clique (i.e., the complete 3-uniform hypergraph with 4 vertices) in a 3-uniform hypergraph, a natural generalization of finding a triangle in an ordinary graph (i.e., a 2-uniform hypergraph). Our initial motivation comes from the complexity-theoretic importance of the problem. Indeed, while it is now well-known that Max-3SAT over n variables is reducible to finding a 4-clique with maximum weight in a 3-uniform hypergraph of $O(2^{n/4})$ vertices (the reduction is similar to the reduction from Max-2SAT to triangle finding mentioned above; we refer to [19] for details), no efficient classical algorithm for 4-clique finding has been discovered so far. Constructing query-efficient algorithms for this problem can be seen as a first step to investigate the possibility of faster (in the time complexity setting) classical or quantum algorithms for Max-3SAT.

Concretely, and more generally, this paper gives a framework based on quantum walks for finding any constant-sized sub-hypergraph in a 3-uniform hypergraph (Theorem 1). This is an extension of the learning-graph-based algorithm in [13] to the hypergraph case in terms of the nested quantum walk [9]. We illustrate this methodology by constructing a quantum algorithm that finds a 4-clique in a 3-uniform hypergraph with query complexity $\tilde{O}(n^{241/128}) = O(n^{1.883})$, while naïve Grover search over the $\binom{n}{4}$ combinations of vertices only gives $O(n^2)$. As another application, we also construct a quantum algorithm that determines if a ternary operator is associative using $\tilde{O}(n^{169/80}) = O(n^{2.113})$ queries, while naïve Grover search needs $O(n^{2.5})$ queries.

In the course of designing the quantum walk framework, we introduce several new technical ideas (outlined below) for analyzing nested quantum walks to cope with difficulties that do not arise in the 2-uniform case (i.e., ordinary graphs), such as the fact that the size of the random subset taken in an inner walk may vary depending on the random subsets taken in outer walks. We believe that these ideas may be applicable to various problems beyond sub-hypergraph finding.

Technical Contribution. Roughly speaking, the subgraph finding algorithm by Lee, Magniez and Santha [13] works as follows. First, for each vertex i in the subgraph H that we want to find, a random subset V_i of vertices of the input graph is taken. This subset V_i represents a set of candidates for the vertex i. Next, for each edge (i, j) in the subgraph H, a random subset of pairs in $V_i \times V_j$ is taken, representing a set of candidates for the edge (i, j). The most effective feature of their algorithm is to introduce a parameter for each ordered pair (V_i, V_j) that controls the average degree of a vertex in the bipartite graph between V_i and V_j. To make the algorithm efficient, it is crucial to keep the degree of every vertex in V_i almost equal to the value specified by the parameter. For this, they carefully devise a procedure for taking pairs from $V_i \times V_j$.

Our basic idea is similar in that we first, for each vertex i in the sub-hypergraph H that we want to find, take a random subset V_i of vertices in the input 3-uniform hypergraph as a set of candidates for the vertex i and then, for each hyperedge $\{i, j, k\}$ of H, take a random subset of triples in $V_i \times V_j \times V_k$. One may think that the remaining task is to fit the pair-taking procedure into the hypergraph case. It, however, turns out to be technically very complicated to generalize the pair-taking procedure from [13] to an efficient triple-taking procedure. Instead we cast the idea into the nested quantum walk of Jeffery, Kothari and Magniez [9] and employ probabilistic arguments. More concretely, we introduce a parameter that specifies the number e_{ijk} of triples to be taken from $V_i \times V_j \times V_k$ for each hyperedge $\{i, j, k\}$ of H. We then argue that, for randomly chosen e_{ijk} triples, the degree of each vertex sharply concentrates around its average, where the degree means the number of triples including the vertex. This makes it substantially easier to analyze the complexity of all involved quantum walks, and enables us to completely analyze the complexity of our approach. Unfortunately, it turns out that this approach (taking the sets V_i first, and then e_{ijk} triples from each $V_i \times V_j \times V_k$) does not lead to any improvement over the naïve $O(n^2)$-query quantum algorithm.

Our key idea is to introduce, for each unordered pair $\{i, j\}$ of vertices in H, a parameter f_{ij}, and modify the approach as follows. After randomly choosing V_i, V_j, V_k, we take three random subsets $F_{ij} \subseteq V_i \times V_j$, $F_{jk} \subseteq V_j \times V_k$, and $F_{ik} \subseteq V_i \times V_k$ of size f_{ij}, f_{jk} and f_{ik}, respectively. We then randomly choose e_{ijk} triples from the set $\Gamma_{ijk} = \{(u, v, w) \,|\, (u, v) \in F_{ij}, (u, w) \in F_{ik} \text{ and } (v, w) \in F_{jk}\}$. The difficulty here is that the size of Γ_{ijk} varies depending on the sets F_{ij}, F_{jk}, F_{ik}. Another problem is that, after taking many quantum-walks (i.e., performing the update operation many times), the distribution of the set of pairs can change. To overcome these difficulties, we carefully define the "marked states" (i.e., "absorbing states") of each level of the nested quantum walk: besides requiring, as usual, that the set (of the form V_i, F_{ij} or Γ_{ijk}) associated to a marked state should contain a part (i.e., a vertex, a pair of vertices or a triple of vertices) of a copy of H, we also require that this set should *satisfy certain regularity conditions*. We then show that the associated sets almost always satisfy the regularity conditions, by using the concentration theorems for hypergeometric distributions. This regularity enables us to effectively bound the complexity of our new approach, giving in particular the claimed $\tilde{O}(n^{241/128})$-query upper bound when H is a 4-clique.

The proofs of some of our results are omitted in this version due to space constraints. We refer to the full version of the present work [11] for all details.

2   Preliminaries

For any $k \geq 2$, an undirected k-uniform hypergraph is a pair (V, E), where V is a finite set (the set of vertices), and E is a set of unordered k-tuples of elements in V (the set of hyperedges). An undirected 2-uniform hypergraph is simply an undirected graph.

In this paper, we use the standard quantum query complexity model formulated in Ref. [2]. We deal with (undirected) 3-uniform hypergraphs $G = (V, E)$ as input, given as a black-box. The operation of the black-box is given as the unitary mapping $|\{u, v, w\}, b\rangle \mapsto |\{u, v, w\}, b \oplus \chi(\{u, v, w\})\rangle$ for $b \in \{0, 1\}$, where the triple $\{u, v, w\}$ is the query to the black-box and $\chi(\{u, v, w\})$ is the answer on whether the triple is a hyperedge of G, namely, $\chi(\{u, v, w\}) = 1$ if $\{u, v, w\} \in E$ and $\chi(\{u, v, w\}) = 0$ otherwise.

Our algorithmic framework is based on the concept of the nested quantum walk introduced by Jeffery, Kothari and Magniez [9]. In the nested quantum walk, for each positive integer t, the walk at level t checks whether the current state is marked or not by invoking the walk at level $t + 1$, and this is iterated recursively until some fixed level m. The data structure of the walk at level t is defined so that it includes the initial state of the walk at level $t + 1$, which means that the setup cost of the walk at level $t \geq 2$ is zero. Jeffery, Kothari and Magniez have shown (in Section 4.1 of [9]) that the overall complexity of such a walk is

$$\tilde{O}\left(\mathsf{S} + \sum_{t=1}^{m} \left(\prod_{r=1}^{t} \frac{1}{\sqrt{\varepsilon_r}}\right) \frac{1}{\sqrt{\delta_t}} \mathsf{U}_t\right)$$

if the checking cost at level m is zero, which will be our case. Here S denotes the setup cost of the whole nested walk, U_t denotes the cost of updating the database of the walk at level t, δ_t denotes the spectral gap of the walk at level t, and ε_r denotes the fraction of marked states for the walk at level r. As in most quantum walk papers, we only consider quantum walks on the Johnson graphs, where the Johnson graph $J(N, K)$ is a graph such that each vertex is a subset with size K

of a set with size N and two vertices corresponding to subsets S and S' are adjacent if and only if $|S\Delta S'| = 2$ (we denote by $S\Delta S'$ the symmetric difference between S and S'). If the walk at level t is on $J(N, K)$, then its spectral gap δ_t is known to be $\Omega(1/K)$.

Consider the update operation of the walk at any level. The update cost may vary depending on the states of the walk we want to update. Assume without loss of generality that the update operation is of the form $U = \sum_i |i\rangle\langle i| \otimes U_i$, where each U_i can be implemented using q_i queries, and the quantum state to be updated is of the form $|s\rangle = \sum_i \alpha_i |i\rangle |s_i\rangle$. Then the following lemma, used in [9], shows that if the magnitude of the states $|i\rangle |s_i\rangle$ that cost much to update (i.e., such that q_i is large) is small enough, we can approximate the update operator U with good precision by replacing all U_i acting on such costly states with the identity operator.

Lemma 1 ([9]). *Let* $U = \sum_i |i\rangle\langle i| \otimes U_i$ *be a controlled unitary operator and let* q_i *be the query complexity of exactly implementing* U_i. *For any fixed integer* T, *define* $\tilde{U}$ *as* $\sum_{i:q_i\leq T} |i\rangle\langle i|\otimes U_i + \sum_{i:q_i>T} |i\rangle\langle i|\otimes \mathbb{I}$, *where* $\mathbb{I}$ *is the identity operator on the space on which* U_i *acts. Then, for any quantum state* $|s\rangle = \sum_i \alpha_i |i\rangle |s_i\rangle$, *the inequality* $\left|\langle s|\tilde{U}U|s\rangle\right| \geq 1 - \epsilon_T$ *holds whenever* $\epsilon_T \geq \sum_{i:q_i>T} |\alpha_i|^2$.

3   Statement of Our Main Result

Let H be a 3-uniform hypergraph with κ vertices. We identify the set of vertices of H with the set $\Sigma_1 = \{1, \ldots, \kappa\}$. We identify the set of hyperedges of H with the set $\Sigma_3 \subseteq \{\{1,2,3\}, \{1,2,4\}, \ldots, \{\kappa - 2, \kappa - 1, \kappa\}\}$. We identify the set of (unordered) pairs of vertices included in at least one hyperedge of H with the set $\Sigma_2 = \{\{i,j\} \mid \{i,j,k\} \in \Sigma_3 \text{ for some } k\}$. By generalizing the definition in [13,9], we define a loading schedule for H as follows.

Definition 1. *A* loading schedule for H of length m *is a list* $S = (s_1, \ldots, s_m)$ *of* m *elements such that the following three properties hold for all* $t \in \{1, \ldots, m\}$: *(i)* $s_t \in \Sigma_1 \cup \Sigma_2 \cup \Sigma_3$; *(ii) if* $s_t = \{i,j\}$, *then there exist* $t_1, t_2 \in \{1, \ldots, t-1\}$ *such that* $s_{t_1} = i$ *and* $s_{t_2} = j$; *(iii) if* $s_t = \{i,j,k\}$, *then there exist* $t_1, t_2, t_3 \in \{1, \ldots, t-1\}$ *such that* $s_{t_1} = \{i,j\}$, $s_{t_2} = \{i,k\}$ *and* $s_{t_3} = \{j,k\}$. *A loading schedule* S *is* valid *if no element of* $\Sigma_1 \cup \Sigma_2 \cup \Sigma_3$ *appears more than once and, for any* $\{i,j,k\} \in \Sigma_3$, *there exists an index* $t \in \{1, \ldots, m\}$ *such that* $s_t = \{i,j,k\}$.

We now introduce the concept of parameters associated to a loading schedule. Formally, these parameters are functions of the variable n representing the number of vertices of the input 3-uniform hypergraphs $G = (V, E)$. We will nevertheless, in a slight abuse of notation, consider that n is fixed, and define them as integers (implicitly depending on n).

Definition 2. *Let* $S = (s_1, \ldots, s_m)$ *be a loading schedule for* H *of length* m. *A* set of parameters for S *is a set of* m *integers defined as follows: for each* $t \in \{1, \ldots, m\}$,

- if $s_t = i$, then the associated parameter is denoted by r_i and satisfies $r_i \in \{1, \ldots, n\}$;
- if $s_t = \{i, j\}$, then the associated parameter is denoted by f_{ij} and satisfies $f_{ij} \in \{1, \ldots, r_i r_j\}$;
- if $s_t = \{i, j, k\}$, then the associated parameter is denoted by e_{ijk} and satisfies $e_{ijk} \in \{1, \ldots, r_i r_j r_k\}$.

The set of parameters is admissible if $r_i \geq 1$, $e_{ijk} \geq 1$, $\frac{r_i r_j}{f_{ij}} \geq 1$, $\frac{f_{ij} f_{ik} f_{jk}/(r_i r_j r_k)}{e_{ijk}} \geq 1$, and the terms $\frac{n}{r_i}$, $\frac{f_{ij}}{r_i}$, $\frac{f_{ij}}{r_j}$, $\frac{f_{ij} f_{ik}}{r_i r_j r_k}$ are larger than n^γ for some constant $\gamma > 0$.

Now we state the main result in terms of loading schedules.

Theorem 1. Let H be a constant-sized 3-uniform hypergraph. Let $S = (s_1, \ldots, s_m)$ be a valid loading schedule for H with an admissible set of parameters. There exists a quantum algorithm that, given as input a 3-uniform hypergraph G with n vertices, finds a sub-hypergraph of G isomorphic to H (and returns "no" if there are no such sub-hypergraphs) with probability at least some constant, and has query complexity

$$\tilde{O}\left(\mathsf{S} + \sum_{t=1}^{m} \left(\prod_{r=1}^{t} \frac{1}{\sqrt{\varepsilon_r}}\right) \frac{1}{\sqrt{\delta_t}} \mathsf{U}_t\right),$$

where S, U_t, δ_t and ε_r are evaluated as follows:

- $\mathsf{S} = \sum_{\{i,j,k\} \in \Sigma_3} e_{ijk}$;
- for $t \in \{1, \ldots, m\}$, (i) if $s_t = \{i\}$, then $\delta_t = \Omega(\frac{1}{r_i})$, $\varepsilon_t = \Omega(\frac{r_i}{n})$ and $\mathsf{U}_t = \tilde{O}\left(1 + \sum_{\{j,k\}:\{i,j,k\} \in \Sigma_3} \frac{e_{ijk}}{r_i}\right)$; (ii) if $s_t = \{i, j\}$, then $\delta_t = \Omega(\frac{1}{f_{ij}})$, $\varepsilon_t = \Omega(\frac{f_{ij}}{r_i r_j})$ and $\mathsf{U}_t = \tilde{O}\left(1 + \sum_{k:\{i,j,k\} \in \Sigma_3} \frac{e_{ijk}}{f_{ij}}\right)$; (iii) if $s_t = \{i, j, k\}$, then $\delta_t = \Omega(\frac{1}{e_{ijk}})$, $\varepsilon_t = \Omega(\frac{e_{ijk} r_i r_j r_k}{f_{ij} f_{ik} f_{jk}})$ and $\mathsf{U}_t = O(1)$.

4   Proof of Theorem 1

In this section, we prove Theorem 1 by constructing an algorithm based on the concept of nested quantum walks, in which the walk at level t will correspond to the element s_t of the loading schedule for each $t \in \{1, \ldots, m\}$. For convenience, we will write $M_{ijk} = 11 \frac{f_{ij} f_{ik} f_{jk}}{r_i r_j r_k}$ for each $\{i, j, k\} \in \Sigma_3$.

4.1   Definition of the Walks

At level $t \in \{1, \ldots, m\}$, the quantum walk will differ according to the nature of s_t, so there are three cases to consider.

Case 1 [$s_t = i$]: The quantum walk will be over the Johnson graph $J(n, r_i)$. The space of the quantum walk will then be $\Omega_t = \{T \subseteq \{1, \ldots, n\} \mid |T| = r_i\}$. A state of this walk is an element $R_t \in \Omega_t$.

Case 2 $[s_t = \{i, j\}]$: The quantum walk will be over $J(r_i r_j, f_{ij})$. The space of the quantum walk will then be $\Omega_t = \{T \subseteq \{1, \ldots, r_i r_j\} \mid |T| = f_{ij}\}$. A state of this walk is an element $R_t \in \Omega_t$.

Case 3 $[s_t = \{i, j, k\}]$: The quantum walk will be over $J(M_{ijk}, e_{ijk})$. The space of the quantum walk will then be $\Omega_t = \{T \subseteq \{1, \ldots, M_{ijk}\} \mid |T| = e_{ijk}\}$. A state of this walk is an element $R_t \in \Omega_t$.

4.2   Definition of the Data Structures of the Walks

Let us fix an arbitrary ordering on the set $V \times V \times V$ of triples of vertices. For any set $\Gamma \subseteq V \times V \times V$ and any $R \subseteq \{1, \ldots, |V|^3\}$, define the set $\mathsf{Y}(R, \Gamma)$ consisting of at most $|R|$ triples of vertices which are taken from Γ by the process below.

- Construct a list Λ of all the triples in $V \times V \times V$ as follows: first, all the triples in Γ are listed in increasing order and, then, all the triples in $(V \times V \times V) \backslash \Gamma$ are listed in increasing order.
- For any $a \in \{1, \ldots, |V|^3\}$, let $\Lambda[a]$ denote the a-th triple of the list.
- Define $\mathsf{Y}(R, \Gamma) = \{\Lambda[a] \mid a \in R\} \cap \Gamma$.

The following lemma will be useful later in this section.

Lemma 2. *Let Γ and Γ' be two subsets of $V \times V \times V$. Let p and r be any parameters such that $1 \leq r \leq p \leq |V|^3$. There exists a permutation π of $\{1, \ldots, p\}$ such that, if R is a subset of $\{1, \ldots, p\}$ of size r taken uniformly at random, then*

$$\Pr_R\left[|\mathsf{Y}(R, \Gamma) \Delta \mathsf{Y}(\pi(R), \Gamma')| \leq \frac{22r|\Gamma \Delta \Gamma'|}{p} + 100 \log n\right] \geq 1 - 2\left(\frac{1}{2}\right)^{\frac{11r|\Gamma \Delta \Gamma'|}{p} + 50 \log n}.$$

Suppose that the states of the walks at levels $1, \ldots, m$ are $R_1, \ldots, R_m$, respectively. Assume that the set of vertices of G is $V = \{v_1, \ldots, v_n\}$. We first interpret the states $R_1, \ldots, R_m$ as sets of vertices, sets of pairs of vertices or sets of triples of vertices in V, as follows. For each $t \in \{1, \ldots, m\}$, there are three cases to consider.

Case 1 $[s_t = i]$: In this case, $R_t = \{a_1, \ldots, a_{r_i}\} \subseteq \{1, \ldots, n\}$. We associate to R_t the set $V_i = \{v_{a_1}, \ldots, v_{a_{r_i}}\}$. For further reference, we will rename the vertices in this set as $V_i = \{v_1^i, \ldots, v_{r_i}^i\}$.

Case 2 $[s_t = \{i, j\}$ **with** $i < j]$: We know that, in this case, there exist $t_1, t_2 \in \{1, \ldots, t - 1\}$ such that $s_{t_1} = i$ and $s_{t_2} = j$. The state R_t represents a set $\{(a_1, b_1), \ldots, (a_{f_{ij}}, b_{f_{ij}})\}$ of f_{ij} pairs in $R_{t_1} \times R_{t_2}$. We associate to it the set $F_{ij} = \{(v_{a_1}^i, v_{b_1}^j), \ldots, (v_{a_{f_{ij}}}^i, v_{b_{f_{ij}}}^j)\}$ of pairs of vertices.

Case 3 $[s_t = \{i, j, k\}$ **with** $i < j < k]$: We know that there exist $t_1, t_2, t_3 \in \{1, \ldots, t - 1\}$ such that $s_{t_1} = \{i, j\}$, $s_{t_2} = \{i, k\}$ and $s_{t_3} = \{j, k\}$, and R_t is a subset of $\{1, \ldots, M_{ijk}\}$ with $|R_t| = e_{ijk}$. Let us define the set

$$\Gamma_{ijk} = \{(u, v, w) \in V_i \times V_j \times V_k \mid (u, v) \in F_{ij}, (u, w) \in F_{ik} \text{ and } (v, w) \in F_{jk}\}.$$

We associate to R_t the set $E_{ijk} = \mathsf{Y}(R_t, \Gamma_{ijk})$.

We are now ready to define the data structures involved in the walks. When the states of the walks at levels $1, \ldots, (m-1)$ are $R_1, \ldots, R_{m-1}$, respectively, and the state of the most inner walk is R_m, the data structure associated with the most inner walk is denoted by $D(R_1, \ldots, R_m)$ and defined as:

$$D(R_1, \ldots, R_m) = \left\{ (\{u, v, w\}, \chi(\{u, v, w\})) \mid (u, v, w) \in \bigcup_{\{i,j,k\} \in \Sigma_3 \,:\, i<j<k} E_{ijk} \right\}.$$

The data structure associated with the walk at level t, for each $t \in \{1, \ldots, m-1\}$, is defined as: $\sum_{R_{t+1} \in \Omega_{t+1}} \cdots \sum_{R_m \in \Omega_m} |R_{t+1}\rangle \cdots |R_m\rangle |D(R_1, \ldots, R_m)\rangle$ (here and hereafter we omit normalization factors).

4.3   Marked States of the Walks

For any $t \in \{1, \ldots, m-1\}$, the purpose of the walk at level $t+1$ is to check if the state of the walk t is marked (for the most inner walk, the state can be checked without running another walk, since all the information necessary is already in the database). In this subsection we define the set of marked states for each walk.

Assume that the hypergraph G contains a (without loss of generality, unique) sub-hypergraph isomorphic to H. Let $\{u_1, \ldots, u_\kappa\}$ denote the vertex set of this sub-hypergraph. For the most outer walk, $s_1 = j$ for some $j \in \{1, \ldots, \kappa\}$ and we say that R_1 is marked if and only if $u_j \in V_j$. Consider a state R_t of the walk at level $t > 1$, and suppose that the states $R_1, \ldots, R_{t-1}$ are all marked. We have again three cases to consider.

Case 1 [$s_t = i$]: R_t corresponds to V_i. We say that R_t is marked if and only if $u_i \in V_i$.

Case 2 [$s_t = \{i, j\}$ with $i < j$]: R_t corresponds to F_{ij}, and we say that R_t is marked if and only if the following four conditions hold: (a) $(u_i, u_j) \in F_{ij}$; (b) for all $u \in V_i$, $\frac{f_{ij}}{2r_i} \leq |\{v \in V_j \mid (u, v) \in F_{ij}\}| \leq 2\frac{f_{ij}}{r_i}$; (c) for all $v \in V_j$, $\frac{f_{ij}}{2r_j} \leq |\{u \in V_i \mid (u, v) \in F_{ij}\}| \leq 2\frac{f_{ij}}{r_j}$; (d) for any k such that there exists $t_1 \in \{1, \ldots, t-1\}$ for which $s_{t_1} = \{i, k\}$, and any $(v, w) \in V_j \times V_k$, $|\{u \in V_i \mid (u, v) \in F_{ij} \text{ and } (u, w) \in F_{ik}\}| \leq 11\frac{f_{ij} f_{ik}}{r_i r_j r_k}$.

Case 3 [$s_t = \{i, j, k\}$ with $i < j < k$]: R_t corresponds to E_{ijk}, and we say that R_t is marked if and only if $(u_i, u_j, u_k) \in E_{ijk}$.

4.4   Analysis of the Algorithm

Our nested quantum walk algorithm finds a marked state in the most inner walk and thus a sub-hypergraph isomorphic to H, with high probability, since, as will be shown below, the ideal nested quantum walks can be approximated with high accuracy. As explained in Section 2, the overall query complexity of the walk is

$$\tilde{O}\left(\mathsf{S} + \sum_{t=1}^{m} \left(\prod_{r=1}^{t} \frac{1}{\sqrt{\varepsilon_r}}\right) \frac{1}{\sqrt{\delta_t}} \mathsf{U}_t\right).$$

We will show below that the values of the terms S, U_t, δ_t and ε_t are as claimed in the statement of Theorem 1.

We first make the following simple observation: when computing U_t and ε_t, we can assume that the state R_{t-1} of the immediately outer walk is marked (and thus, by applying this argument recursively, that the states $R_1, \ldots, R_{t-1}$ of all the outer walks are marked). Indeed, remember that the purpose of the walk at level t is to check if the state R_{t-1} is marked. We first evaluate its complexity under the assumption that R_{t-1} is marked, giving some upper bound T on the complexity. Now, since the checking procedure in our framework has one-sided error, in the case where R_{t-1} is not marked the checking procedure may not terminate after T queries, but we can stop it after T queries anyway and simply output that R_{t-1} is not marked.

The setup cost S for the algorithm is the number of queries needed to construct the superposition $\sum_{R_1} \cdots \sum_{R_m} |R_1\rangle \cdots |R_m\rangle |D(R_1, \ldots, R_m)\rangle$, where each sum $\sum_{R_i}$ is taken over Ω_i. This value is at most $\sum_{\{i,j,k\} \in \Sigma_3} e_{ijk}$.

We next evaluate δ_t and ε_t. The analysis is again divided into three cases.

Case 1 [$s_t = i$]: Since the quantum walk is over $J(n, r_i)$ by the definition in Section 4.1, we have $\delta_t = \Omega(\frac{1}{r_i})$ and $\varepsilon_t = \Omega(\frac{r_i}{n})$.

Case 2 [$s_t = \{i,j\}$ with $i < j$]: Since the quantum walk is over $J(r_i r_j, f_{ij})$, we have $\delta_t = \Omega(\frac{1}{f_{ij}})$. The fraction of states F_{ij} such that $(u_i, u_j) \in F_{ij}$ is $\Omega(\frac{f_{ij}}{r_i r_j})$. While all those states may not be marked, it can be proved (see [11]) that the fraction of those states that are not marked is exponentially small when the set of parameters is admissible. Thus $\varepsilon_t = \Omega(\frac{f_{ij}}{r_i r_j})$.

Case 3 [$s_t = \{i,j,k\}$ with $i < j < k$]: In this case $\delta_t = \Omega(\frac{1}{e_{ijk}})$. Since all the states $R_1, \ldots, R_{t-1}$ of the outer walks are assumed to be marked, by item (d) of the definition of the marked states in Section 4.3, we can upper-bound $|\Gamma_{ijk}| = \sum_{(v,w) \in F_{jk}} |\{u \in V_i \mid (u,v) \in F_{ij} \text{ and } (u,w) \in F_{ik}\}|$ by $|F_{jk}| \frac{11 f_{ij} f_{ik}}{r_i r_j r_k} = M_{ijk}$. Thus, we have $\varepsilon_t = \Omega(\frac{e_{ijk}}{M_{ijk}})$.

Finally, we evaluate the cost U_t, which is the cost of transforming the quantum state

$$\sum_{R_{t+1}} \cdots \sum_{R_m} |R_{t+1}\rangle \cdots |R_m\rangle |D(R_1, \ldots, R_{t-1}, R_t, R_{t+1}, \ldots, R_m)\rangle,$$

to the quantum state

$$\sum_{R_{t+1}} \cdots \sum_{R_m} |R_{t+1}\rangle \cdots |R_m\rangle |D(R_1, \ldots, R_{t-1}, R_t', R_{t+1}, \ldots, R_m)\rangle,$$

for any two states R_t and R_t' adjacent in the corresponding Johnson graph. We again divide the analysis into three cases.

Case 1 [$s_t = i$]: In this case R_t and R_t' are two subsets of $\{1, \ldots, n\}$, both of size r_i, differing by exactly one element. The corresponding subsets V_i and V_i' also differ by exactly one element: let us write $V_i' = (V_i \backslash \{u\}) \cup \{u'\}$. For any

$\{i, j, k\} \in \Sigma_3$, there exist some $t_1, t_2, t_3 \in \{t+1, \ldots, m\}$ such that $s_{t_1} = \{i, j\}$, $s_{t_2} = \{i, k\}$ and $s_{t_3} = \{i, j, k\}$. There also exist some $t_4, t_5, t_6 \in \{1, \ldots, m\}$ such that $s_{t_4} = j$, $s_{t_5} = k$ and $s_{t_6} = \{j, k\}$. Note that t_4, t_5, t_6 can be smaller than t, but we will assume here that they are all larger than t (the other cases are omitted, but they are actually easier to analyze). A state R_{t_4} defines a set V_j of r_j vertices and, for any $R_{t_1} \in \Omega_{t_1}$, the state (R_t, R_{t_1}, R_{t_4}) defines a set of f_{ij} pairs in $V_i \times V_j$, as described in Section 4.3. In the same way, for any $R'_{t_1} \in \Omega_{t_1}$, the state $(R'_t, R'_{t_1}, R_{t_4})$ defines a set of f_{ij} pairs in $V'_i \times V_j$. There exists a permutation π_1 of the elements of Ω_{t_1} such that, for any $R_{t_1} \in \Omega_{t_1}$, the set F_{ij} defined by (R_t, R_{t_1}, R_{t_4}) and the set F'_{ij} defined by $(R'_t, \pi_1(R_{t_1}), R_{t_4})$ are related in the following way:

$$F'_{ij} = (F_{ij} \setminus \{(u, v) \in \{u\} \times V_j \mid (u, v) \in F_{ij}\}) \cup \{(u', v) \in \{u'\} \times V_j \mid (u, v) \in F_{ij}\},$$

which means that each pair of the form (u, v) in F_{ij} is replaced by the pair (u', v) in F'_{ij}, while the other pairs are the same in F_{ij} and in F'_{ij}.

Similarly, there exists a permutation π_2 of the elements of Ω_{t_2} such that, for any $R_{t_2} \in \Omega_{t_2}$, the set F_{ik} defined by (R_t, R_{t_2}, R_{t_5}) and the set F'_{ik} defined by $(R'_t, \pi_2(R_{t_2}), R_{t_5})$ are related in the following way:

$$F'_{ik} = (F_{ik} \setminus \{(u, w) \in \{u\} \times V_k \mid (u, w) \in F_{ik}\}) \cup \{(u', w) \in \{u'\} \times V_k \mid (u, w) \in F_{ik}\}.$$

The states $(R_t, R_{t_1}, R_{t_2}, R_{t_4}, R_{t_5}, R_{t_6})$ define sets $V_i, F_{ij}, F_{ik}, V_j, V_k, F_{jk}, \Gamma_{ijk}$, while the states $(R'_t, \pi_1(R_{t_1}), \pi_2(R_{t_2}), R_{t_4}, R_{t_5}, R_{t_6})$ define sets $V'_i, F'_{ij}, F'_{ik}, V_j, V_k, F_{jk}, \Gamma'_{ijk}$. Given any state R_{t_3}, let $E_{ijk}(R_t, R_{t_1}, R_{t_2}, R_{t_3}, R_{t_4}, R_{t_5}, R_{t_6})$ denote the set of hyperedges to be queried associated with Γ_{ijk} and R_{t_3}, and $E_{ijk}(R'_t, \pi_1(R_{t_1}), \pi_2(R_{t_2}), R_{t_3}, R_{t_4}, R_{t_5}, R_{t_6})$ denote the set of hyperedges to be queried associated with Γ'_{ijk} and R_{t_3}. By Lemmas 1 and 2, the mapping

$$|R_{t_1}\rangle |R_{t_2}\rangle |R_{t_4}\rangle |R_{t_5}\rangle |R_{t_6}\rangle \sum_{R_{t_3}} |R_{t_3}\rangle |E_{ijk}(R_t, R_{t_1}, R_{t_2}, R_{t_3}, \ldots, R_{t_6})\rangle \mapsto$$

$$|\pi_1(R_{t_1})\rangle |\pi_2(R_{t_2})\rangle |R_{t_4}\rangle |R_{t_5}\rangle |R_{t_6}\rangle \sum_{R_{t_3}} |R_{t_3}\rangle |E_{ijk}(R'_t, \pi_1(R_{t_1}), \pi_2(R_{t_2}), R_{t_3}, \ldots, R_{t_6})\rangle,$$

where the sum is over all R_{t_3} in Ω_{t_3}, can be approximated within inverse polynomial precision using $\tilde{O}\left(\frac{e_{ijk} |\Gamma_{ijk} \Delta \Gamma'_{ijk}|}{M_{ijk}} + \log n\right) = \tilde{O}\left(\frac{e_{ijk} |\Gamma_{ijk} \Delta \Gamma'_{ijk}|}{M_{ijk}} + 1\right)$ queries. We will use the following lemma.

Lemma 3. *When R_{t_1}, R_{t_2} and R_{t_6} are taken uniformly at random,*

$$\Pr\left[|\Gamma_{ijk} \Delta \Gamma'_{ijk}| \geq 44 \times \frac{f_{ij} f_{ik} f_{jk}}{r_i^2 r_j r_k}\right] = O\left(\frac{1}{n^{100}}\right).$$

Lemmas 1 and 3 then show that the mapping

$$|R_{t_4}\rangle |R_{t_5}\rangle \sum_{R_{t_1}} \sum_{R_{t_2}} \sum_{R_{t_6}} \sum_{R_{t_3}} |R_{t_1}\rangle |R_{t_2}\rangle |R_{t_6}\rangle |R_{t_3}\rangle |E_{ijk}(R_t, R_{t_1}, \cdots, R_{t_6})\rangle \mapsto$$

$$|R_{t_4}\rangle |R_{t_5}\rangle \sum_{R_{t_1}} \sum_{R_{t_2}} \sum_{R_{t_6}} \sum_{R_{t_3}} |R_{t_1}\rangle |R_{t_2}\rangle |R_{t_6}\rangle |R_{t_3}\rangle |E_{ijk}(R'_t, R_{t_1}, \cdots, R_{t_6})\rangle$$

can be approximated within inverse polynomial precision using $\tilde{O}(e_{ijk}/r_i + 1)$ queries. This argument is true for all $\{i, j, k\} \in \Sigma_3$, so the update cost is

$$\mathsf{U}_t = \tilde{O}\left(1 + \sum_{\{j,k\} \text{ such that } \{i,j,k\} \in \Sigma_3} \frac{e_{ijk}}{r_i}\right).$$

Case 2 $[s_t = \{i, j\}$ with $i < j]$: In this case R_t and R'_t correspond to two subsets F_{ij} and F'_{ij} that also differ by exactly one element. Using an analysis similar to what has been done for Case 1, we can show that the update cost is $\mathsf{U}_t = \tilde{O}\left(1 + \sum_{k \text{ such that } \{i,j,k\} \in \Sigma_3} \frac{e_{ijk}}{f_{ij}}\right)$.

Case 3 $[s_t = \{i, j, k\}$ with $i < j < k]$: R_t and R'_t are two subsets of $\{1, \ldots, M_{ijk}\}$, both of size e_{ijk}, differing by exactly one element. The corresponding E_{ijk} and E'_{ijk} are subsets of the same Γ_{ijk}, and have symmetric difference $|E_{ijk} \Delta E'_{ijk}| \leq 2$, so $\mathsf{U}_t \leq 2$.

Now the proof of Theorem 1 is completed.

5   Applications: 4-clique Detection and Associativity

In this section we describe how to use our method to construct efficient quantum algorithms for 4-clique detection and ternary associativity testing.

First, by applying Theorem 1 to the case where H is a 4-clique, and optimizing both the loading schedule and the parameters, we obtain the following result.

Theorem 2. *There exists a quantum algorithm that detects if a 3-uniform hypergraph on n vertices has a 4-clique, with high probability, using $\tilde{O}(n^{241/128}) = O(n^{1.883})$ queries.*

Next, we consider ternary associativity testing. Let X be a finite set with $|X| = n$. A ternary operator $\mathcal{F}$ from $X \times X \times X$ to X is said to be *associative* if $\mathcal{F}(\mathcal{F}(a, b, c), d, e) = \mathcal{F}(a, \mathcal{F}(b, c, d), e) = \mathcal{F}(a, b, \mathcal{F}(c, d, e))$ holds for every 5-tuple $(a, b, c, d, e) \in X^5$. The function $\mathcal{F}$ is given as a black-box: when we make a query (a, b, c) to $\mathcal{F}$, the answer $\mathcal{F}(a, b, c)$ is returned. We can show that that the property "$\mathcal{F}$ is not associative" has a certificate corresponding to a sub-hypergraph of seven vertices in a 3-uniform directed hypergraph with each edge weighted by an element in X. By applying Theorem 1 with adaptations to directed hypergraphs with non-binary hyperedge weights, we obtain the following result.

Theorem 3. *There exists a quantum algorithm that determines if $\mathcal{F}$ is associative with high probability using $\tilde{O}(n^{169/80}) = \tilde{O}(n^{2.1125})$ queries.*

Acknowledgments. The authors are grateful to Tsuyoshi Ito, Akinori Kawachi, Hirotada Kobayashi, Masaki Nakanishi, Masaki Yamamoto and Shigeru Yamashita for helpful discussions. This work is supported by the Grant-in-Aid for Scientific Research (A) No. 24240001 of the JSPS and the Grant-in-Aid for Scientific Research on Innovative Areas No. 24106009 of the MEXT in Japan.

References

1. Ambainis, A.: Quantum walk algorithm for element distinctness. SIAM J. Comput. 37(1), 210–239 (2007)
2. Beals, R., Buhrman, H., Cleve, R., Mosca, M., de Wolf, R.: Quantum lower bounds by polynomials. J. ACM 48(4), 778–797 (2001)
3. Belovs, A.: Span programs for functions with constant-sized 1-certificates: extended abstract. In: Proceedings of STOC, pp. 77–84 (2012)
4. Belovs, A.: Learning-graph-based quantum algorithm for k-distinctness. In: Proceedings of FOCS, pp. 207–216 (2012)
5. Belovs, A., Childs, A.M., Jeffery, S., Kothari, R., Magniez, F.: Time-efficient quantum walks for 3-distinctness. In: Fomin, F.V., Freivalds, R., Kwiatkowska, M., Peleg, D. (eds.) ICALP 2013, Part I. LNCS, vol. 7965, pp. 105–122. Springer, Heidelberg (2013)
6. Belovs, A., Rosmanis, A.: On the power of non-adaptive learning graphs. In: Proceedings of CCC, pp. 44–55 (2013)
7. Buhrman, H., Dürr, C., Heiligman, M., Høyer, P., Magniez, F., Santha, M., de Wolf, R.: Quantum algorithms for element distinctness. SIAM J. Comput. 34(6), 1324–1330 (2005)
8. Janson, S., Łuczak, T., Ruciński, A.: Random Graphs. Wiley-Interscience Series in Discrete Mathematics and Optimization. John Wiley & Sons (2000)
9. Jeffery, S., Kothari, R., Magniez, F.: Nested quantum walks with quantum data structures. In: Proceedings of SODA, pp. 1474–1485 (2013)
10. Le Gall, F.: Improved output-sensitive quantum algorithms for Boolean matrix multiplication. In: Proceedings of SODA, pp. 1464–1476 (2012)
11. Le Gall, F., Nishimura, H., Tani, S.: Quantum algorithms for finding constant-sized sub-hypergraphs. Full version of the present paper, available as arXiv:1310.4127
12. Lee, T., Magniez, F., Santha, M.: Learning graph based quantum query algorithms for finding constant-size subgraphs. Chicago J. Theor. Comput. Sci., Article 10 (2012)
13. Lee, T., Magniez, F., Santha, M.: Improved quantum query algorithms for triangle finding and associativity testing. In: Proceedings of SODA, pp. 1486–1502 (2013)
14. Magniez, F., Nayak, A., Roland, J., Santha, M.: Search via quantum walk. SIAM J. Comput. 40(1), 142–164 (2011)
15. Magniez, F., Santha, M., Szegedy, M.: Quantum algorithms for the triangle problem. SIAM J. Comput. 37(2), 413–424 (2007)
16. Vassilevska Williams, V., Williams, R.: Finding, minimizing, and counting weighted subgraphs. SIAM J. Comput. 42(3), 831–854 (2013)
17. Vassilevska Williams, V., Williams, R.: Subcubic equivalences between path, matrix and triangle problems. In: Proceedings of FOCS, pp. 645–654 (2010)
18. Williams, R.: A new algorithm for optimal 2-constraint satisfaction and its implications. Theor. Comput. Sci. 348(2-3), 357–365 (2005)
19. Williams, R.: Algorithms and resource requirements for fundamental problems. Ph.D. Thesis, Carnegie Mellon University (2007)
20. Zhu, Y.: Quantum query complexity of constant-sized subgraph containment. Int. J. Quant. Inf. 10(3), 1250019 (2012)

Approximation Algorithm for the Balanced 2-Connected Bipartition Problem

Di Wu[1], Zhao Zhang[1,*], Weili Wu[2], and Xiaohui Huang[1]

[1] College of Mathematics Physics and Information Engineering,
Zhejiang Normal University, Zhejiang, Jinhua, 321004, China
[2] Department of Computer Science, University of Texas at Dallas
Richardson, Texas, 75080, USA
hxhzz@sina.com

Abstract. For two positive integers m, k and a connected graph $G = (V, E)$ with a nonnegative vertex weight function w, the balanced m-connected k-partition problem, denoted as BC_mP_k, is to find a partition of V into k disjoint nonempty vertex subsets $(V_1, V_2, \ldots, V_k)$ such that each $G[V_i]$ (the subgraph of G induced by V_i) is m-connected, and $\min_{1 \le i \le k}\{w(V_i)\}$ is maximized. In this paper, we study the BC_2P_2 problem on 4-connected interval graphs. First, a 3/2-approximation algorithm is given. Then, assuming that w is integral, a fully polynomial time approximation scheme (FPTAS) is obtained. As far as we known, this is the first paper studying balanced connected partition problem with higher connectivity requirement on each part.

Keywords: balanced m-connected k-partition, interval graph, pseudo-polynomial time algorithm, FPTAS.

1 Introduction

Let $G = (V, E, w)$ be an undirected graph, in which w is a nonnegative weight function on vertices. For a vertex subset $U \subseteq V$, denote by $G[U]$ the subgraph of G induced by U, and $w(U) = \sum_{v \in U} w(v)$ the weight of U. For an integer $k \ge 2$, suppose $V = V_1 \cup V_2 \cup \ldots V_k$ such that each V_i is a non-empty subset of V and $V_i \cap V_j = \emptyset$ for $i \ne j$. Then, $(V_1, V_2, \ldots, V_k)$ is called a *k-partition* of V. It is a *connected k-partition* of G if every $G[V_i]$ is connected for $1 \le i \le k$. In many applications such as image processing [9] and clustering [10], it is desirable to find a connected partition whose parts are as balanced as possible. Such a setting can be modeled as a *balanced connected k-partition* (BCP_k) problem, in which one looks for a connected k-partition $(V_1, V_2, \ldots, V_k)$ with $\min_{1 \le i \le k} w(V_i)$ being maximized.

If one is looking for a clustering in which the communities are more compact, then each part should have higher connectivity. As far as we know, we have not seen any work on balanced partition with higher requirement on connectivity, which motivates us to study BC_mP_k problem defined as follows.

---

* Corresponding author.

Z. Cai et al. (Eds.): COCOON 2014, LNCS 8591, pp. 441–452, 2014.
© Springer International Publishing Switzerland 2014

For a connected graph $G = (V, E)$, a *vertex cut* of G is a vertex set $C \subseteq V$ such that $G - C$ is disconnected. In particular, if $G - v$ is disconnected, then vertex v is a *cut vertex* of G. In this paper, we adopt the convention that a connected graph G is *m-connected* if there is no vertex cut of cardinality less than m in G. In particular, a graph G is 2-connected if and only if there is no cut vertex in G. Under this convention, a complete graph on n vertices, denoted as K_n, is regarded as m-connected for any integer m. Furthermore, a graph on n vertices with $n \leq m$ is m-connected if and only if it is K_n.

Definition 1. A k-partition $(V_1, V_2, \ldots, V_k)$ of V is an *m-connected k-partition* of G if every $G[V_i]$ is m-connected for $1 \leq i \leq k$. The *balanced m-connected k-partition* problem $(BC_m P_k)$ looks for an m-connected k-partition $(V_1, V_2, \ldots, V_k)$ of G such that $\min_{1 \leq i} w(V_i)$ is maximized.

In particular, $BC_1 P_k$ is exactly BCP_k. In this paper, we study $BC_2 P_2$ on interval graphs, which is denoted as $IBC_m P_k$. It has been shown in [11] that the BCP_k problem is NP-hard, even when the graph is complete. Under our assumption that a complete graph is m-connected for any m, a k-partition of a complete graph is also an m-connected k-partition. Then, by observing that complete graphs are special interval graphs, the computation of the $IBC_m P_k$ problem is NP-hard.

The contributions of this paper are:

1) As far as we know, this paper is the first one to study $BC_m P_k$ problem for $m \geq 2$. Because of the difficulties brought by the higher connectivity requirement, as a starting point, this paper studies approximation algorithms for $BC_2 P_2$ in interval graphs.

2) A 3/2-approximation algorithm is give for $BC_2 P_2$ on 4-connected interval graphs. As a consequence, a lower bound for the optimal value is obtained. It should be noted that connectivity 4 is necessary because there exists 3-connected interval graph which has no 2-connected bipartition.

3) For the special situation when the weight function is integral, an FPTAS for $BC_2 P_2$ on 4-connected interval graphs is given, which achieves approximation ratio $(1 + \varepsilon)$ in time $O((1 + \frac{1}{\varepsilon})n^5)$, where ε is an arbitrary positive real number.

Although the FPTAS is obtained through the classic dynamic programming method and the classic scaling technique, adapted to suit the new setting, a prerequisite to obtain the desired approximation ratio is a lower bound for the optimal value. For the BCP_2 problem without higher connectivity requirement, it is not difficult to obtain such a lower bound [3], even for a general 2-connected graph. However, with the requirement on higher connectivity of each part, this task becomes much more difficult. We managed to derive such a lower bound for 4-connected interval graphs, which is a corollary of our constant approximation algorithm for this problem.

The paper is organized as follows. In Section 2, we introduce some related works. Notations and preliminary results are presented in Section 3. Section 4 presents a 3/2-approximation algorithm for $BC_2 P_2$ on 4-connected interval graphs. In Section 5, under the assumption that the weight function w is integral, an FPTAS for BC_2P_2 on 4-connected interval graphs is given. Section 6 concludes the paper and proposes some problems for future researches.

2 Related Works

A classic result on partition problem is due to Lovász [8] and Györi [7], saying that for any k-connected graph G and any integers $n_1, n_2, \ldots, n_k$ with $\sum_{i=1}^{k} n_i = |V(G)|$, graph G can be partition into k connected subgraphs of orders $n_1, n_2, \ldots, n_k$. The proof used in [8] is topological and the proof used in [7] is graph theoretical. But neither of them imply a polynomial time algorithm for finding such a partition.

In [4], it was shown that the *unweighted BCP_k* problem is NP-hard for any fixed $k \geq 2$. Polynomial-time algorithms exist for the unweighted BCP_k problem in the special case when the graph has at most two articulation vertices in each block [3,5].

Considering weight, Chebíková showed in [3] that BCP_2 is NP-hard in the strong sense and can not be approximated within an absolute error guarantee of $V^{1-\varepsilon}$, for any $\varepsilon > 0$ unless NP=P. A 4/3-approximation algorithm was given in [3]. For BCP_3 and BCP_4 on 3-connected and 4-connected graphs, 2-approximation algorithms were proposed by Chataigner *et al.* [2].

The BCP_k problem on interval graphs is denoted as $IBCP_k$. For the special case that G is a complete graph, the BCP_2 is at least as hard as PARTITION problem ([SP12] in [6]) and is therefore NP-hard. Since the class of interval graphs includes complete graphs, the $IBCP_k$ problem is also NP-hard for any fixed $k \geq 2$ [11].

In [11], Wu first gave a pseudo-polynomial time algorithm for $IBCP_2$. Then by a scaling technique, a fully polynomial time approximation scheme (FPTAS) was obtained. For any $\varepsilon > 0$, the FPTAS finds a $(1 + \varepsilon)$-approximation in time $O((1 + \frac{1}{\varepsilon})n^3)$, where n is the number of vertices. He also generalized the results to the case $k > 2$, requiring that the interval graph is k-connected.

As far as we know, our paper is the first work on BC_mP_k problem for $m \geq 2$.

3 Preliminaries

Let $G = (V, E, w)$ be the graph with vertex set V, edge set E, and nonnegative vertex weight function w. For convenience, n and W are used to denote $|V|$ and $w(V)$, respectively. The optimum value of BC_2P_k on graph G is denoted as $\beta_2^*(G, k)$, i.e., $\beta_2^*(G, k) = \max \min_{1 \leq i \leq k} w(V_i)$, where the maximum is taken over all 2-connected k-partitions of G.

In an *interval graph*, vertices can be represented by intervals on the real horizontal axis such that there is an edge between two vertices if and only if the two intervals intersect. An interval on the real horizontal axis is typically defined by its left endpoint l and right endpoint r. Hence we use $[l, r]$ to represent the interval and use terminologies "vertex" and "interval" interchangeably. Assume, without loss of generality, that the endpoints of the intervals are all distinct. For a vertex v of an interval graph, denote by $l(v)$ and $r(v)$ the left and the right endpoints of interval v, respectively. For a vertex set U, define $l(U) = \min\{l(v): v \in U\}$ and $r(U) = \max\{r(v): v \in U\}$.

For a vertex subset U, denote by $v_{l,j}(U)$ the vertex of U with the j-th leftmost left endpoint, and by $v_{r,j}(U)$ the vertex of U with the j-th rightmost right endpoint. That is, if $|U| = m$, then vertices in U can be ordered according to the increasing order of their left endpoints such that $l(v_{l,1}(U)) < l(v_{l,2}(U)) < \cdots < l(v_{l,m}(U))$. Vertices in U can also be ordered according to the decreasing order of their right endpoints such that $r(v_{r,1}(U)) > r(v_{r,2}(U)) > \cdots > r(v_{r,m}(U))$. Denote by $l_j(U) = l(v_{l,j}(U))$ the j-th leftmost endpoint, and $r_j(U) = r(v_{r,j}(U))$ the j-th rightmost endpoint. In particular, $l_1(U) = l(U)$ and $r_m(U) = r(U)$. Using Menger's theorem (see [1]), we can obtain the following results.

Lemma 1. *Let $G = (V, E)$ be an m-connected interval graph, and $\overline{v}$ be the vertex of V such that either*
   (i) $r(\overline{v}) = \min\{r(u): u \in V\}$, *or*
   (ii) $l(\overline{v}) = \max\{l(u): u \in V\}$.
Then, graph $G - \overline{v}$ is still m-connected.

Lemma 1 says that removing the interval with the leftmost right endpoint (or the rightmost left endpoint) keeps the m-connectedness. The following corollary says that removing the *set of intervals* with the leftmost right endpoints (or the rightmost left endpoints) keeps the m-connectedness.

Corollary 1. *Suppose $G = (V, E)$ is an m-connected interval graph, U is a subset of V such that $r(U) < \min\{r(v) : v \in V \setminus U\}$ or $l(U) > \max\{l(v) : v \in V \setminus U\}$. Then $G - U$ is still m-connected.*

Lemma 2. *Let G be an m-connected graph on n vertices. Suppose graph G' is obtained from G by adding a vertex v and joining v to at least $\min\{n, m\}$ distinct vertices of G. Then G' is also m-connected.*

4   A 3/2-Approximation for IBC_2P_2

In this section, we present a 3/2-approximation algorithm for the IBC_2P_k problem on 4-connected interval graphs.

Let $G = (V, E)$ be a connected interval graph. For a vertex $v \in V$, denote by $I(v)$ the interval of v, i.e., $I(v) = [l(v), r(v)]$. For a vertex subset $U \subseteq V$, denote by $I(U) = \bigcup_{v \in U} I(v)$. Since G is connected, $I(V)$ is a continuous interval on the real horizontal axis. Let $I_1 = [l_1(V), l_2(V))$ and $I_2 = (r_2(V), r_1(V)]$. Set $MI(V) = I(V) \setminus (I_1 \cup I_2)$. For $|V| \geq 2$, $MI(V) \neq \emptyset$. If a point $p \in MI(V)$ belongs to at most one interval of V, say $p \in I(u)$, then u is a cut vertex of G (see Fig.1). Thus, we have the following observation.

Observation 1. *For any 2-connected interval graph $G = (V, E)$ on at least two vertices, any point in $MI(V)$ belongs to at least two intervals of V.*

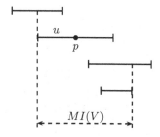

Fig. 1. The definition of $MI(V)$ and an illutration of Observation 1

Theorem 1. *Let $G = (V, E)$ be a 4-connected interval graph, and (V_1, V_2) be a 2-connected bipartition of G with $|V_2| \geq 2$. Then there is a vertex $u \in V_2$ such that $(V_1 \cup \{u\}, V_2 \setminus \{u\})$ is still a 2-connected bipartition of G. Furthermore, there are at least two distinct vertices in V_2 which can serve as u.*

Proof. For simplicity, denote $G_1 = G[V_1]$ and $G_2 = G[V_2]$.

If $|V| \leq 4$, then G is a complete graph, and the result holds trivially. So, suppose $|V| \geq 5$. If $|V_1| = 1$, let v be the only vertex in V_1. Then $G_2 = G - v$ is 3-connected, and thus removing any vertex from G_2 results in a 2-connected subgraph. Since a 4-connected graph on at least five vertices has minimum degree at least four, vertex v has at least four neighbors in V_2, any of such a neighbor can serve as u satisfying the condition of this theorem. In the following, we assume $|V_1| \geq 2$.

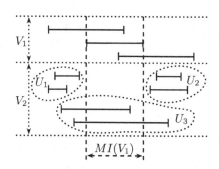

Fig. 2. An illustration for the proof of Theorem 1

Let $U_1 = \{v \in V_2 : r(v) < l(MI(V_1))\}$, $U_2 = \{v \in V_2 : l(v) > r(MI(V_1))\}$, and $U_3 = V_2 \setminus (U_1 \cup U_2)$ (see Fig.2). Since $|V_1| \geq 2$, we have $MI(V_1) \neq \emptyset$. It follows that U_1 and U_2 are well-defined (they might be empty sets) and $I(U_1) \cap I(U_2) = \emptyset$. As a consequence,

$$\text{if } U_1 \neq \emptyset \text{ and } U_2 \neq \emptyset, \text{ then } G[U_1 \cup U_2] = G_2 - U_3 \text{ is disconnected.} \quad (1)$$

By the definition of U_1 and U_2, every vertex in $U_1 \cup U_2$ is adjacent with at most one vertex of V_1. To be more concrete, $v_{l,1}(V_1)$ is the only vertex in V_1 which might be adjacent with vertices in U_1, and $v_{r,1}(V_1)$ is the only vertex in V_1 which might be adjacent with vertices in U_2.

For any vertex $v \in U_3$, since $l(v) \le r(MI(V_1))$ and $r(v) \ge l(MI(V_1))$, interval $I(v)$ contains some point in $MI(V_1)$. By the 2-connectedness of $G[V_1]$ and by Observation 1, this point belongs to at least two intervals of V_1, and thus vertex v is adjacent with at least two vertices of V_1. So, $G_1 + v = G[V_1 \cup \{v\}]$ is 2-connected by Lemma 2. To prove the theorem, we are to find a vertex $u \in U_3$ such that $G_2 - u = G[V_2 \setminus \{u\}]$ is 2-connected.

If both $U_1 = U_2 = \emptyset$ and $|U_3| = 2$, then $G[V_2]$ is a complete graph on two vertices, the removal of any one vertex leaves a complete graph on one vertex, which is 2-connected. In the following, we consider the remaining cases.

Claim 1. $|U_3| \ge 3$.

If $U_1 = U_2 = \emptyset$, then by the remark before this claim, we may assume $|U_3| \ge 3$. If $U_1 \cup U_2 \ne \emptyset$, suppose, without loss of generality, that $U_1 \ne \emptyset$. Recall that $v_{l,1}(V_1)$ is the only vertex in V_1 which may be adjacent with vertices in U_1. Since $|V_1| \ge 2$, we have $V_1 \setminus \{v_{l,1}(V_1)\} \ne \emptyset$. Thus $U_3 \cup \{v_{l,1}(V_1)\}$ is a vertex cut of G, the deletion of which separates U_1 and $V_1 \setminus \{v_{l,1}(V_1)\}$ (property (1) is used here). Since G is 4-connected, we have $|U_3 \cup \{v_{l,1}(V_1)\}| \ge 4$, and thus $|U_3| \ge 3$. Claim 1 is proved.

Claim 2. Suppose u is a vertex in U_3 such that graph $H = G[U_3 \setminus \{u\}]$ is 2-connected. Then $G_2 - u$ is also 2-connected.

Notice that $H = G_2 - u - U_1 - U_2$. To prove this claim, the idea is to show that enlarging H by adding vertices of $U_1 \cup U_2$ one by one, the 2-connectedness is kept.

Recall that vertices in U_1 can be ordered as $v_{r,1}(U_1), v_{r,2}(U_1), \ldots, v_{r,|U_1|}(U_1)$ such that $v_{r,i}(U_1)$ is the vertex of U_1 with the i-th rightmost endpoint. For $1 \le i \le |U_1|$, denote $U_1^{(i)} = \{v_{r,1}(U_1), \ldots, v_{r,i}(U_1)\}$ and $\widetilde{U}_1^{(i)} = U_1 \setminus U_1^{(i)}$. Let $U_1^{(0)} = \emptyset$. By the definition of U_1 and $\widetilde{U}_1^{(i-1)}$, we have $r(\widetilde{U}_1^{(i-1)}) \le r(U_1) < \min\{r(v) : v \in V_1\}$ and $r(\widetilde{U}_1^{(i-1)}) < \min\{r(v) : v \in V_2 \setminus \widetilde{U}_1^{(i-1)}\}$. Hence $r(\widetilde{U}_1^{(i-1)}) < \min\{r(v) : v \in V \setminus \widetilde{U}_1^{(i-1)}\}$. By Corollary 1, Claim 1, and the assumption that $|V_1| \ge 2$,

$$G - \widetilde{U}_1^{(i-1)} \text{ is a 4-connected graph on at least five vertices.} \qquad (2)$$

Next, we shall show that expanding H by adding vertices of U_1 sequentially in the order of $v_{r,1}(U_1), v_{r,2}(U_1), \ldots, v_{r,|U_1|}(U_1)$ keeps the 2-connectedness. For this purpose, denote $H^{(i)} = H + U_1^{(i)}$. We shall show by induction on i that $H^{(i)}$ is 2-connected. This is true for $H^{(0)} = H$. Suppose $H^{(i-1)}$ is 2-connected. We claim that vertex $v_{r,i}(U_1)$ is adjacent with at least two vertices of $H^{(i-1)}$. In fact, by property (2), vertex $v_{r,i}(U_1)$ has degree at least four in $G - \widetilde{U}_1^{(i-1)} = G[V_1 \cup (V_2 - \widetilde{U}_1^{(i-1)})]$. Since $v_{r,i}(U_1)$, being a vertex of U_1, is adjacent with at most one vertex of V_1, it is adjacent with at least three vertices of $V_2 - \widetilde{U}_1^{(i-1)} = U_2 \cup U_3 \cup U_1^{(i-1)}$. By property (1), $v_{r,i}(U_1)$ is not adjacent with any vertex of U_2. Hence, $v_{r,i}(U_1)$ is adjacent with at least three vertices of $U_3 \cup U_1^{(i-1)}$, and thus at least two vertices of $H^{(i-1)}$ (notice that $V(H^{(i-1)}) = U_3 \cup U_1^{(i-1)} - u$). Then by Lemma 2, $H^{(i)}$ is 2-connected. The induction proof is completed. In particular, $H + U_1 = H^{(|U_1|)}$ is 2-connected.

By a symmetric argument, it can be shown that expanding $H + U_1$ by adding vertices of U_2 in the order of $v_{l,1}(U_2), v_{l,2}(U_2), \ldots, v_{l,|U_2|}(U_2)$ preserves the 2-connectedness. Thus $G_2 - u = H + U_1 + U_2$ is 2-connected. Claim 2 is proved.

Claim 3. Let $\hat{u}$ and $\tilde{u}$ be the vertices in U_3 such that $r(\hat{u}) = \min\{r(v): v \in U_3\}$ and $l(\tilde{u}) = \max\{l(v): v \in U_3\}$. Then $G_2 - \hat{u}$ and $G_2 - \tilde{u}$ are both 2-connected.

Since $r(U_1) < \min\{r(v) : v \in V_2 \setminus U_1\}$ and $l(U_2) > \max\{l(v) : v \in V_2 \setminus U_2\}$, by Corollary 1, graph $G_2 - U_1 - U_2 = G[U_3]$ is 2-connected. Since $r(\hat{u}) = \min\{r(v): v \in U_3\}$, graph $H = G[U_3 \setminus \{\hat{u}\}]$ is also 2-connected by Lemma 1. Then by Claim 2, $G_2 - \hat{u}$ is 2-connected. The 2-connectedness of $G_2 - \tilde{u}$ can be proved symmetrically. Claim 3 is proved.

If $\hat{u} \neq \tilde{u}$, then we are done. Now, suppose $\hat{u} = \tilde{u}$. In this case, every vertex $u \in U_3 \setminus \{\hat{u}\}$ has $l(u) < l(\hat{u})$ and $r(u) > r(\hat{u})$. Then, $G[U_3]$ is a complete graph. So, for any vertex $u \in U_3$, $G[U_3 \setminus \{u\}]$ is 2-connected, and thus $G_2 - u$ is 2-connected by Claim 2. That is, any vertex in $U_3 - \hat{u}$ can play the role of u required by the theorem. Notice that there are at least two of such u, since $|U_3| \geq 3$ by Claim 1. The theorem is proved.   □

Based on Theorem 1, we now present Algorithm 1 which returns a 2-connected bipartition of a 4-connected interval graph.

Algorithm 1. A $3/2$-approximation algorithm for IBC_2P_2 on 4-connected interval graph

---

Input: A 4-connected interval graph G and a nonnegative vertex-weight function w.
Output: A 2-connected bipartition (V_1, V_2) of G.

1: Let v_1 be the vertex with the maximum weight. Initialize $V_1 = \{v_1\}$ and $V_2 = V \setminus V_1$.
2: **while** $w(V_1) < \frac{1}{2}W$, **do**
3:      Let $V_0 = \{u \in V_2 : (V_1 \cup \{u\}, V_2 \setminus \{u\})$ is a 2-connected bipartition of $G\}$. Choose $u \in V_0$ such that $w(u) = \min_{v \in V_0} w(v)$.
4:      **if** $w(u) < W - 2w(V_1)$ **then,**
5:          $V_1 = V_1 \cup \{u\}, V_2 = V_2 \setminus \{u\}$;
6:      **else**
7:          Return (V_1, V_2);
8:      **end if**
9: **end while**
10: Return (V_1, V_2).

Theorem 2. *For any 4-connected interval graph $G = (V, E)$ with nonnegative vertex weight function w, suppose v_1 is the vertex with the maximum weight. Algorithm 1 returns a 2-connected bipartition (V_1, V_2) of G with the following property:*

(a) *If $w(v_1) \geq \frac{1}{2}W$, then (V_1, V_2) is an optmal solution.*

(b) *If $w(v_1) < \frac{1}{2}W$, then $\min\{w(V_1), w(V_2)\} \geq W/3$, where $W = w(V)$ is the total weight of the graph.*

Proof. When $w(v_1) \geq W/2$, the optimal solution is clearly $(v_1, V \setminus v_1)$, thus (a) is proved. In the following, suppose $w(v_1) < W/2$.

First, notice that since $v_1 \notin V_2$ and $|V_0| \geq 2$ by Theorem 1, the vertex u chosen in Line 3 has at most the third largest weight. So, $w(u) \leq W/3$.

The algorithm terminates either in Line 7 with $w(V_1) < \frac{1}{2}W$ and vertex u currently chosen in Line 3 satisfies $w(u) \geq W - 2w(V_1)$, or in Line 10 with $w(V_1) \geq \frac{1}{2}W$. In the former case, $\frac{1}{2}W > w(V_1) \geq \frac{1}{2}(W - w(u))$ and $w(V_2) = W - w(V_1) > \frac{1}{2}W$. In the latter case, by Line 4, the last vertex u chosen into V_1 satisfies $w(u) < W - 2w(V_1 \setminus \{u\}) = W - 2w(V_1) + 2w(u)$. So, $\frac{1}{2}W \leq w(V_1) < \frac{1}{2}(W + w(u))$ and thus $\frac{1}{2}W \geq w(V_2) > \frac{1}{2}(W - w(u))$. In both cases, $\min\{w(V_1), w(V_2)\} \geq \frac{1}{2}(W - w(u))$. By $w(u) \leq W/3$, we have $\min\{w(V_1), w(V_2)\} \geq W/3$.   □

As a corollary of Theorem 2, we have the following lower bound for $\beta_2^*(G, 2)$.

Corollary 2. *For any 4-connected interval graph G in which $\max\{w(v): v \in V\} \leq W/2$, $\beta_2^*(G, 2) \geq W/3$.*

Since $W/2$ is a trivial upper bound for $\beta_2^*(G, 2)$, Algorithm 1 is in fact a constant approximation for the IBC_2P_2 problem on 4-connected interval graphs.

Corollary 3. *Algorithm 1 is a polynomial-time $\frac{3}{2}$-approximation for the IBC_2P_2 problem on 4-connected interval graphs.*

5   An FPTAS for the IBC_2P_2 Problem

In this section we assume that the vertex weight function w is integral, and give an FPTAS for the IBC_2P_2 problem on 4-connected interval graphs.

5.1   A Pseudo-polynomial Time Algorithm

For the pseudo-polynomial time algorithm in this subsection, it suffices to require the interval graph to be 2-connected instead of 4-connected.

Order the vertices of the interval graph as $V = \{v_1, v_2, \ldots, v_n\}$ such that $v_i = [l_i, r_i]$ and $0 < l_1 < l_2 < \cdots < l_n$. Let $U_i = \{v_j: 1 \leq j \leq i\}$ and $G_i = G[U_i]$. By Corollary 1, G_i is 2-connected for $i = 1, 2, \ldots, n$. The algorithm will iteratively find 2-connected bipartitions of G_i based on 2-connected bipartitions of G_{i-1}. In our algorithm, $(\emptyset, U_i)$ is also regarded as a 2-connected bipartition of G_i. Notice that an optimal solution (V_1, V_2) must have $\min\{w(V_1), w(V_2)\} > 0$, and thus no part is empty. So, trivial partitions $(\emptyset, U_i)$ only play an auxiliary role in the algorithm and will not affect the correctness of the algorithm.

The next lemma provides a condition under which a 2-connected bipartition of G_{i-1} can be extended to a 2-connected bipartition of G_i.

Lemma 3. *Suppose $(U, U_{i-1} \setminus U)$ is a 2-connected bipartition of G_{i-1}. Then $(U \cup \{v_i\}, U_{i-1} \setminus U)$ is a 2-connected bipartition of G_i if and only if*
   (i) either $U = \emptyset$, or
   (ii) $|U| = 1$ and $l(v_i) < r_1(U)$, or
   (iii) $|U| \geq 2$ and $l(v_i) < r_2(U)$.

Proof. To prove the necessity, it suffices to show that $G[U \cup \{v_i\}]$ is 2-connected. If Case (i) occurs, then $G[U \cup \{v_i\}] = K_1$. If Case (ii) occurs, suppose $U = \{v\}$. Since $l(v) \leq l(U_{i-1}) < l(v_i) < r_1(U) = r(v)$, vertex v_i is adjacent with vertex v, and thus $G[U \cup \{v_i\}] = K_2$. In both cases, $G[U \cup \{v_i\}]$ is 2-connected. If Case (iii) occurs, by $\max\{l(v_{r,2}(U)), l(v_{r,1}(U))\} \leq l(U_{i-1}) < l(v_i) < r_2(U) = r(v_{r,2}(U)) < r(v_{r,1}(U))$, we see that vertex v_i is adjacent with both $v_{r,2}(U)$ and $v_{r,1}(U)$. Hence $G[U \cup \{v_i\}]$ is 2-connected by Lemma 2.

Conversely, if $|U| \geq 1$ and $l(v_i) > r_1(U)$, then v_i is an isolated vertex of $G[U \cup \{v_i\}]$. If $|U| \geq 2$ and $r_2(U) < l(v_i) < r_1(U)$, then $v_{r,1}(U)$ is the only vertex of U which is adjacent with v_i, and thus $G[U \cup \{v_i\}] - v_{r,1}(U)$ is disconnected. In any case, $G[U \cup \{v_i\}]$ is not 2-connected. □

By Lemma 3, we can iteratively generate all possible 2-connected bipartitions of G_i based on those 2-connected bipartitions of G_{i-1}. The optimal solution can be found when the iteration reaches G_n. Notice that such an operation needs exponential time. However, since we only want to find *one* optimal bipartition, it is unnecessary to enumerate all of them. The following is our strategy.

For a 2-connected bipartition (V_1, V_2) of G_i, we only need to record the six parameters $(r_1(V_1), r_2(V_1), w(V_1), r_1(V_2), r_2(V_2), w(V_2))$. Since $w(V_1) + w(V_2) = w(U_i)$ and $\max\{r_1(V_1), r_2(V_2)\} = r(U_i)$, six parameters can be further reduced to four parameters. Define a *configuration* of a 2-connected bipartition (V_1, V_2) of G_i as a quadruple of integers (x_1, x_2, x_3, y), where $x_1 = r_1(V_1)$, $x_2 = r_2(V_1)$, $x_3 = r_2(V_2)$, and $y = w(V_1)$. This notion implicitly assumes that $r_1(V_2) = r(U_i)$ and $w(V_2) = w(U_i) - y$. If $|V_1| = 0$, then $x_1 = x_2 = 0$. If $|V_1| = 1$, then $x_1 \neq 0$, and $x_2 = 0$. A configuration (x_1, x_2, x_3, y) is *feasible* for G_i if there exists a 2-connected bipartition $(V_1^{(i)}, V_2^{(i)})$ of G_i whose configuration is exactly (x_1, x_2, x_3, y).

For the sake of the simplicity, in the following we only show how to compute $\beta_2^*(G, 2)$. If an optimal solution is required instead of just the optimal value, then for each configuration, a corresponding set V_1 is recorded. The time complexity remains the same but more space is needed. Notice that there might be different 2-connected bipartitions corresponding to a same configuration, but only one of them is stored. This is why the time complexity can be lowered.

The algorithm is presented in Algorithm 2. Initially, set $Q_1 = \{(0, 0, 0, 0)\}$ in Line 1, which is the configuration of the only 2-connected bipartition $(\emptyset, \{v_1\})$ of G_1. Each configuration $(x_1, x_2, x_3, y) \in Q_{i-1}$ corresponds to a 2-connected bipartition (V_1, V_2) of G_{i-1}. If $G[V_2 \cup \{v_i\}]$ is 2-connected, then the configuration corresponding to $(V_1, V_2 \cup \{v_i\})$ is inserted into Q_i in Line 6. If $G[V_1 \cup \{v_i\}]$ is 2-connected, then the configuration corresponding to $(V_1 \cup \{v_i\}, V_2)$ or $(V_2, V_1 \cup \{v_i\})$ is inserted into Q_i in Line 10 or in Line 12, respectively, depending on

---

Algorithm 2. $WIBC_2P_2$

Input: A 2-connected interval graph G with vertex set $V = \{v_i = [l_i, r_i] : 1 \le i \le n\}$
such that $0 < l_1 < l_2 < \ldots < l_n$.
Output: $\beta_2^*(G, 2)$.

1: $Q_1 \leftarrow \{(0, 0, 0, 0)\}$.
2: **for** $i = 2$ to n, **do**
3:     $Q_i \leftarrow \emptyset$;
4:     **for** each $(x_1, x_2, x_3, y) \in Q_{i-1}$, **do**
5:         **if** $[l(v_i) < x_3]$ or $[x_3 = 0$ and $l(v_i) < r(U_{i-1})]$, **then**
6:             insert $(x_1, x_2, \min\{\max\{r(v_i), x_3\}, r(U_{i-1})\}, y)$ into Q_i;
7:         **end if**
8:         **if** $[l(v_i) < x_2]$ or $[x_2 = 0$ and $l(v_i) < x_1]$ or $[x_1 = 0]$, **then**
9:             **if** $r(v_i) < r(U_{i-1})$, **then**
10:                 insert $(\max\{x_1, r(v_i)\}, \max\{\min\{x_1, r(v_i)\}, x_2\}, x_3, y + w(v_i))$ into Q_i;
11:             **else**
12:                 insert $(r(U_{i-1}), x_3, x_1, w(U_i) - y - w(v_i))$ into Q_i;
13:             **end if**
14:         **end if**
15:     **end for**
16: **end for**
17: return the maximum of $\min\{y, W - y\}$ for all (x_1, x_2, x_3, y) in Q_n.

---

whether $r(v_i) < r(U_{i-1})$ or $r(v_i) > r(U_{i-1})$ (recall that the two parts are ordered such that the second part has the rightmost endpoint $r(U_i)$).

Theorem 3. *Algorithm 2 computes the optimal value of the IBC_2P_2 problem in $O(n^4W)$ time.*

Proof. Suppose Q_{i-1} has enumerated all feasible configurations of G_{i-1}. We shall show that all feasible configurations of G_i can be inserted into Q_i by Algorithm 2. Suppose $(V_1^{(i)}, V_2^{(i)})$ is a 2-connected bipartition of G_i. In the case that $v_i \in V_1^{(i)}$, let $V_1^{(i-1)} = V_1^{(i)} \setminus \{v_i\}$ and $V_2^{(i-1)} = V_2^{(i)}$. If $G[V_1^{(i-1)}]$ has a cut vertex v_j, let R be the rightmost component of $G[V_1^{(i-1)}] - v_j$ and let L be the remaining components of $G[V_1^{(i-1)}] - v_j$. Then $r(L) < l(R) < l(v_i)$. It follows that v_i is not adjacent with any vertex in L, and thus v_j is also a cut vertex of $G[V_1^{(i)}]$, contradicting that $G[V_1^{(i)}]$ is 2-connected. So $(V_1^{(i-1)}, V_2^{(i-1)})$ is a 2-connected bipartition of G_{i-1}, whose configuration is in Q_{i-1} by the induction hypothesis. It should be noted that such a configuration may perhaps come from a 2-connected bipartition of G_{i-1} which is different from $(V_1^{(i-1)}, V_2^{(i-1)})$ but has the same configuration with $(V_1^{(i-1)}, V_2^{(i-1)})$. Nevertheless, when such a configuration is considered in Line 4 of Algorithm 2, by Lemma 3, the configuration corresponding to $(V_1^{(i)}, V_2^{(i)})$ is inserted into Q_i. The case $v_i \in V_2^{(i)}$ can be argued similarly. Iteratively using this argument for $i = 1, 2, \ldots, n$, Q_n enumerates all feasible configurations of $G_n = G$, and thus the value output by Line 17 is $\beta_2^*(G, 2)$.

The configuration set Q_i can be implemented by an $n \times n \times n \times W$ table, and the computation of Q_i takes $|Q_{i-1}| = O(n^3 W)$ time. Since there are n iterations, the total time complexity is $O(n^4 W)$. $\qquad\square$

5.2   A Fully Polynomial Time Approximation Scheme

In this section, we develop an FPTAS for IBC_2P_2 on 4-connected interval graphs. By Theorem 2 (a), if $\max\{w(v) : v \in V\} \geq \frac{1}{2}w(V)$, then the problem can be solved trivially. Hence, assume $\max\{w(v) : v \in V\} \leq W/2$ in the following.

The FPTAS is obtained by the classic scaling method. For the completeness of this paper, and in order to get some feeling about the importance of a lower bound for $\beta_2^*(G, k)$, we include the proof in the following.

Scaling down the weight function w by a factor $f = \rho W/(3n)$, where ρ is a constant, we obtain an instance of IBC_2P_2 problem on interval graph $G' = (V, E, w')$, where $w'(v) = \lfloor w(v)/f \rfloor$ for all $v \in V$.

Lemma 4. *Let (V_1', V_2') be the optimal solution of instance G' output by Algorithm 2. Then, $\min\{w(V_1'), w(V_2')\} \geq (1 - \rho)\beta_2^*(G, 2)$.*

Proof. Let (V_1, V_2) be an optimal solution of G. Then $\beta_2^*(G, 2) = \min\{w(V_1), w(V_2)\}$. Since (V_1', V_2') is an optimal solution of G', we have

$$\min\{w'(V_1), w'(V_2)\} \leq \min\{w'(V_1'), w'(V_2')\}.$$

By the definition of w', for $i = 1, 2$, we have

$$fw'(V_i) \leq w(V_i) < fw'(V_i) + f|V_i| < fw'(V_i) + fn.$$

So,

$$\min\{w(V_1'), w(V_2')\} \geq f \min\{w'(V_1'), w'(V_2')\} \geq f \min\{w'(V_1), w'(V_2)\}$$
$$> \min\{w(V_1), w(V_2)\} - nf = \beta_2^*(G, 2) - nf.$$

By Corollary 2, $\beta_2^*(G, 2) \geq W/3$. So,

$$\frac{\beta_2^*(G, 2)}{\min\{w(V_1'), w(V_2')\}} < \frac{\beta_2^*(G, 2)}{\beta_2^*(G, 2) - nf} \leq \frac{1}{1 - (nf)/(W/3)} = \frac{1}{1 - \rho}.$$

The lemma is proved. $\qquad\square$

Theorem 4. *Suppose $G = (V, E, w)$ is a 4-connected interval graph and w is an integral vertex weight function. For any $\varepsilon > 0$, there is a $(1 + \varepsilon)$-approximation algorithm for the IBC_2P_2 problem on G which runs in time $O((1 + \frac{1}{\varepsilon})n^5)$.*

Proof. For any $\varepsilon > 0$, let $\rho = \frac{\varepsilon}{1+\varepsilon}$. By lemma 4, the approximation ratio is $\frac{1}{1-\rho} = 1 + \varepsilon$. Let $W' = \sum_{v \in V} w'(v)$. Then $W' \leq W/f = 3n/\rho$. By Theorem 3, the time complexity is $O(n^4 W') = O((1 + \frac{1}{\varepsilon})n^5)$. $\qquad\square$

6   Conclusion

In this paper, we studied the balanced 2-connected bipartition problem (BC_2P_2). A pseudo-polynomial time algorithm, a constant-approximation algorithm, and an FPTAS is given for BC_2P_2 on interval graphs, where the first one requires the graph to be 2-connected, and the latter two require the graph to be 4-connected. Recently, we have found a way to generalize this work to BC_2P_k for $k > 2$. Due to the limited space, the generalization will be given in a future paper.

Acknowledgements. This research is supported by NSFC (61222201), SRFDP (20126501110001), Xingjiang Talent Youth Project (2013711011), and by National Science Foundation of USA under grants CNS0831579 and CCF0728851.

References

1. Bondy, J.A., Murty, U.S.R.: Graph Theory with Applications. The Macmillan Press, London (1976)
2. Chataigner, F., Salgado, L.R.B., Wakabayashi, Y.: Approximation and inaproximability results on balanced connected partitions of graphs. Discrete Mathematics and Theoretical Computer Science 9, 177–192 (2007)
3. Chlebíková, J.: Approximating the maximally balanced connected partition problem in graphs. Information Processing Letters 60, 225–230 (1996)
4. Dyer, M., Frieze, A.: On the complexity of partitioning graphs into connected subgraphs. Discrete Applied Mathematics 10, 139–153 (1985)
5. Galbiati, G., Maffioli, F., Morzenti, A.: On the approximability of some maximum spanning tree problems. Theoretical Computer Science 181(1), 107–118 (1997)
6. Garey, M.R., Johnson, D.S.: Computers and Intractability: A Guide to The Theory of NP-Completeness. Freeman, New York (1979)
7. Györi, E.: On division of graph to connected subgraphs. In: Combinatoris (Proc. Fifth Hungarian Colloq., Koszthely, 1976). Colloq. Math. Soc. János Bolyai, vol. I, 18, pp. 485–494. North-Holland, Amsterdam (1978)
8. Lovász, L.: A homology theory for spanning trees of a graph. Acta Math. Acad. Sci. Hunger. 30, 241–251 (1977)
9. Lucertini, M., Perl, Y., Simeone, B.: Most uniform path partitioning and its use in image processing. Discrete Applied Math. 42, 227–256 (1993)
10. Maravalle, M., Simeone, B., Naldini, R.: Clustering on trees. Comput. Statist. Data Anal. 24, 217–234 (1997)
11. Wu, B.Y.: Fully polynomial time approximation schemes for the max-min connected partition problem on interval graphs. Discrete Math. Algorithm. Appl. 04, 1250005 (2012)

Improved Approximation for Time-Dependent Shortest Paths

Masoud Omran and Jörg-Rüdiger Sack

School of Computer Science, Carleton University, Ottawa, Ontario K1S 5B6, Canada
{mtomran,sack}@scs.carleton.ca

Abstract. We study the approximation of minimum travel time paths in time dependent networks. The travel time on each link of the network is a piecewise linear function of the departure time from the start node of the link. The objective is to find the minimum travel time to a destination node d, for all possible departure times at source node s. Dehne *et al.* proposed an exact output-sensitive algorithm for this problem [6, 7] that improves, in most cases, upon the existing algorithms. They also provide an approximation algorithm. In [10, 11], Foschini *et al.* show that this problem has super-polynomial complexity and present an ϵ-approximation[1] algorithm that runs $O(\frac{\lambda}{\epsilon} \log(\frac{T_{max}}{T_{min}}) \log(\frac{L}{\lambda \epsilon T_{min}}))$ shortest path computations, where λ is the total number of linear pieces in travel time functions on links, L is the horizontal span of the travel time function and T_{min} and T_{max} are the minimum and maximum travel time values, respectively.

In this paper, we present two ϵ-approximation algorithms that improve upon Foschini *et al.*'s result. Our first algorithm runs $O(\frac{\lambda}{\epsilon}(\log(\frac{T_{max}}{T_{min}}) + \log(\frac{L}{\lambda T_{min}})))$ shortest path computations at fixed departure times. In our second algorithm, we reduce the dependency on L, by using only $O(\lambda(\frac{1}{\epsilon} \log(\frac{T_{max}}{T_{min}}) + \log(\frac{L}{\lambda \epsilon T_{min}})))$ total shortest path computations.

1   Introduction

Static shortest path computations arise naturally in areas such as Navigation Systems, Network Routing, Robotics, Social Networks, and VLSI design. However, in many applications, the travel time on links are dynamically changing over time. In this case, the minimum travel time from source s to destination d depends on the departure time at s. E.g., a departure time before, or after, rush hour may lead to a reduced travel time when compared to rush hour travel. The problem of finding shortest paths to d for all possible departure times at s is referred to as *Time-Dependent Shortest Path* problem (TDSP for short). The first variant of TDSP (referred to also as "earliest arrival time" and "minimum travel time" problem in some texts) was introduced in 1966 by Cooke and Halsey [3]. Their result is obtained by discretizing time. To capture the

---

[1] We denote by ϵ-approximation any algorithm whose quality is less than or equal to $(1 + \epsilon) \cdot \mathcal{OPT}$, where $\mathcal{OPT}$ is the optimal achievable quality.

Z. Cai et al. (Eds.): COCOON 2014, LNCS 8591, pp. 453–464, 2014.

dynamically changing nature of travel time on links in many applications, most recent results consider continuous time. E.g., with deteriorating road conditions, say due to snow fall, travel along a road may become progressively harder, i.e., slower. Later, snow may melt or be removed. Piece-wise linear functions are commonly used either to approximate non-linear travel time functions, or to model link travel times when they are changing linearly during specific periods of time. Let λ_{max} be the maximum number of linear pieces on link travel time functions. A solution to TDSP, for a specified departure time, can be obtained by applying a simple modification to Dijkstra's shortest path algorithm [8], using Fredman and Tarjan's implementation [12]. Using this approach on a network $G(V, E)$, with vertices, V, and links (edges), E, the minimum travel time from s to d at a given time instance is computed in $O(|E| \log \lambda_{max} + |V| \log |V|)$ time [2]. Here, $\log \lambda_{max}$ in the first term is the time for a binary search to evaluate arrival times on link functions (to find the right linear piece in piecewise linear functions).

In time-dependent networks, the travel time data could either be obtained in real-time by traffic sensors, polling, or using historic data gathered over a period of time. Some navigation systems companies collect such information on the devices and upload it to their servers when maps are updated. For a given starting time at s, the minimum travel time when real-time data is available is computed by taking a snapshot of the network and running a static shortest path algorithm. Conversely, for link $e = (u, v)$ of the network, historic data is usually presented by a piecewise linear function $T_e(t)$ that returns the minimum travel time to v for any departure time t at u. Unlike in time-dependent networks with real-time data, in this case, the minimum travel time for all links $(u, v) \in E$ depends on the departure time from u and would be obtained using link arrival time functions.

With the advent of sensors, cameras, cell-phones, navigation systems, and other devices that collect traffic data, users are becoming increasingly interested and dependent on route planning that is time-dependent. Real-time traffic information might be available and sometimes navigation systems utilize these to compute fastest routes. However, computed routes often need to be recomputed and users may have to follow redirections to accommodate changes occurring during their trip. Traffic data though exhibit typically predictable pattern based on historic data. For example, rush hour traffic would mostly follow a similar pattern for specific weekdays, e.g., Mondays, in an alike weather condition. In such cases, time dependent shortest path can be computed that already takes the nature of expected congestion etc. into consideration.

Additionally, using historic travel time data, one could obtain the minimum travel time function for each link of the network, which then can be used to compute for all possible departure times at s the minimum travel time to d. For example, in Navigation Systems, minimum travel time functions can help for trip planning by letting users pick a time that suits their schedule. It also enables users to examine the travel times during trip planning, choose an optimal departure time, and obtain a suitable route.

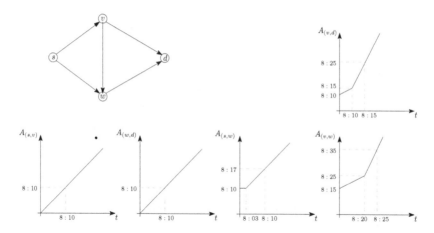

Fig. 1. A time-dependent network and link arrival time functions

Problem Definition. The TDSP problem is formally defined as follows: Let $G(E, V)$ be a graph with edge set E and node set V. Each edge $e = (u, v) \in E$ is assigned a non-decreasing piece-wise linear function $A_e(t)$ denoting the arrival time at v for a given departure time t at u. The travel time function on each edge e denoted by $T_e(t)$ is $A_e(t) - t$. Given a source node s and destination node d, the goal is to approximate $A_{s,d}(t)$ which returns the earliest arrival time at d for any departure time t at s. This is equivalent to approximating the minimum travel time function, i.e., $T_{s,d}(t) = A_{s,d}(t) - t$. We say a function $T'_{s,d}(t)$ is an ϵ–approximation of $T_{s,d}(t)$ if $|T_{s,d}(t) - T'_{s,d}(t)| \leq \epsilon \cdot T_{s,d}(t)$ for all t.

Solving TDSP for all possible times in its general form, where waiting at nodes is not allowed, is known to be NP–hard [17]. In 2004, Dean [5] conjectured that TDSP when waiting at nodes is allowed has super-polynomial complexity. Foschini et al. [11] proved this conjecture. In [16], it has been shown that any non-FIFO network that allows unrestricted waiting on nodes can be converted to a FIFO network with zero waiting on nodes. The FIFO property on links implies that on each link (u, v), a later departure time at u results in a later (or equal) arrival time at v. In other words, link minimum travel time functions can not have a slope less than -1. Since the FIFO property appears to be a natural assumption holding in many applications, it is a common assumption in the literature (see e.g., [5, 11, 13–15]) and one that is also used in this paper.

Figure 1 depicts an instance of a time-dependent network and link arrival time functions. Let $p_1 = \langle s, v, d \rangle$, $p_2 = \langle s, v, w, d \rangle$, and $p_3 = \langle s, w, d \rangle$ be the three possible paths from s to d. Then, the arrival time functions on p_1, p_2, and p_3 are $A_{p_1} = A_{(v,d)}(A_{(s,v)})$, $A_{p_2} = A_{(w,d)}(A_{(v,w)}(A_{(s,v)}))$, and $A_{p_3} = A_{(w,d)}(A_{(s,w)})$, respectively. Figure 2(a) shows the arrival time function on each path as well as the earliest arrival time function from s to d, $A_{s,d}$. For the same network, travel time functions on paths p_1, p_2, and p_3 are $T_{p_1} = A_{p_1} - t$, $T_{p_2} = A_{p_2} - t$, and $T_{p_3} = A_{p_3} - t$, respectively. The travel time function on each path as well as the minimum travel time function from s to d, $T_{s,d}$, is shown in Figure 2(b).

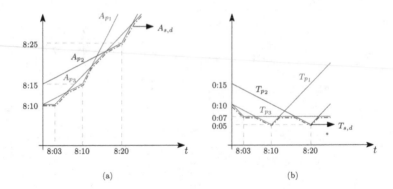

(a)   (b)

Fig. 2. a) The earliest arrival time function for the network shown in Figure 1. b) The minimum travel time function for the network shown in Figure 1.

Related Work. Our goal is to find the minimum travel time to d for all possible departure times t at s on a network, where the travel time on each link is a piecewise linear FIFO function of the departure time from s, i.e., solve TDSP.

Orda and Rom [16] proposed an algorithm for TDSP using a modified version of Bellman-Ford's label-correcting algorithm for shortest paths [1]. For a FIFO network with piece-wise linear link functions, the time complexity of their algorithm is $O(F_{max}|V||E|)$, where F_{max} is the maximum number of linear pieces on the minimum travel time function from s to any node in the network. Dean [5] proposed the first label-setting algorithm (building on Dijkstra's algorithm [8] for the static shortest path problem) that performs a single chronological scan through time and runs, an approach similar to that used in solutions to parametric shortest path problems to establish output functions. The time complexity of this algorithm is $O(|E|F^* \log |V|)$, where F^* is the total number of pieces in all output functions.

Ding et al. [9] presented a simpler label-setting algorithm for TDSP that scans a sequence of time steps. As analyzed in [6], the total number of time steps depends on the values of the arrival time functions. Therefore, the time complexity of their algorithm is $O(\alpha(|E| + |V| \log |V|))$. Note that α depends on travel time values which can be arbitrarily large. See [6] for an example of an instance, where α is independent on $|E|, |V|$ and λ, and can be arbitrarily large.

Dehne et al. [7] proposed an improved label-setting algorithm to solve TDSP that exploits the structural properties of the problem as well as combinatorial properties of the travel time functions. Using these properties, they could discard unnecessary minimum travel time function computations for all intermediate nodes of the network and focus on finding the minimum travel time function for just the destination node, and only at crucial time-points (i.e., function breakpoints). Their algorithm for solving TDSP runs in time $O((F_d + \lambda)(|E| + |V| \log |V|))$, where F_d is the output size (number of linear pieces needed to represent the minimum travel time function from s to d) and λ is the input size (total number of linear pieces in all link travel time functions).

Recently, Foschini *et al.* [11] proved that the minimum travel time function could have super-polynomial complexity (i.e., $O(n^{\Theta(\log n)})$). They also presented an algorithm that uses kinetic data structures and holds a set of certificates that are linear functions of t that guarantee the correctness of the shortest path tree. They also keep a priority queue of events (certificate failure times) and remove an event and update the corresponding certificates. Their algorithm has time-complexity $O((\lambda + F_{max}|E|)\log^2|V|)$, where F_{max} is the maximum of the complexity of travel time functions from s to each nodes in the network.

All algorithms mentioned above provide exact results. Dehne *et al.*'s algorithm outperforms others in most cases and provides a real output sensitive algorithm whose time complexity depends on the complexity of output function, i.e., F_d. The complexity of a minimum travel time function from s to d could be super-polynomial, and in most cases, link travel time functions are approximation of reality. Therefore, efficient approximation algorithms with customizable quality of result are both acceptable and often favorable for TDSP. In the following, we provide an overview of existing approximation algorithms.

Dehne *et al.* [6] presented an additive approximation algorithm for TDSP that approximates the earliest arrival time function from s to d. Their algorithm is based on running a reverse shortest path algorithm at arrival time values at d that are ϵ apart from each other. A reverse shortest path is obtained by running Dijkstra's algorithm on graph G^r obtained by reversing all links of G and inverting all link arrival time functions [4]. This returns, for each arrival time at d, the latest departure time from s. The function obtained by connecting all such sample points is an approximation of the earliest arrival time function from s to d. Their algorithm runs in $O(\frac{\Delta}{\epsilon}(|E| + |V|\log|V|))$ time, where Δ is the vertical span of the earliest arrival time function from s to d.

Foschini *et al.* [11] subsequently proposed an ϵ–approximation algorithm for TDSP that approximates the minimum travel time function from s to d. Note that, unlike earliest arrival time functions, it is not possible to run a reverse shortest path algorithm on minimum travel time functions directly to obtain, for a given travel time from s to d, the starting time at s. This is because it is not possible to evaluate the travel time on each link without knowing the arrival time at d. Therefore, they run a combination of forward (standard) shortest path computations on travel time functions and reverse shortest path computations on arrival time functions to obtain the sample points. Their algorithm requires $O(\frac{\lambda}{\epsilon}\log(\frac{T_{max}}{T_{min}})\log(\frac{L}{\lambda\epsilon T_{min}}))$ shortest path computations, where L is the horizontal span of the travel time function from s to d. T_{min} and T_{max} are the minimum and maximum travel time values of the function, respectively (see Section 2.1 for more details).

Contributions. We propose new algorithms that improve upon Foschini *et al.*'s algorithms by reducing the number of shortest path computations. Our first algorithm runs shortest path computations at fixed departure times, chosen so that the relative ratio is fixed to $(1+\epsilon)$ for any consecutive pair of time instances. The total number of shortest path computations required by our algorithm is $O(\frac{\lambda}{\epsilon}(\log(\frac{T_{max}}{T_{min}}) + \log(\frac{L}{\lambda T_{min}})))$ ("+" instead of "·"in the complexity). Next, we

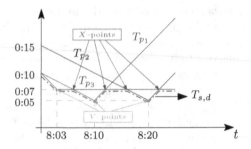

Fig. 3. The V–points and the X–points of $T_{s,d}$ for the network shown in Figure 1

further reduce shortest path computations by increasing the horizontal distance between sample points as the slope function converges to zero. The total number of shortest path computations required for this approach is $O(\lambda(\frac{1}{\epsilon}\log(\frac{T_{max}}{T_{min}}) + \log(\frac{L}{\lambda\epsilon T_{min}})))$ ($\frac{1}{\epsilon}$ factor is removed from the $\log(\frac{L}{\lambda T_{min}})$ term).

Organization. Next, in Section 2, we provide required details of previous algorithms and then present our solutions. Finally, in Section 3 we conclude the paper.

2   Solutions

In this section, we first sketch the previous approximation approaches by Dehne et al. [7] and Foschini *et al.* (Section 2.1). Then, we propose new ϵ–approximation algorithms that improve upon the running time of Foschini *et al.*'s algorithm (Section 2.2 and 2.3).

2.1   Previous Algorithms

Dehne *et al.* [7] showed that $T_{s,d}(t)$ is composed of two types of breakpoints: V–points and X–points. V–points originate from breakpoints in link functions and X–points are the points at which a shortest path from s to d switches to a new path (Figure 3). They also showed that between any consecutive pair of V–points the sub-function is concave and has slope not less than -1. Let λ be the total number of breakpoints on all link functions. The number of V–points in $T_{s,d}(t)$ is at most $O(\lambda)$. Thus, in order to approximate $T_{s,d}(t)$, it suffices to approximate $O(\lambda)$ concave sub-functions. Let $v' = (t', T_{s,d}(t'))$ and $v'' = (t'', T_{s,d}(t''))$ be an arbitrary pair of consecutive V–points. We denote the sub-function of $T_{s,d}(t)$ between v' and v'' by $T_{s,d}[v', v'']$. Because of the concavity of $T_{s,d}[v', v'']$, the minimum value, τ_{min}, is $\min(T_{s,d}(t'), T_{s,d}(t''))$ and its maximum value, τ_{max}, is $2T_{s,d}((t' + t'')/2)$. Let T_{min} and T_{max} be the minimum τ_{min} and maximum τ_{max} among all consecutive V–points, respectively. The general approach used by Foschini et al. to approximate $T_{s,d}[v', v'']$ is to intersect $T_{s,d}(t)$ with the horizontal lines $y = (1 + \epsilon)^k T_{min}$ for each $k \geq 0$ such that $(1 + \epsilon)^k T_{min} \leq T_{max}$. This generates $\approx \frac{2}{\epsilon} \log(T_{max}/T_{min})$ sample points. Connecting consecutive sample points with line-segments results in an ϵ-approximation of $T_{s,d}[v', v'']$.

The above approach requires $O(\frac{\lambda}{\epsilon} \log(\frac{T_{max}}{T_{min}}))$ reverse shortest path computations on $T_{s,d}(t)$. However, reverse shortest path computation is only feasible for $A_{s,d}(t)$, not for $T_{s,d}(t)$. Also note that an ϵ-approximation of $A_{s,d}(t)$ does not necessarily result in an ϵ-approximation for $T_{s,d}(t)$. Foschini et al. resolved the issue by running a combination of forward and reverse shortest path computations. They first partition the range of $T_{s,d}[v', v'']$ using horizontal lines $y = (1 + \epsilon)^{i/2} T_{min}$ for $i \leq 2 \log(T_{max}/T_{min})/\log(1 + \epsilon)$. Starting from $t = t'$, on parts of $T_{s,d}[v', v'']$ with slope at least 1, they compute a reverse shortest path for $A_{s,d}(t) = t + T_{s,d}(t) \cdot \sqrt{1 + \epsilon}$, then set t to the resulting starting time value. They show that this approach performs a constant number of reverse shortest path computations in each horizontal partition of the range of $T_{s,d}[v', v'']$. Therefore, the value of $T_{s,d}(t)$ at consecutive sample points differ by a factor of $1 + \epsilon$. Linearly interpolating between the sample points obtained using the above approach provides an $\epsilon-$ approximation of $T_{s,d}[v', v'']$ for the parts with slope at least 1. In total, their approach performs $O(\frac{\lambda}{\epsilon} \log(\frac{T_{max}}{T_{min}}))$ reverse shortest path computation on $A_{s,d}(t)$.

They use a different approach to find sample points on parts of $T_{s,d}[v', v'']$ with slope at most 1. They perform bisection with forward shortest path computations on $T_{s,d}(t)$ until the approximation error between consecutive sample points is at most $1+\epsilon$. They show that their approach requires $O(\frac{\lambda}{\epsilon} \log(\frac{T_{max}}{T_{min}}) \log(\frac{L}{\lambda \epsilon T_{min}}))$ forward shortest path computations, where L is the horizontal span of $T_{s,d}(t)$. The reader is referred to [11] for more details.

2.2   First Improved Algorithm

In this section, we present an improved algorithm, referred to as Aprx–A, which reduces the number of shortest path probes significantly. For two consecutive V–points, $v' = (t', T_{s,d}(t'))$ and $v'' = (t'', T_{s,d}(t''))$, we approximate $T_{s,d}[v', \bar{v}]$ as in [11] where $\bar{v} = (\bar{t}, T_{s,d}(\bar{t}))$ is the first point on $T_{s,d}[v', v'']$, in increasing order of time, whose slope is less than or equal to 1. However, on $T_{s,d}[\bar{v}, v'']$, we use a different approach to obtain sample points. Our new strategy is illustrated next.

Let $\bar{l}$ be the line, with slope 1, that goes through $\bar{v}$, and l'' be the line, with slope -1, that goes through v'' (Figure 4). Starting from $\bar{v}$, in the ith iteration, we compute $T_{s,d}(\bar{t}_i)$ where $\bar{t}_i$ is the starting time at s that corresponds to the intersection of $\bar{l}$ with the horizontal line $y = (1 + \epsilon)^i T_{s,d}(\bar{t})$. The operation terminates at the smallest k for which $T_{s,d}(\bar{t}_k) \geq T_{s,d}(\bar{t}_{k+1})$. Similarly, starting from v'', in the jth iteration, we compute $T_{s,d}(t''_j)$ where t''_j is the starting time at s that corresponds to the intersection of l'' with line $y = (1+\epsilon)^j T_{s,d}(t'')$. The operation terminates at the smallest z at which $T_{s,d}(t''_z) \geq T_{s,d}(t''_{z+1})$. We then connect the set of sample points obtained using the above approach, in increasing order of the starting times (see Figure 4). The following theorem states that this approach results in an ϵ-approximation of $T_{s,d}(t)$.

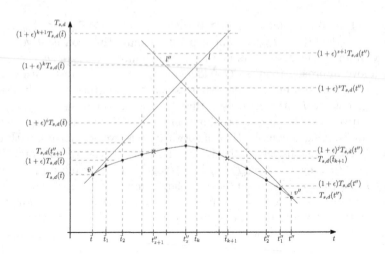

Fig. 4. Sample point placement of Algorithm Aprx–A

Theorem 1. *Algorithm Aprx–A is an ϵ-approximation of $T_{s,d}(t)$.*

Proof. Let $\bar{t}_{i-1}$ and $\bar{t}_i$ be the starting times corresponding to two consecutive sample points of Algorithm Aprx–A. Let $\bar{v}_{i-1} = (\bar{t}_{i-1}, T_{s,d}(\bar{t}_{i-1}))$ be the $i - 1^{\text{th}}$ sample point and $\bar{l}_{i-1}$ be the line that goes through $\bar{v}$ and $\bar{v}_{i-1}$. Also, let $v_c = (\bar{t}_i, T_{s,d}(\bar{t}_{i-1}))$, and v_d be the intersection point of the lines $\bar{l}_{i-1}$ and $x = \bar{t}_i$. We define vertical distances $A, B, C, D,$ and Y as depicted in Figure 5. We show that between $\bar{t}_{i-1}$ and $\bar{t}_i$, the approximation ratio is at most $1 + \epsilon$. We have:

$$m := \frac{B}{D} = \frac{A}{C} = \frac{Slope(\bar{l})}{Slope(\bar{l}_{i-1})} \tag{1}$$

$$B \geq D, \ A \geq C \Rightarrow m \geq 1 \tag{2}$$

$$D \geq C \tag{3}$$

$$\frac{B+Y}{A+Y} = 1 + \epsilon \tag{4}$$

Then:

$$\frac{B+Y}{A+Y} - \frac{D+Y}{C+Y} = \frac{BC + BY + CY + Y^2 - DA - DY - AY - Y^2}{(A+Y)(C+Y)}$$

$$\stackrel{(1)}{=} \frac{(B+C-D-A)Y}{(A+Y)(C+Y)} \stackrel{(1)}{=} \frac{(mD+C-D-mC)Y}{(A+Y)(C+Y)}$$

$$= \frac{((m-1)D - (m-1)C)Y}{(A+Y)(C+Y)} \stackrel{(2)(3)}{\geq} 0$$

$$\Rightarrow \frac{B+Y}{A+Y} - \frac{D+Y}{C+Y} \geq 0 \stackrel{(4)}{\Rightarrow} \frac{D+Y}{C+Y} \leq 1 + \epsilon \tag{5}$$

Since $T_{s,d}(t)$ is a concave function, it lies inside the triangle $\triangle(\bar{v}v_cv_d)$. By Equation 5, the error ratio between $\bar{t}_{i-1}$ and $\bar{t}_i$ is at most $1 + \epsilon$. Due to the symmetry of $\bar{l}$ and l'', a similar argument can be made for sample points along l''. Therefore, Algorithm Aprx–A provides an ϵ–approximation of $T_{s,d}(t)$ which proves the theorem. $\qquad\square$

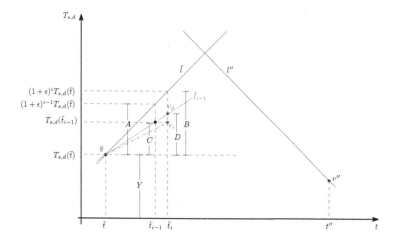

Fig. 5. A sample step of Algorithm Aprx–A

Theorem 2. *Approximation Algorithm Aprx–A requires*
$O(\frac{\lambda}{\epsilon}(\log(\frac{T_{max}}{T_{min}}) + \log(\frac{L}{\lambda T_{min}}))$ *shortest path computations.*

Proof. Each V–point on $T_{s,d}(t)$ corresponds to a breakpoint on a link travel time function. Therefore, the number of V–points is $O(\lambda)$. On parts of $T_{s,d}(t)$ that have slope at least 1, we run a reverse Dijkstra. In [11], it has been shown that there are $O(\frac{1}{\epsilon}\log(\frac{T_{max}}{T_{min}}))$ such points for any two consecutive V–points. Since $\bar{l}$ and l'' have slopes 1 and -1, respectively, the vertical span of $T_{s,d}(t)$ between $\bar{v}$ and v'' is at most $2\mathcal{L}$, where $\mathcal{L}$ is the horizontal distance between $\bar{v}$ and v''. Therefore, our improved method would place $O(\frac{1}{\epsilon}\log(\frac{\mathcal{L}}{T_{min}}))$ sample points along each line for consecutive V–points, v' and v''. Let $\mathcal{L}_i$ be the horizontal distance between the ith pair of V–points. Then, the total horizontal span of $T_{s,d}(t)$, L, is $\sum_{i=1}^{\lambda} \mathcal{L}_i$. The above results can be extended to L using the log-sum inequality, $\sum_{i=1}^{\lambda} \log \frac{\mathcal{L}_i}{T_{min}} \leq \lambda \log \frac{L}{\lambda T_{min}}$. As a result, the total number of shortest path computations by Algorithm Aprx–A is $O(\frac{\lambda}{\epsilon}(\log(\frac{T_{max}}{T_{min}}) + \log(\frac{L}{\lambda T_{min}})))$. □

2.3 Further Improvement

In this section, we present a second algorithm, Aprx–B, for approximating $T_{s,d}(t)$. Similar to Aprx–A, Aprx–B approximates the travel time function between every consecutive pair of V–points. Now, we replace in algorithm Aprx–A the sample point placement on $T_{s,d}(t)$ when the slope is less than or equal to 1. Let $\bar{v} = (\bar{t}, T_{s,d}(\bar{t}))$ be the first point on $T_{s,d}(t)$, in increasing order of time, whose slope is less than or equal to 1. We present a new approach for finding sample points of an ϵ-approximation for $T_{s,d}[\bar{v}, v'']$. Here, our general approach is to partition the range of $T_{s,d}[\bar{v}, v'']$ and find at least one sample point in each partition. We set the ith horizontal partitioning line to $y = (1 + \epsilon)^{i/2} T_{s,d}(T_{min})$, so that the maximum error between any two consecutive sample points is at most $(1 + \epsilon)T_{s,d}(T_{min})$.

We split $T_{s,d}[\bar{v}, v'']$ into two sections: the parts of the function with slopes between 0 and 1; and the remaining parts with slopes between -1 and 0. We now illustrate our approach to find sample points on the first section of $T_{s,d}[\bar{v}, v'']$. Let $\bar{p}_0 = \bar{v}$ be the initial sample point and $\bar{l}_1$ be the line that goes through $\bar{p}_0$ with slope 1. Let $\bar{p}_0$ be in the rth partition of the range. Let $\bar{t}_1$ be the time instance that corresponds to the intersection of $\bar{l}_1$ with the $(r+1)^{\text{st}}$ partitioning line, i.e., $y = (1+\epsilon)^{(r+1)/2}T_{s,d}(T_{min})$. We then obtain sample point $\bar{p}_1 = (\bar{t}_1, T_{s,d}(\bar{t}_1))$ by running a forward Dijkstra at time $\bar{t}_1$. Now, let $\bar{p}_{i-1}$ be in the kth partition of the range and $\bar{l}_i$ be the line segment that goes through $\bar{p}_{i-2}$ and $\bar{p}_{i-1}$. Let $\bar{t}_i$ be the time instance that corresponds to the intersection of $\bar{l}_i$ with the $k+1^{\text{st}}$ partitioning line, i.e., $y = (1+\epsilon)^{(k+1)/2}T_{s,d}(T_{min})$. We then obtain sample point $\bar{p}_i = (\bar{t}_i, T_{s,d}(\bar{t}_i))$ by running a forward Dijkstra at time $\bar{t}_i$. The process stops at step j when $T_{s,d}(\bar{t}_{j-1}) > T_{s,d}(\bar{t}_j)$. To find sample points on parts of $T_{s,d}[\bar{v}, v'']$ with slope between -1 and 0 we run a process similar to above, but with $p_0'' = v''$. Let l_1'' be the line that goes through p_0'' with slope -1. Figure 6 depicts steps of Aprx–B.

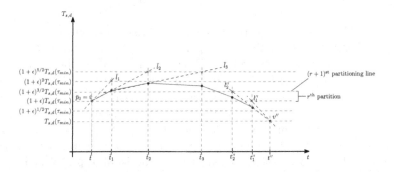

Fig. 6. Sample point placement of Algorithm Aprx–B

Theorem 3. *Algorithm Aprx–B is an ϵ-approximation of $T_{s,d}(t)$.*

Proof. Proof is omitted due to space constraint.

Theorem 4. *Approximation Algorithm Aprx–B requires* $O(\lambda(\frac{1}{\epsilon}\log(\frac{T_{max}}{T_{min}}) + \log(\frac{L}{\lambda\epsilon T_{min}})))$ *shortest path computations.*

Proof. As mentioned earlier, between any two consecutive V–points of $T_{s,d}(t)$, namely v' and v'', Algorithm Aprx–B would place at least a sample point in each horizontal partition of the function range. The total number of partitions of the function range is $O(\frac{\lambda}{\epsilon}\log(\frac{T_{max}}{T_{min}}))$. However, the algorithm may place extra sample points in each partition. We show below that, in total, the algorithm may place $O(\lambda \log(\frac{L}{\lambda \epsilon T_{min}}))$ extra sample points in existing partitions. Between v' and v'', the vertical length of the ith partition is

$$(1 + \epsilon)^{i/2} T_{s,d}(\tau_{min}) - (1 + \epsilon)^{(i-1)/2} T_{s,d}(\tau_{min})$$
$$= ((1 + \epsilon)^{1/2} - 1)(1 + \epsilon)^{(i-1)/2} T_{s,d}(\tau_{min}),$$

where τ_{min} is the minimum travel time value between v' and v''. This increases for higher values of i. Therefore, at the ith step of the algorithm, if p_i is in the same partition as p_{i-1}, then $slope(l_i) < \frac{1}{2} slope(l_{i-1})$. The algorithm terminates if $slope(l_i) < \frac{\epsilon T_{min}}{\mathcal{L}}$ since in that case, the maximum error value would be less than ϵT_{min}. So, the total number of extra sample points can be determined as follows:

$$\frac{1}{2^k} < \frac{\epsilon T_{min}}{\mathcal{L}} \quad \Rightarrow \quad \log 2^k > \log \frac{\mathcal{L}}{\epsilon T_{min}} \quad \Rightarrow \quad k > \log \frac{\mathcal{L}}{\epsilon T_{min}}$$

Let $\mathcal{L}_i$ be the horizontal distance between the ith pair of consecutive V–points and $L = \sum_{i=1}^{\lambda} \mathcal{L}_i$. The above result can be extended to L using the log-sum inequality, $\sum_{i=1}^{\lambda} \log \frac{\mathcal{L}}{\epsilon T_{min}} \leq \lambda \log \frac{L}{\lambda \epsilon T_{min}}$. $\quad\square$

3   Conclusions

We studied the approximation of the time dependent shortest path problem (TDSP), where travel times on links are piecewise linear FIFO functions. We proposed two ϵ-approximation algorithms (referred to as Aprx–A and Aprx–B, respectively) which improve upon the existing approximation algorithm for TDSP by Foschini et al. [11]. The algorithms presented here use simple data structures, are easy to implement and employ as key building block the modification of Dijkstra's algorithm studied in [4].

In [7], it has been shown that, within the same time complexity, for a given departure time at s, not only can one compute the minimum travel time to d, but it is also possible to obtain the slope of the function at that time instance. This could be used in Algorithm Aprx–B as a heuristic improvement to reduce the number of computations by using the function tangent instead of lines connecting consecutive sample points. However, it would not change the worst-case time bound.

References

1. Bertsekas, D.P.: A simple and fast label correcting algorithm for shortest paths. Networks 23(7), 703–709 (1993)
2. Brodal, G.S., Jacob, R.: Time-dependent networks as models to achieve fast exact time-table queries. Electronic Notes in Theoretical Computer Science 92, 3–15 (2004)
3. Cooke, K.L., Halsey, E.: The shortest route through a network with time-dependent internodal transit times. Journal of Mathematical Analysis and Applications 14(3), 493–498 (1966)

4. Daganzo, C.F.: Reversibility of the time-dependent shortest path problem. Transportation Research 36(7), 665–668 (2002)
5. Dean, B.C.: Shortest paths in FIFO time-dependent networks: Theory and algorithms. Technical report, MIT Department of Computer Science (2004)
6. Dehne, F., Omran, M.T., Sack, J.-R.: Shortest paths in time-dependent FIFO networks using edge load forecasts. In: Proceedings of the 2nd International Workshop on Computational Transportation Science, pp. 1–6 (2009)
7. Dehne, F., Omran, M.T., Sack, J.-R.: Shortest paths in time-dependent FIFO networks. Algorithmica 62(1-2), 416–435 (2012)
8. Dijkstra, E.W.: A note on two problems in connexion with graphs. Numerische Mathematik 1(1), 269–271 (1959)
9. Ding, B., Yu, J.X., Qin, L.: Finding time-dependent shortest paths over large graphs. In: EDBT 2008, pp. 205–216 (2008)
10. Foschini, L., Hershberger, J., Suri, S.: On the complexity of time-dependent shortest paths. In: Proceedings of the 22nd Annual ACM-SIAM Symposium on Discrete Algorithms, pp. 327–341 (2011)
11. Foschini, L., Hershberger, J., Suri, S.: On the complexity of time-dependent shortest paths. Algorithmica, 1–23 (2012)
12. Fredman, M.L., Tarjan, R.E.: Fibonacci heaps and their uses in improved network optimization algorithms. J. ACM 34(3), 596–615 (1987)
13. Kanoulas, E., Du, Y., Xia, T., Zhang, D.: Finding fastest paths on a road network with speed patterns. In: Proceedings of the 22nd International Conference on Data Engineering, p. 10 (2006)
14. Kontogiannis, S.C., Zaroliagis, C.D.: Distance oracles for time-dependent networks. CoRR, abs/1309.4973 (2013)
15. Nachtigall, K.: Time depending shortest-path problems with applications to railway networks. European Journal of Operational Research 83(1), 154–166 (1995)
16. Orda, A., Rom, R.: Shortest-path and minimum-delay algorithms in networks with time-dependent edge-length. J. ACM 37(3), 607–625 (1990)
17. Orda, A., Rom, R.: Minimum weight paths in time-dependent networks. Networks 21, 295–319 (1991)

Approximate Sorting of Data Streams
with Limited Storage

Farzad Farnoud (Hassanzadeh), Eitan Yaakobi, and Jehoshua Bruck

California Institute of Technology, Pasadena, CA, USA
{farnoud,yaakobi,bruck}@caltech.edu

Abstract. We consider the problem of approximate sorting of a data stream (in one pass) with limited internal storage where the goal is not to rearrange data but to output a permutation that reflects the ordering of the elements of the data stream as closely as possible. Our main objective is to study the relationship between the quality of the sorting and the amount of available storage. To measure quality, we use permutation distortion metrics, namely the Kendall tau and Chebyshev metrics, as well as mutual information, between the output permutation and the true ordering of data elements. We provide bounds on the performance of algorithms with limited storage and present a simple algorithm that asymptotically requires a constant factor as much storage as an optimal algorithm in terms of mutual information and average Kendall tau distortion.

1   Introduction

In many applications, such as sensor networks, finance, and web applications, data may be available as a transient stream that is not permanently accessible [1]. Often, in these applications, the large volume of data or time constraints prevent storage of the whole stream before processing. Even if data is locally stored, certain storage media only allow sequential access in a time-efficient manner.

In this paper, we study the fundamental problem of sorting a data stream when internal storage is limited. As the nature of the problem makes rearranging the data into a sorted stream impossible, by sorting we mean determining the ordering of the elements of the stream. In our model, the amount of available internal storage limits the number of elements of the data stream that can be stored internally. Furthermore, only elements in internal storage can be compared with each other. Lack of storage capable of holding the whole data stream implies that sorting must be approximate; the goal is to produce a permutation that represents the ordering of the elements of the data stream as faithfully as possible. As in [1], we consider algorithms that make only one pass over the data stream.

To evaluate performance, we measure the distortion between the output permutation and the permutation representing the true ordering of the data. There are many possible distortion measures on permutations [6], among which we consider the Kendall tau metric and the Chebyshev metric. The Kendall tau metric can be viewed as the number of mistakes made by the algorithm, while the

Z. Cai et al. (Eds.): COCOON 2014, LNCS 8591, pp. 465–476, 2014.
© Springer International Publishing Switzerland 2014

Chebyshev metric represents the maximum error in the rank of any element. Another quality measure considered in the paper is the mutual information between the true permutation and the output permutation, which reflects the amount of relevant information present in the output.

We first provide universal bounds on the performance of algorithms with limited storage, namely an upper bound on mutual information, and lower bounds on distortion, between the true permutation and the output. Further, we present a simple algorithm that is asymptotically optimal in terms of mutual information and asymptotically requires a constant factor as much storage as any algorithm with the same average Kendall tau distortion. For the Chebyshev distortion, the algorithm is also asymptotically constant-factor-optimal, provided that normalized distortion, to be defined later, is bounded away from 0.

The problem of sorting a data stream with limited storage goes back to the work of Munro and Paterson [12], where they considered sorting of, and selecting from, data stored on a read-only tape and showed that for exact sorting of a stream of length n in p passes, one requires storage of size $\Theta(n/p)$. While they allowed making multiple passes over the data and considered only exact sorting, in this work we study the quality of *approximate* sorting that can be obtained in *one* pass. Since the work of Munro and Paterson, many papers have studied problems related to *selection* in data streams, such as finding the kth highest value or quantiles, in one or many passes, e.g., [2, 8, 10, 11]. The problem of approximate sorting in one pass, however, to the best of our knowledge, has not been studied.

The rest of this paper is organized as follows. In Section 2, we present the formal problem statement and preliminaries. Section 3 includes universal bounds on the performance of algorithms with limited storage. In Section 4, an algorithm for sorting with limited storage is given and its performance is analyzed.

2   Problem Statement and Preliminaries

For a positive integer n, we let $[n] = \{1, \ldots, n\}$. The set of all permutations of $[n]$ is denoted by $\mathbb{S}_n$. For a permutation $\pi \in \mathbb{S}_n$ and distinct $i, j \in [n]$, we use $i \prec_\pi j$ (resp. $i \succ_\pi j$) to denote that i appears before j (resp. after j) in π. For example, if $\pi = (2, 3, 1)$, we have $2 \prec_\pi 3$ and $1 \succ_\pi 2$. The inverse of π is denoted by π^{-1}. The *rank* of element i in π is its position in π, that is, $\pi^{-1}(i)$.

The data stream is denoted by the sequence $s = s_1, s_2, \ldots, s_n$. We assume there is a permutation $X \in \mathbb{S}_n$ that represents the ordering of the elements of s; if $i \prec_X j$, then $s_i < s_j$ with respect to X. The goal is to approximate X as closely as possible. While X is not directly accessible in our setting, the relationship between every two elements s_i and s_j of s can be queried (or computed) if they are both present in internal storage, and the result of the query is either $i \prec_X j$ or $j \prec_X i$. Throughout the paper, our assumption is that X is chosen uniformly and at random among the permutations of $\mathbb{S}_n$ but we only consider deterministic algorithms.

The elements of s are revealed in a streaming fashion, i.e., one by one. If an element of the stream is not stored internally when revealed, it will not be possible to access it in the future. The storage limitation is that there are m cells each of which can store one element of s and thus any algorithm can only access m elements of the sequence s at any one time. The set of these m cells is termed *stream memory*. When a new element s_i of the stream s arrives, it can only be stored in the stream memory if there is an empty cell or if the contents of a cell is discarded; otherwise, s_i is ignored. To make a query regarding the relative order of s_i and s_j with respect to X, both s_i and s_j should be stored in the stream memory. We do not impose any other type of storage limitation. For example, there is no restriction on the number of integer values that an algorithm can store and access. This assumption is for simplifying the analysis and is also valid when each element of s is much larger than other types of data that an algorithm may require. To avoid trivial cases, we assume $n, m \geq 2$.

The output of the algorithms considered here is a permutation, denoted Y. To measure performance, we evaluate how "close" Y is to X. Closeness between two permutations can be quantified in a variety of ways. We use the Kendall tau and Chebyshev metrics, defined below, as well as the mutual information between X and Y.

The *Kendall tau distance* between two permutations $\pi, \sigma \in \mathbb{S}_n$ is the number of pairs of distinct elements i and j such that $i \prec_\pi j$ and $j \prec_\sigma i$, or equivalently, the number of adjacent transpositions needed to take π to σ. This distance is denoted as $d_\tau(\pi, \sigma)$. The Chebyshev distance between π and σ, denoted $d_C(\pi, \sigma)$, is defined as

$$\max_{i \in [n]} \left| \pi^{-1}(i) - \sigma^{-1}(i) \right| .$$

In other words, the Chebyshev distance is the maximum difference in the rank of any element in the two permutations.

For two functions f_n and g_n of n, the notation $f_n \sim g_n$ is used to denote $\lim_{n \to \infty} f_n / g_n = 1$. Furthermore, we use lg and ln as shorthands for $\log_2$ and $\log_e$, respectively.

3   Universal Bounds

In this section, we present bounds on the performance of any algorithm that can only store m elements of the sequence s. To derive these bounds, we use the fact that to make a query for comparing two elements s_i and s_j, both need to be present in the stream memory and so the amount of information that can be obtained via queries is limited because of the limitation on storage. As mentioned earlier, X is a random element of $\mathbb{S}_n$ but only deterministic algorithms are considered. We first present bounds on the mutual information between X and the output permutation Y and then consider distortion under the Kendall tau and Chebyshev metrics.

We use $H(X)$ and $I(X; Y)$ to refer to the entropy of X and the mutual information between X and Y, respectively. For these functions, logarithms are base 2. Note that as X is a random element of $\mathbb{S}_n$, we have $H(X) = \lg n!$.

Theorem 1. *For any algorithm with stream memory of size m, we have*

$$I(X;Y) \leq n \lg m - m \lg e + O(\lg m) .$$

Furthermore, $I(X;Y^*)/H(X) \sim \lg m/\lg n$ for $m, n \to \infty$, where Y^* is the output of an algorithm that maximizes the mutual information between X and Y.

Proof. Let Z be the set of responses provided to the comparison queries made by the algorithm. Since the algorithm can only have access to X through Z, by the data processing inequality [5], we have $I(Y;X) \leq I(Y;Z)$. Furthermore, $I(Y;Z) \leq H(Z)$.

We now show that $H(Z) \leq \lg m! + (n-m) \lg m$. The first m elements of s can be fully compared and so $m!$ cases arise from their ordering. Let Z'_0 be an integer in $[m!]$ denoting the permutation representing the ordering of the first m elements. Each of the next $n - m$ elements can at most be compared with $m - 1$ elements already present in the stream memory. These $m - 1$ elements define m intervals, into one of which the new element falls. For $i \in \{m+1, \ldots, n\}$, let Z'_i be an integer in $[m]$ denoting the interval in which the ith element of the stream falls. Given the algorithm, Z is a deterministic function of $(Z'_0, Z'_{m+1}, Z'_{m+2}, \ldots, Z'_n)$ and thus

$$H(Z) \leq H\left(Z'_0, Z'_{m+1}, Z'_{m+2}, \ldots, Z'_n\right)$$
$$\leq H(Z'_0) + \sum_{i=m+1}^{n} H(Z'_i)$$
$$\leq \lg m! + (n-m) \lg m .$$

It follows that $I(X;Y) \leq H(Z) \leq \lg m! + (n-m) \lg m$. The first theorem statement then follows from the Stirling approximation: For a positive integer k, we have $\lg k! = k \lg k - k \lg e + O(\lg k)$.

Since $I(X;Y) \leq \lg m! + (n-m) \lg m$ holds for $Y = Y^*$, we have

$$\frac{I(X;Y^*)}{H(X)} \leq \frac{n \lg m + O(m)}{n \lg n + O(n)} = \frac{\lg m}{\lg n}(1 + o(1)) , \qquad (1)$$

where we have used the fact that $\frac{m}{n \lg m} = O(1/\lg n) = o(1)$. In Section 4, we present an algorithm that produces an output Y_1 such that

$$\frac{I(X;Y_1)}{H(X)} \geq \frac{\lg m}{\lg n}(1 + o(1)), \quad m, n \to \infty .$$

Since $I(X;Y^*) \geq I(X;Y_1)$, we have

$$\frac{I(X;Y^*)}{H(X)} \geq \frac{\lg m}{\lg n}(1 + o(1)), \quad m, n \to \infty . \qquad (2)$$

The second statement of the theorem follows from (1) and (2). $\qquad \square$

In particular, if $m = n^\beta + O(1)$ for a constant β, then a β fraction of the information of X can be recovered by an algorithm with stream memory m.

Next, we use the rate-distortion theory to find lower bounds on the average Kendall tau distortion between X and Y, defined as $E[d_\tau(X,Y)]$. We use δ to denote the normalized version of this distortion, that is, $\delta = E[d_\tau(X,Y)]/n$. This choice leads to simpler expressions. Note that since d_τ can be of the order of n^2, δ can take on values in the range $[0,\infty)$.

The following theorem applies to any algorithm with stream memory m. We use W_0 and W_{-1} to respectively denote the principal and the lower branches of the Lambert W function. The Lambert W function $W(x)$ is defined as the function satisfying $W(x)e^{W(x)} = x$ [4].

Theorem 2. *Let* $\mu = \frac{m}{n}$ *and* $\delta = \frac{E[d_\tau(X,Y)]}{n}$. *Suppose* ϵ *is a positive constant. For any algorithm with stream memory* m *and* $\delta > \epsilon$, *we have*

$$\mu \geq -W_0\left(-\frac{\delta^\delta}{e(1+\delta)^{1+\delta}}\right)\left(1 - \frac{K_\epsilon \lg n}{n}\right), \tag{3}$$

where K_ϵ *is a constant that depends on* ϵ, *and*

$$\mu \geq \frac{1}{e^2\delta}\left(1 + O\left(\frac{\lg n}{n}\right) + O\left(\frac{1}{\delta}\right)\right). \tag{4}$$

Proof. Since we only consider deterministic algorithms, the number M of outputs of a given algorithm is bounded from above by $m!m^{n-m}$. This statement can be proven in a similar manner to the upper bound on $H(Z)$ in Theorem 1.

Let $A = \frac{1}{n}\lg\frac{M}{n!}$. We have $\lg M \leq n\lg m - m\lg e + O(\lg m)$ and so

$$A \leq \frac{1}{n}(n\lg m - m\lg e - n\lg n + n\lg e + O(\lg n))$$
$$= \lg\left(\mu e^{1-\mu}\right) + O\left(n^{-1}\lg n\right).$$

Hence, there exists a positive constant K_1 such that $A \leq \lg\left(\mu e^{1-\mu}\right) + K_1\lg n/n$.

The parameter M can be viewed as the size of a rate-distortion code. Hence, from [7, Theorem 5], we have the following relationship between the average distortion $E[d_\tau(X,Y)]$ and M, expressed in terms of δ and A,

$$A \geq \lg\frac{\delta^\delta}{(1+\delta)^{1+\delta}} - \frac{\lg n}{n}.$$

From this and the fact that $A \leq \lg\left(\mu e^{1-\mu}\right) + K_1\lg n/n$, we obtain

$$\mu e^{1-\mu} \geq \frac{\delta^\delta}{(1+\delta)^{1+\delta}}\left(1 - K_2\frac{\lg n}{n}\right),$$

where $K_2 = (1+K_1)\ln 2$, or equivalently,

$$-\mu e^{-\mu} \leq \frac{-\delta^\delta}{e(1+\delta)^{1+\delta}}\left(1 - K_2\frac{\lg n}{n}\right).$$

Hence

$$\mu \geq -W_0 \left(\frac{-\delta^\delta}{e(1+\delta)^{1+\delta}} \left(1 - K_2 \frac{\lg n}{n}\right)\right) . \tag{5}$$

For convenience, let $g(\delta) = \frac{\delta^\delta}{e(1+\delta)^{1+\delta}}$. By taking derivatives, one can show that the function W_0 is concave. Hence,

$$W_0\left(-g(\delta) + g(\delta)K_2 \frac{\lg n}{n}\right) \leq W_0(-g(\delta)) + W_0'(-g(\delta))g(\delta)K_2 \frac{\lg n}{n}$$

$$= W_0(-g(\delta)) + \frac{W_0(-g(\delta))}{-g(\delta)(1+W_0(-g(\delta)))}g(\delta)K_2 \frac{\lg n}{n}$$

$$= W_0(-g(\delta))\left(1 - \frac{K_2}{1+W_0(-g(\delta))}\frac{\lg n}{n}\right)$$

$$\leq W_0(-g(\delta))\left(1 - K_\epsilon \frac{\lg n}{n}\right) ,$$

where $K_\epsilon = \frac{K_2}{1+W_0(-g(\epsilon))}$. Note that for $\delta \geq 0$, the expression $-g(\delta)$ is strictly increasing and we have $-g(\delta) \in [-1/e, 0)$. Since $\epsilon > 0$, we have $-g(\epsilon) > -1/e$ and so $W_0(-g(\epsilon)) > -1$. Thus K_ϵ is well defined. Furthermore, $W_0(-g(\delta)) > W_0(-g(\epsilon))$ for $\delta > \epsilon$ and from this and the fact that $W_0(-g(\delta)) < 0$, the last step of the above derivation follows. We finally have,

$$\mu \geq -W_0\left(-\frac{\delta^\delta}{e(1+\delta)^{1+\delta}}\right)\left(1 - \frac{K_\epsilon \lg n}{n}\right) . \tag{6}$$

To prove the second statement, note that by concavity of W_0 and the facts that $W_0(0) = 0$ and $W_0'(0) = 1$, we have

$$W_0\left(-\frac{\delta^\delta}{e(1+\delta)^{1+\delta}}\right) \leq -\frac{\delta^\delta}{e(1+\delta)^{1+\delta}}$$

$$\leq -\frac{1}{e^2(1+\delta)}$$

$$= -\frac{1+O(1/\delta)}{e^2\delta} . \tag{7}$$

The second statement of the theorem then follows from (6) and (7). □

Finally, we consider the Chebyshev distortion between X and Y. The normalized Chebyshev distortion is $\chi = E\left[d_C(X,Y)\right]/n$. We only consider the case of $\chi \leq 1/2$ which is more important as it represents small distortions.

Theorem 3. *Let* $\mu = \frac{m}{n}$ *and* $\chi = \frac{E[d_C(X,Y)]}{n}$. *Suppose* $2/n \leq \chi \leq 1/2$. *For any algorithm with stream memory* m, *we have*

$$\mu \geq -W_0\left(-\frac{(e/2)^{2\chi}}{2\chi n}\right)\left(1 + O\left(n^{-1}\lg n\right)\right) .$$

Proof. Let M be defined as in the proof of Theorem 2 and let $R = \frac{1}{n} \lg M$. Since $\lg M \leq n \lg m - m \lg e + O(\lg m)$, we have $R \leq \lg m - \frac{m}{n} \lg e + O(n^{-1} \lg m)$. From [7, Theorem 16], we find $R \geq \lg \frac{1}{2\chi} + 2\chi \lg \frac{e}{2} + O(n^{-1} \lg n)$ for $\chi \leq 1/2$. Hence,

$$\lg \frac{1}{2\chi} + 2\chi \lg \frac{e}{2} \leq \lg m - \frac{m}{n} \lg e + O(n^{-1} \lg n)$$

implying that $\mu e^{-\mu} \geq \frac{(e/2)^{2\chi}}{2\chi n} \left(1 + O\left(n^{-1} \lg n\right)\right)$, or equivalently,

$$\mu \geq -W_0\left(-\frac{(e/2)^{2\chi}}{2\chi n}\left(1 + O\left(n^{-1} \lg n\right)\right)\right).$$

Since $\chi \leq 1/2$, we have $(e/2)^{2\chi} \leq e/2$ and since $\chi \geq 2/n$, we have $2\chi n \geq 4$. So $-\frac{(e/2)^{2\chi}}{2\chi n} \geq -\frac{e}{8} > -\frac{1}{e}$. Hence, $W_0\left(-\frac{(e/2)^{2\chi}}{2\chi n}\right)$ is bounded away from -1. We have

$$W_0\left(-\frac{(e/2)^{2\chi}}{2\chi n}\left(1 + O\left(n^{-1} \lg n\right)\right)\right)$$

$$\overset{(a)}{\leq} W_0\left(-\frac{(e/2)^{2\chi}}{2\chi n}\right) + W_0'\left(-\frac{(e/2)^{2\chi}}{2\chi n}\right)\frac{(e/2)^{2\chi} O\left(n^{-1} \lg n\right)}{2\chi n}$$

$$= W_0\left(-\frac{(e/2)^{2\chi}}{2\chi n}\right) + W_0\left(-\frac{(e/2)^{2\chi}}{2\chi n}\right)\frac{O\left(n^{-1} \lg n\right)}{1 + W_0\left(-\frac{(e/2)^{2\chi}}{2\chi n}\right)}$$

$$\overset{(b)}{=} W_0\left(-\frac{(e/2)^{2\chi}}{2\chi n}\right)\left(1 + O\left(n^{-1} \lg n\right)\right).$$

where (a) and (b) follow from the concavity of W_0 and the fact that $1 + W_0\left(-\frac{(e/2)^{2\chi}}{2\chi n}\right)$ is bounded away from 0, respectively. □

4   Algorithm for Limited-Storage Approximate Sorting

We present the following simple algorithm for approximately sorting a stream using storage of size m and then present results regarding its performance. Let $c_1, \ldots, c_m$ denote the m memory cells capable of storing elements of the stream. Recall that $s_i < s_j$ if i appears before j in X, i.e., $i \prec_X j$.

Algorithm 1

1. Store the first $m - 1$ elements of s in memory cells $c_1, \ldots, c_{m-1}$.
2. Find permutation y of $\{1, \ldots, m - 1\}$ such that $s_{y_1} < s_{y_2} < \ldots < s_{y_{m-1}}$.
3. Let $Y_1 \leftarrow y$.
4. For each new element $s_i, i = m, m+1, \ldots, n$, of the stream:
   (a) Store s_i in c_m.
   (b) If there exists j such that $s_{y_{j-1}} < s_i < s_{y_j}$, insert i immediately before y_j in Y_1.
   (c) If $s_i < s_j$ for all $j \in [m - 1]$, insert i immediately before y_1 in Y_1.
   (d) If $s_i > s_j$ for all $j \in [m - 1]$, append i to the end of Y_1.

In this algorithm, the first $m-1$ elements, namely, $s_1, \ldots, s_{m-1}$, are stored in the memory for the duration of the algorithm and every new element is compared with these. An element that is stored in memory, for the purpose that new elements can be compared with it, is called a *pivot*.

Example 1. Suppose $X = (5, 4, 2, 3, 7, 6, 1, 9, 8)$ and $m = 3$. After step 3 of Algorithm 1, we have $y = Y_1 = (2, 1)$. For $i = 3$, Y_1 is updated to $(\mathbf{2}, 3, \mathbf{1})$, where the indices of the pivots are shown in bold. For $i = 4$ and $i = 5$, Y_1 is respectively updated to $(4, \mathbf{2}, 3, \mathbf{1})$ and $(4, 5, \mathbf{2}, 3, \mathbf{1})$. The final output is $Y_1 = (4, 5, \mathbf{2}, 3, 6, 7, \mathbf{1}, 8, 9)$. For the Kendall tau and Chebyshev distortions, we have $d_\tau(X, Y_1) = 3$ and $d_C(X, Y_1) = 1$.

In Algorithm 1, the index set of pivots is $\{1, 2, \ldots, m-1\}$ and they are in correct order in Y_1. However, indices of elements between the pivots, and between the pivots and the boundaries, are sorted in the natural increasing order which may differ from their order in X, e.g., the subsequence $3, 6, 7$ of Y_1 in the preceding example. Let $r_1, \ldots, r_{m-1}$ be an increasing sequence that denotes the positions of the indices of the pivots in X (or equivalently in Y_1). Furthermore, let $r_0 = 0$ and $r_m = n+1$. In Example 1, we have $r_0 = 0$, $r_1 = 3$, $r_2 = 7$, and $r_3 = 10$. For $j \in [m]$, the elements of Y_1 between positions r_{j-1} and r_j can have any order in X. Additionally, all possibilities are equally probable. Given Y_1, the number of possible cases for X is given by $\prod_{j=1}^{m}(r_j - r_{j-1} - 1)!$. We will use this fact to compute the conditional entropy of X given Y_1 in the next theorem.

Theorem 4. *Algorithm 1 is asymptotically optimal for $m, n \to \infty$, with respect to mutual information.*

Proof. Since $I(X; Y_1) = H(X) - H(X|Y_1)$ and $H(X) = \lg n!$, to find $I(X; Y_1)$, it suffices to find $H(X|Y_1)$. We have

$$H(X|Y_1) = \sum_{z \in \mathbb{S}_n} P(Y_1 = z) H(X|Y_1 = z)$$

$$= \binom{n}{m-1}^{-1} \sum_{r_0 < \cdots < r_m} \sum_{j=1}^{m} \lg(r_j - r_{j-1} - 1)! .$$

For given values of $r_0, \ldots, r_m$, the set $[n]$ is divided into m blocks with lengths $r_j - r_{j-1} - 1$. To compute the above sum, we count how many times a block of size k occurs for all possible values of $r_0, \ldots, r_m$. The number of times a block of length k appears starting at position 1 equals $\binom{n-k-1}{m-2}$ since we have $r_1 = k+1$ but must choose the values of $r_2, \ldots, r_{m-1}$ among the $n - (k+1)$ possibilities. The number of times a block of size k ends at position n is the same. A similar argument shows that the number of times a block of length k starts at position i and ends at position $i+k-1$, for each $i \in \{2, \ldots, n-k\}$, is $\binom{n-k-2}{m-3}$. Thus, the total number of blocks of size k is $2\binom{n-k-1}{m-2} + (n-k-1)\binom{n-k-2}{m-3} = m\binom{n-k-1}{m-2}$. Hence,

$$H(X|Y_1) = \binom{n}{m-1}^{-1} m \sum_{k=2}^{n-m+1} \binom{n-k-1}{m-2} \lg k! . \qquad (8)$$

It can be shown that

$$\sum_{k=1}^{n-m+1} \binom{n-k-1}{m-2} k \lg k \le \binom{n}{m} \left(\lg \frac{n}{m} + O(1) \right) . \tag{9}$$

From (8), (9), and the fact that $\lg k! < k \lg k$, we obtain

$$H\left(X|Y_1\right) \le (n-m+1) \left(\lg \frac{n}{m} + O(1) \right) = n \lg \frac{n}{m} + O(n)$$

and so $I(X;Y_1) \ge \lg n! - n \lg \frac{n}{m} + O(n) = n \lg m + O(n)$. Thus $\frac{I(X;Y_1)}{H(X)} \ge \frac{\lg m}{\lg n} (1 + o(1))$ for $m, n \to \infty$. Recall from the proof of Theorem 1 that $\frac{I(X;Y)}{H(X)} \le \frac{\lg m}{\lg n} (1 + o(1))$ for the output Y of any algorithm. Therefore,

$$\frac{I\left(X;Y_1\right)}{H\left(X\right)} = \frac{\lg m}{\lg n} (1 + o(1)), \quad m, n \to \infty ,$$

which is optimal.   □

Next, we discuss the average Kendall tau distortion of Algorithm 1.

Theorem 5. *Suppose Algorithm 1 has stream memory $m = \mu_1 n$ and produces an output with average Kendall tau distortion δn. We have*

$$\mu_1 \le \left(1 + \delta - \sqrt{\delta(\delta+2)}\right) (1 + O(1/n)) .$$

Furthermore, Algorithm 1 asymptotically requires at most a constant factor as much storage as an optimal algorithm with the same average Kendall tau distortion.

Proof. For a random permutation of length k, the average Kendall tau distance from the identity is $\frac{1}{2}\binom{k}{2}$. Hence, from the discussion preceding Theorem 4, we obtain

$$E\left[d_\tau(X,Y_1)\right] = \binom{n}{m-1}^{-1} \sum_{r_0 < \cdots < r_m} \sum_{j=1}^{m} \frac{1}{2} \binom{r_j - r_{j-1} - 1}{2}$$

$$= \frac{1}{2} \binom{n}{m-1}^{-1} m \sum_{k=2}^{n-m+1} \binom{k}{2} \binom{n-k-1}{m-2}$$

$$= \frac{1}{m+1} \binom{n-m+1}{2} , \tag{10}$$

where for the second equality, we have used an argument similar to that of the proof of Theorem 4. We have $\delta n = E\left[d_\tau(X,Y_1)\right] = \frac{1}{\mu_1 n+1} \binom{n-\mu_1 n+1}{2}$ and thus $\delta n \le \frac{(n-\mu_1 n+1)^2}{2\mu_1 n}$. It follows that

$$\mu_1 \le 1 + \delta + \frac{1}{n} - \sqrt{\delta(\delta+2+2/n)}$$

$$= \left(1 + \delta - \sqrt{\delta(\delta+2)}\right) (1 + O(1/n)) .$$

In particular, for large δ, we have $\mu_1 \leq 1/(2\delta)\,(1 + O(1/\delta))\,(1 + O(1/n))$.

Let $\mu^* n$ be the smallest amount of stream memory of any algorithm with average Kendall tau distortion δn. From (4), we have

$$\frac{\mu_1}{\mu^*} \leq \frac{1/(2\delta)}{1/\left(e^2\delta\right)}\,(1 + O(1/\delta))\,(1 + O(\lg n/n)) \ .$$

Thus there is a constant c such that for $\delta, n \geq c$, μ_1/μ^* is bounded.

On the other hand, if $\delta < c$, from (3) and using the fact that μ^* is a decreasing function of δ, we have $\mu^* \geq -W_0\left(-c^c e^{-1}(1+c)^{-1-c}\right)(1 + O(\lg n/n))$. Furthermore $\mu_1 \leq 1$. Hence, if $\delta < c$, then μ_1/μ^* is bounded. □

Remark 1. There is an alternative way to show that $E\,[d_\tau(X, Y_1)] = \frac{1}{m+1}\binom{n-m+1}{2}$. Without loss of generality, assume $1 \prec_X \cdots \prec_X m - 1$. Consider distinct $i, j \in \{m, \ldots, n\}$, with $i < j$. The pair i, j will have incorrect order in Y_1, if and only if $j \prec_X i$ and there is no $p \in \{1, \ldots, m - 1\}$ such that $j \prec_X p \prec_X i$ (in other words, there is no pivot s_p such that $s_j < s_p < s_i$). Since X is random, it is straightforward to see that the probability of this event is $1/(m + 1)$. There are $\binom{n-m+1}{2}$ possible choices for the pair i, j. The desired result then follows by the linearity of expectation.

The next theorem concerns the average Chebyshev distortion of Algorithm 1.

Theorem 6. *Suppose Algorithm 1 has memory m and produces an output with average Chebyshev distortion χn. Furthermore, suppose that $\chi \leq 1/2$ and $m \geq 2$. We have*

$$m \leq -\frac{1}{\chi}W_{-1}\left(-\frac{\chi}{e}\right) \ .$$

Additionally, if χ is bounded away from zero, Algorithm 1 asymptotically requires at most a constant factor as much memory as an optimal algorithm with the same average Chebyshev distortion.

Proof. Consider an element i in Y_1 that is between positions r_{j-1} and r_j. We know that the position of this element in X is also between r_{j-1} and r_j. Thus, $\left|X^{-1}(i) - Y_1^{-1}(i)\right| \leq r_j - r_{j-1} - 1$ and so

$$d_C\,(X, Y_1) \leq \max_j\,(r_j - r_{j-1} - 1) \ .$$

Suppose a stick of length n is randomly broken at $m - 1$ points. Let the length of the longest piece among the m pieces be denoted by S. From [9], we have $E[S] = nE[S']/m$, where S' is the largest random variable among m iid exponential random variables with mean 1. We have $E[S'] = \sum_{i=1}^{m} 1/i \leq \ln m + \gamma_e + \frac{1}{2m}$, where the inequality follows from [3] and γ_e is Euler's constant. Since the positions of the pivots in Algorithm 1 are random, with a coupling argument one can show that the expected length of the longest segment is not more than $E[S]$. That is, $E\,[\max_j\,(r_j - r_{j-1} - 1)] \leq E\,[S]$. Hence,

$$E[d_C(X, Y_1)] = \chi n \leq \frac{n}{m}\left(\ln m + \gamma_e + \frac{1}{2m}\right) \ .$$

Since $m \geq 2$ we have $\gamma_e + \frac{1}{2m} \leq 1$, and thus $\chi \leq \frac{\ln(me)}{m}$. This in turn implies that $-m\chi e^{-m\chi} \leq -\frac{\chi}{e}$, from which it follows that $-(1/\chi)W_0\left(-\chi/e\right) \leq m \leq -(1/\chi)W_{-1}\left(-\chi/e\right)$ and so we have the first statement in the theorem. Note that for $\chi \leq 1$, we have $-(1/\chi)W_0\left(-\chi/e\right) \leq 1$ and hence the inequality $-(1/\chi)W_0\left(-\chi/e\right) \leq m$ does not give us any useful information.

Let μ_1 denote m/n for Algorithm 1 and μ^* denote the smallest amount of storage of any algorithm with Chebyshev distortion χn. From Theorem 3,

$$\frac{\mu_1}{\mu^*} \leq \frac{-\frac{1}{\chi n}W_{-1}\left(-\frac{\chi}{e}\right)}{-W_0\left(-\frac{(e/2)^{2\chi}}{2\chi n}\right)}\left(1 + O\left(\frac{\lg n}{n}\right)\right)$$

$$\sim \frac{-\frac{1}{\chi n}W_{-1}\left(-\frac{\chi}{e}\right)}{\frac{(e/2)^{2\chi}}{2\chi n}}$$

$$\sim \frac{-2W_{-1}\left(-\frac{\chi}{e}\right)}{(e/2)^{2\chi}}. \tag{11}$$

Suppose χ is bounded away from 0. It follows that $-\chi/e$ is also bounded away from 0. This in turn implies that $-W_{-1}(-\chi/e)$ is bounded and so is the right side of (11). This completes the proof of the theorem. $\qquad\square$

Acknowledgements. The authors would like to thank Ryan Gabrys and Yue Li for useful discussions and comments.

References

1. Babcock, B., Babu, S., Datar, M., Motwani, R., Widom, J.: Models and issues in data stream systems. In: Proc. 21st ACM Symp. Principles of Database Systems (PODS), New York, NY, USA (2002)
2. Chakrabarti, A., Jayram, T.S., Pătraşcu, M.: Tight lower bounds for selection in randomly ordered streams. In: ACM-SIAM Symp. Discrete Algorithms (SODA), pp. 720–729. Society for Industrial and Applied Mathematics, Philadelphia (2008), http://dl.acm.org/citation.cfm?id=1347082.1347161
3. Chen, C.P., Qi, F.: The best lower and upper bounds of harmonic sequence. Global Journal of Applied Mathematics and Mathematical Sciences 1(1), 41–49 (2008)
4. Corless, R.M., Gonnet, G.H., Hare, D.E.G., Jeffrey, D.J., Knuth, D.E.: On the Lambert W function. Advances in Computational Mathematics 5(1), 329–359 (1996), http://dx.doi.org/10.1007/BF02124750
5. Cover, T.M., Thomas, J.A.: Elements of information theory. John Wiley & Sons (2006)
6. Diaconis, P.: Group Representations in Probability and Statistics, vol. 11. Institute of Mathematical Statistics (1988)
7. Farnoud, F., Schwartz, M., Bruck, J.: Rate-distortion for ranking with incomplete information. arXiv preprint (2014), http://arxiv.org/abs/1401.3093
8. Greenwald, M., Khanna, S.: Space-efficient online computation of quantile summaries. In: Proc. ACM SIGMOD Int. Conf. Management of Data, pp. 58–66. ACM, New York (2001), http://doi.acm.org/10.1145/375663.375670

476   F. Farnoud, E. Yaakobi, and J. Bruck

9. Holst, L.: On the lengths of the pieces of a stick broken at random. Journal of Applied Probability 17(3), 623–634 (1980)
10. Manku, G.S., Rajagopalan, S., Lindsay, B.G.: Approximate medians and other quantiles in one pass and with limited memory. In: Proc. ACM SIGMOD Int. Conf. Management of Data, pp. 426–435. ACM, New York (1998), http://doi.acm.org/10.1145/276304.276342
11. McGregor, A., Valiant, P.: The shifting sands algorithm. In: ACM-SIAM Symp. Discrete Algorithms (SODA), pp. 453–458. SIAM (2012), http://dl.acm.org/citation.cfm?id=2095116.2095155
12. Munro, J., Paterson, M.: Selection and sorting with limited storage. Theoretical Computer Science 12(3), 315–323 (1980), http://www.sciencedirect.com/science/article/pii/0304397580900614

Simpler Algorithms for Testing Two-Page Book Embedding of Partitioned Graphs*

Seok-Hee Hong[1] and Hiroshi Nagamochi[2]

[1] University of Sydney, Australia
seokhee.hong@sydney.edu.au
[2] Kyoto University, Japan
nag@amp.i.kyoto-u.ac.jp

Abstract. In this paper, we study the problem of testing whether a given graph admits a 2-page book embedding with a fixed edge partition. We first show that finding a 2-page book embedding of a given graph can be reduced to the planarity testing of a graph, which yields a simple linear-time algorithm for solving the problem. We then characterize the graphs that do not admit 2-page book embeddings via forbidden subgraphs, and give a linear-time algorithm for detecting the forbidden subgraph of a given graph.

1 Introduction

For an integer $k \geq 1$, a k-page book embedding (or a k-stack layout) of a graph is to place the vertices linearly on a spine (a line segment) and the edges on k pages (k half planes sharing the spine) so that each edge is embedded in one of the pages without edge crossings. It was shown that a planar graph has a 2-page book embedding if and only if it is sub-Hamiltonian [3]. A planar graph is sub-Hamiltonian if and only if it is Hamiltonian or can be made Hamiltonian by inserting additional edges without violating planarity. Since the problem of testing sub-Hamiltonicity is NP-complete [8], the problem of testing whether a given planar graph G has a 2-page book embedding is NP-complete.

The 2-page book embedding problem contains two combinatorial aspects. One is how to partition an edge set in two edge subsets, each corresponds to one of the two sides along the spine. The other is how to decide an ordering of the vertices on the spine. Note that if an ordering π of all the vertices along the spine is fixed, then we can test whether a given graph admits a 2-page book embedding with π in linear time; the problem can be converted into a planarity testing problem by adding edges between every two consecutive vertices in π (where the last vertex is connected by the first one). However, it was not known whether the problem remains NP-complete or can be solved in polynomial time if a partition of the edge set is prescribed.

In this paper, we consider the problem of testing whether a given graph admits a 2-page book embedding for a fixed edge partition. Based on structural properties of biconnected planar graphs, we show that the problem can be solved in linear time.

---

* This is an extended abstract. For the full version of this paper with omitted proofs, see [7].

Z. Cai et al. (Eds.): COCOON 2014, LNCS 8591, pp. 477–488, 2014.

In a preliminary version of this paper [6], we characterized 2-page book embeddings as "splitter-free" and "disjunctive" plane embeddings (see Section 4 for the definitions), and show that such an embedding, if any, can be constructed in linear time by designing the following three procedures:

1. procedure for detecting *rigid splitters*, a special type of splitters;
2. procedure for computing *splitter-free* plane embeddings;
3. procedure for computing *disjunctive* plane embeddings.

Recently, Angelini et al. [1] implemented the algorithm in [6] with a simplified procedure for computing disjunctive plane embeddings, reducing the hidden constant factor in the time bound.

In this paper, we present simpler algorithms for testing 2-page book embedding with edge partition. More specifically, the new contribution of this paper is the following:

1. We first show that the given instance of 2-page book embedding with edge partition can be converted into another instance with a special structure, called a *canonical instance* (see Theorem 1 in Section 3).
2. We then prove that finding a 2-page book embedding of a given partitioned graph can be reduced to the *planarity testing* of a modified graph, which yields a simple linear-time algorithm for solving the problem without using any of the three procedures in [6] (see Theorem 3 in Section 5).
3. We also characterize the graphs that do not admit 2-page book embeddings via forbidden subgraphs, and present a linear-time algorithm for either detecting a *forbidden subgraph* or constructing a 2-page book embedding of a given graph. This algorithm is obtained by significantly simplifying the first procedure for detecting rigid splitters in [6] utilizing the restricted structure of canonical instances (see Theorem 4 in Section 6).

The 2-page book embedding problem studied in this paper also has two important applications in Graph Drawing: *simultaneous embedding* and *clustered graph planarity*. Angelini et al. [2] showed the relationship between the 2-page book embedding problem with edge partition and the simultaneous embedding with fixed edges (SEFE) problem. They used the algorithm in [6] to solve some specific instance of the SEFE problem.

Previously in [6], we also showed the reduction between the 2-page book embedding problem with edge partitions and the clustered planarity (c-planarity) testing problem with two clusters. The same c-planarity problem can be modeled as the HH-drawing problem, and a testing algorithm was reported in [4]. Note that our new algorithms presented in this paper are simpler than those presented in [1,4,6].

2 Preliminaries

Let $G = (V, E)$ be a graph. The set of edges incident to a vertex $v \in V$ is denoted by $E(v; G)$. A path with endvertices u and v is called a u, v-*path*. The degree of a vertex v in G is denoted by $\deg(v; G)$. For a subset $X \subseteq E$ (resp., $X \subseteq V$), $G - X$ denotes the graph obtained from G by removing the edges in X (resp., the vertices in X together with the edges in $\cup_{v \in X} E(v; G)$). *Subdividing* an edge $e = (u, v)$ is to replace the edge with a u, v-path $u, w_1, w_2, \ldots, w_k, v$ for some $k \geq 1$. A graph H is a subdivision of G if H is obtained by subdividing some edges in G.

A *block* (biconnected component) of a graph is a maximal biconnected subgraph (which possibly consists of a single vertex or a single edge). A graph each of whose blocks is a simple cycle or a single edge is called a *cactus*, i.e., a graph in which any two distinct cycles share at most one vertex. Note that a graph is a cactus if and only if no two vertices are joined by three vertex-disjoint paths.

A planar graph $G = (V, E)$ with a fixed plane embedding F of G is called a *plane* graph. The set of vertices, set of edges and set of facial cycles of a plane graph H are denoted by $V(H)$, $E(H)$ and $F(H)$, respectively. We say that a cycle Q in a plane embedding of a graph *separates* a vertex/edge a_1 and a vertex/edge a_2 (which are not elements of Q) if a_1 and a_2 are respectively contained in the two connected regions R_1 and R_2 of the plane divided by Q.

A planar graph is called *outerplanar* if it admits an *outerplane* embedding, such that all the vertices appear along the outer boundary. We observe that each block B of a simple outerplanar graph is a single edge or a cycle Q of length at least 3 possibly with some chords. Note that such a cycle Q for the block B is uniquely determined, and we call Q the *frame* of B (we let $Q = B$ if B is a single edge). We observe the next result.

Lemma 1. *Let B_i, $i = 1, 2, \ldots, p$ be the blocks of a simple outerplanar graph G, and Q_i be the boundary of B_i. Then a plane embedding Γ_G of G is an outerplane embedding of G if and only if each frame Q_i appears on the outer boundary of Γ_G.*

In this paper, we denote an instance of two-page book embedding problem by a graph $G = (V, E_1 \cup E_2)$ with a partition of its edge set E into E_1 and E_2 (i.e., $E_1 \cup E_2 = E$ and $E_1 \cap E_2 = \emptyset$), where two vertices may be joined by two edges $e \in E_1$ and $e' \in E_2$. We call the edges in E_1 (resp., E_2) *red edges* (resp., *blue edges*). A subgraph H of G is called *red* (resp., *blue*) if $E(H)$ consists of only red (resp., blue) edges. A vertex to which only red (resp., blue) edges are incident is called an *r-vertex* (resp., *b-vertex*). A vertex to which both red and blue edges are incident is called a *br-vertex*.

A *2-page book embedding* (2PB-embedding, for short) π of a graph $G = (V, E_1 \cup E_2)$ is a linear ordering of the vertices such that all vertices are placed in this order on a spine and all red edges are drawn above the spine and all blue edges are drawn below the spine without any edge crossings.

In a 2PB-embedding π of a graph $G = (V, E_1 \cup E_2)$, we can join the first and last vertices on the spine with a new curve so that the spine together with the curve forms a simple closed curve which encloses all red edges but no blue edges. Thus, a 2PB-embedding π can be regarded as a plane embedding Γ of G in which a simple closed curve λ visits each vertex exactly once without intersecting any edge and encloses all red edges but no blue edges. We call λ a *separating curve* of Γ.

Note that the first and last vertices appear on the outer facial cycle in the plane embedding Γ. However by choosing a new outer face, any vertex v can appear along the outer facial cycle. This does not change the combinatorial embedding, and thereby the vertex v can appear as the first vertex on the spine in the 2PB-embedding obtained from the resulting plane embedding Γ'.

3 Canonical Instances

In this section, we give a linear-time algorithm for converting a given 2PB embedding instance into another instance with a special structure, called a "canonical" instance, without changing the 2PB-embeddability.

Clearly an instance $G = (V, E_1 \cup E_2)$ admits a 2PB-embedding only when G is planar, and each E_i induces an outerplanar graph. If $E_1 = \emptyset$ or $E_2 = \emptyset$, then G has a 2PB-embedding if and only if G is outerplanar. In what follows, we assume that $E_1 \neq \emptyset \neq E_2$. Also we assume that G is not a simple cycle (otherwise it always admits a 2PB-embedding). An instance $G = (V, E_1 \cup E_2)$ is *canonical* if:
(i) G is a simple, biconnected planar graph, but G is not a simple cycle;
(ii) Each E_i induces a cactus $(V, E_i \neq \emptyset)$; and
(iii) Each br-vertex of G is of degree 2.

Lemma 2. *A planar graph $G = (V, E_1 \cup E_2)$ admits a 2PB-embedding π with partition E_1 and E_2 if and only if each block H of G has a 2PB-embedding π_H for the partition $E(H) \cap E_1$ and $E(H) \cap E_2$ of $E(H)$.*

Now we can assume that an instance $G = (V, E_1 \cup E_2)$ is biconnected. We next explain why each induced graph (V, E_i) of G can be assumed to be a cactus. Let Γ_i be an outerplane embedding of (V, E_i) (where all vertices in V appear along the outer boundary). In Γ_i, each block of (V, E_i) is either a single edge or a cycle Q with some chords, where each chord is drawn within the cycle. By Lemma 1, in any such embedding of (V, E_i), chords are drawn inside the corresponding cycle and no other edges/vertices are contained in the interior of the cycle.

On the other hand, in any 2PB-embedding of G, such a cycle (frame) C with $E(Q) \subseteq E_i$ is drawn without surrounding any edge in E_j $(j \neq i)$ in its interior. This means that we can remove all the chords in the embedding Γ_i $(i = 1, 2)$ in the sense that we can put them back into any 2PB-embedding of the resulting instance without creating any edge crossings. Hence we can assume that each (V, E_i) is a simple cactus.

We transform a biconnected graph $G = (V, E = E_1 \cup E_2)$ into a canonical instance of 2PB-embedding problem as follows.

Definition 1: Procedure of Transformation

Step 1. For each outerplanar graph (V, E_i), $i = 1, 2$, remove the chords of each block B to obtain a cactus (V, E_i') in the resulting instance $G' = (V, E_1' \cup E_2')$ (see Fig. 1(a) and (b)).

Step 2. Let V_{br} be the set of br-vertices $v \in V$ of degree ≥ 3 in G'.
Replace each $v \in V_{br}$ of degree ≥ 3 with three vertices v_1, w_v and v_2 joined by a new red edge (v_1, w_v) and a new blue edge (v_2, w_v) (see Fig. 1(b) and (c)). Let $G'' = (V'', E_1'' \cup E_2'')$ be the resulting graph.

Theorem 1. *Let G'' be the canonical instance obtained from a biconnected instance $G = (V, E_1 \cup E_2)$ by Definition 1. Then G admits a 2PB-embedding if and only if G'' admits a 2PB-embedding. Furthermore a 2PB-embedding of G'' can be converted into a 2PB-embedding of G in linear time.*

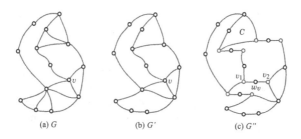

Fig. 1. (a) a biconnected graph $G = (V, E_1 \cup E_2)$; (b) $G' = (V, E_1' \cup E_2')$ with cactus (V, E_i'), $i = 1, 2$; and (c) a canonical instance $G'' = (V'', E_1'' \cup E_2'')$ with cactus (V, E_i''), $i = 1, 2$

A canonical instance G is simple, since G is not a cycle and it has no pair of blue and red multiple edges. Also, if there are a red u, v-path and a blue u, v-path for some vertices $u, v \in V$, then u and v are br-vertices of degree 2. In the following two sections, a given instance is assumed to be canonical.

4   Disjunctive and Splitter-free Plane Embeddings

In a plane embedding Γ of G, a red (resp., blue) cycle Q of G is called a *splitter* if each of the two regions obtained by the cycle Q contains a vertex $v \in V - V(Q)$ or a blue (resp., red) edge. In Γ, a vertex v is called *disjunctive* if for each $i = 1, 2$, all the edges in $E_i(v; G)$ appear consecutively around v. We call Γ *disjunctive* if all vertices in V are disjunctive.

Theorem 2. *Let $G = (V, E_1 \cup E_2)$ be a biconnected planar graph (not necessarily canonical). Then G admits a 2PB-embedding π if and only if G admits a disjunctive and splitter-free plane embedding Γ. Moreover, a 2PB-embedding π of G can be obtained from a disjunctive and splitter-free plane embedding Γ of G in linear time.*

Note that any plane embedding of a canonical instance is disjunctive, since red and blue edges meet at a br-vertex of degree 2.

5   Reduction to Planarity Testing

Lemma 3. *Let $G = (V, E_1 \cup E_2)$ be a canonical instance, and Q be a red (or blue) cycle. In a plane embedding Γ of a canonical instance G, Q is a splitter if and only if Q is not a facial cycle of Γ.*

We call a plane embedding Γ of G *proper* if every red/blue cycle of G appears as a facial cycle of Γ. Detecting a proper embedding Γ of G can be reduced to the standard planarity testing of an augmented graph as follows. For each cycle C in the cactus (V, E_i), $i = 1, 2$, we subdivide each edge e in C with a new vertex w_e, create a new vertex v_C, and add new edges (v_C, w_e), $e \in E(C)$.

Fig. 2(a) shows the graph $\widetilde{G}$ obtained from the canonical instance in Fig. 1(c). Let $\widetilde{G} = (\widetilde{V}, \widetilde{E}_1 \cup \widetilde{E}_2)$ denote the resulting graph, where $\widetilde{E}_i$ is the set of edges obtained by subdividing an edge in E_i or introduced to augment a cycle in the cactus (V, E_i).

Theorem 3. *A canonical instance $G = (V, E_1 \cup E_2)$ admits a proper embedding if and only if $\widetilde{G}$ is planar. A proper embedding of G can be obtained from a plane embedding of $\widetilde{G}$ in linear time.*

Proof. **Only-if part:** Let Γ_G be a proper embedding of G. Then each cycle block C of the cactus (V, E_i) surrounds no edge in E_j $(j \neq i)$. Hence, we can draw the newly added vertices v_C, and w_e, $e \in E(C)$ and edges between them inside the empty region of C without creating edge crossings. Thus the resulting embedding is a plane embedding of $\widetilde{G}$.

If part: Since G is biconnected, $\widetilde{G}$ is also biconnected. Let $\Gamma_{\widetilde{G}}$ be a plane embedding of $\widetilde{G}$, where without loss of generality the outer facial cycle f^o contains a br-vertex z of G. Note that z is not in any cycle of a cactus (V, E_i). Let $\Gamma_{\widetilde{E}_i}$ denote the plane embedding induced from $\Gamma_{\widetilde{G}}$ by the edges in $\Gamma_{\widetilde{E}_i}$.

We first show that for each block C in (V, E) augmented with vertex v_C, the vertex v_C is surrounded by the cycle C (i.e., C separates v_C from f^o) in $\Gamma_{\widetilde{E}_i}$. If for some C, both v_C and f^o are outside C, then by the way of augmentation, only one vertex $u \in V(C)$ can be adjacent to vertices outside C (see Fig. 2(b)), and u would be a cut-vertex separating v_C and z in $\widetilde{G}$, contradicting the biconnectivity of $\widetilde{G}$. Now v_C of each cycle is located inside the subdivided cycle C in $\Gamma_{\widetilde{E}_i}$.

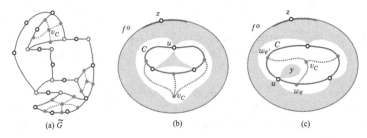

(a) $\widetilde{G}$ (b)   (c)

Fig. 2. Illustration for a plane embedding of the augmented graph $\widetilde{G}$; (a) Graph $\widetilde{G}$ obtained from the graph in Fig. 1(c) by augmenting each cycle in red and blue cacti with stars; (b) f^o and v_C are outside the subdivided cycle C; (c) other vertex y than v_C is inside the subdivided cycle C

Next we show that no other vertex than v_C is located inside the subdivided cycle C in $\Gamma_{\widetilde{G}}$. Assume that for some C, a vertex y $(\neq v_C)$ is inside the subdivided cycle C in $\Gamma_{\widetilde{G}}$ (see Fig. 2(c)). Since G is connected, y is connected to a vertex u by a path in the subdivided C. Let u be the one closest to y among such u. Then in G, $u \in V(C)$, and u is adjacent to two edges $e, e' \in E(C)$. Now in the plane embedding $\Gamma_{\widetilde{G}}$, the cycle $(u, w_e, v_C, w_{e'})$ surrounds the set Y of all vertices reachable from y without passing through u. This, however, implies that u is a cut-vertex separating y and z in $\widetilde{G}$, a contradiction. Hence, each subdivided C encloses no other vertex than v_C in $\Gamma_{\widetilde{G}}$.

Let Γ be the plane embedding of G induced by $\Gamma_{\widetilde{G}}$; i.e., remove the augmented vertices v_C and ignore the introduced vertices w_e for all cycles C in cacti (V, E_1) and (V, E_2). Clearly, each C of a cactus (V, E_i) encloses no edges/vertices in Γ. Since each red (resp., blue) cycle Q in G is a (facial) cycle of (V, E_1) (resp., (V, E_2)), Γ is proper. $\qquad\square$

Since testing planarity and constructing a plane embedding, if any, can be done in linear time, we can find a 2PB-embedding of a given instance, if any, in linear time. When $\Gamma_{\widetilde{G}}$ is not planar, it contains a subdivision of K_5 or $K_{3,3}$. However, such a subgraph is not a direct evidence of a given infeasible instance due to the augmentation. In the next section, we give an algorithm that detects a forbidden subgraph of a given instance.

6   Forbidden Subgraphs in Two-Page Book Embeddings

A graph H is called *pseudo-triconnected* if it is a subdivision of a triconnected graph G. Observe that a graph H is pseudo-triconnected if and only if H has at least three vertices of degree ≥ 3 and every two vertices u and v of degree ≥ 3 in H are connected by three internally disjoint paths. We call a pseudo-triconnected subgraph S of G a *forbidden subgraph* if a cycle Q with edges of E_i in a plane embedding of S separates two edges $e_1, e_2 \in E_j$ $(j \neq i)$. By the uniqueness of plane embedding of S, such a subgraph S (and hence G) cannot admit a 2PB-embedding. In this section, we show that the converse is true establishing the following result.

Theorem 4. *Let $G = (V, E_1 \cup E_2)$ be a planar graph with a partition E_1 and E_2 such that each E_i induces a simple and outerplanar subgraph (V, E_i). Then G admits a 2PB-embedding if and only if there is no forbidden subgraph in G. Furthermore, either a 2PB-embedding or a forbidden subgraph of G can be found in linear time.*

The forbidden subgraph characterization in Theorem 4 is more interesting than merely testing feasibility, since our algorithm can detect some structures of a graph that never (or always) admit a 2PB-embedding. For example, every graph instance such that each of E_1 and E_2 induces a forest always admits a 2PB-embedding because it cannot have any forbidden subgraph in Theorem 4. Hence Theorem 4 implies the next.

Corollary 1. *Let $G = (V, E_1 \cup E_2)$ be a graph with a partition E_1 and E_2 of $E(G)$ such that each E_i induces a forest (V, E_i). Then G admits a 2PB-embedding if and only if G is planar. Furthermore, a 2PB-embedding of G if any can be found in linear time.*

To prove Theorem 4, we first design a linear-time algorithm for detecting a forbidden subgraph of a given canonical instance. It is not difficult see that even if a canonical instance G is obtained from the original instance G^* by Definition 1, a forbidden subgraph S of G gives a forbidden subgraph of G^*.

We then prove that a canonical instance with no forbidden subgraphs admits a proper embedding (and hence a 2PB-embedding) by designing a linear-time algorithm for constructing a proper embedding for such an instance in Section 7.

6.1   SPR Tree Decomposition

To consider all the possible plane embeddings of a biconnected graph, we use the SPR tree, a simplified version of the SPQR tree [5]. The SPR tree $\mathcal{T}$ of a biconnected graph G represents the decomposition of G into triconnected components. Each node ν in $\mathcal{T}$

is associated with a graph $\sigma(\nu) = (V_\nu, E_\nu)$ called the *skeleton* of ν, which corresponds to a triconnected component of G.

Each skeleton consists of *real edges* (i.e., edges of G) and *virtual edges* (i.e., edges newly introduced during the decomposition process). Each skeleton $\sigma(\nu)$ provides an abstract structure of the entire graph G; for each virtual edge $e = (u, v) \in E_\nu$, G has an induced subgraph G_e which shares $\{u, v\}$ as a cut-pair with the complementary part $G - (V(G_e) - \{u, v\})$. Two nodes ν and μ are connected by an edge in $\mathcal{T}$, if their skeletons share the same virtual edge $e = (u, v)$ (i.e., a cut-pair (u, v)).

There are 3 types of nodes in the SPR tree:

1. S-nodes: $\sigma(\nu)$ is a simple cycle with at least 3 vertices.
2. P-nodes: $\sigma(\nu)$ consists of two vertices connected by at least 3 edges.
3. R-nodes: $\sigma(\nu)$ is a simple triconnected graph with at least 4 vertices.

In this paper, a given graph G is assumed to be canonical with no multiple edges. Hence, for each skeleton $\sigma(\nu)$, there is a subgraph S of G that is a subdivision of $\sigma(\nu)$, which is obtained by replacing each virtual edge (u, v) with a u, v-path of G.

For each edge of the skeleton of a node ν of $\mathcal{T}$ of a canonical instance $G = (V, E_1 \cup E_2)$, the graph G_e is called an *r-graph* (resp., *b-graph*) if it has a red (resp., blue) u, v-path. Note that G_e cannot have both a red u, v-path and a blue u, v-path, since otherwise u and v would be br-vertices of degree 2 and could not be adjacent to other vertices in $V - V(G_e)$. A virtual edge e is called an *r-edge* (resp., *b-edge*) if G_e is an *r-graph* (resp., *b-graph*). Note that no virtual edge can be an r-edge and b-edge at the same time. We also treat a red (resp., blue) real edge as an r-edge (resp., b-edge).

For a subgraph H of the skeleton $\sigma(\nu)$ of a node ν, let $E^r(H)$ (resp., $E^b(H)$) denote the set of r-edges (resp., b-edges) in H. Note that each of $E^r(H)$ and $E^b(H)$ induces a cactus, since each (V, E_i) is a cactus. Hence each P-node ν of $\mathcal{T}$ satisfies

$$|E^r(\sigma(\nu))| + |E^b(\sigma(\nu))| \leq 2. \tag{1}$$

6.2   Splitters and Forbidden Subgraphs

Now we examine the structure of red/blue splitters in canonical instances. A cycle Q' in the skeleton $\sigma(\nu)$ of a node ν is called an *r-cycle* (resp., *b-cycle*) if $E(Q') \subseteq E^r(\sigma(\nu))$ (resp., $E(Q') \subseteq E^b(\sigma(\nu))$).

In canonical instances, non-facial r- and b-cycles in the skeleton of an S- or a P-node can be avoided by an adequate choice of plane embeddings (i.e., proper embeddings) of the instances, which will be discussed in Section 7. However, non-facial r- and b-cycles in the skeleton of an R-node need to be checked. In fact, splitters and forbidden subgraphs are equivalent in the following sense.

Lemma 4. *Let $G = (V, E_1 \cup E_2)$ be a canonical instance.*

(i) *Let Q' be a non-facial r-cycle in the skeleton $\sigma(\nu)$ of an R-node ν, and let Q be a red cycle of G corresponding to Q'. Then the subgraph S of G obtained from $\sigma(\nu)$ by replacing each virtual edge $e = (u, v) \in E(Q')$ (resp., $e = (u, v) \in E(\sigma(\nu)) - E(Q')$) with a u, v-path of Q (resp., G) is a forbidden subgraph of G.*

(ii) *Let S be a forbidden subgraph such that a red cycle Q in S separates two blue edges e_1, e_2 of G. Then for the R-node ν such that $V(\nu)$ contains all the vertices* *of degree ≥ 3 of S, the set of r-edges $e \in E^r(\sigma(\nu))$ with $E(G_e) \cap Q \neq \emptyset$ is a non-facial r-cycle Q' in the skeleton $\sigma(\nu)$.*

Lemma 5. *Given $E^r(\sigma(\nu))$ (resp., $E^b(\sigma(\nu))$) for the skeleton $\sigma(\nu)$ of an R-node ν, testing if there is a non-facial r-cycle (b-cycle) in a plane embedding γ_ν of $\sigma(\nu)$ can be done in $O(|E(\sigma(\nu))|)$ time.*

Lemma 5 implies that once we know $E^r(\sigma(\nu))$ (resp., $E^b(\sigma(\nu))$) for all nodes ν in the SPR tree, a non-facial r-/b-cycle in the skeleton of an R-node can be found in linear time. In the next section, we show how to compute the r-edges and b-edges in the skeletons of all nodes in the SPR tree.

6.3   Computing the Color of Edges in the Skeletons of the Rooted SPR Tree

We choose a node in the SPR tree and treat it as a rooted tree. Let μ be a non-root node in $\mathcal{T}$, and ν be the parent of μ. The graph $\sigma(\mu)$ has exactly one virtual edge e in common with $\sigma(\nu)$, called the *parent virtual edge* in $\sigma(\mu)$, and a *child virtual edge* in $\sigma(\nu)$. Let $Ch(\nu)$ denote the set of all children of ν.

We denote the graph formed from $\sigma(\nu)$ by deleting its parent virtual edge pe(ν) as $\sigma^-(\nu)$, if ν is not the root of $\mathcal{T}$. Let $G^-(\nu)$ denote the subgraph of G induced by the set of all vertices in the graphs $\sigma^-(\mu)$ for all descendants μ of ν, including ν itself.

We first give an overview of the algorithm for computing the r-edges in the skeletons of all nodes in the SPR tree of G (computing b-edges can be done symmetrically):

1. First choose a node as the root of the SPR tree of G.
2. By traversing the rooted SPR tree in a bottom-up manner, compute the r-edges in the skeleton $\sigma^-(\nu)$ of each node ν (except for the parent virtual edge pe(ν)), based on the computation of r-edges in the skeletons of children of ν.
3. By traversing the rooted SPR tree in a top-down manner, identify the children $\mu \in Ch(\nu)$ of each node ν such that the parent virtual edge pe(μ) is an r-edge.

Since G is not a simple cycle, the SPR tree of G has an R- or P-node. We choose an R- or P-node ν^* as the root of the SPR tree $\mathcal{T}$. For a leaf node ν in the rooted SPR tree $\mathcal{T}$, we know that $E^r(\sigma^-(\nu))$ is the set of the red edges in the subgraph $G^-(\nu)$. The next lemma says that the r-edges in the skeletons $\sigma^-(\nu)$ of all other nodes in the SPR tree $\mathcal{T}$ can be computed in linear time.

Lemma 6. *Given $E^r(\sigma^-(\mu))$ for all $\mu \in Ch(\nu)$ of a node ν in $\mathcal{T}$ the set $E^r(\sigma^-(\nu))$ of all r-edges in the skeleton $\sigma^-(\nu)$ except pe(ν) can be computed in $O(|E(\sigma(\nu))| + \sum_{\mu \in Ch(\nu)} |E(\sigma(\mu))|)$ time.*

By the lemma, we can compute $E^r(\sigma^-(\nu))$ for all nodes ν in $\mathcal{T}$. The next lemma says that the parent virtual r-edges of the skeletons $\sigma(\nu)$ of all nodes in the SPR tree $\mathcal{T}$ can be computed in linear time.

Lemma 7. *Given $E^r(\sigma^-(\nu))$ for a node ν in $\mathcal{T}$, the set of all children $\mu \in Ch(\nu)$ such that pe(μ) is an r-edge can be computed in $O(|E(\sigma(\nu))|)$ time.*

By Lemma 4, any forbidden subgraph can be found as a non-facial r- or b-cycle in the skeleton of an R-node. After we compute the r-edges in the skeletons of all nodes in the SPR tree in linear time by Lemmas 6 and 7, we test if each R-node has a non-facial r-cycle in its skeleton in time linear of the size of the skeleton by Lemma 5, which takes $O(|V| + |E|)$ time in total over all R-nodes. Symmetrically we can find a non-facial b-cycle in the skeleton of an R-node in linear time. Hence finding a forbidden subgraph of a canonical instance, if any, can be done in linear time.

To prove Theorem 4, the remaining task is to design a linear-time algorithm for constructing a proper embedding for a canonical instance with no forbidden subgraphs.

7 Constructing Proper Embeddings

In this section, we assume that a given canonical instance $G = (V, E_1 \cup E_2)$ has no forbidden subgraph, i.e., no non-facial r- or b-cycle in the skeleton of any R-node, and present a linear-time algorithm for constructing a proper plane embedding.

In what follows, for a graph $H = \sigma^-(\nu)$ or $H = G^-(\nu)$ of each non-root node ν in $\mathcal{T}$, an embedding ψ of H means a plane embedding of the graph such that both end vertices u and v of the parent virtual edge $pe(\nu) = (u, v)$ of ν appear in the boundary of the plane embedding. When we traverse the boundary of ψ clockwise, we denote the path along the boundary from u to v (resp., from v to u) by $B_{u,v}(\psi)$ (resp., $B_{v,u}(\psi)$). The path $B_{u,v}(\psi)$ is called *r-rimmed* if $B_{u,v}(\psi)$ is the unique r-u, v-path (i.e., red u, v-path). We define *b-rimmed* boundaries symmetrically. Fig. 3(a) and (b) show an embedding γ_ν of the skeleton $\sigma(\nu)$ and an embedding Γ_ν of the graph $G^-(\nu)$ for an R-node ν ($B_{u,v}(\gamma_\nu)$ and $B_{u,v}(\Gamma_\nu)$ are r-rimmed; $B_{v,u}(\gamma_\nu)$ and $B_{v,u}(\Gamma_\nu)$ are not r-rimmed).

An embedding ψ of H is called *proper* if (i) every r-cycle/b-cycle in ψ is a facial cycle; and (ii) $B_{u,v}(\psi)$ or $B_{v,u}(\psi)$ is r-rimmed (resp., b-rimmed) when $pe(\nu)$ is an r-edge (resp., b-edge). A plane embedding ψ of $H = G$ or $H = \sigma(\nu)$ for the root node ν is called *proper* if every r-cycle/b-cycle in ψ is a facial cycle.

Assuming that each edge $e \in E(\sigma^-(\nu))$ of a node ν admits a proper embedding Γ_e of G_e, we show that a proper embedding Γ_ν of $G^-(\nu)$ can be obtained from a set of proper embeddings Γ_e, $e \in E(\sigma^-(\nu))$. For this, we first observe that the skeleton of each node admits a proper embedding.

Lemma 8. *Let G be a canonical instance with no forbidden subgraph, and ν be a node in the SPR tree of G. Then:*

(i) *When ν is an S- or R-node or a P-node, with $|E^r(\sigma(\nu))| + |E^b(\sigma(\nu))| \leq 1$, any plane embedding γ_ν of the skeleton $\sigma(\nu)$ of the root ν or of $\sigma^-(\nu)$ of a non-root ν is proper.*

(ii) *When ν is a P-node with $|E^r(\sigma(\nu))| = 2$ (resp., $|E^b(\sigma(\nu))| = 2$), a plane embedding γ_ν of the skeleton $\sigma(\nu)$ of the root ν or of $\sigma^-(\nu)$ of a non-root ν is proper if one of the two edges in $E^r(\sigma(\nu))$ (resp., $E^b(\sigma(\nu))$) appear consecutively (possibly one appears as $B_{u,v}(\gamma_\nu)$ and the other as the parent virtual edge $pe(\nu)$).*

Based on proper embeddings of all nodes in the SPR tree, we show how to construct a proper embedding for $H = G$ traversing the rooted SPR tree $\mathcal{T}$ in a bottom-up manner.

We first construct proper embeddings for leaf nodes of $\mathcal{T}$. Then we construct a proper embedding of a non-leaf node ν by assembling the proper embeddings of all children of ν.

1. Leaf nodes: For each leaf S- or R-node ν of $\mathcal{T}$, any plane embedding of $G^-(\nu) = \sigma^-(\nu)$ is proper by Lemma 8. Note that there is no leaf P-node since G is simple.

2. Internal nodes: Let ν be an internal node of $\mathcal{T}$, $\mathrm{pe}(\nu) = (u,v)$ be the parent virtual edge of ν, and γ_ν be a proper embedding of $\sigma^-(\nu)$ in Lemma 8. For each virtual edge $e \in E(\sigma^-(\nu))$, let Γ_e denote a proper embedding of G_e.
(1) S-node: Let ν be an S-node. In this case, γ_ν is a single path joining u and v. Let Γ_ν be an embedding of $G^-(\nu)$ obtained from γ_ν by replacing each virtual edge $e \in E(\sigma^-(\nu))$ with Γ_e. If $G^-(\nu)$ is not an r-graph or a b-graph, then the resulting embedding Γ_ν of $G^-(\nu)$ is already proper. Assume that $G^-(\nu)$ is an r-graph (the case of a b-graph can be treated symmetrically). Then,

> (α) for each edge e in $B_{u,v}(\xi_\nu)$, we flip each Γ_e if necessary so that the r-rimmed boundary of ψ_e appears along the outer face of γ_ν.

The resulting embedding Γ_ν of $G^-(\nu)$ is now proper.

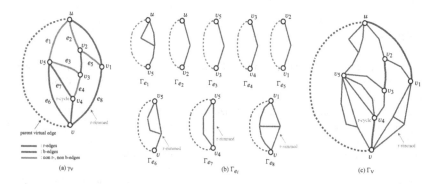

Fig. 3. (a) A proper embedding γ_ν of the skeleton $\sigma(\nu)$ of an R-node ν, where $e_1, \ldots, e_8$ are the virtual edges of $\sigma(\nu)$; (b) A proper embedding Γ_{e_i} of subgraph G_e for each virtual edge e_i, $i = 1, 2, \ldots, 8$; and (c) A proper embedding Γ_ν of subgraph $G^-(\nu)$ obtained from γ_ν by replacing each virtual edge e_i with Γ_{e_i}

(2) R-node: Let ν be an R-node, and γ_ν be a proper embedding of $\sigma^-(\nu)$, where we assume without loss of generality that $B_{u,v}(\gamma_\nu)$ is the unique r-u, v-path of $\sigma^-(\nu)$. See Fig. 3, where (a) shows a proper embedding γ_ν of $\sigma^-(\nu)$ of the R-node ν, and (b) shows a proper embedding Γ_{e_i} of G_{e_i} of each virtual edge $e_i \in E(\sigma(\nu))$. Consider the case where $G^-(\nu)$ is an r-graph and the parent virtual edge of ν is an r-edge (the other case can be treated analogously). We replace each virtual edge e in $\sigma^-(\nu)$ such that

> (α) for each edge e in $B_{u,v}(\gamma_\nu)$, the r-rimmed boundary of Γ_e of appears along the outer face of γ_ν; and

(β) for each edge e in a facial r-cycle Q' of γ_ν, the r-rimmed boundary of Γ_e appears facing the interior of Q' (note that an r-edge cannot belong to two r-faces in a canonical instance).

Let Γ_ν be the resulting embedding of $G^-(\nu)$. See Fig. 3(c) for the resulting embedding Γ_ν obtained from the embeddings in Fig. 3(a) and (b). We see that Γ_ν is proper, since γ_ν and all ψ_e are proper.

(3) P-node: Let ν be a P-node and γ_ν be a proper embedding of $\sigma^-(\nu)$. That is, two edges in $E^r(\sigma(\nu))$ (resp., $E^r(\sigma(\nu))$), if any, appear consecutively (possibly one as $B_{u,v}(\gamma_\nu)$ and the other as the parent edge pe(ν)). We replace each virtual edge e in $\sigma^-(\nu)$ by Γ_e according to the same rules (α) and (β). We easily see that the resulting embedding Γ_ν of $G^-(\nu)$ is proper.

3. Root nodes: For the root R- or P-node ν, we can obtain a proper embedding Γ of G by replacing each virtual edge e in a proper embedding γ_ν of $\sigma(\nu)$ with a proper embedding Γ_e according to the same rule (β).

This completes an inductive proof for the existence of proper embeddings in canonical instances when there is no forbidden subgraph. It is not difficult to see that the above procedure for constructing a proper embedding of G can be implemented to run in linear time. Therefore, this completes the proof of Theorem 4.

References

1. Angelini, P., Di Bartolomeo, M., Di Battista, G.: Implementing a partitioned 2-page book embedding testing algorithm. In: Didimo, W., Patrignani, M. (eds.) GD 2012. LNCS, vol. 7704, pp. 79–89. Springer, Heidelberg (2013)
2. Angelini, P., Di Battista, G., Frati, F., Patrignani, M., Rutter, I.: Testing the simultaneous embeddability of two graphs whose intersection is a biconnected or a connected graph. J. Discrete Algorithms 14, 150–172 (2012)
3. Bernhart, F., Kainen, P.C.: The book thickness of a graph. J. Combin. Theory Ser. B 27(3), 320–331 (1979)
4. Biedl, T.: Drawing planar partitions III: Two constrained embedding problems, Technical Report RRR 12-98, RUTCOR, Rutgers University (1998)
5. Di Battista, G., Tamassia, R.: On-line maintenance of triconnected components with SPQR-trees. Algorithmica 15, 302–318 (1996)
6. Hong, S., Nagamochi, H.: Two-page book embedding and clustered graph planarity, TR[2009-004], Dept. of Applied Mathematics and Physics, University of Kyoto (2009)
7. Hong, S., Nagamochi, H.: Simpler testing for two page book embedding of partitioned graphs, TR[2013-001], Dept. of Applied Mathematics and Physics, University of Kyoto (2013)
8. Wigderson, A.: The complexity of the Hamiltonian circuit problem for maximal planar graphs, Technical Report 298, EECS Department, Princeton University (1982)

Classifying the Clique-Width
of H-Free Bipartite Graphs*

Konrad Kazimierz Dabrowski and Daniël Paulusma

School of Engineering and Computing Sciences, Durham University,
Science Laboratories, South Road, Durham DH1 3LE, United Kingdom
{konrad.dabrowski,daniel.paulusma}@durham.ac.uk

Abstract. Let G be a bipartite graph, and let H be a bipartite graph with a fixed bipartition (B_H, W_H). We consider three different, natural ways of forbidding H as an induced subgraph in G. First, G is H-free if it does not contain H as an induced subgraph. Second, G is strongly H-free if G is H-free or else has no bipartition (B_G, W_G) with $B_H \subseteq B_G$ and $W_H \subseteq W_G$. Third, G is weakly H-free if G is H-free or else has at least one bipartition (B_G, W_G) with $B_H \not\subseteq B_G$ or $W_H \not\subseteq W_G$. Lozin and Volz characterized all bipartite graphs H for which the class of strongly H-free bipartite graphs has bounded clique-width. We extend their result by giving complete classifications for the other two variants of H-freeness.

1 Introduction

The *clique-width* of a graph G, is a well-known graph parameter that has been studied both in a structural and in an algorithmic context. It is the minimum number of labels needed to construct G by using the following four operations:

(i) creating a new graph consisting of a single vertex v with label i;
(ii) taking the disjoint union of two labelled graphs G_1 and G_2;
(iii) joining each vertex with label i to each vertex with label j $(i \neq j)$;
(iv) renaming label i to j.

We refer to the surveys of Gurski [13] and Kamiński, Lozin and Milanič [14] for an in-depth study of the properties of clique-width.

We say that a class of graphs has *bounded* clique-width if every graph from the class has clique-width at most p for some constant p. As many NP-hard graph problems can be solved in polynomial time on graph classes of bounded clique-width [10,15,20,21], it is natural to determine whether a certain graph class has bounded clique-width and to find new graph classes of bounded clique-width. In particular, many papers determined the clique-width of graph classes characterized by one or more forbidden induced subgraphs [1,2,3,4,5,6,7,8,9,11,16,17,18,19].

In this paper we focus on classes of bipartite graphs characterized by a forbidden induced subgraph H. A graph G is H-free if it does not contain H as

---

* The research in this paper was supported by EPSRC (EP/G043434/1 and EP/K025090/1) and ANR (TODO ANR-09-EMER-010).

an induced subgraph. If G is bipartite, then when considering notions for H-freeness, we may assume without loss of generality that H is bipartite as well. For bipartite graphs, the situation is more subtle as one can define the notion of freeness with respect to a fixed bipartition (B_H, W_H) of H. This leads to two other notions (see also Section 2 for formal definitions). We say that a bipartite graph G is strongly H-free if G is H-free or else has no bipartition (B_G, W_G) with $B_H \subseteq B_G$ and $W_H \subseteq W_G$. Strongly H-free graphs have been studied with respect to their clique-width, although under less explicit terminology (see e.g. [14,17,18]). In particular, Lozin and Volz [18] completely determined those bipartite graphs H, for which the class of strongly H-free graphs has bounded clique-width (we give an exact statement of their result in Section 3). If G is H-free or else has at least one bipartition (B_G, W_G) with $B_H \not\subseteq B_G$ or $W_H \not\subseteq W_G$, then G is said to be weakly H-free. As far as we are aware this notion has not been studied with respect to the clique-width of bipartite graphs.

Our Results: We completely classify the classes of H-free bipartite graphs of bounded clique-width. We also introduce the notion of weak H-freeness for bipartite graphs and characterize those classes of weakly H-free bipartite graphs that have bounded clique-width. In this way, we have identified a number of new graph classes of bounded clique-width. Before stating our results precisely in Section 3, we first give some terminology and examples in Section 2. In Section 4 we give the proofs of our results.

2   Terminology and Examples

We first give some terminology on general graphs, followed by terminology for bipartite graphs. We illustrate the definitions of H-freeness, strong H-freeness and weak H-freeness of bipartite graphs with some examples. As we will explain, these examples also make clear that all three notions are different from each other.

General Graphs: Let G and H be graphs. We write $H \subseteq_i G$ to indicate that H is an induced subgraph of G. A bijection of the vertices $f : V_G \to V_H$ is called a *(graph) isomorphism* when $uv \in E_G$ if and only if $f(u)f(v) \in E_H$. If such a bijection exists then G and H are *isomorphic*. Let $\{H_1, \ldots, H_p\}$ be a set of graphs. A graph G is $(H_1, \ldots, H_p)$-*free* if no H_i is an induced subgraph of G. If $p = 1$ we may write H_1-free instead of (H_1)-free. The *disjoint union* $G + H$ of two vertex-disjoint graphs G and H is the graph with vertex set $V_G \cup V_H$ and edge set $E_G \cup E_H$. We denote the disjoint union of r copies of G by rG.

Bipartite Graphs: A graph G is *bipartite* if its vertex set can be partitioned into two (possibly empty) independent sets. Let H be a bipartite graph. We say that H is a *labelled* bipartite graph if we are also given a *black-and-white labelling* ℓ, which is a labelling that assigns either the colour "black" or the colour "white" to each vertex of H in such a way that the two resulting monochromatic colour classes B_H^ℓ and W_H^ℓ form a partition of H into two (possibly empty) independent sets. From now on we denote a graph H with such a labelling ℓ by

$H^\ell = (B_H^\ell, W_H^\ell, E_H)$. Here the pair (B_H^ℓ, W_H^ℓ) is *ordered*, that is, $(B_H^\ell, W_H^\ell, E_H)$ and $(W_H^\ell, B_H^\ell, E_H)$ are different labelled bipartite graphs.

We say that two labelled bipartite graphs H_1^ℓ and $H_2^{\ell^*}$ are *isomorphic* if the (unlabelled) graphs H_1 and H_2 are isomorphic, and if in addition there exists an isomorphism $f : V_{H_1} \to V_{H_2}$ such that for all $u \in V_{H_1}$, $u \in W_{H_1}^\ell$ if and only if $f(u) \in W_{H_2}^{\ell^*}$. Moreover, if $H_1 = H_2$, then ℓ and ℓ^* are said to be *isomorphic* labellings. For example, the bipartite graphs $(\{u, v\}, \emptyset)$ and $(\{x, y\}, \emptyset)$ are isomorphic, and the labelled bipartite graph $(\{u, v\}, \emptyset, \emptyset)$ is isomorphic to the labelled bipartite graph $(\{x, y\}, \emptyset, \emptyset)$. However, $(\{x, y\}, \emptyset, \emptyset)$ is neither isomorphic to $(\emptyset, \{x, y\}, \emptyset)$ nor to $(\{x\}, \{y\}, \emptyset)$ (also see Fig. 1).

We write $H_1^\ell \subseteq_{li} H_2^{\ell^*}$ if $H_1 \subseteq_i H_2$, $B_{H_1}^\ell \subseteq B_{H_2}^{\ell^*}$ and $W_{H_1}^\ell \subseteq W_{H_2}^{\ell^*}$. In this case we say that H_1^ℓ is a *labelled* induced subgraph of $H_2^{\ell^*}$. Note that the two labelled bipartite graphs $H_1^{\ell_1}$ and $H_2^{\ell_2}$ are isomorphic if and only if $H_1^{\ell_1}$ is a labelled induced subgraph of $H_2^{\ell_2}$, and vice versa.

Fig. 1. The graph $2P_1$ partitioned into three ways; none of these three labelled bipartite graphs are isomorphic to each other

Let G be an (unlabelled) bipartite graph, and let H^ℓ be a labelled bipartite graph. We say that G contains H^ℓ as a *strongly labelled* induced subgraph if $H^\ell \subseteq_{li} (B_G, W_G, E_G)$ for some bipartition (B_G, W_G) of G. If not, then G is said to be *strongly H^ℓ-free*. We say that G contains H^ℓ as a *weakly labelled* induced subgraph if $H^\ell \subseteq_{li} (B_G, W_G, E_G)$ for all bipartitions (B_G, W_G) of G. If not, then G is said to be *weakly H^ℓ-free*. Equivalently, G is strongly H^ℓ-free if for every labelling ℓ^* of G, G^{ℓ^*} does not contain H^ℓ as a labelled induced subgraph and G is weakly H^ℓ-free if there is a labelling ℓ^* of G such that G^{ℓ^*} does not contain H^ℓ as a labelled induced subgraph. Note that these two notions of freeness are only defined for (unlabelled) bipartite graphs. Let $\{H_1^{\ell_1}, \ldots, H_p^{\ell_p}\}$ be a set of labelled bipartite graphs. Then a graph G is *strongly (weakly)* $(H_1^{\ell_1}, \ldots, H_p^{\ell_p})$-free if G is strongly (weakly) $H_i^{\ell_i}$-free for $i = 1, \ldots, p$.

The following lemma shows that for all labelled bipartite graphs H^ℓ, the class of H-free graphs is a (possibly proper) subclass of the class of strongly H^ℓ-free bipartite graphs and that the latter graph class is a (possibly proper) subclass of the class of weakly H^ℓ-free bipartite graphs.

Lemma 1. *Let G be a bipartite graph and H^ℓ be a labelled bipartite graph. The following two statements hold:*

(i) If G is H-free, then G is strongly H^ℓ-free.
(ii) If G is strongly H^ℓ-free, then G is weakly H^ℓ-free.

Moreover, the two reverse statements are not necessarily true.

Proof. Statements (i) and (ii) follow by definition. The following two examples, which are also depicted in Fig. 2, show that the reverse statements may not necessarily be true. Let G be isomorphic to $S_{1,1,3}$ with $V_G = \{u_1, \ldots, u_6\}$ and $E_G = \{u_1u_2, u_1u_3, u_1u_4, u_4u_5, u_5u_6\}$. Let $H = K_{1,3} + P_1$. We denote the vertex set and edge set of H by $V_H = \{x_1, x_2, x_3, x_4, x_5\}$ and $E_H = \{x_1x_2, x_1x_3, x_1x_4\}$.

Let $H^\ell = (\{x_2, x_3, x_4\}, \{x_1, x_5\}, E_H)$. We first notice that G is not H-free, because $G[u_1, u_2, u_3, u_4, u_6]$ is isomorphic to $K_{1,3} + P_1$. However, we do have that G is strongly H^ℓ-free, because H^ℓ is neither a labelled induced subgraph of $(\{u_1, u_5\}, \{u_2, u_3, u_4, u_6\}, E_G)$ nor of $(\{u_2, u_3, u_4, u_6\}, \{u_1, u_5\}, E_G)$.

Let $H^{\ell^*} = (\{x_2, x_3, x_4, x_5\}, \{x_1\}, E_H)$. Then G is not strongly H^{ℓ^*}-free, because $(\{u_2, u_3, u_4, u_6\}, \{u_1\}, \{u_1u_2, u_1u_3, u_1u_4\})$ is isomorphic to H^{ℓ^*}. However, G is weakly H^{ℓ^*}-free, because H^{ℓ^*} is not a labelled induced subgraph of $(\{u_1, u_5\}, \{u_2, u_3, u_4, u_6\}, E_G)$). $\qquad\square$

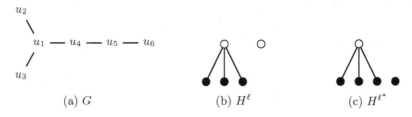

(a) G (b) H^ℓ (c) H^{ℓ^*}

Fig. 2. The graphs G, H^ℓ and H^{ℓ^*} from the proof of Lemma 1

Special Graphs: For $r \geq 1$, the graphs C_r, K_r, P_r denote the cycle, complete graph and path on r vertices, respectively, and the graph $K_{1,r}$ denotes the star on $r+1$ vertices. If $r = 3$, the graph $K_{1,r}$ is also called the *claw*. For $1 \leq h \leq i \leq j$, let $S_{h,i,j}$ denote the tree that has only one vertex x of degree 3 and that has exactly three leaves, which are of distance h, i and j from x, respectively. Observe that $S_{1,1,1} = K_{1,3}$. A graph $S_{h,i,j}$ is called a *subdivided claw*.

Let $H^\ell = (B_H^\ell, W_H^\ell, E_H)$ be a labelled bipartite graph. The *opposite* of H^ℓ is defined as the labelled bipartite graph $H^{\overline{\ell}} = (W_H^\ell, B_H^\ell, E_H)$. We say that $\overline{\ell}$ is the *opposite* black-and-white labelling of ℓ. Suppose that H is a bipartite graph such that among all its black-and-white labellings, all those that maximize the number of black vertices are isomorphic. In this case we pick one of such labelling and call it b.

3 The Classifications

A full classification of the boundedness of the clique-width of strongly H^ℓ-free bipartite graphs was given by Lozin and Voltz [18], except that in their result the trivial case when $H^\ell = (sP_1)^b$ or $H^\ell = (sP_1)^{\overline{b}}$ for some $s \geq 1$ was missing. Their proof is correct except that it overlooked this case, which occurs when one of the colour classes of the labelled graph H^ℓ is empty. However, strongly $(sP_1)^b$-free bipartite graphs can have at most $2s - 2$ vertices, and as such form a class of bounded clique-width. Below we state their result after incorporating this small correction, followed by our results for the other two variants of freeness. We refer to Fig. 3 for pictures of the labelled bipartite graphs used in Theorems 1 and 3.

Theorem 1 ([18]). *Let H^ℓ be a labelled bipartite graph. The class of strongly H^ℓ-free bipartite graphs has bounded clique-width if and only if one of the following cases holds:*

- $H^\ell = (sP_1)^b$ $\qquad\qquad$ *or* $\quad H^\ell = (sP_1)^{\overline{b}}$ $\qquad\qquad$ *for some $s \geq 1$*
- $H^\ell \subseteq_{li} (K_{1,3} + 3P_1)^b$ $\quad$ *or* $\quad H^\ell \subseteq_{li} (K_{1,3} + 3P_1)^{\overline{b}}$
- $H^\ell \subseteq_{li} (K_{1,3} + P_2)^b$ $\quad$ *or* $\quad H^\ell \subseteq_{li} (K_{1,3} + P_2)^{\overline{b}}$
- $H^\ell \subseteq_{li} (P_1 + S_{1,1,3})^b$ $\quad$ *or* $\quad H^\ell \subseteq_{li} (P_1 + S_{1,1,3})^{\overline{b}}$
- $H^\ell \subseteq_{li} (S_{1,2,3})^b$ $\qquad\quad$ *or* $\quad H^\ell \subseteq_{li} (S_{1,2,3})^{\overline{b}}.$

Theorem 2. *Let H be a graph. The class of H-free bipartite graphs has bounded clique-width if and only if one of the following cases holds:*

- $H = sP_1$ *for some $s \geq 1$*
- $H \subseteq_i K_{1,3} + 3P_1$
- $H \subseteq_i K_{1,3} + P_2$
- $H \subseteq_i P_1 + S_{1,1,3}$
- $H \subseteq_i S_{1,2,3}.$

Theorem 3. *Let H^ℓ be a labelled bipartite graph. The class of weakly H^ℓ-free bipartite graphs has bounded clique-width if and only if one of the following cases holds:*

- $H^\ell = (sP_1)^b$ $\qquad\quad$ *or* $\quad H^\ell = (sP_1)^{\overline{b}}$ $\qquad\qquad$ *for some $s \geq 1$*
- $H^\ell \subseteq_{li} (P_1 + P_5)^b$ $\quad$ *or* $\quad H^\ell \subseteq_{li} (P_1 + P_5)^{\overline{b}}$
- $H \subseteq_i P_2 + P_4$
- $H \subseteq_i P_6.$

4 The Proofs of Our Results

We first recall a number of basic facts on clique-width known from the literature. We then state a number of other lemmas which we use to prove Theorems 2 and 3.

4.1 Facts about Clique-width

The *bipartite complement* of a bipartite graph *with respect to* a bipartition (B, W) is the bipartite graph with bipartition (B, W), in which two vertices $u \in B$ and $v \in W$ are adjacent if and only if $uv \notin E$. For instance, the graph $2P_2$ has C_4 as its only bipartite complement, whereas the graph $2P_1$ has $2P_1$ and P_2 as its bipartite complements. For two disjoint vertex subsets X and Y in G, the *bipartite complementation* operation with respect to X and Y acts on G by replacing every edge with one end-vertex in X and the other one in Y by a non-edge and vice versa. The *edge subdivision* operation replaces an edge vw in a graph by a new vertex u with edges uv and uw.

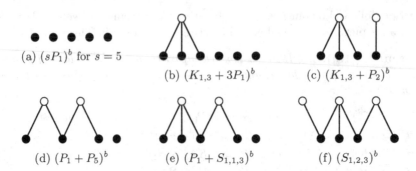

(a) $(sP_1)^b$ for $s = 5$

(b) $(K_{1,3} + 3P_1)^b$

(c) $(K_{1,3} + P_2)^b$

(d) $(P_1 + P_5)^b$

(e) $(P_1 + S_{1,1,3})^b$

(f) $(S_{1,2,3})^b$

Fig. 3. The labelled bipartite graphs used in Theorems 1 and 3

We now state some useful facts for dealing with clique-width. We will use these facts throughout the paper. We will say that a graph operation *preserves* boundedness of clique-width if for every constant k and every graph class $\mathcal{G}$, the graph class $\mathcal{G}_{[k]}$ obtained by performing the operation at most k times on each graph in $\mathcal{G}$ has bounded clique-width if and only if $\mathcal{G}$ has bounded clique-width.

Fact 1. Vertex deletion preserves boundedness of clique-width [16].

Fact 2. Bipartite complementation preserves boundedness of clique-width [14].

Fact 3. For a class of graphs $\mathcal{G}$ of *bounded* degree, let $\mathcal{G}'$ be the class of graphs obtained from $\mathcal{G}$ by applying zero or more edge subdivision operations to each graph in $\mathcal{G}$. Then $\mathcal{G}$ has bounded clique-width if and only if $\mathcal{G}'$ has bounded clique-width [14].

We also use some other elementary results on the clique-width of graphs. In order to do so we need the notion of a *wall*. We do not formally define this notion, but instead refer to Fig. 4, in which three examples of walls of different height are depicted. A k-*subdivided wall* is a graph obtained from a wall after subdividing each edge exactly k times for some constant $k \geq 0$. The next well-known lemma follows from combining Fact 3 with the fact that walls have maximum degree 3 and unbounded clique-width (see e.g. [14]).

Lemma 2. *For every constant k, the class of k-subdivided walls has unbounded clique-width.*

We let $\mathcal{S}$ be the class of graphs each connected component of which is either a subdivided claw $S_{h,i,j}$ for some $1 \leq h \leq i \leq j$ or a path P_r for some $r \geq 1$. This leads to the following lemma, which is well-known and follows from the fact that walls have maximum degree at most 3 and from Lemma 2 by choosing an appropriate value for k (also note that k-subdivided walls are bipartite for all $k \geq 0$).

Lemma 3. *Let $\{H_1, \ldots, H_p\}$ be a finite set of graphs. If $H_i \notin \mathcal{S}$ for $i = 1, \ldots, p$ then the class of $(H_1, \ldots, H_p)$-free bipartite graphs has unbounded clique-width.*

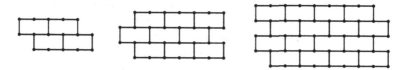

Fig. 4. Walls of height 2, 3, and 4, respectively

4.2   A Number of Other Lemmas

We start with a lemma which is related to Lemma 1 and which follows immediately from the corresponding definitions.

Lemma 4. *Let G and H be bipartite graphs. Then G is H-free if and only if G is strongly H^ℓ-free for all black-and-white labellings ℓ of H.*

A graph G that contains a graph H as an induced subgraph may be weakly H^ℓ-free for all black-and-white labellings ℓ of H; take for instance the graphs G and H from the proof of Lemma 1. However, we can make the following observation, which also follows directly from the corresponding definitions.

Lemma 5. *Let H be a bipartite graph with a unique black-and-white labelling ℓ (up to isomorphism). Then every bipartite graph G is H-free if and only if it is weakly H^ℓ-free.*

Note that there exist both connected bipartite graphs (for example $H = P_6$) and disconnected bipartite graphs (for example $H = 2P_2$) that satisfy the condition of Lemma 5.

Two black-and-white labellings of a bipartite graph H are said to be *equivalent* if they are isomorphic or opposite to each other; otherwise they are said to be *non-equivalent*. The following lemma follows directly from the definitions.

Lemma 6. *Let ℓ and ℓ^* be two equivalent black-and-white labellings of a bipartite graph H. Then the class of strongly (weakly) H^ℓ-free graphs is equal to the class of strongly (weakly) H^{ℓ^*}-free graphs.*

The following lemma is due to Lozin and Rautenbach [17].

Lemma 7 ([17]). *Let $\{H_1^{\ell_1}, \ldots, H_p^{\ell_p}\}$ be a finite set of labelled bipartite graphs. For $i = 1, \ldots, p$, let F_i denote the bipartite complement of H_i with respect to $(B_{H_i}^{\ell_i}, W_{H_i}^{\ell_i})$. If $H_i \notin \mathcal{S}$ for all $1 \le i \le p$ or $F_i \notin \mathcal{S}$ for all $1 \le i \le p$, then the class of strongly $(H_1^{\ell_1}, \ldots, H_p^{\ell_p})$-free bipartite graphs has unbounded clique-width.*

In the next lemma we demonstrate a list of H-free bipartite classes with unbounded clique-width. It is obtained by combining a known result of Lozin and Voltz [18] with a number of new results.

Lemma 8. *The class of H-free bipartite graphs has unbounded clique-width if $H \in \{2P_1 + 2P_2, 2P_1 + P_4, 4P_1 + P_2, 3P_2, 2P_3\}$.*

Proof. Lozin and Voltz [18] showed that $2P_3$-free bipartite graphs have unbounded clique-width. Let $H \in \{2P_1 + 2P_2, 2P_1 + P_4, 4P_1 + P_2, 3P_2\}$, and let $\{H^{\ell_1}, \ldots, H^{\ell_p}\}$ be the set of all non-equivalent labelled bipartite graphs isomorphic to H. For $i = 1, \ldots, p$, let F_i denote the bipartite complement of H with respect to $(B_H^{\ell_i}, W_H^{\ell_i})$. We will show that every F_i does not belong to $\mathcal{S}$. Then, by Lemma 7 the class of strongly $(H_1^{\ell_1}, \ldots, H_p^{\ell_p})$-free bipartite graphs has unbounded clique-width. Because a bipartite graph is H-free if and only if it is strongly $(H_1^{\ell_1}, \ldots, H_p^{\ell_p})$-free (by Lemmas 4 and 6), this means that the class of H-free bipartite graphs has unbounded clique-width.

Suppose $H \in \{2P_1 + 2P_2, 2P_1 + P_4\}$. Let $V_H = \{x_1, \ldots, x_6\}$ with $E_H = \{x_1x_2, x_3x_4\}$ if $H = 2P_1 + 2P_2$ and $E_H = \{x_1x_2, x_2x_3, x_3x_4\}$ if $H = 2P_1 + P_4$. Then H has only two non-equivalent black-and-white labellings. We may assume without loss of generality that one of these two labellings colours x_1, x_3, x_5, x_6 black and x_2, x_4 white, whereas the other one colours x_1, x_3, x_5 black and x_2, x_4, x_6 white. Let F_1 and F_2 be the bipartite complements corresponding to the first and second labellings, respectively. The vertices x_2, x_4, x_5, x_6 induce a C_4 in F_1, whereas the vertices x_1, x_4, x_5, x_6 induce a C_4 in F_2. Hence, F_1 and F_2 do not belong to $\mathcal{S}$.

Suppose $H = 4P_1 + P_2$. Let $V_H = \{x_1, \ldots, x_6\}$ and $E_H = \{x_1x_2\}$. Then H has three non-equivalent black-and-white labellings. We may assume without loss of generality that the first one colours x_1, x_3, x_4, x_5, x_6 black and x_2 white, the second one colours x_1, x_3, x_4, x_5 black and x_2, x_6 white, and the third one colours x_1, x_3, x_4 black and x_2, x_5, x_6 white. Let F_1, F_2, F_3 denote the corresponding bipartite complements. The vertices $x_2, \ldots, x_6$ induce a $K_{1,4}$ in F_1. The vertices x_2, x_3, x_4, x_6 induce a C_4 in F_2 and F_3. Hence, none of F_1, F_2, F_3 belongs to $\mathcal{S}$.

Suppose $H = 3P_2$. Let $V_H = \{x_1, \ldots, x_6\}$ and $E_H = \{x_1x_2, x_3x_4, x_5x_6\}$. Let ℓ be a black-and-white labelling of H that colours x_1, x_3, x_5 black and x_2, x_4, x_6 white. Then every other labelling ℓ^* of H is isomorphic to ℓ. The bipartite complement of H with respect to (B_H^ℓ, W_H^ℓ) is isomorphic to C_6, which does not belong to $\mathcal{S}$.   □

We will also need the following lemma. We omit the proof due to space restrictions.

Lemma 9. *Let $H \in \mathcal{S}$. Then H is $(2P_1 + 2P_2, 2P_1 + P_4, 4P_1 + P_2, 3P_2, 2P_3)$-free if and only if $H = sP_1$ for some integer $s \geq 1$ or H is an induced subgraph of one of the graphs in $\{K_{1,3} + 3P_1, K_{1,3} + P_2, P_1 + S_{1,1,3}, S_{1,2,3}\}$.*

The last lemma we need before proving the main results of this paper is the following one (we use it several times in the proof of Theorem 3).

Lemma 10. *Let H^ℓ be a labelled bipartite graph. The class of weakly H^ℓ-free bipartite graphs has unbounded clique-width in both of the following cases:*

(i) H^ℓ contains a vertex of degree at least 3, or
(ii) H^ℓ contains four independent vertices, not all of the same colour.

Proof. Let b_1 be a black-and-white labelling of $4P_1$ that colours three vertices black and one vertex white. Let b_2 be a black-and-white labelling of $4P_1$ that colours two vertices black and two vertices white. We show below that the class of weakly H^ℓ-free bipartite graphs has unbounded clique-width if $H^\ell \in \{(K_{1,3})^b, (4P_1)^{b_2}, (4P_1)^{b_3}\}$. Then we are done by Lemma 6.

Consider a 1-subdivided wall G' obtained from a wall G. Recall that 1-subdivided walls are bipartite. Moreover, the vertices that were introduced when subdividing every edge of G all have degree 2 and form one class of a bipartition (B, W) of G'. Let this class be B. Then $(K_{1,3})^b$ is not a labelled induced subgraph of $(B, W, E_{G'})$. Hence, G' is weakly $(K_{1,3})^b$-free. This means that the class of weakly $(K_{1,3})^b$-free graphs contains the class of 1-subdivided walls. As such, it has unbounded clique-width by Lemma 2. The bipartite complement G'' of G' with respect to (B, W) is weakly $(4P_1)^{b_1}$-free, as $(K_{1,3})^b$ is the bipartite complement of $(4P_1)^{b_1}$ and $(K_{1,3})^b$ is not a labelled induced subgraph of $(B, W, E_{G'})$. Hence, the class of weakly $(4P_1)^{b_1}$-free graphs has unbounded clique-width by Fact 2. The class of weakly $(4P_1)^{b_2}$-free bipartite graphs has unbounded clique-width by Lemma 1 and Theorem 1.   □

4.3   The Proof of Theorem 2

Proof. We first deal with the bounded cases. First suppose $H = sP_1$ for some $s \geq 1$. Then every H-free bipartite graph G has at most $s - 1$ vertices in each partition class for every bipartition. This means that the clique-width of G is at most $2s - 2$. Now suppose that $H \in \{K_{1,3} + 3P_1, K_{1,3} + P_2, P_1 + S_{1,1,3}, S_{1,2,3}\}$. Then the claim follows from combining Lemma 1 with Theorem 1.

We now deal with the unbounded cases. Suppose $H \neq sP_1$ for some $s \geq 1$ and that H is not an induced subgraph of one of the graphs in $\{K_{1,3} + 3P_1, K_{1,3} + P_2, P_1 + S_{1,1,3}, S_{1,2,3}\}$. Then by Lemma 9, either $H \notin S$ or, H is not $(2P_1 + 2P_2, 2P_1 + P_4, 4P_1 + P_2, 3P_2, 2P_3)$-free. Hence, the clique-width of the class of H-free bipartite graphs is unbounded by Lemmas 3 and 8, respectively.   □

4.4   The Proof of Theorem 3

Proof. We first consider the bounded cases. First suppose $H^\ell = (sP_1)^b$ for some $s \geq 1$ (the $H^\ell = (sP_1)^{\bar{b}}$ case is equivalent). Then every weakly H^ℓ-free bipartite graph has a bipartition (B, W) with $|B| \leq s - 1$. Hence, the clique-width of such a graph is at most $s + 1$ (first introduce the vertices of B by using distinct labels, then use two more labels for the vertices of W, introducing them one-by-one).

Before considering the case $H^\ell = (P_1 + P_5)^b$, we first consider the case where $H \subseteq_i P_2 + P_4$ or $H \subseteq_i P_6$. We first assume that $H = P_2 + P_4$ or $H = P_6$. Then $H \subseteq_i S_{1,2,3}$, which implies that that the class of H-free bipartite graphs has bounded clique-width by Theorem 2. All black-and-white labellings of $P_2 + P_4$ are isomorphic. Similarly, all black-and-white labellings of P_6 are isomorphic. Hence, the class of H-free bipartite graphs coincides with the class of weakly H^ℓ-free graphs by Lemma 5. We therefore conclude that the latter class also has bounded clique-width.

Now let $H \subseteq_i P_2 + P_4$ or $H \subseteq_i P_6$, but $H \notin \{P_2 + P_4, P_6\}$. Note that $P_2 + P_4$ and P_6 have a unique labelling b (up to isomorphism). If H^ℓ is not a labelled induced subgraph of one of $\{(P_2 + P_4)^b, P_6^b\}$ then H must have two non-equivalent black-and-white labellings. Since H is a linear forest, it must have at least two components with an odd number of vertices. Therefore $H \in \{2P_1, 3P_1, P_1 + P_3, 2P_1 + P_2\}$. However, in all these cases, for every labelling ℓ of H, $H^\ell \subseteq_{li} P_6^b$ or $H^\ell \subseteq_{li} (P_2 + P_4)^b$. Therefore, if $H \subseteq_i P_2 + P_4$ or $H \subseteq_i P_6$ then for every labelling ℓ of H, the weakly H^ℓ-free bipartite graphs are a subclass of either the P_6-free or $(P_2 + P_4)$-free bipartite graphs. In particular, this holds for $H^\ell = (P_1 + 2P_2)^b$ (we need this observation for the following case).

Finally, suppose $H^\ell = (P_1 + P_5)^b$. Let G be a weakly H^ℓ-free bipartite graph. Then G has a labelling ℓ^* such that H^ℓ is not a labelled induced subgraph of $(B_G^{\ell^*}, W_G^{\ell^*}, E_G)$. If $|B_G^{\ell^*}|$ is even, then we delete a vertex of $B_G^{\ell^*}$. We may do this by Fact 1. Hence $|B_G^{\ell^*}|$ may be assumed to be odd. Let X be the subset of $W_G^{\ell^*}$ that consists of all vertices that are adjacent to less than half of the vertices of $B_G^{\ell^*}$. We apply the bipartite complementation between X and $B_G^{\ell^*}$. We may do this by Fact 2. Let G_1 be the resulting bipartite graph, with bipartition classes $B_{G_1}^{\ell^*} = B_G^{\ell^*}$ and $W_{G_1}^{\ell^*} = W_G^{\ell^*}$.

Suppose $B_{G_1}^{\ell^*}$ contains three vertices b_1, b_2, b_3 and $W_{G_1}^{\ell^*}$ contains two vertices w_1, w_2 such that $G_1^{\ell^*}[b_1, b_2, b_3, w_1, w_2]$ is isomorphic to $(P_1 + 2P_2)^b$. By construction and because $|B_{G_1}^{\ell^*}| = |B_G^{\ell^*}|$ is odd, w_1 and w_2 have at least one common neighbour $b_4 \in B_{G_1}^{\ell^*}$. Then $G_1^{\ell^*}[b_1, b_2, b_3, b_4, w_1, w_2]$ is isomorphic to $(P_1 + P_5)^b$. However, then $G^{\ell^*}[b_1, b_2, b_3, b_4, w_1, w_2]$ is also isomorphic to $(P_1 + P_5)^b$ (irrespective of whether w_1 or w_2 belong to X), which is a contradiction. We conclude that G_1 is weakly $(P_1 + 2P_2)^b$-free. As observed above, this means that G_1 has bounded clique-width. Hence G has bounded clique-width.

We now consider the unbounded cases. Let H^ℓ be a labelled bipartite graph that is not isomorphic to one of the (bounded) cases considered already. Suppose that H contains a cycle or an induced subgraph isomorphic to $2P_3$. Then the class of weakly H^ℓ-free graphs has unbounded clique-width by combining Lemma 1 with Theorem 2. Suppose that H contains a vertex of degree at least 3. Then the class of weakly H^ℓ-free bipartite graphs has unbounded clique-width by Lemma 10(i). It remains to consider the case when $H = sP_1 + tP_2 + P_r$ for some constants $1 \le r \le 6$, $s \ge 0$ and $t \ge 0$, where $\max\{s, t\} \ge 1$ (as H is not an induced subgraph of P_6).

Suppose $5 \le r \le 6$. Assume without loss of generality that three vertices of the copy of P_r in H^ℓ are coloured black. If $r = 6$ or $t \ge 1$ or some copy P_1 in H^ℓ is coloured white, or two copies of P_1 in H^ℓ are coloured black, then we can apply Lemma 10(ii). Hence, $H^\ell = (P_1 + P_5)^b$, which is not possible by assumption.

Suppose $r = 4$. If two vertices in the induced subgraph of H^ℓ isomorphic to $sP_1 + tP_2$ have the same colour then we can apply Lemma 10(ii). Hence we may assume that $s \le 2$ and $t \le 1$, and moreover that $s = 0$ if $t = 1$. Also we would have $H \subseteq_i P_2 + P_4$ if $s = 0$ and $t = 1$ or if $s = 1$ and $t = 0$. Hence, it remains to consider the case $s = 2$ and $t = 0$, such that one copy of P_1 is coloured black and the other one white. In that case, we may apply Lemma 10(ii).

Suppose $r = 3$. Assume without loss of generality that the two vertices of the copy of P_3 in H^ℓ are coloured black. Recall that $s \geq 1$ or $t \geq 1$. If $t \geq 2$, then we can apply Lemma 10(ii). Suppose $t = 1$. Then $s = 0$ otherwise H^ℓ would contain an induced $4P_1$ in which not all the vertices are the same colour, in which case we could apply Lemma 10(ii). However, this means that H is an induced subgraph of $P_2 + P_4$. Now suppose $t = 0$. Then $s \geq 2$, as otherwise H is an induced subgraph of $P_2 + P_4$. If $s \geq 3$ then H^ℓ contains an induced $4P_1$ in which not all the vertices are the same colour, in which case we apply Lemma 10(ii). Hence, $s = 2$ and both copies are coloured black (otherwise we apply Lemma 10(ii)). However, in this case H^ℓ is a labelled induced subgraph of $(P_1 + P_5)^b$, which is not possible by assumption.

Finally suppose that $r \leq 2$. Then we may write $H = sP_1 + tP_2$ instead. We must have $s + t \geq 4$ or $t \geq 3$, otherwise H would be an induced subgraph of $P_2 + P_4$ or P_6. If $t = 0$ then since $H^\ell \neq (sP_1)^b$ and $H^\ell \neq (sP_1)^{\bar{b}}$ we can find four copies of P_1 in H that are not all of the same colour and apply Lemma 10(ii). If $t \geq 1, s + t \geq 4$, we can also find four copies of P_1 that are not all of the same colour and apply Lemma 10(ii). Finally, suppose $s = 0, t = 3$. In this case we combine Lemmas 1 and 8. This completes the proof. $\square$

5   Conclusions

We have completely determined those bipartite graphs H for which the class of H-free bipartite graphs has bounded clique-width. We also characterized exactly those labelled bipartite graphs H for which the class of weakly H-free bipartite graphs has bounded clique-width. These results complement the known characterization of Lozin and Volz [18] for strongly H-free bipartite graphs. A natural direction for further research would be to characterize, for each of the three notions of H-freeness, the clique-width of classes of $\mathcal{H}$-free bipartite graphs when $\mathcal{H}$ is a set containing at least 2 graphs. In a follow-up paper [12], we apply our results for H-free bipartite graphs to determine classes of (H_1, H_2)-free (general) graphs of bounded and unbounded clique-width.

References

1. Boliac, R., Lozin, V.: On the clique-width of graphs in hereditary classes. In: Bose, P., Morin, P. (eds.) ISAAC 2002. LNCS, vol. 2518, pp. 44–54. Springer, Heidelberg (2002)
2. Bonomo, F., Grippo, L.N., Milanič, M., Safe, M.D.: Graphs of power-bounded clique-width. arXiv, abs/1402.2135 (2014)
3. Brandstädt, A., Engelfriet, J., Le, H.-O., Lozin, V.: Clique-width for 4-vertex forbidden subgraphs. Theory of Computing Systems 39(4), 561–590 (2006)
4. Brandstädt, A., Klembt, T., Mahfud, S.: P_6- and triangle-free graphs revisited: structure and bounded clique-width. Discrete Mathematics and Theoretical Computer Science 8(1), 173–188 (2006)
5. Brandstädt, A., Kratsch, D.: On the structure of (P_5,gem)-free graphs. Discrete Applied Mathematics 145(2), 155–166 (2005)

6. Brandstädt, A., Le, H.-O., Mosca, R.: Gem- and co-gem-free graphs have bounded clique-width. International Journal of Foundations of Computer Science 15(1), 163–185 (2004)

7. Brandstädt, A., Le, H.-O., Mosca, R.: Chordal co-gem-free and (P_5,gem)-free graphs have bounded clique-width. Discrete Applied Mathematics 145(2), 232–241 (2005)

8. Brandstädt, A., Mahfud, S.: Maximum weight stable set on graphs without claw and co-claw (and similar graph classes) can be solved in linear time. Information Processing Letters 84(5), 251–259 (2002)

9. Brandstädt, A., Mosca, R.: On variations of P_4-sparse graphs. Discrete Applied Mathematics 129(2-3), 521–532 (2003)

10. Courcelle, B., Makowsky, J.A., Rotics, U.: Linear time solvable optimization problems on graphs of bounded clique-width. Theory of Computing Systems 33(2), 125–150 (2000)

11. Dabrowski, K.K., Golovach, P.A., Paulusma, D.: Colouring of graphs with Ramsey-type forbidden subgraphs. Theoretical Computer Science 522, 34–43 (2013)

12. Dabrowski, K.K., Paulusma, D.: Clique-width of graph classes defined by two forbidden induced subgraphs. coRR, abs/1405.7092 (2014)

13. Gurski, F.: Graph operations on clique-width bounded graphs. CoRR, abs/cs/0701185 (2007)

14. Kamiński, M., Lozin, V., Milanič, M.: Recent developments on graphs of bounded clique-width. Discrete Applied Mathematics 157(12), 2747–2761 (2009)

15. Kobler, D., Rotics, U.: Edge dominating set and colorings on graphs with fixed clique-width. Discrete Applied Mathematics 126(2-3), 197–221 (2003)

16. Lozin, V., Rautenbach, D.: On the band-, tree-, and clique-width of graphs with bounded vertex degree. SIAM Journal on Discrete Mathematics 18(1), 195–206 (2004)

17. Lozin, V., Rautenbach, D.: The tree- and clique-width of bipartite graphs in special classes. Australasian Journal of Combinatorics 34, 57–67 (2006)

18. Lozin, V., Volz, J.: The clique-width of bipartite graphs in monogenic classes. International Journal of Foundations of Computer Science 19(02), 477–494 (2008)

19. Makowsky, J., Rotics, U.: On the clique-width of graphs with few P_4's. International Journal of Foundations of Computer Science 10(3), 329–348 (1999)

20. Oum, S.-I.: Approximating rank-width and clique-width quickly. ACM Trans. Algorithms 5(1), 10:1–10:20 (2008)

21. Rao, M.: MSOL partitioning problems on graphs of bounded treewidth and clique-width. Theoretical Computer Science 377(1-3), 260–267 (2007)

Displacing Random Sensors
to Avoid Interference*

Evangelos Kranakis[1] and Gennady Shaikhet[2]

[1] School of Computer Science, Carleton University, Ottawa, ON, Canada
[2] School of Mathematics and Statistics, Carleton University, Ottawa, ON, Canada

Abstract. Consider sensors on a line. Assume that for a given parameter $s > 0$ two sensors' signals interfere with each other during communication if their distance is $\leq s$. We are allowed to move the sensors on the line, if needed, so as to avoid interference. We call total movement the sum of displacements that the sensors have to move so that the distance between any two sensors is $> s$. We study the following sensor displacement problem for avoiding interference. Assume that n sensors are thrown randomly and independently with the Poisson distribution having arrival rate $\lambda = n$ in the interval $[0, +\infty)$. What is the expected minimum total distance that the sensors have to move from their initial position to a new destination so that any two sensors are at a distance more than s apart? In this paper we study tradeoffs between the interference distance s and the expected minimum total movement, denoted by $E(s)$. (Clearly, the higher the value of s the more the resulting displacement $E(s)$.)

For the line. we prove the following results. 1) If $s \leq \frac{1}{nt}$ then $E(s) \leq \min\{t^2/(t-1)^3, (n-1)/2t\}$, where $t > 1$. 2) For $s \geq 1/n + \Omega(n^{-\alpha})$ we show that $E(s) \in \Omega(n^{2-\alpha})$, $2 \geq \alpha \geq 0$, while for $|s - 1/n| \in \Theta(n^{-3/2})$, we show that $E(s) \in \Theta(\sqrt{n})$. These results show a critical regime for the expected minimum total displacement $E(s)$, for s in the interval $[1/n - 1/n^{3/2}, 1/n + 1/n^{3/2}]$.

Similar results concerning the expcted optimal sum of displacements are obtained when the sensors are located on the plane and their coordinates are generated by two independent identical Poisson processes. In the critical regime for sensors on the plane, we show that $E(s) \in \Theta(n^{3/4})$ provided that s is in the interval $[1/n^{1/2} - 1/n^{3/4}, 1/n^{1/2} + 1/n^{3/4}]$.

Keywords and Phrases. Displacement, Interference, Line, Plane, Poisson, Random, Sensors.

1 Introduction

It is well known that proximity between neighbouring sensors affects their transmission and reception signals and degrades network performance (see [5]). In fact, the closer their distance the higher the resulting interference and hence performance degradation. Therefore to avoid interference a critical value, say s,

---

* Research supported in part by NSERC grants.

Z. Cai et al. (Eds.): COCOON 2014, LNCS 8591, pp. 501–512, 2014.

is established and unless the sensors maintain this minimum critical distance interference will occur.

Consider n sensors thrown randomly with the Poisson distribution (having arrival rate n) in the half-line $[0, +\infty)$. For a given interference value s, no two consecutive sensors can be at distance less than s. What is the expected minimum total displacement (i.e., sum of distances) that the sensors have to move so as to ensure that in their final position no two sensors are at distance $\leq s$? In particular, what are tradeoffs on the movement of the sensors with respect to the parameter s?

1.1   Preliminaries and Notation

We consider a Poisson process with arrival rate $\lambda = n$ and suppose that the i-th event represents the location of the i-th sensor, for $i = 1, 2, \ldots, n$. Let X_i be the arrival time of the ith event in this Poisson process. If $T_1, T_2, \ldots$ are the interarrival times of the Poisson process then $X_i = T_1 + T_2 + \cdots + T_i$. Recall (see [14]) that for $i = 1, 2, \ldots$ the interarrival times are independent and identically distributed exponential random variables such that

$$\Pr[T_i = s] = \lambda e^{-\lambda s} \text{ and } \Pr[T_i \geq s] = e^{-\lambda s}.$$

In can be shown from this that $E[T_i] = \frac{1}{\lambda}$ and $Var(T_i) = \frac{1}{\lambda^2}$. The random variable $N(0, t]$ which counts the number $N(0, t]$ of points in the interval $(0, t]$ is Poisson with arrival rate λt and $E[N(0, t]] = \lambda t$ and $Var(N(0, t]) = \lambda t$. The law of large numbers for the Poisson distribution (see [9][pages 41-43]) states that $\lim_{t \to \infty} \frac{N(0,t]}{t} = \lambda$, with probability 1. For related work on Poisson distributions and random geometric graphs we refer the reader to [6] and [13], respectively.

Throughout we assume that $s \geq 0$ is a given parameter denoting the *inter-ference value* or *interference distancee*. Two sensors *avoid interference* if their distance is $> s$. A sensor *avoids interference* if its distance from all other sensors is $> s$. Two sensors *overlap* if their distance is $< s$. There is a *gap of length g* between the ith and $i + 1$st sensors if their distance is $> g$. Similarly, we say that there is gap of length g following the ith sensor if the distance between the ith and $i + 1$st sensors is $> g$. The sensors are required to move from their starting to a final position so as to attain an interference distance of at least s. In this context, the *displacement* of a sensor is the distance that it has to move from its starting to its final position and the *total displacement* is the sum of displacements of all sensors.

Definition 1. *Given n, s, we denote by $E_n(s)$ the expected minimum total displacement required so that in their final poisitions every pair of sensors is at distance $> s$.*

Usually, we abbreviate $E_n(s)$ by $E(s)$ when the number n of sensors is clear from the context. Our intention is to study tradeoffs between the interference value s and the expected minimum total distance $E(s)$.

1.2   Related Results

Several papers study interference in relation to network performance degradation [5,8], while [2] proposes connectivity preserving and spanner constructions which are interference optimal. [12] considers the *average interference* problem while maintaining connectivity.

Related to our work, is the waiting time between events which has been analyzed by several researchers. Let $W(s)$ be the random variable measuring the waiting time until two events of the Poisson process occur within time s. According to [17][Equation 15.4] (see also [3][Equation 1.3]), the following identity is valid

$$\Pr[W(s) > t] = e^{-\lambda t} \sum_{k=0}^{\lceil t/s \rceil} \frac{(t - (k-1)s)^k \lambda^k}{k!},$$

i.e., the probability that the Poisson process has no two consecutive events in the interval $[0, t]$ separated by a distance which is shorter than s. Additional related work can be found in [17][Equation 15.4] and [15,16,3].

All the studies mentioned above refer only to static sensor models in that no movement of the sensors is investigated. Our work is closely related to the work of [10] where the authors consider the expected minimum total displacement for establishing full coverage of a unit interval for n sensors placed uniformly at random. In a sense, the coverage problem is opposite to our problem but the analysis required for the solution of the interference problem is different.

Finally, it is worth mentioning the related problem of choosing transmission radii which minimize the maximum interference while maintaining a connected symmetric communication graph; this is shown by [1] to be NP-complete. In addition, [7] gives an algorithm which yields a maximum interference of $O(\sqrt{n})$, while [11] studies the max interference problem for a set of sensors on a line.

1.3   Outline and Results of the Paper

We assume that the arrival rate $\lambda = n$. The tradeoffs derived for the expected minimum total movement depend on the required interferences distance s between consecutive sensors. Table 1 displays results proved in the paper. We study tradeoffs between the interference distance s and the expected minimum total

Table 1. Table displays our results about the expected minimum total movement in the interval $[0, +\infty)$ as a function of the interference distance s. Similar results are obtained for sensors on the plane.

Interference Distance s	Total Displacement $E(s)$	Theorem
$s - \frac{1}{n} \in \Omega\left(n^{-\alpha}\right), 2 \geq \alpha \geq 0$	$\Omega(n^{2-\alpha})$	1
$\left\lvert s - \frac{1}{n} \right\rvert \in O\left(n^{-3/2}\right)$	$\Theta(\sqrt{n})$	2
$s \leq \frac{1}{tn}, t > 1$	$\leq \frac{t^2}{(t-1)^3}$	3

displacement $E(s)$. The results are presented relative to the size of the interference distance s.

Our study indicates the existence of a *critical threshold* around the value $\frac{1}{n}$ as this affects the expected total movement of the sensors to solve the interference problem. First of all, in Section 2.1 we prove a basic $\Omega(n^{2-\alpha})$ lower bound when $s - \frac{1}{n} \in \Theta(n^{-\alpha})$, for $2 \geq \alpha \geq 0$. Next, sensor interference distance $\frac{1}{n}$ is shown to be a critical threshold 1) around which and within the interval $\left[\frac{1}{n} - \frac{1}{n^{3/2}}, \frac{1}{n} + \frac{1}{n^{3/2}}\right]$ we have a critical regime, i.e., the value $E(s)$ (the expected minimum total movement required so that every pair of sensors is at distance at least s) is exactly $\Theta(\sqrt{n})$, 2) below which $E(s)$ declines sharply to a constant $O(1)$, and 3) above which it increases above $\Theta(\sqrt{n})$.

Similar results are obtained when the sensors are located on the plane. Like for the line we determine a critical threshold and show that in the interval $\left[\frac{1}{n^{1/2}} - \frac{1}{n^{3/4}}, \frac{1}{n^{1/2}} + \frac{1}{n^{3/4}}\right]$ we have a critical regime.

Here is an outline of the paper. In Section 2 we consider sensors on the line. In Subsection 2.2 we prove a $\Theta(\sqrt{n})$ bound when s is sufficiently close to $\frac{1}{n}$. Subsection 2.3 investigates interference, displacemnt tradeoffs for $s \leq \frac{1}{tn}$ as a function of the paprameter t. Section 3 considers sensors on the plane. We conclude with some open problems.

2   Sensors on the Line

In this section we analyze sensor interference when the sensors are placed on the half-line $[0, +\infty)$.

2.1   Expected Displacement for $s - \frac{1}{n} \in \Omega(n^{-\alpha})$

In this subsection we study the total displacement when the interference value is sufficiently above $1/n$. Our first theorem gives a lower bound on the expected total displacement when $\left(s - \frac{1}{n}\right) \in \Omega(n^{-\alpha})$, where $2 \geq \alpha \geq 0$.

Theorem 1. *Assume that the interference value between sensors is s. If $s - \frac{1}{n} \in \Omega(n^{-\alpha})$ then $E(s) \in \Omega(n^{2-\alpha})$, where $2 \geq \alpha \geq 0$ is a real number. In particular, if $s - \frac{1}{n} \in \Omega(n^{-3/2})$ then $E(s) \in \Omega(\sqrt{n})$.*

Proof. (Theorem 1) Suppose that the i-th sensor's displacement is equal to D_i. After the sensors move to their final destinations it must be true that $D_{i+1} + X_{i+1} \geq D_i + X_i + s$, for all $1 \leq i \leq n-1$, so as to ensure that the two sensors are at least distance s apart. It follows that $E[D_{i+1}] + E[X_{i+1}] \geq E[D_i] + E[X_i] + s$. However, $E[X_{i+1}] = \frac{i+1}{n}$ and $E[X_i] = \frac{i}{n}$. Therefore $E[D_{i+1}] \geq E[D_i] + s - \frac{1}{n}$, for $1 \leq i \leq n-1$. Repeating this inequality recursively we see that

$$E[D_{i+1}] \geq E[D_1] + \left(s - \frac{1}{n}\right)i,$$

for $1 \leq i \leq n - 1$. In particular, we conclude that the expected minimum total movement must satisfy

$$\sum_{i=1}^{n} E[|D_i|] \geq \sum_{i=1}^{n} E[D_i] \geq nE[D_1] + \left(s - \frac{1}{n}\right)\frac{n(n-1)}{2}. \tag{1}$$

Now we can prove that the expected minimum total movement is in $\Omega(n^{2-\alpha})$, where $2 \geq \alpha \geq 0$ is a real number. There are two cases to consider.

Case 1. If $E[D_1] \geq 0$. Then the first part of the theorem follows immediately from Inequality (1).

Case 2. If $E[D_1] < 0$. Then observe that $\Pr[T_1 \geq s] = e^{-\lambda s}$ and so with with high probability the sensor must fall within the interval $[0, \frac{c \ln n}{n}]$. Therefore the sensor will never need to move more than $O\left(\frac{c \ln n}{n}\right)$. Hence, $|nE[D_1]|$ is in $O(\ln n)$ and therefore the first part of the theorem follows immediately from Inequality (1).

Finally, the proof of the second part of the theorem concerning the expected minimum total movement follows immediately from the first part using the value $\alpha = 3/2$. ∎

2.2   Expected Displacement for $\left|s - \frac{1}{n}\right| \in O\left(n^{-3/2}\right)$

Next we look at the expected minimum total displacement when s is close to $1/n$. As a consequence of our analysis we prove that there is a critical regime when $\left|s - \frac{1}{n}\right| \in O\left(n^{-3/2}\right)$.

First we prove a lemma, which was first stated and proved in [10] for the unit interval using the binomial distribution. Our random placement model uses the Poisson distribution and is on the half-line $[0, +\infty)$. Therefore we must restate and prove the result in this new model; but our proof is simpler

Lemma 1. *The expected sum of displacements of n sensors to move from their current location to anchor locations $a_i := \frac{i}{n} - \frac{1}{2n}$, for $i = 1, \ldots, n$, respectively is given by the formula*

$$-\frac{1}{2} + \frac{1}{n}\sum_{i=1}^{n} e^{-\frac{2i-1}{2}}\sum_{j=0}^{i-1}\frac{\left(\frac{2i-1}{2}\right)^j}{j!} + \frac{2}{n}\sum_{i=1}^{n} ie^{-\frac{2i-1}{2}}\frac{\left(\frac{2i-1}{2}\right)^i}{i!} \tag{2}$$

Moreover, the expected sum of displacements is in $\Theta(\sqrt{n})$.

Proof. (Lemma 1) Let X_i be the arrival time of the ith event in a Poisson process with arrival rate λ. It turns out (see [14]) that X_i obeys the Gamma distribution with parameters i, λ, where $i = 1, 2, \ldots, n$, i.e., $\Pr[X_i = s] = \lambda e^{-\lambda s}\frac{(\lambda s)^{i-1}}{(i-1)!}$. If the arrival rate is $\lambda = n$ then $E[X_i] = \frac{i}{n}$, for $i = 1, 2, \ldots, n$.

Next we calculate the expected displacement of ith sensor. Let the i-th sensor move to position $a_i = \frac{i}{n} - \frac{1}{2n}$. Let $D_i(a_i) := E[|X_i - a_i|]$ denote the expected displacement of the ith sensor. $D_i(a_i)$ is given by the following formulas which are easily proved.

$$D_i(a_i) = \left(a_i - \frac{i}{\lambda}\right)\left(1 - 2e^{-\lambda a_i}\sum_{j=0}^{i-1}\frac{(\lambda a_i)^j}{j!}\right) + \frac{2i}{\lambda}e^{-\lambda a_i}\frac{(\lambda a_i)^i}{i!}. \tag{3}$$

If we assume that the Poisson arrival rate λ satisfies $\lambda = n$ then using Formula (3) we derive the following identity for the displacement of the ith sensor to the anchor $a_i = \frac{2i-1}{2n}$

$$D_i\left(\frac{2i-1}{2n}\right) = -\frac{1}{2n}\left(1 - 2e^{-\frac{2i-1}{2}}\sum_{j=0}^{i-1}\frac{(\frac{2i-1}{2})^j}{j!}\right) + \frac{2i}{n}e^{-\frac{2i-1}{2}}\frac{(\frac{2i-1}{2})^i}{i!}$$

$$= -\frac{1}{2n} + \frac{1}{n}e^{-\frac{2i-1}{2}}\sum_{j=0}^{i-1}\frac{(\frac{2i-1}{2})^j}{j!} + \frac{2i}{n}e^{-\frac{2i-1}{2}}\frac{(\frac{2i-1}{2})^i}{i!}. \tag{4}$$

Using the last Identity (4) above, the expected sum of displacements of all n sensors is given by the formula

$$\sum_{i=1}^{n}D_i\left(\frac{2i-1}{2n}\right) = \sum_{i=1}^{n}E\left[\left|X_i - \frac{2i-1}{2n}\right|\right]$$

$$= -\frac{1}{2} + \frac{1}{n}\sum_{i=1}^{n}e^{-\frac{2i-1}{2}}\sum_{j=0}^{i-1}\frac{(\frac{2i-1}{2})^j}{j!} + \frac{2}{n}\sum_{i=1}^{n}ie^{-\frac{2i-1}{2}}\frac{(\frac{2i-1}{2})^i}{i!}$$

This proves Identity (2) in the first part of the theorem. We now prove the second part. It turns out this sum is in $\Theta(n)$. To this end observe that we can bound the terms in the sum above as follows.

$$e^{-\frac{2i-1}{2}}\sum_{j=0}^{i-1}\frac{(\frac{2i-1}{2})^j}{j!} = \sqrt{e}\cdot e^{-i}\sum_{j=0}^{i-1}\frac{(i-\frac{1}{2})^j}{j!} \le \sqrt{e}\cdot e^{-i}\sum_{j=0}^{i-1}\frac{i^j}{j!} \le \sqrt{e}\cdot ie^{-i}\frac{i^i}{i!}$$

$$ie^{-\frac{2i-1}{2}}\frac{(\frac{2i-1}{2})^i}{i!} = \sqrt{e}\cdot ie^{-i}\frac{(i-\frac{1}{2})^i}{i!}$$

Therefore it follows that the sum (2) is dominated up to a constant factor by the term $\frac{1}{n}\sum_{i=1}^{n}ie^{-i}\cdot\frac{i^i}{i!}$. Recall from Stirling's Inequality (see Feller [4][page 54]) that we have the following bounds on the factorial functions:

$$\sqrt{2\pi}n^{n+\frac{1}{2}}e^{-n+\frac{1}{12n+1}} < n! < \sqrt{2\pi}n^{n+\frac{1}{2}}e^{-n+\frac{1}{12n}} \tag{5}$$

Using Inequalities (5) we see that

$$ie^{-i}\frac{i^i}{i!} > \frac{1}{\sqrt{2\pi}}\frac{ie^{-i}i^i}{i^{i+\frac{1}{2}}e^{-i+\frac{1}{12i}}} = \frac{1}{\sqrt{2\pi}}\frac{i^{\frac{1}{2}}}{e^{\frac{1}{12i}}} = \frac{1}{\sqrt{2\pi}e^{\frac{1}{12}}}\sqrt{i}$$

$$ie^{-i}\frac{i^i}{i!} < \frac{1}{\sqrt{2\pi}}\frac{ie^{-i}i^i}{i^{i+\frac{1}{2}}e^{-i+\frac{1}{12i+1}}} = \frac{1}{\sqrt{2\pi}}\frac{i^{\frac{1}{2}}}{e^{\frac{1}{12i+1}}} = \frac{1}{\sqrt{2\pi}e^{\frac{1}{13}}}\sqrt{i}$$

It is now easy to see that $\frac{1}{n}\sum_{i=1}^{n} ie^{-i} \cdot \frac{i^i}{i!} \in \Theta(n^{1/2})$. This completes the proof of Lemma 1. ∎

Using this lemma we are now in a position to study upper bounds on the expected minimum displacement when the interference value s is sufficiently close to $1/n$, namely $\left|s - \frac{1}{n}\right| \in O\left(n^{-3/2}\right)$.

Theorem 2 (Critical Regime). *Assume the interference value between sensors is s. If $\left|s - \frac{1}{n}\right| \in O\left(n^{-3/2}\right)$ then $E(s) \in \Theta(\sqrt{n})$.*

Proof. (Theorem 2) Assume $\left|s - \frac{1}{n}\right| \in O\left(n^{-3/2}\right)$.

First consider the case $s \geq \frac{1}{n}$. Define the position $b_i := (i-1)s$ and have the i-th sensor move to position b_i, for $i \geq 1$. It is sufficient to show that

$$\sum_{i=1}^{n} E[|X_i - b_i|] \in O(\sqrt{n}). \tag{6}$$

Clearly, if the ith sensor occupies position b_i, for $i = 1, 2, \ldots, n$, then the distance between consecutive sensors is equal to s.

Now consider the anchor positions $a_i = \frac{2i-1}{2n}$, for $i = 1, \ldots, n$, used in the proof of Lemma 1. By Lemma 1 we know that $\sum_{i=1}^{n} E[|X_i - a_i|] \in \Theta(\sqrt{n})$. Now observe from the triangle inequality that

$$|X_i - b_i| \leq |X_i - a_i| + |a_i - b_i|. \tag{7}$$

From the definition of the anchor points it is easy to see that

$$a_i - b_i = \frac{2i-1}{2n} - (i-1)s = \frac{1}{2n} + (i-1)\left(\frac{1}{n} - s\right).$$

Therefore we conclude that

$$\sum_{i=1}^{n} |a_i - b_i| \leq \frac{1}{2} + \left|s - \frac{1}{n}\right| \cdot \sum_{i=1}^{n}(i-1)$$

$$\leq \frac{1}{2} + \left|s - \frac{1}{n}\right| \cdot \frac{n(n-1)}{2}. \tag{8}$$

Taking the sums in Inequality (7), for $i = 1, \ldots, n$, we conclude that

$$\sum_{i=1}^{n} E[|X_i - b_i|] \leq \sum_{i=1}^{n} E[|X_i - a_i|] + \sum_{i=1}^{n} |a_i - b_i|$$

Therefore the theorem follows using Inequality (8) above and Lemma 1.

Next consider the case $s \leq \frac{1}{n}$.. First of all observe that for the positions b_i, $i = 1, 2, \ldots, n$, defined above we have that

$$E(s) \geq \sum_{i=1}^{n} |X_i - b_i|.$$

From the triangle inequality we have that $|X_i - b_i| \geq |X_i - a_i| - |a_i - b_i|$. Therefore using Inequality (8) we see that

$$\sum_{i=1}^{n} E[|X_i - b_i|] \geq \sum_{i=1}^{n} E[|X_i - a_i|] - \sum_{i=1}^{n} |a_i - b_i|$$

$$\geq \sum_{i=1}^{n} E[|X_i - a_i|] - \frac{1}{2} \left| s - \frac{1}{n} \right| \cdot \frac{n(n-1)}{2}.$$

This is sufficient to complete the proof of Theorem 1. ∎

2.3   Expected Displacement for $s \leq \frac{1}{tn}$

In this subsection we prove an upper bound on the expected minimum displacement for $s \leq \frac{1}{tn}$, provided that $t > 1$.

A simple upper bound can be obtained as follows for $t \geq 1$. Consider the locations $X_1, X_2, \ldots, X_n$ of the n sensors in the interval $[0, +\infty)$. Apply a recursive procedure that moves sensors to the right of their current position so as to ensure that in their new position each pair of consecutive sensors is separated by a distance of at least s. An algorithm that accomplishes this task is as follows.

Algorithm 1: Moving Sensors

1. Set $M_1 = 0$;
2. **for** $i = 2$ **to** n **do**
3. move sensor X_i to new position $X_i + M_i$ s.t.
4a. $X_{i-1} + M_{i-1} \leq X_i + M_i$;
4b. $X_i + M_i \leq s + X_{i-1} + M_{i-1}$;

Observe that in this movement the order of the sensors is maintained. However, when a sensor moves to the right, consecutive sensors to its right will also have to move cummulatively. It follows that $X_i + M_i \leq X_1 + M_1 + (i-1)s = X_1 + (i-1)s$ holds for all $i \geq 1$. Therefore the total movement of all the sensors satisfies the inequality

$$E(s) \leq \sum_{i=2}^{n} M_i \leq \sum_{i=2}^{n} (i-1)s \leq \frac{n-1}{2t}. \tag{9}$$

However, we can improve this result by using standard results in queueing theory. We have the following result.

Theorem 3. *Assume the interference distance between sensors is s. If $s \leq \frac{1}{tn}$ then*

$$E(s) \leq \min \left\{ \frac{t^2}{(t-1)^3}, \frac{n-1}{2t} \right\}, \tag{10}$$

where $t > 1$.

Proof. (Theorem 3) The upper bound $\frac{n-1}{2t}$ is immediate from the previously proved Inequality (9). Therefore we only need to prove the upper bound $\frac{t^2}{(t-1)^3}$.

Our model is equivalent to a single-server service station in which customers arrive according to a Poisson process having rate λ. An arriving customer is served immediately if the server is free; and if not, the customer joins the queue (i.e., waits in line). Successive service times are independent with a common distribution and the system will alternate between idle periods (with no customers in the system—server is idle), and busy periods (with customers in the system—server is busy). A busy period begins when an arrival finds the system empty, and because of the memoryless property of the Poisson arrivals it follows that the distribution of the length of a busy period will be the same for each such period. Let the random variable B denote the length of a busy period and S the service time of the first customer in the busy period.

According to [14][pages 348-349] its mean and variance are given by the formulas below

$$E[B] := \frac{E[S]}{1 - \lambda E[S]} \tag{11}$$

$$Var(B) := \frac{Var(S) + \lambda E^3[S]}{(1 - \lambda E[S])^3} \tag{12}$$

In our model we have that $E[S] = s$ and $\lambda = n$. Let the random variable N denote the number of customers in the busy period B. Since B is the length of the busy period, and the service for a customer is equal to s, it follows that the number of customers in the busy period must be exactly equal to $\frac{B}{s}$. It follows from Identities (11) and (12) that

$$E[N] = \frac{E[B]}{s} = \frac{E[S]}{(1 - \lambda E[S])s} = \frac{1}{1 - ns} \leq \frac{t}{t-1} \tag{13}$$

$$Var(N) = \frac{Var(B)}{s^2} = \frac{ns}{(1-ns)^3} \leq \frac{t^2}{(t-1)^3} \tag{14}$$

$$E[N^2] = Var(N) + E[N]^2 \leq \frac{t^2}{(t-1)^3} + \frac{t^2}{(t-1)^2} = \frac{t^3}{(t-1)^3} \tag{15}$$

Let the random variable D denote the sum of displacements of the sensors in the busy period B. It is clear that $D \leq N^2 s$. Therefore using Equation (13) we conclude

$$E[D] \leq E[N^2 s] = E[N^2]s \leq \left(\frac{t}{t-1}\right)^3 s = \frac{t^2}{n(t-1)^3}. \tag{16}$$

Let the random variable T denote the sum of displacements of all the sensors. Recall that a busy period has $\frac{B}{s}$ customers. Let the consecutive busy periods be $B_1, B_2, \ldots, B_m$, where $m \leq n$.

Let D_i denote the sum of displacements of the sensors participating in the busy period B_i. Since the randomm variables D_i are identical, it follows from the above discussion and Inequality (16) that the expected sum of displacements of all the sensors satisfies

$$E[T] = E\left[\sum_{i=1}^{m} D_i\right] = mE[D] \leq \frac{t^2}{(t-1)^3},$$

which proves the theorem.   ∎

3   Sensors on the Plane

Assume that n sensors are placed in the quadrant $[0, +\infty) \times [0, +\infty)$ according to a "double" Poisson process, i.e., there are two identical and independent Poisson processes X_i, Y_i, for $i = 1, 2, \ldots, m$, each with arrival rate m, and $n = m^2$, where X_i (respectively, Y_i) represents the ith arrival in the x-axis (respectively, y-axis). The position of a sensor in the plane is determined by the pair (X_i, Y_j), $i, j = 1, 2, \ldots, m$. To prevent interference sensors must be displaced (in the plane) so that any pair of sensors is at Euclidean distance $> s$, where as before s is a given positive real number called the interference distance. Similarly to the line, let $E(s)$ denote the expected minimum sum of displacements of all sensors in the plane which ensures that in their final position every pair of sensors is at distance $> s$.[1]

We now embark to extend our results to the plane. Arguing as in Theorem 1 we can prove the following result.

Theorem 4. *If $s - \frac{1}{n^{1/2}} \in \Omega(n^{-\beta})$ then $E(s) \in \Omega(n^{3/2-\beta})$, where $1 \geq \beta \geq 0$ is a real number. In particular, for $\beta = 3/4$ we see that if $s - \frac{1}{n^{1/2}} \in \Omega(n^{-3/4})$ then $E(s) \in \Omega(n^{3/4})$.*

Proof. (Theorem 4) Put $\alpha = 2\beta$. Assume $m = n^{1/2}$. By assumption we have that $s - \frac{1}{m} \in \Omega(m^{-\alpha})$. Observe that if a sensor moves from position (X_i, Y_j) to position $(X_i + M_{i,j}, Y_j + N_{i,j})$ then the Euclidean distance between the old and new position is $D_{i,j} := (M_{i,j}^2 + N_{i,j}^2)^{1/2}$ while for the sum of movements of all the sensors we have that the folllowing inequalities are valid

$$\frac{1}{\sqrt{2}} \sum_{i,j=1}^{m} (|M_{i,j}| + |N_{i,j}|) \leq \sum_{i,j=1}^{m} (M_{i,j}^2 + N_{i,j}^2)^{1/2} \leq \sum_{i,j=1}^{m} (|M_{i,j}| + |N_{i,j}|) \quad (17)$$

It is clear that if the displacement $D_{i,j}$ of sensor (X_i, Y_j) is $> s$ then $|M_{i,j}| + |N_{i,j}| > s$. By Theorem 1 applied to the integer $m = n^{1/2}$ we have that $HE^i(s) + VE^j(s) \in \Omega(m^{2-\alpha}) = \Omega(n^{1-\alpha/2})$, for all $i = 1, 2, \ldots, m$, where $2 \geq \alpha \geq 0$ is a real number, and $HE^i(s), VE^j$ are the corresponding minimum sum of horizontal and vertical displacements of the sensors (X_i, Y_j), for $j = 1, 2, \ldots, m$, and of the sensors (X_i, Y_j), for $i = 1, 2, \ldots, m$, respectively. Using the lefthand side of Inequality (17) we see that $\sum_{i,j=1}^{m}(HE^i(s) + VE^j(s)) \leq E(s)$, which yields $E(s) \in \Omega(n^{3/2-\alpha/2})$. The second part of the theorem is easy.   ∎

---

[1] By abuse of notation, we will use the same symbol for sensors in the plane as for sensors in the line. However the precise interpretation will be clear from the context.

Consider the anchor positions $A_{i,j} := (a_i, a_j)$, for $i, j = 1, 2, \ldots, m$, where $a_i := \frac{2i-1}{2n}$. First we prove the following lemma which is the analogue of Lemma 1.

Lemma 2. *The expected minimum sum of displacements of $n = m^2$ sensors to move to the anchor positions $A_{i,j}$, for $i, j = 1, 2, \ldots, m$, is in $\Theta(n^{3/4})$.*

Proof. (Lemma 2) For the $O(n^{3/4})$ upper bound note that the movement of all sensors in each row and column is in $O(m^{1/2}) = O(n^{1/4})$. Hence, the result follows since there are m rows and m columns of sensors. For the $\Omega(n^{3/4})$ lower bound note that if a sensor moves from position (X_i, Y_j) to position $(X_i + M_{i,j}, Y_j + N_{i,j})$ then the Euclidean distance between the old and new position is $D_{i,j} := (M_{i,j}^2 + N_{i,j}^2)^{1/2}$ while for the sum of movements of all the sensors we have that Inequality (17). However, for each given i, j, we have that $\sum_{j=1}^m |M_{i,j}| \in \Omega(m^{1/4})$ and $\sum_{i=1}^m |N_{i,j}| \in \Omega(m^{1/4})$, which proves the lemma. ∎

Theorem 5 (Critical Regime). *Assume the interference distance between sensors is s. If $\left|s - \frac{1}{n^{1/2}}\right| \in O\left(n^{-3/4}\right)$ then $E(s) \in \Theta(n^{3/4})$.*

Proof. (Theorem 5) This is similar to the proof of Theorem 2. ∎

The analogue of Theorem 3 in the plane for $s \leq \frac{1}{tn}$, where $t > 1$, can also be proved as before.

Theorem 6. *Assume the interference value between sensors is s. If $s \leq \frac{1}{tn^{1/2}}$ then*

$$E(s) \leq \min\left\{\frac{t^2}{(t-1)^3}, \frac{n^{1/2}-1}{2t}\right\}, \tag{18}$$

where $t > 1$.

Proof. (Theorem 6) This is an immediate consequence of Theorem 3 applied to $m = n^{1/2}$. By displacing the respective sensors located at $X_1, X_2, \ldots, X_m$ and $Y_1, Y_2, \ldots, Y_m$ to ensure interference distance at least s, respectively, we guarantee that any pair among the sensors (X_i, Y_j) is also at distance at least s. This proves the theorem. ∎

4   Conclusion

In this paper we have considered the sensor displacement problem to avoid sensor interference. An interesting question would be to study the coverage problem simultaneously with the interference problem. Namely, what is the expected minimum total displacement of n sensors each of range r from their initial position to a new destination so that any two consecutive sensors are at a distance more than s (where $r > s$) apart while at the same time there are no coverage gaps from the left endpoint to the last rightmost sensor?

Acknowledgements. Many thanks to Danny Krizanc for useful discussions on the subject of the paper.

References

1. Buchin, K.: Minimizing the maximum interference is hard, arXiv:0802.2134 (February 2008)
2. Burkhart, M., Wattenhofer, R., Zollinger, A.: Does topology control reduce interference? In: Proceedings of the 5th ACM International Symposium on Mobile Ad Hoc Networking and Computing, pp. 9–19. ACM, New York (2004)
3. Covo, S.: On probabilities associated with the minimum distance between events of a poisson process in a finite interval, and erratum/addendum to it. arXiv preprint arXiv:1007.0283 (2010)
4. Feller, W.: An Introduction to Probability Theory and its Applications, vol. 1. John Wiley, NY (1968)
5. Gupta, P., Kumar, P.R.: The capacity of wireless networks. IEEE Transactions on Information Theory 46(2), 388–404 (2000)
6. Haight, F.A.: Handbook of the Poisson distribution. Wiley, New York (1967)
7. Halldórsson, M.M., Tokuyama, T.: Minimizing interference of a wireless ad-hoc network in a plane. Theoretical Computer Science 402(1), 29–42 (2008)
8. Jain, K., Padhye, J., Padmanabhan, V.N., Qiu, L.: Impact of Interference on Multi-Hop Wireless Network Performance. Wireless Networks 11(4), 471–487 (2005)
9. Kingman, J.F.C.: Poisson processes, vol. 3. Oxford University Press (1992)
10. Kranakis, E., Krizanc, D., Morales-Ponce, O., Narayanan, L., Opatrny, J., Shende, S.: Expected sum and maximum of displacement of random sensors for coverage of a domain. In: Proceedings of the 25th SPAA, pp. 73–82. ACM (2013)
11. Kranakis, E., Krizanc, D., Morin, P., Narayanan, L., Stacho, L.: A tight bound on the maximum interference of random sensors in the highway model. arXiv preprint arXiv:1007.2120 (2010)
12. Moscibroda, T., Wattenhofer, R.: Minimizing interference in ad hoc and sensor networks. In: Proceedings of the 2005 Joint Workshop on Foundations of Mobile Computing, pp. 24–33. ACM, New York (2005)
13. Penrose, M.: Random geometric graphs, vol. 5. Oxford University Press, Oxford (2003)
14. Ross, S.: Introduction to probability models, 10th edn. Elsevier (2010)
15. Todinov, M.: Statistics of defects in one-dimensional components. Computational Materials Science 24(4), 430–442 (2002)
16. Todinov, M.: Minimum failure-free operating intervals associated with random failures of non-repairable components. Computers & Industrial Engineering 45(3), 475–491 (2003)
17. Todinov, M.: Reliability and risk models: Setting reliability requirements. Wiley (2005)

On the Performance of Mildly Greedy Players in Cut Games[*]

Vittorio Bilò[1] and Mauro Paladini[2]

[1] Department of Mathematics and Physics "Ennio De Giorgi", University of Salento, Provinciale Lecce-Arnesano, P.O. Box 193, 73100 Lecce, Italy
vittorio.bilo@unisalento.it
[2] Department of Mathematics and Physics "Ennio De Giorgi", University of Salento, Provinciale Lecce-Arnesano, P.O. Box 193, 73100 Lecce, Italy

Abstract. We continue the study of the performance of mildly greedy players in cut games initiated by Christodoulou *et al.* in [14], where a mildly greedy player is a selfish agent who is willing to deviate from a certain strategy profile only if her payoff improves of a factor of more than $1+\epsilon$, for some $\epsilon \geq 0$. Hence, in presence of mildly greedy players, the classical concepts of pure Nash equilibria and best-responses generalize to those of ϵ-approximate pure Nash equilibria and ϵ-approximate best-responses, respectively. We first show that the ϵ-approximate price of anarchy, that is the price of anarchy of ϵ-approximate pure Nash equilibria, is at least $\frac{1}{2+\epsilon}$ and that this bound is tight for any ϵ. Then, we evaluate the approximation ratio of the solutions achieved after an ϵ-approximate one-round walk starting from any initial strategy profile, where an approximate one-round walk is a sequence of ϵ-approximate best-responses, one for each player. We improve the currently known lower bound on this ratio from $\min\left\{\frac{1}{4+2\epsilon}, \frac{\epsilon}{4+2\epsilon}\right\}$ up to $\min\left\{\frac{1}{2+\epsilon}, \frac{2\epsilon}{(1+\epsilon)(2+\epsilon)}\right\}$ and show that this is tight for any ϵ.

1 Introduction

It has been known since the early fifties that the strategic behavior of selfish players in non-cooperative games usually produces suboptimal outcomes with respect to the ones which could be potentially enforced by a dictatorial authority, the Prisoner's Dilemma being the most famous and pragmatic example. Nevertheless, it has been only after the seminal paper of Koutsoupias and Papadimitriou [21] in 1999 that this phenomenon, termed as *price of anarchy*, became object of a thorough analytical scrutiny by the scientific community.

Formally speaking, given a *social function* measuring the overall quality of all the strategy profiles which can be realized in a game, the price of anarchy measures the *worst-case* ratio between the social value of a strategy profile optimizing the social function and the social value of a Nash equilibrium.

[*] This work was partially supported by the PRIN 2010–2011 research project ARS TechnoMedia: "Algorithmics for Social Technological Networks" funded by the Italian Ministry of University.

Z. Cai et al. (Eds.): COCOON 2014, LNCS 8591, pp. 513–524, 2014.

In the last years, however, a ground-breaking sequence of complexity results has provided a strong evidence of the computational intractability of the problem of computing Nash equilibria in several games of interest. In particular, the problem of computing a pure Nash equilibrium has been shown to be PLS-complete in congestion games by Fabrikant et al. [17] and in some of their special cases by Ackermann et al. [1], where congestion games, introduced by Rosenthal in [23], is a well-known and significant class of games represented in succinct form for which any best-response dynamics is always guaranteed to converge to a pure Nash equilibrium in a finite number of steps. Moreover, the problem of computing a (mixed) Nash equilibrium has been shown to be PPAD-complete for any number of players (Chen and Deng [11], Daskalakis et al. [15], Daskalakis and Papadimitriou [16]), even in games represented in standard normal form, i.e., by explicitly listing the utility of each player in any possible strategy profile.

For such a reason, the price of anarchy has to be intended as a theoretical bound of inefficiency to which a system populated by selfish agents may ideally tend to the limit, but which is unlikely to be attained in practice because of computational issues. Because of these limitations, in the last years, quite an attention has been moved to the analysis of the performance of less demanding solution concepts, among which are approximate pure Nash equilibria and best-response dynamics of polynomially bounded length.

Approximate pure Nash equilibria are pure Nash equilibria for mildly greedy players, that is, players who are willing to be part of any strategy profile in which they experience a utility which is "not too far" from the best utility they can get by deviating to another strategy. More formally, given a value $\epsilon \geq 0$, an ϵ-approximate pure Nash equilibrium is a strategy profile σ such that the utility that each player gets when deviating to any other strategy is no more than $1 + \epsilon$ times the utility that she gets in σ. Any 0-approximate pure Nash equilibrium is a pure Nash equilibrium by definition, hence, the set of pure Nash equilibria is a proper subset of that of ϵ-approximate pure Nash equilibria for any $\epsilon > 0$. For sufficiently high values of ϵ, the problem of computing an ϵ-approximate pure Nash equilibrium becomes polynomial time solvable in several games of interest. In particular, there exist polynomial time algorithms for computing one such an equilibrium in several special cases of congestion games (Bhalgat et al. [4], Caragiannis et al. [8,9], Chien and Sinclair [12]).

A best-response dynamics, instead, is an evolutive processes in which, starting from a given strategy profile, the players are processed sequentially and, at each step, each player is allowed to change her current strategy by best-responding to the strategies played by the others. Clearly, when players can compute in polynomial time their best-responses, a best-response dynamics of polynomially bounded length, i.e., with a polynomial number of steps, can be efficiently computed. We speak of an approximate best-response dynamics when it involves mildly greedy players. In particular, an ϵ-approximate best-response dynamics is a dynamics in which each player changes her strategy only when it improves her utility of a factor of more than $1 + \epsilon$. By definition, any fixed point of an ϵ-approximate best-response dynamics is an ϵ-approximate pure Nash equilibrium.

One may define several special cases of best-response dynamics: for instance, Mirrokni and Vetta [22] introduce the notions of covering walks, k-covering walks, one-round walks, k-round walks and random one-round walks. A covering walk is a sequence of best-response dynamics in which each player plays at least once, a k-covering walk is a concatenation of k covering walks, a one-round walk is a covering walk in which each player plays exactly once, a k-round walk is a concatenation of k one-round walks, while a random one-round walk is a one-round walk such that the order in which players are processed is chosen randomly. When considering mildly greedy players, the analogous notions of approximate covering walks, approximate k-covering walks, approximate one-round walks, and so on, may be defined.

Our Contribution. In this paper, we study the performance of mildly greedy players in cut games, a relevant subclass of congestion games. Cut games are naturally defined by an undirected edge weighted graph G. Each vertex of G is owned by a player and has to be placed in one of the two possible sides of a bipartition. Each player has to decide which side to choose so as to maximize the sum of the weights of the edges connecting her node to all the nodes belonging to the opposite side. Thus, each strategy profile induces a cut of G and each player wants to maximize the contribution given to the total weight of the cut by the edges incident to her node. The social function mainly used in the literature to measure the overall quality of a strategy profile is the total weight of the induced cut which is half of the sum of the players' utilities.

Each cut game, being a particular instance of congestion games, always admits pure Nash equilibria; moreover, any best-response dynamics is guaranteed to converge to one such an equilibrium in a finite number of steps. However, the computation of one such an equilibrium, being strongly related to that of a local optimum of the MAXCUT problem, is a PLS-complete problem, hence widely believed to be computationally untractable. This justifies the idea of resorting to mildly greedy players who can give life to solutions having a more permissive computational complexity. To this aim, Bhalgat et al. [4] give a polynomial time algorithm to compute a $(3+\epsilon)$-approximate pure Nash equilibrium, for any $\epsilon > 0$.

Standard arguments from the theory of approximation algorithms imply that the price of anarchy of cut games is $1/2$ and that so is also the approximation ratio of the solutions achieved after a one-round walk starting from the empty strategy profile. Chrisodoulou et al. [14] show that a random one-round walk converges to a $1/8$ approximation of the social optimum, while, on the negative side, they show that there exist k-round walks converging to an $O(k/n)$ approximation of the social optimum and that there are strategy profiles at exponential distance from any pure Nash equilibrium. Such a worst-case poor deterministic convergence, however, does not occur when mildly greedy players come into play, since they prove that any ϵ-approximate one-round walk starting from any initial strategy profile converges to a $\left(\min\left\{\frac{1}{4+2\epsilon}, \frac{\epsilon}{4+2\epsilon}\right\}\right)$-approximation of the social optimum.

We give exact bounds on the worst-case performance guarantee of mildly greedy players in cut games by considering either approximate pure Nash equilibria and approximate one-round walks. In particular, we show that the ϵ-approximate price of anarchy, that is the price of anarchy of ϵ-approximate pure Nash equilibria, is at least $\frac{1}{2+\epsilon}$ and that this bound is tight for any ϵ. We then move to the evaluation of the approximation ratio of the solutions achieved after an ϵ-approximate one-round walk starting from any initial strategy profile. This notion can be seen as an analogy of the price of anarchy for ϵ-approximate one-round walks and is defined as the worst-case ratio between the value of the social optimum and the social value of a strategy profile realized at the end of the walk. We show that this ratio is at least $\min\left\{\frac{1}{2+\epsilon}, \frac{2\epsilon}{(1+\epsilon)(2+\epsilon)}\right\}$, thus significantly improving the previous lower bound of $\min\left\{\frac{1}{4+2\epsilon}, \frac{\epsilon}{4+2\epsilon}\right\}$ given by Christodoulou et al. [14], and prove that also this bound is tight for any ϵ.

Our lower bounds are both obtained by exploiting the primal-dual method introduced by Bilò in [5]. In particular, for the case of approximate one-round walks, a simple but tricky analysis of all the situations which may occur during the walk allows us to exploit the power of the primal-dual method at its full magnitude. In fact, the lower bound that we achieve is much better (at least the double) than the one that could be obtained by Christodoulou et al. [14] by making use of only combinatorial arguments.

Related Work. The study of (approximate) best-response dynamics plays a crucial role in the determination of both (approximate) pure Nash equilibria and solutions of good social value in a variety of congestion games.

As to the computation of (approximate) pure Nash equilibria, Fabrikant et al. [17] design a polynomial time algorithm to compute a pure Nash equilibrium in symmetric network congestion games, Ackermann et al. [1] show that, when the strategy set of each player is the bases of a matroid over the set of resources, any best-response dynamics converges in polynomial time to a pure Nash equilibrium, while Caragiannis et al. [8,9] give polynomial time algorithms for computing an $O(1)$-approximate pure Nash equilibrium either in congestion games with polynomial latency functions and in their generalization with weighted players. Chien and Sinclair [12] show that, under some mild assumptions, any ϵ-approximate best-response dynamics converges to an ϵ-approximate pure Nash equilibrium after a polynomial number of steps in symmetric congestion games. Such a result has been complemented by Skopalik and Vöcking [24] who showed that there exist asymmetric congestion games with strategy profiles being at an exponential distance from any approximate pure Nash equilibrium.

For what concerns the convergence to solutions of good social value, Goemans et al. [20] show that any random best-response dynamics converges in polynomial time to a constant approximation of the social optimum in weighted congestion games with polynomial latency functions, as well as in basic and valid utility games. Awerbuch et al. [2] prove that, under some mild assumptions, any approximate best-response dynamics converges in polynomial time to an approximation of the social optimum which is arbitrarily close to the price of anarchy in weighted congestion games and that such a property does not hold for best-response dynamics. A similar, but independent, result by Bhalgat et *al.* [3] states that any approximate best-response dynamics converges in poly-nomial time to a constant approximation of the social optimum in congestion games whose resources have similar latency functions. Despite the negative result of Awerbuch *et al.* [2], Fanelli and Moscardelli [19] show that, with some additional reasonable assumptions, best-response dynamics are guaranteed to converge in polynomial time to a constant approximation of the social optimum in weighted congestion games with polynomial latency functions. For the special case of congestion games with linear latency functions, Fanelli *et al.* [18] show that $\Theta(n \log \log n)$ best-responses are necessary and sufficient to attain a constant approximation of the social optimum, where n is the number of players. Christodoulou *et al.* [14] prove that any one-round walk starting from the empty strategy profile converges to a $(2 + \sqrt{5})$-approximation of the social optimum and that such a value grows to $4 + 2\sqrt{3}$ in the case of weighted players, while, when starting from any initial strategy profile, the bound becomes $O(n)$ which is asymptotically tight. The $2 + \sqrt{5}$ upper bound has been shown to be tight by Bilò *et al.* [6], while Caragiannis *et al.* [10] give a lower bound of $3 + 2\sqrt{2}$ for the case of weighted players.

The approximate price of anarchy of congestion games with linear latency functions has been characterized by Christodoulou *et al.* [13].

Paper Organization. Next section contains all necessary definitions and notation, while Sections 3 and 4 illustrate the technical contributions of the paper. In particular, in the former, we bound the approximate price of anarchy, while, in the latter, we focus on approximate one-round walks starting from any initial strategy profile. Finally, in the last section, we discuss our results and possible future research directions. Due to space limitations, some proofs have been omitted.

2 Definitions and Notation

A *strategic game* is a triple $\mathsf{SG} = \left([n], (\Sigma_i)_{i \in [n]}, (\mathsf{U}_i)_{i \in [n]}\right)$, where $[n] := \{1, \ldots, n\}$ is a set of *players* and, for each $i \in [n]$, Σ_i is the *set of strategies* for player i and $\mathsf{U}_i : \times_{j \in [n]} \Sigma_j \to \mathbb{R}_{\geq 0}$ is her *utility function*.

A *strategy profile* $\boldsymbol{\sigma} = (\sigma_1, \ldots, \sigma_n)$ is an n-tuple of strategies, one for each player. We denote as $\Sigma = \times_{j \in [n]} \Sigma_j$ the set of all the strategy profiles of SG. Given a strategy profile $\boldsymbol{\sigma}$, a player $i \in [n]$ and a strategy $\tau \in \Sigma_i$, we denote with $\boldsymbol{\sigma}_{-i} \diamond \tau = (\sigma_1, \ldots, \sigma_{i-1}, \tau, \sigma_{i+1}, \ldots, \sigma_n)$ the strategy profile obtained from $\boldsymbol{\sigma}$ when player i unilaterally changes her strategic choice from σ_i to τ. Classical greedy (or selfish) players choose their strategies so as to maximize their utility functions. *Mildly greedy players*, instead, are players who may be willing to be part of some suboptimal profiles in a sense that we make precise in the following.

Definition 1. *Given a strategy profile $\boldsymbol{\sigma}$ and a value $\epsilon \geq 0$, a strategic choice $\tau \in \Sigma_i$ is an ϵ-**approximate improving deviation** for player i in $\boldsymbol{\sigma}$ if it holds $\mathsf{U}_i(\boldsymbol{\sigma}_{-i} \diamond \tau) > (1 + \epsilon)\mathsf{U}_i(\boldsymbol{\sigma})$.*

Let us denote with $\mathsf{ID}_\epsilon(\boldsymbol{\sigma}, i) = \{\tau \in \Sigma_i : \mathsf{U}_i(\boldsymbol{\sigma}_{-i} \diamond \tau) > (1 + \epsilon)\mathsf{U}_i(\boldsymbol{\sigma})\}$ the set of ϵ-approximate improving deviations for player i in $\boldsymbol{\sigma}$.

Definition 2. *Given a strategy profile $\boldsymbol{\sigma}$ and a value $\epsilon \geq 0$, a strategic choice $\tau \in \Sigma_i$ is an ϵ-**approximate best-response** for player i in $\boldsymbol{\sigma}$ if it holds $\tau \in \arg\max_{\tau' \in \mathsf{ID}_\epsilon(\boldsymbol{\sigma}, i) \cup \sigma_i} \{\mathsf{U}_i(\boldsymbol{\sigma}_{-i} \diamond \tau')\}$.*

Thus, an ϵ-approximate best-response for player i in $\boldsymbol{\sigma}$ is one of her ϵ-approximate improving deviations providing the highest improvement when she possesses at least one such a deviation (i.e., $\mathsf{ID}_\epsilon(\boldsymbol{\sigma}, i) \neq \emptyset$), while it is equal to the strategy currently played by i otherwise.

An ϵ-*greedy player* is never willing to be part of a strategy profile for which she has an ϵ-approximate improving deviation, but she accepts strategy profiles for which she has only ϵ'-approximate improving deviations for any $\epsilon' < \epsilon$. This yields the following definitions of approximate equilibrium and approximate one-round walk.

Definition 3. *Given a value $\epsilon \geq 0$, a strategy profile $\boldsymbol{\sigma} \in \Sigma$ is an ϵ-**approximate pure Nash equilibrium** if, for each $i \in [n]$ and for each $\tau \in \Sigma_i$, it holds $\mathsf{U}_i(\boldsymbol{\sigma}_{-i} \diamond \tau) \leq (1 + \epsilon)\mathsf{U}_i(\boldsymbol{\sigma})$.*

Definition 4. *Given a value $\epsilon \geq 0$, an ϵ-**approximate one-round walk** is an $(n + 1)$-tuple of strategy profiles $\boldsymbol{W} = (\boldsymbol{w}^0, \boldsymbol{w}^1, \ldots, \boldsymbol{w}^n)$ such that, for each $0 \leq i < n$, $\boldsymbol{w}^{i+1} = \boldsymbol{w}^i_{-i} \diamond \tau$, where τ is an ϵ-approximate best-response for player i in $\boldsymbol{w}^i$. The strategy profiles $\boldsymbol{w}^0 := s(\boldsymbol{W})$ and $\boldsymbol{w}^n := f(\boldsymbol{W})$ are the **starting** and the **final strategy profile** of $\boldsymbol{W}$, respectively.*

Note that the concepts of ϵ-approximate improving deviation, ϵ-approximate best-response, ϵ-approximate pure Nash equilibrium and ϵ-approximate one-round walk are proper generalizations of their counterparts defined for greedy players since the latter ones can be obtained from the former when $\epsilon = 0$.

Let $G = (V, E, c)$, with $c : E \to \mathbb{R}_{\geq 0}$, be an edge weighted undirected graph. Graph G defines a *cut game* $\mathcal{C}(G)$ as follows. Each node $v_i \in V$ corresponds to a player $i \in [n]$ whose set of strategies is $\Sigma_i = \{\{0\}, \{1\}\}$. For a strategy profile $\boldsymbol{\sigma}$, let $T_{\boldsymbol{\sigma}} = \{v_i \in V(G) : \sigma_i = 0\}$ and $\overline{T}_{\boldsymbol{\sigma}} = V(G) \setminus T_{\boldsymbol{\sigma}}$. Each strategy profile $\boldsymbol{\sigma}$ induces a bipartition $(T_{\boldsymbol{\sigma}}, \overline{T}_{\boldsymbol{\sigma}})$ of the nodes of G, that is, a cut of G. Let $Adj(i) = \{j \in [n] : \{v_i, v_j\} \in E\}$ be the set of players whose corresponding node is adjacent to v_i in G and denote $c_{ij} := c(\{v_i, v_j\})$. The utility of player i in the strategy profile $\boldsymbol{\sigma}$ is defined as $\mathsf{U}_i(\boldsymbol{\sigma}) = \sum_{\{v_i, v_j\} \in E : \sigma_i \neq \sigma_j} c_{ij} = \sum_{j \in Adj(i) : \sigma_i \neq \sigma_j} c_{ij}$. Hence, the utility of player i in $\boldsymbol{\sigma}$ is equal to the total weight of the edges which are incident to v_i in G and belong to the cut $(T_{\boldsymbol{\sigma}}, \overline{T}_{\boldsymbol{\sigma}})$. We use the *social function* $\mathsf{CUT}(\boldsymbol{\sigma}) = \sum_{\{v_i, v_j\} \in E : \sigma_i \neq \sigma_j} c_{ij}$, that is, the total weight of all the edges in the cut $(T_{\boldsymbol{\sigma}}, \overline{T}_{\boldsymbol{\sigma}})$ as a measure of the quality of a given strategy profile. It is easy to see that $\mathsf{CUT}(\boldsymbol{\sigma}) = \frac{1}{2} \sum_{i \in [n]} \mathsf{U}_i(\boldsymbol{\sigma})$, which justifies the use of the function CUT as a measure of the total welfare of the players.

For a cut game $\mathcal{C}(G)$, let $\mathcal{NE}_\epsilon(\mathcal{C}(G))$ be the set of its ϵ-approximate pure Nash equilibria and let $\boldsymbol{\sigma}^*$ be the strategy profile maximizing the social function CUT, that is, a maximum cut in G. The ϵ-**approximate price of anarchy** of $\mathcal{C}(G)$ is defined as

$$\mathsf{PoA}_\epsilon(\mathcal{C}(G)) = \min_{\sigma \in \mathcal{NE}_\epsilon(\mathcal{C}(G))} \frac{\mathsf{CUT}(\sigma)}{\mathsf{CUT}(\sigma^*)}.$$

Moreover, let $\mathcal{W}_\epsilon(\mathcal{C}(G))$ be the set of ϵ-approximate one-round walks starting from any initial strategy profile for $\mathcal{C}(G)$. The **approximation ratio of the solutions achieved after an ϵ-approximate one-round walk starting from any initial strategy profile** of $\mathcal{C}(G)$ is defined as

$$\mathsf{Apx}_\epsilon(\mathcal{C}(G)) = \min_{W \in \mathcal{W}_\epsilon(\mathcal{C}(G))} \frac{\mathsf{CUT}(f(W))}{\mathsf{CUT}(\sigma^*)}.$$

Hence, by definition, both $\mathsf{PoA}_\epsilon(\mathcal{C}(G))$ and $\mathsf{Apx}_\epsilon(\mathcal{C}(G))$ take values in the interval $[0, 1]$. These metrics naturally extend to the whole class of cut games $\mathcal{C}$ as follows:

$$\mathsf{PoA}_\epsilon(\mathcal{C}) = \inf_{\mathcal{C}(G) \in \mathcal{C}} \mathsf{PoA}_\epsilon(\mathcal{C}(G)) \qquad \text{and} \qquad \mathsf{Apx}_\epsilon(\mathcal{C}) = \inf_{\mathcal{C}(G) \in \mathcal{C}} \mathsf{Apx}_\epsilon(\mathcal{C}(G)).$$

3   The ϵ-Approximate Price of Anarchy

In this section, we give an exact characterization of the ϵ-approximate price of anarchy of cut games for any value of ϵ. Our theoretical analysis relies on the application of the primal-dual method introduced by Bilò in [5]. When instantiated to our scenario of investigation, such a method operates as follows.

Given a cut game $\mathcal{C}(G)$, let us denote with σ a generic ϵ-approximate pure Nash equilibrium of $\mathcal{C}(G)$. We formulate the problem of minimizing the ratio $\frac{\mathsf{CUT}(\sigma)}{\mathsf{CUT}(\sigma^*)}$ via linear programming. The two strategy profiles σ and σ^* play the role of fixed constants, while, for each $\{v_i, v_j\} \in E(G)$, the values c_{ij} defining the edge weights are variables that must be suitably chosen so as to satisfy two constraints: the first assures that σ is an ϵ-approximate pure Nash equilibrium of $\mathcal{C}(G)$, while the second normalizes to 1 the value of the social optimum $\mathsf{CUT}(\sigma^*)$. The objective function aims at minimizing the social value $\mathsf{CUT}(\sigma)$ which, being the social optimum normalized to 1, is equivalent to minimizing the ratio $\frac{\mathsf{CUT}(\sigma)}{\mathsf{CUT}(\sigma^*)}$.

Let us denote with $\mathsf{LP}(\sigma, \sigma^*)$ such a linear program. By the Weak Duality Theorem, each feasible solution to the dual program of $\mathsf{LP}(\sigma, \sigma^*)$ yields a lower bound on the optimal solution of $\mathsf{LP}(\sigma, \sigma^*)$. Hence, by providing a feasible dual solution, we obtain a lower bound on the ratio $\frac{\mathsf{CUT}(\sigma)}{\mathsf{CUT}(\sigma^*)}$ for $\mathcal{C}(G)$. Since no assumptions are made on $\mathcal{C}(G)$, if the provided dual solution is independent on the particular choice of σ and σ^*, we obtain a lower bound on the ratio $\frac{\mathsf{CUT}(\sigma)}{\mathsf{CUT}(\sigma^*)}$ for any possible pair of profiles σ and σ^* in any possible cut game $\mathcal{C}(G)$, which means that we obtain a lower bound on the ϵ-approximate price of anarchy of cut games.

Given a strategy profile σ, let us introduce a boolean variable α_{ij}^σ which, for any pair of indexes $i, j \in [n]$, takes the value 1 if and only if the edge $\{v_i, v_j\}$ belongs to the cut $(T_\sigma, \overline{T}_\sigma)$, that is, $\alpha_{ij}^\sigma = 1$ if and only if $\sigma_i \neq \sigma_j$. Moreover, define $\overline{\alpha}_{ij}^\sigma = 1 - \alpha_{ij}^\sigma$ as the complement of α_{ij}^σ.

By the definition of ϵ-approximate pure Nash equilibria, we obtain the following linear program $\mathsf{LP}(\sigma, \sigma^*)$.

$$min \sum_{\{v_i, v_j\} \in E} c_{ij} \alpha_{ij}^\sigma$$

subject to

$$(1+\epsilon) \sum_{j \in Adj(i)} c_{ij} \alpha_{ij}^\sigma - \sum_{j \in Adj(i)} c_{ij} \overline{\alpha}_{ij}^\sigma \geq 0, \quad \forall i \in [n]$$

$$\sum_{\{v_i, v_j\} \in E} c_{ij} \alpha_{ij}^{\sigma^*} = 1,$$

$$c_{ij} \geq 0, \qquad\qquad\qquad\qquad \forall \{v_i, v_j\} \in E$$

The dual program $\mathsf{DLP}(\sigma, \sigma^*)$ is defined as follows.

$$max \; \gamma$$

subject to

$$(1+\epsilon) \alpha_{ij}^\sigma (x_i + x_j) - \overline{\alpha}_{ij}^\sigma (x_i + x_j) + \gamma \alpha_{ij}^{\sigma^*} \leq \alpha_{ij}^\sigma, \quad \forall \{v_i, v_j\} \in E$$

$$x_i \geq 0, \qquad\qquad\qquad\qquad\qquad\qquad \forall i \in [n]$$

We obtain the following lower bound on the ϵ-approximate price of anarchy of cut games.

Theorem 1. *For any cut game* $\mathcal{C}(G)$ *and value* $\epsilon \geq 0$*, it holds* $\mathsf{PoA}_\epsilon(\mathcal{C}(G)) \geq \frac{1}{2+\epsilon}$.

Proof. It suffices to show that the solution in which $\gamma = \frac{1}{2+\epsilon}$ and $x_i = \frac{1}{4+2\epsilon}$ for each $i \in [n]$ is feasible for $\mathsf{DLP}(\sigma, \sigma^*)$. Since all the values x_i are non-negative, it remains to show that, for each $\{v_i, v_j\} \in E$, it holds

$$\frac{1+\epsilon}{2+\epsilon} \alpha_{ij}^\sigma - \frac{1}{2+\epsilon} \overline{\alpha}_{ij}^\sigma + \frac{1}{2+\epsilon} \alpha_{ij}^{\sigma^*} \leq \alpha_{ij}^\sigma. \tag{1}$$

Note that inequality (1) is implied by the following inequality

$$\frac{1+\epsilon}{2+\epsilon} \alpha_{ij}^\sigma - \frac{1}{2+\epsilon} \overline{\alpha}_{ij}^\sigma + \frac{1}{2+\epsilon} \leq \alpha_{ij}^\sigma \tag{2}$$

which we show to be always true.

For the case of $\alpha_{ij}^\sigma = 1$, inequality (2) becomes

$$\frac{1+\epsilon}{2+\epsilon} + \frac{1}{2+\epsilon} \leq 1$$

which is always true. On the other hand, for the case of $\alpha_{ij}^\sigma = 0$, inequality (2) becomes

$$-\frac{1}{2+\epsilon} + \frac{1}{2+\epsilon} \leq 0$$

which is again always true. $\qquad\qquad\qquad\qquad\qquad\qquad\qquad \square$

We now present a matching upper bound.

Theorem 2. *For any* $\epsilon \geq 0$, *there exists a cut game* $\mathcal{C}(G)$ *such that* $\mathsf{PoA}_\epsilon(\mathcal{C}(G)) = \frac{1}{2+\epsilon}$.

Proof. Consider the cut game $\mathcal{C}(C_4)$ defined by the four cycle C_4 in which the weights of the edges $\{v_1, v_2\}$ and $\{v_3, v_4\}$ are equal to $\frac{1}{4+2\epsilon}$, while the weights of the remaining ones are equal to $\frac{1+\epsilon}{4+2\epsilon}$. Being C_4 a bipartite graph, the maximum cut contains all the edges and, thus, has a social value equal to 1. It is easy to verify by inspection that the strategy profile $\boldsymbol{\sigma}$ such that $T_{\boldsymbol{\sigma}} = \{v_1, v_4\}$ is an ϵ-approximate pure Nash equilibrium such that $\mathsf{CUT}(\boldsymbol{\sigma}) = \frac{1}{2+\epsilon}$, which yields the claim. □

As a consequence of Theorems 1 and 2, it follows that, for any $\epsilon \geq 0$, the ϵ-approximate price of anarchy of cut games is equal to $1/(2+\epsilon)$. Note that, for $\epsilon = 0$, we reobtain the known value of $1/2$ for the price of anarchy of pure Nash equilibria.

Corollary 1. *For any* $\epsilon \geq 0$, *it holds* $\mathsf{PoA}_\epsilon(\mathcal{C}) = \frac{1}{2+\epsilon}$.

4 Performance of ϵ-Approximate One-Round Walks

Given a cut game $\mathcal{C}(G)$ and a value $\epsilon \geq 0$, let $\boldsymbol{W}$ be an ϵ-approximate one-round walk for $\mathcal{C}(G)$. For notation purposes, set $\boldsymbol{\sigma}^S := s(\boldsymbol{W})$, $\boldsymbol{\sigma}^F := f(\boldsymbol{W})$ and denote with $\mathcal{CH}$ the set of players changing their strategy during the walk, that is, $\mathcal{CH} = \{i \in [n] : \sigma_i^F \neq \sigma_i^S\}$. By using the same approach of the previous section, we obtain the following linear program $\mathsf{LP}(\boldsymbol{\sigma}^S, \boldsymbol{\sigma}^F, \boldsymbol{\sigma}^*)$.

$$min \sum_{\{v_i, v_j\} \in E} c_{ij} \alpha_{ij}^{\sigma^F}$$

subject to

$$(1+\epsilon) \sum_{j \in Adj(i):j<i} c_{ij} \alpha_{ij}^{\sigma^F} + (1+\epsilon) \sum_{j \in Adj(i):j>i} c_{ij} \alpha_{ij}^{\sigma^S}$$
$$- \sum_{j \in Adj(i):j<i} c_{ij} \overline{\alpha}_{ij}^{\sigma^F} - \sum_{j \in Adj(i):j>i} c_{ij} \overline{\alpha}_{ij}^{\sigma^S} \geq 0, \qquad \forall i \in [n] \setminus \mathcal{CH}$$

$$-(1+\epsilon) \sum_{j \in Adj(i):j<i} c_{ij} \overline{\alpha}_{ij}^{\sigma^F} - (1+\epsilon) \sum_{j \in Adj(i):j>i} c_{ij} \alpha_{ij}^{\sigma^S}$$
$$+ \sum_{j \in Adj(i):j<i} c_{ij} \alpha_{ij}^{\sigma^F} + \sum_{j \in Adj(i):j>i} c_{ij} \overline{\alpha}_{ij}^{\sigma^S} \geq 0, \qquad \forall i \in \mathcal{CH}$$

$$\sum_{\{v_i, v_j\} \in E} c_{ij} \alpha_{ij}^{\sigma^*} = 1,$$

$$c_{ij} \geq 0, \qquad \forall \{v_i, v_j\} \in E$$

The first constraint states that, for each player not changing her strategy in $\boldsymbol{W}$, the utility that she is getting at the time in which she is processed is at least

$1 + \epsilon$ times the one that she can achieve when moving to the opposite side, that is, each player not belonging to $\mathcal{CH}$ is currently playing an ϵ-approximate best response at the time she is processed by $\boldsymbol{W}$. Note that, in order to compute the utility that player i gets at the time she is processed by $\boldsymbol{W}$, we have to consider the fact that all nodes with indexes smaller than i might have changed side with respect to $\boldsymbol{\sigma}^S$; hence, we evaluate their side with respect to $\boldsymbol{\sigma}^F$. For the nodes with indexes greater than i, we evaluate their side with respect to $\boldsymbol{\sigma}^S$, since their associated players have not been processed yet at the time in which player i is. The second constraint states that each player changing her strategy in $\boldsymbol{W}$ has an ϵ-approximate improving deviation, that is, each such a player improves by a factor of at least $1+\epsilon$ by moving to the opposite side. Note that, by the definition of ϵ-approximate improving deviations, the second constraint should have been a strict inequality; nevertheless, this does not affect our analysis since, as by so doing we are expanding the set of the feasible solutions of a minimization problem, we can only obtain less significant lower bounds (our objective, in fact, is to obtain the highest possible lower bound).

The dual program $\mathsf{DLP}(\boldsymbol{\sigma}^S, \boldsymbol{\sigma}^F, \boldsymbol{\sigma}^*)$ is defined as follows.

$$
\begin{aligned}
&max\ \gamma\\
&subject\ to\\
&(1+\epsilon)\left(\alpha_{ij}^{\sigma^F} x_j + \alpha_{ij}^{\sigma^S} x_i\right)\\
&\quad -\overline{\alpha}_{ij}^{\sigma^F} x_j - \overline{\alpha}_{ij}^{\sigma^S} x_i + \gamma\alpha_{ij}^{\sigma^*} \leq \alpha_{ij}^{\sigma^F}, \quad \forall\{v_i, v_j\}\in E: i,j \notin \mathcal{CH}\\
&-(1+\epsilon)\left(\overline{\alpha}_{ij}^{\sigma^F} y_j + \alpha_{ij}^{\sigma^S} y_i\right)\\
&\quad +\alpha_{ij}^{\sigma^F} y_j + \overline{\alpha}_{ij}^{\sigma^S} y_i + \gamma\alpha_{ij}^{\sigma^*} \leq \alpha_{ij}^{\sigma^F}, \quad \forall\{v_i, v_j\}\in E: i,j \in \mathcal{CH}\\
&(1+\epsilon)\left(\alpha_{ij}^{\sigma^F} x_j - \alpha_{ij}^{\sigma^S} y_i\right)\\
&\quad -\overline{\alpha}_{ij}^{\sigma^F} x_j + \overline{\alpha}_{ij}^{\sigma^S} y_i + \gamma\alpha_{ij}^{\sigma^*} \leq \alpha_{ij}^{\sigma^F}, \quad \forall\{v_i, v_j\}\in E: i\in \mathcal{CH}, j \notin \mathcal{CH}\\
&-(1+\epsilon)\left(\overline{\alpha}_{ij}^{\sigma^F} y_j - \alpha_{ij}^{\sigma^S} x_i\right)\\
&\quad +\alpha_{ij}^{\sigma^F} y_j - \overline{\alpha}_{ij}^{\sigma^S} x_i + \gamma\alpha_{ij}^{\sigma^*} \leq \alpha_{ij}^{\sigma^F}, \quad \forall\{v_i, v_j\}\in E: i\notin \mathcal{CH}, j \in \mathcal{CH}\\
&x_i, y_i \geq 0, \quad\quad\quad\quad\quad\quad\quad\quad\quad\quad \forall i \in [n]
\end{aligned}
$$

We obtain the following lower bound of the approximation ratio of the solutions achieved after an approximate one-round walk starting from any initial strategy profile.

Theorem 3. *For any cut game $\mathcal{C}(G)$ and value $\epsilon \geq 0$, it holds $\mathsf{Apx}_\epsilon(\mathcal{C}(G)) \geq \min\left\{\frac{1}{2+\epsilon}, \frac{2\epsilon}{(1+\epsilon)(2+\epsilon)}\right\}$.*

Also in this case, we give a matching upper bound as shown in the following theorem.

Theorem 4. *For any $\epsilon \geq 0$, there exists a cut game $\mathcal{C}(G)$ such that $\mathsf{Apx}_\epsilon(\mathcal{C}(G)) \leq \min\left\{\frac{1}{2+\epsilon}, \frac{2\epsilon}{(1+\epsilon)(2+\epsilon)}\right\}$.*

As a consequence of Theorems 3 and 4, it follows that the approximation ratio of the solutions achieved after an ϵ-approximate one-round walk starting from any initial strategy profile in cut games is equal to $1/(2 + \epsilon)$ for any $\epsilon \geq 1$ and equal to $\frac{2\epsilon}{(\epsilon+1)(\epsilon+2)}$ for any $0 \leq \epsilon < 1$. Note that $\mathsf{Apx}_0(\mathcal{C}) = 0$ models the fact that there are cut games with some strategy profiles starting from which some 0-approximate one-round walk may end up to a solution whose social value is arbitrarily far from the social optimum, as already shown by Christodoulou *et al.* [14].

Corollary 2. *For any $\epsilon \geq 0$, it holds $\mathsf{Apx}_\epsilon(\mathcal{C}) = \min \left\{ \frac{1}{2+\epsilon}, \frac{2\epsilon}{(\epsilon+1)(\epsilon+2)} \right\}$.*

5   Conclusions

Our findings reveal an unexpected and surprisingly good performance of the very simple solutions generated after an ϵ-approximate one-round walk, independently of which is the initial strategy profile, for some values of ϵ. In particular, since for $\epsilon \geq 1$ it holds $\mathsf{PoA}_\epsilon(\mathcal{C}) = \mathsf{Apx}_\epsilon(\mathcal{C})$, any ϵ-approximate one-round walk converges to an approximation of the social optimum which is never worse than the ϵ-approximate price of anarchy. To the best of our knowledge, this is the first evidence of such a quick convergence to such a good quality solution. Moreover, for $\epsilon = 1$, it holds $\mathsf{Apx}_\epsilon(\mathcal{C}) = 1/3$, whereas a recent paper by Bilò *et al.* [7] shows that the approximation ratio of the solutions achieved after a one-round walk starting from any initial strategy profile performed by greedy players who apply a farsighted, rather than a myopic, rationality (termed as *sequential price of anarchy*) is exactly $1/3$. Thus, 1-approximate greedy players perform as well as farsighted greedy ones in the worst-case.

In the light of these interesting situations, determining whether mildly greedy players may exhibit a similar outstanding performance in other contexts as well becomes an intriguing research direction.

References

1. Ackermann, H., Röglin, H., Vöcking, B.: On the impact of combinatorial structure on congestion games. Journal of the ACM 55(6) (2008)
2. Awerbuch, B., Azar, Y., Epstein, A., Mirrokni, V.S., Skopalik, A.: Fast Convergence to Nearly Optimal Solutions in Potential Games. In: Proceedings of the 9th ACM Conference on Electronic Commerce (EC), pp. 264–273. ACM Press (2008)
3. Bhalgat, A., Chakraborty, T., Khanna, S.: Nash Dynamics in Congestion Games with Similar Resources. In: Leonardi, S. (ed.) WINE 2009. LNCS, vol. 5929, pp. 362–373. Springer, Heidelberg (2009)
4. Bhalgat, A., Chakraborty, T., Khanna, S.: Approximating Pure Nash Equilibrium in Cut, Party Affiliation, and Satisfiability Games. In: Proceedings of the 11th ACM Conference on Electronic Commerce (EC), pp. 73–82. ACM Press (2010)
5. Bilò, V.: A Unifying Tool for Bounding the Quality of Non-cooperative Solutions in Weighted Congestion Games. In: Erlebach, T., Persiano, G. (eds.) WAOA 2012. LNCS, vol. 7846, pp. 215–228. Springer, Heidelberg (2013)

6. Bilò, V., Fanelli, A., Flammini, M., Moscardelli, L.: Performance of One-Round Walks in Linear Congestion Games. Theory of Computing Systems 49(1), 24–45 (2011)
7. Bilò, V., Flammini, M., Monaco, G., Moscardelli, L.: Some Anomalies of Farsighted Strategic Behavior. Theory of Computing Systems (to appear)
8. Caragiannis, I., Fanelli, A., Gravin, N., Skopalik, A.: Efficient Computation of Approximate Pure Nash Equilibria in Congestion Games. In: Proceedings of the IEEE 52nd Annual Symposium on Foundations of Computer Science (FOCS), pp. 532–541. IEEE Computer Society (2011)
9. Caragiannis, I., Fanelli, A., Gravin, N., Skopalik, A.: Approximate Pure Nash Equilibria in Weighted Congestion Games: Existence, Efficient Computation, and Structure. In: Proceedings of the ACM Conference on Electronic Commerce (EC), pp. 284–301. ACM Press (2012)
10. Caragiannis, I., Flammini, M., Kaklamanis, C., Kanellopoulos, P., Moscardelli, L.: Tight Bounds for Selfish and Greedy Load Balancing. Algorithmica 61(3), 606–637 (2011)
11. Chen, X., Deng, X., Teng, S.: Settling the Complexity of Computing Two-Player Nash Equilibria. Journal of ACM 56(3) (2009)
12. Chien, S., Sinclair, A.: Convergence to Approximate Nash Equilibria in Congestion Games. Games and Economic Behavior 71(2), 315–327 (2001)
13. Christodoulou, G., Koutsoupias, E., Spirakis, P.G.: On the Performance of Approximate Equilibria in Congestion Games. Algorithmica 61(1), 116–140 (2011)
14. Christodoulou, G., Mirrokni, V.S., Sidiropoulos, A.: Convergence and Approximation in Potential Games. Theoretical Computer Science 438, 13–27 (2012)
15. Daskalakis, K., Goldberg, P.W., Papadimitriou, C.H.: The Complexity of Computing a Nash Equilibrium. Communications of ACM 52(2), 89–97 (2009)
16. Daskalakis, K., Papadimitriou, C.H.: Three-Player Games Are Hard. Electronic Colloquium on Computational Complexity (ECCC) (139) (2005)
17. Fabrikant, A., Papadimitriou, C.H., Talwar, K.: The Complexity of Pure Nash Equilibria. In: Proceedings of the 36th Annual ACM Symposium on Theory of Computing (STOC), pp. 604–612. ACM Press (2004)
18. Fanelli, A., Flammini, M., Moscardelli, L.: The Speed of Convergence in Congestion Games Under Best-response Dynamics. ACM Transactions on Algorithms 8(3), 25 (2012)
19. Fanelli, A., Moscardelli, L.: On Best-response Dynamics in Weighted Congestion Games with Polynomial Delays. Distributed Computing 24(5), 245–254 (2011)
20. Goemans, M.X., Mirrokni, V.S., Vetta, A.: Sink Equilibria and Convergence. In: Proceedings of the 46th Annual IEEE Symposium on Foundations of Computer Science (FOCS), pp. 142–154. IEEE Computer Society (2005)
21. Koutsoupias, E., Papadimitriou, C.: Worst-case equilibria. In: Meinel, C., Tison, S. (eds.) STACS 1999. LNCS, vol. 1563, pp. 404–413. Springer, Heidelberg (1999)
22. Mirrokni, V.S., Vetta, A.: Convergence Issues in Competitive Games. In: Jansen, K., Khanna, S., Rolim, J.D.P., Ron, D. (eds.) APPROX and RANDOM 2004. LNCS, vol. 3122, pp. 183–194. Springer, Heidelberg (2004)
23. Rosenthal, R.W.: A Class of Games Possessing Pure-Strategy Nash Equilibria. International Journal of Game Theory 2, 65–67 (1973)
24. Skopalik, A., Vöcking, B.: Inapproximability of Pure Nash Equilibria. In: Proceedings of the 40th Annual ACM Symposium on Theory of Computing (STOC), pp. 355–364. ACM Press (2008)

Statistical Properties of Short RSA Distribution and Their Cryptographic Applications

Pierre-Alain Fouque[1] and Jean-Christophe Zapalowicz[2]

[1] Université de Rennes 1 and Institut Universitaire de France
Pierre-Alain.Fouque@univ-rennes1.fr
[2] Inria, France
jean-christophe.zapalowicz@inria.fr

Abstract. In this paper, we study some computational security assumptions involved in two cryptographic applications related to the RSA cryptosystem. To this end, we use exponential sums to bound the statistical distances between these distributions and the uniform distribution. We are interested in studying the k least (or most) significant bits of $x^e \bmod N$, where N is an RSA modulus and x only belongs to a small interval of $[0, N)$.

First of all, we provide the first rigorous evidence that the cryptographic pseudo-random generator proposed by Micali and Schnorr is based on firm foundations. This proof is missing in the original paper and does not cover the parameters chosen by the authors. Consequently, we extend the proof to get a new result closer to these parameters using recently new exponential sums results and we show some limitations of our technique. Finally, we look at the semantic security of the RSA padding scheme called PKCS#1 v1.5 which is still used a lot in practice. We show that parts of the ciphertext are indistinguisable from uniform bitstrings.

Keywords: Exponential Sums, Security Proof for Micali-Schnorr pseudorandom generator, semantic security of RSA padding scheme.

1 Introduction

The RSA assumption states that, given a random value y in $\mathbb{Z}/N\mathbb{Z}$ where N is the product of two large primes, it is difficult to compute a e-th root of y, *i.e.* find x such that $y = x^e \bmod N$. The RSA problem has been heavily studied by mathematicians and no attack more efficient than factoring the RSA modulus has been found since its discovery. Usually, it is very difficult to prove a computational assumption such as RSA and cryptographers try to prove that this one is at least as difficult as another one, for instance factoring. However, some evidences for the non-equivalence of these two hard problems has been provided by Boneh et Venkatesan in [7] while the RSA assumption seems to hold.

RSA is a valid cryptographic assumption since on average its difficulty seems to be established thanks to its self-reducibility property. Indeed, it is well-known that if we are able to invert RSA on a non-negligible subset of $\mathbb{Z}/N\mathbb{Z}$, then

Z. Cai et al. (Eds.): COCOON 2014, LNCS 8591, pp. 525–536, 2014.

we can invert nearly all values in $\mathbb{Z}/N\mathbb{Z}$ with high probability. Based on this assumption, cryptographers have proposed and proved that the RSA signature and encryption schemes using adequate padding functions [4,3] are secure in the random oracle model [2]. The security proof of RSA-OAEP for encryption appeared in [11].

Another direction to assess the security of a computational assumption consists in showing that the values we are looking for are computationally or statistically indistinguishable from the uniform distribution on bitstrings of the same size. Consequently, the best the adversary can do is to guess this value until he finds it. For RSA, it is easy to see that the value y is uniformly distributed if x is. In this paper, we will be interested in the short RSA problem: given y and the promise that $x < M \ll N$, find x such that $x^e \bmod N$. Clearly, if $M \leq N^{1/e}$ then it is possible to recover the value x using the well-known Hensel's lifting lemma whose complexity is linear in the size of x. However, we can wonder what is the security of this new assumption when $M \gg N^{1/e}$. It is trivial to see that if $M = N^{\frac{1}{e}+\varepsilon}$ then, by guessing the high order bits of x, in time N^ε times a polynomial in $\log N$, we can invert x. However for larger values of M, the problem seems to be hard. This assumption is made in some standards, such as the standard PKCS#1 v1.5 that is used to protect RSA encryption and we will study it in a special case. To assess the security of this short RSA assumption, the classical technique consists in studying the distribution of values $x^e \bmod N$ when $x < M$. This distribution cannot be uniform in $\mathbb{Z}/N\mathbb{Z}$ when the output is larger than M, but it can be computationally difficult to distinguish it from the uniform distribution in $\mathbb{Z}/N\mathbb{Z}$. We will study the short RSA distribution when we consider only some part of the output bits. In this case, it is possible to prove some mathematical statements on the statistical distance between this distribution and the uniform distribution. We will also study the security of the Micali-Schnorr pseudorandom generator. At COCOON 2013, we have shown some attacks that explain the choice of the parameters proposed by Micali and Schnorr [9]. This generator assumes that the distribution of the least significant bits of $x^e \bmod N$ is indistinguishable from the uniform distribution in $\{0,1\}^k$ if $x < M$ with $M = N^{2/e}$.

Our Contributions. In the first part of this paper, we will prove the following informal theorem for different values of M.

Theorem 1. Let $N = pq$ be a balanced RSA modulus, e the public exponent and $M < N$. Let the function $f : \mathbb{Z}/M\mathbb{Z} \longrightarrow \mathbb{Z}/N\mathbb{Z}$ defined as $f(x) = x^e \bmod N$. The k least significant bits of $f(x)$ for $k < \log N$ are statistically indistinguishable from the uniform distribution on $\{0,1\}^k$.

For $M \gg \sqrt{N}$, we will show it using classical bounds, and for $N^{1/e} \ll M \ll \sqrt{N}$, we will use more recent results proved by Wooley [17]. This last bound is very close to be optimal since for $M \leq N^{1/e}$, it is possible in polynomial-time to recover $x \leq M$ given the $k \geq M$ least significant bits of $f(x)$ as we explain in [9]. Indeed, in this case the function f becomes non modular and the problem of retrieving x is quite easy by using Hensel's lifting.

In a second part, we will show two applications of these theorems. Micali and Schnorr original proof refers to the first bound $M \gg \sqrt{N}$ when $e = 3$, or the second one otherwise. However the proof is missing and they do not give any hint to explain their more aggressive choice of parameters. Indeed, it would be possible to output less bits at each iteration of the generator, but the efficiency of this generator would be less efficient than the Blum-Blum-Shub generator [6] for instance. Micali and Schnorr prefer to output more bits and avoid the previous attack when $M = N^{1/e}$. Last year, we developed some attacks to go beyond this bound using some time/memory tradeoff techniques which require exponential time complexity. Our result allows us to propose parameters ensuring the randomness of the output. We also explain that the parameters proposed by Micali and Schnorr are close to be optimal in the special case of $e = 3$.

Finally, we propose to study the semantic security of the RSA padding called PKCS #1 v1.5 proposed by the RSA Labs. This padding is used in practice in many IETF standards and its security has been studied in [1] under various security notions. Here, we study the semantic security, *i.e.* the most interesting security notion, to assess this assumption as much as we can using mathematical and rigorous statements. This notion means that no bit of the plaintext leaks when we see the ciphertext. In this paper, we will show that no bit of the plaintext leaks when we see some part of the bits of the ciphertext.

2   Some Mathematical Backgrounds

Statistical Distance. We need several results on the regularity of the probability distributions related to the studied problem. Recall that the *statistical distance* between a random variable X on a finite set S and the uniform distribution is defined as:

$$\Delta_1(X) = \frac{1}{2} \cdot \sum_{s \in S} \left| \Pr[X = s] - \frac{1}{|S|} \right|.$$

We say that X is δ-*statistically close to uniform* when $\Delta_1(X) \leq \delta$.

In addition we will consider the *collision probability* of a random variable X on a finite set S, defined as:

$$\mathrm{Col}(X) = \sum_{s \in S} \Pr[X = s]^2.$$

Finally the link between the statistical distance and the collision probability is given by the following lemma proven in [15]:

Lemma 1. *Let X be a random variable with values in a finite set S of size m. If X has collision probability β and distance δ from uniform on S, then:*

$$\delta \leq \frac{1}{2} \sqrt{m\beta - 1}.$$

Exponential Sums. Our proof of Theorem 1 relies on exponential sums in $\mathbb{Z}/m\mathbb{Z}$, so we first fix some notations and recall useful standard results. For any integer m, we denote by $\mathbf{e}_m$ the additive character $\mathbb{Z}/m\mathbb{Z} \rightarrow \mathbb{C}^*$ given by $\mathbf{e}_m(x) = \exp(2i\pi x/m)$. The following results hold.

Proposition 1 (Orthogonality). *For all $x \in \mathbb{Z}/m\mathbb{Z}$, we have:*

$$\sum_{c=0}^{m-1} \mathbf{e}_m(cx) = \begin{cases} 0 & \text{if } c \not\equiv 0 \pmod{m}, \\ m & \text{if } c \equiv 0 \pmod{m}. \end{cases}$$

Lemma 2 ([16, Problem 11.c]). *For any modulus[1] $m \geq 60$ and any non negative integers h, k, we have:*

$$\sum_{c=1}^{m-1} \left| \sum_{x=k}^{k+h} \mathbf{e}_m(cx) \right| \leq (m-1)\log m.$$

Lemma 3 (Weil [13]). *Consider a prime modulus p. For all polynomials $g(X), h(X) \in \mathbb{F}_p[X]$ such that the rational function $f(X) = h(X)/g(X)$ is not constant on $\mathbb{F}_p$, the bound:*

$$\left| \sum_{\substack{x \in \mathbb{F}_p \\ g(x) \neq 0}} \mathbf{e}_p\big(f(x)\big) \right| \leq \big(\max(\deg g, \deg h) + v - 1 \big) \cdot p^{1/2}$$

holds, with v the number of distinct zeros of $g(X)$ in the algebraic closure of $\mathbb{F}_p$.

3   Main Results

To prove Theorem 1, we have to estimate the statistical distance between the function $\mathsf{lsb}_k(x^e \bmod N)$ for x randomly chosen in $\mathbb{Z}/M\mathbb{Z}$ and the uniform distribution modulo 2^k. In function of the values of N, e, k and M, we will be able to show (or not) the statistically indistinguishability of these two probability ensembles. More precisely the ratio between M and N is crucial in our analysis and we propose two different techniques for estimating the statistical distance. The first one is meaningful when $M \gg \sqrt{N}$ (see Theorem 2) and the second one is useful for $M \ll \sqrt{N}$ (see Theorem 3).

3.1   First Bound When $M \gg \sqrt{N}$

The first technique we propose uses mostly the technical lemmas on exponential sums which are recalled in Section 2. We obtain a bound on the statistical distance which is negligible when the parameter M is sufficiently larger than $\sqrt{N}$. With this first analysis we cannot hope to approach the optimal bound, i.e. $M \gg N^{1/e}$ for $e \geq 3$. However its advantage may lie in providing concrete values of M and k for cryptographic sizes of modulus N. It is thus an useful bound for our applications.

---

[1] We assume that this bound on m holds for all moduli involved in our computations below, i.e. $p, q > 60$, which is of course satisfied for all RSA moduli in practice.

Theorem 2. *Let $N = pq$ be a balanced RSA modulus (i.e. $N = pq$ for primes p, q such that $60 < q < p < 2q$), e the public exponent and $M < N$ an integer such that $M \gg \sqrt{N}$. Then the random variable $X = \mathsf{lsb}_k(x^e \bmod N)$ for x randomly chosen in $\mathbb{Z}/M\mathbb{Z}$ is δ-statistically close to uniform with:*

$$\delta = \sqrt{\frac{2^k}{N} + \frac{2^{k/2}e^2\sqrt{N}\log^{3/2} N}{M}}.$$

Proof. The values of the random variable X are taken in $[0, 2^k)$ with the following distribution: x is chosen uniformly at random in $\mathbb{Z}/M\mathbb{Z}$ and we output $f(x) = \mathsf{lsb}_k(x^e \bmod N)$. We are interested in bounding the collision probability of this random variable. By denoting $K = \lfloor \frac{N-1}{2^k} \rfloor$, we can evaluate this probability using the orthogonality property of additive characters (Prop. 1):

$$\mathrm{Col}(X) = \frac{1}{M^2} \times \left|\{(x,y) \in [0, M-1]^2 \mid \exists u \in [\![-K, K]\!], x^e - y^e = 2^k \cdot u \bmod N\}\right|,$$

$$\leq \frac{2}{M^2N} \sum_{x=0}^{M-1} \sum_{y=0}^{M-1} \sum_{u=0}^{K} \sum_{a=0}^{N-1} \mathbf{e}_N(a(x^e - y^e - 2^k \cdot u)).$$

We now define by B the value $\max_a |S(a, M)|$ where $S(a, M) = \sum_{0 \leq x < M} \mathbf{e}_N(ax^e)$ to uniformly bound $\sum_{x=0}^{M-1}\sum_{y=0}^{M-1} \mathbf{e}_N(ax^e - ay^e) = S(a, M)\overline{S(a, M)} = |S(a, M)|^2$. The contribution of $a = 0$ is exactly $2(K+1)/N$ and if we put it aside we get:

$$\mathrm{Col}(X) \leq \frac{2(K+1)}{N} + \frac{2}{M^2N} \sum_{a=1}^{N-1} |S(a, M)|^2 \left| \sum_{u=0}^{K} \mathbf{e}_N(-a2^k \cdot u) \right|.$$

The probability collision can be bounded using B and Lemma 2 since the function $a \in \mathbb{Z}/N\mathbb{Z} \setminus \{0\} \to 2^k a \in \mathbb{Z}/N\mathbb{Z} \setminus \{0\}$ is a bijection:

$$\mathrm{Col}(X) \leq \frac{2(K+1)}{N} + \frac{2}{M^2N} \left(\max_{1 \leq a \leq N-1} |S(a, M)|^2 \right) \sum_{a=1}^{N-1} \left| \sum_{u=0}^{K} \mathbf{e}_N(-a2^k \cdot u) \right|,$$

$$\mathrm{Col}(X) \leq \frac{2(K+1)}{N} + \frac{2}{M^2N}B^2 \cdot N \log N. \tag{1}$$

It remains to bound B for all $a \in \mathbb{Z}/N\mathbb{Z} \setminus \{0\}$. The incomplete exponential sum $S(a, M)$ can be expressed using the complete one as follow:

$$S(a, M) = \sum_{x < M} \mathbf{e}_N(ax^e) = \sum_{x \in \mathbb{Z}/N\mathbb{Z}} \mathbf{e}_N(ax^e) \cdot [\![x \in [\![0, M-1]\!]]\!],$$

$$= \sum_{x \in \mathbb{Z}/N\mathbb{Z}} \mathbf{e}_N(ax^e) \frac{1}{N} \sum_{m=0}^{M-1} \sum_{b \in \mathbb{Z}/N\mathbb{Z}} \mathbf{e}_N(b(x - m)),$$

$$= \frac{1}{N} \sum_{b \in \mathbb{Z}/N\mathbb{Z}} \sum_{m=0}^{M-1} \mathbf{e}_N(-bm) \sum_{x \in \mathbb{Z}/N\mathbb{Z}} \mathbf{e}_N(ax^e + bx).$$

where $[\cdot]$ is the usual Iverson bracket notation: for a statement U, $[U] = 1$ is U is true and 0 otherwise. If we pick integers u, v such that $up + vq = 1$, we see that the sum in x decomposes as:

$$\sum_{x \in \mathbb{Z}/N\mathbb{Z}} \mathbf{e}_N(ax^e + bx) = \sum_{x \in \mathbb{Z}/N\mathbb{Z}} \mathbf{e}_N\big((up + vq) \cdot (ax^e + bx)\big)$$

$$= \sum_{x_p \in \mathbb{F}_p} \mathbf{e}_p\big(vg_{a,b}(x_p) \bmod p\big) \sum_{x_q \in \mathbb{F}_q} \mathbf{e}_q\big(ug_{a,b}(x_q) \bmod q\big)$$

where the function $g_{a,b}$ is given by $g_{a,b}(x) = ax^e + bx$. Now if $a \neq 0 \bmod p$, $vg_{a,b}$ is a non constant function in $\mathbb{F}_p$, so Lemma 3 ensures:

$$|T_p| = \left| \sum_{x_p \in \mathbb{F}_p} \mathbf{e}_p\big(vg_{a,b}(x_p) \bmod p\big) \right| \leq e\sqrt{p}.$$

On the other hand, if $a = 0 \bmod p$, we have:

$$\sum_{x_p \in \mathbb{F}_p} \mathbf{e}_p\big(vg_{a,b}(x_p) \bmod p\big) = \sum_{x_p \in \mathbb{F}_p} \mathbf{e}_p(vbx) = \begin{cases} p & \text{if } b = 0 \bmod p, \\ 0 & \text{otherwise.} \end{cases}$$

A corresponding result holds for $T_q = \sum_{x_q \in \mathbb{F}_q} \mathbf{e}_q(ug_{a,b}(x_q) \bmod q)$ and as a result, to bound $|S(a, M)|$, we have to separate the case when a is invertible modulo N from the case where it is a multiple of p or q. When a is invertible modulo N, we directly have:

$$|S(a, M)| \leq \frac{1}{N} \sum_{b \in \mathbb{Z}/N\mathbb{Z}} \left| \sum_{m=0}^{M-1} \mathbf{e}_N(-bm) \right| \cdot |T_p| \cdot |T_q|,$$

$$\leq \frac{1}{N} \sum_{b \in \mathbb{Z}/N\mathbb{Z}} \left| \sum_{m=0}^{M-1} \mathbf{e}_N(-bm) \right| \cdot e\sqrt{p} \cdot e\sqrt{q} \leq \frac{1}{N} e^2 \sqrt{N} \cdot N \log N$$

by Lemma 2. On the other hand, assume that a is a multiple of p. We have:

$$|S(a, M)| = \left| \frac{p}{N} \sum_{\substack{b \in \mathbb{Z}/N\mathbb{Z} \\ b=0 \bmod p}} \sum_{m=0}^{M-1} \mathbf{e}_N(-bm) \sum_{x_q \in \mathbb{F}_q} \mathbf{e}_q\big(ug_{a,b}(x_q) \bmod q\big) \right|$$

$$\leq \frac{1}{q} \sum_{b'=0}^{q-1} \left| \sum_{m=0}^{M-1} \mathbf{e}_N(-bm) \right| \cdot e\sqrt{q} \leq e\sqrt{q} \log q < e^2 \sqrt{N} \log N,$$

by applying Lemma 2. the same bound holds when a is a multiple of q, so that $B \leq e^2 \sqrt{N} \log N$. Together with (1), we obtain:

$$\text{Col}(X) \leq \frac{2(K+1)}{N} + \frac{2e^4 N \log^3 N}{M^2}.$$

Finally Lemma 1 provides a bound on the statistical distance $\Delta_1(X)$:

$$\Delta_1(X) \leq \sqrt{\frac{2^k(K+1)}{N} - 1} + \frac{2^{k/2}e^2\sqrt{N}\log^{3/2} N}{M},$$

and to conclude the proof, it is easy to show that $\left|\frac{K+1}{N} - 1/2^k\right| \leq 1/N.$ ☐

3.2   Second Bound When $M \ll \sqrt{N}$

Here we treat the case where M is smaller than $\sqrt{N}$, which will be interesting to approach the optimal bound, i.e. $M \gg N^{1/e}$. Even if the following lemma and corollaries do not require anything on the size of M (except that it is less than N obviously), the bounds we find for $\Delta_1(X)$ are only interesting for small values of M, meaning $M \ll \sqrt{N}$.

Theorem 3. *Let Let N, e, X be as in Theorem 2 and $M < N$ an integer such that $M \ll \sqrt{N}$. Then X is δ-statistically close to uniform with:*

$$\delta = \sqrt{\frac{2^k}{N} + 2^{k/2}\log^{3/2} N \left(\frac{1}{M} + \frac{N}{M^e}\right)^{\frac{1}{2e(e-1)+1}}} + \frac{2^{k/2}e\log^{3/2} N}{M}.$$

Proof. This result is based on a more recent result proved by Wooley, which provides another evaluation of the exponential sum $S(a, M)$. We give here a specific version adapted to our case:

Theorem 4 (Wooley, [17]). *Let e be an integer with $e \geq 2$, and let $a/N \in \mathbb{R}$. Suppose that, for some $c \in \mathbb{Z}$ and $N \in \mathbb{N}$ with $\gcd(c, N) = 1$, one has $|a/N - c/N| \leq N^{-2}$ and $N \leq M^e$. Then one has:*

$$\sum_{1 \leq x \leq M} \mathbf{e}_N(ax^e) \ll M^{1+\varepsilon}(N^{-1}+M^{-1}+N \cdot M^{-e})^{\sigma(e)} \quad \text{where} \quad \sigma(e)^{-1} = 2e(e-1).$$

According to [17], the factor M^ε may be replaced by $\log(2M)$, if one increases $\sigma(e)^{-1}$ from $2e(e-1)$ to $2e^2 - 2e + 1$. For sake of simplicity, we bound $\log(2M)$ by $\log N$ (with the weak assumption that $M \leq N/2$) and we neglect the term $1/N$ since it is negligible compared to $\min(M^{-1}, N \cdot M^{-e})$. Thus we obtain:

$$|S(a, M)| \ll M \log N(M^{-1} + N \cdot M^{-e})^{\frac{1}{2e(e-1)+1}}.$$

Note that this bound is correct for $a \in (\mathbb{Z}/N\mathbb{Z})^*$ and because it is meaningful when $M \ll \sqrt{N}$, one cannot bound as before the value $|S(a, M)|$ for $a \notin (\mathbb{Z}/N\mathbb{Z})^*$. Starting from:

$$\text{Col}(X) \leq \frac{2(K+1)}{N} + \frac{2}{M^2N} \sum_{a=1}^{N-1} |S(a, M)|^2 \left|\sum_{u=0}^{K} \mathbf{e}_N(-a2^k \cdot u)\right|,$$

we decompose the sum in a as:

$$\sum_{a=1}^{N-1} |S(a, M)|^2 \left|\sum_{u=0}^{K} \mathbf{e}_N(-a2^k \cdot u)\right| = S^* + S_p + S_q$$

with S^* the sum in $a \in (\mathbb{Z}/N\mathbb{Z})^*$ and S_p (resp. S_q) the one in $a \in \mathbb{Z}/N\mathbb{Z}\backslash\{0\}$ such that $a = 0 \bmod p$ (resp. $a = 0 \bmod q$). The sum S^* is bounded using Theorem 4 and Lemma 2, whereas we treat S_p (and similarly S_q) using an intermediate result from the previous proof and Lemma 2, i.e.:

$$S^* \leq M^2 \log^2 N (M^{-1} + N \cdot M^{-e})^{\frac{2}{2e(e-1)+1}} \cdot N \log N,$$

$$S_p \leq e^2 q \log^2 q \cdot \sum_{a_p \in \mathbb{F}_q^*} \left| \sum_{u=0}^{K} e_q(-a_p 2^k \cdot u) \right| \leq e^2 q^2 \log^3 q \leq e^2 N \log^3 N.$$

We thus have:

$$\mathrm{Col}(X) \leq \frac{2(K+1)}{N} + 2\log^3 N (M^{-1} + N \cdot M^{-e})^{\frac{2}{2e(e-1)+1}} + \frac{4e^2 \log^3 N}{M^2}$$

$$\Delta_1(X) \leq \sqrt{\frac{2^k}{N} + 2^{k/2} \log^{3/2} N \left(\frac{1}{M} + \frac{N}{M^e}\right)^{\frac{1}{2e(e-1)+1}}} + \frac{2^{k/2} e \log^{3/2} N}{M}$$

$\square$

Since Theorem 3 does not require any assumption of size on M, we want to define this parameter using the bound $N^{1/e}$. In other words, we write M as $M = N^{1/e} \log^c N$ with $c > 1$ and we propose two corollaries, proved in the full version of the paper, which treat the case $1/M \leq N/M^e$ (see Corollary 1) and $1/M \geq N/M^e$ (see Corollary 2). This distinction is done for simplifying the bound of Theorem 3 and implies a bound of the value c.

Corollary 1. *Let N, e and X be as in Theorem 2. Let $M = N^{1/e} \log^c N$ and $c \leq \frac{1}{e(e-1)} \frac{\log N}{\log \log N}$. Then X is δ-statistically close to uniform with:*

$$\delta = \frac{2^{k/2}}{(\log N)^{ce\sigma(e)-3/2}} + o(1) \quad \text{where} \quad \sigma^{-1}(e) = 2e(e-1) + 1.$$

Note that the statistical distance will be negligible if $ce\sigma(e) - 3/2 \gg 0$, meaning if $c \gg 3(e-1) + 3/2e$.

Corollary 2. *Let N, e and X be as in Theorem 2. Let $M = N^{1/e} \log^c N$ and $\frac{1}{e(e-1)} \frac{\log N}{\log \log N} \leq c < \frac{e-1}{e} \frac{\log N}{\log \log N}$. Then X is δ-statistically close to uniform with:*

$$\delta = \frac{2^{k/2}}{N^{\frac{\sigma(e)}{e}} (\log N)^{c\sigma(e)-3/2}} + o(1) \quad \text{where} \quad \sigma^{-1}(e) = 2e(e-1) + 1.$$

An interesting value of c is $c = (\frac{1}{2} - \frac{1}{e})\frac{\log N}{\log \log N}$ which represents the case $M = \sqrt{N}$. For $e = 3$, this is the lower bound of c in Corollary 2 and for $e > 3$ it is included in the defined interval.

Let us give a numerical example for Corollary 1, the most interesting corollary since it treats values of M as close as possible to the optimal bound $N^{1/e}$. We consider classical cryptographic parameters for the upper bound of δ, i.e. 2^{-80}, and we put $e = 3$. Suppose that we want to have a negligible statistical distance for $k = 160$, then a modulus of 4096 bits leads to an impossibility. Indeed, Corollary 1 requires that $7 \ll c < 56$ and the result on δ is true for $c \geq 65$. However, with a modulus of 8192 bits one obtains the condition $60 \leq c < 105$. In other words, with an input of at least 3511 bits for the function f, the 160 least significant bits of $f(x)$ are statistically indistinguishable from the uniform distribution on $\{0,1\}^{160}$.

To conclude, we extend ours theorems and corollaries to another model. More precisely we prove in the full version that all our results are still valid when we study the indistinguishability of the most significant bits of the function $x^e \bmod N$. That is the following corollary:

Corollary 3. *Let N, e be as in Theorem 2 and $M < N$. The results from Theorem 2 and Theorem 3 on the statistical distance between $\mathsf{lsb}_k(x^e \bmod N)$ for x randomly chosen in $\mathbb{Z}/M\mathbb{Z}$ and the uniform distribution modulo 2^k are still valid for $\mathsf{msb}_k(x^e \bmod N)$.*

4   Applications of These Bounds to Cryptographic Cases

4.1   Security of Micali-Schnorr Pseudo-Random Number Generator

Micali-Schnorr PRNG. A pseudorandom generator is a deterministic polynomial time algorithm that expands short seeds (made of truly random bits) into longer bit sequences, whose distribution cannot be distinguished from uniformly random bits by a computationally bounded algorithm.

Let (e, N) a RSA public key with e small compared to $\log N$ and $x_0 \in [0, 2^r)$ with $2^r \ll N$ a secret seed of size r. The Micali-Schnorr pseudorandom generator proposed in [14] is defined as follows:

$$v_i = x_{i-1}^e \mod N \quad \text{and} \quad v_i = 2^k x_i + w_i \quad \text{for} \quad i \geq 1.$$

At each iteration, this generator outputs the k least significant bits of v_i, denoted by w_i. In addition, denoting n the size of the modulus N, only x_i of size $r = n - k$, unknown, is reused for the next iteration. Since the generator outputs $O(k/\log e)$ bits per multiplication, one wants k to be as large as possible and e to be as small as possible to be efficient. This pseudorandom generator is proven secure under the following strong assumption:

Assumption 1. *The distribution of $x^e \bmod N$ for random r-bit integers x is indistinguishable by all polynomial-time statistical tests from the uniform distribution of elements of $(\mathbb{Z}/N\mathbb{Z})^*$.*

Clearly this assumption cannot be true if one does not restrain the tests to polynomial-time ones only because of the lack of entropy in input. Micali and Schnorr have proposed the parameters $r = 2n/e$ and thus $k = n(1 - 2/e)$ which are very aggressive parameters in order to increase the efficiency of the generator.

Our Result. We do not contradict this assumption since our theorems give upper bounds on δ. However Theorem 2 is useful to define the sizes of parameters k and r such that the statistical distance is bounded as desired. Corollary 4, which is proven in the full version, consists in determining the minimal size of the input in order to have an indistinguishable output from the uniform distribution. This is really interesting to note that when N tends to infinity, this bound tends to $2/3$. In other words we cannot expect to have a positive result of indistinguishability according to our results if one outputs more than $(\log N)/3$ of the least significant bits asymptotically.

Corollary 4. *Let (e, N) a RSA public key with e small compared to $\log N$ and d a security parameter such that $\delta < 2^{-d}$. Let $\alpha \in (0, 1)$ such that Micali-Schnorr pseudo-random number generator outputs the $(1 - \alpha) \log N$ least significant bits at one iteration. This output is indistinguishable from the uniform distribution on $\{0, 1\}^{(1-\alpha) \log N}$ if*

$$\alpha > 2/3 + \frac{2d + 4 \log e}{3 \log N} + \frac{\log \log N}{\log N}.$$

As a concrete example, for $N = 2^{1024}, e = 3$ and $d = 80$, that gives an input greater than 747 bits (and thus an output lesser than 277 bits). Note that we study a single iteration of the generator as in [10] for example, the consideration of two or more consecutive outputs is a more difficult task. Finally remark that Theorem 3 is useless for this application because of the necessarily link between k and M: if $M = N^\alpha$ then $2^k \simeq N^{1-\alpha}$. For $M \ll \sqrt{N}$, there is not enough entropy to prove the indistinguishability.

4.2   Semantic Security of PKCS #1 v1.5 Encryption

RSA is a well-known asymmetric cryptosystem which was first publicized in 1977. Even nowadays it is frequently used in applications where security of digital data is a concern. However the basic RSA encryption process, meaning without any padding of the plaintext, is vulnerable to quite simple or clever attacks (see for example [8,12]). The standard PKCS #1 v1.5 proposes a padding which avoids a part of these attacks, nevertheless it has been defeated by Bleichenbacher in [5].

PKCS #1 v1.5 Encryption. We recall the encryption scheme proposed in the standard PKCS #1 v1.5. By denoting (e, N) the RSA public key with n the size of the modulus N, a message m of size at most $\ell = n - 88$ is padded as $m_{padded} = 00||02||PS||00||m$ with PS the padding string of size $r = n - \ell - 24$. Then, m_{padded} is encrypted using the RSA encryption process.

Semantic Security. The semantic security requires that the adversary should not gain any advantage or information from having seen the ciphertext resulting from an encryption algorithm. This can be formalized by this concrete definition:

Definition 1. *An encryption scheme* (Enc, Dec) *is* (t, o, ε) *semantically secure if for every distribution* X *over messages, every functions* $I : \{0,1\}^m \to \{0,1\}^*$ *and* $f : \{0,1\}^m \to \{0,1\}^*$ *(of arbitrary complexity) and every function* A *of complexity* $t_A \leq t$, *there is a function* A' *of complexity* $\leq t_a + o$ *such that*

$$\left| \Pr[A(Enc(K, M), I(m)) = f(M)] - \Pr[A'(I(m)) = f(M)] \right| \leq \varepsilon.$$

$I(m)$ can be seen as the knowledge of the adversary on the message M, whereas $f(M)$ represents the knowledge the adversary would learn.

This notion is equivalent to the notion of indistinguishability. Informally, consider that a challenger chooses m_0, m_1 two messages of same length and s the state information (possibly including the public key). Now suppose that a random one of m_0 and m_1 is selected and encrypted as y. With the knowledge of s and y, the goal of the challenger consists in determining if y was selected as the encryption of m_0 or m_1. If the ciphertext is indistinguishable from the uniform distribution, the advantage of the challenger is negligible.

In the case of PKCS #1 v1.5, the indistinguishability can only be proven for a message of one bit. When it is larger, the security of this scheme is based on a computational assumption because of a lack of entropy. To propose a positive result of indistinguishability we thus consider only some bits of the ciphertext.

Our Result. We are interested by bounding the statistical distance between the function $\mathsf{lsb}_k(Pad(x)^e \bmod N)$ for x randomly chosen in $\mathbb{Z}/2^r\mathbb{Z}$ and the uniform distribution on $\{0,1\}^k$, x being an integer whose binary representation is PS. More precisely we define $Pad(x)$ as $Pad(x) = 2^{n-16}b + 2^{l+8}x + m$ with b representing 02 and m of size ℓ. Note that the standard adds another requirement for PS: this random value should not have any null byte. For simplicity we omit this property but we assume that it would not change the results in the following corollary.

Corollary 5. *Let* N, e *be as in Theorem 3 and* $M < N$. *Let* $Pad(x)$ *a function defined same as above. The results from Theorem 2 and Theorem 3 on the statistical distance between* $\mathsf{lsb}_k(x^e \bmod N)$ *for* x *randomly chosen in* $\mathbb{Z}/M\mathbb{Z}$ *and the uniform distribution on* $\{0,1\}^k$ *are still valid for* $\mathsf{lsb}_k(Pad(x)^e \bmod N)$.

With a security parameter $d = 80$, $\log N = 1024$, $e = 3$ and $\log M = 872$ (we consider the encryption of a symmetric cryptosystem key of size 128), we obtain the condition $k < 524$.

References

1. Bauer, A., Coron, J.-S., Naccache, D., Tibouchi, M., Vergnaud, D.: On the Broadcast and Validity-Checking Security of PKCS#1 v1.5 Encryption. In: Zhou, J., Yung, M. (eds.) ACNS 2010. LNCS, vol. 6123, pp. 1–18. Springer, Heidelberg (2010)
2. Bellare, M., Rogaway, P.: Random Oracles are Practical: A Paradigm for Designing Efficient Protocols. In: ACM Conference on Computer and Communications Security, pp. 62–73 (1993)

3. Bellare, M., Rogaway, P.: Optimal Asymmetric Encryption. In: De Santis, A. (ed.) EUROCRYPT 1994. LNCS, vol. 950, pp. 92–111. Springer, Heidelberg (1995)
4. Bellare, M., Rogaway, P.: The Exact Security of Digital Signatures - How to Sign with RSA and Rabin. In: Maurer, U. (ed.) EUROCRYPT 1996. LNCS, vol. 1070, pp. 399–416. Springer, Heidelberg (1996)
5. Bleichenbacher, D.: Chosen Ciphertext Attacks against Protocols Based on the RSA Encryption Standard PKCS #1. In: Krawczyk, H. (ed.) CRYPTO 1998. LNCS, vol. 1462, pp. 1–12. Springer, Heidelberg (1998)
6. Blum, L., Blum, M., Shub, M.: A Simple Unpredictable Pseudo-Random Number Generator. SIAM J. Comput. 15(2), 364–383 (1986)
7. Boneh, D., Venkatesan, R.: Breaking RSA May Not Be Equivalent to Factoring. In: Nyberg, K. (ed.) EUROCRYPT 1998. LNCS, vol. 1403, pp. 59–71. Springer, Heidelberg (1998)
8. Coppersmith, D., Franklin, M.K., Patarin, J., Reiter, M.K.: Low-Exponent RSA with Related Messages. In: Maurer, U. (ed.) EUROCRYPT 1996. LNCS, vol. 1070, pp. 1–9. Springer, Heidelberg (1996)
9. Fouque, P.-A., Vergnaud, D., Zapalowicz, J.-C.: Time/Memory/Data Tradeoffs for Variants of the RSA Problem. In: Du, D.-Z., Zhang, G. (eds.) COCOON 2013. LNCS, vol. 7936, pp. 651–662. Springer, Heidelberg (2013)
10. Friedlander, J., Shparlinski, I.: On the distribution of the power generator. Math. Comput. 70(236), 1575–1589 (2001)
11. Fujisaki, E., Okamoto, T., Pointcheval, D., Stern, J.: RSA-OAEP Is Secure under the RSA Assumption. In: Kilian, J. (ed.) CRYPTO 2001. LNCS, vol. 2139, pp. 260–274. Springer, Heidelberg (2001)
12. Håstad, J.: Solving Simultaneous Modular Equations of Low Degree. SIAM J. Comput. 17(2), 336–341 (1988)
13. Lidl, R., Niederreiter, H.: Finite Fields. Cambridge University Press (1996)
14. Micali, S., Schnorr, C.-P.: Efficient, Perfect Polynomial Random Number Generators. J. Cryptology 3(3), 157–172 (1991)
15. Shoup, V.: A computational introduction to number theory and algebra. Cambridge University Press (2006)
16. Vinogradov, I.M.: Elements of number theory. Dover (1954)
17. Wooley, T.D.: Vinogradov's mean value theorem via efficient congruencing. Annals of Mathematics 175(3), 1575–1627 (2012)

Numerical Tic-Tac-Toe on the 4 × 4 Board

Bryce Sandlund[1], Kerrick Staley[2], Michael Dixon[2], and Steve Butler[2]

[1] University of Wisconsin–Madison, Madison, WI 53706, USA
bcsandlund@gmail.com
[2] Iowa State University, Ames, IA 50011, USA
kerrick@kerrickstaley.com,
{medixon,butler}@iastate.edu

Abstract. Numerical Tic-Tac-Toe on the $n \times n$ board is a two player game where the numbers $\{1, 2, \ldots, n^2\}$ are divided between the two players (usually as odds and evens) and then players alternately play by placing one of their numbers on the board. The first player to complete a line of n numbers (played by either player) that add up to $n(n^2 + 1)/2$ is the winner. The original 3×3 game was created and analyzed by Ron Graham nearly fifty years ago and it has been shown that the first player has a winning strategy. In this paper we consider the 4×4 game and determine that in fact the second player has a winning strategy.

Keywords: Tic-Tac-Toe, games, symmetry, backtracking, pruning.

1 Introduction

Tic-Tac-Toe is a classic game that is familiar to people of all ages. Because of its familiarity this game is often used as a starting example of how to mathematically analyze a game, and it is well known that in optimal play by both players the game will always end in a tie.[1] One of the best known Tic-Tac-Toe strategy guides was created by Randall Munroe in an XKCD posting [6].

Nearly fifty years ago Ron Graham created a variation of Tic-Tac-Toe which has come to be known as "Numerical Tic-Tac-Toe". The game is still played on the same 3×3 board but now instead of using ×'s and ○'s the two players are given the numbers $\{1, 3, 5, 7, 9\}$ and $\{2, 4, 6, 8\}$, respectively. The players take turns (with the odd player going first) and at each round the players put one of their unused numbers on an open square on the board. The first player to create any three numbers in a line that sum to 15 wins the game.

Numerical Tic-Tac-Toe is easily played by two players and the reader is encouraged to give it a try to help get a sense of the game. It should be noted that there are many implementations of this game online as well as apps for portable devices.

---

[1] Even modest training can result in a "good" player. At one time people would compete against chickens playing Tic-Tac-Toe in casinos in Atlantic City (though the chickens themselves responded more to lighting cues than strategy).

Z. Cai et al. (Eds.): COCOON 2014, LNCS 8591, pp. 537–546, 2014.

By extensive case analysis carried out by hand, Graham [1] determined that remarkably the *first* player has a strategy that can guarantee a win. An exhaustive computer analysis by Markowsky [4,5] thirty years later verified the result of Graham, while Orr and Cooper [7] subsequently gave a compact strategy for the game.

This game can be generalized to be played on any size board. Usually, on the $n \times n$ board the numbers $\{1, 2, \ldots, n^2\}$ are divided into the odd's and even's. The players take turns, starting with the odd player, and at each round the players put one of their unused numbers on the board. The first player to complete a line of any n numbers that sum to $n(n^2+1)/2$ wins the game. (The value $n(n^2+1)/2$ comes from the average value of $1, 2, \ldots, n^2$ being $(n^2+1)/2$ and then having n of them.)

In this paper we will outline the computation that was done to carry out the analysis for $n = 4$. Our exhaustive computation shows that in optimal play the *second* player has a winning strategy (though we do not have a compact description of such a strategy).

We mention in passing that there are other variations that could be considered for this game. One variation that we also explored was altering the initial division of $1, 2, \ldots, 9$ between the two players for the 3×3 game. It turns out that regardless of which five numbers the first player has they will always be able to win. This was independently confirmed by Bennett Hansen [2]. For the 4×4 case one could also create variations by changing the initial division; or giving each player the numbers $\{1, 2, \ldots, 8\}$ and then the winner would be the first person to get four numbers in a line that total to 18. We have not considered these variations here but the technique we will give can be used to analyze these and other situations as well.

One popular approach to solve perfect information games that has proven effective is using retrograde analysis, i.e., starting at the finishing positions and then working backwards. This works well when there are relatively few finishing positions, e.g., chess endgames [9]. This technique was also used to solve end game for checkers when there were relatively few pieces (later Schaeffer et al. [8] gave a complete solution of checkers combining various ideas). While this technique could work in this setting we will opt instead to use symmetry and efficient pruning of the game tree to determine the result. A survey of games that have been solved, including a discussion of various techniques to solve them, is given by van den Herik et al. [3].

2   Symmetries of the 4×4 Game

One naïve approach to this problem is to determine every possible state of the board and then form a graded poset with the unique maximal entry corresponding to an empty board and then as we go down we look at all possible ways to legally insert one number until either there is a win or the game results in a tie. Given such a poset we could then easily determine the winner of the game by working from the bottom to the top. However an approximation for the number of boards at depth k is

$$\binom{16}{k}\binom{8}{\lfloor k/2 \rfloor}\binom{8}{\lceil k/2 \rceil}k!.$$

That is, choose k positions out of 16 possible positions, then choose which odds are to be played, which evens are to be played and put them on the board in all possible ways. Summing up over all possible k then gives us 2.7×10^{15} boards. For comparison the 3×3 version of Numerical Tic-Tac-Toe has 9.3×10^6 and classical Tic-Tac-Toe has $6,046$. (This count gives all possible boards, but some of these boards would not occur in gameplay as they contain within them two or more disjoint winning lines which would indicate play has already stopped.)

Even with advances in computing power and memory storage this is still prohibitive to approach an analysis of the 4×4 board. We will employ several techniques to reduce the size of this problem to a point where the computation can be carried out efficiently.

One of the most important tools that we have is to use the symmetry of the board. That is a bijection from the board to itself which preserves lines. To be more precise when we have the following board:

A	B	C	D
E	F	G	H
I	J	K	L
M	N	O	P

then the lines are:

$$\{A,B,C,D\}, \{E,F,G,H\}, \{I,J,K,L\}, \{M,N,O,P\}, \{A,E,I,M\},$$
$$\{B,F,J,N\}, \{C,G,K,O\}, \{D,H,L,P\}, \{A,F,K,P\}, \{D,G,J,M\}.$$

Proposition 1. *The bijections of the board consist of compositions of the following maps: rotations, reflections, and the two maps shown below.*

For the two maps given above we will call the one on the left the "cross-symmetry" and the one on the right the "X-symmetry". When all of these are combined we end up with 32 different bijections that preserve the lines of the 4×4 board.

Proof. A simple check will verify that each one of those maps will preserve lines, so it suffices to show that if we have preserved lines that we must be a composition of these maps.

Next we note that each map we have outlined is reversible (i.e., for rotation we reverse the direction and for the other maps we simply apply the map a second time). Therefore to show that we have all possible bijections by combining these maps, it suffices to show how we can apply these maps to get back to the starting board.

So suppose we have a board that has preserved lines. Then by applying rotations we can place the A in the upper left corner. Note that A, D, M and P will always have to form the corners of a square and so if needed we can apply reflection to place D in the upper right corner. Reflection will achieve this because A and D cannot be on opposite corners, since that would force B and C to be one of the four center squares. The center squares are involved in three lines, but B and C are only involved in two, meaning they can never be center tiles. Therefore we are in one of the following two situations:

Similarly, F, G, J, and K will form another square and their positioning must agree with the above boards in preserving diagonal lines. So in the first case we now have the following four possibilities:

A			D
	F	G	
	J	K	
M			P

A			D
	F	J	
	G	K	
M			P

A			D
	K	G	
	J	F	
M			P

A			D
	K	J	
	G	F	
M			P

Every remaining unfilled square is contained in two lines and we can determine what value (if possible) must go in the square by taking the intersection of the two lines. The first possibility reduces to the identity, the fourth gives the X-symmetry, and the other two are impossible.

For the second case we have the following four possibilities:

F			G
	A	D	
	M	P	
J			K

F			J
	A	D	
	M	P	
G			K

K			G
	A	D	
	M	P	
J			F

K			J
	A	D	
	M	P	
G			F

Again proceeding as before we can fill in any remaining squares by looking at the intersection of the lines. The first possibility reduces to cross-symmetry, the fourth possibility is the result of composition of X-symmetry and cross-symmetry maps, and the other two are impossible.

Thus using only our given maps we have accounted for each valid bijection of the board. ∎

There is one other natural candidate for symmetry involving manipulation of the numbers themselves rather than the location of each entry. The symmetry was used by Markowsky [4] in the 3 × 3 version and was found due to the fact:

$$a + b + c = 15 \leftrightarrow (10 - a) + (10 - b) + (10 - c) = 15.$$

Simple algebra takes the original sum $n(n^2 + 1)/2$ and finds that in the general case, subtracting each filled entry from $n^2 + 1$ presents a possible symmetry. For the 4 × 4 version, each filled cell q would then be replaced by $17 - q$. An example is given below:

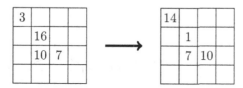

However, due to the fact replacing q by $17 - q$ changes the parity of an entry, this symmetry is invalid for two reasons. First, if we apply this after an odd number of plays then we will be in an impossible board, i.e., one with more even numbers than odd numbers. Second, if we apply this after an even number of plays then situations can change dramatically. As an example of this latter case, consider the above scenario where it is now the first player's turn. In the original board the first player can block but cannot win with the next move, while in the second board the first player can place 9 in the lower right corner and win.

We note that this last symmetry does work for the 3 × 3 board and more generally any $n \times n$ board when n is odd, due to the fact $n^2 + 1$ is even whenever n is odd.

3 Splitting the Computation

As already mentioned, we can form a graded poset consisting of all boards where connections go between consecutive levels between boards that differ by a legal move. In this poset we then identify each "winning move" and break all connections below such boards and work from the bottom up. Each board can be labeled with one of three possibilities $P1$, $P2$, or T for "player one wins", "player two wins" and "neither player can guarantee a win". Working from the leaf nodes upwards we visit each board and identify the labels of all boards below it which it connects to and then determine the label of the board by the following rule: If it is player one's turn then $P1 \succ T \succ P2$ (i.e., $P1$ is preferred to T which is preferred to $P2$ and the player takes the best labeling of all boards immediately below); while if it is player two's turn then $P2 \succ T \succ P1$. Finally, the label of the empty board indicates the outcome of optimal play on the part of both players.

Instead we will opt for starting at the root and working our way down the tree and filling in this information as we go along. The implemented program is a version of the minimax algorithm, working in a depth-first, backtracking manner. One advantage of this is that we can use alpha-beta pruning to eliminate the need to compute significant parts of the tree (i.e., avoid having to consider some board configurations). Suppose that below is part of the tree for us to consider (where on the left we have indicated which player is playing):

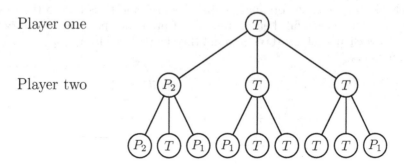

Given that at each board we will scan its children from left to right, we can then trim off parts of the tree that we guarantee are not necessary to visit, giving us the following:

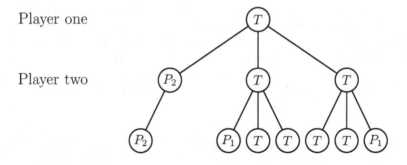

Since this technique is applied at each level of the tree, this pruning has a dramatic effect on execution time. But we can prune even more if we run the computation twice and instead of looking at $P1$, $P2$ and T we ask the following two questions:

- Can player one force a win?
- Can player two force a win?

While forming the tree for each of these two situations we will use Y and N for "Yes" and "No" respectively. For the first question we note that $Y = \{P_1\}$ and $N = \{P_2, T\}$, further when it is player one's turn $Y \succ N$ and for player two $N \succ Y$. For the second question we note that $Y = \{P_2\}$ and $N = \{P_1, T\}$, further when it is player one's turn $N \succ Y$ and for player two $Y \succ N$.

Applying this to the above tree and trimming as we are wont will result in the following two trees:

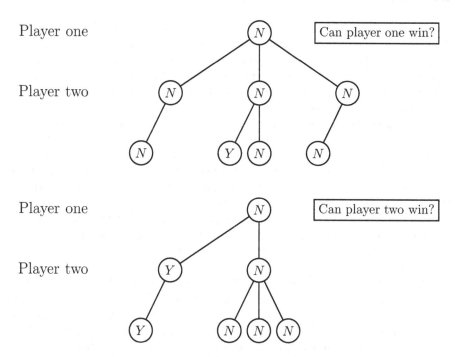

In this particular problem space, we found that examining only two utility values instead of three sped up computation exponentially due to the fact many tie boards exist in the game space. With this technique, only one player must search to find a winning configuration, but the other can return as soon as a tie board is found. In theory, this could reduce runtime from $O(b^d)$ to $O(b^{d/2})$ (where b is the branching factor and d is the depth), due to the repeating pattern of $1 * b * 1 * b...$ corresponding to an immediate return vs. searching every child. In practice, we found the strategy more effective for player one's computation than player two's computation (sensible considering our result) and encourage the reader to look at the number of boards visited at each depth in the attached appendix, which is explained in more detail in the following section.

The authors comment that this general technique is probably known in AI literature and can be seen as similar to other techniques such as proof-number search [10]. For the record, however, we note that the above strategy could be easily translated to any general problem space with k utility values to turn the original question into $O(log(k))$ separate questions by binary searching for the best achievable utility value.

4   Results

In the previous two sections we have outlined our basic approach. First we run two computations to ask starting from an empty board whether each player can win. To further help prune we store all boards up through some preset depth and the corresponding answers for those boards. Symmetries were only used in the first five depths, a cutoff found by trial and error to avoid the expensive cost of calculating symmetries but keep the gain of avoiding duplicate computation. Boards were still stored after this point since it is quite possible to arrive at the same board from two or more different paths. Finally, due to memory constraints, to answer player two's computation, periodically we removed configurations from memory at the deeper depths of the memory storage.

In the appendix is the information regarding the outcome when we ran the program for the question: Can player one force a win? For each depth (i.e., number of rounds played in the game) it records the number of boards that were visited, how many were determined because they were already calculated in memory (stored up through depth 9), and how many times the answer was "Yes". Since our program looked one move ahead to determine if the opponent can win, all recorded "Yes's" represent a win at least one depth away from the winning board. Similarly, in the appendix is the same information when we ran the program for the question: Can player two force a win? (Where we stored boards up through depth 10.)

In particular, we see that for the 4×4 Numerical Tic-Tac-Toe game that player two can force a win. Comparing the amount of computation required to determine the answers to our original questions, it appears that in random play that it is much easier to stop player one from winning than it is to make player two win. This is easy to convince yourself of in the 3×3 case where it is not so hard to stop player two from winning but far from obvious how to have player one win.

By using an appropriately pruned tree for the question "Can player two force a win?" we would have a perfect strategy for the game. Unfortunately this requires immense amounts of storage and so is likely impractical to program. Therefore we expect that for most players (even most low-powered computer players with limited time to act in each move), games will likely end in a tie.

There are of course many interesting questions left to ask for numerical tic-tac-toe, including whether or not one of the players always has a winning strategy. One can imagine that we divide the numbers up arbitrarily or play in larger boards. If one player did always have a winning strategy is there an easy explanation for which player it is and what strategy they should pursue? Another question that we did not answer here is whether player two could force an early win, i.e., is it possible to finish the game in the fourteenth round? We do not yet have the answers to these questions, but look forward to the next move in this area.

Implementation

The program for carrying out the computation was written in Java, and is available online: https://github.com/brycesandlund/4x4TicTacToe/

References

1. Graham, R.: Personal communication
2. Hansen, B.: Personal communication
3. can den Herik, H.J., Uiterwijk, J.W.H.M., van Rijswijck, J.: Games solved: Now and in the future. Artificial Intelligence 134, 277–311 (2002)
4. Markowsky, G.: Numerical tic-tac-toe–I. J. of Recreational Math. 22, 114–123 (1990)
5. Markowsky, G.: Numerical tic-tac-toe–II. J. of Recreational Math. 22, 192–200 (1990)
6. Munroe, R.: xkcd: Tic-Tac-Toe, http://xkcd.com/832/
7. Orr, S., Cooper, C.: A compact strategy for numerical tic-tac-toe. J. of Recreational Math. 27, 161–171 (1995)
8. Schaeffer, J., Burch, N., Björnsson, Y., Kishimoto, A., Müller, M., Lake, R., Lu, P., Sutphen, S.: Checkers is solved. Science 317, 1518–1522 (2007)
9. Thompson, K.: Retrograde analysis of certain endgames. ICCA Journal 9, 131–139 (1986)
10. Allis, L.V., van der Meulen, M., van den Herik, H.J.: Proof-number search. Artificial Intelligence 66(1), 91–124 (1994)

Appendix

For the question "Can player one force a win?" we have the following information in regards to the computation:

Depth	Boards Visited	Found in Memory	How many "Yes"'s
0	1	0	0
1	128	112	0
2	16	0	0
3	1,440	136	0
4	1,304	0	0
5	88,430	33,554	0
6	57,295	0	2,419
7	2,158,685	1,131,526	2,419
8	1,334,445	6,328	309,061
9	23,654,937	9,012,830	306,283
10	18,237,546	0	3,896,978
11	215,581,273	0	3,896,978
12	221,312,077	0	9,627,782
13	1,462,159,978	0	9,627,782
14	1,452,532,196	0	0
15	2,818,792,528	0	0
16	2,818,792,528	0	0

For the question "Can player two force a win?" we have the following information in regards to the computation:

Depth	Boards Visited	Found in Memory	How many "Yes"'s
0	1	0	1
1	128	112	128
2	72	0	16
3	1,467	19	1,411
4	17,358	329	1,396
5	155,748	5,215	140,006
6	2,822,029	150,020	135,188
7	14,434,502	833,734	11,892,556
8	185,959,992	16,012,248	11,122,103
9	572,733,507	45,970,492	413,409,269
10	5,686,646,311	699,206,459	374,906,758
11	10,231,334,279	0	5,593,772,898
12	82,553,255,357	0	5,593,772,898
13	109,945,222,392	0	32,985,739,933
14	521,913,263,546	0	32,985,739,933
15	488,927,523,613	0	0
16	488,927,523,613	0	0

On Linear Congestion Games
with Altruistic Social Context⋆

Vittorio Bilò

Dipartimento di Matematica e Fisica "Ennio De Giorgi", Università del Salento
Provinciale Lecce-Arnesano, P.O. Box 193, 73100 Lecce, Italy
vittorio.bilo@unisalento.it

Abstract. We study the issues of existence and inefficiency of pure Nash equilibria in linear congestion games with altruistic social context, in the spirit of the model recently proposed by de Keijzer *et al.* [14]. In such a framework, given a real matrix $\Gamma = (\gamma_{ij})$ specifying a particular social context, each player i aims at optimizing a linear combination of the payoffs of all the players in the game, where, for each player j, the multiplicative coefficient is given by the value γ_{ij}. We give a broad characterization of the social contexts for which pure Nash equilibria are always guaranteed to exist and provide tight or almost tight bounds on their prices of anarchy and stability. In some of the considered cases, our achievements either improve or extend results previously known in the literature.

1 Introduction

Congestion games are, perhaps, the most famous class of non-cooperative games due to their capability to model several interesting competitive scenarios, while maintaining some nice properties. In these games there is a set of players sharing a set of *resources*, where each resource has an associated *latency function* which depends on the number of players using it (the so-called *congestion*). Each player has an available set of strategies, where each strategy is a non-empty subset of resources, and aims at choosing a strategy minimizing her cost which is defined as the sum of the latencies experienced on all the selected resources.

Congestion games have been introduced by Rosenthal [21]. He proved that each such a game admits a bounded *potential function* whose set of local minima coincides with the set of *pure Nash equilibria* of the game, that is, strategy profiles in which no player can decrease her cost by unilaterally changing her strategic choice. This existence result makes congestion games particularly appealing especially in all those applications in which pure Nash equilibria are elected as the ideal solution concept.

In these contexts, the study of the inefficiency of pure Nash equilibria, usually measured by the sum of the costs experienced by all players, has affirmed as a

---

⋆ This work was partially supported by the PRIN 2010–2011 research project ARS TechnoMedia: "Algorithmics for Social Technological Networks" funded by the Italian Ministry of University.

fervent research direction. To this aim, the notions of *price of anarchy* (Koutsoupias and Papadimitriou [18]) and *price of stability* (Anshelevich et al. [2]) are widely adopted. The price of anarchy (resp. stability) compares the performance of the worst (resp. best) pure Nash equilibrium with that of an optimal cooperative solution.

Congestion games with unrestricted latency functions are general enough to model the Prisoner's Dilemma game, whose unique pure Nash equilibrium is known to perform arbitrarily bad with respect to the solution in which all players cooperate. Hence, in order to deal with significative bounds on the prices of anarchy and stability, some kind of regularity needs to be imposed on the latency functions associated with the resources. To this aim, lot of research attention has been devoted to the case of polynomial latency functions. In particular, Awerbuch et al. [4] and Christodoulou and Koutsoupias [12] proved that the price of anarchy of congestion games is $5/2$ for linear latency functions and $d^{\Theta(d)}$ for polynomial latency functions of degree d. Subsequently, Aland et al. [1] obtained exact bounds on the price of anarchy for congestion games with polynomial latency functions. Still for linear latencies, Caragiannis et al. [7] proved that the same bounds hold for *load balancing* games as well, that is, for the restriction in which all possible strategies are singleton sets, while for *symmetric* load balancing games, that is load balancing games in which the players share the same set of strategies, Lücking et al. [19] proved that the price of anarchy is $4/3$. Moreover, the works of Caragiannis et al. [7] and Christodoulou and Koutsoupias [13] show that the price of stability of congestion games with linear latency functions is $1 + 1/\sqrt{3}$, while an exact characterization for the case of polynomial latency functions of degree d has been recently given by Christodoulou and Gairing [11].

Motivations and Previous Related Works. To the best of our knowledge, Chen and Kempe [10] were the first to study the effects of altruistic (and spiteful) behavior on the existence and inefficiency of pure Nash equilibria in some well-understood non-cooperative games. They focus on the class of non-atomic congestion games, where there are infinitely many players each contributing for a negligible amount of congestion, and show that price of anarchy decreases as the degree of altruism of the players increases. Hoefer and Skopalik [16] consider (atomic) linear congestion games with γ_i-altruistic players, where $\gamma_i \in [0,1]$, for each player i. According to their model, player i aims at minimizing a function defined as $1 - \gamma_i$ times her cost plus γ_i times the sum of the costs of all the players in the game (*also counting* player i). They show that pure Nash equilibria are always guaranteed to exist via a potential function argument, while, in all the other cases in which existence is not guaranteed, they study the complexity of the problem of deciding whether a pure Nash equilibrium exists in a given game. In such a context, Chen et al. [9] show that the price of anarchy of coarse correlated equilibria is upper bounded by $\frac{5+2\hat{\gamma}+2\check{\gamma}}{2-\hat{\gamma}+2\check{\gamma}}$, where $\hat{\gamma} = \max_i \gamma_i$ and $\check{\gamma} = \min_i \gamma_i$. Such a result implies the same upper bound also on the price of anarchy of correlated equilibria, mixed Nash equilibria and pure Nash equilibria.

Caragiannis et al. [8] consider a more general model of altruistic behavior: in fact, for a parameter $\gamma_i \in [0,1]$, they model a γ_i-altruistic player i as a player who aims at minimizing a function defined as $1 - \gamma_i$ times her cost plus γ_i times the sum of the costs of all the players in the game *other than* i[1]. In such a way, the more γ_i increases, the more γ_i-altruistic players tend to favor the interests of the others to their own ones, with 1-altruistic and 0-altruistic players being the two opposite extremal situations in which players behave in a completely altruistic or in a completely selfish way, respectively. Caragiannis *et al.* [8] study the basic case of $\gamma_i = \gamma$ for each player i and show that the price of anarchy is $\frac{5-\gamma}{2-\gamma}$ for $\gamma \in [0, 1/2]$ and $\frac{2-\gamma}{1-\gamma}$ for $\gamma \in [1/2, 1]$ and that these bounds hold also for load balancing games. This result appears quite surprising, because it shows that altruism can only have a harmful effect on the efficiency of linear congestion games, since the price of anarchy increases from 5/2 up to an unbounded value as the degree of altruism goes from 0 to 1. On the positive side, they prove that, for the special case of symmetric load balancing games, the price of anarchy is $\frac{4(1-\gamma)}{3-2\gamma}$ for $\gamma \in [0, 1/2]$ and $\frac{3-2\gamma}{4(1-\gamma)}$ for $\gamma \in [1/2, 1]$, which shows that altruism has a beneficial effect as long as $\gamma \in [0, 0.7]$. Note that, that for $\gamma = 1/2$, that is when selfishness and altruism are perfectly balanced, the price of anarchy drops to 1 (i.e., all pure Nash equilibria correspond to socially optimal solutions), while, as soon as γ approaches 1, the price of anarchy again grows up to an unbounded value.

Recently, de Keijzer *et al.* [14] proposed a model for altruistic and spiteful behavior further generalizing the one of Caragiannis *et al.* [8]. According to their definition, each non-cooperative game with n players is coupled with a real matrix $\Gamma = (\gamma_{ij}) \in \mathbb{R}^{n \times n}$, where γ_{ij} expresses how much player i cares about player j. In such a framework, player i wants to minimize the sum, for each player j in the game (thus also counting i), of the cost of player j multiplied by γ_{ij}. Thus, a positive (resp. negative) value γ_{ij} expresses an altruistic (resp. spiteful) attitude of player i towards player j[2]. When considering linear congestion games with altruistic players, along the lines of the negative results of Caragiannis *et al.* [8], as soon as there are two players i, j such that $\gamma_{ij} > \gamma_{ii}$, i.e., player i cares more about player j than about herself, the price of anarchy may become unbounded. Therefore, de Keijzer *et al.* [14] focus on the scenario, which they call *restricted altruistic social context*, in which $\gamma_{ii} \geq \gamma_{ij}$ for each pair of players i and j. They show an upper bound of 7 on the price of anarchy of coarse correlated equilibria. Moreover, they prove that, when restricting to load balancing games with identical resources, such an upper bound decreases to $2 + \sqrt{5} \approx 4.236$. Very recently, Rahn and Scäfer [20] improved the 7 upper bound to 17/3 and showed that this is tight even for pure Nash equilibria.

Noting that matrix Γ implicitly represents the *social context* (for instance, a social network) in which the players operate, the model of de Keijzer *et al.* [14] falls within the scope of the so-called *social context games*. In these games, the

---

[1] Note that each game with γ_i-altruistic players, where $\gamma_i \in [0, 1]$, in the model of Hoefer and Skopalik [16] maps to a game with γ_i'-altruistic players, where $\gamma_i' \in [0, 1/2]$, in the model of Caragiannis *et al.* [8].

[2] Nevertheless, a model in which $\gamma_{ii} = 1$ and $\gamma_{ij} = \gamma_{ji}$ for each $i, j \in [n]$ had been previously considered by Hoefer and Skopalik in [17].

payoff of each player is redefined as a function, called *aggregating function*, of her cost and of those of her neighbors in a given *social context graph*.

Social context games have been introduced and studied by Ashlagi, Krysta, and Tennenholtz [3] for the class of load balancing games, in the case in which the aggregating function is one among the minimum, maximum, sum and ranking functions, for which they gave an almost complete characterization of the cases in which existence of pure Nash equilibria is guaranteed. The model of de Keijzer *et al.* [14], hence, coincides with a social context game in which the social context graph has weighted edges and the aggregating function is a weighted sum. The issues of existence and inefficiency of pure Nash equilibria for the case of social context linear congestion games have been considered by Bilò *et al.* [6]. In particular, for the aggregating function sum, pure Nash equilibria are shown to exist for each social context graph via an exact potential function argument and the price of anarchy is shown to fall within the interval [5; 17/3].

Finally, the particular case of social context games in which the social context graph is a partition into cliques coincide with games in which static coalitions among players are allowed. These games have been considered by Fotakis, Kontogiannis and Spirakis [15] who focus on weighted congestion game defined on a parallel link graph when the aggregating function is the maximum function (i.e, the coalitional generalization of the KP-model of Koutsoupias and Papadimitriou [18]).

Our Contribution. We consider the issues of existence and inefficiency of pure Nash equilibria in linear congestion games with social context as defined by de Keijzer *et al.* [14]. In particular, we restrict our attention to the case of altruistic players, that is, the case in which the matrix Γ has only non-negative entries. Hoefer and Skopalik [17] had shown that pure Nash equilibria are always guaranteed to exist via an exact potential function argument when either the altruistic social context is restricted and Γ is symmetric. We prove that both properties are essential to guarantee an existential result, by providing instances with three players not admitting pure Nash equilibria as soon as exactly one of them is not satisfied.

We then show that, in the restricted altruistic social context, the price of anarchy of coarse correlated equilibria remains 17/3 even in the special case of load balancing games. Such a result, which improves the one given by Rahn and Schäfer [20], proves that the assumption of having identical resources is essential in the upper bound of $2 + \sqrt{5}$ given by de Keijzer *et al.* [14] for the case of load balancing games. As to the price of stability, we give an upper bound of 2 holding for each symmetric matrix Γ and a lower bound of $1 + 1/\sqrt{2} \approx 1.707$ holding for the case in which Γ is a boolean symmetric matrix.

Finally, we also consider the special case in which Γ is such that $\gamma_{ij} = \gamma_i$ for each pair of indexes i, j with $i \neq j$, which coincides with the general model of γ_i-altruistic players of Caragiannis *et al.* [8]. We show that pure Nash equilibria are always guaranteed to exist in any case via an exact potential function argument (this slightly improves the existential result by Hoefer and Skopalik [16] since they only proved the existence of a weighted potential function). Moreover, we

give an upper bound on the price of anarchy equal to $\frac{5+2\hat{\gamma}-3\hat{\gamma}}{2-\hat{\gamma}}$ when $\hat{\gamma} \in [0, 1/2]$
and to $\frac{2-\hat{\gamma}}{1-\hat{\gamma}}$ when $\hat{\gamma} \in [1/2, 1]$. We stress that these upper bounds improve on the
one shown by Chen et al. [9] which, however, holds for the more general setting
of the price of anarchy of coarse correlated equilibria. Finally, we show that, for
the basic case of $\gamma_i = \gamma$ for each player i, the price of stability is $\frac{(\sqrt{3}+1)(1-\gamma)}{\sqrt{3}-\gamma(\sqrt{3}-1)}$
when $\gamma \in [0, 1/2]$ and $\frac{3-\sqrt{3}-2\gamma(2-\sqrt{3})}{2(1-\gamma)}$ when $\gamma \in [1/2, 1]$.

Paper Organization. In the next section, we introduce all formal definitions
and notation; in Section 3, we address the issue of existence of pure Nash equilibria for general altruistic social contexts; in Section 4, we bound the inefficiency
of pure Nash equilibria for games with restricted altruistic social contexts, while,
in Section 5, we consider the special case of γ_i-altruistic players of Caragiannis
et al. [8]. Due to space limitations, some of the proofs have been omitted.

2 Preliminaries

A congestion game is a tuple $\mathcal{G} = \langle [n], E, S_{i\in[n]}, \ell_{e\in E} \rangle$, where $[n] := \{1, \ldots, n\}$
is a set of $n \geq 2$ players, E is a set of resources, $\emptyset \neq S_i \subseteq 2^E$ is the set of
strategies of player i, and $\ell_e : \mathbb{N} \to \mathbb{R}_{\geq 0}$ is the latency function of resource e.
The special case in which, for each $i \in [n]$ and each $s \in S_i$, it holds $|s| = 1$
is called load balancing congestion game. Denoted by $\mathcal{S} := \times_{i\in[n]} S_i$ the set of
strategy profiles in $\mathcal{G}$, that is, the set of outcomes of $\mathcal{G}$ in which each player selects
a single strategy, the cost of player i in the strategy profile $S = (s_1, \ldots, s_n) \in \mathcal{S}$
is defined as $c_i(S) = \sum_{e \in s_i} \ell_e(n_e(S))$, where $n_e(S) := |\{j \in [n] : e \in s_j\}|$ is the
congestion of resource e in S, that is, the number of players using e in S.

Given a strategy profile $S = (s_1, \ldots, s_n)$ and a strategy $t \in S_i$ for a player
$i \in [n]$, we denote with $S_{-i} \diamond t$ the strategy profile obtained from S by replacing
the strategy played by i in S with t. A pure Nash equilibrium is a strategy
profile S such that, for any player $i \in [n]$ and for any strategy $t \in S_i$, it holds
$c_i(S_{-i} \diamond t) \geq c_i(S)$.

The function $\mathsf{SUM} : \mathcal{S} \to \mathbb{R}_{\geq 0}$ such that $\mathsf{SUM}(S) = \sum_{i\in[n]} c_i(S)$, called
the social function, measures the social welfare of a game. Given a congestion
game $\mathcal{G}$, let $\mathcal{NE}(\mathcal{G})$ denote the set of its pure Nash equilibria (such a set has
been shown to be non-empty by Rosenthal [21]) and S^* be the strategy profile
minimizing the social function. The price of anarchy (PoA) of $\mathcal{G}$ is defined as
$\max_{S\in\mathcal{NE}(\mathcal{G})} \left\{ \frac{\mathsf{SUM}(S)}{\mathsf{SUM}(S^*)} \right\}$, while the price of stability (PoS) of $\mathcal{G}$ is defined as
$\min_{S\in\mathcal{NE}(\mathcal{G})} \left\{ \frac{\mathsf{SUM}(S)}{\mathsf{SUM}(S^*)} \right\}$.

A linear congestion game is a congestion game such that, for each $e \in E$, it
holds $\ell_e(x) = \alpha_e x + \beta_e$, with $\alpha_e, \beta_e \geq 0$. For these games, the cost of player i
in the strategy profile $S = (s_1, \ldots, s_n)$ becomes $c_i(S) = \sum_{e \in s_i} (\alpha_e n_e(S) + \beta_e)$,
while the social value of S becomes $\mathsf{SUM}(S) = \sum_{i\in[n]} \sum_{e \in s_i} (\alpha_e n_e(S) + \beta_e) = \sum_{e \in E} (\alpha_e n_e(S)^2 + \beta_e n_e(S))$.

A linear congestion game with an altruistic social context is a pair $(\mathcal{G}, \Gamma)$
such that $\mathcal{G}$ is a linear congestion game with n players and $\Gamma = (\gamma_{ij}) \in \mathbb{R}^{n \times n}$

is a real matrix such that $\gamma_{ij} \geq 0$ for each $i, j \in [n]$. The set of players and strategies is defined as in the underlying linear congestion game $\mathcal{G}$, while, for any strategy profile S, the cost of player i is $S = (s_1, \ldots, s_n)$ is defined as $\widehat{c}_i(S) = \sum_{j \in [n]} (\gamma_{ij} \cdot c_j(S)) = \sum_{j \in [n]} (\gamma_{ij} (\alpha_e n_e(S) + \beta_e)) = \sum_{e \in E} \left((\alpha_e n_e(S) + \beta_e) \sum_{j \in [n]: e \in s_j} \gamma_{ij} \right)$, where $c_j(S)$ is the cost of player j in S in the underlying linear congestion game $\mathcal{G}$. For a strategy profile S, a player $i \in [n]$ and a strategy $t \in S_i$, for the sake of brevity, let us denote with $x_e := n_e(S)$. It holds

$$
\widehat{c}_i(S) - \widehat{c}_i(S_{-i} \diamond t) = \sum_{e \in s_i \setminus t} \left(\gamma_{ii} (\alpha_e x_e + \beta_e) + \alpha_e \sum_{j \neq i: e \in s_j} \gamma_{ij} \right)
$$
$$
- \sum_{e \in t \setminus s_i} \left(\gamma_{ii} (\alpha_e (x_e + 1) + \beta_e) + \alpha_e \sum_{j: e \in s_j} \gamma_{ij} \right). \tag{1}
$$

The special case in which $\gamma_{ii} \geq \gamma_{ij}$ for each $i, j \in [n]$, is called restricted altruistic social context. Note that, in such a case, as pointed out by de Keijzer et al. [14], it is possible to assume without loss of generality that $\gamma_{ii} = 1$ for each $i \in [n]$[3].

3   Existence of Pure Nash Equilibria

In this section, we provide a complete characterization of the social contexts for which pure Nash equilibria are guaranteed to exist, independently of which is the underlying linear congestion game.

Theorem 1. *[17] Each linear congestion game with restricted altruistic social context $(\mathcal{G}, \Gamma)$ such that Γ is symmetric admits an exact potential function*

$$
\Phi(S) = \frac{1}{2} \sum_{e \in E} \left(\alpha_e \left(n_e(S)(n_e(S) + 1) + \sum_{(i,j) \in P_e(S)} \gamma_{ij} \right) + 2\beta_e n_e(S) \right),
$$

where $P_e(S) = \{(i, j) \in [n] \times [n] : i \neq j \wedge e \in s_i \cap s_j\}$.

In order to prove that the characterization given in Theorem 1 is tight, we provide the following two non-existential results. In the first one, although preserving the property that Γ is symmetric, we relax the constraint that the game is played in a restricted altruistic social context: in particular, we allow $\gamma_{ii} = 0$ for some player $i \in [n]$.

Theorem 2. *There exists a three-player linear congestion game $\mathcal{G}$ and a symmetric matrix $\Gamma \in \mathbb{R}^{3 \times 3}$ such that the linear congestion game with altruistic social context $(\mathcal{G}, \Gamma)$ does not admit pure Nash equilibria.*

---

[3] This claim follows from the fact that both the set of pure Nash equilibria and the social value of any strategy profile do not change when dividing all the entries in row i of Γ by the value γ_{ii}.

In the second result, although preserving the property that the game is played in a restricted altruistic social context, we relax the constraint that Γ is symmetric.

Theorem 3. *There exists a three-player linear congestion game $\mathcal{G}$ and a matrix $\Gamma \in \mathbb{R}^{3 \times 3}$ with a unitary main diagonal such that $(\mathcal{G}, \Gamma)$ does not admit pure Nash equilibria.*

4   Inefficiency of Pure Nash Equilibria

In this section, we give bounds on the prices of anarchy and stability of linear congestion games with restricted social context.

Rahn and Schäfer [20] show that the price of anarchy of coarse correlated equilibria is 17/3 and that this bound is tight even for pure Nash equilibria. We improve the lower bound so as to hold even in the special case in which the underlying linear congestion game is a load balancing one. The basic idea of our construction, suitably extended to comply with our altruistic scenario, is borrowed from Caragiannis et al. [7].

Theorem 4. *For any $\epsilon > 0$, there exists a linear congestion game with restricted altruistic social context $(\mathcal{G}, \Gamma)$, such that $\mathcal{G}$ is a load balancing game and Γ is a symmetric boolean matrix, for which $\mathsf{PoA}(\mathcal{G}, \Gamma) \geq \frac{17}{3} - \epsilon$.*

Note that such a lower bound implies that the the assumption of identical resources in crucial in the upper bound of $2 + \sqrt{5}$ given by de Keijzer et al. [14] for load balancing games with restricted altruistic social context.

We now turn our attention to the study of the price of stability. To this aim, we recall that it is possible to assume without loss of generality that $\beta_e = 0$ for each $e \in E$ as long as we are not interested in load balancing games. For a given linear congestion game with altruistic social context $(\mathcal{G}, \Gamma)$, we denote with $K = (k_1, \ldots, k_n)$ and $O = (o_1, \ldots, o_n)$, respectively, a Nash equilibrium and a social optimum of $(\mathcal{G}, \Gamma)$ and we use $K_e := n_e(K)$ and $O_e := n_e(O)$ to denote the congestion of resource e in K and O, respectively. By exploiting the potential function defined in Theorem 1 and the fact that there exists a pure Nash equilibrium K such that $\Phi(K) \leq \Phi(O)$, we easily obtain the following upper bound.

Theorem 5. *For any linear congestion game with restricted altruistic social context $(\mathcal{G}, \Gamma)$ such that Γ is symmetric, it holds $\mathsf{PoS}(\mathcal{G}, \Gamma) \leq 2$.*

Proof. Let K be a pure Nash equilibrium obtained after a sequence of improving deviation starting from O. The existence of K is guaranteed by the existence of the potential function Φ. Moreover, it holds $\Phi(K) \leq \Phi(O)$. Hence, it follows that

$$\mathsf{SUM}(K) = \sum_{e \in E} \left(\alpha_e K_e^2 \right) \leq \sum_{e \in E} \left(\alpha_e \left(K_e(K_e + 1) + \sum_{(i,j) \in P_e(K)} \gamma_{ij} \right) \right) =$$

$$= \Phi(K) \le \Phi(O) = \sum_{e \in E} \left(\alpha_e \left(O_e(O_e + 1) + \sum_{(i,j) \in P_e(O)} \gamma_{ij} \right) \right) \le$$

$$\le \sum_{e \in E} (\alpha_e \left(O_e(O_e + 1) + O_e(O_e - 1) \right)) = 2 \sum_{e \in E} (\alpha_e O_e^2) = 2\mathsf{SUM}(O),$$

where the last inequality follows from the fact that $\gamma_{ij} \in [0,1]$ for each $i, j \in [n]$ and $|P_e(O)| = O_e(O_e - 1)$. $\qquad\square$

In this case, we are only able to provide a lower bound of $1 + \frac{1}{\sqrt{2}} \approx 1.707$.

Theorem 6. *For any $\epsilon > 0$, there exists a linear congestion game with restricted altruistic social context $(\mathcal{G}, \Gamma)$, such that Γ is a symmetric boolean matrix, for which* $\mathsf{PoS}(\mathcal{G}, \Gamma) \ge 1 + \frac{1}{\sqrt{2}} - \epsilon$.

5   Results for Simple Social Contexts

In this section, we focus on the special case given by model of Caragiannis *et al.* [8] in which, for each $i \in [n]$, it holds $\gamma_{ii} = 1 - \gamma_i$ and $\gamma_{ij} = \gamma_i$ for each $j \ne i \in [n]$, where $\gamma_i \in [0,1]$. In such a model, the restricted altruistic social context coincides with the case in which, for each $i \in [n]$, it holds $\gamma_i \le 1/2$. Caragiannis *et al.* [8] show that, when $\gamma_i = \gamma$ for each $i \in [n]$, the price of anarchy is exactly $\frac{2-\gamma}{1-\gamma}$ for general altruistic social contexts (that is, for $\gamma > 1/2$) and $\frac{5-\gamma}{2-\gamma}$ in the restricted one.

First of all, we prove that pure Nash equilibria are always guaranteed to exist via an exact potential function argument. An existential result had already been given by Hoefer and Skopalik [16], nevertheless, their proof makes use of a weighted potential function. So, our result is slightly stronger and, more importantly, provides a better potential function to be subsequently exploited in the derivation of an upper bound on the price of stability of these games.

Let $\Lambda_n := [0,1]^n$ be the set of n-dimensional vectors whose entries belong to the interval $[0,1]$. Given a vector $\gamma = (\gamma_1, \ldots, \gamma_n) \in \Lambda_n$, denote with Γ_γ the $n \times n$ matrix Γ such that, for each $i \in [n]$, it holds $\gamma_{ii} = 1 - \gamma_i$ and $\gamma_{ij} = \gamma_i$ for each $j \ne i \in [n]$.

Theorem 7. *Each n-player linear congestion game with altruistic social context $(\mathcal{G}, \Gamma)$ such that $\Gamma = \Gamma_\gamma$ for some $\gamma \in \Lambda_n$ admits an exact potential function.*

Proof. Consider a strategy profile $S = (s_1, \ldots, s_n)$, a player $i \in [n]$ and a strategy $t \in S_i$, and again denote with $x_e := n_e(S)$. From equation (1), since $\sum_{j \ne i : e \in s_j} \gamma_{ij} = (x_e - 1)\gamma_i$ and $\sum_{j : e \in s_j} \gamma_{ij} = x_e \gamma_i$, it follows that

$$\widehat{c}_i(S) - \widehat{c}_i(S_{-i} \diamond t)$$
$$= \sum_{e \in s_i \setminus t} (\alpha_e (x_e - \gamma_i) + (1 - \gamma_i)\beta_e) - \sum_{e \in t \setminus s_i} (\alpha_e (x_e + 1 - \gamma_i) + (1 - \gamma_i)\beta_e). \quad (2)$$

Consider, now, the following potential function

$$\Phi(S) = \frac{1}{2} \sum_{e \in E} \left(\alpha_e \left(x_e(x_e + 1) - 2 \sum_{j:e \in s_j} \gamma_j \right) + 2\beta_e \sum_{j:e \in s_j} (1 - \gamma_j) \right).$$

It holds

$$\Phi(S) - \Phi(S_{-i} \diamond t)$$
$$= \frac{1}{2} \sum_{e \in s_i \setminus t} (\alpha_e (2x_e - 2\gamma_i) + 2(1 - \gamma_i)\beta_e)$$
$$- \frac{1}{2} \sum_{e \in t \setminus s_i} (\alpha_e (2(x_e + 1) - 2\gamma_i) + 2(1 - \gamma_i)\beta_e)$$
$$= \sum_{e \in s_i \setminus t} (\alpha_e (x_e - \gamma_i) + (1 - \gamma_i)\beta_e) - \sum_{e \in t \setminus s_i} (\alpha_e (x_e + 1 - \gamma_i) + (1 - \gamma_i)\beta_e)$$

which shows that Φ is an exact potential function for $(\mathcal{G}, \Gamma)$.   □

By exploiting the potential function defined above, we obtain an upper bound on the price of stability for the case in which $\gamma_i = \gamma$ for each $i \in [n]$ as follows.

The fact that there exists a pure Nash equilibrium K such that $\Phi(K) \leq \Phi(O)$ easily implies the following inequality (where, as usual, we have removed the terms β_e from the latency functions):

$$\sum_{e \in E} (\alpha_e (K_e(K_e + 1) - 2\gamma K_e - O_e(O_e + 1) + 2\gamma O_e)) \leq 0, \qquad (3)$$

where we have used the equalities $\sum_{j:e \in k_j} \gamma_j = \gamma K_e$ and $\sum_{j:e \in o_j} \gamma_j = \gamma O_e$.

By exploiting the inequality $\widehat{c}_i(K) - \widehat{c}_i(K_{-i} \diamond o_i) \leq 0$, we obtain that, for each $i \in [n]$, it holds

$$\sum_{e \in k_i \setminus o_i} (\alpha_e (K_e - \gamma)) - \sum_{e \in o_i \setminus k_i} (\alpha_e (K_e + 1 - \gamma)) \leq 0,$$

which implies

$$\sum_{e \in k_i} (\alpha_e (K_e - \gamma)) - \sum_{e \in o_i} (\alpha_e (K_e + 1 - \gamma)) \leq 0. \qquad (4)$$

We now apply the primal-dual technique that we introduced in [5]. This method aims at formulating the problem of maximizing the ratio $\frac{\text{SUM}(K)}{\text{SUM}(O)}$ via linear programming. The two strategy profiles K and O play the role of fixed constants, while, for each $e \in E$, the values α_e defining the latency functions are variables that must be suitably chosen so as to satisfy a set of constraints: some of them, assures that K is a pure Nash equilibrium with some desired properties, while the last one normalizes to 1 the value of the social optimum $\text{SUM}(O)$. The objective function aims at maximizing the social value $\text{SUM}(K)$ which, being the social optimum normalized to 1, is equivalent to maximize the ratio $\frac{\text{SUM}(K)}{\text{SUM}(O)}$. Let us denote with $LP(K,O)$ such a linear program, which, in our scenario of investigation, by making use of inequalities (3) and (4), becomes the following one:

$$maximize \sum_{e \in E} \left(\alpha_e K_e^2\right)$$

subject to

$$\sum_{e \in E} \left(\alpha_e \left(K_e(K_e + 1) - 2\gamma K_e - O_e(O_e + 1) + 2\gamma O_e\right)\right) \leq 0$$

$$\sum_{e \in k_i} \left(\alpha_e \left(K_e - \gamma\right)\right) - \sum_{e \in o_i} \left(\alpha_e \left(K_e + 1 - \gamma\right)\right) \leq 0, \qquad \forall i \in [n]$$

$$\sum_{e \in E} \left(\alpha_e O_e^2\right) = 1,$$

$$\alpha_e \geq 0, \qquad \forall e \in E$$

Let $DLP(K,O)$ be the dual program of $LP(K,O)$. By the Weak Duality Theorem, each feasible solution to $DLP(K,O)$ provides an upper bound on the optimal solution of $LP(K,O)$. Hence, by providing a feasible dual solution, we obtain an upper bound on the ratio $\frac{\text{SUM}(K)}{\text{SUM}(O)}$. Anyway, if the provided dual solution is independent on the particular choice of K and O, we obtain an upper bound on the ratio $\frac{\text{SUM}(K)}{\text{SUM}(O)}$ for any possible pair of profiles K and O, which means that we obtain an upper bound on the price of stability of pure Nash equilibria. The dual program $DLP(K,O)$ is

$$minimize \ \theta$$

subject to

$$x \left(K_e(K_e + 1) - 2\gamma K_e - O_e(O_e + 1) + 2\gamma O_e\right)$$
$$+ \sum_{i:e \in k_i} \left(y_i \left(K_e - \gamma\right)\right) - \sum_{i:e \in o_i} \left(y_i \left(K_e + 1 - \gamma\right)\right) + \theta O_e^2 \geq K_e^2, \quad \forall e \in E$$

$$x \geq 0,$$

$$y_i \geq 0, \qquad \forall i \in [n]$$

Theorem 8. *For any n-player linear congestion game with altruistic social context $(\mathcal{G}, \Gamma)$ such that $\Gamma = \Gamma_\gamma$ for some $\gamma \in \Lambda_n$ with $\gamma_i = \gamma$ for each $i \in [n]$, it holds $\mathsf{PoS}(\mathcal{G}, \Gamma) \leq \frac{(\sqrt{3}+1)(1-\gamma)}{\sqrt{3}-\gamma(\sqrt{3}-1)}$ when $\gamma \in [0, 1/2]$ and $\mathsf{PoS}(\mathcal{G}, \Gamma) \leq \frac{3-\sqrt{3}-2\gamma(2-\sqrt{3})}{2(1-\gamma)}$ when $\gamma \in [1/2, 1]$.*

We now show matching lower bounds.

Theorem 9. *For any $\epsilon > 0$, there exists an n-player linear congestion game with altruistic social context $(\mathcal{G}, \Gamma)$ such that $\Gamma = \Gamma_\gamma$ for some $\gamma \in \Lambda_n$ with $\gamma_i = \gamma \in [0, 1/2]$ for each $i \in [n]$ for which it holds $\mathsf{PoS}(\mathcal{G}, \Gamma) \geq \frac{(\sqrt{3}+1)(1-\gamma)}{\sqrt{3}-\gamma(\sqrt{3}-1)} - \epsilon$*

and an n-player linear congestion game with altruistic social context $(\mathcal{G}', \Gamma')$ *such that* $\Gamma' = \Gamma'_\gamma$ *for some* $\gamma \in \Lambda_n$ *with* $\gamma_i = \gamma \in [1/2, 1]$ *for each* $i \in [n]$ *for which it holds* $\mathsf{PoS}(\mathcal{G}', \Gamma') \geq \frac{3 - \sqrt{3} - 2\gamma(2 - \sqrt{3})}{2(1-\gamma)} - \epsilon.$

Note that for $\gamma = 1/2$, the price of stability is 1 which means that, when players are half selfish and half altruistic, there always exists a social optimal solution which is also a pure Nash equilibrium. For $\gamma = 0$, that is, when players are totally selfish, we reobtain the well-known bound of $1 + 1/\sqrt{3}$ on the price of stability of linear congestion games proven by Caragiannis *et al.* [7]. For $\gamma = 1$, the price of stability goes to infinity, i.e., all Nash equilibria may perform extremely bad with respect to the social optimal solution. This implies that totally altruistic players are tremendously harmful in a non-cooperative system, since they yield games in which even the price of stability may be unbounded. Finally, in the restricted altruistic social context, i.e., $\gamma \in [0, 1/2]$, when γ goes from 0 to 1/2, the price of anarchy increases from 5/2 to 3, while the price of stability decreases from $1 + 1/\sqrt{3}$ to 1. In particular, the increase in the price of anarchy is always compensated by a slightly higher decrease in the price of stability.

We now conclude by considering the general case in which the players have a different degree of altruism. For a vector $\gamma \in \Lambda_n$, denote with $\hat{\gamma}$ and $\check{\gamma}$ the maximum and minimum entry in γ, respectively. For the price of anarchy, by simply exploiting inequality (2), we get the following dual program

$$minimize\ \theta$$
$$subject\ to$$
$$\sum_{i:e \in k_i} (x_i (K_e - \hat{\gamma})) - \sum_{i:e \in o_i} (x_i (K_e + 1 - \check{\gamma})) + \theta O_e^2 \geq K_e^2, \quad \forall e \in E$$
$$x_i \geq 0, \qquad\qquad\qquad\qquad\qquad\qquad\qquad\qquad\qquad \forall i \in [n]$$

Theorem 10. *For any n-player linear congestion game with altruistic social context* $(\mathcal{G}, \Gamma)$ *such that* $\Gamma = \Gamma_\gamma$ *for some* $\gamma \in \Lambda_n$, *it holds* $\mathsf{PoA}(\mathcal{G}, \Gamma) \leq \frac{2 - \check{\gamma}}{1 - \hat{\gamma}}$ *when* $\hat{\gamma} \in [1/2, 1]$ *and* $\mathsf{PoA}(\mathcal{G}, \Gamma) \leq \frac{5 + 2\hat{\gamma} - 3\check{\gamma}}{2 - \hat{\gamma}}$ *when* $\hat{\gamma} \in [0, 1/2]$.

Note that, for $\hat{\gamma} = \check{\gamma}$, we reobtain the upper bounds already proved by Caragiannis *et al.* [8]. Moreover, the bound for the case $\hat{\gamma} \in [0, 1/2]$, improves on the one proved by Chen *et al.* [9] which however holds for the more general setting of coarse correlated equilibria.

References

1. Aland, S., Dumrauf, D., Gairing, M., Monien, B., Schoppmann, F.: Exact Price of Anarchy for Polynomial Congestion Games. SIAM Journal on Computing 40(5), 1211–1233 (2011)
2. Anshelevich, E., Dasgupta, A., Kleinberg, J., Tardos, E., Wexler, T., Roughgarden, T.: The Price of Stability for Network Design with Fair Cost Allocation. SIAM Journal on Computing 38(4), 1602–1623 (2008)

3. Ashlagi, I., Krysta, P., Tennenholtz, M.: Social Context Games. In: Papadimitriou, C., Zhang, S. (eds.) WINE 2008. LNCS, vol. 5385, pp. 675–683. Springer, Heidelberg (2008)

4. Awerbuch, B., Azar, Y., Epstein, A.: The Price of Routing Unsplittable Flow. In: Proceedings of STOC 2005, pp. 57–66. ACM Press (2005)

5. Bilò, V.: A Unifying Tool for Bounding the Quality of Non-cooperative Solutions in Weighted Congestion Games. In: Erlebach, T., Persiano, G. (eds.) WAOA 2012. LNCS, vol. 7846, pp. 215–228. Springer, Heidelberg (2013)

6. Bilò, V., Celi, A., Flammini, M., Gallotti, V.: Social Context Congestion Games. Theoretical Computer Science 514, 21–35 (2013)

7. Caragiannis, I., Flammini, M., Kaklamanis, C., Kanellopoulos, P., Moscardelli, L.: Tight Bounds for Selfish and Greedy Load Balancing. Algorithmica 61(3), 606–637 (2011)

8. Caragiannis, I., Kaklamanis, C., Kanellopoulos, P., Kyropoulou, M., Papaioannou, E.: The Impact of Altruism on the Efficiency of Atomic Congestion Games. In: Wirsing, M., Hofmann, M., Rauschmayer, A. (eds.) TGC 2010, LNCS, vol. 6084, pp. 172–188. Springer, Heidelberg (2010)

9. Chen, P.-A., de Keijzer, B., Kempe, D., Schäfer, G.: The Robust Price of Anarchy of Altruistic Games. In: Chen, N., Elkind, E., Koutsoupias, E. (eds.) WINE 2011. LNCS, vol. 7090, pp. 383–390. Springer, Heidelberg (2011)

10. Chen, P.A., Kempe, D.: Altruism, Selfishness and Spite in Traffic Routing. In: Proceedings of EC 2008, pp. 140–149. ACM Press (2008)

11. Christodoulou, G., Gairing, M.: Price of Stability in Polynomial Congestion Games. In: Fomin, F.V., Freivalds, R., Kwiatkowska, M., Peleg, D. (eds.) ICALP 2013, Part II. LNCS, vol. 7966, pp. 496–507. Springer, Heidelberg (2013)

12. Christodoulou, G., Koutsoupias, E.: The Price of Anarchy of Finite Congestion Games. In: Proceedings of STOC 2005, pp. 67–73. ACM Press (2005)

13. Christodoulou, G., Koutsoupias, E.: On the Price of Anarchy and Stability of Correlated Equilibria of Linear Congestion Games. In: Brodal, G.S., Leonardi, S. (eds.) ESA 2005. LNCS, vol. 3669, pp. 59–70. Springer, Heidelberg (2005)

14. Anagnostopoulos, A., Becchetti, L., de Keijzer, B., Schäfer, G.: Inefficiency of Games with Social Context. In: Vöcking, B. (ed.) SAGT 2013. LNCS, vol. 8146, pp. 219–230. Springer, Heidelberg (2013)

15. Fotakis, D., Kontogiannis, S., Spirakis, P.G.: Atomic Congestion Games among Coalitions. ACM Transactions on Algorithms 4(4) (2008)

16. Hoefer, M., Skopalik, A.: Altruism in Atomic Congestion Games. In: Fiat, A., Sanders, P. (eds.) ESA 2009. LNCS, vol. 5757, pp. 179–189. Springer, Heidelberg (2009)

17. Hoefer, M., Skopalik, A.: Social Context in Potential Games. In: Goldberg, P.W. (ed.) WINE 2012. LNCS, vol. 7695, pp. 364–377. Springer, Heidelberg (2012)

18. Koutsoupias, E., Papadimitriou, C.: Worst-case equilibria. In: Meinel, C., Tison, S. (eds.) STACS 1999. LNCS, vol. 1563, pp. 404–413. Springer, Heidelberg (1999)

19. Lücking, T., Mavronicolas, M., Monien, B., Rode, M.: A New Model for Selfish Routing. Theoretical Computer Science 406(2), 187–206 (2008)

20. Rahn, M., Schäfer, G.: Bounding the Inefficiency of Altruism through Social Contribution Games. In: Chen, Y., Immorlica, N. (eds.) WINE 2013. LNCS, vol. 8289, pp. 391–404. Springer, Heidelberg (2013)

21. Rosenthal, R.W.: A Class of Games Possessing Pure-Strategy Nash Equilibria. International Journal of Game Theory 2, 65–67 (1973)

Scheduling over Scenarios on Two Machines*

Esteban Feuerstein[1], Alberto Marchetti-Spaccamela[2], Frans Schalekamp[3],
René Sitters[4,5], Suzanne van der Ster[4], Leen Stougie[4,5], and Anke van Zuylen[3]

[1] Departamento de Computación, FCEyN, UBA, Buenos Aires, Argentina
efeuerst@dc.uba.ar
[2] Sapienza University of Rome, Italy
alberto@dis.uniroma1.it
[3] College of William and Mary, Department of Mathematics,
Williamsburg, VA 23185, USA
{frans,anke}@wm.edu
[4] Vrije Universiteit Amsterdam, The Netherlands
{r.a.sitters,suzanne.vander.ster,l.stougie}@vu.nl
[5] CWI Amsterdam, The Netherlands
{r.a.sitters,stougie}@cwi.nl

Abstract. We consider scheduling problems over scenarios where the
goal is to find a single assignment of the jobs to the machines which performs well over all possible scenarios. Each scenario is a subset of jobs
that must be executed in that scenario and all scenarios are given explicitly. The two objectives that we consider are minimizing the maximum
makespan over all scenarios and minimizing the sum of the makespans
of all scenarios. For both versions, we give several approximation algorithms and lower bounds on their approximability. With this research
into optimization problems over scenarios, we have opened a new and
rich field of interesting problems.

Keywords: job scheduling, makespan minimization, scenarios, approximation.

1 Introduction

We consider optimization problems *over scenarios* where the goal is to find a
single solution that performs well for each scenario in a given set of scenarios.
In particular, we consider the scheduling problem where the objective function
is the makespan: we are given a set J of jobs, each with a processing time,
and a set of scenarios; each scenario is specified by a subset of jobs in J that
must be executed in that scenario. Our goal is to find an assignment of jobs
to machines that *is the same for all scenarios* and optimizes a function of the
makespan, i.e., the completion time of the last completed job, *over all scenarios*.

---

* This work was partially supported by EU-IRSES grant EUSACOU and Tinbergen
Institute. EF was partially supported by projects PRH PICT 2009-119 and UBA-
CYT 20020090100149. AvZ was partially supported by Suzann Wilson Matthews
summer research award.

Z. Cai et al. (Eds.): COCOON 2014, LNCS 8591, pp. 559–571, 2014.
© Springer International Publishing Switzerland 2014

The two objectives that we consider are minimizing the *maximum* makespan over all scenarios and minimizing the *sum* of the makespan of all scenarios. We note that when the input contains only a single scenario, both versions of the problem reduce to the usual makespan minimization problem.

As an example, suppose that J contains three jobs, numbered 1, 2, and 3, that must be executed on two machines; the processing time of job 1 is 2 while the processing time of jobs 2 and 3 is 1. There are three scenarios $S_1 = \{1, 2, 3\}$ and $S_2 = S_3 = \{2, 3\}$. Assigning job 1 to the first machine and jobs 2 and 3 to the second machine minimizes the maximum makespan over all scenarios, while assigning jobs 1 and 2 to the first machine and job 3 to the second one minimizes the sum of the makespans of all scenarios.

The more egalitarian objective function of minimizing the maximum makespan over all scenarios fits in the framework of *robust* optimization, where usually not so much a finite set of scenarios is explicitly given, as in our problem, but ranges for values of input parameters (see [3]). We will refer to this objective as the Min-Max objective. The more *utilitarian* objective function of minimizing the sum of the makespans of all scenarios fits in the framework of *a priori* optimization, though a priori optimization has so far only been introduced as a problem where the scenarios are random objects and the objective is to minimize the expected objective value. In that sense, minimizing the sum of makespans could be seen as the a priori problem with a uniform discrete distribution over a finite set of fully specified scenarios. In general, the deterministic problem of optimizing over a finite set of scenarios can be seen as an alternative to the stochastic a priori setting [13], in case a limited number of likely scenarios exists. We refer to this objective as the MinSum objective.

In an indirect way, combinatorial optimization problems over scenarios with the MinSum objective have appeared as the first-stage problem in a boosted sampling approach to two-stage stochastic optimization problems [11]. In [11], scenarios are defined within a so-called black box, meaning that they can only be learnt by sampling. From the black box, a finite set of scenarios is sampled, giving rise to a deterministic optimization problem over the drawn set of scenarios, in which a single solution needs to be found, that minimizes the sum of the objective values for the drawn scenarios. In this sense some results on combinatorial optimization problems over scenarios have appeared, like VERTEX COVER, STEINER TREE and UNCAPACITATED FACILITY LOCATION [11].

Modeling optimization problems over a finite set of given scenarios yields a rich source of interesting new combinatorial optimization problems, which are in general harder than their single-scenario versions. Specifically, almost any single-scenario scheduling problem has an interesting multi-scenario variation. As mentioned before, in this paper we focus, as a first example, on minimizing the maximum makespan over all scenarios and minimizing the sum (or, equivalently, the average) of the makespan of all scenarios.

The specific setting of the scheduling problem over scenarios appears in situations where jobs have to be performed by skilled machines (workers), and some investment is required to attain the skill for a particular job. In such situations, one should decide on an assignment of all possible jobs to the workers, such that the workers can train for the jobs assigned to them ahead of time. The problem then is to assign jobs (specializations) to machines (workers), so that the workload of a machine for any scenario of jobs, from a set of scenarios likely to occur, is minimized. Examples of such a setting are assignment of clients to lawyers, households to power sources, compile-time assignment of computational tasks to processors. In most of such situations, the robust version of the problem with the MinMax objective is rather plausible, especially in situations where a set of likely scenarios to hedge against can be specified upfront.

Another motivation, though a bit indirect, comes from distributed information retrieval: in a term-partitioned index, it is good to allocate to the same processor terms appearing frequently together in queries, so as to minimize the communication cost (to solve an intersection query between two terms that reside in different processors, one of the posting lists must be sent to the processor holding the other). But this goal must be complemented with that of balancing the load, as it is not viable to put all the terms in the same processor. Therefore, it is necessary to divide "clusters" of commonly co-occurring terms among the processors, trying to balance the load. Naturally, queries appear sequentially over time and are not known a priori. One could, as an approximation, optimize considering as input the more likely scenarios. The partition must indeed be done a priori, because lists must be assigned to processors a priori.

To the best of our knowledge, this problem has not been considered in the literature. An a priori version of scheduling with stochastic scenarios has been studied in [4,5], albeit not from an approximation theory point of view, but merely presenting experimental results, and with the scheduling objective of minimizing the sum of completion times of all the jobs per scenario.

We now give a formal definition of the two problems we consider. We restrict ourselves to the case of two machines. We are given a set of jobs J with for each job $j \in J$ a processing times p_j, and a set of k scenarios $\mathcal{S} = \{S_1, S_2, \ldots, S_k\}$, where each scenario $S_i \in \mathcal{S}$ is a subset of J. In each scenario, we are interested in minimizing the makespan, but we are restricted to finding a solution, i.e., an assignment of the jobs to the machines, that applies to every one of the scenarios. Clearly, a solution that is good for one scenario may be bad for another. This gives rise to specifying objectives that reflect the trade-off between the various scenarios. In this paper we define the following two versions of the problem.

- **MM2** Assign the jobs in J to two machines in such a way that the maximum makespan over the given scenarios is minimized. In other words, if we denote the makespan of a subset $S \subseteq J$ of jobs by $p(S) = \sum_{j \in S} p_j$, we are looking for a partition $A, \bar{A}$ of J, that minimizes $\max_{i=1,2,\ldots,k} \max\{p(A \cap S_i), p(\bar{A} \cap S_i)\}$.

- **SM2** Assign the jobs to the machines such that the sum of the makespans of the given scenarios is minimized. Using the notation just introduced, we are seeking a partition $A, \bar{A}$ of J, that minimizes $\sum_{i=1}^{k} \max\{p(A \cap S_i), p(\bar{A} \cap S_i)\}$.

For both objective functions, the problems are NP-hard, since the single-scenario version is NP-hard. However, the single-scenario version is only weakly NP-hard for 2 machines and an FPTAS exists [1], whereas the problems defined here are strongly NP-hard. We will give various approximability and inapproximability results for several different versions of the problem depending on restrictions of the input. In particular, the special cases that we consider are the following:

1. $p_j = 1 \; \forall j \in J$, that is, the case where all processing times are unitary;
2. $|S_i| \leq r \; \forall S_i \in \mathcal{S}$, that is, the case where the number of jobs in each scenario is bounded by a constant;
3. $k = |\mathcal{S}|$ is constant, that is, the case that the number of scenarios is a constant.

In Section 2, we study the problem MM2; we show that the problem cannot be approximated to within a ratio of $2 - \varepsilon$ already in the case where $p_j = 1$ and a ratio of $3/2$ if $|S_i| \leq 3$ and $p_j = 1$. On the positive side, we give a polynomial-time algorithm for the version in which every scenario contains 2 jobs. If k, the number of scenarios, is constant then there exists a PTAS; for an arbitrary number of scenarios, a $O(\log^2 k)$ approximation ratio exists. The latter two results are a consequence of an observed direct relation to the so-called VECTOR SCHEDULING problem (see Section 2 for its definition) and results of [6].

In Section 3, we study problem SM2. We prove inapproximability within 1.0196 assuming P$\neq$NP, and within 1.0404 under the *Unique Games Conjecture* [15]. On the positive side, we present a $3/2$-approximate randomized algorithm. For instances with scenarios of size at most 3, we use a reduction to MAX CUT to obtain a 1.12144-approximation algorithm. For scenarios of size at most r, we present a reduction to WEIGHTED MAX NOT-ALL-EQUAL r-SAT and use this to obtain better-than-$3/2$- approximations for problems where the scenario sizes are not larger than 4.

Some thoughts about related problems, and ideas for future research are contained in a concluding section.

2   Minimizing Maximum Makespan

We obtain inapproximability of MM2 using a recent result [2] on the hardness of HYPERGRAPH BALANCING: given a hypergraph find a 2-coloring of the vertices such as to minimize over all hyperedges the discrepancy between the number of vertices of the two colors.

Theorem 1. *It is NP-hard to approximate MM2 with unitary jobs within ratio $2 - \epsilon$.*

This is remarkable since, trivially, any solution, for any job sizes, is 2-approximate (since we consider the problem for two machines only). In the full version of this paper [8] we prove a hardness bound of $3/2$ when $|S_i| \leq 3$ and $p_j = 1$. This result completes the hardness characterization.

We now show that, if the number of jobs per scenario is 2, then the problem is solvable in polynomial time.

Theorem 2. *MM2 with $|S_i| = 2$ for all $S_i \in \mathcal{S}$ can be solved in time $O(|\mathcal{S}| \log |\mathcal{S}|)$.*

Proof. We create a graph with a vertex for each job and connect by an edge the jobs that appear together in a scenario. We define the weight of edge (j, k) to be $p_j + p_k$, i.e., the sum of the processing times of the jobs associated to the incident vertices. Note that a solution for the a priori scheduling problem is a partitioning of the job set, and can be associated with a coloring of the vertices in this graph problem with two color classes. The objective value is then equal to the maximum of the highest weight of any monochromatic edge and the largest processing time of any job.

In other words, we should find a 2-coloring of the vertices of this graph, such that the maximum weight of a monochromatic edge is minimized. A lowest weight edge in any odd cycle gives a lower bound on the objective value.

Consider the following algorithm. Starting with all vertices being part of their own singleton component, and having color 1, we grow components by inserting edges, and label the vertices with the component they belong to, and with a color that can assume two values; 1 and 2. A color inversion of a vertex changes the color of the vertex (i.e., if it is colored 1, the color is changed to 2, and vice versa). We consider the edges in order of descending weight. When considering the next edge, say (j, k), the following 3 cases can occur.

Case 1. Vertices j and k have the same color, and are in the same component. We end the algorithm. An optimal partitioning of the job set is given by the two color classes, where jobs that have color 1 (respectively, 2) are assigned to machine 1 (respectively, 2) and the objective is equal to the weight of edge (j, k).

Case 2. Vertices j and k have different colors. If the vertices are in different components, then we update the component label for all nodes of the smaller component (breaking ties arbitrarily), so that all vertices have the same label. We then proceed to the next edge.

Case 3. Vertices j and k have the same color, and are in different components. In this case we invert the color of all nodes in the smaller component (breaking ties arbitrarily), and then proceed as in Case 2.

By construction, two vertices of the same color in the same component are joined by an even-length path. Therefore, when the algorithm terminates in Case 1, we have found an odd cycle in the graph, of which this last edge has lowest weight. Note that the assignment of jobs with the same color to the same machine implies that the makespan of the scenario is bounded by the weight of the last considered edge. Since the weight of any such edge is a lower bound on the objective value, we have found an optimal solution. Its value is given by the maximum of the weight of a monochromatic edge and $p_{\max} = \max_{j \in J} p_j$.

The running time follows from the observation that any time we invert the color and/or update the label of a vertex, it ends up in a component of at least twice the size of the component it belonged to before. Hence, the label of a vertex can be updated at most $\log |J|$ times. The total time can thus be bounded by $|\mathcal{S}| \log |\mathcal{S}|$ time for sorting the edges by weight, plus $|J| \log |J|$ time for updating the vertex colors and labels. Finally, we may assume without loss of generality that each job appears in at least one scenario, so $|\mathcal{S}| \geq |J|/2$.   □

Another sharp characterization w.r.t. the number of scenarios, is obtained for the case of a constant number of scenarios. For jobs with unit processing times, the problem can be solved exactly: given that the number of scenarios is constant, there is only a constant number of job types, where the type of a job is the set of scenarios it is in. Then, the number of jobs on machine 1 of each type can be guessed. There are only a polynomial number of choices;, an extension can also accommodate a constant number of machines in polynomial time. We notice that this also solves SM2 under the same restrictions in polynomial time.

Theorem 3. *MM2 and SM2 having jobs with unitary processing times can be solved in polynomial time if the number of scenarios is constant.*

A similar idea, with guessing the optimal value and rounding, leads to a PTAS in the general case under a constant number of scenarios, but this is also implied by the following result.

We conclude this section by noticing that if we consider any number of machines, the problem of minimizing the maximum makespan reduces to the VECTOR SCHEDULING problem, where each coordinate corresponds to a scenario.

Definition 1. *In the VECTOR SCHEDULING problem we are given a set V of n rational d-dimensional vectors $v_1, \ldots, v_n$ from $[0, \infty)^d$ and a number m. A valid solution is a partition of V into m sets $A_1, \ldots, A_m$. The objective is to minimize $\max_{1 \leq i \leq m} \| \sum_{v_j \in A_i} v_j \|_\infty$.*

This problem is a d-dimensional generalization of the makespan minimization problem, where each job is a d-dimensional vector and the machines are d-dimensional objects as well. In our setting, the dimension d equals the number of scenarios $|\mathcal{S}|$. Each coordinate of job j equals its processing time in the corresponding scenario (either 0 or p_j). Results of Chekuri et al. [6] on VECTOR SCHEDULING can directly be translated into our setting.

Theorem 4 ([6]). *For the problem of minimizing the maximum makespan over scenarios $S_i \in \mathcal{S}$ on m machines,*

1. *there exists a PTAS for the case that $k = |\mathcal{S}|$ is constant*
2. *there exists a polynomial-time $O(\log^2 k)$-approximation for k scenarios;*
3. *there exists no c-approximation algorithm for any $c > 1$, when dealing with any number of scenarios.*

3   Minimizing Sum of Makespans

We now turn our attention to SM2, the problem of minimizing the sum of the makespans over all scenarios, in the case of 2 machines.

We start this section by noting that SM2 is MAX SNP-hard even with unitary processing times and scenarios containing two jobs each.

Theorem 5. *SM2 is NP-hard to approximate to within a factor of* 1.0196 *and UGC-hard to approximation to within a factor of* 1.0404, *even if all jobs have length 1, and all scenarios contain two jobs.*

The proof is through a reduction from MAX CUT [9], and the hardness of approximation results shown by Håstadt [12] and Khot et al. [16]. The details are given in the full version of this paper [8].

In the remainder of this section, we will give approximation results for SM2. As for MM2 in the previous section, we notice that also for this problem any solution is a trivial 2-approximation. In the remainder of this section, we will first show that the algorithm that randomly assigns the jobs to the two machines independently with equal probability gives a 3/2-approximation. We then show two deterministic approximation algorithms, which give good approximation guarantees if the number of jobs per scenario is small.

3.1   A Randomized Approximation Algorithm

Lemma 1. *Consider a scenario S, and let $A, \bar{A}$ be any partitioning of the jobs in S. When assigning each job of S to the two machines independently with equal probability, the expected load of the least loaded machine is at least $\frac{1}{2} \min\{p(A), p(\bar{A})\}$.*

Proof. An assignment of jobs to the two machines induces a partition of A into sets A', A'', and a partition of $\bar{A}$ into sets $\bar{A}', \bar{A}''$ where the jobs in the same set of the partition are assigned to the same machine. The sets $A', A'', \bar{A}', \bar{A}''$ are not necessarily all non-empty. We will prove the lemma by showing that, conditioned on the sets $A', A'', \bar{A}', \bar{A}''$, the machine load of the least loaded machine is at least $\frac{1}{2} \min\{p(A), p(\bar{A})\}$, which implies that the statement also holds unconditionally.

Conditioned on the sets $A', A'', \bar{A}', \bar{A}''$, the least loaded machine has a load of $\min\{p(A') + p(\bar{A}'), p(A'') + p(\bar{A}'')\}$ with probability $\frac{1}{2}$ (namely, if $A', \bar{A}'$ are assigned to one machine, and $A'', \bar{A}''$ to the other machine), and $\min\{p(A') + p(\bar{A}''), p(A'') + p(\bar{A}')\}$ with probability $\frac{1}{2}$ (namely, if $A', \bar{A}''$ are assigned to one machine, and $A'', \bar{A}'$ are assigned to the other machine). Hence, conditioned on the partition of A into A', A'' and of $\bar{A}$ into $\bar{A}', \bar{A}''$, the expected load of the least loaded machine is

$$\tfrac{1}{2} \min\{p(A') + p(\bar{A}'), p(A'') + p(\bar{A}'')\} + \tfrac{1}{2} \min\{p(A') + p(\bar{A}''), p(A'') + p(\bar{A}')\}.$$

Note that a simple case analysis shows that the sum of the two terms is either at least $\frac{1}{2}(p(A') + p(A'')) = \frac{1}{2}p(A)$ or at least $\frac{1}{2}(p(\bar{A}') + p(\bar{A}'')) = \frac{1}{2}p(\bar{A})$. So the load of the least loaded machine is at least $\frac{1}{2} \min\{p(A), p(\bar{A})\}$. □

Theorem 6. *Randomly assigning each job to the two machines independently with equal probability is a 3/2-approximation for SM2.*

Proof. Consider a scenario S, and let A be the set of jobs processed on machine 1, and $\bar{A} = S \setminus A$ the set of jobs processed on machine 2 in a schedule with minimum makespan. Hence, the optimal makespan for S is $\max\{p(A), p(\bar{A})\}$. By Lemma 1, the load of the least loaded machine in scenario S, if the jobs are randomly assigned to the machines with equal probability, is at least $\frac{1}{2}\min\{p(A),p(\bar{A})\}$. Hence, the load of the machine with the highest load is at most $p(A)+p(\bar{A})-\frac{1}{2}\min\{p(A),p(\bar{A})\}$ $= \max\{p(A),p(\bar{A})\} + \frac{1}{2}\min\{p(A),p(\bar{A})\} \le \frac{3}{2}\max\{p(A),p(\bar{A})\}$.

Hence, the expected makespan for scenario S is at most $\frac{3}{2}$ times the optimal makespan for scenario S, which implies that the sum over all scenarios of the expected makespans is at most $\frac{3}{2}$ times the optimal summed makespan of all scenarios.   □

We remark that the proof of the previous lemma bounds the objective value by comparing the load on a machine in a given scenario to the load for the optimal schedule *for that scenario*, rather than the optimal schedule for our problem.

It is easy to see that the analysis of the simple randomized algorithm is tight, by considering an instance of two jobs $\{1,2\}$ with unitary execution time and one scenario $S_1 = \{1,2\}$. The optimal solution is to assign one job to each machine, whereas the randomized algorithm either assigns both jobs to the same machine with probability $\frac{1}{2}$, or one job to each machine with probability $\frac{1}{2}$.

3.2   Deterministic Approximation Algorithms

To obtain a deterministic approximation algorithm, we show that the SM2 problem can be reduced to the WEIGHTED MAX NOT-ALL-EQUAL SATISFIABILITY problem, that we will abbreviate as MAX-NAE SAT.

Definition 2. *In* MAX-NAE SAT, *a boolean expression is given, and a weight for each clause. A clause in the expression is satisfied if it contains both true and false literals. The problem is to find an assignment of true/false values to the variables, such as to maximize the total weight of the clauses satisfied.*

Note that if r is such that $|S_i| \le r$ for all $S_i \in \mathcal{S}$, then by adding dummy jobs of processing time 0, we can assume that every scenario contains exactly the same number of jobs, i.e., $|S_i| = r$ for all $S_i \in \mathcal{S}$. We will reduce the SM2 problem with scenarios of size at most r to the MAX-NAE SAT problem with clauses of length r (MAX-NAE r-SAT).

Theorem 7. *A $(1 - \gamma_r)$-approximation for* MAX-NAE r-SAT *implies a $(1 + 2^{r-2}\gamma_r)$-approximation for the SM2 problem with $|S| \le r$ for all scenarios $S \in \mathcal{S}$.*

Proof. We start by formulating the SM2 problem as a MAX-NAE SAT problem. Each job j corresponds to a variable x_j in the MAX-NAE SAT instance. An assignment of the variables in the MAX-NAE SAT instance corresponds to an assignment in SM2 as follows: machine 1 is assigned all jobs for which the corresponding variable is set to true, and machine 2 processes all jobs for which the corresponding variable is set to false.

We now construct a set of weighted clauses for each scenario such that the weight of the satisfied clauses for a given assigment is equal to the load of the least loaded machine in the scenario. Hence, maximizing the weight of the satisfied clauses will maximize the weight of the least loaded machine, and it will thus minimize the weight of the machine with the heaviest load, i.e., the makespan.

For a given scenario S of SM2 with r jobs, we construct 2^{r-1} clauses of length r as follows. For each partitioning of S into two sets A and $\bar{A}$, we create a clause denoted by $C_S(\{A, \bar{A}\})$. In clause $C_S(\{A, \bar{A}\})$, all variables corresponding to jobs in one set appear negated, all variables corresponding to the other set appear non-negated. Note that $C_S(\{A, \bar{A}\})$ has the same truth table as $C_S(\{\bar{A}, A\})$ (namely, a clause is false if and only if all its literals are false, or all its literals are true). Note that this means that if A is assigned to the first machine and $\bar{A}$ is assigned to the second machine, then all clauses *except* $C_S(\{A, \bar{A}\})$ are satisfied.

Denote by $w_S(\{A, \bar{A}\})$ the weight on the clause $C_S(\{A, \bar{A}\})$. To ensure the weight of the satisfied clauses is equal to the weight of the least loaded machine in SM2, we define weights on the clauses to be so that

$$\sum_{B,\bar{B}:B\cup\bar{B}=S,B\cap\bar{B}=\emptyset} w_S(\{B, \bar{B}\}) - w_S(\{A, \bar{A}\}) = \min\{p(A), p(\bar{A})\}.$$

Let $N = 2^{r-1}$, i.e., N is the number of clauses corresponding to scenario S. The solution to this system of equations is to set

$$w_S(\{A, \bar{A}\}) = \frac{1}{N-1} \sum_{B,\bar{B}:B\cup\bar{B}=S,B\cap\bar{B}=\emptyset} \min\{p(B), p(\bar{B})\} - \min\{p(A), p(\bar{A})\}.$$

The weights thus defined are not necessarily non-negative: consider a scenario S that contains $r = 4$ jobs of unit length. There are four ways of partitioning S into one set of size one and one set of size three, and there are $\binom{4}{2}/2 = 3$ ways of partitioning S into two sets of size two. Therefore $\sum_{B,\bar{B}} \min\{p(B), p(\bar{B})\} = 10$, but that means that for a partitioning into sets $A, \bar{A}$ of size two $w_S(\{A, \bar{A}\}) = \frac{1}{7}(10) - 2 < 0$.

To use approximation algorithms for MAX-NAE SAT, we need to make sure that all weights are non-negative. We accomplish this by adding a constant $K(S)$ to all weights of clauses corresponding to scenario S, where we set $-K(S)$ equal to a lower bound on the weights. We derive a lower bound on the weights by noting that (1) $\frac{1}{N}\sum_{B,\bar{B}} \min\{p(B), p(\bar{B})\}$ is the expected value of the least loaded machine when all jobs are assigned to a machine with probability $\frac{1}{2}$ independently, hence, by Lemma 1, its value is lower bounded by $\frac{1}{2}\max_{B,\bar{B}} \min\{p(B), p(\bar{B})\}$; and (2) trivially, $\max_{B,\bar{B}} \min\{p(B), p(\bar{B})\} \leq \frac{1}{2}p(S)$. Therefore

$$w_S(\{A, \bar{A}\}) = \frac{1}{N-1} \sum_{B,\bar{B}:B\cup\bar{B}=S, B\cap\bar{B}=\emptyset} \min\{p(B), p(\bar{B})\} - \min\{p(A), p(\bar{A})\}$$

$$= \frac{N}{N-1} \frac{1}{N} \sum_{B,\bar{B}:B\cup\bar{B}=S, B\cap\bar{B}=\emptyset} \min\{p(B), p(\bar{B})\} - \min\{p(A), p(\bar{A})\}$$

$$\geq \frac{N}{N-1} \frac{1}{2} \max_{B,\bar{B}} \min\{p(B), p(\bar{B})\} - \min\{p(A), p(\bar{A})\}$$

$$\geq \frac{\frac{1}{2}N - (N-1)}{N-1} \max_{B,\bar{B}} \min\{p(B), p(\bar{B})\}$$

$$= -\frac{1}{2} \frac{N-2}{N-1} \max_{B,\bar{B}} \min\{p(B), p(\bar{B})\}$$

$$\geq -\frac{1}{4} \frac{N-2}{N-1} p(S).$$

Thus, we set $K(S) = \frac{1}{4} \frac{N-2}{N-1} p(S)$, such that $\tilde{w}_S(\{A, \bar{A}\}) = w_S(\{A, \bar{A}\}) + K(S) \geq 0$ for all partitionings $A, \bar{A}$ of S into two sets.

A solution to the MAX-NAE SAT instance is now mapped to a solution of SM2, by assigning the jobs for which the variable is set to true to machine 1, and scheduling the other jobs on machine 2. We note that the w-weights of the clauses corresponding to scenario S were chosen so that the sum of the weights of the clauses that are satisfied is exactly equal to the load on the least loaded machine in scenario S. Also, $N - 1$ clauses of scenario S are satisfied in any solution to the MAX-NAE SAT instance. Therefore the total $\tilde{w}$-weight of the clauses for scenario S that are satisfied in any MAX-NAE SAT solution is equal to the load on the least loaded machine in scenario S plus an additional $(N - 1)K(S)$.

We let $L = \sum_S p(S)$, and denote by $L^*_{\min}$ the sum over all scenarios of the load of the least loaded machine in an optimal solution, and by $L^*_{\max}$ the sum over all scenarios of the load of the most loaded machine in an optimal solution, so that $L^*_{\min} + L^*_{\max} = L$. Note that the additional term $K(S)$ in the $\tilde{w}$-weights of the MAX-NAE SAT solution causes an increase of the objective value with respect to the w-weights solution by adding an additional $\sum_S (N - 1)K(S) = \sum_S \frac{1}{4}(N - 2)p(S) = \frac{1}{4}(N - 2)L$ to each solution.

In particular, an optimal solution to the MAX-NAE SAT instance, has objective value $L^*_{\min} + \frac{1}{4}(N - 2)L$, and a $(1 - \gamma)$-approximation algorithm for the MAX-NAE SAT instance, therefore, has objective value at least $(1 - \gamma)(L^*_{\min} + \frac{1}{4}(N - 2)L)$. Let us denote by $ALG(L_{\min})$ and $ALG(L_{\max})$ the sum over all scenarios of the least and most loaded machines in the corresponding job assignment. Note that $ALG(L_{\min}) \geq (1 - \gamma)\left((L^*_{\min} + \frac{1}{4}(N - 2)L) - \frac{1}{4}(N - 2)L = (1 - \gamma)L^*_{\min} - \frac{1}{4}\gamma(N - 2)L$. Therefore,

$$ALG(L_{\max}) = L - ALG(L_{\min}) \leq L - ((1 - \gamma)L^*_{\min} - \frac{1}{4}\gamma(N - 2)L)$$

$$= (1 - \gamma)(L - L^*_{\min}) + \gamma L + \frac{1}{4}\gamma(N - 2)L$$

$$= (1 - \gamma)L^*_{\max} + \frac{1}{4}\gamma(N + 2)L.$$

Noting that $L \leq 2L^*_{\max}$ gives $ALG(L_{\max}) \leq (1 - \gamma)L^*_{\max} + \frac{1}{2}\gamma(N + 2)L^*_{\max} = (1 + \frac{1}{2}\gamma N)L^*_{\max}$ which proves the theorem, since $N = 2^{r-1}$. □

For $r = 3$, Zwick [18] gives a 0.90871-approximation for MAX-NAE 3-SAT. By the previous lemma, this gives a 1.18258-approximation for SM2 with scenarios of length at most three. For $r = 4$, Karloff et al. [14] give a $\frac{7}{8}$-approximation for MAX-NAE 4-SAT. By our lemma, this implies a $\frac{3}{2}$-approximation for SM2 with scenarios of size 4. Note that this matches the guarantee we proved for the algorithm that randomly assigns each job to one of the two machines. For general r, the best approximation factor known for MAX-NAE SAT is 0.74996 due to Zhang, et al. [17], and the implied approximation guarantees for our problem are worse than the guarantee for the random assignment.

If every scenario has exactly two jobs, then we can obtain a better approximation guarantee by reducing SM2 to MAX CUT as follows: we create a vertex for every job, and add an edge between i and j of weight $\min\{p_i, p_j\}$ for every scenario that contains jobs i and j. For any cut, the weight of the edges crossing the cut is then exactly the sum over all scenarios of the load of the least loaded machine. Since the makespan for a scenario S is $p(S)$ minus the load of the least loaded machine, maximizing the load of the least loaded machine, summed over all scenarios, is equivalent to minimizing the sum of the makespans.

If every scenario has at most three jobs, we can also reduce SM2 to MAX CUT, but the reduction, given in the full version [8], is slightly more involved.

Theorem 8. *There exists a $(1 + \gamma)$-approximation algorithm for the SM2 problem with scenarios containing at most three jobs, where $1 - \gamma$ is equal to the approximation ratio for MAX CUT.*

The 0.87856-approximation for MAX CUT of Goemans et al. [10] gives us the following corollary.

Corollary 1. *There exists a 1.12144-approximation algorithm for the SM2 problem with scenarios containing at most three jobs.*

4 Epilogue

This paper presents some first results on a basic scheduling problem under a set of scenarios. The objective is to find a single solution that is applied to all the scenarios specified. We studied this problem for scheduling with two different objectives: minimizing the maximum objective value over all scenarios, the Min-Max version, and minimizing the sum of the objective values of all scenarios, the MinSum version.

To the best of our knowledge, combinatorial optimization problems under a set of fully explicitly specified scenarios has hardly been studied in the literature. Apart from posing theoretically interesting questions as we hope to have shown with this paper, it enhances our ability to model decisions problems where a learning aspect for performing jobs prohibits that job assignments can be adjusted on a day-by-day basis, but merely require a fixed assignment whose quality then necessarily differs over the various instances.

In relation to the MinMax version of the problem, we also like to mention a version of combinatorial optimization which has become known under the name universal optimization. E.g., [7] study a universal scheduling problem. In such a problem, the scenarios are not explicitly specified, but can be seen to be chosen by an adversary. The quality of an algorithm is then measured by comparing its solution to the optimal solution when the adversarial choices are known beforehand.

For future research, anyone can choose her or his favorite combinatorial optimization problem and study its multiple-scenario version.

We finish with the some questions emerging from our multiple-scenario scheduling problem. The result in [2] suggests a 3/2-approximation for MM2 with 4 jobs per scenario and unitary jobs. Can this be extended to any job sizes? For the SM2 version the question is to close the gap between the 3/2-approximate randomized algorithm for the general case and the 1.0404 lower bound under the Unique Games Conjecture. It would also be interesting to find out if our randomized algorithm can be derandomized.

References

1. Ausiello, G., Crescenzi, P., Gambosi, G., Kann, V., Marchetti-Spaccamela, A., Protasi, M.: Complexity and approximation: Combinatorial optimization problems and their approximability properties. Springer (1999)
2. Austrin, P., Guruswami, V., Håstad, J.: $(2 + \epsilon)$-SAT is NP-hard. Electronic Colloquium on Computational Complexity, TR13-159 (2013)
3. Ben-Tal, A., Nemirovski, A.: Robust optimization–methodology and applications. Mathematical Programming 92(3), 453–480 (2002)
4. Bouyahia, Z., Bellalouna, M., Ghedira, K.: Load balancing a priori strategy for the probabilistic weighted flowtime problem. Comput. Ind. Eng. 64(1), 1–10 (2013)
5. Bouyahia, Z., Bellalouna, M., Jaillet, P., Ghedira, K.: A priori parallel machines scheduling. Comput. Ind. Eng. 58(3), 488–500 (2010), Supply, Production and Distribution Systems
6. Chekuri, C., Khanna, S.: On multidimensional packing problems. SIAM Journal on Computing 33(4), 837–851 (2004)
7. Epstein, L., Levin, A., Marchetti-Spaccamela, A., Megow, N., Mestre, J., Skutella, M., Stougie, L.: Universal sequencing on an unreliable machine. SIAM Journal on Computing 41, 565–586 (2012)
8. Feuerstein, E., Marchetti-Spaccamela, A., Schalekamp, F., Sitters, R., van der Ster, S., Stougie, L., van Zuylen, A.: Scheduling over scenarios on two machines. CoRR, abs/1404.4766 (2014)
9. Garey, M.R., Johnson, D.S.: Computers and Intractability; A Guide to the Theory of NP-Completeness. W. H. Freeman & Co., New York (1990)
10. Goemans, M.X., Williamson, D.P.: Improved approximation algorithms for maximum cut and satisfiability problems using semidefinite programming. J. ACM 42(6), 1115–1145 (1995)
11. Gupta, A., Pál, M., Ravi, R., Sinha, A.: Sampling and cost-sharing: Approximation algorithms for stochastic optimization problems. SIAM Journal on Computing 40(5), 1361–1401 (2011)
12. Håstad, J.: Some optimal inapproximability results. J. ACM 48(4), 798–859 (2001)
13. Jaillet, P.: A priori solution of a traveling salesman problem in which a random subset of the customers are visited. Oper. Res. 36(6), 929–936 (1988)

14. Karloff, H.J., Zwick, U.: A 7/8-approximation algorithm for MAX 3SAT? In: FOCS, pp. 406–415. IEEE Computer Society (1997)
15. Khot, S.: On the power of unique 2-prover 1-round games. In: Proceedings of 34th Annual ACM Symposium on Theory of Computing, pp. 767–775 (2002)
16. Khot, S., Kindler, G., Mossel, E., O'Donnell, R.: Optimal inapproximability results for MAX-CUT and other 2-variable CSPs? SIAM J. Comput. 37(1), 319–357 (2007)
17. Zhang, J., Ye, Y., Han, Q.: Improved approximations for max set splitting and max NAE SAT. Discrete Applied Mathematics 142(1-3), 133–149 (2004)
18. Zwick, U.: Outward rotations: A tool for rounding solutions of semidefinite programming relaxations, with applications to MAX CUT and other problems. In: STOC, pp. 679–687 (1999)

The Complexity of Bounded Register and Skew Arithmetic Computation

Vikraman Arvind and S. Raja

The Institute of Mathematical Sciences (IMSc), Chennai, India
{arvind,rajas}@imsc.res.in

Abstract. We study two register arithmetic computation and skew arithmetic circuits. Our main results are the following:

- For commutative computations, we present an exponential circuit size lower bound for a model of 2-register straight-line programs (SLPs) which is a universal model of computation (unlike width-2 algebraic branching programs that are not universal [AW11]).
- For noncommutative computations, we show that Coppersmith's 2-register SLP model [BOC88], which can efficiently simulate arithmetic formulas in the commutative setting, is not universal. However, assuming the underlying noncommutative ring has quaternions, Coppersmith's 2-register model can simulate noncommutative formulas efficiently.
- We consider skew noncommutative arithmetic circuits and show:
  - An exponential separation between noncommutative monotone circuits and noncommutative monotone skew circuits.
  - We define k-regular skew circuits and show that $(k+1)$-regular skew circuits are strictly powerful than k-regular skew circuits, where $k \leq \frac{n}{\omega(\log n)}$.

1 Two Register Arithmetic Computations

An *arithmetic circuit* over a field $\mathbb{F}$ and indeterminates $X = \{x_1, x_2, \cdots, x_n\}$ is a directed acyclic graph with each node of indegree zero labeled by a variable or a scalar constant. Each internal node g of the DAG is labeled by $+$ or $\times$ (i.e. it is a plus or multiply gate) and is of indegree two. A node of the DAG is designated as the output gate. Each gate of the arithmetic circuit computes a polynomial in the commutative ring $\mathbb{F}[X]$, by adding or multiplying its input polynomials. The polynomial computed at the output gate is the polynomial computed by the circuit.

If the indeterminates $X = \{x_1, x_2, \cdots, x_n\}$ are noncommuting with no relations between them, then the circuit is called a *noncommutative circuit* and it computes a polynomial in the free *noncommutative ring* $\mathbb{F}\langle X\rangle$.

We can view an arithmetic circuit C as a *straight-line program* (called an SLP, for short), which prescribes an order of gate evaluation for C. More precisely, a straight-line program corresponding to circuit C is a topologically sorted listing

Z. Cai et al. (Eds.): COCOON 2014, LNCS 8591, pp. 572–583, 2014.

of the nodes of the DAG defining C. For each internal gate g_i we have an instruction like $g_i := g_j \circ g_k$, where g_j and g_k are the inputs to g_i and $\circ \in \{+, *\}$. Thus, a straight-line program is a list of input variables and scalars followed by a list of assignment statements of the form $x := y \circ z$, where y and z are already computed.

The above notion of straight-line programs (SLPs) can be refined by introducing *registers*. Each instruction $x := y \circ z$ entails that x be stored in a register. The input variables and scalars (which occur only on the right-hand side) of the assignments are freely accessible by the program. This naturally leads to the notion of *bounded register* SLPs, where the bound is the number of registers used by the program. We give a general definition of bounded register SLPs.

Definition 1. *A straight-line program using w registers $R_1, R_2, \ldots, R_w$ over the field $\mathbb{F}$ and indeterminates from $X = \{x_1, x_2, \ldots, x_n\}$ consists of a sequence of instructions. Let $R_i^{(t)}$ denote the contents of register R_i at stage t. Then the tth instruction in the sequence transforms the tuple of register contents $(R_1^{(t-1)}, R_2^{(t-1)}, \ldots, R_w^{(t-1)})$ to $(R_1^{(t)}, R_2^{(t)}, \ldots, R_w^{(t)})$, where each $R_w^{(t)} = f(R_1^{(t-1)}, R_2^{(t-1)}, \ldots, R_w^{(t-1)}, x_1, \ldots, x_n)$, where the function f comes from a fixed set of polynomials.*

In an important paper in the area of arithmetic complexity, Ben-Or and Cleve [BOC92] showed that *three* registers suffice to efficiently simulate arithmetic formulas. The instructions they use in their SLPs are of the form: $R_i^{(t)} = l_1 R_j^{(t-1)} + l_2 R_k^{(t-1)} + R_i$ where l_1, l_2 are affine linear forms in $x_1, ..., x_n$. In general, when we allow instructions of the form

$$R_i^{(t)} = l_1 R_j^{(t-1)} + l_2 R_k^{(t-1)} + l_3 R_i,$$

where the l_i are affine linear forms, the polynomial in $\mathbb{F}[X]$ computed by such 3-register SLPs can be computed as the $(1,1)^{th}$ entry of the product of a sequence of 3×3 matrices, one for each SLP instruction, where these matrices have affine linear forms in the x_i's as entries. This is precisely the width-3 algebraic branching program (ABP) model of arithmetic computation which, by the Ben-Or and Cleve result is equivalent to arithmetic formulas (upto polynomial size).

A natural question is about the power of 2-register computations. Allender and Wang [AW11] have shown that 2-register ABPs (defined similarly via the iterated product of 2×2 matrices with affine linear form entries) are not even universal. Indeed, they show that the quadratic polynomial $x_1 x_2 + x_2 x_3 + ... + x_{n-1} x_n$ cannot be computed by width-2 ABPs. However, there are interesting 2-register SLP models that merit further investigation. Coppersmith (as mentioned in [BOC88]) has already observed that 2-register SLPs, with instructions of the form: $R_i = R_i + \alpha R_j^2$, $\alpha \in \mathbb{F}$ and $R_i = R_i + l$ where l is a affine linear form, can simulate formulas efficiently if char $\mathbb{F} \neq 2$.

Notice that Definition 1 allows for SLPs that are more powerful than the Ben-Or Cleve model because the general SLPs allow for multiplication of registers and hence can compute polynomials of degree exponential in their size. If we restrict ourselves to polynomials in $\mathbb{F}[X]$ of polynomially bounded degree, then arbitrary SLPs, indeed arbitrary arithmetic circuits of polynomial degree, can be transformed to $n^{O(\log n)}$ size formulas and hence width-3 ABPs.

This motivates the study of other *universal 2-register* SLP models of computation. It turns out that these models are universal because we allow a *reset instruction* $R_i = c$ for any $c \in \mathbb{F}$. Apart from the reset instruction, even if we restrict ourselves to only the ABP-like "skew" instructions $R_i = \ell_1 R_i + \ell_2 R_j$ for affine linear forms ℓ_1 and ℓ_2, the model is surprisingly quite powerful. It can efficiently simulate $\Sigma\Pi\Sigma$ arithmetic circuits and it would be interesting to either prove lower bounds for such a model or even to separate it from $\Sigma\Pi\Sigma$ arithmetic circuits. As we are unable to show such results, we consider a somewhat weaker universal 2-register SLP model and show exponential size lower bounds for a polynomial that has $\Sigma\Pi\Sigma$ circuits of polynomial size. Due to page limit, we omit several proofs in this conference version and we refer to ECCC version of the paper [AR14] for the full proofs.

Model 1

At each time instant t, we allow instructions of the form

1. $R_i^{(t)} = \alpha R_i^{(t-1)} + \beta R_j^{(t-1)}$ where $i, j \in \{1, 2\}$ and $\alpha, \beta \in \mathbb{F}$.
2. $R_i^{(t)} = x R_i^{(t-1)}$ where x is a variable.
3. $R_i^{(t)} = \alpha R_i^{(t-1)} + l$ where l is any affine linear form and $\alpha \in \mathbb{F}$.

Here $R_i^{(t)}$ denotes the contents of register R_i at time t. At each time instant t, we can apply any instruction to R_1 and R_2 simultaneously. Notice that the instruction (3) gives the power of resetting contents of a register to a nonzero constant. This is a crucial difference from width-2 ABPs (which cannot do such a reset). However, width-3 ABPs can simulate Model 1 efficiently.

Lemma 1. *Model 1 is universal.*

In Section 2 an exponential lower bound for Model 1. As noted above, the variant in which we allow $R_i^{(t)} = l R_i^{(t-1)}$ for an affine linear form l is at least as powerful as $\Sigma\Pi\Sigma$ circuits. We refer to this variant as Model 2.

Lemma 2. *Model 2 is at least as powerful as $\Sigma\Pi\Sigma$ circuits. I.e., for any $\Sigma\Pi\Sigma$ circuit of size s there is a size $O(s)$ SLP of Model 2 type.*

2   A Lower Bound for 2-Register SLP of Model 1

In this section we show an exponential size lower bound for 2-register SLPs (of Model 1) computing an explicit multivariate polynomial over any field $\mathbb{F}$. The instructions allowed in Model 1 are: (1) $R_i = \alpha R_i + \beta R_j$, where $\alpha, \beta \in \mathbb{F}$, (2) $R_i = x R_i$ where x is a variable, and (3) $R_i = \alpha R_i + l$, where l is an affine linear form and $\alpha \in \mathbb{F}$.

Model 1 is universal by Lemma 1. In the absence of instruction (3) (which allows resets $R_i := c$ for any $c \in \mathbb{F}$), the model can be simulated by width-2 arithmetic branching programs (ABP), which are not universal, as shown by Allender and Wang [AW11]. As mentioned earlier, they give a sparse polynomial that is not width-2 ABP computable. In Model 1 there is an SLPs of size $O(td)$ to compute any t-sparse polynomial of degree d. The idea is to compute the monomials, one at a time, in the first register, add it to the second register, and reset the first register to continue.

Consider the polynomial $Q = P_1 + P_2 + \ldots + P_k$ where $P_i = \Pi_{j=1}^{n}(x_{ij} + y_{ij})$ for each i. We will show an exponential size lower bound for SLPs in Model 1 computing the polynomial Q. Let $V = \{x_{ij}, y_{ij}\}_{i \in [k], j \in [n]}$ denote the set of indeterminates of polynomial Q. Note that $|V| = 2nk$. Each P_i is a homogeneous multilinear polynomial of degree n with 2^n distinct monomials. Furthermore, each P_i has an $O(n)$ size $\Pi\Sigma$ arithmetic circuit, and Q has an $O(nk)$ size $\Sigma\Pi\Sigma$ arithmetic circuit.

Let R_1, R_2 be the two registers used by an SLP computing polynomial Q, and let R_1 be the output register. A rough outline of the argument is as follows: based on the structure of the SLP, we pick some indeterminates of the polynomial Q and set them to 0. We then analyze the resulting SLP to establish the lower bound.

Theorem 1. *The polynomial* $Q = P_1 + P_2 + \cdots + P_k$, *defined above, for a suitably large constant* k, *requires* $2^{\Omega(n)}$ *size SLPs in the 2-register SLP of Model 1.*

Proof. Consider an SLP of Model 1 computing polynomial Q that is of *minimal* size s. Let $V = \{x_{ij} | i \in [k], j \in [n]\} \cup \{y_{ij} | i \in [k], j \in [n]\}$ denote the variable set of Q. The SLP consists of a sequence of instructions (of type (1) – (3)). At time $t, 0 \le t \le s$ the SLP computes the register contents $R_1^{(t)}$ and $R_2^{(t)}$ simultaneously from $R_1^{(t-1)}$ and $R_2^{(t-1)}$ by permissible instructions. We can assume that the contents of the registers R_1 and R_2 at time 0 is $R_1^{(0)} = R_2^{(0)} = 0$. The final contents of the two registers is $R_1^{(s)}$ and $R_2^{(s)}$ at time instant s, and let $R_1^{(s)}$ be the output of the SLP. Our aim is to lower bound s. We assume to the contrary that $s = 2^{o(n)}$ and derive a contradiction.

Clearly, at any time t both $R_1^{(t)}$ and $R_2^{(t)}$ contain polynomials of degree at most t in indeterminates from V over $\mathbb{F}$. We define the set of all *good monomials* as $GOOD = \{m \mid m$ has nonzero coefficient in $Q\}$. Notice that good monomials have degree n and there are $k.2^n$ good monomials. A monomial m is said to be in $R_i^{(t)}$ if it has nonzero coefficient in the polynomial $R_i^{(t)}$. Since the SLP is not monotone, the set of good monomials in $R_i^{(t)}$ need not monotonically increase in cardinality with t. But the SLP satisfies the following two properties:

- At time s the number of good monomials in $R_1^{(s)}$ is $k2^n$.
- The number of good monomials in R_1 and R_2 together, at time $t' > t$, can exceed the number at time t only if we apply the instruction $R_i = xR_i$ for some variable $x \in V$ to a register R_i and then $R_j = \alpha R_1 + \beta R_2$. Clearly, taking only linear combinations of R_1 and R_2 cannot increase the set of good

monomials. But any given variable x occurs in exactly 2^{n-1} good monomials in Q. Consequently, at each time step the total number of good monomials in the two registers can increase by at most 2^{n-1}.

Therefore, there is a time instant $\hat{t} \leq s$ such that: (i) $l2^n \leq |GOOD_{\hat{t}}| \leq (l+1)2^n$ and (ii) $\forall t \geq \hat{t}$ we have $|GOOD_t| \geq l2^n$. Choose l such that $k/8 < l < k/4$. For $t \geq \hat{t}$, for our convenience, we rename the registers at each time instant so that $R_1^{(t)}$ has at least as many good monomials as $R_2^{(t)}$.

Hence, without loss of generality, we can assume that $\forall t \geq \hat{t}$, $R_1^{(t)}$ has at least $\frac{l}{2}.2^n$ many good monomials. This means the SLP cannot apply the instruction $R_1^{(t+1)} = xR_1^{(t)}$ for all $\forall t \geq \hat{t}$. Otherwise, $R_1^{(t+1)}$ will have fewer than 2^{n-1} many good monomials, contradicting the above property of $\hat{t}$.

The number of good monomials at time $\hat{t}$ is bounded by $(l+1)2^n$. Hence, there are at least $(k - l - 1)2^n$ good monomials to be still included in the registers. The only way to generate new good monomials is to apply the instruction $R_2 = xR_2$ for $x \in V$. Let $t_0 > \hat{t}$ be the first such time instant. We set this variable $x = 0$. As a result $R_2^{(t_0)} = 0$. Without loss of generality, assume x occurs in the polynomial P_k. We can set all variables in P_k to zero and the polynomial Q becomes $P_1 + P_2 + ... + P_{k-1}$ which the SLP must still compute. Notice that the number of good monomials in $R_1^{(t_0)}$ is still at least $(\frac{l}{2} - 1)2^n$ and at most $(l + 1).2^n$.

Since $\frac{k}{8} < l < \frac{k}{4}$, there are at least $k - 2l - 1$ polynomials P_i in $\{P_1, P_2, ..., P_{k-1}\}$ such that $R_i^{(t_0)}$ contains fewer than 2^{n-1} good monomials of P_i. Without loss of generality, let these polynomials be $\{P_1, P_2, ..., P_q\}$ where $q \geq k - 1 - 2l$. Let $Var(P_j)$ denote the variable set of the polynomial P_j.

We analyze the SLP for time instants after t_0 in segments. For any polynomial $P_i \in \{P_1, P_2, ..., P_q\}$, a P_i-segment is a maximal time interval $[t', t'']$ such that in this time interval, all variable multiplication instructions $R_2 = xR_2$ are for $x \in Var(P_i)$. As a consequence, the last variable multiplication $R_2 = yR_2$ that occurs before time instant t' is either for some $y \notin Var(P_i)$ or $t' = t_0$.

A P_i-segment $[t', t'']$ is *nontrivial* if $2^{\Omega(n)}$ good monomials of P_i, not present in $R_1^{(t')}$ are included in $R_1^{(t'')}$.

Clearly, for each $P_i \in \{P_1, P_2, ..., P_q\}$ there must be at least one nontrivial P_i segment in the SLP. Otherwise, the SLP cannot compute Q because $s = 2^{o(n)}$.

Without loss of generality suppose the first nontrivial segment is P_1-segment and it occurs in the time interval $[t', t'']$. Clearly, the first variable multiplication to R_2 after time instant t'' is of the form $R_2 = zR_2$ for $z \notin Var(P_1)$, because the SLP must have nontrivial segments for all polynomials in $\{P_1, P_2, ..., P_q\}$. Likewise, either $t' = 0$ or the last variable multiplication to R_2 before time instant t' is of the form $R_2 = yR_2$ for $y \notin Var(P_1)$. In any case, by setting $z = 0$ and $y = 0$ we enforce the condition that at time t' the register $R_2 = 0$ and R_1 has at least $(\frac{\ell}{2} - 3)2^n$ good monomials. Let G denote the set of these good monomials in $R_1^{(t')}$. Also at time t'' the register R_2 is zero.

The rest of the proof will proceed as follows. We will first argue that at time t'', for every good monomial $g \in G$, the register R_1 has a monomial of the form

$m_g g$, where m_g is a nontrivial monomial in $Var(P_1)$. Furthermore, we shall argue that the rest of the SLP cannot remove these "bad" monomials if $s = 2^{o(n)}$. The main steps of the argument are in the following claims.

Claim. In the SLP segment $[t', t'']$, if we replace $R_1^{(t')}$ by a fresh variable u, the polynomial computed in R_1 at time t'' can be uniquely expressed as $Q_1 \cdot u + Q_2$. Hence, in the original SLP segment $R_1^{(t'')}$ can be expressed as $R_1^{(t'')} = Q_1 \cdot R_1^{(t')} + Q_2$, for the uniquely defined polynomials Q_1 and Q_2.

Likewise, at any time instant $t \in [t', t'']$ we can uniquely express the contents of $R_1^{(t)}$ as $Q_1^{(t)} \cdot R_1^{(t')} + Q_2^{(t)}$, where the polynomials $Q_i^{(t)}$ are uniquely defined in the same way. The above claim is clearly true on inspection of the SLP segment $[t', t'']$. We can similarly express the contents of $R_2^{(t)}$ as $P_1^{(t)} \cdot R_1^{(t')} + P_2^{(t)}$, for uniquely defined polynomials $P_i^{(t)}$.

Claim. The polynomial Q_1, where $R_1^{(t'')} = Q_1 \cdot R_1^{(t')} + Q_2$, is a nonconstant polynomial over the variable set $Var(P_1)$.

Proof. The proof is by an inductive argument. We divide the SLP segment into blocks. More precisely, there are time instants $t' < t_1 < t_2 < \cdots < t_r \leq t''$ when a linear combination of R_1 and R_2 is computed and stored in R_1.

Assume to the contrary that Q_1 is a constant. Let $j_1 \in (t', t'')$ be the least index such that $R_1^{(t_{j_1})} = T_{j_1} . R_1^{(t')} + Q_2^{(j_1)}$, where $T_{j_1} = \sum_l \alpha_l m_l$ is a nonconstant polynomial and m_l strictly divides m_{l+1} for each l. In the sequel, by the u-degree of polynomial $R_1^{(t)}$, we will mean the degree of the polynomial $Q_1^{(t)}$.

The u-degree of polynomial $R_1^{(t_{j_1})}$ is the degree of T_{j_1}. By minimality of j_1, $R_1^{(t_{j_1}-1)} = c_{j_1-1} . R_1^{(t')} + Q_2^{(j_1-1)}$. It is easy to see that $Q_2^{(j_1-1)}$ and $Q_2^{(j_1)}$ are both s-sparse polynomials. Let $j_2 \in (j_1, t'']$ be the least index such that $R_1^{(t_{j_2})} = c_{j_2} . R_1^{(t')} + Q_2^{(j_2)}$, where $c_{j_2} \in \mathbb{F}$. Intuitively, in the interval $(j_1, j_2]$, the SLP is removing the nonconstant multiplicative factors of $R_1^{(t')}$ at time j_2 in register $R_1^{(j_2)}$. We note that in the SLP segment (j_1, j_2), R_1 can not be used by R_2. Otherwise, the u-degree of R_2 will equal the u-degree of R_1 at that time, and in subsequent computation the nonconstant factors cannot be removed. This forces that the polynomial $Q_2^{(j_2)}$ remains an s-sparse polynomial. The rest of the proof proceeds inductively by considering the least time instant $j_3 \in (j_2, t'']$ such that the u-degree of $R_1^{(j_3)}$ is zero. We can similarly argue that the polynomial $Q_1^{(j_3)}$ will remain s-sparse. Continuing thus, it follows that if Q_1 is constant then Q_2 is s-sparse contradicting the assumption that the SLP adds $2^{\Omega(n)}$ monomials of P_1 in the $[t', t'']$ segment.

Claim. If $R_1^{(t'')} = Q_1 R_1^{(t')} + Q_2$ and $deg(Q_1) \geq 1$ then $R_1^{(t'')}$ has exponentially many monomials of the form $m_g g$ whose degree is $> n$, where $g \in G$ and m_g is a monomial over $Var(P_1)$.

Proof. Let π be the lexicographic ordering on the variable set $Var(P_1)$. Note that π induces a monomial ordering on all monomials over $Var(P_1)$. Let $B = \{m \cdot g \in R_1^{t''} \mid g \in G, Var(m) \subseteq Var(P_1), \ \& \ m \text{ is the largest such monomial under } \pi\}$. Here G is the set of $(\frac{\ell}{2} - 3)2^n$ good monomials in $R_1^{(t')}$. Clearly, $|B| = |G| \geq (\frac{\ell}{2} - 3)2^n$. Let $\hat{m}$ be the largest monomial under π in the polynomial Q_1. Clearly, each monomial in the set $\hat{m} \cdot B = \{\hat{m} \cdot mg \mid mg \in B\}$ is nonzero in the polynomial $Q_1 \cdot R_1^{t'}$. These monomials will also occur with nonzero coefficient in $R_1^{(t'')} = Q_1 R_1^{(t')} + Q_2$ because each monomial in Q_2 has all but at most one of its variables from $Var(P_1)$. On the other hand, each monomial in $\hat{m} \cdot B$ has exactly n variables not in $Var(P_1)$. This proves the claim.

Putting the claims together, we have shown that at time $t'' + 1$ the SLP has an exponential sized error set $\hat{m}B$ in R_1 and R_2 is 0. This error set can only be shifted and cannot be removed in the subsequent SLP computation assuming $s = 2^{o(n)}$. This completes the proof.   □

2.1   Noncommutative 2-register Arithmetic Computations

We now briefly consider 2-register noncommutative arithmetic computations. Here we are working in the free noncommutative ring $\mathbb{F}\langle X \rangle$ where $X = \{x_1, ..., x_n\}$ is a set of noncommuting free variables and $\mathbb{F}$ is any field. Thus monomials are words in the noncommutative free monoid X^*, and polynomials are $\mathbb{F}$-linear combinations of monomials. Nisan [Nis91] has shown that noncommutative ABPs can simulate noncommutative formulas efficiently. An examination of Ben-or and Cleve [BOC92] result shows that width-3 *noncommutative* ABPs can efficiently simulate noncommutative arithmetic formulas and are, in fact, equivalent to them. This has been observed before (e.g. see [AJS09]). Therefore, it is interesting to examine the power of 2-register noncommutative arithmetic computations. Width-2 noncommutative ABPs are also not universal [AW11]. However, the noncommutative version of Model 1 is universal. We can consider a noncommutative generalization of Model 1 in which we allow both left and right multiplication by an indeterminate: $R_i^t = xR_i^{t-1}$ and $R_i^t = R_i^{t-1}x$.

Lemma 3. *Noncommutative 2-register SLPs of Model 1 type have $O(n)$ size SLPs for the Palindrome polynomial $P_n(x_0, x_1) = \sum_{w \in \{x_0, x_1\}^n} w.w^R$.*

We omit the easy proof of the above lemma. In the noncommutative setting, $Q = P_1 + P_2 + ... + P_k$, where $P_i = \Pi_{j=1}^n (x_{ij} + y_{ij})$, has linear size $\Sigma \Pi \Sigma$ circuits. But by Theorem 1 any 2-register SLP of Model 1 for Q requires exponential size. The palindrome polynomial cannot be computed by polynomial size noncommutative $\Sigma \Pi \Sigma$ circuits [Nis91] but we have linear sized 2-register SLP of Model 1 by Lemma 3.

Corollary 1. *In the noncommutative setting, 2-register SLPs of Model 1 are incomparable to $\Sigma \Pi \Sigma$ circuits (or even ABPs).*

Next, we show that Coppersmith's model [BOC88] is not universal for non-commutative polynomials. Recall that the model is $R_i = R_i + \alpha R_j^2$, $R_i = R_i + l$ where $\alpha \in \mathbb{F}$ and l is a affine linear form.

Proposition 1. *The Coppersmith model is not universal in the noncommutative ring of polynomials $\mathbb{F}x, y$ for any field $\mathbb{F}$.*

However, if we assume the presence of quaternions in the ring $\mathbb{F}\langle x_1, ..., x_n \rangle$ we can show that Coppersmith's 2-register model is universal and even efficiently simulates noncommutative arithmetic formulas.

Lemma 4. *Let $R = \mathbb{F}\langle x_1, ..., x_n, i, j, k \rangle$ be a noncommutative ring where $char\mathbb{F} \neq 2$, and $x_1, ..., x_n$ are free noncommuting variables and i, j, k satisfy the relations: $i^2 = j^2 = k^2 = -1$, $ij = k, ji = -k, jk = i, kj = -i, ki = j, ik = -j, \forall y \in \{i, j, k\}$ $\forall x \in \{x_1, ..., x_n\}$ $yx = xy$. For any noncommuting arithmetic formula of size s we can give an equivalent 2-register SLP of size $s^{O(1)}$.*

3   Skew Noncommutative Computation

An arithmetic circuit is *skew* if for every multiplication gate one of its inputs is a scalar or an indeterminate $x_i \in X$. This model of arithmetic circuits has been studied in [AJMV98], especially in connection with depth reduction for noncommutative circuits. Although commutative skew circuits are equivalent to ABPs, in the noncommutative setting, skew circuits are known to be strictly more powerful than noncommutative ABPs [Nis91, AJMV98]. This is because multiplications could be either to the left or right. For instance, consider the palindrome polynomial $P(x_0, x_1) = \sum_{w \in \{x_0, x_1\}^n} w w^R$, where w^R denotes the reverse of w can be computed by an $O(n)$ size noncommutative skew circuit (using both left and right skew multiplications), but requires $2^{\Omega(n)}$ size ABPs as show by Nisan [Nis91].

Remark 1. It is an interesting question whether we can prove superpolynomial size lower bounds for noncommutative skew circuits computing the noncommutative Permanent or Determinant. A lower bound argument given in [AJMV98, Theorem 7.12] is unfortunately not correct. The idea there was to convert a given skew circuit into a left skew circuit (which is just a noncommutative ABP) by moving all right skew multiplications to the left, and then to apply Nisan's rank argument, to the resulting ABP. However, the modified circuit, in general, does not compute a polynomial weakly equivalent (in the sense of [Nis91]) to the one computed by the original circuit. Recently Limaye, Malod, and Srinivasan [LMS14] have shown a $2^{\Omega(n)}$ lower bound for general noncommutative skew circuits.

Although our results fall short of proving a lower bound for general noncommutative skew circuits, we show some exponential lower bounds separations. Specifically, we define a natural subclass of skew circuits, which we call k-regular skew circuits, and show exponential separations between k-regular and $k + 1$-regular skew circuits for each k, resulting in an infinite hierarchy of separations above noncommutative algebraic branching programs. Indeed, noncommutative ABPs form a proper subclass of 1-regular skew circuits.

We also compare the power of monotone noncommutative skew circuits with unrestricted noncommutative monotone circuits. Exponential size lower bounds for arbitrary monotone circuits computing the Permanent in both commutative and noncommutative settings are already well known (e.g. see [Nis91] for one proof). Here, we show an exponential separation between noncommutative monotone circuits and noncommutative monotone skew circuits.

Lower Bounds for k-regular Skew Circuits

Let C be a noncommutative skew circuit of size s computing a homogeneous polynomial in $\mathbb{F}\langle X \rangle$, where $X = \{x_1, x_2, \ldots, x_n\}$ are noncommuting free variables. We can first convert C into a *layered* circuit of size $\text{poly}(n, s)$ such that (i) all gates at layer i compute polynomials of degree i, (ii) all edges are between gates at layer i and $i + 1$ for each i, (iii) An edge from a gate u in layer i to a gate v layer $i + 1$ is labeled by a homogeneous linear form and the symbol l or r (indicating whether the linear form is to be multiplied to the left or right of the polynomial produced at gate u). This product of the linear form and the polynomial at gate u is the contribution of u to v. The polynomial computed at gate v is the sum of contributions over all incoming edges to v from layer i.

If the polynomial f computed by C is not homogeneous we can compute each homogeneous part of f by a layered skew circuit as described above. Let C is a layered skew circuit that computes a homogeneous polynomial $f \in \mathbb{F}\langle X \rangle$ of degree d.

A path ρ from a gate g in layer i to the output gate is said to be an (a, b)-*type* if there are exactly a left-skew multiplications and b right-skew multiplications in ρ, where $a + b = d - i$.

Definition 2. *A layered skew circuit C is said to be k-regular if for each multiplication layer i, we can associate a set of types $S_i = \{(c, d) \mid c, d \geq 0 \text{ and } c + d = d - i\}$ such that $|S_i| \leq k$ and for each gate g in layer i, if there is a path of (a, b)-type from gate g, where $a + b = d - i$, then $(a, b) \in S_i$.*

Note that in a k-regular skew circuit, the number of gates in each layer can be unbounded. Furthermore, the type sets S_i are not fixed and can depend on the computed polynomial. Of course the set S_i and the circuit structure between the ith and $i + 1^{st}$ multiplication layers will determine S_{i+1}. Given a layered skew circuit C we can check if it is k-regular for a given k, and also efficiently compute the minimum k for which C is k-regular.

Remark 2. We note here that even 1-regular skew circuits are strictly more powerful than ABPs. For, ABPs are just 1-regular skew circuits in which all multiplications are right skew, and there is an $O(n)$ size 1-regular skew circuit (which is also monotone) for the palindrome polynomial which requires exponential size ABPs [Nis91].

It is convenient to first convert a given layered noncommutative skew circuit into one which is homogeneous in a specific sense that we call *type homogeneous*. Something more general is known for general noncommutative circuits [HWY10]. This conversion can be carried out in deterministic polynomial time.

A gate g in layer i is said to be an (a, b)-*type gate* if all paths from g to the output are of (a, b)-type. Clearly, the output gate is always type homogeneous. Starting from the output gate and proceedings downward in a given layered skew circuit C of size s we can convert it, layer by layer, into a type homogeneous layered skew circuit of size poly(d, s) size.

Lemma 5. *A layered skew circuit C of size s computing a homogeneous polynomial of degree d in $\mathbb{F}\langle X \rangle$ can be transformed in polynomial time (in s and n) into a layered type homogeneous skew circuit C^T of size poly(d, s) computing f.*

In general the gates in layer i of a type homogeneous skew circuit C can have any of the $d - i + 1$ possible types. Notice that, crucially, type homogenization of skew circuit does not alter the set of types of the paths at any layer. Specifically, if C is k-regular then C^T remains k-regular.

Now, we show that for each fixed $k > 0$, an exponential size separation between k-regular and $(k+1)$-regular skew circuits. We define the following homogeneous degree $2(k + 1)n$ polynomial $P_{i,k}(X, Y, Z)$ where the $2(k + 1)n$ variables are partitioned as $X \cup Y \cup Z$ such that $|X| = |Y| = n$ and $|Z| = 2kn$.

$P_{i,k}(X, Y, Z) = \sum_{w \in S} z_1 z_2 \dots z_{2(i-1)n} . w . w^R . z_{2(i-1)n+1} . z_{2(i-1)n+2} \dots z_{2kn}$, where $S = \{x_1, y_1\} \times \{x_2, y_2\} \times \dots \times \{x_n, y_n\}$.

Theorem 2. *Let $P = \sum_{i \in [k+1]} P_{i,k}(X_i, Y_i, Z_i)$ where variable sets X_i, Y_i, Z_i are disjoint $\forall i \in [k + 1]$. Any k-regular skew circuit for P requires size at least 2^n. But there is polynomial sized $(k + 1)$-regular skew circuit for P.*

Examining the proof of Theorem 2, we note that in the definition of polynomial P if we set $k = n$ (hence k is not constant) then the polynomial P can be computed by a polynomial size layered skew circuit but any $\frac{n}{\omega(\log n)}$-regular skew circuit that computes P is of exponential size. Putting it together we have the following.

Corollary 2. *There is an explicit noncommutative polynomial in n variables that has polynomial-size skew circuits but any $\frac{n}{\omega(\log n)}$-regular skew circuit that computes it is of size $n^{\omega(1)}$.*

Remark 3. Chien and Sinclair [CS07] considered the question of lower bounds for the Determinant polynomial whose entries are 2×2 matrices over a field $\mathbb{F}$, where char $\mathbb{F} \neq 2$. Based on Nisan's rank argument they showed that any ABP computing the order n Determinant, DET_n, whose entries are 2×2 matrices requires $2^{\Omega(n)}$ size.

As in the proof of Theorem 2, we can show that any k-regular skew circuit computing DET_n can be transformed into an ABP computing the $\frac{n}{k+1} \times \frac{n}{k+1}$ determinant. It now follows from Chien and Sinclair's result that for any constant k, k-regular skew circuits computing DET_n over 2×2 matrices are of size $2^{\Omega(n)}$.

A Rank Based Approach to Lower Bounds for Skew Circuits

Let $P \in \mathbb{F}\langle X \rangle$ be a homogeneous noncommutative polynomial of degree d on the variables $X = \{x_1, x_2, \ldots, x_n\}$. For each $0 \leq k \leq d$, we can associate a matrix $M_k(P)$ over the field $\mathbb{F}$ with n^k rows and $(d - k + 1)n^{(d-k)}$ columns. Each row of $M_k(P)$ is labeled by a distinct monomial m of degree k. Each column is labeled by a pair of monomials (m_1, m_2) such that the sum of their degrees is $d - k$. A monomial $\hat{m}$ of degree d can be factored as $\hat{m} = m_1 m m_2$ in $d - k + 1$ different ways, where m is of degree k. The property we demand of the matrix $M_k(P)$ is that the coefficient $P(\hat{m})$ of the monomial $\hat{m}$ in P can be written as

$$P(\hat{m}) = \sum_{\hat{m} = m_1 m m_2} M_K(m, (m_1, m_2)), \tag{1}$$

where m is a degree k monomial.

Remark 4. Note that unlike for ABPs, the matrix $M_k(P)$ is not uniquely defined for the polynomial P. In a skew circuit a monomial $\hat{m}$ can occupy more than one entry in $M_k(P)$, but we require that these entries sum to the coefficient of $\hat{m}$. In particular, it is clear that in a skew circuit computing a homogeneous degree d polynomial, each monomial can occupy $O(d)$ nonzero entries in $M_k(P)$.

However, we can still relate the minimum size of a skew circuit computing a homogeneous degree d polynomial P to the rank of $M_k(P)$. Let $S(P)$ denote the minimum size of a layered skew circuit computing the polynomial P. Similar to Nisan's [Nis91] ABP size characterization we have the following.

For $0 \leq k \leq d$, let $rank_k(P)$ denote the *minimum rank* attained by a matrix $M_k(P)$ satisfying Equation 1.

Theorem 3. *For homogeneous polynomials P of degree d*

$$S(P) \geq \sum_{k=0}^{d} rank_k(P).$$

Theorem 3 seems difficult to use for general skew circuits as we need to lower bound $rank_k(P)$. However, we can use it for the monotone case for the following reason. Suppose $P \in \mathbb{R}\langle X \rangle$ is a monotone noncommutative homogeneous polynomial of degree d and C is a monotone layered skew circuit computing P. As the circuit C is monotone, the matrices L_k and R_k in the proof of Theorem 3 will both be monotone. Let $mrank_k(P)$ denote the minimum rank of a *nonnegative* matrix $M_k(P)$ corresponding to P and let $mS(P)$ denote the minimum size of a monotone skew circuit for P. It follows from the proof of Theorem 3 that $mS(P) \geq \sum_{k=0}^{d} mrank_k(P)$.

Let $Q = P_1(x_1, x_2)P_2(x_3, x_4)$ where P_i are the palindrome polynomials of degree $2n$. Clearly, Q has polynomial size monotone noncommutative circuits: we can compute P_1 and P_2 with $O(n)$ size monotone skew circuits and multiply their outputs. We will show that Q requires exponential size monotone skew circuits. Note that Q is a homogeneous degree $4n$ polynomial with 2^{2n} monomials.

Let $M_k(Q)$ be a *nonnegative* matrix corresponding to polynomial Q with rows labeled by degree k monomials. By $M_k^{a,b}(Q)$ we mean the submatrix of $M_k(Q)$ with all columns (m_1, m_2) of $M_k(Q)$ such that $|m_1| = a, |m_2| = b$.

Theorem 4. *Let $Q = P_1(x_1, x_2)P_2(x_3, x_4)$, where P_i are palindrome polynomials of degree $2n$ each. If $M_{3n}(Q)$ is nonnegative then $rank(M_{3n}(Q)) = 2^{\Omega(n)}$ and hence $mS(Q) = 2^{\Omega(n)}$.*

Examining the proof of Theorem 2, we note that the upper bounds are obtained by monotone skew circuits. Thus we also have an exponential separation between the size of monotone k-regular skew circuits and monotone $k+1$-regular skew circuits.

Corollary 3. *There is an exponential size separation between the following monotone circuit classes via explicit monotone noncommutative polynomials: (i) monotone circuits and monotone skew circuits, (ii) monotone skew circuits and k-regular monotone skew circuits, (iii) $k + 1$-regular and k-regular monotone skew circuits.*

Acknowledgments. We thank Meena Mahajan and Yadu Vasudev for discussions and comments. We are grateful to the referees for their comments.

References

[AJMV98] Allender, E., Jiao, J., Mahajan, M., Vinay, V.: Non-commutative arithmetic circuits: Depth reduction and size lower bounds. Theor. Comput. Sci. 209(1-2), 47–86 (1998)

[AW11]   Allender, E., Wang, F.: On the power of algebraic branching programs of width two. In: Aceto, L., Henzinger, M., Sgall, J. (eds.) ICALP 2011, Part I. LNCS, vol. 6755, pp. 736–747. Springer, Heidelberg (2011)

[AJS09]   Arvind, V., Joglekar, P.S., Srinivasan, S.: Arithmetic circuits and the hadamard product of polynomials. In: FSTTCS, pp. 25–36 (2009)

[AR14]   Arvind, V., Raja, S.: The Complexity of Two Register and Skew Arithmetic Computation. Electronic Colloquium on Computational Complexity (ECCC) 21, 28 (2014)

[BOC88]   Ben-Or, M., Cleve, R.: Computing algebraic formulas with a constant number of registers. In: Proceedings of the Twentieth Annual ACM Symposium on Theory of Computing (STOC), pp. 254–257. ACM (1988)

[BOC92]   Ben-Or, M., Cleve, R.: Computing algebraic formulas using a constant number of registers. SIAM Journal on Computing 21(1), 54–58 (1992)

[CS07]   Chien, S., Sinclair, A.: Algebras with polynomial identities and computing the determinant. SIAM Journal on Computing 37(1), 252–266 (2007)

[Nis91]   Nisan, N.: Lower bounds for non-commutative computation (extended abstract). In: STOC, pp. 410–418 (1991)

[RS05]   Raz, R., Shpilka, A.: Deterministic polynomial identity testing in noncommutative models. Computational Complexity 14(1), 1–19 (2005)

[HWY10]   Hrubes, P., Wigderson, A., Yehudayoff, A.: Noncommutative arithmetic circuits and the sum-of-squares problem. In: STOC 2010 (2010)

[LMS14]   Limaye, N., Malod, G., Srinivasan, S.: Talk at an arithmetic circuits workshop (February 2014)

Minimizing Average Flow-Time
under Knapsack Constraint

Suman Kalyan Bera[1], Syamantak Das[2], and Amit Kumar[2]

[1] Dartmouth College, USA
suman.k.bera.gr@dartmouth.edu
[2] Indian Institute of Technology Delhi, India
{sdas,amitk}@cse.iitd.ac.in

Abstract. We give the first logarithmic approximation for minimizing average flow-time of jobs in the subset parallel machine setting (also called the restricted assignment setting) under a single knapsack constraint. In a knapsack constraint setting, each job has a profit, and the set of jobs which get scheduled must have a total profit of at least a quantity Π. Our result extends the work of Gupta, Krishnaswamy, Kumar and Segev (APPROX 2009) who considered the special case where the profit of each job is unity. Our algorithm is based on rounding a natural LP relaxation for this problem. In fact, we show that one can use techniques based on iterative rounding.

1 Introduction

Classical scheduling problems that minimize an underlying objective function require that all the jobs in the input get processed. However, for many applications, one might require only a subset of the jobs to be scheduled so as to meet a pre-specified hard profit. In this paper, we consider the well studied objective of minimizing the average flow time of jobs in a multiprocessor environment under such a hard profit constraint which we call the *knapsack* constraint. Formally, there is an associated profit π_j with each job. The goal is to schedule a subset of jobs whose sum of profits is at least a fixed quantity Π. Equivalently, the sum of profit (weights) of the rejected jobs should fit in a knapsack of size B. Note that the special case where each job has unit profit corresponds to the problem where there is a lower bound on the number of jobs to be processed.

Charikar et al. [3] initiated the study of this model where there is a lower bound on the number of jobs which need to be scheduled. Gupta et al. [8] extended this model to a wide variety of scheduling problems. In particular, they consider the problem of minimizing average flow-time on identical parallel machines where jobs have unit profit. In this paper, we generalize this result in two directions as described below.

1.1 Our Results and Techniques

We give the first $O(\log P)$-approximation algorithm for the problem of minimizing average flow-time in the restricted assignment setting (each job

Z. Cai et al. (Eds.): COCOON 2014, LNCS 8591, pp. 584–595, 2014.
© Springer International Publishing Switzerland 2014

can only be processed by a subset of machines) under a single knapsack constraint(FlowKnap). Here P is the ratio of the largest to the smallest job size. Previously, such a result was only known for the parallel machines setting and when each job had unit profit [8].

Our technique is based on a standard LP relaxations for these problems. Directly working with such a relaxation turns out to be involved and tricky. Instead, we show that one can often extract some interesting properties of a fractional solution, and we can write a second simpler LP relaxation based on these properties. This simpler LP relaxation can then be iteratively rounded.

As has been remarked in [8], the presence of knapsack constraint makes the problem significantly harder than the non-profit version even in the simple setting of a *single machine*. In fact, this is the version we shall consider in greater details and later on show how to extend it to the case of subset parallel assignments.

As is standard for algorithms for flow-time, we first divide the jobs into *classes* based on their sizes p_j – a job is of class k if its processing time lies in the range $[2^k, 2^{k+1}]$. Using established techniques, it turns out that it is enough to get a constant factor approximation for the special case when all the jobs are of the same class *provided* our rounding algorithm schedules these jobs in the same timeslots in which class k jobs were scheduled by the fractional solution (upto some constant number of slots violations). Thus, we consider the case when all jobs are of roughly the same size and they are only allowed to be processed in some given region of the timeline. We call this the Scheduling with Forbidden Regions(ForbidFlow) problem. We first write a time-indexed LP relaxation for this problem which schedules each job to a fractional extent between 0 and 1. Given a fractional solution, we perform the following steps:

- We create a modified instance and its corresponding fractional solution by compressing all the *gaps* in the LP schedule, i.e., we move the jobs back in time (and decrease their release dates) to fill all available space. This considerably simplifies the subsequent rounding steps. Note that in the algorithm of Gupta et al. [8], considerable technical work is required to get around this problem. The high level reason is that one would want to change the fractional assignment of jobs such that these become 0 or 1. Naive ways to achieve this may end up charging to the entire length of the schedule including the gaps in between. So one needs to do a finer analysis to prevent this. However, once we move the jobs back, this issue goes away.

- It turns out we can get rid of the time-indexed LP relaxation, and just work with the fraction to which a job is processed. We write a simpler LP relaxation to capture this and show that it can be rounded using iterative rounding.

- Finally, we show how to *expand* this schedule to obey the earlier release dates without incurring much in the flow-time.

The analysis essentially shows that we do not incur a lot of cost compared to LP in each of these steps.

1.2 Related Work

Scheduling jobs on machines subject to various constraints is one of the central problems in combinatorial optimization. See [9, 11] for earlier work on the problems of minimizing makespan and GAP in multiple machines environment. Garg and Kumar [5, 6] gave upper and lower bounds for minimizing average flow-time of jobs under various settings.

More recently, there has been a growing interest in the study of *scheduling with rejections*. Several works ([1, 4, 2]) have studied the prize collecting model where one has to pay a penalty for the jobs that are rejected. On the other hand, Charikar et al. [3] considered scheduling under the knapsack constraint where the objective is to schedule jobs so as to achieve a target profit. Guha et al. [7] presented an LP rounding based algorithm for minimizing average completion-time with outliers where the outlier constraint is violated by a constant factor. Gupta et al. [8] gave approximation algorithms for scheduling with outliers under various settings. They give a constant factor approximation for GAP under the knapsack constraint(approximation factor improved by Saha and Srinivasan [10]), an $O(\log k)$ approximation for average completion time on unrelated machines minimization under k knapsack constraints and a logarithmic approximation for minimizing total(average) flow time of jobs on identical parallel machines under knapsack constraint where each job has *unit profit*.

2 Preliminaries

In this section, we formally describe the scheduling problems that will be considered in this paper. We will be given a set of jobs J and a set of machines M. In the subset parallel machines setting, each job j specifies a subset S_j of machines on which it can be processed and its processing time on these machines is p_j. Recall that in the *knapsack* constraint, each job j also comes with a profit π_j, and we are given a lower bound Π. Given a release date r_j for each job j, The flow-time of a j is the difference between its completion time and release date. The goal is to schedule a subset of jobs of total profit at least Π while minimizing the total flow-time of scheduled jobs. We allow jobs to be pre-empted but we do not allow migration across machines. However, our approximation ratio holds with respect to a migratory optimum as well.

We need to solve a related problem, which we call scheduling with forbidden regions(ForbidFlow). Here, we are given a single machine, and the setting is same as minimizing total flow-time with knapsack constraint. Further, all processing times are within a factor of 2 of each other. We are also given a quantity $z(t)$ for each timeslot $[t, t + 1]$. Any schedule can only use $1 - z(t)$ amount of space in this timeslot. In Section 3, we give an $O(1)$-approximation algorithm with an additive $\sum_t z(t)$ factor for this problem. We use this result as a subroutine for the FlowKnap problem (Section 4).

3 Scheduling with Forbidden Regions on a Single Machine

In this section, we give an approximation algorithm for the ForbidFlow problem. An instance $\mathcal{I} = \text{ForbidFlow}(J, Q, \Pi, z)$ is similar to an instance of FlowKnap – each job $j \in J$ has associated parameters r_j, p_j and π_j as before while Π is a hard profit. Processing times of the jobs are bounded in a range $[Q, 2Q]$ for some positive integer $Q, Q \geq 1$. Let T be a guess for the time at which the optimal solution completes processing all jobs. Moreover, z is a vector in $[0, 1]^T$. In any time slot $[t, t+1]$, the schedule can do at most $1 - z_t$ amount of processing. For a time slot t we refer to the quantity z_t as *forbidden space* while the rest is called *available space*. We write the following LP relaxation for this problem. Here x_{jt} refers to the amount of processing of job j done during the timeslot $[t, t + 1]$. This variable is defined only for $t \geq r_j$. For a job j, y_j refers to the fraction of job j which gets processed. It is well-known that the objective function is a lower bound on the value of the flow-time of the corresponding schedule [5]. The first term in the objective function refers to *fractional flow-time* and the second term refers to half of the processing time of the schedule. Constraint (1) states that a job should be processed to the extent of p_j if y_j is 1. Constraint (2) says that a timeslot $[t, t+1]$ can only be used to the extent of $1 - z_t$. Finally, constraint (3) states that the total profit of jobs selected for processing must be at least Π. Thus, this is a relaxation for the ForbidFlow problem.

$$\min \sum_{j=1}^{n} \sum_{t} \frac{x_{jt}(t - r_j)}{2Q} + \frac{1}{2} \sum_{j} y_j p_j \qquad \text{(LP1)}$$

$$\sum_{t \geq r_j} x_{jt} = y_j p_j, \forall j \qquad (1)$$

$$\sum_{j=1}^{n} x_{jt} \leq 1 - z_t, \forall t \qquad (2)$$

$$\sum_{j=1}^{n} y_j \pi_j \geq \Pi \qquad (3)$$

$$0 \leq y_j \leq 1, \forall j \qquad (4)$$

Let $(x^\star, y^\star)$ be an optimal solution to LP1 and $\text{flow}(x^\star, y^\star)$ denote the LP objective. We state the main theorem of this section.

Theorem 1. *There is a polynomial time algorithm to schedule jobs of total profit at least Π such that total flow-time of the scheduled jobs is at most $O(\text{flow}(x^\star, y^\star) + \sum_t z_t)$.*

3.1 The Rounding Algorithm

Our algorithm will only use the $y_j^\star$ variables – this is so because there is an optimal fractional solution which schedules jobs in order of their release dates(proof deferred to full version of paper). Hence, given $y_j^\star$ values, we can deduce the $x_{jt}^\star$ values. We now describe the details of each of the conceptual steps of our algorithm:

CloseGaps :

1. Initialize $\bar{y}_j = 0$ for all j and $\bar{x}_{jt} = 0$ for all j, t.
2. For $l = 1, 2, \ldots, n$
   Let t_l be the first time such that $z(t_l) + \sum_{s=1}^{l-1} \bar{x}_{j_s t_l} < 1$
   (i.e., there is some more processing which can be done at this time slot)
   In the instance $\mathcal{I}'$, define $\bar{s}_{j_l} = t_l$.
   Process the job j_l to an extent of $y_{j_l}^\star$ from time t_l onwards, i.e.,
   for $t = t_l$, set $\bar{x}_{jt} = \min(1 - z_t - \sum_{s=1}^{l-1} \bar{x}_{j_s t}, p_j y_j^\star)$ and
   for $t > t_l$, iteratively set $\bar{x}_{jt} = \max(0, \min(1 - z_t, p_j y_j^\star - \sum_{t' < t} \bar{x}_{jt}))$.
   Set $\bar{y}_{j_l} = y_{j_l}^\star, \bar{r}_j = r_j - (s_j^\star - \bar{s}_j)$.

Fig. 1. Algorithm for modifying $\mathcal{I}$ to $\mathcal{I}'$

Closing the Gaps: Given the instance $\mathcal{I}$ and the solution $(x^\star, y^\star)$, our first step is to create a new instance $\mathcal{I}'$ and a corresponding fractional solution $(\bar{x}, \bar{y})$. The idea is that in the solution $(x^\star, y^\star)$ there may be gaps in the fractional schedule – a gap is unused timeslot which appears in the middle of a schedule. The reason why we cannot process a job during the gap is that all subsequent jobs have release dates beyond this timeslot. In $\mathcal{I}'$ we would like to move jobs to the left so that these gaps go away. $\mathcal{I}'$ will be identical to $\mathcal{I}$ except that the release date of a job in $\mathcal{I}'$ will appear before that in $\mathcal{I}$.

We give some notation first. Let $j_1, \ldots, j_n$ be an ordering of the input jobs according to increasing release dates. For a job j, let $s_j^\star$, the starting time of j, denote the first timeslot $[t, t+1]$ in which the solution $(x^\star, y^\star)$ processes j, i.e., the smallest t such that $x_{jt}^\star > 0$ – we can assume that $y_j^\star > 0$ for all jobs j because we may not consider any job for which $y_j^\star = 0$. When we construct $\mathcal{I}'$, $\bar{r}_j$ will denote the release date of job j in this instance, and $\bar{s}_j$ will denote the first timeslot $[t, t+1]$ for which $\bar{x}_{jt} > 0$. We iteratively modify the instance $\mathcal{I}$ to $\mathcal{I}'$ and construct a fractional schedule simultaneously. The procedure is described in Fig. 1. It essentially moves each job back so that all gaps get filled. The starting time of the job moves back accordingly. The release dates also move back by the same amount as the starting time. It is not difficult to see that the relative ordering of the release dates in $\mathcal{I}'$ are same as that in $\mathcal{I}$.

Since we are only moving the processing of a job back in time, it is easy to prove by induction that for each job j, $\bar{s}_j \leq s_j^\star$ and so, $\bar{r}_j \leq r_j$. For indices $t_1 < t_2$, let $\texttt{avail}(t_1, t_2)$ denote the total available space in the timeslots $[t_1, t_1+1], \ldots, [t_2, t_2+1]$, i.e., $\sum_{t=t_1}^{t_2} (1 - z_t)$. Let $\texttt{forbid}(t_1, t_2)$ be the total forbidden space in the corresponding timeslots, i.e., $\sum_{t=t_1}^{t_2} z_t$. Clearly, $\texttt{avail}(t_1, t_2) + \texttt{forbid}(t_1, t_2) = t_2 - t_1 + 1$.

We note one simple fact which will be useful later. Again, the reason why this is true is because in $(x^\star, y^\star)$, we could have had gaps between the execution of two jobs j and j', but these will not be present in $(\bar{x}, \bar{y})$.

Claim. Let j, j' be jobs with $r_j \leq r_{j'}$. Then $\texttt{avail}[s_j^\star, s_{j'}^\star] \geq \texttt{avail}[\bar{s}_j, \bar{s}_{j'}]$.

The first observation is that the cost of the solution $(\bar{x}, \bar{y})$ is close to that of $(x^\star, y^\star)$.

Lemma 1

$$\sum_{j,t} \frac{\bar{x}_{jt}(t - \bar{r}_j)}{2Q} \leq \sum_{j,t} \frac{x_{jt}^\star(t - r_j)}{2Q} + \sum_t z_t + \sum_j y_j^\star p_j.$$

Proof. Let $\bar{C}_j$ be the completion time of j in the schedule $(\bar{x}, \bar{y})$. For a job j, we will show that

$$(5) \qquad \sum_{t \geq \bar{s}_j} \frac{\bar{x}_{jt}(t - \bar{s}_j)}{2Q} \leq \sum_{t \geq s_j^\star} \frac{x_{jt}^\star(t - s_j^\star)}{2Q} + \sum_{t = \bar{s}_j + 1}^{\bar{C}_j} z_t + y_j^\star p_j$$

For the sake of argument, we divide the processing of job j into very small pieces such that each piece "fits" in one timeslot. More formally, suppose ε is a small enough positive quantity such that all positive $x_{jt}^\star$ and $\bar{x}_{jt}$ values are integral multiples of ε. We can think of $N_j = \frac{y_j p_j}{\varepsilon}$ such contiguous pieces of j, which get processed in both schedules in this order. Now, such a piece c getting processed at time t in $x^\star$ will contribute $\frac{\varepsilon(t - r_j)}{2Q}$ to the sum $\sum_{t \geq s_j^\star} \frac{x_{jt}^\star(t - s_j^\star)}{2Q}$. We claim that this piece will finish processing in the schedule $(\bar{x}, \bar{y})$ by time $t' = \bar{s}_j + 1 + (t - s_j^\star) + \texttt{forbid}(\bar{s}_j + 1, \bar{C}_j)$. Indeed, if $t' \geq \bar{C}_j$, then there is nothing to prove because by definition of $\bar{C}_j$, j will finish by $\bar{C}_j \leq t'$. So assume $t' < \bar{C}_j$. Now, $\texttt{forbid}(\bar{s}_j + 1, t')$ is at most $\texttt{forbid}(\bar{s}_j + 1, \bar{C}_j)$. So,

$$\texttt{avail}(\bar{s}_j + 1, t') \geq t' - \bar{s}_j - \texttt{forbid}(\bar{s}_j + 1, \bar{C}_j) \geq t - s_j^\star + 1.$$

But then the piece c should finish processing by time t' in $(\bar{x}, \bar{y})$ because the total processing requirement of pieces of j coming before c (including c) is at most $t - s_j^\star + 1$. Thus, this piece will contribute at most $\frac{\varepsilon(1 + (t - s_j^\star) + \texttt{forbid}(\bar{s}_j + 1, \bar{C}_j))}{2Q}$ to the sum $\sum_{t \geq \bar{s}_j} \frac{\bar{x}_{jt}(t - \bar{s}_j)}{2Q}$. Summing over all the pieces of j, we get

$$\sum_{t \geq \bar{s}_j} \frac{\bar{x}_{jt}(t - \bar{s}_j)}{2Q} \leq \sum_{t \geq s_j^\star} \frac{x_{jt}^\star(1 + (t - s_j^\star) + \texttt{forbid}(\bar{s}_j + 1, \bar{C}_j))}{2Q}$$

$$\leq \sum_{t \geq s_j^\star} \frac{x_{jt}^\star(t - s_j^\star)}{2Q} + y_j^\star p_j + \texttt{forbid}(\bar{s}_j + 1, \bar{C}_j),$$

because $p_j \geq 1$ and $\sum_t x_{jt}^\star = y_j^\star p_j \leq 2 y_j^\star Q$. This proves inequality (5). Summing this over all jobs and noting that the closed intervals $[(\bar{s}_j + 1, \bar{C}_j]$ are disjoint for different jobs, we get

(6) $$\sum_{j,t} \frac{\bar{x}_{jt}(t - \bar{s}_j)}{2Q} \leq \sum_{j,t} \frac{x_{jt}^{\star}(t - s_j^{\star})}{2Q} + \sum_t z_t + \sum_j y_j^{\star} p_j.$$

This implies the desired result because

$$\sum_{j,t} \frac{\bar{x}_{jt}(t - \bar{r}_j)}{2Q} - \sum_{j,t} \frac{x_{jt}^{\star}(t - r_j)}{2Q}$$

$$= \sum_{j,t} \frac{\bar{x}_{jt}(t - \bar{s}_j)}{2Q} + \sum_{j,t} \frac{\bar{x}_{jt}(\bar{s}_j - \bar{r}_j)}{2Q} - \sum_{j,t} \frac{x_{jt}^{\star}(t - r_j)}{2Q} - \sum_{j,t} \frac{x_{jt}^{\star}(r_j - s_j^{\star})}{2Q}$$

$$= \sum_{j,t} \frac{\bar{x}_{jt}(t - \bar{s}_j)}{2Q} - \sum_{j,t} \frac{x_{jt}^{\star}(t - s_j^{\star})}{2Q}$$

because $r_j - s_j^{\star} = \bar{r}_j - \bar{s}_j$ and $\sum_{j,t} \bar{x}_{j,t} = \bar{y}_j = y_j^{\star} = \sum_{j,t} x_{j,t}^{\star}$. $\quad\square$

Iterative Rounding: Now we show how to round the solution $(\bar{x}, \bar{y})$ for the instance $\mathcal{I}'$. Again, our rounding algorithm will only use the $\bar{y}$ values. We achieve this by an iterative rounding procedure which works with a simple LP relaxation – this simpler LP relaxation only looks at the $\bar{y}_j$ values. We first motivate the new LP relaxation. It is not hard to see that the objective function for the solution $(\bar{x}, \bar{y})$ can be written as $\sum_t \frac{tu(t)}{2Q} - \sum_j \frac{\bar{r}_j \bar{y}_j p_j}{2Q} + \frac{1}{2} \sum_j p_j \bar{y}_j$, where $u(t)$ is the amount of processing done in timeslot $[t, t+1]$. It will turn out that the quantity $u(t)$ will not change. Hence, we can treat the objective function as $-\sum_j \frac{\bar{r}_j \bar{y}_j p_j}{2Q} + \frac{1}{2} \sum_j p_j \bar{y}_j$.

Our algorithm will iteratively reject or select some jobs. Hence, at any point of time during our algorithm, we will work with a residual budget Π' – this is the remaining profit we need to recover from jobs for which we have not made a decision. Our algorithm will also maintain a set of jobs J' initialized to the set of all jobs. These will be the set of jobs about which our algorithm has not decided whether they will be scheduled or not. For a set of jobs J', let $F(J')$ be the first three jobs in J' (according to release dates).

Given jobs j, j' we say that $j \prec j'$ if $\bar{r}_j \leq \bar{r}_{j'}$. For a job j, let $\bar{V}_j$ denote the amount of processing done by the solution $(\bar{x}, \bar{y})$ on jobs released before j (including j), i.e., $\bar{V}_j = \sum_{j':j' \prec j} \sum_t \bar{x}_{j't} = \sum_{j':j' \prec j} \bar{y}_{j'} p_{j'}$. Our new LP relaxation is shown below.

$$\min -\sum_{j \in J'} \frac{p_j y_j \bar{r}_j}{2Q} + \frac{1}{2} \sum_{j \in J'} y_j p_j \qquad\qquad \text{(LP2)}$$

$$\sum_{j \in J'} y_j \pi_j = \Pi' \qquad\qquad (7)$$

$$\sum_{j' \in J':j \prec j'} y_{j'} p_{j'} = V_j \qquad\qquad \forall j \in J' - F(J') \qquad (8)$$

$$0 \leq y_j \leq 1 \qquad\qquad \forall j \in J' \qquad (9)$$

Constraint (7) ensures the total profit remains unchanged. Constraints (8) imposes that total volume of scheduled jobs which come before j remain restricted to a certain volume V_j (which will be initially $\bar{V}_j$). We write these constraints for $j \in J' - F(J')$.

IterRound:

1. Initialize $J' \leftarrow J$, $V_j \leftarrow \bar{V}_j$ for all jobs j. Set $S = \emptyset$, $\Pi' = \Pi$.
2. While $|J'| > 3$
   - (i) Find a vertex solution y to LP2.
   - (ii) If there is a job $j \in F(J')$ with $y_j = 0$, set $J' \leftarrow J' \setminus \{j\}$.
   - (iii) Else if there is a job $j \in F(J')$ with $y_j = 1$, set $S \leftarrow S \cup \{j\}$ and
     - a. $J' \leftarrow J' \setminus \{j\}$.
     - b. $\Pi' \leftarrow \Pi' - \pi_j$.
     - c. $\forall j' \prec j, j' \in J'$, $V_{j'} \leftarrow V_{j'} - p_j$.
3. $S \leftarrow S \cup J'$
4. Return S.

Fig. 2. Algorithm for rounding LP relaxation for $\mathcal{I}'$

Now we present the iterative rounding algorithm for converting the fractional schedule into an integral one (Algorithm **IterRound**). Note that the LP relaxation does not write the constraints (8) for the first three jobs in J'. This will ensure that a vertex solution will have at least one variable which is 0 or 1. In the end, we shall return the jobs which were added to the set S and the remaining three jobs in J'. This is the set of jobs chosen by our algorithm. We will prove that either step 2(ii) or step 2(iii) will be executed in each iteration of this algorithm. Let S be the set of jobs returned by the above algorithm. Finally, we show how to schedule these jobs feasibly with respect to the original release dates.

We first show that the iterative rounding algorithm will terminate.

Lemma 2. *Consider a vertex solution y to (LP2). Assuming $|J'| \geq 4$, there exists a job $j \in F(J')$ for which y_j is 0 or 1.*

Proof. Suppose not. Let j_4 be the fourth job (according to release date) in J'. Note that constraints (8) is written for j_4 and jobs released after j_4. By subtracting the constraint for all jobs j' released after j_4 from the constraint for job j_4, we get an equivalent LP. However, in this LP, the variables $y_{j_1}, y_{j_2}, y_{j_3}$, where j_1, j_2, j_3 are the first three jobs in J', appear in only two tight constraints – constraint (8) for j_4 and constraint (7). $\qquad \square$

We give some more notations. We index the iterations of the `while` loop in the algorithm starting from n downwards. Thus, at the beginning of iteration k, $|J'| = k$. Let LP2(k) denote the corresponding LP, and $y^{(k)}$ be the vertex solution found by the Algorithm **IterRound** for this LP. So the first iteration finds the solution $y^{(n)}$, then $y^{(n-1)}$ and so on till $y^{(4)}$. Let $y^{(3)}$ be a vertex solution to LP2(3) when $|J'|$ is 3 (and V_j, Π' values are given by the end of iteration indexed 4) – even though our algorithm will not use this LP solution, it will be useful for analysis. Again, let $V_j^{(k)}$ and $\Pi^{(k)}$ to be the values of V_j and Π' when $|J'| = k$. Let $S^{(k)}$ be the set S when $|J'| = k$. So, $S^{(n)}$ is $\emptyset$.

We now give a procedure which given values y_j for all jobs $j \in J$, outputs a corresponding schedule x_{jt}. This is similar to the algorithm in Figure 1. The procedure arranges the jobs in ascending order of release dates and schedules them in this order. The procedure is described in Figure 3. Note that the schedule itself does not care about the release dates of the jobs and so may not even respect the release dates.

ScheduleJobs :

   Input : Values $y_j \in [0, 1]$ for all jobs $j \in J$.
   Output : A schedule of the jobs $\bar{x}_{jt}$, such that $\sum_t \bar{x}_{jt} = y_j$.
   1. For $l = 1, 2, \ldots, n$
      Let t_l be the first time such that $z(t_l) + \sum_{s=1}^{l-1} \bar{x}_{j_s t_l} < 1$
      (i.e., there is some more processing which can be done at this time slot)
      Process the job j_l to an extent of y_{j_l} from time t_l onwards, i.e.,
      for $t = t_l$, set $\bar{x}_{j_l t} = 1 - z_t - \sum_{s=1}^{l-1} \bar{x}_{j_s t_l}$ and
      for $t > t_l$, iteratively set $\bar{x}_{j_l t} = \max(0, \min(1 - z_t, p_{j_l} y_{j_l}^* - \sum_{t' < t} \bar{x}_{j_l t'}))$.

Fig. 3. Algorithm for building schedule from a solution y

Given the solution $y^{(k)}$ for a subset of jobs J', we extend it to the set of all jobs in J by setting $y_j^{(k)} = 1$ for all $j \in S^{(k)}$, and 0 for jobs in $J - J' - S^{(k)}$. We shall call this the *extended solution* $y^{(k)}$. Let $x^{(k)}$ be the corresponding schedule of these jobs given by calling **ScheduleJobs** on the extended solution $y^{(k)}$. We shall use $\texttt{flow}(x^{(k)}, y^{(k)})$ to denote the objective function value if we view this as a solution to (LP1) with respect to release dates $\bar{r}$, i.e.,

$$\texttt{flow}(x^{(k)}, y^{(k)}) = \sum_j \sum_t \frac{x_{jt}^{(k)}(t - \bar{r}_j)}{2Q} + \frac{1}{2} \sum_j y_j^{(k)} p_j.$$

Lemma 3. *The extended solution $y^{(k)}$ is feasible to (LP2) during every iteration k of the algorithm* **IterRound**. *Further, each of these solutions processes jobs of total volume $\sum_j \bar{y}_j p_j$, and*

$$\texttt{flow}(x^{(3)}, y^{(3)}) \leq \texttt{flow}(\bar{x}, \bar{y}).$$

Proof. Clearly, $\bar{y}$ is a feasible solution to LP2(n). Hence, LP2(n) has non-empty set of feasible solutions, and $y^{(n)}$ is well-defined. Assume (by induction) that LP2(k) has a non-empty set of feasible solutions, and hence $y^{(k)}$ is well-defined. Given the solution $y^{(k)}$, suppose we select a job j with $y_j^{(k)} = 1$ in iteration k (the other case is similar). It is easy to see that $y^{(k)}$ is also a feasible solution to

LP2$(k-1)$ in the next iteration if we do not consider the variables corresponding to j. Further, the constraint (8) for the last job (according to release date) ensures that the total volume of the processed jobs in the solution $y^{(k-1)}$ is same as that in the solution $(x^{(k)}, y^{(k)})$ minus p_j. Hence, if we consider the extended solution $y^{(k-1)}$ to all the jobs, then the volume processed by it does not change.

Since all the solutions $y^{(k)}$ process the same amount of volume, and do not have any gaps, they will occupy each slot to the same extent. So

$$\texttt{flow}(x^{(k)}, y^{(k)}) = \sum_t \frac{u(t)t}{2Q} - \sum_j \frac{\bar{r}_j y_j^{(k)} p_j}{2Q} + \frac{1}{2}\sum_j y_j^{(k)} p_j.$$

The quantity $u(t)$ denotes the amount of processing done in $[t, t+1]$, and will not depend on k. Since LP2 treats the other terms above in the objective, it is easy to show that $\texttt{flow}(x^{(k-1)}, y^{(k-1)}) \leq \texttt{flow}(x^{(k)}, y^{(k)})$. □

We now consider the schedule corresponding to S. Define a solution $\tilde{y}$ as $\tilde{y}_j = 1$ iff $j \in S$. Let $(\tilde{x}, \tilde{y})$ be the corresponding schedule obtained by calling **ScheduleJobs** on $\tilde{y}$.

Lemma 4. *For the schedule* $(\tilde{x}, \tilde{y})$,

$$\texttt{flow}(\tilde{x}, \tilde{y}) \leq \texttt{flow}(\bar{x}, \bar{y}) + 12\sum_j \bar{y}_j p_j + 6\sum_t z_t.$$

Proof. The schedule for S is obtained from $(x^{(3)}, y^{(3)})$ by adding 3 jobs in the beginning. This will require *shifting* the jobs scheduled in $(x^{(3)}, y^{(3)})$ to the right by at most $6Q$ amount of available space so that these 3 jobs can be accommodated. The processing time of these 3 jobs is at most $6Q$. Now, if we look at any timeslot $[t, t+1]$, the number of alive jobs at this time can increase by at most 6 (because these many new jobs which were being processed before t could now be getting processed after t). Hence the increase in flow-time is at most 6 times the makespan of the schedule, which is at most 6 times $\sum_t z_t$ plus the total processing volume. Assuming that there are at least 4 jobs in optimal solution (otherwise we could just enumerate), this would mean that the processing time of the 3 new jobs can be charged to the processing volume of $(x^{(3)}, y^{(3)})$. This proves the desired result. □

Corollary 1. *The total profit of the jobs* S *returned by the algorithm* **Iter-Round** *is at least* Π.

Proof. We prove by induction on k that the total profit of the jobs in $S - S^{(k)}$ is at least $\Pi^{(k)}$. Base case is $k = 3$. We know that there is a feasible solution to (LP2) when there are just 3 jobs remaining in J' (Lemma 3). In this fractional solution, we get a profit of at least $\Pi^{(3)}$. Our algorithm picks all the three jobs and hence its profit must be at least $\Pi^{(3)}$ as well. Assuming that this is true for some k, it is easy to see that the statement holds true for $k-1$ as well (we are updating $\Pi^{(k)}$ accordingly). When $k = n$, $S^{(k)}$ is $\emptyset$, and so we are done. □

Lemma 5. *For a job* $j \in J$, *let* $p(S_{\prec j})$ *be the total processing time of jobs in* S *which are released before* j. *Then* $p(S_{\prec j})$ *lies in the interval* $[\bar{V}_j - 6Q, \bar{V}_j + 6Q]$.

Proof. Consider the first iteration of LP2 when j is among the first 3 jobs of J' – say this is iteration k. So far, we have chosen a set of jobs $S^{(k)}$. Since we are always writing the constraint (8) for j, we know that $V_j^{(k)} = \bar{V}_j - p(S^{(k)})$. Since there is a feasible solution to this LP (Lemma 3), we know that $V_j^{(k)} \geq 0$, and so the jobs selected in S have processing time at least $\bar{V}_j - V_j^{(k)}$. But now $V_j^{(k)}$ involves just 3 jobs, and so it cannot be larger than $6Q$. This proves the lower bound on $p(S_{\prec j})$. The upper bound follows similarly – the only eligible jobs in $p(S_{\prec j})$ are either those in $S^{(k)}$ or the first three jobs of this iteration.   □

Corollary 2. *Let* j, j' *be jobs in* J, $j \prec j'$. *Let* $S_{j,j'}$ *be the of jobs in* S *which lie between* j *and* j' *(excluding* j *and* j'*) with respect to the order* $\prec$. *Then* $p(S_{j,j'}) \leq 12Q + \sum_{j'':j \prec j'' \prec j'} p_{j''} y_{j''}^{\star}$.

Using the above lemma, we now show that it is possible to modify the solution $(\tilde{x}, \tilde{y})$ such that it obeys the release dates $\bar{r}_j$ for the jobs. But we will not need such a result in our analysis because we can directly use the above result. We now give proofs for the final schedule constructed by our algorithm.

Opening the Gaps: Recall that $(\tilde{x}, \tilde{y})$ is the schedule obtained for S where we just start from the beginning and fill available space without looking at the release dates (Figure 3). As a final step in our algorithm, we convert this into a feasible schedule that respects all release dates. We consider these jobs in the ascending order of release dates r_j (which is same as the ordering with respect to $\bar{r}$). We schedule j in the earliest available slots after r_j.

Let $\hat{C}_j$ be the completion time of job $j \in S$ in the final schedule. We need to account for $\sum_{j \in S}(\hat{C}_j - r_j^{\star})$. We split this sum into two parts and give bounds on them separately. The proofs of the following lemma can be found in full version of the paper.

Lemma 6
$$\sum_{j \in S}(s_j^{\star} - r_j^{\star}) \leq 8\texttt{flow}(\tilde{x}, \tilde{y}) + 7\sum_t z_t.$$

Lemma 7
$$\sum_{j \in S}(\hat{C}_j - s_j^{\star}) \leq 14\texttt{flow}(\tilde{x}, \tilde{y}) + 14\sum_t z_t.$$

Combining Lemma 6, Lemma 7, Lemma 4 and Lemma 1, we get Theorem 1.

4   Flow Time Minimization on Multiple Machines under Knapsack Constraint

We use our algorithm for ForbidFlow to design an algorithm for Flowknap in the subset parallel setting. Recall that each job has a release date r_j, a processing requirement p_j and a profit π_j while a target profit Π has to be attained by the scheduled jobs. The machines are identical, but a job can go only to a subset of machines S_j. The objective is to minimize total flow time of the scheduled jobs. The proof of the following theorem can be found in full version of the paper.

Theorem 2 *There is a polynomial time $O(\log P)$-approximation algorithm for* Flowknap *in the subset parallel setting under a single knapsack constraint. Here, P is the ratio between largest and the smallest processing time of a job.*

We write a natural LP relaxation for this problem, which is an extension of the LP relaxation used in [6]. We divide jobs into classes – a job is of class k if its processing time lies in the interval $[2^{k-1}, 2^k)$. Our algorithm runs over several iterations – in each iteration it schedules jobs of a particular class, say class k. Using the optimal solution to the LP relaxation, we figure out the extent to which each timeslot processes jobs of class k. We now use the ForbidFlow algorithm to schedule jobs of class k where we are allowed to use a timeslot to this extent only. Combining the solutions for all classes gives Theorem 2.

References

[1] Bansal, N., Blum, A., Chawla, S., Dhamdhere, K.: Scheduling for flow-time with admission control. In: Di Battista, G., Zwick, U. (eds.) ESA 2003. LNCS, vol. 2832, pp. 43–54. Springer, Heidelberg (2003)

[2] Bartal, Y., Leonardi, S., Marchetti-Spaccamela, A., Sgall, J., Stougie, L.: Multiprocessor scheduling with rejection. In: Proc. ACM-SIAM SODA (1996)

[3] Charikar, M., Khuller, S.: A robust maximum completion time measure for scheduling. In: Proc. ACM-SIAM SODA (2006)

[4] Engels, D.W., Karger, D.R., Kolliopoulos, S.G., Sengupta, S., Uma, R.N., Wein, J.: Techniques for scheduling with rejection. In: Bilardi, G., Pietracaprina, A., Italiano, G.F., Pucci, G. (eds.) ESA 1998. LNCS, vol. 1461, pp. 490–501. Springer, Heidelberg (1998)

[5] Garg, N., Kumar, A.: Better algorithms for minimizing average flow-time on related machines. In: Bugliesi, M., Preneel, B., Sassone, V., Wegener, I. (eds.) ICALP 2006, Part I. LNCS, vol. 4051, pp. 181–190. Springer, Heidelberg (2006)

[6] Garg, N., Kumar, A.: Minimizing average flow-time: Upper and lower bounds. In: Proc. IEEE FOCS (2007)

[7] Guha, S., Munagala, K.: Model-driven optimization using adaptive probes. In: Proc. ACM-SIAM SODA (2007)

[8] Gupta, A., Krishnaswamy, R., Kumar, A., Segev, D.: Scheduling with outliers. In: Dinur, I., Jansen, K., Naor, J., Rolim, J. (eds.) APPROX and RANDOM 2009. LNCS, vol. 5687, pp. 149–162. Springer, Heidelberg (2009)

[9] Lenstra, J.K., Shmoys, D.B., Tardos, É.: Approximation algorithms for scheduling unrelated parallel machines. Math. Program. (1990)

[10] Saha, B., Srinivasan, A.: A new approximation technique for resource-allocation problems. In: Proc. ICS (2010)

[11] Shmoys, D.B., Tardos, É.: Scheduling unrelated machines with costs. In: Proc. ACM-SIAM SODA (1993)

Depth Lower Bounds
against Circuits with Sparse Orientation

Sajin Koroth and Jayalal Sarma

Department of Computer Science and Engineering,
Indian Institute of Technology Madras, Chennai 600036, India
{sajin,jayalal}@cse.iitm.ac.in

Abstract. We study depth lower bounds against non-monotone circuits, parametrized by a new measure of non-monotonicity: the orientation[1] of a function f is the characteristic vector of the minimum sized set of negated variables needed in any DeMorgan[2] circuit computing f. We prove trade-off results between the depth and the weight/structure of the orientation vectors in any circuit C computing the **CLIQUE** function on an n vertex graph. We prove that if C is of depth d and each gate computes a Boolean function with orientation of weight at most w (in terms of the inputs to C), then $d \times w$ must be $\Omega(n)$. In particular, if the weights are $o(\frac{n}{\log^k n})$, then C must be of depth $\omega(\log^k n)$. We prove a barrier for our general technique. However, using specific properties of the **CLIQUE** function (used in [4]) and the Karchmer-Wigderson framework [11], we go beyond the limitations and obtain lower bounds when the weight restrictions are less stringent.

We then study the depth lower bounds when the structure of the orientation vector is restricted. We demonstrate that this approach reaches out to the limits in terms of depth lower bounds by showing that slight improvements to our results separates NP from NC.

As our main tool, we generalize Karchmer-Wigderson game [11] for monotone functions to work for non-monotone circuits parametrized by the weight/structure of the orientation. We also prove structural results about orientation and prove connections between number of negations and weight of orientations required to compute a function.

1 Introduction

Deriving size/depth lower bounds for Boolean circuits computing NP-complete problems has been one of the main goals in circuit complexity. Attempts to prove size lower bounds against constant depth circuits has yielded useful results (see survey [1,2] and textbook [10]). However, despite many efforts, for computing explicit functions, the best size lower bound known against general circuits is

---

[1] A generalization of monotone functions are studied under the name *unate functions*(cf. [7]). We inherit the terminology of *orientation* from that setting. We remark that our definition is universal unlike the case of unate functions.

[2] Circuits where negations appear only at the leaves.

Z. Cai et al. (Eds.): COCOON 2014, LNCS 8591, pp. 596–607, 2014.

still a constant factor on the number of inputs [8], and the best depth lower bound known against general bounded fan-in circuits is (derived from formula size lower bound due to Håstad [17]) less than $3 \log n$.

Notable progress has been made in proving lower bounds against monotone circuits. Razborov [15] proved a super-polynomial size lower bound against monotone circuits computing the **CLIQUE** function which is NP-hard. This was further strengthened to exponential lower bounds by Alon and Boppana [3]. A super polynomial monotone size lower bound is also known [16] for the **PMATCH** problem. The latter result also showed that non-monotonicity helps in size restricted settings as **PMATCH** is known to be in P[5].

Moving in the direction of non-monotonicity, Amano and Maruoka [4] established super-polynomial lower bounds against circuits with at most $\frac{1}{6} \log \log n$ negations computing the **CLIQUE** function. A chasm was already known at the $\log n$ negations; Fisher [6] proved that any circuit of polynomial size can be converted to a circuit of polynomial size having only $\log n$ negations. In particular, this implies that if we are able to extend the lower bounds to the case of circuits having $O(\log n)$ negations, then it separates P from NP. The gap was further tightened by Jukna [9] (for multi-output functions), where he showed a super-polynomial size lower bound against circuits with $\log n - 16 \log \log n$ negations.

In terms of depth lower bounds, it is known that **CLIQUE** function and the **PMATCH** function on graphs of n vertices require $\Omega(n)$ depth for any bounded fan-in monotone circuit computing them [14]. Thus, non-monotonicity is useful in the depth restricted setting also as **PMATCH** is known to be in non-uniform NC^2 [12]. One main technique involved in [14] is a characterization of circuit depth using a communication game defined between two players. Raz and Wigderson [13] used this framework to obtain a lower bound of $\Omega(n^2)$ on the number of negations at the leaves for any $O(\log n)$ depth DeMorgan circuit solving the s-t connectivity problem. However, we do not know[3] depth lower bounds against circuits where there are negations at arbitrary locations using the Karchmer-Wigderson framework.

Our Results : We study an alternative way of limiting the non-monotonicity in the circuit. To arrive at our restriction, we define a new measure called *orientation* of a Boolean function. A function $f : \{0, 1\}^n \to \{0, 1\}$ is said to have *orientation* $\beta \in \{0, 1\}^n$ if there is a monotone function $h : \{0, 1\}^{2n} \to \{0, 1\}$ such that : $\forall x \in \{0, 1\}^n, f(x) = h(x, (x \oplus \beta))$. The orientation of a Boolean function is simply the indicator vector of the set of inputs which are required to be negated in any DeMorgan circuit computing the function f. Indeed, if f itself is monotone, the orientation is simply the all-0s vector. The *weight of the orientation* is simply the number of 1s in β, and can be thought of as a parameter indicating how "close" f is to a monotone function.

---

[3] Indeed, size lower bounds against bounded fan-in circuits in the presence of negations [4] also imply depth lower bounds against them. In particular, [4] implies that any circuit with $\frac{1}{6} \log \log n$ negation gates computing **CLIQUE**$(n, (\log n)^{\sqrt{\log n}})$ requires depth $\Omega((\log n)^{\sqrt{\log n}})$.

The same definitions can be extended to circuits as well. We consider circuits where the function computed at each gate can be non-monotone, but the corresponding orientation (with respected to the inputs to the circuit) must be of limited weight. We say a circuit C is weight w oriented if every internal gate of C computes a function which has an orientation β with $|\beta| \leq w$. The semantic restriction we study limits the weight of the orientation of the function computed at each gate of the circuit (in terms of the original inputs of the circuit). We prove the following theorem which presents a depth vs weight trade-off.

Theorem 1. *If C is a Boolean circuit of depth d and weight of the orientation w ($w > 0$), computing* **CLIQUE** *then, $d \times w$ must be $\Omega(n)$.*

In particular, if the weights are $o(\frac{n}{\log^k n})$, the **CLIQUE** function requires $\omega(\log^k n)$ depth. By contrast, any circuit computing **CLIQUE** has weight of the orientation at each gate at most n^2. We prove the above theorem by extending the Karchmer-Wigderson framework to the case of non-monotone sparsely oriented circuits. The proof critically requires the route via Karchmer-Wigderson games since it is unclear how to directly simulate the above non-monotone circuit model using a monotone circuit. We remark that the above theorem applies even to circuits computing **PMATCH**.

The difficulty in extending the above lower bound to more general lower bounds is the potential presence of gates computing densely oriented functions. In this context, we explore the usefulness of having gates with non-zero orientation in the circuit. We argue that allowing even a constant number of non-zero (but dense) oriented gates makes the circuit more powerful in the limited depth setting. In particular, we show (see Theorem 7) that *there exists a monotone function f which cannot be computed by poly-log depth monotone circuits, but there is a poly-log depth circuit computing it such that there are at most two internal gates which has a non-zero orientation β.*

The above theorem indicates that the densely oriented gates are indeed useful, and that Theorem 1 cannot be improved in terms of the number of densely oriented gates it can handle, without using specific properties about the function(for example, **CLIQUE**) being computed.

Going beyond the above limitations, we exploit the known properties of the **CLIQUE** function and the generalized Karchmer-Wigderson games to prove lower bounds against less stringent weight restrictions (in particular, we can restrict the weight restrictions to only negation gates and their inputs).

Theorem 2. *For any circuit family $\mathcal{C} = \{C_m\}$ (where $m = \binom{n}{2}$) computing* **CLIQUE**$(n, n^{\frac{1}{6\alpha}})$ *where there are $\ell + k$ negation gates, with $\ell \leq 1/6 \log \log n$ and the k negation gates in C_m are computing functions which are sensitive only on w inputs (i.e., the orientation of their input as well as their output is at most w) and the remaining ℓ negations compute functions of arbitrary orientation:*
$$\text{Depth}(C_m) \geq n^{\frac{1}{2^{\ell+8}}} - kw - \ell$$

This theorem implies that **CLIQUE** cannot be computed by circuits with depth $n^{o(1)}$ even if we allow only constant number of gates to have non-zero (even dense) orientation - thus going beyond the earlier hurdle presented for **PMATCH**. We remark that the above theorem also generalizes the case of circuits with negations at the leaves ($\ell = 0$, and $w = 1$).

We now turn to circuits where the structure of the orientation is restricted. The restriction is on the number of vertices of the input graph involved in edges indexed by β.

Theorem 3. *If C is a circuit computing the **CLIQUE** function and for each gate g of C, the number of vertices of the input graph involved in edges indexed by β_g (the orientation vector of gate g) is at most w, then $d \times w$ must be $\Omega(\frac{n}{\log n})$.*

We also study a sub-class of the above circuits for which we prove lower bound results very close to the required ones. A circuit is said to be of *uniform orientation* if there exists a $\beta \in \{0,1\}^n$ such that every gate in it computes a function which has orientation β.

Theorem 4. *Let C be a circuit computing the **CLIQUE** function, with uniform orientation $\beta \in \{0,1\}^n$ such that there is a subset of vertices U, $|U| \geq \log^{k+\epsilon} n$ for which $\beta_e = 0$ for all edges e within U, then C must have depth $\omega(\log^k n)$.*

We remark that a DeMorgan circuit has an orientation of weight exactly equal to the number of negated variables. However, this result is incomparable with that of [13] against DeMorgan circuits for two reasons : (1) this is for the **CLIQUE** function. (2) the lower bounds and the class of circuits are different.

In contrast to the above theorem, we show that an arbitrary circuit can be transformed into one having our structural restriction on the orientation with $|U| = O(\log^k n)$.

Theorem 5. *If there is a circuit C computing **CLIQUE** with depth d then for any set of $c \log n$ vertices U, there is an equivalent circuit C' of depth $d + c \log n$ with orientation β such that none of the edges $e(u, v)$, $u, v \in U$ has $\beta_{e(u,v)} = 1$.*

Thus if either Theorem 4 is extended to $|U| = O(\log^k n)$ or the transformation in Theorem 5 can be modified to give $|U| = O(\log^{k+\epsilon} n)$ for some constant $\epsilon > 0$, then a depth lower bound for **CLIQUE** function against general circuits of depth $O(\log^k n)$ will be implied.

2   Preliminaries

For $x, y \in \{0,1\}^n$, $x \leq y$ if and only if for all $i \in [n]$, $x_i \leq y_i$. A Boolean function f is said to be monotone if for all $x \leq y$, $f(x) \leq f(y)$. In other words value of a monotone function does not decrease when input bits are changed from 0 to 1.

For a set U, we denote by $\binom{U}{2}$ the set $\{\{u, v\} \,|\, u, v \in U\}$. In an undirected graph $G = (V, E)$, a clique is a set $S \subseteq V$ such that $\binom{S}{2} \subseteq E(G)$. **CLIQUE**$(n, k)$ is a Boolean function $f : \{0,1\}^{\binom{n}{2}} \to \{0,1\}$ such that for any $x \in \{0,1\}^{\binom{n}{2}}$, $f(x) = 1$ if G_x, the undirected graph represented by the undirected adjacency matrix x has a clique of size k. **CLIQUE**(n, k) is a monotone function as adding edges (equivalent to turning 0 to 1 in adjacency matrix) cannot remove a k-clique, if one already exists. By **CLIQUE**, we denote **CLIQUE**$(n, \frac{n}{2})$. A perfect matching of an undirected graph $G = (V, E)$ is a $M \subseteq E(G)$ such that no two edges in M share an end vertex and it is such that every vertex $v \in V$ is contained as an end vertex of some edge in M. Corresponding Boolean function **PMATCH** $: \{0, 1\}^{\binom{n}{2}} \to \{0, 1\}$ is defined as **PMATCH**$(x) = 1$ if G_x contains a perfect matching. It is easy to note that **PMATCH** is also a monotone function.

A circuit is a directed acyclic graph whose internal nodes are labeled with $\wedge$, $\vee$ and $\neg$ gates, and leaf nodes are labeled with inputs. The function computed by the circuit is the function computed by a designated "root" node. All our circuits are of bounded fan-in. The depth of a circuit C, denoted by **Depth**(C) is the length of the longest path from root to any leaf, and **Depth**(f) denotes the minimum possible depth of a circuit computing f. By **Depth**$_t(f)$ we denote the minimum possible depth of a circuit computing f with at most t negations. Size of a circuit is simply the number of internal gates in the circuit, and is denoted by **Size**(C). **Size**(f), **Size**$_t(f)$ are defined analogous to **Depth**(f), **Depth**$_t(f)$ respectively. We refer the reader to a standard textbook (cf. [18]) for more details.

We now review the Karchmer-Wigderson games and the related lower bound framework. The technique is a strong connection between circuit depth and communication complexity of a specific two player game where the players say Alice and Bob are given inputs $x \in f^{-1}(1)$ and $y \in f^{-1}(0)$, respectively. In the case of general circuits, the game is denoted by **KW**(f) and the goal is to find an index i such that $x_i \neq y_i$. In the case of monotone circuits, the game is denoted by **KW**$^+(f)$ and the goal is to find an index i such that $x_i = 1$ and $y_i = 0$. We abuse the notation and use **KW**(f) and **KW**$^+(f)$ to denote the number of bits exchanged in the worst case for the best protocols for the corresponding communication games. Karchmer and Wigderson [11] proved that for any function f depth of the best circuit computing f, denoted by **Depth**(f) is equal to **KW**(f). For any monotone function f the depth of the best monotone circuit computing f, denoted by **Depth**$^+(f)$ is equal **KW**$^+(f)$. Raz and Wigderson [14] showed that **KW**$^+$(**CLIQUE**) and **KW**$^+$(**PMATCH**) are both $\Omega(n)$.

Characterization of Orientation: If $\beta \in \{0, 1\}^n$ is an orientation for a function f, then any $\beta' \geq \beta$ is also an orientation for f by definition of orientation. For any general circuit C computing a function $f : \{0, 1\}^n \to \{0, 1\}$ there is a circuit C' of at most twice the size of C, such that it has a uniform β for some $\beta \in \{0, 1\}^n$. This can be seen by simply converting the circuit into a DeMorgan circuit by pushing down the negations applying De-Morgan's laws. Also, for any function f whose orientation is β, there is a circuit C of uniform orientation β. We prove a sufficient condition for the β_i to be 1 in the orientation.

Proposition 1. *For any function f, if there exists a pair (u, v) such that $u_i = 0, v_i = 1, u_{[n]\setminus\{i\}} = v_{[n]\setminus\{i\}}$ and $f(u) = 1, f(v) = 0$ then any orientation β of the function must have $\beta_i = 1$.*

It is not a priori clear that the minimal orientation for a function f is unique. We defer the proof of the following proposition to the full version.

Proposition 2. *Minimal orientation for a function* $f : \{0,1\}^n \to \{0,1\}$ *is well defined and it is* $\beta \in \{0,1\}^n$ *such that* $\beta_i = 1$ *if and only if there exists a pair* (u,v) *such that* $u_i = 0, v_i = 1,$ $u_{[n]\setminus\{i\}} = v_{[n]\setminus\{i\}}$ *and* $f(u) = 1, f(v) = 0.$

3   Lower Bound Argument for Sparsely Oriented Circuits

In this section, we prove Theorem 1 which shows the trade-off between depth and weight of orientation of the internal gates of a circuit. We prove the following main lemma of our paper.

Lemma 1. *If C is a circuit of depth d such that each internal gate computes a Boolean function whose orientation has weight at most w and C is computing a monotone function $f : \{0,1\}^n \to \{0,1\}$ which is sensitive on all its inputs, then $d \times (4w + 1) \geq \mathbf{KW}^+(f).$*

Proof. The proof idea is to devise a protocol for $\mathbf{KW}^+(f)$ using C having **Depth**(C) rounds and each round having a communication cost of $4w + 1$.

Alice is given $x \in f^{-1}(1)$ and Bob is given $y \in f^{-1}(0)$. The goal is to find an index i such that $x_i = 1, y_i = 0$. The protocol is described in Algorithm 1.

We now prove that the protocol(Algorithm 1) solves $\mathbf{KW}^+(f)$. The following invariant which is maintained during the run of the protocol is crucial for the proof.

Invariant: When the protocol is at a node which computes a function f with orientation vector β it is guaranteed a priori that the inputs held by Alice and Bob, x' and y' are equal on the indices where $\beta_i = 1$, $f(x') = 1, f(y') = 0$ and restriction of f obtained by fixing variables where $\beta_i = 1$ to $x_i'(= y_i')$ is a monotone function.

Assuming that the invariant is maintained, we claim that when the protocol stops at an input node of the circuit computing a function f with $f(x') = 1$ and $f(y') = 0$ then $f = x_i$ for some $i \in [n]$. If the input node is a negative literal, say $\bar{x}_i$ then by Proposition 1, orientation of $\bar{x}_i$ has $\beta_i = 1$. By the guarantee that $x'_\beta = y'_\beta$, $x'_i = y'_i$, contradicting $f(x') \neq f(y')$. Hence whenever the protocol stops at leaf node it is guaranteed that the leaf is labeled by a positive literal. And when input node is labeled by a positive literal x_i, then a valid solution is output as $f(x') = 1, f(y') = 0$ implies $x'_i = 1$ and $y'_i = 0$. Note that during the run of the protocol we only changed x, y at some indices $i, x_i \neq y_i$ to $x'_i = y'_i$. Hence, any index where $x'_i \neq y'_i$ it is the case that $x_i = x'_i$ and $y_i = y'_i$.

Now we prove the invariant. Note that it is vacuously true at the root gate as f is a monotone function implying $\beta = 0^n$, and in the standard $\mathbf{KW}^+(f)$ game $x \in f^{-1}(1)$ and $y \in f^{-1}(0)$. We argue that, while descending down to one of the children of the current node the invariant is maintained. To begin with, we show that the protocol does not get stuck in step 8 (and similarly for step 17). To prove this, we claim that at an $\wedge$ gate $f = f_1 \wedge f_2$, if the protocol failed to find an i in step 4 such that $x'_i = 1, y'_i = 0$ then on the modified input y'' at least one of $f_1(y'')$ or $f_2(y'')$ is guaranteed to be zero. Since the protocol failed to output an i such that $x'_i = 1, y'_i = 0$, it must be the case that $x'_i \leq y'_i$ for

---

Algorithm 1. Modified Karchmer-Wigderson Protocol

---

1. {Let x' and y' be the current inputs. At the current gate g computing f, with the input gates g_1 and g_2, f_1 and f_2 be the corresponding sub-functions and β_1, β_2 be the corresponding orientations (and are known to both Alice and Bob). If g_1 or g_2 is a negation gate, let γ_1 and γ_2 be the orientation vectors of input functions to g_1 and g_2, otherwise they are 0-vectors. Let $\alpha = \beta_1 \vee \beta_2 \vee \gamma_1 \vee \gamma_2$. Let $S = \{i : \alpha_i = 1\}$, x_S is the substring of x indexed by S. }

2. **if** g is $\wedge$ **then**

3.   Alice sends x'_S to Bob. Bob compares x'_S with y'_S.

4.   **if** there is an index $i \in S$ such that $x'_i = 1$ and $y'_i = 0$ **then**

5.   Output i.

6.   **else**

7.   Define $y'' \in \{0,1\}^n$: $y''_S = x'_S$ and $y''_{[n]\setminus S} = y'_{[n]\setminus S}$.

8.   Bob sends $i \in \{1,2\}$ such that $f_i(y'') = 0$ to Alice. They recursively run the protocol on g_i with $x' = x'$ and $y' = y''$.

9.   **end if**

10. **end if**

11. **if** g is $\vee$ **then**

12.   Bob sends y'_S to Alice. Alice compares y'_S with x'_S.

13.   **if** there is an index $i \in S$ such that $x'_i = 1$ and $y'_i = 0$ **then**

14.   Output i.

15.   **else**

16.   Define $x'' \in \{0,1\}^n$: $x''_S = y'_S$ and $x''_{[n]\setminus S} = x'_{[n]\setminus S}$.

17.   Alice sends $i \in \{1,2\}$ such that $f_i(x'') = 1$ to Bob. They recursively run the protocol on g_i with $x' = x''$ and $y' = y'$.

18.   **end if**

19. **end if**

---

indices indexed by β_1, β_2. Let U be the subset of indices indexed by β_1 and β_2 where $x_i = 0$ and $y_i = 1$. Bob obtains y'' from y' by setting $y''_i = 0$ for all $i \in U$. Thus we have made sure that x' and y'' are the same on the variables whose negations are required to compute f, f_1 and f_2.

Consider the functions $f', f'' : \{0,1\}^{n-|\beta_1 \vee \beta_2|} \to \{0,1\}$ which are obtained by restricting the variables indexed by orientation vectors of f_1 and f_2 to the value of those variables in x'. Both f' and f'' are monotone as they are obtained by restricting all negated input variables of the DeMorgan circuits computing f_1 and f_2 for orientations β_1 and β_2 respectively. The changes made to x', y' were only at places where they differed. Thus at all the indices where x', y' were same, x', y'' is also same. Hence monotone restriction $f_{x'_\beta}$ of f obtained by setting variables indexed by β to their values in x' is a consistent restriction for y'' also. It is easy to note that $y'' \leq y'$. Hence $f(y'') = 0$ because y'' agrees with y' on variables indexed by β (as x'' agrees with y' and y'' on variables indexed by β) implying $f_{x'_\beta}(y''_{[n]\setminus\beta}) \leq f_{x'_\beta}(y'_{[n]\setminus\beta}) = 0$. Since $f(y'') = 0$, it is guaranteed that one of

$f_1(y''), f_2(y'')$ is equal to 0. Bob sets $y' = y''$ and sends 0 if it is $f_1(y'') = 0$ or 1 otherwise, indicating Alice which node to descend to. Note that $x'_{\beta_1} = y''_{\beta_1}, x'_{\beta_2} = y''_{\beta_2}$ and restriction of f_1, f_2 to $x'_{\beta_1}, x'_{\beta_2}$ respectively gives monotone functions f', f'' thus maintaining the invariant for both f_1 and f_2.

We claim that if any of the input gates g_1, g_2 to the current $\wedge$ gate g is a $\neg$ gate then the protocol will not take the path through the negation gate. To argue this, we use the following Lemma.

Lemma 2. *If $\ell, \bar{\ell}$ are functions with orientations β, γ, then for all $x, y \in \{0, 1\}^n$ such that $x_{\beta \vee \gamma} = y_{\beta \vee \gamma}, \ell(x) = \ell(y)$.*

Proof. We know that for a function ℓ, if there exists a pair $(u, v) \in \{0, 1\}^n \times \{0, 1\}^n$ with $u \leq v$, $u_i \neq v_i$, $u_{[n] \backslash \{i\}} = v_{[n] \backslash \{i\}}$ and $\ell(u) = 1, \ell(v) = 0$ then by Proposition 1 for every orientation β, $\beta_i = 1$. Let i be an index on which ℓ is sensitive, i.e., there exists $(u, v) \in \{0, 1\}^n \times \{0, 1\}^n$ with $u \leq v$, $u_i \neq v_i$, $u_{[n] \backslash \{i\}} = v_{[n] \backslash \{i\}}$ and $\ell(u) \neq \ell(v)$. Note that l is sensitive on i need not force $\beta_i = 1$, as it could be that $\ell(u) = 0$ and $\ell(v) = 1$. But in this case $\bar{\ell}(u) = 1$ and $\bar{\ell}(v) = 0$, hence $\gamma_i = 1$ for $\bar{\ell}$. Hence, ℓ is sensitive only on indices in $\beta \vee \gamma$.   □

The lemma establishes that at every negation gate in weight w oriented circuit, a function which is sensitive on at most $2w$ indices is computed. Hence, the root gate cannot be a negation gate for a function sensitive on all inputs if $2w < n$. Suppose only one child is a negation gate, say f_1. Since we ensure $x'_{\beta_1 \vee \gamma_1} = y''_{\beta_1 \vee \gamma_1}$, the above lemma implies $f_1(x') = f_1(y'')$. But the protocol does not descend down a path where x', y'' are not separated. Hence the claim.

This also proves that when the protocol reaches an $\wedge$ node with both children negated, at the round for that node protocol outputs an index i and stops. Otherwise, since we ensure $x'_S = y''_S$, $f_1(y'') = f_1(x') = 1$ and $f_2(y'') = f_2(x') = 1$. But this contradicts the fact that at a node $f = f_1 \wedge f_2$ either $f_1(y'') = 0$ or $f_2(y'') = 0$ (or both).

Proof of equivalent claims for an $\vee$ gate is similar except for the fact that Alice modifies her input.

Thus, using the above protocol we are guaranteed to solve $\mathbf{KW^+}(f)$. Communication complexity of the protocol is upper bounded by $\mathbf{Depth}(C) \times (4w + 1)$. Communication cost of a round is $4w + 1$. Because if any of the children is a negation gate then we have to send its orientation along with the orientation of its complement. The protocol clearly stops after $\mathbf{Depth}(C)$ many rounds.   □

4   Dense Orientation

Currently our depth lower bound technique cannot handle orientations of weight $\frac{n}{\log^k n}$ or more for obtaining $\omega(\log^k n)$ lower bounds. In light of this, we explore the usefulness of densely oriented gates in a circuit. First we prove that any polynomial sized circuit can be transformed into an equivalent circuit of polynomial size but having only $O(n \log n)$ gates of non-zero orientation by studying the connection between orientations and negations. Next we present a limitation of our technique in a circuit having only two gates of non-zero (but dense) orientation. Thus, strengthening of our technique will have to use some property of the function being computed. Finally we show how to use a property of **CLIQUE** function to slightly get around the limitation.

4.1 From Negation Gates to Orientation

Since weight of the orientation can be thought of as a measure of non-monotonicity in a circuit, a natural question to explore is the connection between the number of negations and number of non-zero orientations required to compute a function f. By analyzing the sub-functions produced at each negation gate and carefully combining them, we show the following (detailed proof is deferred to the full version):

Theorem 6. *For any function $f : \{0,1\}^n \to \{0,1\}$, if there is a circuit family $\{C_n\}$ computing f with $t(n)$ negations then there is also a circuit family $\{C'_n\}$ computing f such that $\mathbf{Size}(C'_n) \leq 2^t \times (\mathbf{Size}(C_n) + 2^t) + 2^t$, and there are at most $2^{t-1}(t+2) - 1$ internal gates whose orientation is non-zero.*

In particular, in conjunction with the result of Fisher [6] mentioned before, this implies that it is enough to prove lower bounds against circuits with at most $O(n \log n)$ internal nodes of dense orientations to obtain lower bounds against the general circuits.

4.2 Power of Dense Orientation

We show that even as few as two densely oriented internal gates can help to reduce the depth from super poly-log to poly-log for some functions. We defer the proof to the full version.

Theorem 7. *There exists a monotone Boolean function f such that it cannot be computed by poly-log depth monotone circuits, but there is a poly-log depth circuit computing it such that at most two internal gates have non-zero orientation β.*

This theorem combined with the sparse orientation protocol implies that the two non-zero orientations β_1, β_2 is such that $|\beta_1| + |\beta_2|$ is not only non-zero but is super poly-log. Because our protocol will spend $|\beta_1| + |\beta_2|$ for handling these two gates, and on the remaining gates in the circuit it will spend 1 bit each. Hence the cost of the sparse orientation protocol will be at most $|\beta_1| + |\beta_2| + \mathbf{Depth}(C)$, thus $|\beta_1| + |\beta_2|$ is at least $\mathbf{KW}^+(f) - \mathbf{Depth}(C)$ which is super poly-log as $\mathbf{Depth}(C)$ is poly-log and $\mathbf{KW}^+(f)$ is super poly-log.

By Theorem 7 we get a function which has an NC^2 circuit with two non-zero orientation gates which has no monotone circuit of poly-log depth. Thus our bounds cannot be strengthened to handle higher weight without incorporating the specifics of the function being computed. In section 4.3, we rescue the situation slightly using the specific properties of the **CLIQUE** function.

4.3 Lower Bounds for CLIQUE Function

The number of gates with high orientations can be arbitrary in general. In this subsection we give a proof for Theorem 2. We first extend our technique to handle the low weight negations efficiently so that we get a circuit on high weight negations (see Lemma 3 below). To complete the proof of Theorem 2, we appeal to depth lower bounds against negation-limited circuits computing f. We refer the reader to the full version for the details.

Lemma 3. *For any circuit family $\mathcal{C} = \{C_n\}$ computing a monotone function f where there are k negations in C_n computing functions which are sensitive only on w inputs (i.e., the orientation of their input as well as their output is at most w) and the remaining ℓ negations compute functions of arbitrary orientation:*
$$\mathbf{Depth}(C_n) \geq \mathbf{Depth}_{2^\ell}(f) - kw - \ell$$

5   Structural Restrictions on Orientation

In this section we study structural restrictions on the orientation and prove stronger lower bounds.

5.1   Restricting the Vertex Set Indexed by the Orientation

We first consider restrictions on the set of vertices[4] indexed by the orientation - in order to prove Theorem 3 stated in the introduction. As in the other case, we argue the following lemma, which establishes the trade-off result. By using the lower bound for $\mathbf{KW}^+$ games for **CLIQUE** function, the theorem follows.

Lemma 4. *Let C be a circuit of depth d computing **CLIQUE**, with each gate computing a function whose orientation is such that the number of vertices of the input graph indexed by the orientation β is at most $\frac{w}{\log n}$, then d is $\Omega\left(\frac{\mathbf{KW}^+(f)}{4w+1}\right)$.*

Proof. It is enough to solve the $\mathbf{KW}^+(f)$ on the min-term, max-term pairs which in case of **CLIQUE**(n, k) is a k-clique and a complete $k - 1$-partite graph. We play the same game as in the proof of Theorem 1, but instead of sending edges we send vertices included in the edge set indexed by β with some additional information. If it is Alice's turn, then x'_β defines an edge sub-graph of her clique. Both Alice and Bob know β and hence knows which vertices are spanned by edges $e_{u,v}$ such that $\beta_{e(u,v)} = 1$. So Alice can send a bit vector of length at most w (in the case of Alice we can handle up to w), indicating which of these vertices are part of her clique. This information is enough for Bob to deduce whether any $e_{u,v}$ indexed by β is present in Alice's graph or not. Since Bob makes sure that $x'_\beta = y'_\beta$ by modifying his input, and Alice keeps her input unchanged, Alice knows what modifications Bob has done to his graph.

Similarly on Bob's turn, he sends the vertices in the partition induced by y_β and the partition number each vertex belongs to (hence the $\log n$ overhead for Bob) to Alice. With this information Alice can deduce whether any $e_{u,v} \in \beta$ is present in Bob's graph or not. Inductively they maintain that they know of the changes made to other parties input in each round. Hence the game proceeds as earlier. This completes the proof of the theorem.

---

[4] Notice that the input variables to the **CLIQUE** function represents the edges. This makes the results of this section incomparable with the depth lower bounds of [13].

5.2   Restricting the Orientation to be Uniform

In this section, we consider the circuits where the orientation is uniform and study its structural restrictions. We proceed to the proof of Theorem 4.

Theorem 4: *Let C be a circuit computing the* **CLIQUE** *function with uniform orientation $\beta \in \{0,1\}^n$ such that there is subset of vertices U and $\epsilon > 0$ such that $|U| \geq \log^{k+\epsilon} n$ for which $\beta_e = 0$ for all edges e within U, then C must have depth $\omega(\log^k n)$.*

Proof. We prove by contradiction. Suppose there is a circuit C of depth $c\log^k n$. In the argument below we assume $c = 1$ for simplicity. Without loss of generality, we assume that $|U| = \log^{k+\epsilon} n$. Fix inputs to circuit C in the following way: Choose an arbitrary $K_{\frac{n}{2} - \frac{|U|}{2}}$ comprising of vertices from $[n] \setminus U$ and set those edges to 1. For every edge in $\binom{[n]\setminus U}{2}$ which is not in the clique chosen earlier, set to 0. For every edge between $[n] \setminus U$ and U set it to 1. Since every edge $e(x,y)$ which has $\beta_e = 1$ has at least one of the end points in $[n] \setminus U$, by above setting, all those edges are turned to constants. Thus we obtain a monotone circuit C'' computing **CLIQUE**$(|U|, \frac{|U|}{2})$ of depth at most $(\log n)^k$. In terms of the new input, $(\log n)^k = ((\log n)^{k+\epsilon})^{\frac{k}{k+\epsilon}} = (|U|)^{\frac{k}{k+\epsilon}}$, this contradicts the Raz-Wigderson [14] lower bound of $\Omega(|U|)$, as $\frac{k}{k+\epsilon} < 1$ for $\epsilon > 0$.   □

A Contrasting Picture: Any function has a circuit with a uniform orientation $\beta = 1^n$ $(|\beta| = n)$. We show that the weight of the orientation can be reduced at the expense of depth, when the circuit is computing the **CLIQUE** function.

Theorem 5: *If there is a circuit C computing **CLIQUE** with depth d then for any set of $c\log n$ vertices U, there is an equivalent circuit C' of depth $d + c\log n$ with orientation β such that none of the edges $e(u,v)$, $u, v \in U$ has $\beta_{e(u,v)} = 1$.*

Proof. We modify the KW protocol on circuit C as follows: Alice chooses an arbitrary clique $K_{\frac{n}{2}} \in G_x$ (which she is guaranteed to find as $x \in f^{-1}(1)$). She then obtains x' by deleting edges $e(x,y)$ from $\binom{U}{2}$ which are outside the chosen clique $K_{\frac{n}{2}}$. Note that since $K_{\frac{n}{2}} \in G_{x'}$, $f(x') = 1$. Alice then sends the characteristic vector of vertices in $K_{\frac{n}{2}} \cap U$ which is of length at most $c\log n$, to Bob. Bob then obtains y' from y by removing edges in $\binom{U}{2}$ which are outside the clique formed by $K_{\frac{n}{2}} \cap \binom{U}{2}$. By monotonicity of **CLIQUE**, $f(y') = 0$. If there is an edge $e(u,v) \in K_{\frac{n}{2}} \cap \binom{U}{2}$ which is missing from y' Bob outputs the index $e(u,v)$. Otherwise they run the standard Karchmer-Wigderson game on x', y' using the circuit C to obtain an $e(x,y)$ such that $x_{e(u,v)} = 1$ and $y_{e(u,v)} = 0$. The correctness of the protocol is easy to see The cost of the above protocol is $d + c\log n$. By the connection between **KW**(f) and circuit depth, we get a circuit of desired properties.   □

Thus, if there is a circuit $C \in \mathsf{NC}^k$ computing **CLIQUE**(n, k), then there is a circuit $C' \in \mathsf{NC}^k$ of uniform orientation β computing **CLIQUE**(n, k) such that there are $(c\log n)^k$ vertices V' with none of the edges $e(u, v)$ having $\beta_{e(u,v)} = 1$. In other words, if we improve Theorem 3 to the case when the orientation "avoids" a set of $\log n$ vertices (instead of $(\log n)^{(1+\epsilon)}$ as done), it will imply $\mathsf{NC}^1 \neq \mathsf{NP}$.

References

1. Allender, E.: Circuit Complexity before the Dawn of the New Millennium. In: Chandru, V., Vinay, V. (eds.) FSTTCS 1996. LNCS, vol. 1180, pp. 1–18. Springer, Heidelberg (1996)
2. Allender, E.: Cracks in the defenses: Scouting out approaches on circuit lower bounds. In: Hirsch, E.A., Razborov, A.A., Semenov, A., Slissenko, A. (eds.) CSR 2008. LNCS, vol. 5010, pp. 3–10. Springer, Heidelberg (2008)
3. Alon, N., Boppana, R.B.: The Monotone Circuit Complexity of Boolean Functions. Combinatorica 7(1), 1–22 (1987)
4. Amano, K., Maruoka, A.: A superpolynomial lower bound for a circuit computing the clique function with at most $(1/6) \log \log n$ negation gates. SIAM Journal on Computing 35(1), 201–216 (2005)
5. Edmonds, J.: Paths, trees, and flowers. Canad. J. Math. 17, 449–467 (1965)
6. Fischer, M.: The Complexity of Negation-limited Networks — A Brief Survey. In: Brakhage, H. (ed.) GI-Fachtagung 1975. LNCS, vol. 33, pp. 71–82. Springer, Heidelberg (1975)
7. Impagliazzo, R., Paturi, R., Saks, M.E.: Size-depth tradeoffs for threshold circuits. SIAM Journal of Computing 26(3), 693–707 (1997)
8. Iwama, K., Morizumi, H.: An explicit lower bound of $5n-o(n)$ for boolean circuits. In: Diks, K., Rytter, W. (eds.) MFCS 2002. LNCS, vol. 2420, pp. 353–364. Springer, Heidelberg (2002)
9. Jukna, S.: On the minimum number of negations leading to super-polynomial savings. Information Processing Letters 89(2), 71–74 (2004)
10. Jukna, S.: Boolean Function Complexity: Advances and Frontiers. Algorithms and Combinatorics, vol. 27. Springer New York Inc. (2012)
11. Karchmer, M., Wigderson, A.: Monotone Circuits for Connectivity Require Super-logarithmic Depth. In: STOC, pp. 539–550 (1988)
12. Lovász, L.: On determinants, matchings, and random algorithms. In: Symposium on Fundamentals of Computation Theory (FCT), pp. 565–574 (1979)
13. Raz, R., Wigderson, A.: Probabilistic communication complexity of boolean relations. In: Proc. of the 30th FOCS, pp. 562–567 (1989)
14. Raz, R., Wigderson, A.: Monotone circuits for matching require linear depth. Journal of ACM 39(3), 736–744 (1992)
15. Razborov, A.A.: Lower Bounds for Monotone Complexity of Some Boolean Functions. Soviet Math. Doklady, 354–357 (1985)
16. Razborov, A.A.: Lower bounds on monotone complexity of the logical permanent. Mathematical Notes 37(6), 485–493 (1985)
17. Håstad, J.: The shrinkage exponent of de morgan formulas is 2. SIAM Journal on Computing (1998)
18. Vollmer, H.: Introduction to Circuit Complexity: A Uniform Approach. Springer New York Inc. (1999)

Efficient Respondents Selection for Biased Survey Using Online Social Networks*

Donghyun Kim[1], Jiaofei Zhong[2], Minhyuk Lee[1], Deying Li[3],
and Alade O. Tokuta[1]

[1] Dept. of Math. and Physics, North Carolina Central University,
Durham, NC 27707, USA
{donghyun.kim,atokuta}@nccu.edu, mlee28@eagles.nccu.edu
[2] Dept. of Math. and Computer Science, California State University, East Bay,
Hayward, CA 94542, USA
fayzhong08@gmail.com
[3] School of Information, Renmin University of China, Beijing 100872, China
deyingli@ruc.edu.cn

Abstract. Online social networks are getting lots of attentions from the research communities since they are rich sources of data to learn about the members of our society as well as the relationship among them. With the advances of Internet related technologies, online surveys are established as an essential tool for a wide range of applications. One significant issue of online survey is how to select a good respondent group so that the survey result is reliable. This paper investigates the use of online social network to form a biased survey respondent group which is useful for certain applications. We formally introduce a new optimization problem called the *minimum inverse k-core dominating set problem (MIkCDSP)* for this purpose, show its NP-hardness, and finally and mostly importantly introduce a greedy approximation algorithm for it.

1 Introduction

Recently, online social networks are receiving lots of attentions from the research communities due to the growing popularity of social networking web sites such as Facebook, Twitter, Google+, etc. It is widely recognized that the online social networks are rich sources of data to learn about the interest of each user as well as the relationship among them. Due to the reason, online social networks are investigated for a wide range of applications such as shared interest discovery among users [4], information propagation [6,1], online advertising [5], efficient information propagation [7], community clustering [2], and so on.

* This work was supported in part by US National Science Foundation (NSF) CREST No. HRD-1345219. This research was jointly supported by National Natural Science Foundation of China under grant 91124001, the Fundamental Research Funds for the Central Universities, and the Research Funds of Renmin University of China 10XNJ032.

Z. Cai et al. (Eds.): COCOON 2014, LNCS 8591, pp. 608–615, 2014.

These days, online surveys are established as an essential tool for a wide range of applications such as marketing and political decision making. It is known that in 2006, around 20% of global data-collection expenditure was spent for online survey research [8]. In 2012, US spent more than $1.8 billion for all survey research spending [9]. There are a number of reasons, not to mention its low cost (than the traditional methods), that online survey becomes so popular [10]. In online survey researches, how to find a right sample group of respondents is a long lasting conundrum since this is directly related to the reliability of the survey. Frequently, a biased respondent group is considered to be lack of its reliability. This is because the result from the surveys are mainly used to obtain a statistical information about the general public by consulting with a sample group from the public, the survey result from a sample group which lost its representative is not reliable for this purpose. Due to the reason, many efforts are made to find a representative and unbiased respondent group [9].

Interestingly enough, however, we observe the bias in the survey is not always something to avoid. Consider a product quality manager of a new smartphone, e.g. iPhone 5c, who wants to collect the feedback via an online survey from users so that he/she can improve the quality of the product. Also, suppose most of the customers using the product are happy with it. Then, while the manager is only interested in hearing complaints from the users, it is likely that the online survey result from the respondents selected by the methods whose common goal is to make the result representative and unbiased, would be mostly about their satisfaction about the new product. As a result, such survey is quite wasteful in practice to the manager who is only interested in complaints. Therefore, it would be helpful to form a biased respondent group so that it includes more unsatisfied users.

In this paper, we investigate the use of online social networks to compute a biased but representative respondent group such that the rate of the minority opinion group (e.g. those who are not satisfied with the product) in the respondent group can be magnified. To the best of our knowledge, this is the first effort in the literature which exploits online social network to enhance to the quality of online survey. The rest of this paper is organized as follows: In Section 2, we introduce several important notations and definitions. Especially, we introduce the formal definition of our problem of interest, the *minimum inverse k-core dominating set problem (MIkCDSP)*, corresponding justification, and its NP-hardness result. Section 3 proposes a new greedy approximation algorithm for MIkCDSP. Finally, we conclude this paper in Section 4.

2   Notations, Definitions, and Problem Statement

2.1   Notations and Definitions

In this paper, $G = (V, E)$ represents an online social network graph with a node set $V = V(G)$ and an edge set $E = E(G)$. We assume the relationship between the members are symmetric and thus the edges in E are bidirectional. Also, we use n to denote the number of nodes in V, i.e. $n = |V|$. For any subset $D \subseteq V$,

$G[D]$ is a subgraph of G induced by D. For each node $v \in V$, $N_{v,V}(G)$ is the set of nodes in V neighboring to v in G. Now, we introduce some important definitions.

Definition 1 (DS). *Given a graph G, a subset $D \subseteq V$ is a dominating set (DS) of G if for each node $u \in V \setminus D$, $\exists v \in D$ such that $(v, u) \in E$.*

Definition 2 (MDSP). *Given a graph G, the goal of the minimum dominating set problem (MDSP) is to find a minimum size DS of G.*

Definition 3 (Inverse k-core). *Given a graph G, a subset $D \subseteq V$, and a positive integer k such that $0 \leq k \leq \Delta$, where Δ is the degree of G, D is an inverse k-core in G if for each $v \in D$, $|N_{v,D}(G)| \leq k$.*

2.2   Problem Statement

In this paper, we study an online survey sample (respondents) selection problem such that the rate of people with minority opinion in the sample can be higher than their rate in the overall group.

We claim that people who share similar opinions have better chance to be a friend with each other in the online social networking, which is frequently true in the professional social networks. Suppose for a survey, there exists an online social network relevant to the survey topic. For instance, for the survey on a new smartphone, we assume the existence of some online social network among the technicians. Then, based on our assumption, even though we do not know the opinion of each user in the social network regarding the smartphone, we can assume that two neighboring users in the social network have a smaller chance to have two drastically opposite opinion on the product.

A randomly selected DS of the whole group might be one approach to compute a good representative group since any node in the whole group either is a member of the DS or has a close friend who share similar opinion in the DS. However, note that while the DS has a representativeness, it is quite hard to tell if the DS is biased or not, and if biased, how much it is biased.

Based on our previous discussion, we claim the bias of the DS can be observed by checking its cohesiveness. That is, if there exists a clear majority opinion group in the overall group and the DS is completely randomly selected, then it is likely that the rate of the majority opinion group in the overall group is similar to the rate of the majority group in the DS. Furthermore, they will be appeared as a well-connected subgraph with relatively larger size in the social network graph induced by the DS. Meanwhile, there can be one or more well-connected subgraph with relatively smaller size in the graph, each of which represents a unique minority opinion group in the DS.

This observation implies that when we select a DS for the respondents if the degree of the induced graph by the DS is limited, then, the DS will include more amount of non-majority opinion group members. Formally, such a DS can be defined as the inverse k-core dominating set (IkCDS) showed below, where k is the degree of bias (with higher k, the DS is less biased, and with k equivalent to the degree of the social network, the DS is completely unbiased).

Definition 4 (IkCDS). *Given a graph G, a subset $D \subseteq V$, and a positive integer k, D is an* inverse k-core dominating set (IkCDS) *of G if (a) D is a DS of G and (b) for each $v \in D$, $|N_{v,D}(G)| \leq k$.*

It is noteworthy that there are a number of ways to compute IkCDS of a social network. Apparently, it is more desirable to reduce the size of IkCDS since it will cost less for the actual survey. As a result, the problem of computing a biased online survey respondent group can be formulated as MIkCDSP shown below.

Definition 5 (MIkCDSP). *Given a graph G and a positive integer k, the goal of the* minimum inverse k-core dominating set problem (MIkCDSP) *is to find a minimum size IkCDS of G.*

Remark 1. It is noteworthy that as k decreases, the DS will be more biased and the rate of minority opinion in the survey will increase. At the same time, the size of the IkCDS will decrease. This means that with very small k value, the survey respondent set can be very small and less practical given that the usual degree of social networks is not small. On the other hand, with very high k value, the survey respondent group can be negligibly biased, which also may not be desirable for our application. While selecting proper k value is very significant, it is also application dependent. Since this question is the out of this paper, we assume that k value is given as a part of the inputs of the problem.

The below theorem shows our problem is NP-hard.

Theorem 1. *MIkCDSP is NP-hard.*

Proof. A special case of MIkCDSP with $k = n$ is equivalent to the minimum dominating set problem, the problem of computing a minimum size dominating set of G, which is proven to be NP-hard [3]. As a result, MIkCDSP is NP-hard.

Remark 2. Given any graph G and a non-negative integer k, there exists a feasible solution of MIkCDSP in G. This claim is true since (a) a feasible solution of MIkCDSP with $k = 0$ is clearly a feasible solution of MIkCDSP with any $k \geq 1$, and (b) the following coloring strategy can be used to compute an independent set of G, the subset of nodes in G which are pairwise disjoint with each other, which is a feasible solution of MIkCDSP with $k = 0$: (i) initially color all nodes white, (ii) pick each white node black and its neighbors in gray until there is not white node left, and (iii) return the set of black nodes. Clearly, the set of black nodes is a dominating set and each pair of black nodes are not neighboring from each other.

3   Greedy-MIkCDSA: A Simple Greedy Approximation for MIkCDSP

In this section, we introduce Greedy-MIkCDSA, a simple greedy strategy for MIkCDSP and show that its performance ratio is $(1 + \Delta)$, where Δ is the degree of

---

Algorithm 1. Greedy-MIkCDSA $(G = (V, E), k)$

1: Prepare an empty set D, i.e. $D \leftarrow \emptyset$.
2: For each $v_i \in V$, prepare a counter n_i which is initialized to 0, i.e. $n_i \leftarrow 0$.
3: Suppose $X_j = \{v_i | v_i \in V \text{ and } n_i = j\}$.
4: **while** $X_0 \neq \emptyset$ **do**
5:    Find $v_i \in V \setminus \left((\bigcup_{j \geq k} X_j) \bigcup D \bigcup Q \right)$ so that $|N_{v_i, X_0}(G)|$ is maximized, where
      $Q = \{w_1, \cdots, w_q\}$ such that $w_l \in Q$ has at least one neighbor in $(\bigcup_{j \geq k} X_j)$ and
      $w_l \in D$ is true. A tie can be broken arbitrarily.
6:    Set $D \leftarrow D \cup \{v_i\}$.
7:    **for** each node $v_j \in N_{v_i, V}(G)$ **do**
8:       $n_j \leftarrow n_j + 1$.
9:    **end for**
10: **end while**
11: Output D.

---

the input online social network graph. The formal description of Greedy-MIkCDSA is Algorithm 1. Given an MIkCDSP instance $\langle G, k \rangle$, Greedy-MIkCDSA first prepares an empty set D (Line 1), which will eventually include the output, an inverse k-core dominating set (IkCDS) of G. For each node $v_i \in V$, we create a counter n_i which is initialized to 0 (Line 2). The counter will be used to track the number of neighbors of v_i in D. Depending on the counter, we create a partition of the nodes in V, $X_0, X_1, \cdots$, where X_j is the subset of nodes in V whose counter is j (Line 3). This means that initially X_0 is equal to V and each of the rest is empty. Clearly, the number of the subsets is bounded by n. From Lines 4-10, we iteratively pick a node v_i from $\left((\bigcup_{j \geq k} X_j) \bigcup D \bigcup Q \right)$, i.e. v_i is a node which is

- **Condition 1**: with a counter n_i whose value is less than k (i.e. has less than k neighbors in DS),
- **Condition 2**: not selected as a DS node yet, and
- **Condition 3**: without any neighboring node w_l which is in D and, at the same time, in X_j for some $j \geq k$,

such that the number of neighbors of v_i in X_0 is the maximum. Any tie can be broken arbitrarily. This loop is repeated until all nodes in V is either in D or dominated by some node in D while maintaining $G[D]$ as an inverse k-core.

Clearly, Algorithm 1 produces a feasible solution of MIkCDSP since the algorithm repeatedly constructs D until X_0 becomes empty (which means D is a DS of G) and by Line 5, the degree of $G[D]$, the graph induced by D in G, will be bounded by k (which means D is an inverse k-core). One may wonder if there is a situation in which some node x, which has to be included in D to dominate some other node y, cannot be included in D since it has already k neighbors in D. However, this never becomes a problem since if x cannot be selected, then y itself will be included in D by our algorithm, which means that D is always a valid output.

Now, we show Algorithm 1 is a $(1+\Delta)$-approximation algorithm for MIkCDSP.

Lemma 1. *Given a graph* $G = (V, E)$, *let* OPT_{MDSP} *and* OPT_{MIkCDS} *be an optimal solution of MDSP and an optimal solution of MIkCDSP defined over* $\langle G, k \rangle$ *for some* $k \geq 1$, *respectively. Then,*

$$|OPT_{MDSP}| \leq |OPT_{MIkCDS}|.$$

Proof. By definitions, the goal of MDSP is to find a DS of G with minimum cardinality and the goal of MIkCDSP is to find a DS of G with minimum cardinality such that for each node in the DS, the node is allowed to be adjacent with at most k other nodes in the DS. Therefore, in any given G, an IkCDS of G is also a DS of G, but our choice of IkCDS is more limited than that of DS. As a result, this lemma is true.

Lemma 2. *Given a graph* $G = (V, E)$, *suppose we have an* α-*approximation algorithm of MDSP such that the output* O *of the algorithm is also a feasible solution of MIkCDSP. Then, we have* $|O| \leq \alpha|OPT_{MIkCDS}|$.

Proof. By the definition of an α-approximation algorithm of MDSP, we have $|O| \leq \alpha|OPT_{MDSP}|$. By combining this with Lemma 1, we have $|O| \leq \alpha|OPT_{MDSP}| \leq \alpha|OPT_{MIkCDS}|$, and thus this lemma is true.

Recall that Algorithm 1 produces a feasible solution of MIkCDSP defined over $\langle G, k \rangle$ which is also a feasible solution of MDSP defined over G. Therefore, by Lemma 2, we can obtain the performance ratio of Algorithm 1 for MIkCDSP by bounding the ratio between the size of an output of Algorithm 1 and the size of an optimal DS.

Theorem 2. *The performance ratio of Algorithm 1 for MIkCDSP is* $1 + \ln \Delta$, *where* Δ *is the maximum degree of* G.

Proof. Given $G = (V, E)$ and k, consider $OPT_{MDSP} = \{o_1, o_2, \cdots, o_l\}$ be a minimum DS of G. Then, for each $o_i \in OPT_{MDSP}$ in the increasing order of i, we compute

$$P_1 = \{o_1\} \bigcup N_{o_1, V \setminus OPT_{MDSP}}(G), \text{ and}$$
$$P_i = \left(\{o_i\} \bigcup N_{o_i, V \setminus OPT_{MDSP}}(G)\right) \setminus \left(\bigcup_{1 \leq j \leq i-1} P_j\right) \text{ for } i \neq 1.$$

Then, V is partitioned into $\mathcal{P} = \{P_1, P_2, \cdots, P_l\}$ such that each $P_i \in \mathcal{P}$ exactly includes one $o_i \in OPT_{DS}$.

Suppose Algorithm 1 is applied to $\langle G, k \rangle$ and outputs D. Then, each P_i can include some nodes in D. During the rest of this proof, we will try to find the upper bound of the size (i.e. the number of nodes) of $P_i \cap D$. If we can bound this size by α, we have

$$|D| \leq \max_{1 \leq i \leq l} |P_i \cap D| \cdot |OPT_{MDSP}| = \alpha \cdot |OPT_{MDSP}|.$$

Remember that D is also an IkCDS. Therefore, by Lemma 2, we have

$$|D| \leq \alpha \cdot |OPT_{MDSP}| \leq \alpha \cdot |OPT_{MIkCDS}|,$$

which will complete this proof.

To obtain the upper bound of $|P_i \cap D|$, we consider the following strategy: whenever a node $v \in P_i$ is selected as a member of D by Algorithm 1, we assume each neighbor $u \in (P_i \cap X_0)$ of v immediately (before updating its counter) receives an additional weight $w(u)$, which is equivalent to one divided by the number of neighbors of v in $(P_i \cap X_0)$, i.e.

$$w(u) \leftarrow w(u) + \frac{1}{N_{v,(P_i \cap X_0)}(G)}.$$

Clearly, $\sum_{v \in P_i} w(v) = |P_i \cap D|$.

Next, we show that $\sum_{v \in P_i} w(v) = 1 + \ln \Delta$. If $P_i \cap D = \emptyset$, then this proof is trivial, and thus we assume $P_i \cap D \neq \emptyset$. Let $P_i \cap D = \{z_1, z_2, \cdots, z_p\}$. Also, let X_0 be the set of nodes in 'P_i' whose counter is 0, i.e. has no neighbor in D, yet. Note that each time, a node is selected by Algorithm 1 using the greedy strategy and added to D, there will be less number of nodes left in X_0. Let us use $X_0^{(0)}, X_0^{(1)}, \cdots, X_0^{(p)}$, where $X_0^{(i)}$ is the remaining nodes in X_0 after ith iteration of while loop (Line 4-9 in Algorithm 1). Then, we have

$$|X_0^{(0)}| \geq |X_0^{(1)}| \geq \cdots \geq |X_0^{(p)}|. \tag{1}$$

Note that for any j, $|X_0^{(j-1)}| - |X_0^{(j)}|$ is the number of nodes removed from $X_0^{(j-1)}$ after jth iteration. In other word, $|X_0^{(j-1)}| - |X_0^{(j)}|$ is the number of nodes in $X_0^{(j-1)}$, which are not adjacent to any node in $\{z_1, z_2, \cdots, z_{j-1}\}$ yet, and at the moment that z_j is selected, they are adjacent to z_j.

Suppose the initial iteration is executed and z_1 is selected and added to D. Then, the weight added to each neighbor of z_1 in $P_i \cap X_0^{(0)}$ is $1/N_{z_1,(P_i \cap X_0^{(0)})}$ and the number of such nodes is $|N_{z_1,(P_i \cap X_0^{(0)})}|$. In general, after jth iteration, the weight added to each neighbor of v_j in $P_i \cap X_0^{(j-1)}$ is $1/N_{v_j,(P_i \cap X_0^{(j-1)})}$ and the number of such nodes is $|N_{v_j,(P_i \cap X_0^{(j-1)})}|$. Since we are using a greedy strategy, z_j is always neighboring more nodes in $P_i \cap X_0^{(j-1)}$ than $o_i \in OPT_{DS}$. Therefore, we have

$$|N_{z_j,(P_i \cap X_0^{(j-1)})}| \geq |N_{o_i,(P_i \cap X_0^{(j-1)})}|,$$

which implies

$$\frac{1}{|N_{z_j,(P_i \cap X_0^{(j-1)})}|} \leq \frac{1}{|N_{o_i,(P_i \cap X_0^{(j-1)})}|}.$$

Since o_i is adjacent to all nodes in P_i, $N_{o_i,(P_i \cap X_0^{(j-1)})} = X_0^{(j-1)}$. As a result, after the iteration is repeated for p times. we have

$$\sum_{v \in P_i} w(v) \leq \sum_{1 \leq j \leq p} \frac{|X_0^{(j-1)}| - |X_0^{(j)}|}{|X_0^{(j-1)}|}. \tag{2}$$

By Eq. (1), we have $|X_0^{(j-1)}| - |X_0^{(j)}| > 0$ for all j. Finally, p can be bounded by Δ since all nodes in P_i has to be adjacent to o_i. As a result, the second term of the right side of Eq. (2) can be bound by $H(\Delta)$, where H is a harmonic function. As a result, we have

$$\sum_{v \in P_i} w(v) \leq 1 + H(\Delta) \simeq 1 + \ln \Delta,$$

and this theorem is true.

4   Conclusion

In this paper, we introduce a new approach to use the information from an online social network to enhance the result of online survey. To perform this task efficiently, we introduce to solve a new NP-hard optimization problem, propose a new greedy heuristic algorithm for it, and show the algorithm in fact has a theoretical performance guarantee. To the best of our knowledge, this is the first attempt to use online social network to improve the result of online survey. We plan to further investigate the use of social network to improve the reliability of online voting systems. In this paper, we assume the existence of a single social network for survey. However, in reality, there could be more than one social networks which can be used for this kind of computation. Also, it would be very interesting to consider a social network with weighted edges.

References

1. Zhang, H., Dinh, T.N., Thai, M.T.: Maximizing the Spread of Positive Influence in Online Social Networks. In: Proc. of the IEEE Int. Conference on Distributed Computing Systems, ICDCS (2013)
2. Kim, D., Li, D., Asgari, O., Li, Y., Tokuta, A.O.: A Dominating Set Based Approach to Identify Effective Leader Group of Social Network. In: Du, D.-Z., Zhang, G. (eds.) COCOON 2013. LNCS, vol. 7936, pp. 841–848. Springer, Heidelberg (2013)
3. Garey, M.R., Johnson, D.S.: Computers and Intractability: A Guide to the Theory of NP-completeness. Freeman, San Francisco (1978)
4. Wang, F., Xu, K., Wang, H.: Discovering Shared Interests in Online Social Networks. In: International Workshop on Hot Topics in Peer-to-peer Computing and Online Social Networking (July 2012)
5. Kahl, C., Crane, S., Tschersich, M., Rannenberg, K.: Privacy Respecting Targeted Advertising for Social Networks. In: Ardagna, C.A., Zhou, J. (eds.) WISTP 2011. LNCS, vol. 6633, pp. 361–370. Springer, Heidelberg (2011)
6. Zhang, W., Wu, W., Wang, F., Xu, K.: Positive Influence Dominating Sets in Power-Law Graphs. Social Network Analysis and Mining 2(1), 31–37 (2012)
7. Cha, M., Mislove, A., Gummadi, K.: A Measurement-driven Analysis of Information Propagation in the Flickr Social Network. In: Proc. of the 18th International Conference on World Wide Web (WWW), pp. 721–730 (2009)
8. Vehovar, V., Manfreda, K.L.: Overview: Online Surveys. In: Fielding, N.G., Lee, R.M., Blank, G. (eds.) The SAGE Handbook of Online Research Methods, pp. 177–194. SAGE, London (2008)
9. Terhanian, G., Bremer, J.: A Smarter Way to Select Respondents for Surveys? International Journal of Market Research 54(6), 751–780 (2012)
10. Duffy, B., Smith, K., Terhanian, G., Bremer, J.: Comparing Data from Online and Face-to-face Surveys. International Journal of Market Research 47(6), 615–639 (2005)

Explaining Snapshots of Network Diffusions: Structural and Hardness Results

Georgios Askalidis, Randall A. Berry, and Vijay G. Subramanian

Northwestern University
Evanston, IL, USA

Abstract. Much research has been done on studying the diffusion of ideas or technologies on social networks including the *Influence Maximization* problem and many of its variations. Here, we investigate a type of inverse problem. Given a snapshot of the diffusion process, we seek to understand if the snapshot is feasible for a given dynamic, i.e., whether there is a limited number of nodes whose initial adoption can result in the snapshot in finite time. While similar questions have been considered for epidemic dynamics, here, we consider this problem for variations of the deterministic Linear Threshold Model, which is more appropriate for modeling strategic agents. Specifically, we consider both sequential and simultaneous dynamics when deactivations are allowed and when they are not. Even though we show hardness results for all variations we consider, we show that the case of sequential dynamics with deactivations allowed is significantly harder than all others. In contrast, sequential dynamics make the problem trivial on cliques even though it's complexity for simultaneous dynamics is unknown. We complement our hardness results with structural insights that can lead to better understanding of diffusions on social networks under various dynamics.

Keywords: social influence, linear threshold model, NP-hardness.

1 Introduction

Diffusion processes have been widely studied both theoretically and empirically. One of the main theoretical frameworks is based on modeling a diffusion as the result of a network game, i.e., a model in which rational agents make decisions to maximize a pay-off that depends on the actions of other agents in way that depends in part on an underlying network structure. The network game that we assume in this paper is one where agents are called upon at each discrete time point to make a rational decision based on previous decisions of other neighboring agents. We can think of the action to be concerning the adoption or not of a new technology and assume that all agents start the game with the status quo technology (we will also refer to this state as "deactivated" through out the paper). Moreover, we assume that each agent has a non-negative integer threshold which represents the number of her neighbors in the network that need to adopt the new technology (or "activate" as we refer to that action throughout the paper) in order for her utility to be maximized by her also choosing to adopt.

Z. Cai et al. (Eds.): COCOON 2014, LNCS 8591, pp. 616–625, 2014.

In an influence maximization setting this model translates to the widely used *Linear Threshold Model*.

The problem that we study in this paper is a generalization of the TARGET SET problem introduced in [4], since alongside the network graph G, integer budget k and thresholds $t_1, t_2, \ldots, t_n$, we are also given a subset $S \subseteq V(G)$ that we call a *Snapshot*. We seek to find an initial seed set of size at most k that leads, in finite time, to the activation of *exactly* S. For example, this could model a scenario where a snapshot is observed and one seeks to determine the set of nodes that could have started the underlying diffusion. We call this problem the SNAPSHOT problem, and we study four variations of it. When $S = V(G)$, then the SNAPSHOT problem becomes the TARGET SET problem since we are looking to activate the whole graph.

We consider two order-dynamics for our network game setting. In the *simultaneous* (or *parallel*) best-response process, at each point in time, all agents best respond to the state of the network simultaneously while in the *sequential* best-response process we chose only one agent to best respond at each point in time.

In addition to the linear threshold model, other widely used models for diffusions in social networks are the *Independent Cascade (IC), Susceptible-Infected (SI), Susceptible-Infected-Susceptible (SIS)* and *Susceptible-Infected-Recovered (SIR)* models. We refer to [6] for more information on these models.

In the same spirit as the SI and SIR models, for each of the two order-dynamics (simultaneous and sequential), we consider two variations of our problem: one that forces the agents we choose in the seed set to commit to remain activated forever and one that allows them to deactivate at a later stage if such an action maximizes their utility. Note that this restriction concerns only the nodes in the initial seed set and that all other nodes always best-respond and so are allowed to deactivate at any point in time in both settings. When we force the seed set to commit to remain activated, the set of activated nodes can only grow (weakly) larger at each time step and so we call this case *monotone*. Hence we get four variations of the SNAPSHOT problem: MONOTONE SIMULTANEOUS SNAPSHOT, SIMULTANEOUS SNAPSHOT, MONOTONE SEQUENTIAL SNAPSHOT and SEQUENTIAL SNAPSHOT.

In this work, we start by exploring the connections between feasible snapshots under various dynamics and then show that when we are looking for a single initial adopter we can restrict our attention to the closed neighborhood of the given snapshot. Moreover, in the same case, when trying to find an ordering that produces a given snapshots we can ignore all nodes that are not in the snapshot. We then provide various hardness results for all four variations of the problem, most notably that SEQUENTIAL SNAPSHOT is NP-hard even for $k = 1$. Finally, we take an interest in the special case of cliques, a graph structure not studied as much in related literature, and show that even though SEQUENTIAL SNAPSHOT problem becomes easy to solve, the situation is much more complicated under simultaneous dynamics.

As noted previously, one branch of related work is on variations of the IN-FLUENCE MAXIMIZATION PROBLEM [9]: Given a graph G, threshold vector t and a budget k, choose k nodes to activate, in order to maximize the number of infected nodes. Such problems can be motivated by marketing scenarios where one tries to target specific influential persons by, e.g., giving them some kind of an offer or a free product, with the goal of making the product as popular as possible. Strong hardness and inapproximability results have been shown for even the special case when all agents have threshold 2 [4], [11]. Other related problems have been studied as well. For example, [1], defines a notion of influence for bloggers in the web and studies the problem of identifying the most influential bloggers. Similarly, in [12] the authors define the notions of "starters" and "followers" in social media and try to identify agents from each set. In a different spirit, [13] and [18] study the problem of determining the edges of the network given the activation times of the agents.

The other branch of related work seeks to find the source of a diffusion modeled as arising from a probabilistic epidemic process. Shah and Zaman, [19], use the SI model and propose a measure they call rumor-centrality to find the single source of a rumor spread. Prakash et al. in [17] study the same problem as us but under the probabilistic SI model and they provide experimentally tested heuristics. Similar work, under the IC model in the context of finding users suspected of providing misinformations, has been done in [14]. Lappas et al. in [10] study the problem of finding the initial set that best explains a given snapshot in a network. For each set of nodes, they define a cost function that represents the difference of the expected final set of activated nodes and the actual observed snapshot, and try to minimize that function. Finally, assuming that information propagates in a social network following the IC model, Gundecha et al. in [8] study the problem of finding initial sources as well as other recipients of some information given only a small fraction of the recipients of the information. Even though the problem is NP-hard, they provide an efficient heuristic algorithm that they test with real social media datasets. The main difference of our work from this second body of work is the use of the deterministic Linear Threshold Model in contrast to the stochastic IC and SI ones. This can result in significantly different dynamics.

2   Model

We call the general problem we study the SNAPSHOT problem. The input is a tuple (G, S, t, k) where $G = (V, E)$ is an undirected network graph, $S \subseteq V(G)$ is a set of nodes we call the *snapshot*, $t = (t_1, t_2 \ldots, t_n)$, for $n = |V(G)|$, is a vector of non-negative integer *thresholds*, and k is a positive integer that we call the *budget*. The goal is to find a set $S_0 \subseteq V(G)$ of size at most[1] k, that we will call the *initial activated set* or *seed set*, whose activation will, in finite time, cause the

---

[1] Note that the existence of a seed set of size $\leq k$ does not necessarily imply the existence of a seed set of size exactly k, since here we take care to activate only S and nothing else.

activation of *exactly* S for some valid sequence of best responses.[2] . If such S_0 exists for S, we will call S a *valid* snapshot. Depending the order dynamics used we get the SIMULTANEOUS SNAPSHOT and SEQUENTIAL SNAPSHOT problems. In these versions we don't force the agents in the seed set to commit to remain activated forever and hence they can best respond by deactivating at any time point. When we do force the nodes in S_0 to remain activated forever we get the *monotone* version of each of the two problems, which we will call MONOTONE SIMULTANEOUS SNAPSHOT and MONOTONE SEQUENTIAL SNAPSHOT, respectively.

Of particular interest will be the case where $k = 1$ and hence $S_0 = \{u_0\}$. We will then just say that u_0 is an *initial adopter* for S. It's important to clarify a point here. We do not need the snapshot S to be the *final* state of the activation triggered by S_0. Any S_0 that in finite time t will produce *exactly* S is considered to be an initial seed set for S *even if at time $t + 1$ more nodes will be added to or removed from S*.

Example 1. Suppose that our input graph and thresholds are as shown in Figure 1, our budget is $k = 2$, and we use monotone simultaneous dynamics. It can be seen that snapshot $S_1 = \{u_1, u_2, u_3\}$ is feasible since we can activate $\{u_1, u_3\}$ at time 0, which will activate node u_2 at time 1. We don't mind that at time 2 node u_4 will be activated as well. In contrast, snapshot $S_2 = \{u_1, u_3, u_4\}$ is not feasible for $k = 2$.

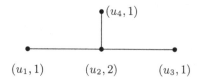

Fig. 1. Example of a network and thresholds. The notation (u_i, t_i), which is used throughout the paper, denotes that node u_i has threshold t_i.

Given that $S = V(G)$, it's implicit from Kempe et al., [9], that finding an influential set of initial adopters of minimum size is NP-hard and later Chen, [4], showed that the problem is hard to approximate within a polylogarithmic factor even when all the thresholds are equal to 2. These will be our starting point towards our hardness results in Section 4.

3   Structural Results

In this section we present various results concerning the structure of the snapshots and seed sets under the various dynamics. Due to space limitations, all proofs are deferred to the full version [2].

---

[2] Note in the case of simultaneous dynamics the sequence of best responses is unique, while for sequential dynamics there are multiple possibilities; we only require that S be activated under one such sequence.

3.1   Feasible Snapshots of Sequential and Simultaneous Dynamics

We start by understanding the relationship between the feasible snapshots under sequential and simultaneous dynamics. We then show similar results between monotone and non-monotone dynamics.

Lemma 1. *Let (G, S, t, k) be an instance of the* SNAPSHOT *problem. If S is feasible for (G, t, k) under monotone simultaneous dynamics then it's also feasible for (G, t, k) under monotone sequential dynamics.*

As the following example shows, the reverse is not true. There are snapshots that are feasible under sequential dynamics (monotone or non-monotone) that cannot be created under simultaneous (monotone or non-monotone) dynamics.

Example 2. Consider the graph and thresholds shown in Figure 1 and assume we have $k = 1$, i.e., we are allowed only one initial adopter.

It can be seen that the snapshot $S_1 = \{u_2, u_3\}$ is feasible under monotone or non-monotone sequential dynamics since we can activate u_2 at time 0 and at time 1 choose u_3 to best respond but S_1 is not feasible under monotone nor non-monotone simultaneous dynamics.

Even though the above example shows that the set of feasible snapshots for sequential and simultaneous dynamics are not the same, the next lemma shows that when we are looking to activate the whole graph, i.e., when $S = V(G)$, and there are no deactivations allowed, then the dynamics are indeed equivalent.

Lemma 2. *Let $(G, t, S = V(G), k)$ be an instance of the* SNAPSHOT *problem. Then S is feasible under monotone sequential dynamics if and only if it is feasible under monotone simultaneous dynamics.*

The key difference that makes Lemma 2 work for $S = V(G)$ but not in general is that when $S = V(G)$ we don't have the issue of *over-activating*, i.e., activating more nodes than are in the snapshot. As can be seen by chosing u_2 as an initial adopter in Example 2, this can occur when S is a strict subset of $V(G)$. Under sequential dynamics we can ensure that a node is not activated by simply never selecting it to best respond, while with simultaneous dynamics we do not have that freedom.

No containment relation holds between the sets of feasible snapshots under simultaneous and sequential dynamics when we don't require monotonicity, as shown in the next example.

Example 3. Assume we have the same graph and thresholds as shown in Figure 1 and $k = 1$. As shown in Example 2, $S_1 = \{u_2, u_3\}$ is feasible under sequential dynamics but not under simultaneous.

In the same example, it can be seen that $S_4 = \{u_1, u_3, u_4\}$ is feasible under non-monotone simultaneous dynamics since we can choose u_2 as our seed. In the next round, u_1, u_3 and u_4 will activate since they had the appropriate number of activated neighbors in the previous round, and u_2 will deactivate since it had

0 activated neighbors in the previous round. In contrast, it can be seen that there is no seed set of size 1 that can produce S_4 under sequential dynamics (monotone or non-monotone).

As Lemma 3 in [2] shows, if S is feasible under monotone sequential dynamics then it's also feasible under non-monotone sequential dynamics. The reverse is not true though, as shown in Example 4 in [2]. No containment relation holds between the feasible snapshots of monotone and non-monotone simultaneous dynamics, as shown in Examples 5 and 6 in [2].

3.2   Distance between the Seed Set and the Snapshot

We study next the distance that a seed set can have from an observed snapshot. Clearly, when we allow for no deactivations the seed set must be part of the observed snapshot and hence the distance is zero. When we have non-monotone sequential dynamics, we show below that for the case of $k = 1$, the seed set can have distance at most 1 from the observed snapshot. We then show that this is not true when we have $k \geq 2$ or non-monotone simultaneous dynamics.

Lemma 3. *Let $I = (G, S, t, 1)$ be an instance of the SNAPSHOT problem. If S is valid under sequential dynamics, then there exists an initial adopter u_0 for S in $N[S]$.*

The main idea behind the proof is that we follow the activation path from a node in S back to u_0 and then argue that this path cannot be deactivated because every node u in that path has at least one 'down stream' neighbor v that it is responsible for activating, i.e., v was activated after u and it's activation required u to be active. Therefore, even if u_0 deactivates, v will ensure that u still has the appropriate number of activated neighbors. Examples 7 and 8 in [2] show that the assumptions of sequential dynamics and $k = 1$ respectively, are necessary for Lemma 3 to hold.

3.3   The Clearing Lemma

Lemma 3 states that when we are looking for a single initial adopter, we can just look in the neighborhood of the snapshot S. We show next, that we can reduce the search space further by simply ignoring all nodes except u_0 that are not in S. This shows that in the case of sequential dynamics and $k = 1$, the activation cannot 'pass through' a node, i.e., use a node v to activate another node $u \in S$ and then leave v deactivated. Again, the result is trivially true when we have monotone dynamics and hence we concentrate on the non-monotone case.

Lemma 4. *Let $I = (G, t, S, 1)$ be an instance of SEQUENTIAL SNAPSHOT and $u_0 \in V(G)$. Then there is an ordering of $V(G)$ that produces S with u_0 as the initial adopter if and only if there is an ordering of $S \cup u_0$, that produces S with u_0 as the initial adopter. Moreover, no node in that ordering other than (possibly) u_0 ever gets deactivated.*

As Example 9 in [2] shows, Lemma 4 does not extend to the case where $k \geq 2$.

4 Hardness

In this section we study the computational complexity of the various versions of the SNAPSHOT problem discussed in this paper. This work extends the already rich literature on the hardness of the Influence Maximization problem, which was first formulated and proved to be NP-hard by Kempe et al. in [9], and TARGET SET (which is the minimization variant of the Influence Maximization problem) that was proved to be APX-hard even in cases of restricted threshold values by Chen in [4]. The decision versions of these two problems coincide and hence we have that TARGET SET is NP-hard. Some tractable cases have also been studied for the TARGET SET problem. Chen in [4] gives a linear-time algorithm for trees and Ben-Zwi et al. in [3] generalized the result by solving the problem in graphs of constant treewidth. Moreover, even though for any constant k the TARGET SET problem can be solved in $O(n^{k+1})$ time, the problem is W[2]-hard for undirected graphs [15] and W[P]-hard for directed graphs [7]. For more on parameterized complexity we refer to [16].

We use the NP-hardness of the TARGET SET problem, [9], [4], as our starting point for proving the following theorem.

Theorem 1. SIMULTANEOUS SNAPSHOT, MONOTONE SIMULTANEOUS SNAPSHOT *and* MONOTONE SEQUENTIAL SNAPSHOT *are all* NP-*hard, even for the case that all thresholds are less than or equal to 2.*

The result follows from Lemmas 6, 7 and 8 in [2] and show the individual NP-hardness results for each of the three problems. All three reductions are *parameter preserving* and as such they carry over the W[2]-hardness result shown in [15] for the TARGET SET. Hence we get the following theorem as well.

Theorem 2. SIMULTANEOUS SNAPSHOT, MONOTONE SIMULTANEOUS SNAPSHOT *and* MONOTONE SEQUENTIAL SNAPSHOT *are all* W[2]-*hard when parameterized by the size of the solution, k.*

Finally, all three reductions are also *approximation preserving*, [5], and as such they carry over the approximation hardness shown by Chen, [4], for the TARGET SET.

Theorem 3. *The optimization version of* SIMULTANEOUS SNAPSHOT, MONOTONE SIMULTANEOUS SNAPSHOT, SEQUENTIAL SNAPSHOT *and* MONOTONE SEQUENTIAL SNAPSHOT *even for the case when all thresholds are less than or equal to 2, cannot be approximated within the ratio of* $O(2^{\log^{1-\epsilon} n})$, *for any fixed constant* $\epsilon > 0$, *unless* NP$\subseteq$DTIME$(n^{polylog(n)})$.

Nevertheless, for constant k, all three problems discussed so far are solvable in time $O(n^{k+1})$ by a brute force search, and hence are polynomial time solvable for $k = 1$. We show next that when we have non-monotone sequential dynamics the problem becomes NP-hard even for $k = 1$.

Theorem 4. *The* SEQUENTIAL SNAPSHOT *problem is* NP-*hard even for* $k = 1$.

Due to space limitations, we defer the proof to the full version [2].

5 SNAPSHOT Problem on Cliques

We briefly discuss here the case of the SNAPSHOT problem restricted on cliques, a special case that unlike others (especially trees) has received little attention in the literature. It's easy to notice that the TARGET SET problem can be solved efficiently on cliques. When trying to activate the whole clique, and because of the strong symmetry of the graph, the best way we can use our budget of k nodes is on the set of k nodes with the highest threshold.

In the SNAPSHOT problem though, we are interested in activating a specific $S \subseteq V(G)$ and nothing more. When we are under monotone sequential dynamics, the problem is still easy since we get to choose which nodes to best respond at each time period and hence we can just ignore all nodes that are not in the snapshot and then solve the problem on $G[S]$, the subgraph induced by S, which is easy to do.

However under monotone simultaneous dynamics, we need to be careful to not over-activate, and hence choosing the strongest seed set (i.e., the k nodes of highest thresholds) may not be optimal, as shown next.

Example 4. Suppose we have a clique of size 10 with thresholds as follows (in increasing order): $t_{u_1} = t_{u_2} = 1, t_{u_3} = t_{u_4} = 2, t_{u_5} = 3, t_{u_6} = 4, t_{u_7} = 5, t_{u_8} = 6, t_{u_9} = 7$ and $t_{u_{10}} = 8$. Suppose we are given the snapshot $S = \{u_1, u_2, \ldots, u_7\}$ and $k = 2$. Then activating the two nodes in S with the highest thresholds, i.e. u_6 and u_7, will activate four nodes, u_1, u_2, u_3, u_4, bringing the total number of activated nodes to 6. After that the remainder of S will be activated, but so will u_8 causing the snapshot to be overshot. If instead we activated u_5 and u_4, we would be able to activate exactly S. The intuition behind this is that we need to keep a balance between the nodes that we need to activate and the nodes that we should not.

We present here some properties that can be used to make the MONOTONE SIMULTANEOUS PROBLEM on cliques simpler. We leave as an open question if these properties can be used to provide provable guarantees on the size of the resulting instance. The proofs are deferred to the full version, [2].

Property 1 *If there exists a node $u \in S$ such that $t_u \geq |S|$, then u must be in the seed set if S is feasible.*

Property 2 *If there is a node $u \notin S$ such that $t_u \leq k$, then S is not feasible, unless $|S| = k$.*

Property 3 *If there are nodes $u \in S$ and $v \notin S$ such that $t_u = t_v$, then u must be part of the seed set if S is feasible.*

Property 4 *Let $t = min\{t_u | u \notin S\}$. Then we can remove all nodes from $V(G) \setminus S$ that have threshold higher than t.*

Property 5 *Let $t = min\{t_u | u \notin S\}$. If $|S| < t$, then S is feasible if and only if the k highest threshold nodes in S can activate S.*

6   Conclusions and Open Problems

In this paper we studied the problem of explaining given snapshots of a network diffusion, i.e., finding a small seed set whose activation will cause the activation of the snapshot. Although we presented strong hardness results for all variations we studied, we also presented a variety of structural results that can help better our understanding of this important problem. These structural results could potentially be used as part of heuristics and/or approximation algorithms. We leave several interesting directions open. One being the complexity of the MONOTONE SNAPSHOT on cliques. If it's proven to be hard, then the question of polynomial size kernels and approximation algorithms arises, and our results from Section 5 could potentially be used towards those directions. Another interesting question is how far can a seed set be from a given snapshot? Can the distance be as large as the diameter of the graph or is it upper bounded by a function of t_{max}, the maximum threshold of the graph, and/or the budget k.

Acknowledgements. We thank Nicole Immorlica and Ming-Yang Kao for useful discussions as well as two anonymous referees for their comments.

References

1. Agarwal, N., Liu, H., Tang, L., Yu, P.S.: Identifying the influential bloggers in a community. In: Proceedings of the 2008 International Conference on Web Search and Data Mining, pp. 207–218. ACM (2008)
2. Askalidis, G., Berry, R.A., Subramanian, V.G.: Explaining snapshots of network diffusions: Structural and hardness results. arXiv preprint arXiv:1402.6273 (2014)
3. Ben-Zwi, O., Hermelin, D., Lokshtanov, D., Newman, I.: An exact almost optimal algorithm for target set selection in social networks. In: Proceedings of the 10th ACM Conference on Electronic Commerce, pp. 355–362. ACM (2009)
4. Chen, N.: On the approximability of influence in social networks. SIAM Journal on Discrete Mathematics 23(3), 1400–1415 (2009)
5. Crescenzi, P.: A short guide to approximation preserving reductions. In: Proceedings of the Twelfth Annual IEEE Conference on Computational Complexity (Formerly: Structure in Complexity Theory Conference), pp. 262–273. IEEE (1997)
6. Easley, D., Kleinberg, J.: Networks, crowds, and markets, vol. 6(1). Cambridge Univ. Press (2010)
7. Eickmeyer, K., Grohe, M., Gruber, M.: Approximation of natural w [p]-complete minimisation problems is hard. In: 23rd Annual IEEE Conference on Computational Complexity, CCC 2008, pp. 8–18. IEEE (2008)
8. Gundecha, P., Feng, Z., Liu, H.: Seeking provenance of information using social media. In: Proceedings of the 22nd ACM International Conference on Information & Knowledge Management, pp. 1691–1696. ACM (2013)
9. Kempe, D., Kleinberg, J., Tardos, É.: Maximizing the spread of influence through a social network. In: Proceedings of the Ninth ACM SIGKDD International Conference on Knowledge Discovery and Data Mining, pp. 137–146. ACM (2003)
10. Lappas, T., Terzi, E., Gunopulos, D., Mannila, H.: Finding effectors in social networks. In: Proceedings of the 16th ACM SIGKDD International Conference on Knowledge Discovery and Data Mining, pp. 1059–1068. ACM (2010)

11. Lu, Z., Zhang, W., Wu, W., Fu, B., Du, D.: Approximation and inapproximation for the influence maximization problem in social networks under deterministic linear threshold model. In: 2011 31st International Conference on Distributed Computing Systems Workshops (ICDCSW), pp. 160–165. IEEE (2011)

12. Mathioudakis, M., Koudas, N.: Efficient identification of starters and followers in social media. In: Proceedings of the 12th International Conference on Extending Database Technology: Advances in Database Technology, pp. 708–719. ACM (2009)

13. Netrapalli, P., Sanghavi, S.: Learning the graph of epidemic cascades. ACM SIG-METRICS Performance Evaluation Review 40, 211–222 (2012)

14. Nguyen, D.T., Nguyen, N.P., Thai, M.T.: Sources of misinformation in online social networks: Who to suspect? In: Military Communications Conference, MILCOM 2012, pp. 1–6. IEEE (2012)

15. Nichterlein, A., Niedermeier, R., Uhlmann, J., Weller, M.: On tractable cases of target set selection. In: Cheong, O., Chwa, K.-Y., Park, K. (eds.) ISAAC 2010, Part I. LNCS, vol. 6506, pp. 378–389. Springer, Heidelberg (2010)

16. Niedermeier, R.: Invitation to fixed-parameter algorithms, vol. 3. Oxford University Press, Oxford (2006)

17. Aditya Prakash, B., Vreeken, J., Faloutsos, C.: Efficiently spotting the starting points of an epidemic in a large graph. Knowledge and Information Systems 38(1), 35–59 (2014)

18. Rodriguez, M.G., Balduzzi, D., Schölkopf, B.: Uncovering the temporal dynamics of diffusion networks. arXiv preprint arXiv:1105.0697 (2011)

19. Shah, D., Zaman, T.: Rumors in a network: Who's the culprit? IEEE Transactions on Information Theory 57(8), 5163–5181 (2011)

Pioneers of Influence Propagation in Social Networks

Kumar Gaurav[1], Bartłomiej Błaszczyszyn[2], and Paul Holger Keeler[3]

[1] UPMC/Inria/ENS, 23 av. d'Italie 75214 Paris, France
Kumar.Gaurav@inria.fr
[2] Inria/ENS, 23 av. d'Italie 75214 Paris, France
Bartek.Blaszczyszyn@ens.fr
[3] Inria/ENS, 23 av. d'Italie 75214 Paris, France
Holger.Keeler@inria.fr

Abstract. In this paper, we present a diffusion model developed by enriching the generalized random graph (a.k.a. configuration model), motivated by the phenomenon of viral marketing in social networks. The main results on this model are rigorously proved in [3], and in this paper we focus on applications. Specifically, we consider random networks having Poisson and Power Law degree distributions where the nodes are assumed to have varying attitudes towards influence propagation, which we encode in the model by their transmitter degrees. We link a condition involving total degree and transmitter degree distributions to the effectiveness of a marketing campaign. This suggests a novel approach to decision-making by a firm in the context of viral marketing which does not depend on the detailed information of the network structure.

1 Introduction

The penetration of internet and the emergence of huge online social networks in the last decade has lead to a decline of the conventional channels of communication and consequently, marketing through them. This has given the firms an opportunity to reach a large subset of their customers through innovative viral marketing campaigns. But the wild uncertainty inherent in whether a marketing campaign goes viral or not, makes it markedly different from conventional marketing and calls for a fundamentally different approach to decision-making.

1.1 Results

In this paper, we introduce a generalized diffusion dynamic on configuration model. Configuration model, while lacking the *community structure* of real-world social networks, approximates the degree distribution of these networks quite well. The diffusion dynamic that we consider can be intuitively described in the following way: an influenced individual in the network influences a random subset of its neighbours, the distribution of which depends on the effectiveness of the marketing campaign.

Z. Cai et al. (Eds.): COCOON 2014, LNCS 8591, pp. 626–636, 2014.

We illustrate large-network-limit results on this model, rigorously proved in [3]. We present a condition involving the total degree and transmitter degree distribution of a uniformly chosen node which, if satisfied, will allow, with a non-negligible probability, the campaign to go viral when started from this particular node. Given this condition, we present an estimate of the fraction of the population that is reached when the campaign does go viral. We then state that under the same condition, the fraction of good pioneers in the network, i.e., the individuals who if targeted initially will lead the campaign to go viral, is non-negligible as well, and we give an estimate of this fraction. We analyze in detail the process of influence propagation on configuration model having two types of degree-distribution: Poisson and Power Law. Three examples illustrating the dynamic of influence propagation on these two networks are considered: (1) Bernoulli transmissions; (2) Node percolation; (3) Coupon-collector transmissions.

Based on the above analysis, we suggest what statistical data a firm should collect from the pioneers of its marketing campaign, and based on these, how to estimate the effectiveness of the campaign and make a cost-benefit analysis.

1.2   Related Work

The diffusion-on-random-graph models have been previously studied in the context of the spread of epidemics in population ([1]), and more recently, to understand the propagation of social and economic behavior through a social network ([7], [2], [9]).

In the context of viral marketing, one approach is to use the detailed information about the network structure and the past instances of influence propagation to come up with a predictor of the most influential individuals who should be targeted for future campaigns ([6]). Another approach is to take into account the current campaign's effectiveness based on the detailed temporal and structural information regarding the ongoing diffusion in the network ([4]). Our analysis follows the latter approach, but differs in that we require much less information regarding the network and the ongoing diffusion in it.

2   Model and Theoretical Claims

In this section, we introduce our model and informally describe the results which are rigorously proved in [3].

2.1   Model

Consider that the only information available to you about an online social network is the number of friends that a subset of network members have. We will work with a uniform random network which agrees with the statistics that you can obtain from the available information. Such a uniform random network is obtained by constructing what is known as *configuration model* (CM); cf [8]. This random network is realized by attaching half-edges to each vertex corresponding to its degree (which represents here, the number of friends) and then uniformly pair-wise matching them to create edges. We assume this model of the social network throughout the paper and will use interchangeably the terms "social graph" and "random network" meaning precisely the CM. We call the vertices of this graph "nodes" or "users" and graph neighbours "friends".

We consider a marketing campaign started from some initial target called *pioneer* in this network. A person influences a subset of its friends who further propagate the campaign in the same manner. The number of friends that a person influences depends on a particular campaign. To model this dynamic, we enhance the configuration model by partitioning the half-edges into *transmitter* half-edges, those through which the influence can flow and *receiver* half-edges which can only receive influence. So, if a person A influences his friend B in the network, then in our representation, A has a transmitter half-edge matched to the transmitter or receiver half-edge of B.

Let D and $D^{(t)}$ denote the random variables corresponding to the empirically observed distributions of total degree and transmitter degree respectively. For notational convenience, we will interchangeably use random variables and their distributions to mean the same thing. Empirical receiver degree distribution, $D^{(r)}$, is $D - D^{(t)}$. Then we have the following large-network-limit results, rigorously proved in [3], but only informally stated here.

2.2   Theoretical Claims

Claim 1. *Starting from a randomly selected pioneer, the campaign can go viral, i.e., reach a strictly positive fraction of the population, with a strictly positive probability if and only if*

$$\mathbb{E}[D^{(t)}D] > \mathbb{E}[D^{(t)} + D]. \tag{1}$$

Note that $\mathbb{E}[D^{(t)}D] > \mathbb{E}[D^{(t)} + D]$ implies

$$\mathbb{E}[D(D-2)] > 0 \tag{2}$$

and recall that this latter condition is necessary and sufficient [1] for the existence of a (unique) connected component of the underlying social graph, called *big component*, encompassing a strictly positive fraction of its population; cf [5]. Obviously, our campaign can go viral only within this big component.

Call *good pioneers* the pioneers from which the campaign can go viral.

Claim 2. *If (1) is satisfied, then the population reached is, more or less, the same, irrespective of the good pioneer chosen initially.*

Let C^* denote the population reached by the campaign when started from a good pioneer and $\overline{C}^*$ the set of good pioneers.

---

[1] Under a few additional technical assumptions, as $0 < \mathbb{E}[D] < \infty$, $\mathbb{P}\{D = 1\} > 0$, which we tacitly assume throughout the paper.

Claim 3. *If (1) is satisfied, then the set of good pioneers $\overline{C}^*$ also forms a strictly positive fraction of the population.*

The next claim gives the estimates on the size of C^* and $\overline{C}^*$. Let

$$H(x) := \mathbb{E}[D]x^2 - \mathbb{E}[D^{(r)}]x - \mathbb{E}[D^{(t)}x^D] \tag{3}$$

and

$$\overline{H}(x) := \mathbb{E}[D]x^2 - \mathbb{E}[D^{(t)}x^{D^{(t)}}] - \mathbb{E}[D^{(r)}x^{D^{(t)}}]x. \tag{4}$$

If condition (1) is satisfied, then $H(x)$ and $\overline{H}(x)$ have unique zeros in $(0,1)$. Call them ξ and $\overline{\xi}$ respectively. Denote also by $G_D(x) = \mathbb{E}[x^D]$ and $G_{D^{(t)}}(x) = \mathbb{E}[x^{D^{(t)}}]$, the probability generating function (pgf) of D and $D^{(t)}$, respectively.

Claim 4. *If (1) is satisfied and n denotes the size of network population, then for n large,*

$$\frac{|C^*|}{n} \approx 1 - G_D(\xi) =: \alpha > 0 \tag{5}$$

and

$$\frac{|\overline{C}^*|}{n} \approx 1 - G_{D^{(t)}}(\overline{\xi}) =: \overline{\alpha} > 0. \tag{6}$$

Note that $\overline{\alpha}$ can be interpreted as the probability that the campaign goes viral when started from a randomly chosen pioneer.

See [3] for formal statements and proofs of the above claims. Recall also from [5] that under assumption (2), the size, $|C_0|$, of the big network component, C_0, satisfies for n large, $\frac{|C_0|}{n} \approx 1 - G_D(\xi_0) =: \alpha_0 > 0$, where ξ_0 is the unique zero of $H_0(x) := \mathbb{E}[D]x^2 - xG'_D(x)$ in $(0,1)$, with $G'_D(x)$ denoting the derivative of the pgf of D.

3   Examples

Let us consider the results of Section 2 in the context of a few illustrative network examples.

3.1   Bernoulli Transmissions

Let us assume some arbitrary distribution of the degree D satisfying (2) (to guarantee the existence of the big component of the social graph). Suppose that each user decides independently for each of its friends with probability $p \in [0,1]$ whether to transmit the influence to him or not. We call this model, *CM with Bernoulli transmissions*, and p, the *transmission probability*. Note that given the total degree D, the transmitter degree D^t is Binomial(D,p) random variable.

Proposition 1. *In the CM with a general degree distribution D satisfying (2) and Bernoulli transmissions, the campaign can go viral if and only if the transmission probability p satisfies*

$$p > \frac{\mathbb{E}[D]}{\mathbb{E}[D^2] - \mathbb{E}[D]}. \tag{7}$$

In this latter case, the fraction of the influenced population and the fraction of good pioneers are approximately equal to each other, i.e., $|C^*|/n \approx |\bar{C}^*|/n =: \alpha$, for large n, and satisfy

$$\alpha = 1 - G_D(\xi), \tag{8}$$

where ξ is the unique zero of the function $\mathbb{E}[D]((x-1)/p+1) - G'_D(x)$ in $(0,1)$.

Proof. Bernoulli transmissions with (3) and (4) imply $H(x) = \mathbb{E}[D]x^2 - (1-p)\mathbb{E}[D]x - pxG'_D(x)$ and $\overline{H}(x) = \mathbb{E}[D]x^2 - G'_D(1-p(1-x))$. Moreover $\overline{G}_{D^{(t)}}(x) = G_D(1-p(1-x))$. Dividing $H(x)$ by px and substituting $y := 1 - p(1-x)$ in $\overline{H}(x)$ and $\overline{G}_{D^{(t)}}(x)$ completes the proof.

Consider two specific network degree examples.

Example 1 (Poisson degree). When D has Poisson distribution of parameter λ (in which case the CM is asymptotically equivalent to the Erdös-Rényi model) the condition (7) reduces to $\lambda p > 1$ and the fraction of the influenced population and good pioneers (8) is equal to $\alpha = (1 - \xi)/p$, where ξ is the unique zero of the function $(x-1)/p + 1 - \exp(\lambda(x-1))$ in $(0,1)$.

More commonly observed degree-distributions in social networks have power-law tails.

Example 2 (Power-Law ("zipf") degree). Assume D having distribution $\mathbb{P}\{D = k\} = k^{-\beta}/\zeta(\beta)$ $k = 1, 2, \ldots$, with $\beta > 2$, where $\zeta(\beta)$ is the zeta function. Recall that the pgf of D is equal to $G_D(x) = \mathrm{Li}_\beta(x)/\zeta(\beta)$, where $\mathrm{Li}_\beta(x) = \sum_{k=1}^{\infty} k^{-\beta}x^k$ is the so-called poly-logarithmic function. Condition (2) for the existence of the big component is equivalent to $\zeta(\beta - 2) - 2\zeta(\beta - 1) > 0$, which is approximately $\beta < 3.48$. Condition (7) reduces to $p > \zeta(\beta - 1)/(\zeta(\beta - 2) - \zeta(\beta - 1))$ and the fraction of the influenced population and good pioneers (8) is equal to $\alpha = 1 - \mathrm{Li}_\beta(\xi)$, where ξ is the unique zero of the function $x\zeta(\beta - 1)((x-1)/p+1) - \mathrm{Li}_{\beta-1}(x)$ in $(0,1)$.

Recall from Proposition 1, that Bernoulli transmissions lead to the model where the fraction of the influenced population and the fraction of good pioneers are asymptotically equal to each other. In what follows we present two scenarios where the set of good pioneers and the influenced population have different size.

3.2   Enthusiastic and Apathetic Users or Node Percolation

Consider CM with a general degree distribution D satisfying (1), whose nodes either transmit the influence to all their friends (these are "enthusiastic" nodes) or do not transmit to any of their friends ("apathetic" ones). Let p denote the fraction of nodes in the network which are enthusiastic. Note that this model corresponds to the *node-percolation* [2] on the CM. Thus, in this model, given D, $D^{(t)} = D$ with probability p and $D^{(t)} = 0$ with probability $1 - p$.

Proposition 2. *Consider node-percolation on the CM with a general degree distribution D satisfying (2). The campaign can go viral if and only if the fraction p of enthusiastic users satisfies condition (1); the same as for the Bernoulli model. Moreover, in this case, the fraction α of reached population is also the same as in the network with Bernoulli transmissions, i.e., equal to (8) with ξ as in Claim 1. However, the fraction $\overline{\alpha}$ of good pioneers is equal to $\overline{\alpha} = p\alpha$.*

The proof follows easily from the general results of Section 2.2. Note that the *campaign on the network with enthusiastic and apathetic users can reach the same population as in the Bernoulli transmissions, however there are less good pioneers.*

3.3   Absentminded Users or Coupon-Collector Transmissions

Consider again CM with a general degree distribution D satisfying (1). Suppose that each user is willing (or allowed) to transmit K messages of influence. In this regard, it randomly selects, K times, one of his friends *with replacement* (as if it forgets its previous choices). An equivalent dynamic of the influence propagation can be formulated as follows: every influenced user, at all times, keeps choosing one of its friends uniformly at random and transmits the influence to him; it stops forwarding the influence after K transmissions.

In this model, the transmission degree, $D^{(t)}$, corresponds to the number of collected coupons in the classical coupon collector problem with the number of coupons being the vertex degree, D, and the number of trials, K. The conditional distribution of $D^{(t)}$, given D, can be expressed as follows: $\mathbb{P}\{ D^{(t)} = k \,|\, D \} = \frac{D!}{(D-k)!D^{-K}} \left\{ \begin{matrix} K \\ k \end{matrix} \right\}$, where $\left\{ \begin{matrix} K \\ k \end{matrix} \right\} = 1/k! \sum_{i=0}^{K}(-1)^i \binom{k}{i}(k-i)^K$ is the Stirling number of the second kind.

Calculating the pgf for this distribution is tedious and we do not present analytical results regarding this model but only simulations and estimation. As we shall see in Section 3.4, in this model, *the influenced population is smaller than the population of good pioneers.*

3.4   Numerical Examples

We will present now a few numerical examples of networks and diffusion models presented above.

---

[2] Different than edge-percolation.

Simulations. In all our examples, we simulate the enhanced configuration model on $N = 1000$ nodes assuming some particular node degree distribution, D, and influence propagation mechanism modeled by the conditional distribution of the transmitter degree, $D^{(t)}$. More precisely, we sample the individual node degrees and transmitter degrees $(D_i, D_i^{(t)})$ $i = 1 \ldots N$ independently from the joint distribution of $(D, D^{(t)})$ and use these values to construct an instance of our enhanced CM by uniform pairwise matching of the half-edges. We calculate the relative size of the influenced population and the set of good pioneers through the exploration of the influenced components for all nodes. [3]

Estimation. We adopt also the following "semi-analytic" approach: Using the sample $(D_i, D_i^{(t)})$, $i = 1, \ldots, N$ used to construct the CM, we consider estimators, $\hat{G}_D(x) := \frac{1}{N} \sum_{i=1}^{N} x^{D_i}$, $\hat{G}_D^{(t)}(x) := \frac{1}{N} \sum_{i=1}^{N} x^{D_i^{(t)}}$, $\hat{H}(x) := \frac{1}{N} \sum_{i=1}^{N} \left(D_i x^2 - (D_i - D_i^{(t)})x - D_i^{(t)} x^{D_i} \right)$, $\hat{\overline{H}}(x) := \sum_{i=1}^{N} \left(D_i x^2 - D_i^{(t)} x^{D_i^{(t)}} - (D_i - D_i^{(t)})x^{D_i^{(t)}+1} \right)$, of the functions, $G_D(x)$, $G_{D^{(t)}}(x)$, $H(x)$ and $\overline{H}(x)$, respectively. We calculate estimators $\hat{\alpha}$ and $\hat{\overline{\alpha}}$ of the fraction of the influenced population, α, and of good pioneers, $\overline{\alpha}$, using Claim 5 and the estimated functions, $\hat{G}_D(x)$, $\hat{G}_{D^{(t)}}(x)$, $\hat{H}(x)$ and $\hat{\overline{H}}(x)$. (That is, we find numerically, zeros, $\hat{\xi}$ and $\hat{\overline{\xi}}$ of $\hat{H}(x)$ and $\hat{\overline{H}}(x)$, respectively, and plug them into (5) and (6), with $\hat{G}_D(x)$ and $\hat{G}_{D^{(t)}}(x)$ replacing $G_D(x)$ and $G_{D^{(t)}}(x)$.)

Note that in the semi-analytic approach, we do not need to know/construct the realization of the underlying model. This observation is a basis of a *campaign evaluation method* that we propose in Section 4. In fact, in reality one usually does not have the complete insight into the network structure and needs to rely on statistics collected from the initially contacted pioneers.

Analytic Evaluation. Finally, for all models, except the "coupon-collector" one of Section 3.3, we calculate numerically, the values of α and $\overline{\alpha}$ using the explicit forms of all the involved functions. (For the coupon-collector model, we obtained the "true" values of α and $\overline{\alpha}$ from a sample of $(D_i, D_i^{(t)})$ of a larger size N.)

When comparing these analytic solutions to the simulation and semi-analytic estimates, we see that in some cases, $N = 1000$ is not big enough to match the theoretical values. One can easily consider larger samples, however, we decided to stay with $N = 1000$ to show how the quality of the estimation varies over different model assumptions. Also, $N = 1000$ seems to be near the lower range of the number of initial pioneers one needs to contact to produce a reasonable prognosis for the development of the campaign.

---

[3] The simulations are run in *python* by modifying code from the *networkx* package. Remark that the *directed configuration model* in networkx package is a completely different model despite superficial similarity.

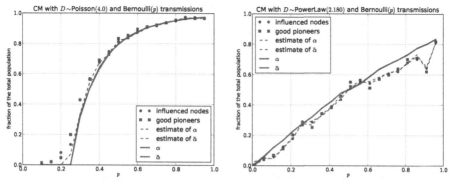

Fig. 1. CM with Poisson degree of mean $\lambda = 4$ and Power-Law degree of parameter $\beta = 2.180$ (corresponding to $\mathbb{E}[D] \approx 4$), both with Bernoulli transmissions with probability p. The set of good pioneers and the influenced population are of the same size. In the Poisson case their fraction is strictly positive for $p > 1/\lambda$ while in the Power-Law case it is so for all $p > 0$ whenever $\beta \leq 3$.

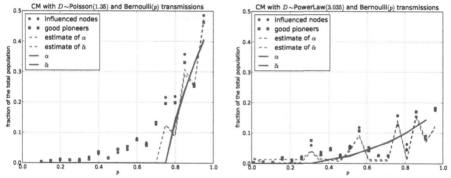

Fig. 2. CM with Poisson and Power-Law degree of mean $\mathbb{E}[D] \approx 1.35$ ($\lambda = 1.35$ and $\beta = 3.035$) and Bernoulli transmissions. The set of good pioneers and the influenced population are of the same size for each model. One observes the phase transition in both models, at $p = 1/\lambda$ and $p = \zeta(\beta - 1)/(\zeta(\beta - 2) - \zeta(\beta - 1)$, respectively.

Case Study. Figure 1 presents Bernoulli influence propagation on the CM with Poisson and Power-Law degree distribution of mean $\mathbb{E}[D] = 2$. Bernoulli transmissions imply that the sets of good pioneers and influenced population are of the same size. The Power-Law degree with $\beta < 3$ leads to a positive fraction of good pioneers and influenced component for all $p > 0$, while for the Poisson degree distribution one observes the phase transition at $p = 1/\lambda$. That is, the fractions of good pioneers and the influenced component are strictly positive, if and only if, $p > 1/\lambda$.

Figure 2 shows again the model with Bernoulli transmissions on CM with Poisson and Power-Law degree distribution, this time for $\mathbb{E}[D] \approx 1.35$ for which both models exhibit the phase transition in p.

A general observation is that the Power-Law degree distribution gives smaller critical values of p for the existence of a positive fraction of influenced popula-

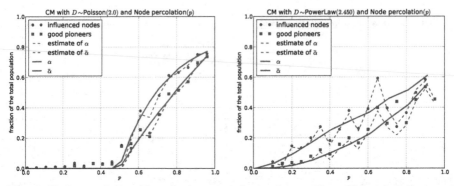

Fig. 3. Node percolation ("apathetic and enthusiastic users") on CM with Poisson and Power-Law degree of mean $\mathbb{E}[D] \approx 2$ ($\lambda = 2$ and $\beta = 2.45$). The influenced component and the critical values for p are equal to these for the CM with Bernoulli transmissions. The set of good pioneers is smaller than the influenced population. We do not observe the phase transition for the Power-Law model since $\beta < 3$.

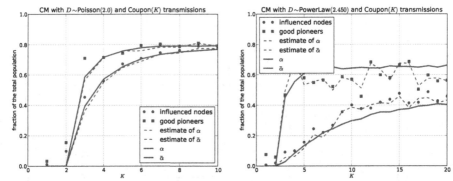

Fig. 4. Coupon collector dynamics "absentminded users") on CM with Poisson and Power-Law degree of mean $\mathbb{E}[D] \approx 2$ ($\lambda = 2$ and $\beta = 2.45$). The set of good pioneers is bigger than the influenced population.

tion and good pioneers, however for these, the size of these sets increase more slowly with the transmission probability, p, in the Bernoulli model. Obviously, the values of $\alpha = \overline{\alpha}$ at $p = 1$ correspond to the size of the biggest connected component of the underlying CM.

Figure 3 shows the node percolation (or the case of "apathetic and enthusiastic users") on CM with Poisson and Power-Law degree distribution of mean $\mathbb{E}[D] \approx 4$. Note that the influenced components have the same size as for Bernoulli transmissions, however good components are smaller. The critical values of p for the phase transition are also the same as for Bernoulli transitions. Note that the estimation of the node percolation model is more difficult than the Bernoulli transmissions because of higher variance of the estimators.

Finally, Figure 4 shows that the coupon collector dynamic (the case of "absentminded users") on CM produces bigger sets of good pioneers than the influenced population.

4   Application to Viral Campaign Evaluation

From the point of view of a firm which has no prior information about the network structure and the campaign effectiveness, it could be useful to assume that the network is a uniform random network, with the total degree and transmitter degree distributions estimated using the information collected from the initial set of pioneers targeted. The collected information, denote it by $(D_i, D_i^{(t)})$, $i = 1, \ldots, N$, allows to estimate various quantities relevant to the potential development of the ongoing campaign, as we did in 3.4. Particularly relevant are the estimates for the following.

Network fragmentation If the value of the estimator, $\mathbb{E}[D^2 - 2D] \approx \frac{1}{N} \sum_{i=1}^{N} \left(D_i^2 - 2D_i \right)$, is not sharply larger than zero, then the firm must assume that the network is too fragmented to allow for viral marketing (condition (2)).

Effectiveness of the campaign If one estimates that the network is not too fragmented, then the firm can evaluate the effectiveness of the ongoing campaign using an estimate of $\mathbb{E}[DD^{(t)} - D^{(t)} - D] \approx \frac{1}{N} \sum_{i=1}^{N} \left(D_i D_i^{(t)} - D_i - D_i^{(t)} \right)$ (condition (1)). If the value of this estimator is sharply larger than zero, then the firm can assume that there is a realistic chance of picking a good pioneer via random sampling and make the campaign go viral.

The estimates of the fraction of good pioneers and the vulnerable population can also be considered to make a cost-benefit analysis of the marketing campaign.

5   Conclusion

Diffusion studies on networks generally tend to focus on the component that can be reached starting from an initial target. Our work in [3], over and above, focuses on the set of good pioneers based on a new approach which consists of identifying this subset as the big component of a *reverse dynamic* in which an "acknowledgment" message is sent in the reversed direction on every edge thus allowing to trace all the possible sources of influence of a given vertex.

In this paper, we consider what insight the graph-theoretical results obtained through this approach provide about the phenomenon of viral marketing on online social networks, particularly to a firm trying to decide how much to spend on a marketing campaign which might go viral or not.

References

1. Bailey, N.: The Mathematical Theory of Infectious Diseases. Books on cognate subjects. Griffin (1975)
2. Banerjee, A.V.: A simple model of herd behavior. The Quarterly Journal of Economics 107(3), 797–817 (1992)

3. Błaszczyszyn, B., Gaurav, K.: Viral marketing on configuration model. arxiv 1309.5779 (2013) (submitted)
4. Cheng, J., Adamic, L., Dow, P.A., Kleinberg, J.M., Leskovec, J.: Can cascades be predicted? In: Proc. of WWW, pp. 925–936 (2014)
5. Janson, S., Luczak, M.J.: A new approach to the giant component problem. Random Structures and Algorithms 34(2), 197–216 (2008)
6. Kempe, D., Kleinberg, J., Tardos, E.: Maximizing the spread of influence through a social network. In: Proc. of ACM SIGKDD, KDD 2003, pp. 137–146. ACM, New York (2003)
7. Moore, C., Newman, M.E.J.: Epidemics and percolation in small-world networks 61, 5678–5682 (May 2000)
8. Van Der Hofstad, R.: Random graphs and complex networks (2009), http://www.win.tue.nl/rhofstad/NotesRGCN.pdf
9. Yagan, O., Qian, D., Zhang, J., Cochran, D.: Conjoining speeds up information diffusion in overlaying social-physical networks. IEEE JSAC 31(6), 1038–1048 (2013)

Empirical Models for Complex Network Dynamics: A Preliminary Study

Douglas Oliveira and Marco Carvalho

Florida Institute of Technology
Department of Computer Science
Melbourne, Florida, USA
doliveira2011@my.fit.edu, mcarvalho@cs.fit.edu

Abstract. Network analysis has draw a considerable amount of attention in the last decade, especially after the discovery of common topological characteristics such as Small World or a Power Law degree distribution. Recently our understanding of complex networks has been augmented with the inclusion of a local view of patterns of connectivity, such patterns that are present more often in real networks than in randomized ones have been called motifs. These global and local perspectives equip us with powerful tools to understand the behavior of many real networks. However, an important aspect of complex network analysis is often neglected: the dynamics of the information flows. The structural elements of the network topology are very important, but to fully understand the dynamics of these networks we need to take a closer look at the dynamics of the information flow in a self-regulation perspective. For example, we know that the performance and reliability of a compute network is likely influenced by the dynamics of the packet flows, as much as it is influenced by the network topology. In a biological regulatory network we need to understand the dynamics that control the excitation and the suppression of gene activity and other transcription factors. In this work we introduce a preliminary simulation study of the flow of information in networks with different topological properties and activation functions. The goal is to approach the analysis of network dynamics from a data-driven approach, using simulations to capture, understand, and possibly model the overall dynamics of the network in a self-regulated perspective.

1 Introduction

The study of networks is not recent, it dates back from the eighteenth century, from the studies of Euler in the problem of seven bridges [1]. He used a graph, a mathematical structure that consists of nodes and edges connecting them, to proof that there was no solution to such problem. Many theories, algorithms and concepts have been developed ever since, however what has been called of the new network science is quite different from the old network science [2]. The new network science is concerned with the structure of naturally occurring network, rather than theoretical ones, like social networks [3], biological networks [4] and communications networks [5].

Z. Cai et al. (Eds.): COCOON 2014, LNCS 8591, pp. 637–646, 2014.

Among the concepts presented with this new network science we can mention Small World networks [6] and Scale Free networks [7]. In a Small World network the distance between any two node is short, precisely it should be O(log n), where n is the number of nodes of the network. The distance between any two nodes is normally calculated using a shortest path algorithm [8]. Not only the average shortest path must be short but also network must be clustered, which holds the percentage of possible cliques of size three in the network. When the network is called Scale Free, its degree distribution follows a Power Law distribution, formally the probability of a node having degree k is: $p(k) = k^{-\lambda}$.

Moving beyond global properties, the work of [9] presents a concept called motifs. Motifs are basic structural elements, subgraphs, that are present more often in real networks rather than in randomized ones. Depending of the kind of network different motifs arise, for example in information-processing network motifs tend to create multiple paths among different nodes in order to improve robustness of the network. Motifs provide a local view of the network, each node only observes its own connections and the connections of its neighbors to uncover the motifs in its neighborhood. Depending of the types of motifs and their frequency in the network many other properties can be deducted [10].

However, all these properties, either global or local ones, are not enough to propertly understand the behavior of many real networks. For example, in a cell regulatory network, that can change its function rapidly depending of the part of the cell cycle that it is in or when there are unexpected changes in the environment due to an stress factor. In these cases, the static properties of the networks remain the same and yet its function changes, this happens because of changes in the information flowing through the network. Such changes can be in the amount of information, in the type of information or in the direction of the information flow. Another characteristic inherent to the dynamics is that each node has its own state that can vary in time, thus any dynamic analysis needs to consider the evolution of the network in time.

According to [11] we are far from comprehend how the collective behavior of thousand of nodes interacting locally contribute to the dynamic behavior that we observe in many real systems. Once we understand how all these nonlinear interactions on the dynamics of real world networks we will have a better understanding of their behavior, even in stressful situations (in a cellular network it could be a cancer state). For this purpose in this work we simulate the flow of information that will be processed by the network, and we measure the various states of the network along the time in many different topological structures. We expect to find the conditions necessary to the self-regulation of the network, in other words, the parameters of the network that will maintain its information flowing without any external interference.

2   Related Work

Several research efforts in the literature focus on the dynamics of the networks from a growth perspective, trying to understand which processes lead to the appearance of certain characteristics in real networks [12]. For example, the work of [13] proposes a model to generate Scale Free networks based on two main principles: that real networks are constantly evolving by the addition of new nodes and new vertices attach preferentially to already well connected nodes.

Another dynamic process commonly found in literature is the study of diffusion process, like the one that happens in disease spread[14]. In this work the authors claim that the robustness present in Scale Free networks is a disadvantage in this situation once it favors the spreading of viruses. A simulation of the spread of the pandemic disease H1N1 was made by [15] that focused in the role of travel restrictions in halting and delaying the spread of disease. Similar works analyzed diffusion processes in Small World networks [16]. The work of [17] analyze the diffusion pattern of websites visitation and claims that the timing of the browsing process is non-poisson contradicting previous works [18].

The authors of [19] propose a realistic model for dynamics in social networks that takes into account digital and social aspects of online social networks. Their work deals with information propagation, that can be either truthful or not (rumor), and how people reacts by receiving such information, if they believe it or not and if they decide to pass such information to its neighbors or not. This works differs considerably from ours in the kind of information that its propagated in the network. In our work we deal with the amount of information that each nodes passes to its neighbors, thus it is represented as a numerical value. In the work of [19] the information has a dual type, positive or negative, and depending of its type it will propagate more or less.

In the work of [20] the authors present the a statistical analysis of the dynamics in a biological network on a genomics scale by combining gene-expression data with transcriptional regulatory information in many different scenarios. Their results shown that less than one percent of the interactions are retained across four or more conditions and that half of the nodes are uniquely expressed in only one condition. Such findings can be easily incorporated in our model.

In this work each node in the network receives an input and following the behavior of a specific function produces an output that is broadcasted to its neighbors. An analogy can be made with artificial neural networks [21], which also receive inputs and produce outputs based on a given function, however the similarities stop there. In artificial neural networks the structure is organized in layers, which have the purpose of learning a pattern from previous data aiming to adjust several parameters to solve a particular problem. Artificial neural networks have been used in many predicting tasks, like electric load forecasting [22] and in pharmaceutical research [23]. The goal of the model used in this work is to analyze how the information flows in the network, it does not have any direction that the information must flow or try to solve any problem.

3 The Simulation Model

In order to simulate dynamic processes of information flow on a network we create a model where, at each time step t, each node i in the network receives input signals from each of its incoming neighbor r. The input signal is denoted by $\varphi_r(t)$ and can be either excitatory or inhibitory. Depending of the strength of these signals and the function (γ) used to calculate the output, the node broadcast an output signal to all its output neighbors, which is denoted by $\omega_i(t)$.

At each time step t each node i receives input signals from all its neighbors, the sum of all these inputs, that will be computed, is denoted by $v_i(t)$ and is formally defined as:

$$v_i(t) = \sum_{0<r<k} \varphi_r(t) \tag{1}$$

Where k is the number of neighbors of node i. The Figure 1 presents an example of a node in the model. The node i receives input signals $\varphi_1(t)$, $\varphi_2(t)$ and $\varphi_3(t)$ from each of its three incoming neighbors (that will compose $v_i(t)$) and thus will produce an output $\omega_i(t)$ that will be broadcast to its outgoing neighbors.

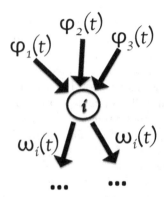

Fig. 1. Example of information flow in a node of the network

To calculate the output, each node uses a function γ on the input $v_i(t)$, we used two well known functions. The first is called Hill function [24]. The Hill function is a common sigmoid function presented in many biological processes, like tumor growth [25]. It consits of a curve that rises from zero and approaches a maximal saturated level δ. Formally, it is defined as:

$$\gamma(v_i(t)) = \frac{\delta}{1 + b^{-v_i(t)}} \tag{2}$$

Where b is the base of calculation, as the value of b increases the function converges faster towards the saturation point δ. At the beginning of the simulation each node defines randomly its own value of b, this variation was made to incorporate the variations of output of different nodes that might receive the same amount of input signals, $v_i(t)$. The second function used is a Poisson function which is widely used and defined by the mean and standard deviation [26].

4   Results

As a result of this model the network's ability to maintain itself 'alive' without any external input is characterized by three main factors: the function (γ) used to calculate the output, the topology of the network and the initial input given to start the simulation. We first analyze the relationship between the topology and γ.

In this work we assume that the network is in a "stable" state if a certain amount of the nodes remains in an state that is neither dead or saturated withou any external input. Nodes in a saturated state receive continuously the maximum input, thus producing its maximum output. This situation may occur due to a loop between two nodes that the output of one is the input of the other and the same happens with the other node. A dead state is a state where the node does not receive any input, thus it cannot produce any output, perhaps due to the lack of input in its incoming neighbors. Such "stable" state is an indicative that the network will not reach a point where no node receives any input (in a biological network this means that the cell died) or all the nodes receive the maximum input, which is not realistic. In order to investigate if such behavior occurs in a large network with real world properties we defined the following metric:

$$\Theta(t) = 1 - \frac{(\eta(v_i^0(t)) + \eta(v_i^\delta(t)))}{n} \tag{3}$$

Where the function η computes the number of nodes at time t that have $v_i(t)$ equals to a specific value. In the case of saturation this value is δ and in a 'dead' state this value is zero. Thereby, $\Theta(t)$ holds the percentage of nodes that are in neither states at each time t. From a global perspective of the network, in order to maintain itself functional the network aims to maximize the value of $\Theta(t)$.

We create a network with the same topological characteristics of the regulatory network of E. coli [27] that has 126 nodes and 326 edges, which can be seen in Figure 2. In Figure 2 the nodes were colored accordingly to their degree, the majority of nodes only have one or two edges and few nodes connect the entire network, like 'crp' and 'h-ns'. The topological properties of this network were analyzed in more details in [28]. Thus we simulate the flow of transcription factors through the network that can excite or inhibit the production of other transcription factor in other nodes.

We measured the variation of $\Theta(t)$ over the time in order to analyze the collective behavior of the network maintain itself in a 'stable' state without any external interference. As can be seen in Figure 3, when we use the Hill function (left) the nodes soon converges to a dead or a saturated state, the percentage of nodes that died is around 30% and the percentage of nodes saturated is around 70%, which leads to $\Theta(t)$ remain close of zero, during the rest of the simulation with almost no change. When we use the Poisson function we perceive a more dynamical behavior of the network, where a few percentage of nodes are dead, less than 10%, and the percentual of node in a saturated state oscillates around 50 and 70%.

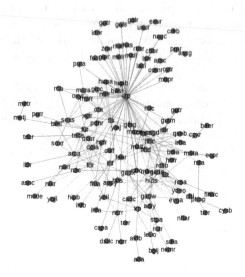

Fig. 2. Regulatory network of *E. coli*

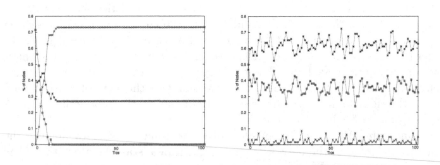

Fig. 3. Variation of the percentage of nodes saturated (X's), dead (asterisk) and $\Theta(t)$ (squares) over time. In the scenario on the left we used the hill function and in the scenario on the right we used the poisson function.

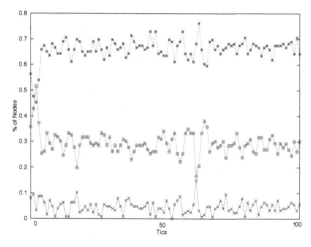

Fig. 4. Variation of the percentage of nodes saturated (X's), dead (asterisk) and $\Theta(t)$ (squares) over time in a Small World network

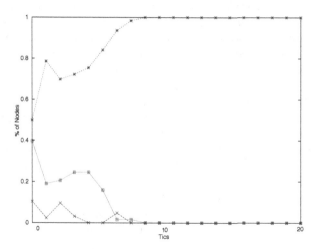

Fig. 5. Variation of the percentage of nodes saturated (X's), dead (asterisk) and $\Theta(t)$ (squares) over time in a Scale Free network

In this case we constructed the network randomly, accordingly to [29]. In order compare the results with a Small-World network, we create an equivalent network (with the same number of nodes and edges) with such characteristics, following the algorithm proposed in [30]. In the Figure 4 we present the results of the simulation of our model in such network that uses a Poisson function. We notice that the results are very similar to the results of the random network Figure 3 (right), although there is a larger percentage of nodes that saturates.

We performed a similar analysis in a Scale Free network. Preserving the dimensions of the network (number of nodes and edges), we rewired the edges aiming that its degree distribution follow a power law distribution with $\lambda=3$, accordingly to [13]. In Figure 5 we see that although the network uses a Poisson function it exhibits a very similar behavior of the randomly created network in Figure 3 (left) where $\Theta(t)$ soon converges to zero. The main difference between these results is that in the Scale Free network the majority of nodes dies and in the random network the majority of nodes saturates.

5   Conclusions and Future Work

In the last few years many advances has been made in network science, including the discovery of Small World properties and its numerous applications, and the characterization of Scale Free networks. Recurrent patterns of local connectivity were also shown to play important role in complex networks, and several algorithms and techniques have been proposed to find and characterize such structures. In this paper we focus on the analysis and understanding of network flow dyanmics. While preliminary, the work proposes a simulation-based model to represent the flow of information. At each time step, each node receives an input from its incoming neighbors and accordingly to a function produces an output that is broadcasted to its output neighbors.

We first introduce two different functions that for node activation: a Hill function and a Poisson function. The Hill function rapidly leads the network to an unrealistic state where almost the totality of nodes are either dead or saturated. In the other hand the Poisson function presented more dynamic results where the percentage of nodes saturated is low and the percentage of nodes dead is not so high, thus the network maintain itself, without any external interference, in such stable state.

We discussed the flow analysis for a Small World network, in contrast to a comparable random network. The results showed, apparently, a similar behavior between the two topologies. The major difference is an increase in the percentage of nodes that saturated, perhaps due to the clustering characteristic of the network that lead the information reach faster between any pair of nodes of the network. We intend to deeper investigate these results aiming to provide a more solid comparison. We also compared the results of such randomly created network with a Scale Free that has the same dimensions. We perceived that due to the majority of nodes only have a couple edges the network rapidly evolves to a completely dead state.

As part of our future work, we intend to further investigate the specific phenomena in Small World networks and Scale Free networks that lead to such behavior presented here. For example, if we change the exponent λ of the power law distribution in Scale Free network the results will remain the same? If the network is more clustered in our Small World version the value of $\Theta(t)$ will increase ou decrease? We also aim to include different motifs in the network and see how if affect the dynamic of the network locally. The possibilities are very broad, thus we also need to come up with other ways to measure the dynamic flow of information of the network

References

1. West, D.B.: Introduction to graph theory. Prentiss Hall, Upper Saddle River (1996)
2. Newman, M., Barabási, A.-L., Watts, D.J.: The structure and dynamics of networks
3. Rapoport, A., Horvath, W.J.: A study of a large sociogram. Behavioral Science 6(4), 279–291 (1961)
4. Jeong, H., Tombor, B., Albert, R., Oltvai, Z.N., Barabási, A.-L.: The large-scale organization of metabolic networks. Nature 407(6804), 651–654 (2000)
5. Faloutsos, M., Faloutsos, P., Faloutsos, C.: On power-law relationships of the internet topology. SIGCOMM Comput. Commun. Rev. 29(4), 251–262 (1999)
6. Watts, D., Strogatz, S.: The small world problem. Collective Dynamics of Small-World Networks 393, 440–442 (1998)
7. Albert, R., Barabási, A.-L.: Statistical mechanics of complex networks. Rev. Mod. Phys. 74, 47–97 (2002)
8. Dijkstra, E.W.: A note on two problems in connexion with graphs. Numerische Mathematik 1(1), 269–271 (1959)
9. Milo, R., Shen-Orr, S., Itzkovitz, S., Kashtan, N., Chklovskii, D., Alon, U.: Network motifs: simple building blocks of complex networks. Science 298(5594), 824–827 (2002)
10. Kashtan, N., Itzkovitz, S., Milo, R., Alon, U.: Topological generalizations of network motifs. Physical Review E 70(3), 031909 (2004)
11. Argollo de Menezes, M., Barabási, A.-L.: Fluctuations in network dynamics. Phys. Rev. Lett. 92, 028701 (2004)
12. Newman, M., Barabási, A.-L., Watts, D.J.: The structure and dynamics of networks. Princeton University Press (2006)
13. Barabási, A.-L., Albert, R.: Emergence of scaling in random networks. Science 286(5439), 509–512 (1999)
14. Pastor-Satorras, R., Vespignani, A.: Epidemic spreading in scale-free networks. Physical Review Letters 86(14), 3200 (2001)
15. Bajardi, P., Poletto, C., Ramasco, J.J., Tizzoni, M., Colizza, V., Vespignani, A.: Human mobility networks, travel restrictions, and the global spread of 2009 h1n1 pandemic. PloS One 6(1) (2011)
16. Monasson, R.: Diffusion, localization and dispersion relations on small-world lattices. The European Physical Journal B-Condensed Matter and Complex Systems 12(4), 555–567 (1999)
17. Jeong, H., Néda, Z., Barabási, A.-L.: Measuring preferential attachment in evolving networks. EPL (Europhysics Letters) 61(4), 567 (2003)
18. Kingman, J.F.C.: Poisson processes, vol. 3. Oxford University Press (1992)
19. Wen, S., Haghighi, M.S., Chen, C., Xiang, Y., Zhou, W., Jia, W.: A sword with two edges: Propagation studies on both positive and negative information in online social networks. IEEE Transactions on Computers (2013)
20. Luscombe, N.M., Babu, M.M., Yu, H., Snyder, M., Teichmann, S.A., Gerstein, M.: Genomic analysis of regulatory network dynamics reveals large topological changes. Nature 431(7006), 308–312 (2004)
21. Yegnanarayana, B.: Artificial neural networks. PHI Learning Pvt. Ltd. (2004)

22. Park, D.C., El-Sharkawi, M.A., Marks, R.J., Atlas, L.E., Damborg, M.J., et al.: Electric load forecasting using an artificial neural network. IEEE Transactions on Power Systems 6(2), 442–449 (1991)
23. Agatonovic-Kustrin, S., Beresford, R.: Basic concepts of artificial neural network (ann) modeling and its application in pharmaceutical research. Journal of Pharmaceutical and Biomedical Analysis 22(5), 717–727 (2000)
24. Alon, U.: An introduction to systems biology: Design principles of biological circuits, vol. 10. CRC Press (2007)
25. Ai, B.-Q., Wang, X.-J., Liu, G.-T., Liu, L.-G.: Correlated noise in a logistic growth model. Physical Review (2003)
26. Marx, M.L., Larsen, R.J.: Introduction to mathematical statistics and its applications. Pearson/Prentice Hall (2006)
27. Salgado, H., Peralta-Gil, M., Gama-Castro, S., Santos-Zavaleta, A., Muñiz-Rascado, L., García-Sotelo, J.S., Weiss, V., Solano-Lira, H., Martínez-Flores, I., Medina-Rivera, A., et al.: Regulondb v8. 0: omics data sets, evolutionary conservation, regulatory phrases, cross-validated gold standards and more. Nucleic Acids Research 41(D1), D203–D213 (2013)
28. Oliveira, D., Carvalho, M.: A comparison of community identification algorithms for regulatory network motifs. In: IEEE International Conference on BioInformatics and BioEngineering (2013)
29. Erdos, P., Rényi, A.: On the evolution of random graphs. Publ. Math. Inst. Hung. Acad. Sci. 5, 17–61 (1960)
30. Watts, D.J., Strogatz, S.H.: Collective dynamics of small-world networks. Nature 393(6684), 440–442 (1998)

Partner Matching Applications of Social Networks

Chunyu Ai[1], Wei Zhong[1], Mingyuan Yan[2], and Feng Gu[3]

[1] Math & Computer Science Division, University of South Carolina Upstate, Spartanburg, SC, USA
[2] Department of Computer Science, Georgia State University, Atlanta, USA
[3] Department of Computer Science, College of Staten Island, The City University of New York, New York, USA

Abstract. People always need to find partners to engage in a lot of daily activities. Therefore, applications of partner matching are significant to help people to find good partners easily. In this paper, we proposed a framework which can match partners for an activity community. In order to improve the matching performance, all users are divided into groups based on a specific classification tree that is built for a specific activity. Maintaining as many stable partnerships as possible in the community is the optimization goal. To achieve the goal, various factors are considered to design matching functions. The simulation results show that the proposed framework can help most of people find stale partners quickly.

Keywords: Social networks, Partner matching, Stable partnership.

1 Introduction

A lot of daily activities require two or more people to collaborate. For instance, playing tennis, squash, rock climbing, and ballroom dancing. However, a lot of people liking these activities suffer from finding good partners. Some people have partners; however, it is still impossible to engage in these activities at the time they preferred since their partners might not be available at that time. Also, partners with different levels usually don't enjoy playing with each other. Everyone wants to have good partners since good partners can help each other improve their performance and skills efficiently. Lacking of matched partners really can ice people's enthusiasm for these activities. An application which can find partners for people has high practical utility. It encourages more people to engage in these activities.

Nowadays, social networks have integrated into our daily life [1–5]; therefore, using social networks to find partners are feasible. However, randomly finding partners through social networks are not efficient. It is difficult and takes time for users to find partners who meet their requirements via social networks. Most importantly, people take various risks to be partners with strangers. Therefore, people rarely try to find partners via social networks. This motivates us to design an application to find partners for people.

Z. Cai et al. (Eds.): COCOON 2014, LNCS 8591, pp. 647–656, 2014.

Usually, people engage in these activities at places such as a gym, fitness center, or classroom. Obviously, people don't want to change their activity locations for a new partner. Therefore, the searching pool should be composed of people who engage in the same activity at the same place. Partner matching applications can be used by these places to help their members to have better experiences and also attract more people.

The challenges of designing a partner matching framework are as follows:

1. How to find a good partner for a user? A good partner for one may be bad for another since users have different definitions of good partners. Therefore, we cannot use the same criteria for every user.
2. How to make the utmost possible effort to benefit all users? It is very difficult to satisfy every user. Then, how to match partners to benefit as many users as possible is a complicated optimization issue.
3. How to protect every user's privacy? In order to match partners very well, users need to provide their personal information such as age, gender, years of experience, and contact information.

In this paper, we proposed a partner matching framework to address this issue. The rest of the paper is organized as follows. Section 2 reviews related work. The proposed mechanism is introduced in Section 3. Section 4 shows the simulation results. Section 5 concludes our work.

2   Related Work

Gale and Shapley proposed an algorithm to solve the college admissions and the stability of marriage problem [6]. The stable marriage problem is the problem of finding a stable match between two sets of elements given a set of preferences for each element. A matching is a mapping from the elements of one set to the elements of the other set. David Gale and Lloyd Shapley proved that, for any equal number of men and women, it is always possible to solve the stable marriage problem and make all marriages stable. The algorithm in [6] is not suitable for the partner matching problem since unlike the marriage relationship, we cannot divide people into two sets. In [7], the preferential partner selection in an evolutionary study of prisoner's dilemma was studied. Marriage and employment relationship matching problem is addressed in [8]. [9] studied the network partner selection.

To the best of our knowledge, existing algorithms are not suitable for solving partner matching problem efficiently. Obviously, randomly matching partners or round-robin will not find stable partnerships quickly since people will keep requesting a new partner when they are not satisfied with the current one. In this paper, we proposed a mechanism to matching partners for an activity community.

3   Partner Matching Framework

In this section, we introduce our proposed partner matching framework. People who need to find partners can send a request. The partner searching pool includes people who request partners and all members of a certain place where the activity is held. Even though some members have no intention to send a partner matching request, they might accept a partner request from others. Therefore, we also add them to the searching pool.

A profile is created for every member in the searching pool. An example profile for an indoor rock climbing gym is shown in Table 1 [10]. For different activities, the profile is designed specifically according to the features of the activity. When a climber uses a harness and rope as protection from falling, he/she needs a belayer to operate these belaying devices to ensure a falling climber does not fall very far. It is important for the belayer to closely monitor the climber's situation, as the belayer's role is crucial to the climber's safety. Therefore, a rock climber usually carefully picks a partner. Usually, an indoor rock climbing gym requires climbers to pass a test to get a belay certification. A climber without a certification is not allowed to belay other climbers. There are 2 belay certification levels, top rope and lead climbing. A climber who can do lead climbing also can do top rope, but not vice versa. The climb route level is from 5.6 to 5.13 usually for an indoor gym. Climbers like to be partners with someone who has similar or more advanced climbing skills than themselves, thus they will improve more quickly.

Table 1. Member profile for a rock climbing gym

Name	Rachel
Phone	555-555-5555
Email	rachel@partnerfinding.com
Gender	female
Age	35
Year of experience	2
Belay certification level	Top rope
Climbing level	5.9-5.10
Climbing time	weekday evenings

For request senders, we can obtain their profiles easily when they send the requests. For the regular members of the place, we use their membership information to fill the profile, thus some information might be missing such as years of experience, level, and time schedule. A time schedule can be summarized according to check-in records of the past few months. If someone does not have a regular time schedule, we use the term *random* to fill it. Also, we can use their years of membership as their estimated years of experience. Then, use the average values of other members with the same years of experience to fill other missing items. For these members with estimated items in their profiles, we put an *estimated* mark on them.

When a user signs in to use the partner matching application, he/she needs to complete a form to describe their requirements to potential partners. An example request form is shown in Table 2. The request form is designed according to features of activities. Requirements are divided into two categories, strong requirements and weak requirements. Strong requirements are more important for users; thus strong requirements must be totally or partially met. Weak requirements don't affect collaboration experience of partners obviously, so these are used for ranking search results when multiple candidates are found. For the climbing gym example, belay certification level, climbing level, and scheduling are strong requirements. Age and gender are weak requirements.

For the strong requirement items, users must specify a value or choose a range. For each requirement item, there are a few options from which to choose. In Table 2, we use semicolons to separate options. For some requirement items, users can choose more than one option. For instance, in Table 2, users can choose more than one *Schedule* and *Climbing level* option. But users can only choose one option at *Belay certification level*. The domain of a requirement item which has values within a certain range are divided into several continuous intervals.

Table 2. Partner request

Requirement	Option	Weight
Belay certification level	Top rope ; Lead climb	0.4
Climbing level	5.6-5.7 ; 5.8-5.9 ; 5.10-15.11 ; 5.12 or above	0.3
Schedule	weekday mornings ; weekday afternoons ; weekday evenings ; weekends	0.2
Preferred gender	male ; female ; none	0.05
Preferred age	under 30 ; 31-40 ; 40 or above ; none	0.05

3.1   Problem Definition

The optimization goal of the partner matching application is to construct as many stable partner relationships as possible. A partner relationship is stable when each party thinks he/she found the best partner already or he/she cannot find a better partner. The rating system in Table 3 is used for rating partners.

A set P is created including everyone who has either sent a partner request or accepted a partner request. For a person p_i in P, R_{p_i} is the rate he/she gave to the current partner or the previous partner if he/she does not have a partner currently. If a person p_i sent a request, but never found a partner, $R_{p_i} = 0$. The goal of matching partners is to

$$Maximize \sum_{i=1}^{|P|} R_{p_i} \qquad (1)$$

3.2   Classification

If a user requests a partner, the system recommends a list of members who meet or almost meet requirements of the user. In order to efficiently find the members who meet the requirements, we use classification techniques to divide

Table 3. Partnership rating

Score	Rate	Action	Possibility of accepting a new partner
4	Excellent	None	0%
3	Good	Little chance of accepting new partners	25%
2	Fair	Big chance of accepting new partners	50%
1	Poor	Request a new partner	75%

all members in the searching pool into groups. SP is the searching pool. $SP = \{m_1, m_2, \cdots, m_{|SP|}\}$ where m_i is the profile of member i in the searching pool.

Assume there are n requirements, $r_1, r_2, \cdots, r_n$, in the partner request form. For every requirement r_i, we assign a weight w_i for it.

$$\sum_{i=1}^{n} w_i = 1 \tag{2}$$

The more important the requirement is, the bigger weight is assigned. Opinions of experts of the activity are necessary for weight assignment. Table 2 shows the weight assignment for the rock climbing gym example. Usually, the total weight of weak requirements are not more than 0.1 since they are not so important compared to strong requirements.

Assume there are s strong requirements. We sort strong requirements in descending order by weights, and store in the list $R_S = r_1, r_2, \cdots, r_s$, where $w_i > w_j$ if $i < j$. Each requirement r_i in R_S has o_i options. r_{i_k} indicates the kth option of requirement r_i. These strong requirements are used to divide all members into groups. The requirement with bigger weight is applied first. The purpose is to reduce the searching range when processing a user's request. A classification guide tree is used to help the group division process.

Initially, we create an empty classification tree T. A tree node t has the structure $\{r_{i_k}, member\_list, child\_list\}$. r_{i_k} is the classification attribute used for the current tree node. $member\_list$ includes all members who fall into the group that the current tree node represents. $child\_list$ contains links to child nodes of the current tree node. In the classification tree, all members are in the leaf nodes; therefore, for any tree node, either $member\_list$ or $child\_list$ is empty.

Algorithm 1 describes the basic process of building a classification tree. First, the root node is created. its $member\_list$ initially includes all members in the searching pool. Then, all options of the requirement r_1 are used as grouping criteria to divide all members into separate groups. For each option, a new group is generated. We create a new tree node for each group to store the members in this group. Also, all these new created nodes are children of the current node. Then, for these new nodes, options of the next requirement in the list R_S are used to partition their members to new groups. This process is repeated until the last strong requirement is applied. However, if members are not evenly distributed based on these options of requirements or there is no sufficient members, a lot of sparse nodes (groups without member or just with few members) will be

---

Algorithm 1. Classification

Input: SP, R_S, OS, and *classification_threshold*
Output: Classification tree T

1: create a queue Q, set Q empty initially
2: create a tree T with only the root node *root* where $root = \{null, SP, \emptyset\}$
3: add the root node *root* to Q
4: **while** Q is not empty **do**
5:　　$t = dequeue(Q)$ {return the first node in the queue}
6:　　**if** $|member\_list|$ of t is greater than *classification_threshold* **then**
7:　　　　**if** the first element classification attribute r_{i_k} of t is *null* **then**
8:　　　　　　*classification_requirement* $= r_1$
9:　　　　**else if** i of r_{i_k} is less than s {did not apply the last strong requirement yet} **then**
10:　　　　　　*classification_requirement* $= r_{i+1}$
11:　　　　**end if**
12:　　　　**for** each option r_{j_k} of *classification_requirement* **do**
13:　　　　　　create a tree node $c = \{r_{j_k}, \emptyset, \emptyset\}$
14:　　　　　　remove every member who meet r_{j_k} in *member_list* of t and add to the *member_list* of c
15:　　　　　　**if** *member_list* of c is not empty **then**
16:　　　　　　　　add c to the *child_list* of t
17:　　　　　　　　add c to Q
18:　　　　　　**end if**
19:　　　　**end for**
20:　　**end if**
21: **end while**

---

generated. The tree with many sparse nodes is not desired since it leads to poor searching performance. In order to avoid generating too many sparse nodes, a *classification threshold* is predefined. When the number of members of the current processing node is less than *classification threshold*, it is not necessary to continue the partition process.

The classification tree of the rock climbing gym example is shown in Fig. 1. Usually, there is no 5.12 and 5.13 top rope routes in an indoor rock climbing gym, so there is no child node for level 5.12-5.13 for top rope. Also, lead climbers don't climb easy routes like level 5.6 and 5.7, so there is no node for level 5.6-5.7 for lead climb. The node for level ≥ 5.12 of lead climb is not partitioned further since there are not enough climbers in the group to support further partition on schedule [10].

After the classification tree is generated, we can search members who meet certain requirements easily via visiting from the top to the corresponding leaf node. If a new member just joins in, we can use his/her profile to search the tree and find the matched leaf node to insert the member. Each member in the searching pool is stored in one leaf node of the classification tree T. If a member's certain attributes in the profile are updated, we can relocate the member to the proper leaf node by removing the member and then reinserting into the tree.

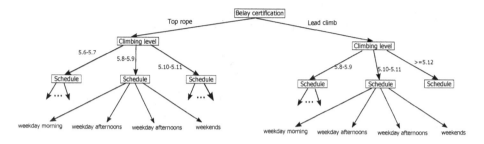

Fig. 1. Classification tree of rock climbing gym

3.3   Partner Matching

Maintaining as many stable partnerships as possible is the optimization goal of the partner matching application. When a partner request pr is received, we will search the classification tree to find all members who meet the strong requirements of pr and add them to candidate set C_{pr}. If there is no more than 10 candidates found in the corresponding leaf node, add left and right direct sibling's members into the candidate set. If a user does not think his/her current partner is the best partner he/she can find, he/she will keep trying to send partners requests or accept requests until the best match is found. In order to make each user find their matched partner as early as possible, the method of ranking candidates becomes significant. For a member m in C_{pr}, function $f(m)$ is designed to measure how well the member m matches the request pr.

$$f(m) = \sum_{1}^{n} w_i * meet_i \tag{3}$$

w_i is the weight of requirement i. $meet_i$ is 1 if member m meets requirement i, otherwise, $meet_i$ is 0.

How well a pair of partners match relies on both parties. The candidate with the largest $f(m)$ is the best candidate for the requester. However, the requester might not meet the expectation of the candidate. Therefore, only considering the benefits of the requester is not sufficient. Benefits of candidates are also important for maintaining stable partnerships.

$$f(m) = (\sum_{1}^{n} w_i * meet_i) * (\sum_{1}^{n} w_i * meet_{i_m}) \tag{4}$$

$meet_{i_m}$ is used to indicate wether the requester meets the candidate m's requirement i. $f(m)$ in Formula 4 are suitable to measure how well the requester and the candidate m match each other.

If the best candidate is satisfied with the current partner, the possibility he/she accepts the request would be low. If the request sender was rejected by the best candidate, he/she also possibly misses the chance to win the second or third best candidate. Thus, the possibility of being accepted by a candidate also need to be considered when we rank all candidates.

$$f(m) = (\sum_1^n w_i * meet_i) * (\sum_1^n w_i * meet_{i_m}) * rate_m \qquad (5)$$

As shown in Formula 5, we add one more factor $rate_m$ which is the possibility candidate m accepts a new request. The estimated possibility of accepting a request is related to the rating of the current partner. Table 3 shows our estimated possibilities. If a person never has a partner, we set the $rate_m$ to 50%.

$f(m)$ is used to rank all candidates of the request pr. Then, the system will send a request to the top 1 candidate. The request includes the profile of the request sender. To protect privacy of members, personal information such as name and contact information are not included. Age is sent as a range format instead of using exact age. If the request is accepted, a new partnership is established. Otherwise, send a request to the next candidate in the ranked list. The system repeatedly sends the request out until someone accepts the request or all candidates are probed already. A reply waiting time is set up. If the receiver did not reply in time, the request is withdrawn and continues to probe the next candidate. If no one in the candidate set accepts the request finally, the partner request sender can file another request. Candidates who made a rejection prior might accept the request from the same sender since they are not satisfied with the current partner.

Since a member changes a partner when there is a better choice, most members will have a stable partnership over time. A few members with improper behaviors are not capable of maintaining a long term partnership. Therefore, they have to keep changing partners.

4   Simulation

We use the rock climbing gym example to evaluate the proposed framework. In our simulation, there are total 500 members. 100 of them request partners initially. If a member gave the current partner a poor rating, he/she definitely sends a new request. Whenever someone accepts the request, he/she will break the partnership with the current partner. The member abandoned by the partner will send a new partner request to find a new partner.

We use the ranking function $f(m)$ in Formula (3), (4), and (5) respectively to evaluate the performance of our partner matching mechanism. The matching program is run by rounds. In each round, a person can only send or receive one request. Fig. 2 shows the percentage of matched partners within 20 rounds. The percentage of matched partners is calculated by the following Formula 6.

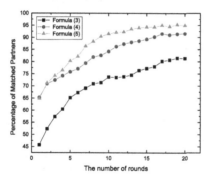

Fig. 2. Percentage of matched partners

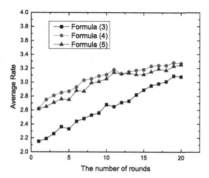

Fig. 3. Average rate to partners

$$\frac{the\ number\ of\ partner\ requesters\ who\ currently\ have\ a\ partner}{the\ total\ number\ of\ partner\ requesters} \qquad (6)$$

As seen in Fig. 2, the effectiveness of Formula (3) is the worst among these 3 formulas. The reason is that only the requester's need is considered; however, a request might not satisfy his/her best candidate's requirements. Formula (4) and (5) performs better since requirements of both requesters and candidates are considered. Overall, Formula (5) achieves the fastest matching progress since it also considers the possibility of the candidates accept the request.

The average rate to partners is shown in Fig. 3. Formula (4) and (5) perform better than Formula (3) since Formula (3) does not consider the need of candidates. Thus, if the chosen candidate accepted the request, the possibility of satisfaction is not high. Formula (4) has higher average rate to partners than

Formula (5) because it always chooses the best matched partners for each other. However, Formula (5) sacrifices the best match to achieve a higher request acceptance rate. Overall, Formula (5) has the fastest matching progress and gains acceptable partner rate scores.

5   Conclusion

In this paper, a partner matching mechanism is proposed to help people find partners in an activity social community. A classification tree is used to partition users into groups to reduce the candidate searching time complexity. To achieve the optimization goal to maintain stable partnerships in the community, we design 3 matching functions. The simulation results show that the proposed framework gains good partner matching performance. In our future work, we will consider partnerships which require more than two parties.

References

1. Adamic, L., Adar, E.: How to search a social network. Social Networks 27(3), 187–203 (2005)
2. Kumar, R., Novak, J., Tomkins, A.: Structure and evolution of online social networks. In: Yu, P.S., Han, J., Faloutsos, C. (eds.) Link Mining: Models, Algorithms, and Applications, pp. 337–357. Springer New York (2010)
3. Cho, E., Myers, S.A., Leskovec, J.: Friendship and mobility: User movement in location-based social networks. In: Proceedings of the 17th ACM SIGKDD International Conference on Knowledge Discovery and Data Mining, KDD 2011, pp. 1082–1090. ACM, New York (2011)
4. Xiang, R., Neville, J., Rogati, M.: Modeling relationship strength in online social networks. In: Proceedings of the 19th International Conference on World Wide Web, WWW 2010, pp. 981–990. ACM, New York (2010)
5. Szell, M., Lambiotte, R., Thurner, S.: Multirelational organization of large-scale social networks in an online world. Proceedings of the National Academy of Sciences 107(31), 13636–13641 (2010)
6. Gale, D., Shapley, L.S.: College admissions and the stability of marriage. The American Mathematical Monthly 69(1), 9–15 (1962)
7. Mortensen, D.T.: Matching: Finding a partner for life or otherwise. American Journal of Sociology 94 (1988)
8. Ashlock, D., Smucker, M.D., Stanley, E., Tesfatsion, L.: Preferential partner selection in an evolutionary study of prisoner's dilemma. Biosystems 37(1-2), 99–125 (1996)
9. Beckman, C.M., Haunschild, P.R., Phillips, D.J.: Friends or strangers? firm-specific uncertainty, market uncertainty, and network partner selection. Organization Science 15(3), 259–275 (2004)
10. Rock climbing ratings from 5.0 to 5.15, http://outdoorswithdave.com/climbing/climbing-ratings

YASCA: An Ensemble-Based Approach for Community Detection in Complex Networks

Rushed Kanawati

University Sorbonne Paris Cité, Paris Nord,
LIPN, CNRS UMR 7030, Villetaneuse, France
rushed.kanawati@lipn.univ-paris13.fr
http://www-lipn.univ-paris13.fr/~kanawati

Abstract. In this paper we present an original approach for community detection in complex networks. The approach belongs to the family of seed-centric algorithms. However, instead of expanding communities around selected seeds as most of existing approaches do, we explore here applying an ensemble clustering approach to different network partitions derived from ego-centered communities computed for each selected seed. Ego-centered communities are themselves computed applying a recently proposed ensemble ranking based approach that allow to efficiently combine various local modularities used to guide a greedy optimization process. Results of first experiments on real world networks for which a ground truth decomposition into communities are known, argue for the validity of our approach.

Keywords: Community detection, complex networks, seed-centric algorithms.

1 Introduction

Complex networks are frequently used for modeling interactions in real-world systems in diverse areas, such as sociology, biology, information spreading and exchanging and many other different areas. One key topological feature of real-work complex networks is that nodes are arranged in tightly knit groups that are loosely connected one to each other. Such groups are called *communities*. Nodes composing a community are generally admitted to share common proprieties and/or be involved in a same function and/or having a same role. Hence, unfolding the community structure of a network could give us much insights about the overall structure a complex network. Works in this field can be roughly divided into two main classes:

- Computing a network partition into communities [12,36], or possibly detecting overlapping communities [27,38].
- Computing a local community centered on a given node [6,4,17].

Recently, an increasing number of work has been proposed with the idea of merging both kind of approaches. The basic idea is to identify some particular

Z. Cai et al. (Eds.): COCOON 2014, LNCS 8591, pp. 657–666, 2014.

nodes in the target network, called *seed nodes*, around which local communities can be computed [16,29,33]. The interest in seed-centric approaches has been boosted in the recent years following the demonstration of serious limitations of *modularity optimization* based approaches considered till lately as the most efficient approaches [14,20]. In this paper we propose an original seed-centric community detection algorithm called: Yasca. Instead of expanding communities around selected seeds as most of existing approaches do, we apply an ensemble clustering approach to different network partitions derived from local communities computed for each selected seed. Local communities are themselves computed applying a recently proposed ensemble ranking based approach that allow to efficiently combine various local modularities used to guide a greedy optimization process [17].

The reminder of this paper is organized as follows. Next in section 2, we review briefly the field of seed-centric approches for community detection. The proposed algoithm is detailed in section 3. First evaluation on benchmark networks are reported and commented in section 4. Comparaison with top algorithms of the state of the art is also made in the same section. Conclusions are given in section 5.

2   Seed-Centric Community Detection Algorithms

Seed-centric approaches constitute an emerging trend in the field of community detection in complex networks. The underlaying idea of these approaches is to select a set of nodes (i.e. seeds) around which communities are constructed. Being based on local computations, these approaches are very attractive to deal with large-scale and/or dynamic networks. A quick review study of existing approaches allow to identify the following criteria for classifying seed-centric algorithms:

- *Seed nature* : A seed can be single node [18,33], a set of nodes [16] , or a connected subgraph [29].
- *Seed number* : The number of seed nodes can be pre-determined [18,7] or computed by the approach itself [16,33].
- *Seed selection policy* : The seed selection process can be : random [18] or informed [16,33,29].
- *Seed community computation* : The community construction can be made applying consensus techniques [32,26,37,8], expansion techniques [4,29,28] or agglomeration techniques [16,33,18].

Next, we present YASCA an original approach that apply an ensemble clustering approach that aggregate different bi-partitions of the whole network inferred from different local communities computed around a set of seeds selected in an informed way.

3   YASCA: The Proposed Algorithm

3.1   General Description

In this section we give the general outlines of the proposed algorithm: YASCA[1] algorithm. Algorithm 1 sketchs the outlines of proposed approach. This is structured into three main steps:

Algorithm 1 The Yasca community detection algorithm

Require: $G = < V, E >$ a connected graph,
1. $\mathcal{C} \leftarrow \emptyset$
2. $S \leftarrow$ **compute_seeds(G)**
3. **for** $s \in S$ **do**
4. $C_s \leftarrow$ **compute_local_com(s,G)**
5. $\mathcal{C} \leftarrow \mathcal{C} + (C_s, \overline{C_s})$
6. **end for**
7. **return Ensemble_Clustering($\mathcal{C}$)**

---

1. The first step is to compute a set of seed nodes $S \subset V$. This is the role of the **compute_seeds()** function (line 2 in algorithm 1). Different selection strategies can be applied for seed election as mentioned in previous section. However, while most of existing approaches seek for seeds that are likely to be at the core of computed communities, we search here to locate seeds as nodes having various positions in the graph. This will be detailed further in section 3.2
2. For each seed node $s \in S$ we compute its local community C_s. This is the role of **compute_local_com()** function (line 4 in algorithm 1). Different algorithms can be applied for local community detection [4,6,2]. We mainly apply here a recent algorithm proposed in [17] that apply a multi-objective greedy optimization approach. The set of vertices V can then be partitioned into two disjoint sets : $P_v = \{C_s, \overline{C_s}\}$ where $\overline{C_s}$ denotes the complement of set C_s.
3. Finally, we apply an ensemble clustering approach [34] in order to merge the different bi-partitions obtained in step 2. This is the role of the **Ensemble_Clutering()** function (line 7 in algorithm 1). The output of this process is the taken to be the final decomposition of the graph into communities.

The overall complexity of the algorithm is determined by the highest complexity of the three above described steps. This depends on specific algorithms applied for implementing each step. However, the ensemble clustering step is usually the most expensive step, computationally speaking. In this work, we apply a classical cluster-based similarity partitioning algorithm [34] that have the following complexity in our case $\mathcal{O}(n^2 \times 2 \times |S|) \sim \mathcal{O}(n^2)$, where n is the number of nodes of the graph, 2 is the number of clusters in each clustering and $|S|$ is the number of different partitions to merge.

---

[1] Yet Another Seed-centric Community detection Algorithm

3.2   Seed Selection

Most existing seed-centric approches search for seeds that are likely to be at the core of communities to be detected. This is mainly the case of leader-based approaches [16,33,18] and set-seeds based approaches [30]. However, the basic idea of our algorithm is to compute a set of seed nodes that occupy diverse positions in the network. Each node will provide a bi-partition of the network from its own point of view. Our intuition is that merging these diverse bi-partitions can provide a good partition of whole network into communities.

Our seeding strategy is inspired form the work of [21] that show that real complex networks are often structured in one huge *bi-connected core* linked to number of small-sized sub-graphs named *whiskers* by a set of *bridges*. Before defining formally, the three mentioned concepts, we recall first the definition of a *bi-connected component* in a graph.

Definition 1 Given a graph $G =< V, E >$ a biconnected component is a maximal induced subgraph $G' =< V', E' >$ that remains connected after removing any node and its adjacent edges in G'.

The size of a biconnected component is defined as the number of edges. We can now define the three above mentioned concepts:

Definition 2 A biconnected core is a maximum size connected graph of G after removing all bi-connected components of size one.

Definition 3 A bridge is a biconnected component of size one which is directly connected to the biconnected core.

Definition 4 A whisker is a maximal subgraph of G that can be detached from the biconnected core by removing a bridge.

We call *articulation nodes*, nodes that are at the intersection of bridges and the biconnected core. Our seeding strategy consists on composing the seed set by the set of articulation nodes to which we add the top high central nodes in the biconnected core. This allows to select nodes playing central role in the major part of the network (the biconnected core) and nodes controlling the periphery of the network (articulation nodes). The computation of the biconnected core and the articulation node is done using a variant of the depth-first traversal of a graph [15] in a nearly linear time complexity.

3.3   Local Community Detection

A main stream in the area of local community detection consists on applying greedy optimization algorithm that starts to explore the network from the query node v_q. Let D be the set of current explored nodes. We can classify nodes in V at any time during the exploration process, into three disjoint sets:

- *The core set (denoted by C)* : is composed of explored nodes whose all neighbors are also explored. In a formal way. We have $C = \{x \in V \, s.t. \, \widehat{\Gamma}(x) \subseteq D\}$.

- *The border set (denoted B):* is composed of explored nodes that have at least one unexplored neighbor node. Formally, $B = \{x \in D \; : \; \exists v \in \Gamma(x) : v \notin D\}$.
- *The shell set (denoted by S):* is composed of nodes partially explored. These are nodes that have some neighbors in the set B. Formally, $S = \{x \in \overline{D} : \Gamma(x) \cap D \neq \phi\}$.
- *The set of unexplored nodes (denoted by U):* This is the set of nodes in V that are not explored at all. Formally, $U = \{x \in \overline{D \cup S}\}$.

Notice that D is equal to $B \cup C$. Figure 1. illustrates the different sets of nodes at a given time t during the exploration process.

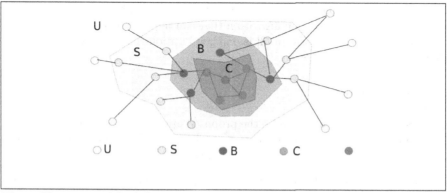

Fig. 1. Illustration of the definitions of the different sets of nodes during the exploration the neighborhood of a query node v_q

A greedy optimization is an iterative process: Initially, C is set to the singleton $\{v_c\}$, B is initialized to the empty set and S is set to $\Gamma(v_c)$ the direct neighbors of v_c. At each iteration, nodes in S are ranked in function of an objective function Q called also local modularity function. The top ranked node is added to the set B. Then all three sets C, B and S are updated. The algorithm iterates while S is not empty and if the local modularity induced by the selected node increases. Different algorithms apply different objective functions [6,22,4]. We apply here an algorithme proposed in [17] that consists apply a multi-objective optimization strategy using ensemble-ranking approaches: Let $\mathcal{Q} = (Q_1, \dots, Q_n)$ be the set of applied objective functions. Let S^{Q_i} be the ranked list of elements of S in function of Q_i. At each iteration of the greedy optimization algorithm, we select the node that is ranked first after fusion of the obtained ranks $\{S^{Q_1}, \dots, S^{Q_k}\}$. Ensemble ranking (a.k.a rank aggregation, rank fuse or social choice algorithms) approaches can be used to obtain the final ranking of elements in S [1,10,5,35].

Let $v_w \in S$ be the winner node: the node that is ranked first after the rank merging process. Let $Q_i(v_w)$ be the ith modularity obtained from adding v_w to B. The algorithm iterates if there exist at least one local modularity that is enhanced or equal to the same modularity computed for the previously winner node selected in the previous iteration.

3.4 Ensemble Clustering

The goal of an ensemble clustering approach is to compute a clustering (here a partition) that combine the different obtained partitions. One widely applied method is based on constructing a **consensus graph** out of the set of partitions to be combined [11,34]. The consensus graph G_{cons} is defined over the same set of nodes of the initial graph G. Two nodes $v_i, v_j \in V$ are linked in G_{cons} if there is at least one partition $P^y_{Q_x}$ where both nodes are in a same cluster. Each link (v_i, v_j) is weighted by the frequency of instances that nodes v_i, v_j are placed in the same cluster. Different approaches have been proposed to detect communities in the consensus graph [34,9]. In this work we propose applying to the consensus graph a community detection algorithm that can handle unconnected, weighted graphs. The Louvain algorithm [3] is one good option that we have adopted for that purpose.

4 Experiments

In a first experiment, we evaluate the proposed approach on a set of four widley used benchmark networks for which a ground-truth decomposition into communities are known. These networks are the following:

Zachary's karate club. This network is a social network of friendships between 34 members of a karate club at a US university in 1970 [39]. Following a dispute the network was divided into 2 groups between the club's administrator and the club's instructor. The dispute ended in the instructor creating his own club and taking about half of the initial club with him. The network can hence be divided into two main communities.

Dolphins social network. : This network is an undirected social network resulting from observations of a community of 62 dolphins over a period of 7 years [23]. Nodes represent dolphins and edges represent frequent associations between dolphin pairs occurring more often than expected by chance. Analysis of the data revealed two main groups.

American political books. This is a political books co-purchasing network. Nodes represent books about US politics sold by the online bookseller *Amazon.com*. Edges represent frequent co-purchasing of books by the same buyers, as indicated by the "customers who bought this book also bought these other books" feature on Amazon. Books are classified into three disjoint classes: liberal, neutral or conservative. The classification was made separately by Mark Newman based on a reading of the descriptions and reviews of the books posted on Amazon.

Next figure shows the structure of the selected networks with real communities indicated by the color code. In table 1 we summarize basic characteristics of selected benchmark real networks.

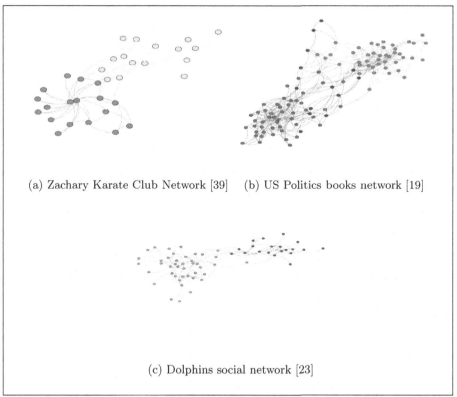

(a) Zachary Karate Club Network [39]   (b) US Politics books network [19]

(c) Dolphins social network [23]

Fig. 2. Real community structure of the selected benchmark networks

Table 1. Characteristics of some well-known benchmark networks

Network	n	m	# com	reference
Zachary club	34	78	2	[39]
Political books	100	441	3	[19]
Dolphins	62	159	2	[23]

In the seeding phase we selected in addition to the articulation nodes the top 15% central nodes in the biconnected core (using the degree centrality). For the consensus graph we keep a link if the associated frequency is equal or greater than 0.5 (these are the best parameters when using the degree centrality for selecting seeds). Next figure shows the obtained results on the three datasets compared to state of the art algorithms : Louvain [3], Infomap [24], Walktrap [31] and edge-betweenness based modularity optimization algorithm (denoted Girvan algorithm in the figure) [13]. Evaluation is made in function of the normalized mutual information (NMI) indice that measures the similarity between computed partition and ground-truth partition [25]. Results show that YASCA yields better results than other algorithms on these small benchmark networks.

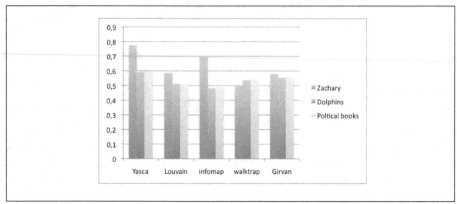

Fig. 3. Comparative results on the three selected dataset in terms of NMI

These results needs confirmation on large-scale datasets. The problem to cope with is to find large-scale networks with reliable known ground-truth partitions.

5 Conclusion

A new seed centric algorithm for community detection is proposed. First results on small networks show the potential of this algorithm compared to the state of the art algorithms. Further investigations about the effects of different parameters of the algorithm are considered (seed selection strategy, the local community algorithm to be used, etc.). Validations on large-scale graphs are also scheduled. This requires to parallelise the step of local community identification of each seed node.

References

1. Arrow, K.: Social choice and individual values, 2nd edn. Cowles Foundation, New Haven (1963)
2. Bagrow, J.P., Bollt, E.M.: A local method for detecting communities. Phys. Rev. E 72, 046108 (2005)
3. Blondel, V.D., Guillaume, J.L., Lefebvre, E.: Fast unfolding of communities in large networks, pp. 1–12 (2008)
4. Chen, J., Zaïane, O.R., Goebel, R.: Local community identification in social networks. In: Memon, N., Alhajj, R. (eds.) ASONAM, pp. 237–242. IEEE Computer Society (2009)
5. Chevaleyre, Y., Endriss, U., Lang, J., Maudet, N.: A short introduction to computational social choice. In: van Leeuwen, J., Italiano, G.F., van der Hoek, W., Meinel, C., Sack, H., Plášil, F. (eds.) SOFSEM 2007. LNCS, vol. 4362, pp. 51–69. Springer, Heidelberg (2007)
6. Clauset, A.: Finding local community structure in networks. Physical Review E (2005)

7. Cordasco, G., Gargano, L., Capocelli, A.R.M.: Community detection via semi-synchronous label propagation algorithms. Informatica (2010)
8. Cupertino, T.H., Huertas, J., Zhao, L.: Data clustering using controlled consensus in complex networks. Neurocomputing (2013), http://www.sciencedirect.com/science/article/pii/S0925231213003160
9. Dahlin, J., Svenson, P.: Ensemble approaches for improving community detection methods. CoRR abs/1309.0242 (2013)
10. Dwork, C., Kumar, R., Naor, M., Sivakumar, D.: Rank aggregation methods for the Web. In: WWW, pp. 613–622 (2001)
11. Fern, X.Z., Brodley, C.E.: Solving cluster ensemble problems by bipartite graph partitioning. In: Brodley, C.E. (ed.) ICML. ACM International Conference Proceeding Series, vol. 69, ACM (2004)
12. Fortunato, S.: Community detection in graphs. Physics Reports 486(3-5), 75–174 (2010)
13. Girvan, M., Newman, M.E.J.: Community structure in social and biological networks. PNAS 99(12), 7821–7826 (2002)
14. Good, B.H., de Montjoye, Y.A., Clauset, A.: The performance of modularity maximization in practical contexts. Physical Review E(81), 046106 (2010)
15. Hopcroft, J., Tarjan, R.: Efficient algorithms for graph manipulation. Communications of the ACM 16(6), 372–378 (1973)
16. Kanawati, R.: Licod: Leaders identification for community detection in complex networks. In: SocialCom/PASSAT, pp. 577–582 (2011)
17. Kanawati, R.: Empirical evaluation of applying ensemble ranking to ego-centered communities identification in complex networks. In: Zaz, Y. (ed.) 4th International Conference on Multimedia Computing and Systems. IEEE, Marrakech (2014)
18. Khorasgani, R.R., Chen, J., Zaiane, O.R.: Top leaders community detection approach in information networks. In: 4th SNA-KDD Workshop on Social Network Mining and Analysis, Washington D.C (2010)
19. Krebs, V.: Political books network, http://www.orgnet.com/
20. Lancichinetti, A., Fortunato, S.: Limits of modularity maximization in community detection. CoRR abs/1107.1 (2011)
21. Leskovec, J., Lang, K.J., Dasgupta, A., Mahoney, M.W.: Community structure in large networks: Natural cluster sizes and the absence of large well-defined clusters. Internet Mathematics 6(1), 29–123 (2009)
22. Luo, F., Wang, J.Z., Promislow, E.: Exploring local community structures in large networks. Web Intelligence and Agent Systems 6(4), 387–400 (2008)
23. Lusseau, D., Schneider, K., Boisseau, O.J., Haase, P., Slooten, E., Dawson, S.M.: The bottlenose dolphin community of doubtful sound features a large proportion of long-lasting associations. Behavioral Ecology and Sociobiology 54, 396–405 (2003)
24. Rosvall, M., Axelsson, D., Bergstrom, C.T.: The map equation. Eur. Phys. J. Special Topics 13, 178 (2009)
25. Meilă, M.: Comparing clusterings by the variation of information. In: Schölkopf, B., Warmuth, M.K. (eds.) COLT/Kernel 2003. LNCS (LNAI), vol. 2777, pp. 173–187. Springer, Heidelberg (2003)
26. de Oliveira, T.B.S., Zhao, L., Faceli, K., de Carvalho, A.C.P.L.F.: Data clustering based on complex network community detection. In: IEEE Congress on Evolutionary Computation, pp. 2121–2126. IEEE (2008)
27. Palla, G., Derônyi, I., Farkas, I., Vicsek, T.: Uncovering the overlapping modular structure of protein interaction networks. FEBS Journal 272, 434 (2005)

28. Pan, L., Dai, C., Chongjun, W., Junyuan, X., Liu, M.: Overlapping community detection via leaders based local expnsion in social networks. In: Proceddings of the 24th IEEE Conferenceon Tools with Artficial Intelligence, ICTA 2012 (2012)
29. Papadopoulos, S., Kompatsiaris, Y., Vakali, A.: A graph-based clustering scheme for identifying related tags in folksonomies. In: Bach Pedersen, T., Mohania, M.K., Tjoa, A.M. (eds.) DAWAK 2010. LNCS, vol. 6263, pp. 65–76. Springer, Heidelberg (2010)
30. Papadopoulos, S., Kompatsiaris, Y., Vakali, A., Spyridonos, P.: Community detection in social media - performance and application considerations. Data Min. Knowl. Discov. 24(3), 515–554 (2012)
31. Pons, P., Latapy, M.: Computing communities in large networks using random walks. J. Graph Algorithms Appl. 10(2), 191–218 (2006)
32. Raghavan, U.N., Albert, R., Kumara, S.: Near linear time algorithm to detect community structures in large-scale networks. Physical Review E 76, 1–12 (2007)
33. Shah, D., Zaman, T.: Community Detection in Networks: The Leader-Follower Algorithm. In: Workshop on Networks Across Disciplines in Theory and Applications, NIPS (2010)
34. Strehl, A., Ghosh, J.: Cluster ensembles: a knowledge reuse framework for combining multiple partitions. The Journal of Machine Learning Research 3, 583–617 (2003)
35. Subbian, K., Melville, P.: Supervised rank aggregation for predicting influencers in twitter. In: SocialCom/PASSAT, pp. 661–665. IEEE (2011)
36. Tang, L., Liu, H.: Community Detection and Mining in Social Media. Synthesis Lectures on Data Mining and Knowledge Discovery. Morgan & Claypool Publishers (2010)
37. Wei, F., Qian, W., Fei, Z., Zhou, A.: Identifying community structures in networks with seed expansion. In: Kitagawa, H., Ishikawa, Y., Li, Q., Watanabe, C. (eds.) DASFAA 2010. LNCS, vol. 5981, pp. 627–634. Springer, Heidelberg (2010)
38. Xie, J., Kelley, S., Szymanski, B.K.: Overlapping community detection in networks: The state-of-the-art and comparative study. ACM Comput. Surv. 45(4), 43 (2013)
39. Zachary, W.W.: An information flow model for conflict and fission in small groups. Journal of Anthropological Research 33, 452–473 (1977)

Mining the Key Structure
of the Information Diffusion Network

Jingzong Yang[1], Li Wang[1,2], and Weili Wu[3]

[1] School of Computer Science and Technology, Taiyuan University of Technology,
Shanxi, China
[2] Institute of Computing Technology, Chinese Academy of Science, Beijing, China
[3] Department of Computer Science, University of Texas at Dallas, USA
yjzong870204@sohu.com, wangli@tyut.edu.cn, weiliwu@utdallas.edu

Abstract. With the development of the online social network (OSN),
huge number pieces of information are propagating over the OSN all
the time which has formed the information diffusion network. We find
that during the process of information spreading, there exists not only
the significant spreaders who play important role in the process of information transmission, but also some special structure that we call it
key structure. In this paper, we define the problem in the large social
network and propose an algorithm to mining the key structure (abbre-
viation as MKS). We evaluate our algorithm on the $SINA$ microblog
datasets and compare it with the classical algorithm $PageRank$. Empirical results indicate that our proposed method can yield out better
performance.

Keywords: Social Network, Influential, Spreading Process, Key Structure.

1 Introduction

Online Social Network (OSN) has become a major service of Internet for people
to communicate with each other, such as Twitter, WeChat. Microblogging has
become an extremely fashionable form of social media over the past year or so.
Similar to weblogs, people most record and share interesting information through
following networks. Compared to blogs, microblogging encourages fast updating
by limiting post size, restricting the content format to text, and by supporting
easy mobile updating. The differences of function potentially are creating new
ways for people to accumulate and share information. Social influence can be
described as power – the ability of a single to influence the actions or thoughts
of others.

Information and influence propagation in social network has been actively
studied for decades in the fields of psychology, sociology, communication, marketing, and political science. For online social networks, researchers has summarized the social structures into three categories: Pyramid, Circular, and Hybrid.
An example of the pyramid structure is Microblogging. Influences such as CNN

Z. Cai et al. (Eds.): COCOON 2014, LNCS 8591, pp. 667–675, 2014.

have millions of followers, while the influencer does not follow back. Face-book is an example of the circular social structure, because Face-book users become friends only by selecting from a number of people or brands. The hybrid social structure combines the circular and pyramid-shaped community frameworks.

The rest of the paper is organized as follows. Section 2 gives the description of the related work. Section 3 presents the problem definition and our proposed algorithm in detail. In section 4, we gives experimental results by using a real dataset - online social network $SINA$ microblog. Section 5 summarizes our work and presents the future work.

2   Related Work

Complex network are pervasive to natural and social sciences, ranging from social and information networks to technological and biological networks[1,2]. The spreading processes of epidemic information attract increasing attention in complex network studies[3], and researchers tried to find the reason that why information spread so quickly and influential[4,5]. Spreading is a ubiquitous process, which describes many important network activities[6,7]. And how to control the information spreading process is of particular interests. In the spreading process, the identification of influential nodes is a crucial issue according to the assumption because that highly influential nodes are more likely to be infected and to infect a large number of nodes[7]. At the same time, this paper[7] also shows that the best spreaders are not necessarily the most connected people in the network and the most efficient spreaders are those who located within the core of the network as identified by the $k - core$ decomposition analysis[8].

Basically, the principle of the $k - core$ decomposition is to assign a core index k to each node such that the nodes with the lowest values are located at the periphery of the network while the nodes with the highest values are located in the center of the network. Thus, the innermost nodes forms the core of the network. Brown et al.[9] observed that the results of the $k - shell$ decomposition on Twitter network are highly skewed. Therefore they proposed a modified algorithm which uses a logarithmic mapping, in order to produce fewer and more meaningful $k - shell$ values. Cataldi et al.[10] proposed to use the well-known $PageRank$ algorithm[11] to assess the distribution of influence throughout the network. The $PageRank$ value of a given node is proportional to the probability of visiting this node in a random walk of the social network, where the set of states of the random walk is the set of nodes. The methods we have just described only exploit the topology of the network, and ignore other important properties, such as nodes' features and the way they process information.

According to the observation, we can find that most $OSNs$ members are passive information consumers. Romero et al.[12] developed a graph-based approach similar to the well-known $HITS$ algorithm, which assigns a relative influence and a passivity score to every users based on the ratio of which they forward information. However, no individual can be a universal influencer, and influential members of the network tend to be influential only in one or some specific domains of knowledge. Therefore, Pal et al.[13] developed a non-graph based, topic-sensitive method. To do so, they defined a set of nodal and topical features for characterizing the network members. Using probabilistic clustering over this feature space, they ranked nodes with a within-cluster ranking procedure to identify the most influential and authoritative people for a given topic.

However, most of the theoretic researches above are all about the identification of influential nodes which are to be unrelated in the set of nodes. In this study, we propose a new algorithm which combines the cluster detection and structure hole to mine the key structure of the spreading process. The key structure is organized by a set of nodes which are connected with each other.

3   An Algorithm to Mining the Key Structure in Large Social Network

Giving a network consisting of N nodes and M links. The goal of our study is to mine the key structure that ensure the influence of information maximization of the diffusion network which formed by the information spreading process. As mentioned above in Section 2, our algorithm combine both the algorithm of cluster detection and the structure hole. And in this paper we use the $(\alpha, \beta) - cluster$[14] to find the overlapping communities and HIS model to identify the structure holes[15].

3.1   The $(\alpha, \beta) - cluster$ Algorithm

What is a good cluster in a social network? Different from the cut-based graph clustering algorithm producing a strict partition of the graph, the example in Figure 1 motivates a new formulation of the graph clustering problem which does not stipulate that each vertex belongs to exactly only one cluster. The objective of the $(\alpha, \beta) - cluster$ algorithm is to identify clusters that are internally dense, i.e., each vertex in the cluster is adjacent to at least a $\beta - fraction$ of the cluster, and externally sparse, i.e., any vertex outside of the cluster is adjacent to at most an $\alpha - fraction$ of the vertices in the cluster.

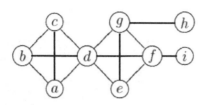

Fig. 1. Overlapping clusters[14]

Definition1. Given a graph, $G = (V, E)$, where every vertex has a self-loop $C \subset V$ is an $(\alpha, \beta) - cluster$ if

Internally Dense: $\forall v \in C, |E(v,C) \geq \beta|C|$

Externally Sparse: $\forall v \in (V - C), |e(u,C) \leq \alpha|C|$, Given $0 \leq \alpha \leq \beta \leq 1$, the $(\alpha, \beta) - cluster$ problem is to find all $(\alpha, \beta) - clusters$.

The new clustering criterion does not seek a strict partitioning of the data. From Figure 1, you can see why clusters can overlap. In Definition 1, we can notice that when $\beta \to 1$, the cluster C approaches a clique, and when $\alpha \to 0$, an $(\alpha, \beta) - cluster$ tends to a disconnected component. We want $\alpha < \beta$ because nodes outside of a cluster should own fewer neighbors in the cluster than node that belong to the cluster. Based on the above analysis, we take $(0.3, 0.5)$ as the parameters values of (α, β) which could achieve a good effect. The details of $(\alpha, \beta) - clustering$ can be get from the literature[14].

3.2   The *HIS* Model

According to the theoretical analysis[15], we assume a setting in which the set V which is consisted of n distinct users form l groups $C = \{C_1, \cdots, C_l\}$, where C is the set of communities. A utility function $Q = (v, C)$ is defined for each node to measure its degree to span structural holes. Formally, we have the following definition:

Definition2. Top-k Structural Hole Spanners[15]. Let $G = (V, E)$ denote a social network, where $V = \{v_1, v_2, \cdots, v_n\}$ is a set of n nodes, and $E \in V \times V$ is a set of undirected social relationships between users. Further that the nodes of the social network can be grouped into l (overlapping) communities $C = \{C_1, \cdots, C_l\}$, with $V = C_1 \cup \cdots C_l$. Then, the $top-k$ structure hole spanners are defined as a subset of k nodes, denoted as V_{SH} in the network, which maximizes the following utility function:

$$\max Q(V_{SH}, C), with |V_{SH}| = k \tag{1}$$

Please note in the definition2, we just focus on the network information but not the content information. Figure 2 shows an example of the structural holes with two communities. Generally speaking, v_6 and v_{12} can be viewed as the structural holes spanners between the two communities. But the HIS model consideres that the v_6 is the only structural holes spanner.

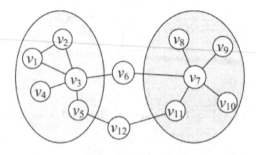

Fig. 2. Illustration of structural holes[15]

3.3   The *MKS* Algorithm

In this section, we propose the *MKS* algorithm. For our study in this paper, we define the key structure of information diffusion network as follows: 1) it is a set of nodes which includes not only the structural holes spanners but also the leader nodes in each cluster; 2) each node in the set can be connected via some hops.

Note we just consider the undirected network. Therefore, for each node in these clusters, we adopt the node degree as the measure of its importance. Besides, we have verified that nodes degree in the top 20% could covers the vast majority of connections in the corresponding cluster. Algorithm 1 gives the pseudocode of *MKS* algorithm.

Algorithm 1. The pseudocode of *MKS* algorithm

1: /*Phase 1 starts here */
2: Input network $N = (V, E)$.
3: Build a graph $G = (V, E)$ based on the network N.
4: Apply the $(\alpha, \beta) - clustering$ algorithm to G and obtain the clusters $C = \{C_1, \cdots, C_l\}$.
5: Apply the *HIS* model to both G and C to obtain the set of the structural hole spanners V_{SH}.
6: /*Phase 2 starts here */
7: for each $C_l \in C$ do
8:   for each node in C_l do
9:     calculate the degree of node Deg_{li}.
10: end for
11: Choose the nodes with top 20% Deg_{li} and put into these nodes into node set N_l
12: end for
13: repeat each N_l do
14:   if exist path from nodes in V_{SH} and N_l do
15:     add the node into new set D.
16: end repeat and output D.

---

4   Experiments

In this section, we evaluate the coverage ratio R_{cov} of our algorithm proposed in Section 3 and compare the efficiency with *PageRank* algorithm. We define the coverage ration R_{cov} as follows:

$$R_{cov} = \frac{L_{cov}}{L_{all}} \tag{2}$$

In the formula, L_{cov} denote the covered links which obtained by algorithms and L_{all} denotes all the links in the information diffusion network. According to the definition, the bigger value of R_{cov} indicates that the algorithm will be more efficiency.

4.1   Data Sets

Experiments are carried out on the real online social network. We consider three different types of networks for researching our problem: Sina micorblog and Coauthor.

The Sina microblog datasets are from the $WISE$2012 Challenge (http://www. wise2012.cs.ucy.ac.cy/challenge.html)which was based on a dataset collected from one of the most popular micro-blog service (http://weibo.com) in China. Since we focus on the information diffusion network and in order to compare the coverage ratio correctly, we collect the datasets by events. Note that we just pay attention to the diffusion network structure and ignore the event content. Therefore, we use the letters A, B, C, D to denote the four events which are used in our experiments. And based on the four events we extract four diffusion networks with different size respectively. The datasets are described as follows $Table$ 1 and $Table$ 2.

Table 1. Dataset 1 - the dataset of the events

Event	#Users	#Relationships
$EventA$	13946	49834
$EventB$	9583	36042
$EventC$	11375	39586
$EventD$	8647	33204

Table 2. Dataset 2 - the dataset with smaller size of the events

Event	#Users	#Relationships
$EventA$	6641	23906
$EventB$	4354	16116
$EventC$	5416	19500
$EventD$	3393	12577

Coauthor is network of authors and the dataset is gained from [16]. The network includes $815,946$ authors and $2,792,833$ coauthorships and for the evaluation purpose, we extract part of them which consists of 52146 authors and 134539 coauthorships from papers published at 28 major computer science conferences. These conferences cover six research areas: Artificial Intelligence (AI), Databases (DB), Data Mining (DM), Distributed Parallel works, Communitications and Performance (NC).The description of the dataset is listed as follows $Table$ 3.

Table 3. Description of Coauthor dataset

Network	#Authors	#Coauthoships
$Coauthor$	52146	134539

4.2   Experiment Results

It is well known that the $PageRank$ algorithm is a classical algorithm about evaluating the nodes importance. We apply the $PageRank$ on the same datasets to estimate the importance of each node and then select those nodes with the highest *PageRank* scores as the influential nodes.

In Figure 3, the $X - axis$ denotes different $top - k$ of *PageRank* and the $Y - axis$ denotes the the R_{cov}. The red line and the blue line respectively represent the *MKS* algorithm and the *PageRank* algorithm. For the *PageRank* algorithm, we can select different $top - k$ values as the most influential nodes and find that the R_{cov} also grows with the increment of value of $top - k$. This tendency of increasing could be easy to understand for the reason that the bigger value of $top - k$, the more nodes are chosen as the influential nodes. However, for the information diffusion network, not all the nodes can be consider as influencial nodes. Therefore, it will make no sense if the value of $top - k$ becomes more bigger. From the four sub-figures, the R_{cov} of the MKS algorithm still remains at about 0.85 because the algorithm both consider the cluster leaders and the structure holes. The cluster leaders would guarantee the the coverage ration at a high level in each cluster and the existence of the structure holes make the clusters be connected which play a significant role in information propagating. In the MKS algorithm, we take the $top - 20$ as the cluster leaders and the R_{cov} is about 0.85 as well as for the *PageRank* algorithm, the R_{cov} is about 0.75 with $top - 20$.

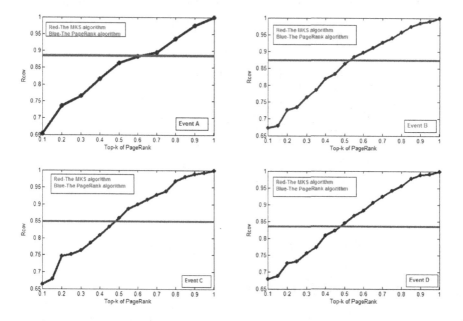

Fig. 3. Comparison about R_{cov} between two algorithms on Sina microblog dataset 1

We also apply the *MKS* algorithm on both the two datasets which include the same events but different size. From Figure 4, conclusion can be obtained that for different network size, our algorithm can produce the same results approximately, which means our method has a good scalability.

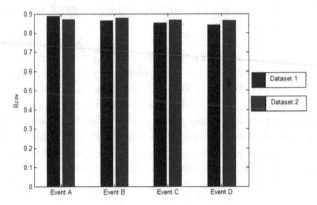

Fig. 4. The MKS algorithm on Sina microblog dataset 2

Besides, the evaluation of the two algorithms are also deployed on the Coauthor dataset. In Figure 5, the $Y-axis$ denotes the R_{cov} and the $X-axis$ denotes the different values of $top-k$ of the $PageRank$ algorithm. The $PageRank$ algorithm makes a better performance when the value of $top-k$ is up to 0.7 but for the smaller values of $top-k$ that the MKS algorithm has much advantages.

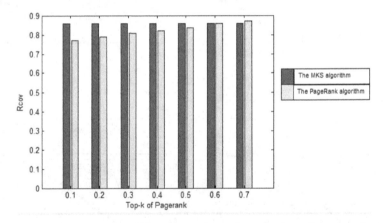

Fig. 5. Comparison about R_{cov} between two algorithms on Coauthor dataset

5   Conclusion and Future Work

With the development of $Web2.0$, the Online Social Network has become a significant component of the Internet. And information propagation also attracted more and more researchers' interest. Different from the obvious works which are with emphasis on identifying the separate influential spreaders, we focus on the structure of the information spreaders in information diffusion network. In order to deal with this problem, we combine the theoretical of clusters and structure holes and propose a new method based on them. Experiments are implemented on the real network and the results show our method performs better than *PageRank*. Also we compare the coverage ratio on different size of event diffusion network. However, the *MKS* algorithm is still remain to improve in some aspects, such as the choosing clusters leaders and improving the coverage ratio accuracy. In the future, we will study these these problems deeply.

Acknowledgments. Partially supported by the Major State Basic Research Development Program of China(973 Program) (Grant No. 2013CB329602), the National High Technology Research and Development Program of China (863 Program) Grant No. 2014AA015204 ,the International Collaborative Project of Shanxi Province of China (Grant No. 2011081034), the National Natural Science Foundation of China (Grant No. 61202215, 61100175, 61232010). China postdoctoral funding (Grant No. 2013M530738).

References

1. Strogatz, S.H.: Exploring complex networks. Nature, 268–276 (2001)
2. Newman, M.E.J.: The structure and function of complex networks. Siam Review, 167–256 (2003)
3. Zhou, T., Fu, Z., Wang, B.: Epidemic dynamics on complex networks. Progress in Natural Science, 452–457 (2006)
4. Lu, L., Chen, D., Zhou, T.: Small world yields the most effective information spreading. CoRR (2011)
5. Doerr, B., Fouz, M., Friedrich, T.: Why Rumors Spread So Quickly in Social Networks, Commun. ACM, 70–75 (2012)
6. Kempe, D., Kleinberg, J., Tardos, E.: Maximizing the Spread of Influence Through a Social Network, pp. 137–146. ACM (2003)
7. Kitsak, M., Gallos, L.K., Havlin, S., Liljeros, F., Muchnik, L., Stanley, H.E., Makse, H.A.: Identification of influential spreaders in complex networks. Nature Physics, 888–893 (2010)
8. Seidman, S.B.: Network Structure and Minimum Degree. SocNet, 269–287 (1983)
9. Brown, P., Feng, J.: Measuring user influence on Twitter using modified k-shell decomposition (2011)
10. Cataldi, M., Di Caro, L., Schifanella, C.: Emerging topic detection on Twitter based on temporal and social terms evaluation (2010)
11. Page, L., Brin, S., Motwani, R., Winograd, T.: The PageRank Citation Ranking: Bringing Order to the Web (1999)
12. Romero, D.M., Galuba, W., Asur, S., Huberman, B.A.: Influence and Passivity in Social Media, pp. 113–114. ACM (2011)
13. Pal, A., Counts, S.: Identifying Topical Authorities in Microblogs, pp. 45–54. ACM (2011)
14. Mishra, N., Schreiber, R., Stanton, I., Tarjan, R.E.: Finding Strongly Knit Clusters in Social Networks. Internet Mathematics, 155–174 (2008)
15. Lou, T., Tang, J.: Mining Structural Hole Spanners Through Information Diffusion in Social Networks. In: International World Wide Web Conferences Steering Committee, pp. 825–836 (2013)
16. Wang, L., Lou, T., Tang, J., Hopcroft, J.E.: Detecting Community Kernels in Large Social Networks, pp. 784–793. IEEE Computer Society (2011)

Handling Big Data of Online Social Networks on a Small Machine

Ming Jia and Jie Wang

Department of Computer Science
University of Massachusetts
Lowell, MA 01854, USA
{mjia,wang}@cs.uml.edu

Abstract. Dealing with big data in computational social networks may require powerful machines, big storage, and high bandwidth, which may seem beyond the capacity of small labs. We demonstrate that researchers with limited resources may still be able to conduct big-data research by focusing on a specific type of data. In particular, we present a system called MPT (Microblog Processing Toolkit) for handling big volume of microblog posts with commodity computers, which can handle tens of millions of micro posts a day. MPT supports fast search on multiple keywords and returns statistical results. We describe in this paper the architecture of MPT for data collection and stat search for returning search results with statistical analysis. We then present different indexing mechanisms and compare them on the micro posts we collected from popular social network sites in China.

Keywords: commodity, indexing, Mongo DB.

1 Introduction

Dealing with big data in computational social networks may require big machines and big storage. This may seem that only large companies or organizations with lucrative budgets can afford big-data research in online social networks (OSNs). We show that, by focusing on a specific type of data, it is possible to carry out big-data research in OSNs using commodity computers in a small lab environment with limited resources. In particular, we present a system for handling big volume of microblog posts (MBPs), and we call the system MPT, which stands for Microblog Processing Toolkit.

Our goals are collecting MBPs from popular OSN sites in China, identifying interesting topics from MBPs, and carrying out statistical analysis on each topic, including gender and location distributions, and discovering hot words and trends. We collected on average approximately 4.5 million (sometimes over 10 million) MBPs a day. We stored these MBPs in a database running Mongo DB on a commodity computer.

To make use of these data, MPT supports, among other things, statistical search that will quickly return, on a set of words entered by the user, the set of

Z. Cai et al. (Eds.): COCOON 2014, LNCS 8591, pp. 676–685, 2014.

MBPs that contain these words and the statistical results of these posts displayed in various graphs. For this purpose we would need to create an appropriate indexing mechanism and update the indexing content regularly. In addition, we also want to retrieve MBPs in real time while we are collecting them, so that we may detect unexpected social events and perform other tasks.

In this paper, we present three indexing methods deployed on commodity computers. Without using clusters of computers, we were able to build a system suitable for implementing a fast search engine and carrying out topic modeling and statistical analysis on large volume of MBPs.

The rest of the paper is organized as follows: In Section II we will describe the data source and the API we use to collect MBPs. In Section III we will introduce the database we use to store the data and describe some of the problems we encountered when storing MBPs. In Section IV we will describe a number of indexing mechanisms, including the default Mongo DB queries using regular expressions, our own implementation of the nextword indexing [1], and a system we built based on Lucene [2]. In particular, we will describe the structures of the systems for indexing, searching, and carrying out statistical analysis. In Section V we will compare the speed of querying, the speed of performing statistical analysis, and the accuracy of each method on real data sets. We conclude the paper in Section VI.

2   Data Collection

We have collected MBPs continuously for over a year from popular OSN sites in China, including Sina, Tencent, and Renren. Both Sina and Tencent provided open API (Application Programming Interface) to developers. They provided different interfaces for different purposes. Since we were focusing on discovering topics from daily MBPs with statistical analysis, rather than on a specific user or a specific topic, we used the public-timeline interface [3] to collect MBPs (see Fig. 1 for an example of such interface).

Under the restriction of the user privilege given to us, we collected a total of about 4.5 million (sometimes over 10 million) MBPs a day from these OSN sites. These MBPs were semi-structured JSON style records.

3   Database

To handle unstructured MBPs in large quantity, we would need a high-performance, steady, and flexible database system. Because MBPs are unstructured, such a database system should be non-relational. We chose Mongo DB [4] for this purpose, which is a common choice for storing unstructured data.

Different MBPs from different sources use different data structures. With Mongo DB, we can store data in different data structures in the same collection. This makes it convenient to manage the data. Moreover, Mongo DB is a database scheme with high performance on the operations of both read and write, which meets our need of intensive writing and querying MBPs.

{
  "statuses": [
    {
      "created_at": "Tue May 31 17:46:55 +0800 2011",
      "id": 11488050248,
      "text": "梦里江.",
      "source": "<a href='http://weibo.com' rel='nofollow'>新浪微博</a>",
      "favorited": false,
      "truncated": false,
      "in_reply_to_status_id": "",
      "in_reply_to_user_id": "",
      "in_reply_to_screen_name": "",
      "geo": null,
      "mid": "5612814510344515491",
      "reposts_count": 8,
      "comments_count": 9,
      "annotations": [],
      "user": {
        "id": 1404376560,
        "screen_name": "saku",
        "name": "saku",
        "province": "11",
        "city": "5",
        "location": "北京 朝阳区",
        "description": "人生五十年，九如梦如幻：有生者有死，壮士复何憾。",
        "url": "http://blog.sina.com.cn/saku",
        "profile_image_url": "http://tp1.sinaimg.cn/1404376560/50/0/1",
        "domain": "saku",
        "gender": "m",
        "followers_count": 1204,
        "friends_count": 447,
        "statuses_count": 2908,
        "favourites_count": 0,
        "created_at": "Fri Aug 28 00:00:00 +0800 2009",
        "following": false,
        "allow_all_act_msg": false,
        "remark": "",
        "geo_enabled": true,
        "verified": false,
        "allow_all_comment": true,
        "avatar_large": "http://tp1.sinaimg.cn/1404376560/180/0/1",
        "verified_reason": "",
        "follow_me": false,
        "online_status": 0,
        "bi_followers_count": 215
      }
    },
    ...
  ],
  "previous_cursor": 0,
  "next_cursor": 11489013766,
  "total_number": 81655
}

Fig. 1. JSON example of public_timeline API

Initially we stored the MBPs posted on the same date in one collection, and stored all the collections in the database. After running it for a few weeks, we experienced unexpected system crashes. The reason was that Mongo DB would write data in the same database into the same file and load all the files that are accessed frequently into the main memory. As more MBPs were stored in the same database, the file grew larger quickly, causing Mongo DB to consume almost all the RAM and crashing the system. To solve this problem, we divide the collections according to a fixed interval of one week of the MBP postings into different databases. Because writing to the database was the main operation of the system and the system only collected real-time data, Mongo DB would load the file of the most recent week into RAM, consuming much less RAM than before. The system has never crashed after we made this change.

4   Data Retrieval

Our Microblo Processing Toolkit performs the following three types of search.

1) Given a keyword, retrieve all MBPs that contain the keyword.

2) Given a set of keywords, retrieve all MBPs that contain at least one of these keywords (this is the logical OR operation)

3) Given a set of keywords, retrieve all MBPs that contain all of the keywords (this is the logical AND operation).

We approached these tasks using the following three methods.

Mongo DB Regular Expressions. Mongo DB provides a built-in regular expression searching method. Given the regular expression we want the text to match, Mongo DB returns all the records that match the regular expression. This search method, however, is inefficient and does not meet our needs.

To speed up the search process, we developed an indexing system based on the nextword indexing scheme [5].

Nextword Indexing. The nextword index consists of a vocabulary of distinct words and, for each word w, a nextword list and a position list. The nextword list consists of each word s that succeeds w anywhere in the database, interleaved with pointers into the position list. Fig. 2 depicts an example of the nextword structure of MBPs.

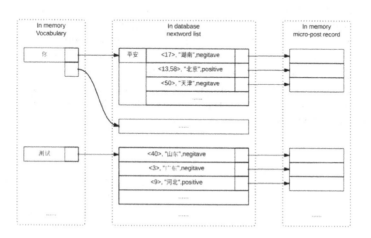

Fig. 2. Nextword index structure

Based on the nextword indexing, we built a system consisting of two parts: (1) index server and (2) search server. The index server indexes in real time the MBPs we collected. In particular, we store the word pair list in the main memory and the information of each word pair in the database. This list contains the position of the word pair in the text and the information of the microblo posts.

Our early version of the system recorded each word pair and stored it in RAM. This method, however, took up too much RAM. We observed that, for less-frequent words, if we just stored the word position and not the word pair positions, then the search speed would only be mildly affected, while the consumption of RAM would drop tremendously. To balance between the search speed and the RAM occupancy by the nextword indexing, we set a threshold on word-pair counts, so that the system only stores the nextword list of words with frequencies over the threshold. This measure cuts down the RAM usage significantly.

After several weeks of running the system, we encountered another problem of data explosion: The number of MBPs provided through the APIs suddenly increased significantly, more than twice the size of the data we collected in one day. Likewise, the index size also increased to occupy about 5GB of RAM. Since we deployed the system on a commodity computer, RAM was a precious and limited resource, and we could not afford such RAM usage. Thus, we were motivated to devise a low-cost RAM solution and we accomplished this using Apache Lucene.

Lucene-Based Indexing. Apache Lucene is a high-performance, full-featured text search engine library, which requires small RAM for indexing and searching, and generates an index file with reasonable size. We customized the Lucene core and built a data retrieval system that solved the problem of memory blowup. Our system consisted of two parts: (1) real-time index server and (2) search server. The real-time index server refreshes the database frequently and indexes dynamically the newly collected MBPs on the fly. The search server returns the MBPs that contain all the keywords entered by the user. These two parts may work simultaneously without conflicting each other, and so the search server can return real-time posts the system collects. Fig. 3 shows the structure of our Lucene-based indexing system.

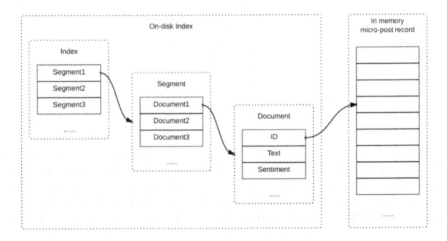

Fig. 3. Lucene-based index structure

5   Statistical Analysis

To carry out statistical analysis on the MBPs we collected, we group the microblog posts according to the following three attributes: 1) posting time, 2) poster's gender, and 3) poster's location.

Using the built-in regular expressions of Mongo DB, we would need to traverse all the MBPs the database returned and then carry out the statistical analysis. This process is extremely time consuming.

Using the nextword indexing, we could complete the statistical analysis in just a few steps without traversing the data.

Using the Lucene-based indexing, we need to customize our own functions. For each MBP, we connect its poster's location, gender, and posting time as one string. We set this string as the key of the post. This means for two posts that are posted in the same hour, the same location and by the posters of the same gender will share the same key. When indexing the MBP, we include this key in the index. When searching for the data, we first carry out the group search by the key. After obtaining the groups with the same key, we traverse each group and perform the statistical analysis in the group (details of statistical analysis are not included in this paper).

6   Experiments

We designed an experiment to test the efficiency and accuracy of the three methods for data retrieval and statistical analysis mentioned in the previous sections. For convenience, we will refer to these methods of Mongo DB's built-in regular expressions, nextword indexing, and Lucene-based indexing as, respectively, Mongo, Nextword, and Lucene. We used the MBPs we collected in one day (the day was randomly chosen) from Sina and Tencent as the data set for comparing the three different methods. There were about 4,250,000 MBPs in this one-day collection. All experiments were executed on a commodity computer with a quad-core CPU and 16GB RAM.

The experiment consisted of two parts. In the first part, we examined how fast each method responded to user queries, as well as how fast the method carried out statistical analysis and returned the results. In the second part, we compared the accuracy of each method.

To run the experiment, we pre-counted all the keyword phrases in the data set and randomly selected a number of keyword phrases as our testing phrases, such that these phrases were made up of two or three keywords and appeared in the test data set for more than 100 times.

For each phrase we selected, we executed a query with each method. We recorded the time that each method incurred to respond to the query and finished up the statistical analysis. We tested Mongo, Nextword, and Lucene separately. For each test, we repeated the process twice and calculated the average time. The results are shown in Fig. 4 and Fig. 5.

First, we compared the response speed of each method. The horizontal axis represents the actual frequency of keywords, and the vertical axis represents the running time of returning the MBPs that contain the keywords.

From Fig. 4 we can see that 1) Lucene offers the fastest responding time, and the responding time will increase as the phrase frequency increases. 2) Nextword is faster, which is slower but close to Lucene; but its time complexity is not as steady as Lucene. 3) Mongo is the slowest, which does not meet the real-time search requirement.

We then compared the speed of carrying out statistical analysis for each method.

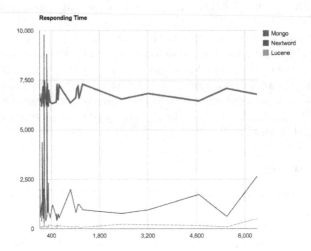

Fig. 4. Responding Time

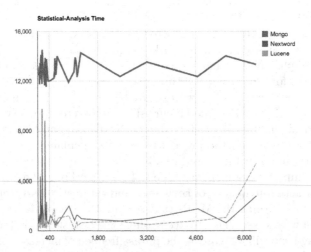

Fig. 5. Statistical-Analysis Time

From Fig. 5 we can see that Lucene is still the fastest, Nextword is slower than Lucene, while Mongo is substantially slower. The time interval between responding to the search query and finishing up statistical analysis is quite short for Lucene, and is much longer for the other two methods. We note that the mechanism for carrying out the statistical analysis of each method is different. By querying Mongo, we would need to traverse all the MBPs returned on the query, which would incur significant computing time when the returning data set is large. The nextword indexing is a well-structured indexing mechanism, where no traverse is needed to perform statistical analysis, and so it can finish the statistical analysis quickly. Lucene finishes statistical analysis by traversing all the groups instead of all the posts.

We can see from Fig. 5 that the time complexity of Mongo search does not increase much when the frequency of the keywords is increased, while the time complexity of Nextword search and the Lucene search is clearly related to the frequency of the keywords.

Finally, we compared the accuracy of each method (see Fig. 6). The horizontal axis represents the actual frequency of the phrase, and the vertical axis represents the counts of the returning results.

From Fig. 6 we can see that the accuracy of each method differs from each other, where Mongo is 100% accurate. In other words, its precision and recall rates are both equal to 1.

Nextword may miss some MBPs. The main reason of missing MBPs is due to the segmentation error of the keywords in the Chinese language. The keyword segmentation in the Chinese language is different from that of English, for the standard Chinese writing contains no space between characters. Thus, different segmentation tools may return different segmentation results. Even for the same keyword using the same segmentation tool, the results may still be different with different text. For the keywords we queried in the experiments, the segmentation result in the MBPs could differ from that in the search query. The MBPs with different segmentation results would be missed.

Lucene, on the other hand, seems to have the worst precision and recall rates, where the number of returned MBPs is usually larger than the actual number that contains the phrase. This is caused by the Lucene indexing structure, where all the MBPs that contain a subset of the search keywords are returned. For example, if phrase X is made up of words A **and** B, when querying X, Lucene will return all the posts that contain A **or** B. So the returning result usually has a larger-than-actual count.

Table 1 shows the precision and recall rates of the search results using the three different methods. We can see from the table that Mongo performs well on accuracy. Both Mongo and Nextword have a value of 1 on the precision rate, which means that the MBPs Mongo returned were exactly those that contain the search phrase. Since Nextword may miss some MBPs, this affected negatively its recall value. Lucene suffers precision loss compared to the other two methods. But it offers better performance in the recall value than Nextword, which means that it may miss fewer MBPs than Nextword.

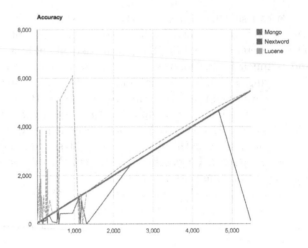

Fig. 6. Accuracy

Table 1. Experiment Results on Precision and Recall Rates

Method	Precision	Recall
Mongo	1	1
Nextword	1	0.72
Lucene	0.53	0.81

From our experiment we can see that each method has its pros and cons. In particular, Mongo provides the regular expression searching method with the perfect accuracy. But it is too slow to meet the needs of real-time querying and stat analysis. Nextword has the best performance in statistical analysis, and the responding time can meet the need of real-time search. But its memory consumption is high, which cannot meet the increasing data requirement. Lucene, on the other hand, has the fastest responding time and the fastest statistical-analysis time, but it incurs low accuracy. This method is good for building a fast search engine but may not meet the requirement of high accuracy.

7   Conclusion

In this paper, we described and compared three methods we used to analyze big volume of microblog posts. We conclude the paper by summarizing our findings as follows:

1) Mongo is good for storing data but not suitable for carrying out search on big data.

2) Nextword offers good performance on real-time search with fast response time and statistical-analysis time. But it would take up too much RAM. Nextword

would be a good choice for analyzing moderate-size data (e.g., less than 3 million) with high requirement on statistical analysis.

3) Lucene is a low RAM-consumption and stable system. But it has the worst performance on precision and recall. For those who need to analyze big volume of data but do not require exact statistical results, Lucene would be a better choice.

Acknowledgment. The authors were supported in part by the NSF under grants CNS-1018422 and CNS-1247875. Any opinions, findings and conclusions or recommendations expressed in this material are those of the authors and do not necessarily reflect the views of the NSF.

References

1. Williams, H.E., Zobel, J., Anderson, P.: What's Next? Index Structures for Efficient Phrase Querying. In: Australasian Database Conference (1999)
2. Apache Lucene, `https://lucene.apache.org/`
3. Open Document for Sina Micro-blog API,
   `http://open.weibo.com/wiki/2/statuses/publictimeline/en`
4. MongoDB, `http://www.mongodb.org/`
5. Bahle, D., Williams, H.E., Zobel, J.: Compaction Techniques for Nextword Indexes. In: SPIRE (2001)